INTRODUCTION TO PHYSICAL POLYMER SCIENCE

THIRD EDITION

L. H. Sperling

Lehigh University
Bethlehem, Pennsylvania

WILEY-INTERSCIENCE

A John Wiley & Sons, Inc., Publication

New York • Chichester • Weinheim • Brisbane • Singapore • Toronto

For ordering and customer service, call 1-800-CALL-WILEY.

Library of Congress Cataloging-in-Publication Data:

Sperling, L. H. (Leslie Howard), 1932–
    Introduction to physical polymer science / L. H. Sperling.—3rd ed.
        p.   cm.
    Includes bibliographical references and index.
    ISBN 0-471-32921-5 (cloth : alk. paper)
    1. Polymers.   2. Polymerization.   I. Title.
QD381.S635   2001
668.9—dc21   00-051316

Printed in the United States of America.

10 9 8 7 6 5 4 3 2

This book is dedicated to my grandchildren,

Ryland Sweigard (b. 10/13/88), who dreams of becoming a professional soccer player, wanting to kick a soccer ball so hard that it will explode at the tip of his toe; and

Tresten Sweigard (b. 2/18/91), an aspiring mechanical engineer, who wants to make the soccer ball material so tough that Ryland can't possibly break it; and

Aubrey Johnson (b. 10/31/95), who plays with plastic doll houses and constantly searches for better clothing materials for her dollies; and

Lyra Johnson (b. 10/20/99), who loves her pacifier and dreams of improvements in the theory of rubber elasticity that will make it still more satisfying.

Dear Reader:
You now know what the better half of this book is about. Please read on. . . .

CONTENTS

4   Concentrated Solutions and Phase Separation Behavior   **123**

PREFACE

Today, there are two main divisions in polymer science: polymer synthesis and polymer physics. While these are often integrated in one textbook to make a one-semester course, more and more colleges and universities are teaching polymer science in two semesters, making textbooks devoted to one-half or the other highly desirable. From another point of view, students interested primarily in chemical engineering, materials science and engineering, mechanical engineering, or physics often have little preparation or interest in synthesis, but a course emphasizing theory and properties is their meat. This book is dedicated to all of those students and faculty who are interested in the physical side of polymer science.

For a long time now, between one-third and one-half of the chemists, chemical engineers, and materials science and engineers graduates of all degree levels have been engaged in some aspects of polymer science and engineering. While this has been commented on many times, the reader is directed to *C&EN*, **77** (26), 57 (1999) for recent statistics. The question of who is really specializing in polymers hinges on who asks and who answers. Take a chemical engineer in charge of a certain factory. The question is asked, "What do you consider your professional specialty? They may answer, "I'm a process engineer." The questioner continues, "And what do you manufacture?" "Why, we make polyester fibers, of course!" Is such an individual actually a process engineer, a polymer engineer, or some of both? Similar questions and responses could involve a physical chemist researching the interfacial bonding forces in polymer blends. I have become conceited enough to believe that such individuals would benefit professionally by at least a course or two in polymer science and engineering.

Actually a new breed of scientist/engineer is emerging: those individuals who are being degreed directly in polymer science and engineering. There are now several departments around America offering this program. In addition there are over 30 interdisciplinary polymer degree programs in America, focused on various centers and institutes. For example, at Lehigh University, we have the Center for Polymer Science and Engineering, offering the M.S., M.E., and Ph.D. degrees, with five departments participating: chemical engineering, chemistry, materials science and engineering, mechanical engineering, and physics. The departments of chemical engineering, chemistry, and materials science and engineering also offer an undergraduate minor in polymer science and engineering. Now, approximately one-half of all of the chemical engineering, chemistry, and materials departments around America offer at least one polymer course. Those interested in further statistics should contact the POLYED committee of the American Chemical Society.

A significant amount of new polymer research has been published since the second edition was finished in 1990, reflecting many key advances. The third edition has been given a general overhaul to incorporate these new findings, while deleting some outdated material. Many new example problems have been added throughout. Two new chapters have been added, polymer surfaces and interfaces, and multicomponent polymeric materials. With the added material, the instructor has several options. While it still is possible to teach substantially the whole book in one regular semester, the instructor may opt to emphasize one portion or the other to suit the student's interests and needs. Alternately, there is now ample material to provide two quarters worth of instruction.

Chapters 1 to 3 continue to be basic polymer science. Chapters 4 to 11 provide the bulk of physical polymer science fundamentals. Chapters 12 and 13, entirely new material, emphasize the new topics of polymer interfaces and multicomponent polymer materials. Chapter 14, about two-thirds new material, focuses on a range of modern topics, with some emphasis on applications. While this book emphasizes the subject of polymer physics, it is broader in outlook, covering aspects of polymer engineering and processing, materials science, and polymer history, while holding synthesis and kinetics to a minimum.

I want to take this opportunity to thank the many students who helped in proofreading the various manuscripts. If I had to pay them even a nickel a correction, I would be a poor man! The result in each case, of course, is one fewer error that future students will have to endure. Many thanks must also be given to the Department of Chemical Engineering and the Department of Materials Science and Engineering, as well as the Materials Research Center and the Center for Polymer Science and Engineering. Special thanks must be given to Ms. Andrea Pressler, photographer par excellence, who provided many of the figures for the book. Special thanks are also due Ms. Gail Kriebel and her staff at the E. W. Fairchild–Martindale Library, who helped with literature searching, and provided me with a carrel.

L. H. SPERLING

BETHLEHEM, PENNSYLVANIA

PREFACE TO THE
SECOND EDITION

When I was small, the books and songs I knew just *were*, with no particular authorship. I got my lesson on the rights and duties of authorship from my mother one day, discussing the Bible. "If we don't think that a given passage is worded in the best way for today, why don't we just rewrite it?' I asked.

My mother was horrified. "That Book was written thousands of years ago! It was handed down from generation to generation as it was set down on paper by its authors!" Books were sacred to her.

So it was that I learned that all books, songs, plays, and operas are written by particular men and women who have a strong and clear message to be expressed. Only they have the privilege of revising their works. If someone wasn't satisfied, they could write their own book!

Now, after a period of still less than a decade, I find that many new ideas have come to the fore in physical polymer science. A few important concepts somehow got left out of the first edition of this book. A handful of pages and paragraphs wandered in that should not have been there. I, the author of this work, joyfully take the privilege that is only mine to prepare this second edition. Thank you for giving me the opportunity.

Major specific changes include the following: three new chapters were added. The first one resulted from a split of the chapter titled, "The Bulk State," into two chapters, "The Amorphous State" and "The Crystalline State." The second one resulted from the vast increase in importance of the liquid crystalline state, resulting in a chapter entitled "Polymers in the Liquid Crystalline State." This chapter absorbed some of the material originally in the chapter entitled "Mechanical Behavior of Polymers." The third new chapter, at the end of the book, is entitled, "Modern Topics."

Other major changes include the addition of the macromolecular hypothesis and historical development to Chapter 1, photophysics and fluorescence to Chapter 2, movement of the sections on thermodynamics of mixing and phase separation and fractionation from Chapter 4 to Chapter 3, ahead of molecular weight determination and additional material on thermodynamics of blending polymers and polymer–polymer phase diagrams to Chapter 4.

The following changes refer to the new chapter numbers. Additional material on self-diffusion of polymers was added to Chapter 5, and fiber spinning and structure were added to Chapter 6. As mentioned previously, Chapter 7 is an entirely new chapter on polymer liquid crystals. Greater discussion of dynamic mechanical behavior is given in Chapter 8. A discussion of gelatinous materials was added to Chapter 9. To Chapter 10, a discussion of rheology and a new demonstration were added (in the appendix). Chapter 11 was given a complete reorganization, because of the new material on stress–strain behavior and crack healing. Chapter 12 is an entirely new chapter, concerned with a variety of topics, emphasizing surface and interfacial behavior, electrical behavior of polymers, nonlinear optics, and high-temperature materials.

Again, I want to take this opportunity to thank they many students who helped in proofreading the manuscript. In proofing the first edition, a student said to me: "Dr. Sperling! You always taught us to write and rewrite manuscripts until they read like a thousand violins playing in the night! This paragraph here, however, reads like 50 violins playing in the day, and two of them are squeaking!" In preparing the second edition, a student again said to me: "Look at this equation! All three terms must always be positive, yet you have equated the sum of them to zero!" The result in each case, of course, is one fewer error that future students will have to endure.

Many thanks must also be given to the Department of Chemical Engineering and the Department of Materials Science and Engineering, as well as the Materials Research Center and the Center for Polymer Science and Engineering at Lehigh University. Dr. D. A. Thomas, long my research partner, contributed some juicy study problems. Special thanks must be given to Ms. Gail Kriebel and her staff at the E. W. Fairchild–Martindale Library, Ms. Andrea Pressler, photographer, and Ms. Virginia Newhard, my secretary.

L. H. Sperling

Bethlehem, Pennsylvania
June 1992

PREFACE TO THE
FIRST EDITION

Research in polymer science continues to mushroom, producing a plethora of new elastomers, plastics, adhesives, coatings, and fibers. All of this new information is gradually being codified and unified with important new theories about the interrelationships among polymer structure, physical properties, and useful behavior. Thus the ideas of thermodynamics, kinetics, and polymer chain structure work together to strengthen the field of polymer science.

Following suit, the teaching of polymer science in colleges and universities around the world has continued to evolve. Where once a single introductory course was taught, now several different courses may be offered. The polymer science and engineering courses at Lehigh University include physical polymer science, organic polymer science, and polymer laboratory for interested seniors and first-year graduate students, and graduate courses in emulsion polymerization, polymer blends and composites, and engineering behavior of polymers. There is also a broad-based introductory course at the senior level for students of chemical engineering and chemistry. The students may earn degrees in chemistry, chemical engineering, metallurgy and materials engineering, or polymer science and engineering, the courses being both interdisciplinary and cross-listed.

The physical polymer science course is usually the first course a polymer-interested student would take at Lehigh, and as such there are no special prerequisites except upper-class or graduate standing in the areas mentioned above. This book was written for such a course.

The present book emphasizes the role of molecular conformation and configuration in determining the physical behavior of polymers. Two relatively new ideas are integrated into the text. Small-angle neutron scattering is doing for polymers in the 1980s what NMR did in the 1970s, by providing an entirely new perspective of molecular structure. Polymer blend science now offers thermodynamics as well as unique morphologies.

Chapter 1 covers most of the important aspects of the rest of the text in a qualitative way. Thus the student can see where the text will lead him or her, having a glimpse of the whole. Chapter 2 describes the configuration of polymer chains, and Chapter 3 describes their molecular weight. Chapter 4 shows the interactions between solvent molecules and polymer molecules. Chapters 5–7 cover important aspects of the bulk state, both amorphous and crystalline, the glass transition phenomenon, and rubber elasticity. These three chapters offer the greatest depth. Chapter 8 describes creep and stress relaxation, and Chapter 9 covers the mechanical behavior of polymers, emphasizing failure, fracture, and fatigure.

Several of the chapters offer classroom demonstrations, particularly Chapters 6 and

7. Each of these demonstrations can be carried out inside a 50-minute class and are easily managed by the students themselves. In fact, all of these demonstrations have been tested by generations of Lehigh students, and they are often presented to the class with a bit of showmanship. Each chapter is also accompanied by a problem set.

The author thanks the armies of students who studied from this book in manuscript form during its preparation and repeatedly offer suggestions relative to clarity, organization, and grammar. Many researchers from around the world contributed important figures. Dr. J. A. Manson gave much helpful advice and served as a Who's Who in highlighting people, ideas, and history.

The Department of Chemical Engineering, the Materials Research Center, and the Vice-President for Research's Office at Lehigh each contributed significant assistance in the development of this book. The Lehigh University Library provided one of their carrels during much of the actual writing. In particular, the author thanks Sharon Siegler and Victoria Dow and the staff at Mart Library for patient literature searching and photocopying. The author also thanks Andrea Weiss, who carefully photographed many of the figures in this book.

Secretaries Jone Susski, Catherine Hildenberger, and Jeanne Loosbrock each contributed their skills. Lastly, the person who learned the most from the writing of this book was . . .

L. H. SPERLING

BETHLEHEM, PENNSYLVANIA
NOVEMBER 1985

SYMBOLS AND DEFINITIONS

SYMBOL	DEFINITION	SECTION
English Alphabet		
A	A_2 = second virial coefficient	3.3.2
	A_1 = first virial coefficient	3.5.3.3
	A_3 = third virial coefficient	3.5.3.3
	A_4 = fourth virial coefficient	3.5.3.3
	A (with various subscripts) = area under a Bragg diffraction line	6.5.4
	Angular amplitude	8.3.3
	A_T = reduced variables shift factor	8.6.1.2
	Surface area (with various subscripts)	12.2.3
B	Bulk modulus	8.1.1.2
C	C^* = Chiral center, optically active carbon	2.3.2, 2.4.1
	C_m = constant	3.3.2
	C_N = neutron scatting equivalent of H	5.2.2.1
	C_∞ = Characteristic ratio	5.3.1.1
	ΔC_p = change in heat capacity	6.1
	CI = crystallinity index	6.5.4
	C_p = heat capacity	8.2.9
	C_1, C_2 = Mooney–Rivlin constants	9.9.1
	C_{100}, C_{010}, C_{200}, C_{400}, C, C', C'' = generalized strain energy constants	9.9.2
	C_1', C_2' = WLF constants	10.4.1
	C_A = concentration of A	A10.1
	C_p, C_v = capacitance of polymer and vacuum	14.7.1
D	Diffusion coefficient	3.6.6, 4.4.2, 5.4.2.1
	D_e = Deborah number	10.2.4
	D' = fractal dimension	12.7.3
	D_2 = IPN phase domain size	13.5.4

SYMBOL	DEFINITION	SECTION
E	Young's modulus	1.3, 8.1.1.1
	ΔE = change in energy	2.2.4
	E_{act} = energy of activation	2.8, 8.6.1.2
	E^* = complex Young's modulus	8.1.8
	E' = storage modulus	8.1.8
	E'' = loss modulus	8.1.8
	E_1, E_2, etc. = spring moduli	10.1.2.1
F	Helmholtz free energy	9.5
G	Gibbs's free energy	3.2
	Group molar attraction constant	3.2.3
	ΔG_M = change in free energy on mixing	3.2
	$G_N{}^0$ = steady-state rubbery shear modulus	5.4.2.1
	Radial growth rate of crystal	6.6.2.2
	Shear modulus	8.1.1.1
	G^* = complex shear modulus	8.1.8
	G' = shear storage modulus	8.2.9
	G'' = shear loss modulus	8.2.9
	$\mathscr{G}$ = fracture energy	11.1.2
	$\mathscr{G}_c$ = critical energy of crack growth	11.1.2
	$\mathscr{G}_{1c}$ = critical energy of crack growth on extension	11.5.2.4
	G^s = surface free energy	12.2.1
H	H_0 = magnetic field	2.2.4
	Optical constant	3.6
	ΔH_f = enthalpy of fusion	6.1
	δH = NMR absorption line width	8.3.4
	Heat energy per unit volume per cycle	8.12
	Wool's general function	11.5.3
	H^s = surface enthalpy	12.2.1
I	I_D = dimer emission intensity	2.10.3.1
	I_M = single mer emission intensity	2.10.3.1
	I_1, I_2, I_3 = strain invarients	9.9.2
	Current	14.7.1
J	Flux	4.4.2
	J_n = de Gennes defect current	5.4.2.1
	Compliance (with various subscripts)	5.4.2.1, 8.1.1.2
	J^* = complex compliance	8.1.8
	J', J'' = storage and loss compliance	10.2.4
K	$\tilde{K}$ = constant	3.6.1
	Wave vector	3.6.1, 5.2.2.1, 12.3.8.1

SYMBOL	DEFINITION	SECTION
	Equilibrium constant of polymerization	3.7.2.1
	Constant in the Mark–Houwink–Sakurada equation	3.8.3
	K_d = distribution coefficient	3.9.2
	$\bar{K}$ = constant relating end-to-end distance to molecular weight	4.3.9
	K_1, K_2 = measures of free volume	8.6.1.1
	K_L, K_H = constants in melt viscosity	10.4.2.1
	K = stress intensity factor	11.2.4.1, 11.3.2
	K_{1c}, K_{2c}, K_{3c} = critical stress intensity factor in the extension, shear, and tearing modes	11.2.4.1, 11.3.2
	ΔK = stress intensity factor range	11.4.2
L	Sample length	9.4
	$L(x), L(\beta)$ = inverse Langevin function	9.10.1
	L_1, L_2 = transverse lengths	12.3.8.1
	$2L_0$ = separation length	12.6.1
M	Molecular weight	
	M_n = number-average molecular weight	1.2.1, 3.4
	M_w = weight-average molecular weight	3.4
	M_z = z-average molecular weight	3.4
	M_v = viscosity-average molecular weight	3.4
	M_1, M_2 = mass fractions	8.8.1
	M_c = number-average molecular weight between cross-links	9.4
	M_e = molecular weight between entanglements	9.4
	M_c' = entanglement molecular weight	9.4
	Mass	11.1.2
N	N_i = number of molecules of molecular weight M_i	1.2.2
	Number of cells	3.3.1.2
	N_A = Avogadro's number	3.3.2
	N_c = number of molecules in 1 cm^3	4.3.2
	N_e = number of mers between entanglements	11.5.2.2
O		
P	$P(\theta)$ = single-chain form factor	3.6.1
	P_c = critical extent of reaction at the gel point	3.7.4
	$\tilde{P}$ = reduced pressure	4.3
	P^* = characteristic value	4.3
	Permeability coefficient	4.4.2
	P_1 = probability of a chain arm folding back on itself	5.4.3
	Probability of barriers being surmounted	8.6.1.2
	P_i = induced polarization	14.8.1

SYMBOL	DEFINITION	SECTION
Q	Partition function	8.6.3.1
	Q_I, Q_{II} = amounts of heat released	9.8.3
R	Gas constant	1.3
	$R\bullet$ = free radical	1.4.1.2
	R = generalized organic group	1.4.1.2
	$R(\theta)$ = Rayleigh's ratio	3.6
	R_g = radius of gyration	3.6.1
	R_i = rate of initiation	3.7.2.2
	R_p = rate of propagation	3.7.2.2
	R_t = rate of termination	3.7.2.2
	R_e = hydrodynamic sphere equivalent radius	3.8.2
	$\mathbf{R}$ = ratio of radii of gyration	9.10.6
	$\bar{R}$ = fracture resistance	11.1.2
	Resistance in ohms	14.7.1
S	Entropy	3.2
	ΔS_M = change in entropy on mixing	3.2
	Solubility coefficient	4.4.2
	Scaling variable	A4.1
	S_k = mean separation distance	6.6.2.5
	Disclination strength	7.6
	S^s = surface entropy	12.2.1
	S_{th} = interphase surface thickness	12.3.7.2
T	Absolute temperature	1.3
	T_f = fusion or melting temperature	1.1, 6.1
	T_g = glass transition temperature	1.3
	ΔT_b = boiling point elevation	3.5.2
	ΔT_f = freezing point depression	3.5.2
	$\tilde{T}$ = reduced temperature	4.3
	T^* = characteristic temperature	4.3
	$\mathbf{T}$-Fraction of light transmitted	5.2.1
	T_f^* = equilibrium melting temperature of crystals	6.8.5
	T_{ll} = liquid–liquid transition	8.4
	T_0 = generalized transition temperature	8.6.1.2
	T_s = arbitrary WLF temperature	8.6.1.2
	T_2 = unifying treatment of the second-order glass transition temperature	8.6.3.4
	T_e = fraction of trapped entanglements	9.10.5.1
	T_e, T_R, T_d = relaxation times	10.2.5
	T_r = reptation time	10.2.5

SYMBOL	DEFINITION	SECTION
U	U_{max} = maximum in scattering intensity in the radial direction	6.5.1
	Internal energy	9.5
	$\delta U_1, \delta U_2, \delta U_3, \delta U_4,$ = energies related to fracture	11.1.1
V	Molar volume	3.2
	V_s = scattering volume	3.6
	V_e = hydrodynamic sphere volume	3.8.2
	V = reduced volume	4.3
	V^* = characteristic volume	4.3
	V_r = volume of one cell	4.3.2
	V_0 = occupied volume	8.6.2.1
	V_t = specific volume	8.6.2.1
	V_1 = molar volume of solvent	9.12
	Voltage	14.7.1
W	Work on elongation	9.7.2
	W_a = work of adhesion	12.3.7.2
X	Brownian motion average distance traversed	5.2.2.1
	X_t = degree of crystallinity	6.6.2.1
	X_B = mole fraction of impurity	6.8.1
	X_1, X_2 = mole fractions	8.8.1
	$X(t)$ = average mer interpenetration depth function of time	10.2.6
	$X(\infty)$ = interpenetration depth constant	10.2.6
Y		
Z	Avrami constant	6.6.2.1
	Constant in cross-link density calculations	8.6.3.2
	Number of carbon atoms in a chain's backbone	10.4.2.1
a	Exponent in the Mark–Houwink–Sakurada equation	3.8.3
	a_H, a_D = scattering lengths	5.2.2.1
	End-to-end distance of a Rouse–Bueche segment	5.4.1
	Cell axis distance	6.1.1
	Van der Wall's constant	9.6
	Half the crack length	11.3.1
	Correlation distance	12.3.8.1
b	b_1, b_2 = polarizabilities	5.2.1
	Kuhn segment length	5.3.1.2
	Defect stored length	5.4.2.1
	Cell axis distance	6.1.1
	Van der Wall's constant	9.6
	Statistical segment step length	12.3.7.2

SYMBOL	**DEFINITION**	**SECTION**
c	Solute concentration	3.5.2
	Cell axis distance	6.1.1
d	Thickness	5.2.1
	Bragg distance	6.2.2
	Domain period	13.6.2.1
e		
f	Functionality of branch units	3.7.4
	$f_0 =$ frictional coefficient	3.8.2
	Function	A4.1
	$f_1 =$ orientation function	5.2.1
	Restoring force	5.4.1
	Shear stress	8.1.1.2
	Fractional free volume	8.6.1.2
	$f_0 =$ fractional free volume at T_g	8.6.1.2
	Retractive force	9.5
	$f_e =$ energetic portion of the retractive force	9.5
	$f_s =$ entropic portion of the retractive force	9.5
	Force on a chain	9.7.1
	$f^* =$ network functionality	9.10.2
	$f_{xx}, f_{yy}, f_{zz} =$ stress components	10.5.2
g	Gauche	2.1.2
h	Planck's constant	2.10.2
i	$i_e =$ Thomson scattering factor	3.6.1
	Square root of minus one	8.1.8
j		
k	Boltzmann's constant	2.8
	$k_i =$ rate constant of initiation	3.7.2.2
	$k_p =$ rate constant of propagation	3.7.2.2
	$k_t =$ rate constant of termination	3.7.2.2
	$k' =$ Huggins's constant	3.8.4
	$k'' =$ Kraemer's constant	3.8.4
l	Length of a link or mer	5.3.1.1
	$\ell =$ crystal thickness	6.4.2.1
m	Meso, same side	2.3.3
	Mass of a polymer chain	3.10
n	Number of mers in the chain	1.1
	Number of network chains per unit volume	1.3
	Mole fraction	3.3.1.1
	Refractive index	3.6, 5.2.1

SYMBOL	DEFINITION	SECTION
	Any whole number	6.2.2
	Avrami constant	6.6.2.1
	n_c, n_p = chemical and physical cross-links	9.10.5.1
	n_{tot} = total number of effective cross-links	9.10.5.1
	Number of stress cycles	11.4.2
	$n(t)$, n_∞ = number of chains intersecting a unit area of interface at t and at infinite time	11.5.3
o		
p	Partial vapor pressure	3.3.1.1
	Fractional conversion	3.7.2.3
	Persistence length	4.2
	Number of pitches	6.3.2
	Probability of Avrami crystal fronts crossing	6.6.2.1
	$1/p^2$ = measure of stiffness	8.3.3
	p_1 = probability of finding a molecule	9.6
q	q_1, q_2 = heat absorbed and released	9.8.3
r	Racmic-opposite side	2.3.3
	End-to-end distance	3.6.2, 9.7.1, 10.2.7
	$\tilde{r}_0{}^2$ = root-mean square end-to-end distance of a chain	3.9.7
	Reptation rate	6.6.2.5
	$\overline{r_i}^2$, $\overline{r_0}^2$ = mean square end-to-end distances of swollen and relaxed chains	9.10.4
	r_y = crack-tip plastic zone radius	11.3.2
	Exponent in interface theory	11.5.3
s	Shear strain	8.1.1.2
t	*Trans*	2.1.2
	Time	3.7.2.4
	Exponent in interface theory	11.5.3
u	Intermolecular excluded volume	3.3.2
	$\bar{u}$ = Stokes terminal velocity	10.5.4
v	Volume fraction	3.2
	v_2 = volume fraction of polymer	4.1.2, 9.10.4
	Excluded volume parameter	4.2
	$v_2{}^*$ = critical volume concentration	7.5.1
	v_f = specific free volume	8.6.1.1
	v_0 = occupied volume	8.6.1.2
	Velocity of chain pullout	11.5.2.2
w	Distance from source	3.6

SYMBOL	DEFINITION	SECTION
x	Mole fraction	4.3.6
	General parameter	A4.1
	Axial ratio of liquid crystalline molecule	7.5.1
y		
z	Charge on the polymer	3.10

Greek Alphabet

A		
B		
Γ		
Δ	Logarithmic decrement	8.12
E		
Z		
H		
Θ		
I		
K		
Λ		
M		
N		
Ξ		
O		
Π		
P		
Σ	$d\Sigma/d\Omega$ = scattering cross section	5.2.2.1
T		
Y		
Φ	Universal constant in intrinsic viscosity	3.8.3
X		
Ψ	Entropic factor	3.3.2
	Ψ_1 = constant	4.1.2
Ω	Number of possible arrangements in space	3.3.1.2
	Solid angle	5.2.2.1
	Probability of finding all the molecules	9.6
	Ω_1 = angular velocity	10.5.4
α	α_x = mechanically induced peak frequency shift	2.9
	Expansion of a polymer coil in a good solvent	3.8.2
	$\alpha_{A/B}$ = gas selectivity ratio	4.4.6.2

SYMBOL	DEFINITION	SECTION
	Volumetric coefficient of expansion	8.3
	α_R = cubic expansion coefficient in the rubbery state	8.6.1.1
	α_G = cubic expansion coefficient in the glassy state	8.6.1.1
	α_f = expansion coefficient of the free volume	8.6.1.2
	Extension ratio	9.4
β	β_1 = lattice constant	3.3.2
	Compressibility	8.1.4
	β_f = compressibility free volume	8.11
	Gaussian distribution term	9.7.1
γ	Number of flexible bonds per mer	8.6.3.2
	$\dot{\gamma}$ = shear rate	10.5.1
	γ_s = surface tension (intrinsic surface energy)	11.3.1
	γ_p = plastic deformation energy	11.3.2
	Surface tension	12.2.1
	$\gamma(r)$ = Debye correlation function	12.3.8.1
δ	Solubility parameter	3.2
	Measure of internal structure	6.6.2.2
	$\tan \delta$ = loss tangent	8.2.9
ε	ε^* = van der Waals energy of interaction	4.3.4
	Tensile strain	8.1.1.1
	ε', ε'' = dielectric storage and loss constants	8.3.4, 14.7.1
ζ		
η	Viscosity of a solution	3.8.1
	η_0 = viscosity of the solvent	3.8.1
	η_{rel} = relative viscosity	3.8.1
	η_{sp} = specific viscosity	3.8.1
	$[\eta]$ = intrinsic viscosity	3.8.1
	Viscosity (of a polymer melt)	8.1.1.2
	η_g = melt viscosity at T_g	8.6.1.2
	η_2, η_3, etc. = dashpot viscosities	10.1.2.2
	η', η'' = storage and loss viscosities	10.5.3
	η^* = complex viscosity	10.5.3
θ	Flory θ-temperature	3.3.2
	Angle of scatter	3.6.1
ι		
κ		
λ	Wavelength	3.6, 6.2.2
	λ^* = chain deformation, phantom network	9.10.6
	Volume element in the Takayanagi model	10.1.2.3

SYMBOL	DEFINITION	SECTION
μ	Magnetic moment	2.2.4
	Number of network junctions	9.10.2
	μ_{tube} = tube mobility of a chain	10.4.2.4
	μ_1 = constant	10.4.2.4
	μ_0 = molecular friction coefficient	11.5.2.2
ν	Frequency	2.2.4
	Kinetic chain length	3.7.2.2
	Poisson's ratio	8.1.1.2
	$\bar{\nu}$ = kinematic viscosity	10.5.4
ξ	Screening length	4.2
	Mesh size	12.7.1
o		
π	Osmotic pressure	3.5.2, 3.5.3
	$\pi_1 = 3.1416$	3.6
ρ	Density	3.2.2, 6.5.4
	ρ_e = electron density	3.6.1
	ρ = reduced density	4.3
σ	Stress	5.2.1, 8.1.1.1
	σ_b = tensile stress to break	1.2.1
	Surface free energy of crystals	6.6.2.3
	σ_R, σ_P = stress of the rubber and plastic components of the Takayanagi model	10.1.2.3
	σ_1, σ_2, σ_3 = components of triaxial stress	11.2.3
τ	NMR scale	2.3.4
	Turbidity	3.6.1
	Relaxation time (various subscripts)	5.4.1, 10.2.1, 11.5.3
υ		
ϕ	Volume fraction of polymer	4.2
	ϕ' = stability parameter	6.8.5
	Volume element in the Takayanagi model	10.1.2.3
	Phase volume fraction	12.3.8.1
χ	Flory–Huggins heat of mixing term (Sometimes written χ_1 or χ_{12})	3.3.2
	χ' = number of cross-links per gram	8.6.3.2
ψ		
ω	Angular frequency in radians	8.12, 10.2.4
	ω_g, ω_e = Carnot cycle work	9.8.3
	Sample-detector distance	5.2.2.1

INTRODUCTION
TO PHYSICAL
POLYMER SCIENCE

THIRD EDITION

1

INTRODUCTION TO
POLYMER SCIENCE

Polymer science was born in the great industrial laboratories of the world of the need to make and understand new kinds of plastics, rubber, adhesives, fibers, and paints. Only much later did polymer science come to academic life. Perhaps because of its origins, polymer science tends to be more interdisciplinary than most sciences, combining chemistry, chemical engineering, materials, and other fields as well.

Chemically, polymers are long-chain molecules of very high molecular weight, often measured in the hundreds of thousands. For this reason, the term "macromolecules" is frequently used when referring to polymeric materials. The trade literature sometimes refers to polymers as resins, an old term that goes back before the chemical structure of the long chains was understood.

The first polymers used were natural products, especially cotton, starch, proteins, and wool. Beginning early in the twentieth century, synthetic polymers were made. The first polymers of importance, Bakelite and nylon, showed the tremendous possibilities of the new materials. However, the scientists of that day realized that they did not understand many of the relationships between the chemical structures and the physical properties that resulted. The research that ensued forms the basis for physical polymer science.

This book develops the subject of physical polymer science, describing the interrelationships among polymer structure, morphology, and physical and mechanical behavior. Key aspects include molecular weight and molecular weight distribution, and the organization of the atoms down the polymer chain. Many polymers crystallize, and the size, shape, and organization of the crystallites depend on how the polymer was crystallized. Such effects as annealing are very important, as they have a profound influence on the final state of molecular organization.

Other polymers are amorphous, often because their chains are too irregular to permit regular packing. The onset of chain molecular motion heralds the glass transition and softening of the polymer from the glassy (plastic) state to the rubbery state. Mechanical behavior includes such basic aspects as modulus, stress relaxation, and elongation to break. Each of these is relatable to the polymer's basic molecular structure and history.

This chapter provides the student with a brief introduction to the broader field of polymer science. Although physical polymer science does not include polymer synthesis, some knowledge of how polymers are made is helpful in understanding configurational aspects, such as tacticity, which are concerned with how the atoms are organized along the chain. Similarly polymer molecular weights and distributions are controlled by the synthetic detail. This chapter starts at the beginning of polymer science, and it assumes no prior knowledge of the field.

1.1   FROM LITTLE MOLECULES TO BIG MOLECULES

The behavior of polymers represents a continuation of the behavior of smaller molecules at the limit of very high molecular weight. As a simple example, consider the normal alkane hydrocarbon series

$$
\begin{array}{ccc}
\underset{\text{Methane}}{\begin{array}{c} \mathrm{H} \\ | \\ \mathrm{H-C-H} \\ | \\ \mathrm{H} \end{array}} &
\underset{\text{Ethane}}{\begin{array}{c} \mathrm{H\ \ H} \\ |\ \ | \\ \mathrm{H-C-C-H} \\ |\ \ | \\ \mathrm{H\ \ H} \end{array}} &
\underset{\text{Propane}}{\begin{array}{c} \mathrm{H\ \ H\ \ H} \\ |\ \ |\ \ | \\ \mathrm{H-C-C-C-H} \\ |\ \ |\ \ | \\ \mathrm{H\ \ H\ \ H} \end{array}}
\end{array}
\tag{1.1}
$$

These compounds have the general structure

$$
\mathrm{H(\!-CH_2\!-)_{\!n}H} \tag{1.2}
$$

where the number of $-CH_2-$ groups, n, is allowed to increase up to several thousand. The progression of their state and properties is shown in Table 1.1.

At room temperature, the first four members of the series are gases. n-Pentane boils at 36.1°C and is a low-viscosity liquid. As the molecular weight of the series increases, the viscosity of the members increases. Although commercial gasolines contain many branched-chain materials and aromatics as well as straight-chain alkanes, the viscosity of gasoline is markedly lower than that of kerosene, motor oil, and grease because of its lower average chain length.

These latter materials are usually mixtures of several molecular species, although they are easily separable and identifiable. This point is important because most polymers are

Table 1.1   Properties of the alkane/polyethylene series

Number of Carbons in Chain	State and Properties of Material	Applications
1–4	Simple gas	Bottled gas for cooking
5–11	Simple liquid	Gasoline
9–16	Medium-viscosity liquid	Kerosene
16–25	High-viscosity liquid	Oil and grease
25–50	Crystalline solid	Paraffin wax candles
50–1000	Semicrystalline solid	Milk carton adhesives and coatings
1000–5000	Tough plastic solid	Polyethylene bottles and containers
$3-6 \times 10^5$	Fibers	Surgical gloves, bullet-proof vests

also "mixtures"; that is, they have a molecular weight distribution. In high polymers, however, it becomes difficult to separate each of the molecular species, and people talk about molecular weight averages.

Compositions of normal alkanes averaging more than about 20 to 25 carbon atoms are crystalline at room temperature. These are simple solids known as wax. It must be emphasized that at up to 50 carbon atoms the material is far from being polymeric in the ordinary sense of the term.

The polymeric alkanes with no side groups that contain 1000 to 3000 carbon atoms are known as polyethylenes. Polyethylene has the chemical structure

$$\mathrm{\left(CH_2 - CH_2\right)_n} \tag{1.3}$$

which originates from the structure of the monomer ethylene, $CH_2 = CH_2$. The quantity n is the number of mers—or monomeric units in the chain. In some places the structure is written

$$\mathrm{\left(CH_2\right)_{n'}} \tag{1.4}$$

or polymethylene. (Then $n' = 2n$.) The relationship of the latter structure to the alkane series is clearer. While true alkanes have CH_3 — as end groups, most polyethylenes have initiator residues.

Even at a chain length of thousands of carbons, the melting point of polyethylene is still slightly molecular-weight-dependent, but most linear polyethylenes have melting or fusion temperatures, T_f, near 140°C. The approach to the theoretical asymptote of about 145°C at infinite molecular weight (1) is illustrated schematically in Figure 1.1.

The greatest differences between polyethylene and wax lie in their mechanical behavior, however. While wax is a brittle solid, polyethylene is a tough plastic. Comparing resistance to break of a child's birthday candle with a wash bottle tip, both of about the same diameter, shows that the wash bottle tip can be repeatedly bent whereas the candle breaks on the first deformation.

Polyethylene is a tough plastic solid because its chains are long enough to connect individual stems together within a lamellar crystallite by chain folding (see Figure 1.2). The chains also wander between lamellae, connecting several of them together. These effects add strong covalent bond connections both within the lamellae and between

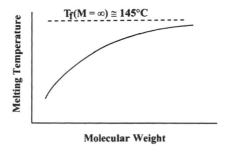

Molecular Weight

Figure 1.1 The molecular weight-melting temperature relationship for the alkane series. An asymptotic value of about 145°C is reached for very high molecular weight linear polyethylenes.

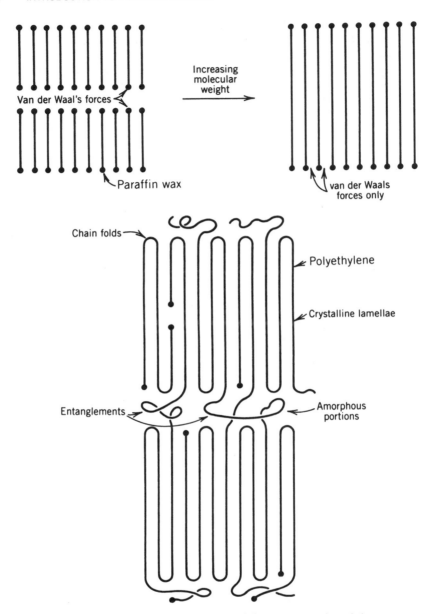

Figure 1.2 Comparison of wax and polyethylene structure and morphology.

them. On the other hand, only weak van der Waals forces hold the chains together in wax.

In addition a certain portion of polyethylene is amorphous. The chains in this portion are rubbery, imparting flexibility to the entire material. Wax is 100% crystalline, by difference.

The long chain length allows for entanglement (see Figure 1.3). The entanglements

Figure 1.3 Entanglement of polymer chains. (*a*) Low molecular weight, no entanglement. (*b*) High molecular weight, chains are entangled. The transition between the two is often at about 600 backbone chain atoms.

help hold the whole material together under stress. In the melt state, chain entanglements cause the viscosity to be raised very significantly also.

The long chains shown in Figure 1.3 also illustrate the coiling of polymer chains in the amorphous state. One of the most powerful theories in polymer science (2) states that the conformations of amorphous chains in space are random coils; that is, the directions of the chain portions are statistically determined.

1.2 MOLECULAR WEIGHT AND MOLECULAR WEIGHT DISTRIBUTIONS

While the exact molecular weight required for a substance to be called a polymer is a subject of continued debate, often polymer scientists put the number at about 25,000 g/mol. This is the minimum molecular weight required for good physical and mechanical properties for many important polymers. This molecular weight is also near the onset of entanglement.

1.2.1 Effect on Tensile Strength

The tensile strength of any material is defined as the stress at break during elongation, where stress has the units of Pa, dyn/cm^2, or lb/in^2; see Chapter 11. The effect of molecular weight on the tensile strength of polymers is illustrated in Figure 1.4. At very low molecular weights the tensile stress to break, σ_b, is near zero. As the molecular weight increases, the tensile strength increases rapidly, and then gradually levels off. Since a major point of weakness at the molecular level involves the chain ends, which do not transmit the covalent bond strength, it is predicted that the tensile strength reaches an asymptotic value at infinite molecular weight. A large part of the curve in Figure 1.4 can be expressed (3,4)

$$\sigma_b = A - \frac{B}{M_n} \tag{1.5}$$

where M_n is the number-average molecular weight (see below) and A and B are constants. Newer theories by Wool (3) and others suggest that more than 90% of tensile strength and other mechanical properties are attained when the chain reaches eight entanglements in length.

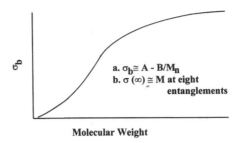

Figure 1.4 Effect of polymer molecular weight on tensile strength.

1.2.2 Molecular Weight Averages

The same polymer from different sources may have different molecular weights. Thus polyethylene from source A may have a molecular weight of 150,000 g/mol, whereas polyethylene from source B may have a molecular weight of 400,000 g/mol (see Figure 1.5). To compound the difficulty, all common synthetic polymers and most natural polymers (except proteins) have a distribution in molecular weights. That is, some molecules in a given sample of polyethylene are larger than others. The differences result directly from the kinetics of polymerization.

However, these facts led to much confusion for chemists early in the twentieth century. At that time chemists were able to understand and characterize small molecules. Compounds such as hexane all have six carbon atoms. If polyethylene with 2430 carbon atoms were declared to be "polyethylene," how could that component having 5280 carbon atoms also be polyethylene? How could two sources of the material having different average molecular weights both be polyethylene, noting A and B in Figure 1.5?

The answer to these questions lies in defining average molecular weights and molecular weight distributions (5,6). The two most important molecular weight averages are the number-average molecular weight, M_n,

$$M_n = \frac{\sum_i N_i M_i}{\sum_i N_i} \tag{1.6}$$

where N_i is the number of molecules of molecular weight M_i, and the weight-average

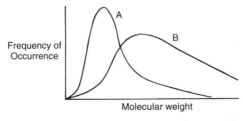

Figure 1.5 Molecular weight distributions of the same polymer from two different sources, A and B.

molecular weight, M_w)

$$M_w = \frac{\sum_i N_i M_i^2}{\sum_i N_i M_i} \tag{1.7}$$

For single-peaked distributions, M_n is usually near the peak. The weight-average molecular weight is always larger. For simple distributions, M_w may be 1.5 to 2.0 times M_n. The ratio M_w/M_n, sometimes called the polydispersity index, provides a simple definition of the molecular weight distribution. Thus all compositions of $+CH_2-CH_2+_n$ are called polyethylene, the molecular weights being specified for each specimen.

For many polymers a narrower molecular distribution yields better properties. The low end of the distribution may act as a plasticizer, softening the material. Certainly it does not contribute as much to the tensile strength. The high-molecular-weight tail increases processing difficulties, because of its enormous contribution to the melt viscosity. For these reasons, great emphasis is placed on characterizing polymer molecular weights.

1.3 MAJOR POLYMER TRANSITIONS

Polymer crystallinity and melting were discussed previously. Crystallization is an example of a first-order transition, in this case liquid to solid. Most small molecules crystallize, an example being water to ice. Thus this transition is very familiar.

A less classical transition is the glass–rubber transition in polymers. At the glass transition temperature, T_g, the amorphous portions of a polymer soften. The most familiar example is ordinary window glass, which softens and flows at elevated temperatures. Yet glass is not crystalline, but rather it is an amorphous solid. It should be pointed out that many polymers are totally amorphous. Carried out under ideal conditions, the glass transition is a type of second-order transition.

The basis for the glass transition is the onset of coordinated molecular motion in the polymer chain. At low temperatures, only vibrational motions are possible, and the polymer is hard and glassy (Figure 1.6, region 1) (7). In the glass transition region, region 2, the polymer softens, the modulus drops three orders of magnitude, and the material becomes rubbery. Regions 3, 4, and 5 are called the rubbery plateau, the rubbery flow, and the viscous flow regions, respectively. Examples of each region are shown in Table 1.2.

Depending on the region of viscoelastic behavior, the mechanical properties of polymers differ greatly. Model stress–strain behavior is illustrated in Figure 1.7 for regions 1, 2, and 3. Glassy polymers are stiff and often brittle, breaking after only a few percent extension. Polymers in the glass transition region are more extensible, sometimes exhibiting a yield point (the hump in the tough plastic stress–strain curve). If the polymer is above its brittle–ductile transition, Section 11.2.3, rubber-toughened, Chapter 13, or semicrystalline with its amorphous portions above T_g, tough plastic behavior will also be observed. Polymers in the rubbery plateau region are highly elastic, often stretching to 500% or more. Regions 1, 2, and 3 will be discussed further in Chapters 8 and 9. Regions 4 and 5 flow to increasing extents under stress; see Chapter 10.

Crosslinked amorphous polymers above their glass transition temperature behave rubbery. Examples are rubber bands and automotive tire rubber. In general, Young's modulus of elastomers in the rubbery-plateau region is higher than the corresponding

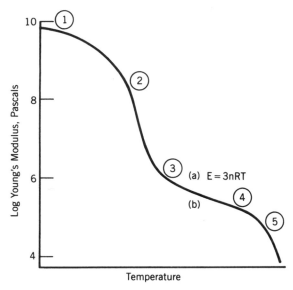

Figure 1.6 Idealized modulus–temperature behavior of an amorphous polymer. Young's modulus, stress/strain, is a measure of stiffness.

Table 1.2 Typical polymer viscoelastic behavior at room temperature (7a)

Region	Polymer	Application
Glassy	Poly(methyl methacrylate)	Plastic
Glass transition	Poly(vinyl acetate)	Latex paint
Rubbery plateau	*Cross*-poly(butadiene–*stat*–styrene)	Rubber bands
Rubbery flow	Chicle[a]	Chewing gum
Viscous flow	Poly(dimethyl siloxane)	Lubricant

[a]From the latex of *Achras sapota*, a mixture of *cis*- and *trans*-polyisoprene plus polysaccharides.

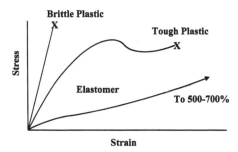

Figure 1.7 Stress-strain behavior of various polymers. While the initial slope yields the modulus, the area under the curve provides the energy to fracture.

Table 1.3   Examples of polymers at room temperature by transition behavior

	Crystalline	Amorphous
Above T_g	Polyethylene	Natural rubber
Below T_g	Cellulose	Poly(methyl methacrylate)

linear polymers, and is governed by the relation $E = 3nRT$, in Figure 1.6 (line not shown); the linear polymer behavior is illustrated by the line (*b*). Here, *n* represents the number of chain segments bound at both ends in a network, per unit volume. The quantities *R* and *T* are the gas constant and the absolute temperature, respectively.

Polymers may also be partly crystalline. The remaining portion of the polymer, the amorphous material, may be above or below its glass transition temperature, creating four subclasses of materials. Table 1.3 gives a common example of each. While polyethylene and natural rubber need no further introduction, common names for processed cellulose are rayon and cellophane. Cotton is nearly pure cellulose, and wood pulp for paper is 80–90% cellulose. A well-known trade name for poly(methyl methacrylate) is Plexiglas®. The modulus–temperature behavior of polymers in either the rubbery-plateau region or in the semicrystalline region are illustrated further in Figure 8.2, Chapter 8.

Actually there are two regions of modulus for semicrystalline polymers. If the amorphous portion is above T_g, then the modulus is generally between rubbery and glassy. If the amorphous portion is glassy, then the polymer may actually be a bit stiffer than expected for a 100% glassy polymer.

1.4   POLYMER SYNTHESIS AND STRUCTURE

1.4.1   Chain Polymerization

Polymers may be synthesized by two major kinetic schemes, chain and stepwise polymerization. The most important of the chain polymerization methods is called free radical polymerization.

1.4.1.1   Free Radical Polymerization   The synthesis of poly(ethyl acrylate) will be used as an example of free radical polymerization. Benzoyl peroxide is a common initiator. Free radical polymerization has three major kinetic steps—initiation, propagation, and termination.

1.4.1.2   Initiation   On heating, benzoyl peroxide decomposes to give two free radicals:

Benzoyl peroxide   Free radical, R ·

(1.8)

In this reaction the electrons in the oxygen–oxygen bond are unpaired and become the active site. With R representing a generalized organic chemical group, the free radical can be written R· . (It should be pointed out that hydrogen peroxide undergoes the same reaction on a wound, giving a burning sensation as the free radicals "kill the germs.")

The initiation step usually includes the addition of the first monomer molecule:

$$
\underset{\text{Free radical}}{R\cdot} + \underset{\text{Ethyl acrylate}}{CH_2=\overset{\displaystyle H}{\underset{\displaystyle O=C-O-C_2H_5}{C}}} \longrightarrow \underset{\text{Growing chain}}{R-CH_2-\overset{\displaystyle H}{\underset{\displaystyle O=C-O-C_2H_5}{C\cdot}}}
\tag{1.9}
$$

In this reaction the free radical attacks the monomer and adds to it. The double bond is broken open, and the free radical reappears at the far end.

1.4.1.3 Propagation After initiation reactions (1.8) and (1.9), many monomer molecules are added rapidly, perhaps in a fraction of a second:

$$
R-CH_2-\overset{\displaystyle H}{\underset{\displaystyle O=C-O-C_2H_5}{C\cdot}} + nCH_2=\overset{\displaystyle H}{\underset{\displaystyle O=C-O-C_2H_5}{C}} \longrightarrow
$$

$$
R-CH_2-\overset{\displaystyle H}{\underset{\displaystyle \underset{\displaystyle H_5C_2-O}{O=C}}{C}} \left(CH_2-\overset{\displaystyle H}{\underset{\displaystyle \underset{\displaystyle O-C_2H_5}{O=C}}{C}}\right)_{n-1} CH_2-\overset{\displaystyle H}{\underset{\displaystyle \underset{\displaystyle O-C_2H_5}{O=C}}{C\cdot}}
\tag{1.10}
$$

On the addition of each monomer, the free radical moves to the end of the chain.

1.4.1.4 Termination In the termination reaction, two free radicals react with each other. Termination is either by combination,

$$
2R-CH_2-\overset{\displaystyle H}{\underset{\displaystyle O=C-O-C_2H_5}{C\cdot}} \longrightarrow R-CH_2-\overset{\displaystyle H}{\underset{\displaystyle \underset{\displaystyle O-C_2H_5}{O=C}}{C}}-\overset{\displaystyle H}{\underset{\displaystyle O=C-O-C_2H_5}{C}}-CH_2-R
\tag{1.11}
$$

where R now represents a long-chain portion, or by disproportionation, where a hydrogen is transferred from one chain to the other. This latter result produces in two final chains. While the normal mode of addition is a head-to-tail reaction (1.10), this termination step is normally head-to-head.

As a homopolymer, poly(ethyl acrylate) is widely used as an elastomer or adhesive, being a polymer with a low T_g, $-22°C$. As a copolymer with other acrylics it is used as a latex paint.

1.4.1.5 Structure and Nomenclature The principal method of polymerizing monomers by the chain kinetic scheme involves the opening of double bonds to form a linear molecule. In a reacting mixture, monomer, fully reacted polymer, and only a small amount of rapidly reacting species are present. Once the polymer terminates, it is "dead" and cannot react further by the synthesis scheme outlined previously.

Polymers are named by rules laid out by the IUPAC Nomenclature Committee (8,9). For many simple polymers the source-based name utilizes the monomer name prefixed by "poly." If the monomer name has two or more words, parentheses are placed around the monomer name. Thus, in the above, the monomer ethyl acrylate is polymerized to make poly(ethyl acrylate).

Table 1.4 provides a selected list of common chain polymer structures and names along with comments as to how the polymers are used. The "vinyl" monomers are characterized by the general structure $CH_2=CHR$, where R represents any side group. One of the best-known vinyl polymers is poly(vinyl chloride), where R is $-Cl$.

Polyethylene and polypropylene are the major members of the class of polymers known as *polyolefins*; see Section 14.1. The term *olefin* derives from the double-bond characteristic of the alkene series.

A slight dichotomy exists in the writing of vinyl polymer structures. From a correct nomenclature point of view, the pendant moiety appears on the left-hand carbon. Thus poly(vinyl chloride) should be written $+CHCl-CH_2+_n$. However, from a synthesis point of view, the structure is written $+CH_2-CHCl+_n$, because the free radical is borne on the pendant moiety carbon. Thus both forms appear in the literature.

The diene monomer has the general structure $CH_2=CR-CH=CH_2$, where on polymerization one of the double bonds forms the chain bonds, and the other goes to the central position. The vinylidenes have two groups on one carbon. Table 1.4 also lists some common copolymers, which are formed by reacting two or more monomers together. In general, the polymer structure most closely resembling the monomer structure will be presented herein.

Today, recycling of plastics has become paramount in preserving the environment. On the bottom of plastic bottles and other plastic items is an identification number and letters; see Table 1.5. This information serves to help in separation of the plastics prior to recycling. Observation of the properties of the plastic such as modulus, together with the identification, will help the student understand the kinds and properties of the plastics in common service.

1.4.2 Step Polymerization

1.4.2.1 A Polyester Condensation Reaction The second important kinetic scheme is step polymerization. As an example of a step polymerization, the synthesis of a polyester is given.

The general reaction to form esters starts with an acid and an alcohol:

$$CH_3-CH_2OH + CH_3-\overset{\overset{\displaystyle O}{\|}}{C}-OH \rightarrow CH_3-CH_2-O-\overset{\overset{\displaystyle O}{\|}}{C}-CH_3 + H_2O \tag{1.12}$$

Ethyl alcohol   Acetic acid   Ethyl acetate   Water

where the ester group is $-O-\overset{\overset{\displaystyle O}{\|}}{C}-$, and water is eliminated.

Table 1.4 Selected chain polymer structures and nomenclature

Structure	Name	Where Used
$\left(\!\!-CH_2-\!\!\underset{\underset{R}{\displaystyle\vert}}{CH}\!-\!\right)_{\!n}$	"Vinyl" class	
$R = -H$	Polyethylene	Plastic
$R = -CH_3$	Polypropylene	Rope
$R = -\langle\bigcirc\rangle$	Polystyrene	Drinking cups
$R = -Cl$	Poly(vinyl chloride)	"Vinyl", water pipes
$R = -O-\overset{\overset{\displaystyle O}{\|}}{C}-CH_3$	Poly(vinyl acetate)	Latex paints
$R = -OH$	Poly(vinyl alcohol)	Fiber
$\left(\!\!-CH_2-\!\!\underset{\underset{O=C-O-R}{\displaystyle\vert}}{\overset{\overset{\displaystyle X}{\vert}}{C}}\!-\!\right)_{\!n}$	$X = -H$, acrylics $X = -CH_3$, methacrylics	
$X = -H, R = -C_2H_5$	Poly(ethyl acrylate)	Latex paints
$X = -CH_3, R = -CH_3$	Poly(methyl methacrylate)	Plexiglas®
$X = -CH_3, R = C_2H_5$	Poly(ethyl methacrylate)	Adhesives
$\left(\!\!-CH_2-\!\!\underset{\underset{C\equiv N}{\displaystyle\vert}}{\overset{\overset{\displaystyle H}{\vert}}{C}}\!-\!\right)_{\!n}$	Polyacrylonitrile[a]	Orlon®
$\left(\!\!-CH_2-\!\!\underset{\underset{R}{\displaystyle\vert}}{C}\!=\!CH-CH_2\!-\!\right)_{\!n}^{a'}$	"Diene" class	
$R = -H$	Polybutadiene	Tires
$R = -CH_3$	Polyisoprene	Natural rubber
$R = -Cl$	Polychloroprene	Neoprene
$\left(\!\!-CX_2-CR_2\!-\!\right)_{\!n}$	Vinylidenes	
$X = -H, R = -F$	Poly(vinylidene fluoride)	Plastic
$X = -F, R = -F$	Polytetrafluoroethylene	Teflon®
$X = -H, R = -CH_3$	Polyisobutene[b]	Elastomer

Common Copolymers

EPDM	Ethylene–propylene–diene–monomer	Elastomer
SBR	Styrene–butadiene–rubber Poly(styrene–*stat*–butadiene)[c]	Tire rubber
NBR	Acrylonitrile–butadiene–rubber Poly(acrylonitrile–*stat*–butadiene)	Elastomer
ABS	Acrylonitrile–butadiene–styrene[d]	Plastic

[a] Polyacrylonitrile is technically a number of the acrylic class because it forms acrylic acid on hydrolysis.

[a'] IUPAC recommends $\left(\!\!-\underset{\underset{R}{\displaystyle\vert}}{C}\!=\!CH-CH_2-CH_2\!-\!\right)_{\!n}$

[b] Also called polyisobutylene. The 2% copolymer with isoprene, after vulcanization, is called butyl rubber.

[c] The term–*stat*–means statistical, as explained in Chapter 2.

[d] ABS is actually a blend or graft of two random copolymers, poly(acrylonitrile–*stat*–butadiene), and poly(acrylonitrile–*stat*–styrene).

Table 1.6 The plastics identification code

Code	Letter I.D.	Polymer Name
♲1	PETE	Poly(ethylene terephthalate)
♲2	HDPE	High-density polyethylene
♲3	V	Poly(vinyl chloride)
♲4	LDPE	Low-density polyethylene
♲5	PP	Polypropylene
♲6	PS	Polystyrene
♲7	Other	Different polymers

Source: From the *Plastic Container Code System*, The Plastic Bottle Information Bureau, Washington, DC.

The chemicals above cannot form a polyester because they have only one functional group each. When the two reactants each have bifunctionality, a linear polymer is formed:

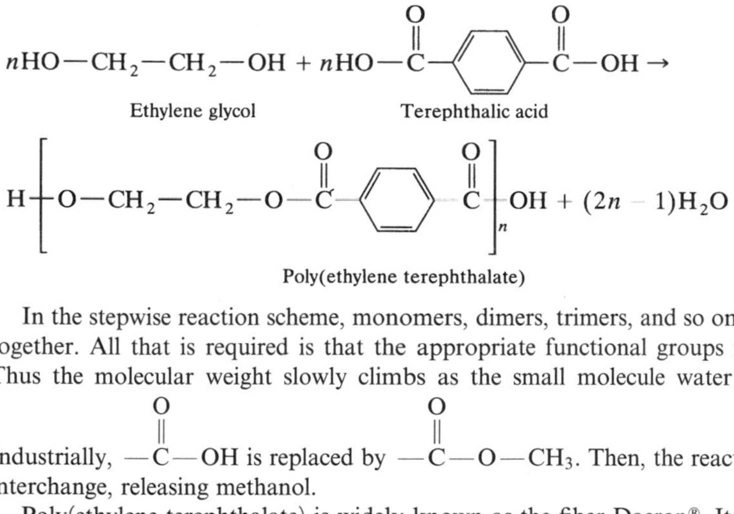

$$n\,HO-CH_2-CH_2-OH + n\,HO-\overset{\overset{O}{\|}}{C}-\langle\bigcirc\rangle-\overset{\overset{O}{\|}}{C}-OH \rightarrow$$

Ethylene glycol   Terephthalic acid

(1.13)

$$H\left[O-CH_2-CH_2-O-\overset{\overset{O}{\|}}{C}-\langle\bigcirc\rangle-\overset{\overset{O}{\|}}{C}\right]_n OH + (2n-1)H_2O$$

Poly(ethylene terephthalate)

In the stepwise reaction scheme, monomers, dimers, trimers, and so on, may all react together. All that is required is that the appropriate functional groups meet in space. Thus the molecular weight slowly climbs as the small molecule water is eliminated. Industrially, $-\overset{\overset{O}{\|}}{C}-OH$ is replaced by $-\overset{\overset{O}{\|}}{C}-O-CH_3$. Then, the reaction is an ester interchange, releasing methanol.

Poly(ethylene terephthalate) is widely known as the fiber Dacron®. It is highly crystalline, with a melting temperature of about +265°C.

Another well-known series of polymers made by step polymerization reactions is the polyamides, known widely as the nylons. In fact there are two series of nylons. In the first series, the monomer has an amine at one end of the molecule and a carboxyl at the other. For example,

$$n\,H_2N-CH_2-CH_2-CH_2-\overset{\overset{O}{\|}}{C}-OH \rightarrow$$

(1.14)

$$H\left(\overset{\overset{H}{|}}{N}-CH_2-CH_2-CH_2-\overset{\overset{O}{\|}}{C}\right)_n OH + (n-1)H_2O$$

which is known as nylon 4. The number 4 indicates the number of carbon atoms in the mer.
In the second series, a dicarboxylic acid is reacted with a diamine:

$$n\text{H}_2\text{N}(\text{CH}_2)_4\text{NH}_2 + n\text{H}-\text{O}-\overset{\overset{\text{O}}{\|}}{\text{C}}(\text{CH}_2)_6\overset{\overset{\text{O}}{\|}}{\text{C}}-\text{OH} \longrightarrow$$

$$\text{H}\overset{}{\left[\!\!\!\begin{array}{c}\overset{\text{H}}{\underset{|}{\text{N}}}-(\text{CH}_2)_4-\overset{\text{H}}{\underset{|}{\text{N}}}-\overset{\overset{\text{O}}{\|}}{\text{C}}-(\text{CH}_2)_6\overset{\overset{\text{O}}{\|}}{\text{C}}\end{array}\!\!\!\right]_n}\text{OH} + (2n-1)\text{H}_2\text{O}$$

(1.15)

which is named nylon 48. Note that the amine carbon number is written first, and the
acid carbon number second. For reaction purposes, acyl chlorides are frequently substituted for the carboxyl groups. An excellent demonstration experiment is described by
Morgan and Kwolek (10), called the nylon rope trick.

1.4.2.2 Stepwise Nomenclature and Structures   Table 1.6 names some of the
more important stepwise polymers. The polyesters have already been mentioned. The
nylons are known technically as polyamides. There are two important subseries of
nylons, where amine and the carboxylic acid are on different monomer molecules (thus
requiring both monomers to make the polymer) or one each on the ends of the same
monomer molecule. These are numbered by the number of carbons present in the
monomer species. It must be mentioned that the proteins are also polyamides.

Other classes of polymers mentioned in Table 1.6 include the polyurethanes, widely
used as elastomers; the silicones, also elastomeric; and the cellulosics, used in fibers and
plastics. Cellulose is a natural product.

Another class of polymers are the polyethers, prepared by ring-opening reactions.
The most important member of this series is poly(ethylene oxide),

$$+\!\text{CH}_2-\text{CH}_2-\text{O}\!+_n$$

Because of the oxygen atom, poly(ethylene oxide) is water soluble.

To summarize the material in Table 1.6, the major stepwise polymer classes contain
the following identifying groups:

Polyesters $\quad\quad -\overset{\overset{\text{O}}{\|}}{\text{C}}-\text{O}-$

Polyamides $\quad\quad -\overset{\text{H}}{\underset{|}{\text{N}}}-\overset{\overset{\text{O}}{\|}}{\text{C}}-$

Polyurethanes $\quad -\overset{\text{H}}{\underset{|}{\text{N}}}-\overset{\overset{\text{O}}{\|}}{\text{C}}-\text{O}-$

Silicones $\quad\quad -\overset{\text{CH}_3}{\underset{\underset{\text{CH}_3}{|}}{\overset{|}{\text{Si}}}}-\text{O}-$

Epoxy resins $\quad -\overset{\text{H}\quad\text{H}}{\underset{\underset{\text{O}}{\diagdown\diagup}}{\text{C}-\text{C}}}- \longrightarrow -\text{CH}_2-\text{CH}_2-\text{O}-\text{R}\diagup^{\diagdown}$

Polyethers $\quad\quad -\text{O}-$

Table 1.6   Selected stepwise structures and nomenclature

Structure[a]	Name	Where Known
$\left[O-CH_2-CH_2-O-\overset{O}{\overset{\|}{C}}-\bigcirc-\overset{O}{\overset{\|}{C}}\right]_n$	Poly(ethylene terephthalate)	Dacron®
$\left[\overset{H}{N}(CH_2)_6\overset{H}{N}-\overset{O}{\overset{\|}{C}}(CH_2)_8\overset{O}{\overset{\|}{C}}\right]_n$	Poly(hexamethylene sebacamide)	Polyamide 610[b]
$\left[\overset{H}{N}-\overset{O}{\overset{\|}{C}}(CH_2)_{5,}\right]_n$	Polycaprolactam	Polyamide 6
$\left[O(CH_2)_4\right]_n$	Polytetrahydrofuran	Polyether
$\left[\left(O(CH_2)_4\right)_m\overset{}{\underset{H}{N}}-\overset{O}{\overset{\|}{C}}\right]_n$	Polyurethane[c]	Spandex Lycra®
$\left(O-\overset{CH_3}{\underset{CH_3}{Si}}\right)_n$	Poly(dimethyl siloxane)	Silicone rubber
$\left(O-\bigcirc-\overset{CH_3}{\underset{CH_3}{C}}-\bigcirc-O-\overset{O}{\overset{\|}{C}}\right)_n$	Polycarbonate	Lexan®
(cellulose chair structure)	Cellulose	Cotton
$\overset{O}{\overset{/\backslash}{H_2C-CH}}-R-\overset{O}{\overset{/\backslash}{CH-CH_2}} \longrightarrow$ $-R'-O\left(CH_2-\overset{OH}{\underset{}{CH}}-R-CH_2-CH_2-O\right)_n R''-$	Epoxy resins	Epon®

[a]Some people see the mer structure in the third row more clearly with

$$\left(\overset{H}{N}(CH_2)_5\overset{O}{\overset{\|}{C}}\right)_n$$

Some other step polymerization mers can also be drawn in two or more different ways. The student should learn to recognize the structures in different ways.

[b]The "6" refers to the number of carbons in the diamine portion, and the "10" to the number of carbons in the diacid. An old name is nylon 610.

[c]The urethane group usually links polyether or polyester low molecular weight polymers together.

Table 1.7   Some natural product polymers

Name	Source	Application
Cellulose	Wood, cotton	Paper, clothing, rayon, cellophane
Starch	Potatoes, corn	Food, thickener
Wool	Sheep	Clothing
Silk	Silkworm	Clothing
Natural rubber	Rubber tree	Tires
Pitch	Oil deposits	Coating, roads

1.4.2.3   *Natural Product Polymers*   Living organisms make many polymers, nature's best. Most such natural polymers strongly resemble step-polymerized materials. However, living organisms make their polymers enzymatically, the structure ultimately being controlled by DNA, itself a polymer.

Some of the more important commercial natural polymers are shown in Table 1.7. People sometimes refer to these polymers as natural products or renewable resources.

Wool and silk are both proteins. All proteins are actually copolymers of polyamide-2 (or nylon-2, old terminology). As made by plants and animals, however, the copolymers are highly ordered, and they have monodisperse molecular weights, meaning that all the chains have the same molecular weights.

Cellulose and starch are both polysaccharides, being composed of chains of glucose-based rings but bonded differently. Their structures are discussed further in Appendix 2.1.

Natural rubber, the hydrocarbon polyisoprene, more closely resembles chain polymerized materials. In fact synthetic polyisoprene can be made either by free radical polymerization or anionic polymerization. The natural and synthetic products compete commercially with each other.

Pitch, a decomposition product, usually contains a variety of aliphatic and aromatic hydrocarbons, some of very high molecular weight.

1.5   CROSS-LINKING, PLASTICIZERS, AND FILLERS

The above provides a brief introduction to simple homopolymers, as made pure. Only a few of these are finally sold as "pure" polymers, such as polystyrene drinking cups and polyethylene films. Much more often, polymers are sold with various additives. That the student may better recognize the polymers, the most important additives are briefly discussed.

On heating, linear polymers flow and are termed thermoplastics. To prevent flow, polymers are sometimes cross-linked (●):

$$(1.16)$$

The cross-linking of rubber with sulfur is called vulcanization. Cross-linking bonds the chains together to form a network. The resulting product is called a thermoset, because it does not flow on heating.

Plasticizers are small molecules added to soften a polymer by lowering its glass transition temperature or reducing its crystallinity or melting temperature. The most widely plasticized polymer is poly(vinyl chloride). The distinctive odor of new "vinyl" shower curtains is caused by the plasticizer, for example.

Fillers may be of two types, reinforcing and nonreinforcing. Common reinforcing fillers are the silicas and carbon blacks. The latter are most widely used in automotive tires to improve wear characteristics such as abrasion resistance. Nonreinforcing fillers, such as calcium carbonate, may provide color or opacity or may merely lower the price of the final product.

1.6 THE MACROMOLECULAR HYPOTHESIS

In the nineteenth century, the structure of polymers was almost entirely unknown. The Germans called it *Schmierenchemie*, meaning grease chemistry (11), but a better translation might be "the gunk at the bottom of the flask," that portion of an organic reaction that did not result in characterizable products. In the nineteenth century and early twentieth century the field polymers and the field of colloids were considered integral parts of the same field. Wolfgang Ostwald declared in 1917 (12):

> All those sticky, mucilaginous, resinous, tarry masses which refuse to crystallize, and which are the abomination of the normal organic chemist; those substances which he carefully sets toward the back of his cupboard . . . , just these are the substances which are the delight of the colloid chemist.

Indeed, those old organic colloids (now polymers) and inorganic colloids such as soap micelles and silver or sulfur sols have much in common (11):

1. Both types of particles are relatively large, 10^{-6} to 10^{-4} mm, and visible via ultramicroscopy[†] as dancing light flashes, that is, Brownian motion.
2. The elemental composition does not change with the size of the particle.

Thus, soap micelles (true aggregates) and polymer chains (which repeat the same structure but are covalently bonded) appeared the same in those days. Partial valences (see Section 6.12) seemed to explain the bonding in both types.

In 1920 Herman Staudinger (13,14) enunciated the *Macromolecular Hypothesis*. It states that certain kinds of these colloids actually consist of very long-chained molecules. These came to be called polymers because many (but not all) were composed of the same repeating unit, or mer. In 1953 Staudinger won the Nobel prize in chemistry for his discoveries in the chemistry of macromolecular substances (15). The Macromolecular Hypothesis is the origin of modern polymer science, leading to our current understanding of how and why such materials as plastics and rubber have the properties they do.

[†] Ultramicroscopy is an old method used to study very small particles dispersed in a fluid for examination, and below normal resolution. Although invisible in ordinary light, colloidal particles become visible when intensely side-illuminated against a dark background.

1.7   HISTORICAL DEVELOPMENT OF INDUSTRIAL POLYMERS

Like most other technological developments, polymers were first used on an empirical basis, with only a very incomplete understanding of the relationships between structure and properties. The first polymers used were natural products that date back to antiquity, including wood, leather, cotton, various grasses for fibers, papermaking, and construction, wool, and protein animal products boiled down to make glues and related material.

Then came several semisynthetic polymers, which were natural polymers modified in some way. One of the first to attain commercial importance was cellulose nitrate plasticized with camphor, popular around 1885 for stiff collars and cuffs as celluloid, later most notably used in Thomas Edison's motion picture film (11). Cellulose nitrates were also sold as lacquers, used to coat wooden staircases, and so on. The problem was the terrible fire hazard existing with the nitrates, which were later replaced by the acetates.

Other early polymer materials included Chardonnet's artificial silk, made by regenerating and spinning cellulose nitrate solutions, eventually leading to the viscose process for making rayon (see Section 6.10) still in use today.

The first truly synthetic polymer was a densely cross-linked material based on the reaction of phenol and formaldehyde; see Section 14.2. The product, called Bakelite, was manufactured from 1910 onward for applications ranging from electrical appliances to phonograph records (16,17). Another early material was the General Electric Company's Glyptal, based on the condensation reaction of glycerol and phthalic anhydride (18), which followed shortly after Bakelite. However, very little was known about the actual chemical structure of these polymers until after Staudinger enunciated the Macromolecular Hypothesis in 1920.

All of these materials were made on a more or less empirical basis; trial and error have been the basis for very many advances in history, including polymers. However, in the late 1920s and 1930s, a DuPont chemist by the name of Wallace Carothers succeeded in establishing the reality of the Macromolecular Hypothesis by bringing the organic-structural approach back to the study of polymers, resulting in the discovery of nylon and neoprene. Actually the first polymers that Carothers discovered were polyesters (19). He reasoned that if the Macromolecular Hypothesis was correct, then if one mixed a molecule with dihydroxide end groups with a another molecule with diacid end groups and allowed them to react, a long, linear chain should result if the stoichiometry was one-to-one.

The problem with the aliphatic polyesters made at that time was their low melting point, making them unsuitable for clothing fibers because of hot water washes and ironing. When the ester groups were replaced with the higher melting amide groups, the nylon series was born. In the same time frame, Carothers discovered neoprene, which was chain-polymerized product of an isoprene-like monomer with a chlorine replacing the methyl group.

Bakelite was a thermoset; that is, it did not flow after the synthesis was complete (20). The first synthetic thermoplastics, materials that could flow on heating, were poly(vinyl chloride), poly(styrene–*stat*–butadiene), polystyrene, and polyamide 66; see Table 1.8 (20). Other breakthrough polymers have included the very high modulus aromatic polyamides, known as Kevlar™ (see Section 7.4), and a host of high temperature polymers.

Further items on the history of polymer science can be found in Appendix 5.1, and Sections 6.1.1 and 6.12.

Table 1.8 Commercialization dates of selected synthetic polymers (20)

Year	Polymer	Producer
1909	Poly(phenol–co–formaldehyde)	General Bakelite Corporation
1927	Poly(vinyl chloride)	B.F. Goodrich
1929	Poly(styrene–stat–butadiene)	I.G. Farben
1930	Polystyrene	I.G. Farben/Dow
1936	Poly(methyl methacrylate)	Rohm and Haas
1936	Nylon 66 (Polyamide 66)	DuPont
1936	Neoprene (chloroprene)	DuPont
1939	Polyethylene	ICI
1943	Poly(dimethyl siloxane)	Dow Corning
1954	Poly(ethylene terephthalate)	ICI
1960	Poly(p-phenylene terephthalamide)[a]	DuPont
1982	Polyetherimide	GEC

[a] Kevlar; see Chapter 7.

1.8 MOLECULAR ENGINEERING

The discussion above shows that polymer science is an admixture of pure and applied science. The structure, molecular weight, and shape of the polymer molecule are all closely tied to the physical and mechanical properties of the final material.

This book emphasizes physical polymer science, the science of the interrelationships between polymer structure and properties. Although much of the material (except the polymer syntheses) is developed in greater detail in the remaining chapters, the intent of this chapter is to provide an overview of the subject and a simple recognition of polymers as encountered in everyday life. In addition to the books in the General Reading section, a listing of handbooks and encyclopedias is given at the end of this chapter.

REFERENCES

1. L. Mandelkern and G. M. Stack, *Macromolecules*, **17**, 87 (1984).

2. P. J. Flory, *Principles of Polymer Chemistry*, Cornell University, Ithaca, NY, 1953.

3. R. P. Wool, *Polymer Interfaces: Structure and Strength*, Hanser, Munich, 1995.

4. L. E. Nielsen and R. F. Landel, *Mechanical Properties of Polymers*, Reinhold, New York, 1994.

5. H. Pasch and B. Trathnigg, *HPLC of Polymers*, Springer, Berlin, 1997.

6. T. C. Ward, *J. Chem. Ed.*, **58**, 867 (1981).

7. L. H. Sperling et al., *J. Chem. Ed.*, **62**, 780, 1030 (1985).

7a. M. S. Alger, *Polymer Science Dictionary*, Elsevier, New York, 1989.

8. A. D. Jenkins, in *Chemical Nomenclature*, K. J. Thurlow, ed., Kluwer Academic Publishers, Dordrecht, 1998.

9. (a) E. S. Wilks, *Polym. Prepr.*, **40**(2), 6 (1999); (b) A Classification of Linear Single-Strand Polymers, *Pure Appl. Chem.*, **61**, 243 (1989).

10. P. W. Morgan and S. L. Kwolek, *J. Chem. Ed.*, **36**, 182, 530 (1959).

11. Y. Furukawa, *Inventing Polymer Science*, University of Pennsylvania Press, Philadelphia, 1998.

12. W. Ostwald, *An Introduction to Theoretical and Applied Colloid Chemistry: The World of Neglected Dimensions*, Dresden and Leipzig, Verlag von Theodor Steinkopff, 1917.

13. H. Staudinger, *Ber.*, **53**, 1073 (1920).

14. H. Staudinger, *Die Hochmolecular Organischen Verbindung*, Springer, Berlin, 1932; reprinted 1960.

15. E. Farber, *Nobel Prize Winners in Chemistry, 1901–1961*, rev. ed., Abelard-Schuman, London, 1963.

16. H. Morawitz, *Polymers: The Origins and Growth of a Science*, Wiley-Interscience, New York, 1985.

17. L. H. Sperling, *Polymer News*, **132**, 332 (1987).

18. R. H. Kienle and C. S. Ferguson, *Ind. Eng. Chem.*, **21**, 349 (1929).

19. D. A. Hounshell and J. K. Smith, *Science and Corporate Strategy: DuPont R&D, 1902–1980*, Cambridge University Press, Cambridge, 1988.

20. L. A. Utracki, *Polymer Alloys and Blends*, Hanser, New York, 1990.

GENERAL READING

H. R. Allcock and F. W. Lampe, *Contemporary Polymer Chemistry*, 2nd ed., Prentice-Hall, Englewood Cliffs, NJ, 1990.

F. W. Billmeyer Jr., *Textbook of Polymer Science*, 3rd ed., Wiley, New York, 1984.

C. E. Carraher Jr., *Seymour/Carraher's Polymer Chemistry: An Introduction*, 5th ed., Dekker, New York, 2000.

M. Doi, *Introduction to Polymer Physics*, Oxford Science, Clarendon Press, Wiley, New York, 1996.

U. Eisele, *Introduction to Polymer Physics*, Springer, Berlin, 1990.

H. G. Elias, *An Introduction to Polymer Science*, VCH, Weinheim, 1997.

J. R. Fried, *Polymer Science and Technology*, Prentice-Hall, Englewood Cliffs, NJ 1995.

U. W. Gedde, *Polymer Physics*, Chapman and Hall, London, 1995.

A. Yu. Grosberg and A. R. Khokhlov, *Giant Molecules*, Academic Press, San Diego, 1997.

A. Kumar and R. K. Gupta, *Fundamentals of Polymers*, McGraw-Hill, New York, 1998.

N. G. McCrum, C. P. Buckley, and C. B. Bucknall, *Principles of Polymer Engineering*, 2nd ed., Oxford Science, Oxford, England, 1997.

J. E. Mark, H. R. Allcock, and R. West, *Inorganic Polymers*, Prentice-Hall, Englewood Cliffs, NJ, 1992.

J. E. Mark, A. Eisenberg, W. W. Graessley, L. Mandelkern, E. T. Samulski, J. L. Koenig, and G. D. Wignall, *Physical Properties of Polymers*, 2nd ed., American Chemical Society, Washington, DC, 1993.

P. Munk, *Introduction to Macromolecular Science*, Wiley-Interscience, New York, 1989.

P. C. Painter and M. M. Coleman, *Fundamentals of Polymer Science: An Introductory Text*, 2nd ed., Technomic, Lancaster, 1997.

J. Perez, *Physics and Mechanics of Amorphous Polymers*, Balkema, Rotterdam, 1998.

A. Ram, *Fundamentals of Polymer Engineering*, Plenum Press, New York, 1997.

F. Rodriguez, *Principles of Polymer Systems*, 4th ed., Taylor and Francis, Washington, DC, 1996.

A. Rudin, *The Elements of Polymer Science and Engineering*, 2nd ed., Academic Press, San Diego, 1999.

M. P. Stevens, *Polymer Chemistry: An Introduction*, 3rd ed., Oxford University Press, New York, 1999.

G. R. Strobl, *The Physics of Polymers*, 2nd ed., Springer, Berlin, 1997.

A. B. Strong, *Plastics Materials and Processing*, 2nd ed., Prentice Hall, Upper Saddle River, NJ, 2000.

S. F. Sun, *Physical Chemistry of Macromolecules: Basic Principles and Issues*, Wiley-Interscience, New York, 1994.

HANDBOOKS, ENCYCLOPEDIAS, AND DICTIONARIES

M. Alger, *Polymer Science Dictionary*, 2nd ed., Chapman and Hall, London, 1997.

G. Allen, ed., *Comprehensive Polymer Science*, Pergamon, Oxford, 1989.

Compendium of Macromolecular Nomenclature, IUPAC, CRC Press, Boca Raton, FL, 1991.

ASM, *Engineered Materials Handbook, Volume 2: Engineering Plastics*, ASM International, Metals Park, OH, 1988.

J. Brandrup and E. H. Immergut, eds., *Polymer Handbook*, 4th ed., Wiley-Interscience, New York, 1998.

N. P. Cheremisnoff, *Handbook of Polymer Science and Technology*, Dekker, New York, 1989.

S. H. Goodman, *Handbook of Thermoset Plastics*, 2nd ed., Noyes Publishers, Westwood, NJ, 1999.

J. I. Kroschwitz, ed., *Encyclopedia of Polymer Science and Engineering*, 2nd ed., Wiley, New York, 1985–1990.

W. A. Kaplan, ed., *Modern Plastics Encyclopedia*, McGraw-Hill, New York, 1998 (published annually).

J. E. Mark, ed., *Polymer Data Handbook*, Oxford University Press, New York, 1999.

J. E. Mark, ed., *Physical Properties of Polymers Handbook*, Springer, New York, 1996.

D. V. Rosato, *Rosato's Plastics Encyclopedia and Dictionary*, Hanser Publishers, Munich, 1993.

H. Saechtling, *International Plastics Handbook*, 2nd ed., English trans., Hanser-Macmillan, New York, 1987.

J. C. Salamone, ed., *Polymer Materials Encyclopedia*, CRC Press, Boca Raton, FL, 1996.

D. W. Van Krevelen, *Properties of Polymers*, 3rd ed., Elsevier, Amsterdam, 1997.

T. Whelen, *Polymer Technology Dictionary*, Chapman and Hall, London, 1992.

G. Wypych, *Handbook of Fillers*, 2nd ed., William Anderson, Norwich, NY, 1999.

WEB SITES

Chemical Abstracts: *http://www.cas.org/EO/polymers.pdf*

Case-Western Reserve University, Department of Macromolecular Chemistry: *http://abalone.cwru.edu/tutorial/enhanced/main.htm*

Pennsylvania College of Technology, Pennsylvania State University, and University of Massachusetts at Lowell: *http://www.pct.edu/prep/*

University of Southern Mississippi, Dept. of Polymer Science, *The Macrogalleria*: *http://www.psrc.usm.edu/macrog/index.html*

STUDY PROBLEMS

1. Polymers are obviously different from small molecules. How does polyethylene differ from oil, grease, and wax, all of these materials being essentially $-CH_2-$?

2. Write chemical structures for polyethylene, polypropylene, poly(vinyl chloride), polystyrene, and polyamide 66.

3. Name the following polymers:

$$+CH_2-\underset{\underset{O=C-O-CH_3}{|}}{\overset{\overset{H}{|}}{C}}\rightarrow_n \qquad +CH_2-\underset{\underset{O=C-O-C_2H_5}{|}}{\overset{\overset{CH_3}{|}}{C}}\rightarrow_n$$

(a)   (b)

$$+CH_2-\underset{\underset{O-\underset{\underset{O}{\|}}{C}-CH_3}{|}}{\overset{\overset{H}{|}}{C}}\rightarrow_n \qquad +CH_2-CF_2\rightarrow_n$$

(c)   (d)

4. What molecular characteristics are required for good mechanical properties? Distinguish between amorphous and crystalline polymers.

5. Show the synthesis of polyamide 610 from the monomers.

6. Name some commercial polymer materials by chemical name that are (a) amorphous, cross-linked, and above T_g; (b) crystalline at ambient temperatures.

7. Take any 10 books off a shelf and note the last page number. What are the number-average and weight-average number of pages of these books? Why is the weight-average number of pages greater than the number-average? What is the poly-dispersity index? Can it ever be unity?

8. Draw a log modulus–temperature plot for an amorphous polymer. What are the five regions of viscoelasticity, and where do they fit? To which regions do the following belong at room temperature: chewing gum, rubber bands, Plexiglas®?

9. Define the terms: Young's modulus, tensile strength, chain entanglements, and glass–rubber transition.

10. A cube 1 cm on a side is made up of one giant polyethylene molecule, having a density of 1.0 g/cm³. (a) What is the molecular weight of this molecule? (b) Assuming an all trans conformation, what is the contour length of the chain (length of the chain stretched out)? Hint: The mer length is 0.254 nm.

APPENDIX 1.1 NAMES FOR POLYMERS

The IUPAC Macromolecular Nomenclature Commission has developed a systematic nomenclature for polymers (9). The Commission recognized, however, that a number of common polymers have semisystematic or trivial names that are well established by usage. For the reader's convenience, the recommended trivial name (or the source-based name) of the polymer is given under the polymer structure, and then the structure-based name is given. For example, the trivial name, polystyrene, is a source-based name, literally "the polymer made from styrene." The structure-based name, poly(1-phenylethylene), is useful both in addressing people who may not be familiar with the structure of polystyrene and in cases where the polymer is not well known. This book uses a source-based nomenclature, unless otherwise specified. The following structures are IUPAC recommended.

$+CH_2CH_2\,\rlap{$-$}{}_n$

polyethylene
poly(methylene)

$+CH=CHCH_2CH_2\,\rlap{$-$}{}_n$

polybutadiene[a]
poly(1-butenylene)

$+\underset{\underset{CH_3}{|}}{CHCH_2}\,\rlap{$-$}{}_n$

polypropylene
poly(1-methylethylene)

$+\underset{\underset{CH_3}{|}}{C}=CHCH_2CH_2\,\rlap{$-$}{}_n$

polyisoprene[b]
poly(1-methyl-1-butenylene)

$+CH_2-\overset{\overset{CH_3}{|}}{\underset{\underset{CH_3}{|}}{C}}\,\rlap{$-$}{}_n$

polyisobutylene
poly(1,1-dimethylethylene)

$+\underset{|}{CHCH_2}\,\rlap{$-$}{}_n$

(phenyl ring)

polystyrene
poly(1-phenylethylene)

$+\underset{\underset{CN}{|}}{CHCH_2}\,\rlap{$-$}{}_n$

polyacrylonitrile
poly(1-cyanoethylene)

$+\underset{\underset{OH}{|}}{CHCH_2}\,\rlap{$-$}{}_n$

poly(vinyl alcohol)
poly(1-hydroxyethylene)

$+\underset{\underset{OOCCH_3}{|}}{CHCH_2}\,\rlap{$-$}{}_n$

poly(vinyl acetate)
poly(1-acetoxyethylene)

$+\underset{\underset{Cl}{|}}{CHCH_2}\,\rlap{$-$}{}_n$

poly(vinyl chloride)
poly(1-chloroethylene)

$+\overset{\overset{F}{|}}{\underset{\underset{F}{|}}{C}}CH_2\,\rlap{$-$}{}_n$

poly(vinylidene fluoride)
poly(1,1-difluoroethylene)

---

[a] Polybutadiene is usually written $+CH_2CH=CHCH_2\rlap{$-$}{}_n$, that is, with the double bond in the center. The structure-based name is given.

[b] Polyisoprene is usually written $+CH_2\overset{\overset{CH_3}{|}}{C}=CHCH_2+$.

$+CF_2CF_2 +_n$

poly(tetrafluoroethylene)
poly(difluoromethylene)

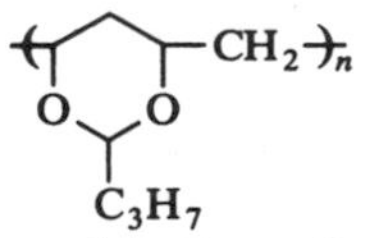

$+\!\!\!\begin{array}{c}\text{CHCH}_2\\|\\\text{COOCH}_3\end{array}\!\!\!+_n$

poly(methyl acrylate)
poly[1-(methoxycarbonyl)ethylene]

$+OCH_2 +_n$

polyformaldehyde
poly(oxymethylene)

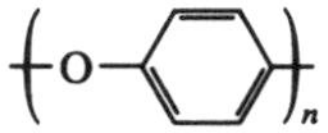

$+NH(CH_2)_6NHCO(CH_2)_4CO +_n$
polyamide 66[a]
poly(hexamethylene adipamide)
poly(iminohexamethyleneiminoadipoyl)

$+OCH_2CH_2OOC-\!\!\!\langle\bigcirc\rangle\!\!\!-CO +_n$

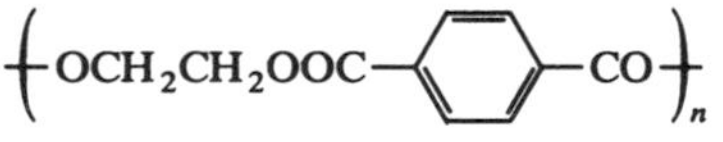

poly(ethylene terephthalate)
poly(oxyethyleneoxyterephthaloyl)

$+\begin{array}{c}\\O\diagdown\diagup O\\|\\C_3H_7\end{array}CH_2 +_n$

poly(vinyl butyral)
poly[(2-propyl-1,3-dioxane-4,
6-diyl)methylene]

$+\!\!\!\begin{array}{c}\text{CH}_3\\|\\\text{C}-\text{CH}_2\\|\\\text{COOCH}_3\end{array}\!\!\!+_n$

poly(methyl methacrylate)
poly[1-(methoxycarbonyl)-
1-methylethylene]

$+O-\!\!\!\langle\bigcirc\rangle\!\!\!+_n$

poly(phenylene oxide)
poly(oxy-1,4-phenylene)

$+OCH_2CH_2 +_n$

poly(ethylene oxide)
poly(oxyethylene)

$+NHCO(CH_2)_5 +_n$

polyamide 6[b]
poly(ε-caprolactam)
poly[imino(1-oxohexamethylene)]

---

[a] Common name. Other ways this is named include nylon 6,6, 66-nylon, 6,6-nylon, and nylon 66.
[b] Common name.

2

CHAIN STRUCTURE
AND CONFIGURATION

In the teaching of physical polymer science, a natural progression of material begins with chain structure, proceeds through morphology, and leads on to physical and mechanical behavior. To a significant measure, one step determines the properties of the next (1). Polymer chains have three basic properties:

1. The molecular weight and molecular weight distribution of the molecules. These properties are discussed in Chapter 3.
2. The conformation of the chains in space. The term conformation refers to the different arrangements of atoms and substituents of the polymer chain brought about by rotations about single bonds. Examples of different polymer conformations include the fully extended planar zigzag, helical, folded chain, and random coils. Some conformations of a random coil might be

$$\text{(2.1)}$$

The methods of determining polymer chain conformation are discussed in Chapters 3 and 5.
3. The configuration of the chain. The term "configuration" refers to the organization of the atoms along the chain. Some authors prefer the term "microstructure" rather than configuration. Configurational isomerism involves the different arrangements of the atoms and substituents in a chain, which can be interconverted only by the breakage and reformation of primary chemical bonds. The configuration of polymer chains constitutes the principal subject of this chapter.

25

2.1   EXAMPLES OF CONFIGURATIONS AND CONFORMATIONS

2.1.1   Head-to-Head and Head-to-Tail Configurations

Before proceeding with the development of theory and instruments, a simple but important example of chain configuration is given. This involves the difference between head-to-head and head-to-tail placement of the monomeric units, or mers, during polymerization. The head-to-tail structure of polystyrene may be written

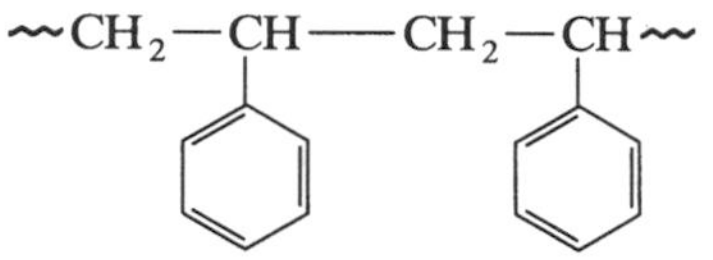

$$\sim CH_2 - CH - CH_2 - CH \sim$$

(2.2)

and its head-to-head structure may be written

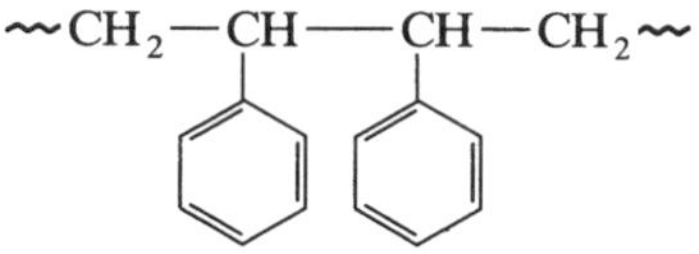

$$\sim CH_2 - CH - CH - CH_2 \sim$$

(2.3)

The thermodynamically and spatially preferred structure is usually the head-to-tail configuration, although most addition polymers contain a small percentage of head-to-head placements. If the synthesis is deliberately arranged so that the head-to-head configuration is obtained, the properties of the polymer are far different. Using polyisobutylene as an example, Malanga and Vogl (2) showed that the melting temperature of the head-to-head configuration was 187°C, whereas the head-to-tail configuration could only be crystallized under stress, and then with a melting temperature of 5°C. The head-to-head and head-to-tail configurations cannot be interchanged without breaking primary chemical bonds.

2.1.2   *Trans*[†]–*Gauche* Conformations

The *trans–gauche* conformations of a polymer chain can be interchanged by simple rotation about the single bond linking the moieties. The *trans* and *gauche* states are defined as follows: Consider a sequence of carbon–carbon single bonds delineated $\ldots i-1, i, i+1, \ldots$. The rotational angle of bond i is defined as the angle between two planes, the first plane defined by bonds $i-1$ and i and the second by bonds i and $i+1$. The planar zigzag conformation, where the angle between the two planes is zero, is called the *trans* state, t. When the angle between the two planes is $\pm 120°$, the *gauche* plus, g^+, and *gauche* minus, g^-, states are defined:

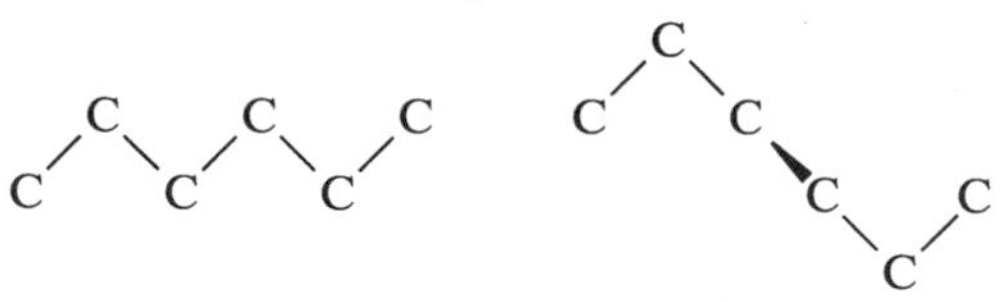

$$\text{(2.4)}$$

$t\ t\ t$ $g^+ t g^-$

[†] *Trans* is also called the anti form.

The bond symbol ◄ means pointed out of the paper toward the reader. The trans and gauche positions are located 120° apart on an imaginary cone of rotation, where the preferred positioning avoids groups on the neighboring chain carbon atoms, the *trans* being the more extended conformation. The *trans* and *gauche* conformations are discussed further in Section 2.8. For C—C bond lengths of 0.154 nm, and C—C—C bond angles of 109°, the mer length of many addition polymers may be taken as 0.254 nm. See Section 6.3.

2.2 THEORY AND INSTRUMENTS

Koenig (3) defines the microstructure of a polymer in terms of its conformation and configuration. The term conformation has taken on two separate meanings: (a) the long-range shape of the entire chain, which is discussed in Chapter 5, and (b) the several possibilities of rotating atoms or short segments of chain relative to one another, to be discussed later. The term configuration includes its composition, sequence distribution, steric configuration, geometric and substitutional isomerism, and so on, and is the major concern of this chapter.

The several aspects of polymer chain microstructure have been studied by both chemical and physical methods. Koenig (3) describes several of these methods, which are summarized in Tables 2.1 and 2.2.

2.2.1 Chemical Methods of Determining Microstructure

The most basic method of characterizing any material uses elemental analysis (Table 2.1). Elemental analysis helps identify unknowns, confirms new syntheses, and yields information on the purity of the polymer.

Table 2.1   Chemical methods of determining polymer chain microstructure (3)

Method	Application	Reference
Elemental analysis	Gross composition of polymers and copolymers, yielding the percent composition of each element; C, H, N, O, S, and so on.	(a)
Functional group analysis	Reaction of a specific group with a known reagent. Acids, bases, and oxidizing and reducing agents are common. Example: titration of carboxyl groups.	(b, c)
Selective degradation	Selective scissions of particular bonds, frequently by oxidation or hydrolysis. Example: ozonalysis of polymers containing double bonds.	(d)
Cyclization reactions	Sequence analysis through formation of lactones, lactams, imides, α-tetralenes, and endone rings.	(e)
Cooperative reactions	Sequence analysis using reactions of one group with a neighboring group.	(f)

References: (a) F. E. Critchfield and D. P. Johnson, *Anal. Chem.*, **33**, 1834 (1961). (b) S. Siggia, *Quantitative Organic Analysis via Functional Groups*, 3rd ed., Wiley, New York, 1963. (c) N. Bikales, *Characterization of Polymers*, Encyclopedia of Polymer Science and Technology, Wiley-Interscience, New York, 1971, p. 91 (d) R. Hill, J. R. Lewis, and J. Simonsen, *Trans. Faraday Soc.*, **35**, 1073 (1939). (e) M. Tanaka, F. Nishimura, and T. Shono, *Anal. Chim. Acta*, **74**, 119 (1975). (f) J. J. Gonzales and P. C. Hammer, *Polym. Lett.* **14**, 645 (1976).

Table 2.2   Physical methods of determining polymer chain microstructure (3)

Method	Application	Reference
Nuclear magnetic resonance	Determination of steric configuration in homopolymers; composition of copolymers, including proteins; chemical functionality, including oxidation products; determination of structural and geometric and substitutional isomerism, conformation, and copolymer microstructure.	(a–c)
Infrared and Raman spectroscopy (considered together)	Molecular identification: determination of chemical functionality; chain and sequence length; quantitative analysis; stereochemical configuration; chain conformation.	(d, e)
Ultraviolet and visible light spectroscopy	Sequence length; conformation and spatial analysis.	(f)
Mass Spectroscopy	Polymer degradation mechanisms; order and randomness of block copolymers, side groups, impurities.	(g)
Electron spectroscopy (ESCA)	Microstructure of polymers, particularly surfaces.	(h)
X-ray and electron diffraction (considered together)	Identification of repeat unit in crystalline polymers; inter- and intramolecular spacings; chain conformation and configuration.	(i)

References: (a) F. Bovey, *High Resolution NMR of Macromolecules*, Academic Press, New York, 1972. (b) C. C. McDonald, W. D. Phillips, and J. D. Glickson, *J. Am. Chem. Soc.*, **93**, 235 (1971). (c) J. C. Randall, *J. Polym. Sci. Polym. Phys. Ed.*, **13**, 889 (1975). (d) J. Haslam, H. A. Willis, and M. Squirrell, *Identification and Analysis of Plastics*, 2nd ed., Ileffe, London, 1972. (e) J. L. Koenig, *Appl. Spectrosc. Rev.*, **4**, 233 (1971). (f) Y. C. Wang and M. A. Winnik, *Macromolecules*, **23**, 4731 (1990). (g) J. L. Koenig, *Spectroscopy of Polymers*, 2nd ed., Elsevier, Amsterdam, 1999. (h) D. T. Clark and W. J. Feast, *J. Macromol. Sci.*, **C12**, 191 (1975). (i) G. Natta, *Makromol. Chem.*, **35**, 94 (1960).

Functional group analysis relates to those reactions that polymers undergo, either intentionally or accidentally. Selective degradation refers to those chemical reactions that a polymer undergoes which cut particular bonds. These may be main chain or side chain. Similarly cyclization reactions and cooperative reactions enable particular sequences to be identified. It must be emphasized that all these methods of characterization are widely used throughout the field of chemistry for big and little molecules alike. This last statement holds also for the physical methods.

2.2.2   General Physical Methods

The more important physical methods of characterizing the microstructure of a polymer are summarized in Table 2.2. Nuclear magnetic resonance and infrared and Raman spectroscopy are considered in the following sections (4).

Ultraviolet and visible light spectroscopy makes use of the quantized nature of the electronic structure of molecules. One example that is commonly observed by eye is the yellow color of polymeric materials that have been slightly degraded by heat or oxidation. Frequently this is due to the appearance of conjugated double bonds (5). For example, the 10-polyene conjugated structure absorbs light at 473 nm in the blue region.

Mass spectroscopy makes use of polymer degradation, and particular masses emerging are identified. For example, polymers having higher alkane side groups usually have a mass peak at 43 g/mol, oxygen as alcohol or ether at 31, 45, or 59 g/mol, and so on (6). Mass spectrometry also provides a powerful method of identifying residual volatile chemicals, which is becoming increasingly important in reducing air pollutants. Mass spectrometry further provides a newer method of determining polymer molecular weights; see Section 3.10. Electron spectroscopy for chemical applications (ESCA) is a relatively new method useful for surface analysis of polymers; see Section 12.3.

X-ray (7) and electron diffraction methods are most useful for determining the structure of polymers in the crystalline state and are discussed in Chapter 6. These methods do, however, provide a wealth of information relative to the inter- and intramolecular spacings, which can be interpreted in terms of conformations and configurations.

2.2.3 Infrared and Raman Spectroscopic Characterization

The total energy of a molecule, big or small, consists of contributions from the rotational, vibrational, electronic, and electromagnetic spin energies. These states define the temperature of the system. Specific energies may be increased or decreased by interaction with electromagnetic radiation of a specified wavelength. In the following discussion, it is important to remember that all such interactions are quantized; that is, only specific energy levels are permitted.

Infrared spectra are obtained by passing infrared radiation through the sample of interest and observing the wavelength of absorption peaks. These peaks are caused by the absorption of the electromagnetic radiation and its conversion into specific molecular motions, such as C—H stretching.

The older, conventional instruments are known as dispersive spectrometers, where the infrared radiation is divided into frequency elements by the use of a monochromator and slit system. Although these instruments are still in use today, the recent introduction of Fourier transform infrared (FT-IR) spectrometers has revitalized the field (4). The FT-IR system is based on the Michelson interferometer. The total spectral information is contained in an interferogram from a single scan of a movable mirror. There are no slits, and the amount of infrared energy falling on the detector is greatly enhanced. Together with the use of modern computer techniques, an entirely new breed of instrument has been created.

Raman spectra (8) are obtained by a variation of a light-scattering technique whereby visible light is passed into the sample. In addition to light of the same wavelength being scattered, there is an inelastic component. The physical cause relates to the light's exchanging energy with the molecule. This inelastic scattering causes light of slightly longer or shorter wavelengths to be scattered. As above, there is an increase or decrease in a specific molecular motion.

Raman and infrared spectroscopy are complementary because they are governed by different selection rules (4,7,9). In order for Raman scattering to occur, the electric field of the light must induce a dipole moment by changing the polarizability of the molecule. By contrast, infrared requires an intrinsic dipole moment to exist, which must change with molecular vibration.

The fields have advanced way beyond the simple determination of spectra and correlating particular bands with particular chemical groups. Today, specific motions are calculated. For an example, see Figure 2.1 (4). Here, two conformational displacements of isotactic polystyrene[†] are shown—one near 550 cm^{-1} in the infrared spectrum, and

---

[†] The term isotactic refers to a specific configuration to be defined later.

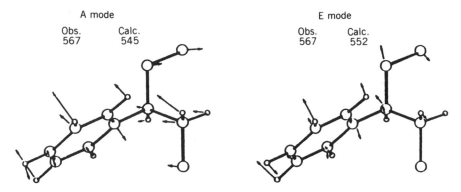

Figure 2.1   Motions associated with the 567 cm^{-1} infrared peak. Quite different motions are associated with the Raman peak at 225 cm^{-1}, not shown. Isotactic polystyrene.

one near 225 cm^{-1} in the Raman spectrum. These motions illustrate a degree of coupling between the ring and backbone vibrations.

2.2.4   Nuclear Magnetic Resonance Methods

Although X-ray (7), Raman spectroscopy (8), and infrared methods (9) are at the disposal of the polymer scientist for stereochemical configuration analyses, by far the most powerful method is nuclear magnetic resonance (NMR). Briefly, when the spin quantum number of a nucleus is $\frac{1}{2}$ or greater, it possesses a magnetic moment. A proton has a spin of $\frac{1}{2}$ and is widely used in NMR studies. When placed in a magnetic field H_0, it can occupy either of two energy levels, which corresponds to its magnetic moment, μ, being aligned with or against the field. The energy differences in the two orientations are given by Bovey et al. (10):

$$\Delta E = h\nu = 2\mu H_0 \tag{2.5}$$

The quantity ΔE indicates the energy that must be absorbed to raise the nuclei in the lower state up to the higher level and is emitted in the reverse process. The separation of energy levels is propositional to the magnetic field strength. In a field of 9400 gauss, the resonant frequency for protons is about 40 MHz. For field strengths of the order of 10,000 gauss and up, the frequency, ν, is in the microwave region.

In a molecule containing many atoms, the field on any one of these is altered by the presence of the others:

$$\Delta E = 2\mu(H_0 + H_L) \tag{2.6}$$

where H_L is the local field with a strength of 5 to 10 gauss. It is these changes that are important in NMR characterization.

Other nuclei besides hydrogen (^{1}H) that have a spin of $\frac{1}{2}$ or greater, and are used in NMR studies, include deuterium (^{2}H), fluorine (^{19}F), carbon-13 (^{13}C), nitrogen-14 and -15 (^{14}N and ^{15}N), and phosphorus-31 (^{31}P). Much higher resolutions are often possible with these nuclei, allowing exact sequences of structures to be determined along the chain.

A new technique for ^{13}C NMR is the so-called magic angle method, which uses oriented specimens spun around an axis at $\theta = 54.7°$ to reduce line broadening due to anisotropic contributions. This particular angle arises because the broadening component is proportional to the quantity $3\cos^2 \theta - 1$, where θ is the angle between the line connecting the nuclei and the direction of the magnetic field in isotropic compositions. At 54.7° this quantity is zero.

While the fundamental unit for determining shifts in NMR peaks is the change in frequency in hertz, an important practical scale is based on the position of the tetramethylsilane peak, leading to the τ scale (see Figure 2.6). This scale is now out moded, and being replaced by scales based on shifts of parts per million, ppm (see Figure 2.8).

2.3 STEREOCHEMISTRY OF REPEATING UNITS

2.3.1 Chiral Centers

Early in the history of organic compounds, two substances were sometimes found that appeared to be chemically identical except that they rotated plane-polarized light equally, but in opposite directions. With the development of the tetrahedral carbon bond model, it gradually became clear that the two isomers were mirror images of each other. As an illustration, the two possible spatial configurations of the compound

$$
\begin{array}{c}
\text{H} \\
| \\
\text{Br}-\text{C}-\text{Cl} \\
| \\
\text{CH}_3
\end{array}
\qquad (2.7)
$$

form the mirror images shown in Figure 2.2 (11). The cause of the optical activity is the asymmetric carbon in the center, known as a chiral center. The two different compounds are known as enantiomorphs, or enantiomers. The important point is that the two mirror images are nonsuperimposable, and the two compounds are really different.

2.3.2 Tacticity in Polymers

The polymerization of a monosubstituted ethylene, such as a vinyl compound, leads to polymers in which every other carbon atom is a chiral center. This is often marked with an asterisk for emphasis:

$$
\begin{array}{c}
\text{H} \\
| \\
\text{CH}_2{=}\text{C} \rightarrow -\text{CH}_2-\text{C}^*- \\
| \qquad\qquad | \\
\text{R} \qquad\qquad \text{R}
\end{array}
\qquad (2.8)
$$

Such carbon atoms are referred to as pseudochiral centers in long-chain polymers because the polymers do not in fact exhibit optical activity (12). The reason for the lack of optical activity can be seen through a closer examination of the substituents on such a pseudochiral center:

$$
\begin{array}{c}
\text{H} \\
| \\
\sim\sim\text{C}^*\sim\sim\sim \\
| \\
\text{R}
\end{array}
\qquad (2.9)
$$

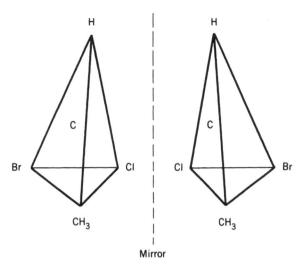

Figure 2.2 Two optical isomers of the same compound as mirror images (11).

The two chain segments are indicated by ∿ and ∿∿ and in general will be of unequal length. The first few atoms of the two chain segments attached to C* are responsible for the optical activity, not those farther away. These near atoms are seen to be the same, and hence the polymer is optically inactive.

The two mirror image configurations remain distinguishable, however. The different possible spatial arrangements are called the tacticity of the polymer. If the R groups on successive pseudochiral carbons all have the same configuration, the polymer is called isotactic (see Figure 2.3) (12). When the pseudochiral centers alternate in configuration from one repeating unit to the next, the polymer is called syndiotactic. If the pseudochiral centers do not have any particular order, but in fact are statistical arrangements, the polymer is said to be atactic.

Figure 2.3 portrays the general three-dimensional structure of the polymer (12). Using Fischer–Hirshfelder or similar models, the actual differences between isotactic and syndiotactic poly(vinyl chloride) can be illustrated (see Figure 2.4). A two-dimensional analogue can be made using what are known as Fisher projections. In these projections the R groups are placed either up or down. All up (or all down) indicates the isotactic structure:

$$-\overset{\displaystyle H}{\underset{\displaystyle H}{C}}-\overset{\displaystyle H}{\underset{\displaystyle R}{C}}-\overset{\displaystyle H}{\underset{\displaystyle H}{C}}-\overset{\displaystyle H}{\underset{\displaystyle R}{C}}-\overset{\displaystyle H}{\underset{\displaystyle H}{C}}-\overset{\displaystyle H}{\underset{\displaystyle R}{C}}-\overset{\displaystyle H}{\underset{\displaystyle H}{C}}-\overset{\displaystyle H}{\underset{\displaystyle R}{C}}- \qquad (2.10)$$

Alternating up and down indicates syndiotactic:

$$-\overset{\displaystyle H}{\underset{\displaystyle H}{C}}-\overset{\displaystyle H}{\underset{\displaystyle R}{C}}-\overset{\displaystyle H}{\underset{\displaystyle H}{C}}-\overset{\displaystyle R}{\underset{\displaystyle H}{C}}-\overset{\displaystyle H}{\underset{\displaystyle H}{C}}-\overset{\displaystyle H}{\underset{\displaystyle R}{C}}-\overset{\displaystyle H}{\underset{\displaystyle H}{C}}-\overset{\displaystyle R}{\underset{\displaystyle H}{C}}- \qquad (2.11)$$

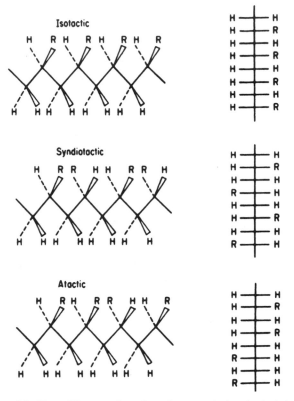

Figure 2.3 Three different configurations of a monosubstituted polyethylene,

$$-\!\!-\!\!\left(\!\text{CH}_2 - \text{CHR}\!\right)_{\!\!n}\!\!-\!\!$$

The dotted and triangular lines represent bonds to substitutents below and above the plane of the carbon–carbon backbone chain, respectively (12).

and random up and down indicates atactic:

$$\begin{array}{c} \text{H} \quad \text{H} \quad \text{H} \quad \text{H} \quad \text{H} \quad \text{R} \quad \text{H} \quad \text{H} \\ | \quad | \quad | \quad | \quad | \quad | \quad | \quad | \\ -\text{C}-\text{C}-\text{C}-\text{C}-\text{C}-\text{C}-\text{C}-\text{C}- \\ | \quad | \quad | \quad | \quad | \quad | \quad | \quad | \\ \text{H} \quad \text{R} \quad \text{H} \quad \text{R} \quad \text{H} \quad \text{H} \quad \text{H} \quad \text{R} \end{array} \qquad (2.12)$$

In specifying the tacticity of the polymer, the prefixes *it* and *st* are placed before the name or structure to indicate isotactic and syndiotactic structures, respectively. For example, *it*-polystyrene means that the polystyrene is isotactic. Such polymers are known as stereoregular polymers. The absence of these terms denotes the corresponding atactic structure.

The structures shown in equations (2.10) to (2.12) result in profoundly different physical and mechanical behavior. The isotactic and syndiotactic structures are both crystallizable because of their regularity along the chain. However, their unit cells and

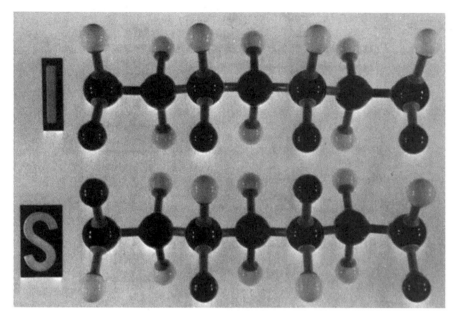

Figure 2.4   Isotactic and syndiotactic structures of poly(vinyl chloride). Allyn and Bacon Molecular Model Set for Organic Chemistry.

melting temperatures are not the same. Atactic polymers, on the other hand, are usually completely amorphous unless the side group is so small or so polar as to permit some crystallinity.[†]

2.3.3   Meso- and Racemic Placements

The Fisher projection in equation (2.10) shows that the placement of the groups corresponds to a meso- (same) or *m* placement of a pair of consecutive pseudochiral centers. The syndiotactic structure in equation (2.11) corresponds to a racemic (opposite) or *r* placement of the corresponding pair of pseudochiral centers. It must be emphasized that the *m* or *r* notation refers to the configuration of one pseudochiral center relative to its neighbor. Several possible configurational sequences are illustrated in Figure 2.5 (13). Each of these, and even more complicated combinations, can be distinguished through NMR studies, as described later.

2.3.4   Proton Spectra by NMR

The 40-MHz [¹]H spectrum of two samples of poly(methyl methacrylate) are illustrated in Figure 2.6 (13). The sample marked (*a*) was prepared via free radical polymerization methods (see Chapter 1). The sample marked (*b*) was synthesized by a then new method, anionic polymerization. The anionic polymerization method was thought to make samples predominantly isotactic, whereas free radical methods resulted in atactic polymers.

[†] One such crystalline atactic polymer is poly(vinylalcohol). Atactic poly(vinylchloride) is slightly crystalline because of syndiotactic "runs."

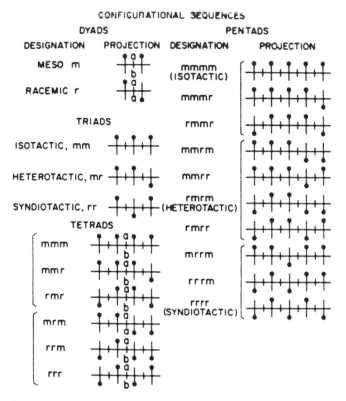

Figure 2.5 Configurational sequences in monosubstituted ethylenes projected in two dimensions (13).

The peaks in Figure 2.6 were assigned as follows (14). The large peak at the left of both (*a*) and (*b*) is that of the chloroform solvent. The methyl ester group appears at 6.40 τ in both spectra and is unchanged by the chain configuration. At 8.78 τ, 8.95 τ, and 9.09 τ are three α-methyl peaks whose relative heights vary greatly with the method of synthesis. Note that the peak at 8.78 τ is much larger in (*b*) than in (*a*), and the peak at 9.09 τ is the more prominent in (*a*). (The τ values and the τ scale refer to a system in which the tetramethylsilane peak is assigned the arbitrary value of +10.000 ppm by definition and is shown on the extreme right in Figure 2.6. These τ values, measured for thousands of organic compounds in carbon tetrachloride solution, are widely used for comparison and identification.)

The peak at 8.78 τ was assigned to the configuration wherein the α-methyl groups of the monomer residues are flanked on both sides by mers of the same configuration—that is, all *m*-placement. The most prominent peak in the free radical polymerized polymers, at 9.09 τ, is attributed to α-methyl groups of central monomer units in syndiotactic, *r*-placement configuration. The peak at 8.95 τ is assigned to α-methyl groups in heterotactic configurations, meaning that the central mer in a triad has opposite configurations at either end. On noting that free radical polymerizations, especially those at low temperatures, tend to be predominantly syndiotactic, the assignment becomes clear. In these

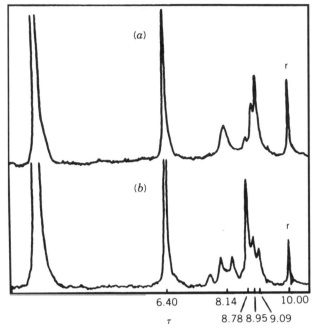

Figure 2.6   The 40-MHz NMR proton spectra of poly(methyl methacrylate) in chloroform. (*a*) Atactic polymer prepared using a free radical initiator. (*b*) Isotactic polymer prepared using *n*-butyllithium initiator by anionic polymerization (14).

early materials, however, the structures were not all one configuration, especially the anionically prepared polymer.

Returning to Figure 2.5, the relative frequencies of the several possibilities have certain necessary relationships (see Table 2.3) (15). For example, considering the triad relationships,

$$mm + mr + rr = 1 \tag{2.13}$$

if the polymer is entirely isotactic, the terms *mr* and *rr* are both zero. These algebraic relationships provide a quantitative basis for determining the probability of certain sequences occurring.

There are, of course, other types of stereoregular polymers. Some of these are briefly described in Appendix 2.1.

2.4   REPEATING UNIT ISOMERISM

2.4.1   Optical Isomerism

There is one important class of polymers that do exhibit strong optical activity, as opposed to the above tactic structures. These are the polymers in which the chiral center is surrounded by different atoms or groups, and a true local center of asymmetry exists.

Table 2.3 Algebraic relations among sequence frequencies (15)

Dyad	$(m) + (r) = 1$
Triad	$(mm) + (mr) + (rr) = 1$
Dyad–triad	$(m) = (mm) + \frac{1}{2}(mr)$
	$(r) = (rr) + \frac{1}{2}(mr)$
Triad–tetrad	$(mm) = (mmm) + \frac{1}{2}(mmr)$
	$(mr) = (mmr) + 2(rmr) = (mrr) + 2(mrm)$
	$(rr) = (rrr) + \frac{1}{2}(mrr)$
Tetrad–tetrad	sum $= 1$
	$(mmr) + 2(rmr) = 2(mrm) + (mrr)$
Pentad–pentad	sum $= 1$
	$(mmmr) + 2(rmmr) = (mmrm) + (mmrr)$
	$(mrrr) + 2(mrrm) = (rrmr) + (rrmm)$
Tetrad–pentad	$(mmm) = (mmmm) + \frac{1}{2}(mmmr)$
	$(mmr) = (mmmr) + 2(rmmr) = (mmrm) + (mmrr)$
	$(rmr) = \frac{1}{2}(mrmr) + \frac{1}{2}(rmrr)$
	$(mrm) = \frac{1}{2}(mrmr) + \frac{1}{2}(mmrm)$
	$(rrm) = 2(mrrm) + (mrrr) = (mmrr) + (rmrr)$
	$(rrr) = (rrrr) + \frac{1}{2}(mrrr)$

An example is poly(propylene oxide),

$$
\left[\begin{array}{c} H \\ | \\ O-C^*-CH_2 \\ | \\ CH_3 \end{array}\right]_n
\tag{2.14}
$$

where the chiral center is surrounded by $-H$, $-CH_3$, $-CH_2-$, and $-O-$.

2.4.2 Geometric Isomerism

The most important examples in this class are the *cis* and *trans* isomerism about double bonds. Take polybutadiene as an example,

$$
\left(\begin{array}{c} CH_2 \qquad CH_2 \\ \diagdown \qquad \diagup \\ C=C \\ \diagup \qquad \diagdown \\ H \qquad\quad H \end{array}\right)_n
\qquad
\left(\begin{array}{c} CH_2 \qquad H \\ \diagdown \qquad \diagup \\ C=C \\ \diagup \qquad \diagdown \\ H \qquad CH_2 \end{array}\right)_n
\tag{2.15}
$$

cis-polybutadiene *trans*-polybutadiene

The *cis–trans* isomerism arises because rotation about the double bond is impossible without disrupting the structure. Thus the formula on the left of equation (2.15) is written *cis*-polybutadiene. The reader should note that the *cis–trans* isomerism is entirely different from the *trans–gauche* structures written in equation (2.4).

The *cis* and *trans* formulas are both crystallizable when appearing in pure form, but with different melting temperatures. If a mixture of *cis* and *trans* isomers occurs, crystallization may be suppressed, similar to the atactic polymers.

2.4.3 Substitutional Isomerism

In the synthesis of diene type polymers, yet another type of isomerism may occur, that of 1,2 versus 1,4 addition:

$$
\begin{array}{cc}
\underset{\substack{| \\ H}}{\overset{\substack{H \\ |}}{-C}}-\underset{\substack{| \\ \underset{\substack{\| \\ CH_2}}{C-H}}}{\overset{\substack{H \\ |}}{C}}- & \underset{\substack{| \\ H}}{\overset{\substack{H \\ |}}{-C}}-\underset{}{\overset{\substack{H \\ |}}{C}}=\underset{}{\overset{\substack{H \\ |}}{C}}-\underset{\substack{| \\ H}}{\overset{\substack{H \\ |}}{C}}- \\
\text{1,2-addition} & \text{1,4-addition}
\end{array}
\tag{2.16}
$$

In the case of 1,2 addition, polymerization is similar to that of vinyl structures. Note that if the diene is substituted, as in isoprene,

$$
\underset{\substack{| \\ H}}{\overset{\substack{H \\ |}}{C}}=\underset{\substack{| \\ CH_3}}{\overset{}{C}}-\underset{}{\overset{\substack{H \\ |}}{C}}=\underset{\substack{| \\ H}}{\overset{\substack{H \\ |}}{C}}
\tag{2.17}
$$

1,2, 1,4, and 3,4 polymerizations are each distinguished. Of course, the 1,4 polymerizations also exhibit the *cis–trans* isomerism simultaneously. All these may appear together in various percentages in a given preparation.

2.4.4 Infrared and Raman Spectroscopic Characterization

Historically studies of the selective absorption of infrared radiation preceded the Raman effect, although the latter has played a critical role in the analysis of chemical structures. A very large number of polymers have now been characterized by infrared (16) and Raman (17).

For infrared, the most important region has been the medium infrared region, stretching from 2.5 to 50 μm (4000 to 200 cm^{-1}). For example, the infrared spectra of bisphenol A polycarbonate, an engineering plastic, is shown in Figure 2.7. The structure of the mer is

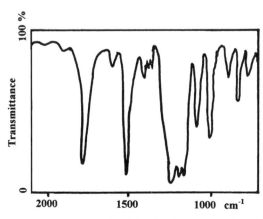

Figure 2.7 Medium infrared spectra of bisphenol A polycarbonate. Based on Ref. 16, vol. 1, p. 241.

Its principal absorption bands include those at 823 cm^{-1} due to ring C—H bending, at 1164 and 1231 cm^{-1} due to C—O stretching, 1506 cm^{-1} due to skeletal ring vibrations, and 1776 cm^{-1} due to C=O stretching (18).

By examining the area under each curve, quantitative analyses can be made. Since the physical and mechanical properties of polymers composed of various mers and their isomers differ, careful determination of the configuration of each preparation is made daily in many chemical industries. Of course, for new polymer syntheses infrared and Raman spectra provide primary evidence as to the exact structure prepared.

Much additional information can be obtained from infrared and Raman spectra. When a polymer is in the crystalline state, the chains are aligned. If the polymer is melted, the chain conformation becomes disordered. The spectra change accordingly. Many new frequencies appear arising from the new conformations in the melt, and some of the frequencies characteristic of the crystalline state disappear.

For example, in the solid state one structure predominates for poly(ethylene oxide), the *tgt* conformation. The O—C bond is *trans*, the C—C bond is gauche, and the C—O bond is *trans*. As illustrated in Table 2.4 (18), all possible conformations exist in

Table 2.4 IR and Raman spectra of molten poly(ethylene oxide) (18)

IR Frequency (cm^{-1})	Raman Frequency (cm^{-1})	Assignment	Form	Models (Tentative)
1485 sh		CH$_2$ scissor	*t*	*ttt, ttg, gtg*
1460 m	1470 s	CH$_2$ scissor	*g*	*tgt, tgg, ggg*
	1448 sh	CH$_2$ scissor	*g*	*tgg, ggg*
1352 m	1352 m	CH$_2$ wag	*g*	*ggg*
1326 (m)	1326 w	CH$_2$ wag	*t*	*ttt, ttg, gtg*
1296 m	1292 m	CH$_2$ twist	*g, t*	All
	1283 s	CH$_2$ twist	*t*	*ttt, ttg, gtg*
1249 m		CH$_2$ twist	*g*	*tgg, ggg*
	1239 m	CH$_2$ twist	*t*	
1140 sh	1134 s	C—O, C—C	*g, t*	All
1107 (s)		C—O, C—C, CH$_2$ rock	*g, t*	All
	1052 m(P)	C—O, C—C	*t*	*tgg*
1038 (m)		C—O, C—C, CH$_2$ rock	*t*	*ttt*
992 w		C—O, C—C, CH$_2$ rock	*t*	*ttt, ttg*
945 (m)		C—O, C—C, CH$_2$ rock	*g*	*tgt*
~915	919 sh	CH$_2$ rock, C—O, C—C	*g*	*tgg, ggg*
886	884 (mw)(P)	CH$_2$ rock	*g*	*ggg*
855 (m)	—	CH$_2$ rock	*g*	*tgg, ggg*
	834 m(P)	CH$_2$ rock	*t*	*ttt, ttg, gtg*
~810 (sh)	807 m(P)	CH$_2$ rock	*g*	*tgg*
	556 w			
	524 w			
	261 (P)			

Table 2.5   Some copolymer terminology (19,20)

Type	Connective	Example
	Short Sequences	
Unspecified	*–co–*	poly(*A–co–B*)
Statistical	*–stat–*	poly(*A–stat–B*)
Random	*–ran–*	poly(*A–ran–B*)
Alternating	*–alt–*	poly(*A–alt–B*)
Periodic	*–per–*	poly(*A–per–B–per–C*)
	Long Sequences	
Block	*–block–*	poly *A–block–*poly *B*
Graft	*–graft–*	poly *A–graft–*poly *B*
Star	*–star–*	*star–*poly *A*
Blend	*–blend–*	poly *A–blend–*poly *B*
Starblock	*–star–····–block–*	*star–*poly *A–block–*poly *B*
	Networks	
Cross-linked	*–cross–*	*cross–*poly *A*
Interpenetrating	*–inter–*	*cross–*poly *A–inter–cross–*poly *B*
AB-crosslinked	*–cross–*	poly *A–cross–*poly B

the molten state—*tgt*, *tgg*, *ggg*, *ttt*, *ttg*, and *gtg*. The intensities of the Raman line at 807 cm^{-1} indicate that the *tgg* isomer predominates.

2.5   COMMON TYPES OF COPOLYMERS

In the discussion above, polymers made from only one kind of monomeric unit, or mer, were considered. Many kinds of polymers contain two kinds of mers. These can be combined in various ways to obtain interesting and often highly useful materials. Some of the basic copolymer nomenclature is presented in Table 2.5 (19, 20). If three mers—*A*, *B*, and *C*—are considered, some of the possible copolymers are also named in Table 2.5. The connectives in copolymer nomenclature will be defined below.

2.5.1   Unspecified Copolymers

An unspecified sequence arrangement of different monomeric units in a polymer is represented by

$$\text{poly}(A–co–B) \tag{2.18}$$

Thus an unspecified copolymer of styrene and methyl methacrylate is named

$$\text{poly[styrene–}co\text{–(methyl methacrylate)]} \tag{2.19}$$

In the older literature, *–co–* was used to indicate a random copolymer, where the mers were added in random order, or perhaps addition preference was dictated by thermodynamic or spatial considerations. These are now distinguished from one another. Dendrimers, polycatenanes, and other novel structures are described in Section 14.5.

2.5.2 Statistical Copolymers

Statistical copolymers are copolymers in which the sequential distribution of the monomeric units obeys known statistical laws. The term *–stat–* embraces a large proportion of those copolymers that are prepared by simultaneous polymerization of two or more monomers in admixture. Thus the term *–stat–* is now preferred over *–co–* for most usage.

The arrangement of mers in a statistical copolymer of A and B might appear as follows:

$$\cdots -A-A-B-A-B-B-B-A-B-A-A-B-A-\cdots \qquad (2.20)$$

The statistical arrangement of mers A and B is indicated by

$$\mathrm{poly}(A-stat-B) \qquad (2.21)$$

See Table 2.5.

2.5.3 Random Copolymers

A random copolymer is a statistical copolymer in which the probability of finding a given monomeric unit at any given site in the chain is independent of the nature of the neighboring units at that position. Stated mathematically, the probability of finding a sequence $\ldots ABC \ldots$ of monomeric units $A, B, C, \ldots, P(\ldots ABC \ldots)$ is

$$P(\ldots ABC \ldots) = P(A) \cdot P(B) \cdot P(C) \cdots = \prod_i P(i), \qquad i = A, B, C \ldots \quad (2.22)$$

where $P(A)$, $P(B)$, $P(C)$, and so on, are the unconditional probabilities of the occurrence of the various monomeric units.

2.5.4 Alternating Copolymers

In the discussion above, various degrees of randomness were assumed. An alternating copolymer is just the opposite, comprising two species of monomeric units distributed in alternating sequence:

$$\cdots -A-B-A-B-A-B-A-B-A-B-\cdots \qquad (2.23)$$

Alternating copolymerization is caused either by A or B being unable to add itself, or the rate of addition of the other monomer being much faster than the addition of itself. An important example of an alternating copolymer is

$$\mathrm{poly}[\mathrm{styrene}-alt-(\mathrm{maleic\ anhydride})] \qquad (2.24)$$

2.5.5 Periodic Copolymers

The alternating copolymer is the simplest case of a periodic copolymer. For three mers,

$$\cdots A-B-C-A-B-C-A-B-C-\cdots \qquad (2.25)$$

the structure is indicated by

$$\text{poly}(A\text{-}per\text{-}B\text{-}per\text{-}C) \qquad (2.26)$$

2.6 NMR IN MODERN RESEARCH

2.6.1 Dilute Solution Studies: Mer Distribution

The definitions above of statistical and random copolymers are idealized. In reality, significant nonrandomness may exist. Since the physical and mechanical behavior of polymers sometimes depends critically on the exact order or lack of order in the copolymer structure, this demands special attention. Randall and Hsieh (21) studied the ^{13}C NMR spectrum of a series of copolymers of ethylene and 1-hexene,

$$CH_2 = CH - CH_2 - CH_2 - CH_2 - CH_3$$

(see Figure 2.8) in dilute solution (21). These data were analyzed in the form of sequence distributions, where E and H represent the two mers, respectively. The resulting triad distributions from this copolymer and another are shown in Table 2.6 (21). Since both of these polymers are rich in ethylene, it is not unexpected that the triad sequence EEE predominates. More interesting are the other triad concentrations, which describe the statistical arrangements of the two mers along the chain.

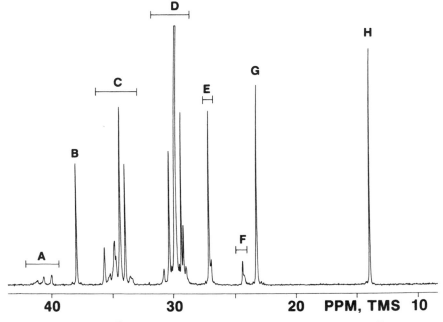

Figure 2.8 The 50.3 MHz ^{13}C NMR spectrum for an 83/17 ethylene/1-hexene copolymer. The temperature was 125 °C, and the concentration was 15% by weight in 1,2,4-trichlorobenzene (21). $\tau(8.14) = 1.86$ PPM.

Table 2.6   Triad distributions in two ethylene/1-hexene copolymers (21)

	83/17 Copolymer	97/3 Copolymer
(*EHE*)	0.098	0.031
(*EHH*)	0.053	0.000
(*HHH*)	0.022	0.000
(*HEH*)	0.043	0.000
(*HEE*)	0.164	0.061
(*EEE*)	0.620	0.908

Note: These copolymers are typical of the "linear low density polyethylenes," LLDPE, used to make shopping bags, etc.

With the information given in Table 2.6, a "run number" may be calculated. A run number, first introduced by Harwood and Ritchey (22), is defined as the average number of like mer sequences or "runs" occurring in a copolymer per 100 mers. This is calculated as follows:[†]

$$(H) = (HHH) + (EHH) + (EHE) \tag{2.27}$$

$$(E) = (EEE) + (HEE) + (HEH) \tag{2.28}$$

$$\text{Run number} = (\tfrac{1}{2})(HE)$$

$$= (EHE) + \tfrac{1}{2}(EHH)$$

$$= (HEH) + \tfrac{1}{2}(HEE) \tag{2.29}$$

The average sequence lengths can then be calculated as follows:

$$\text{Average "}E\text{" sequence length} = (E)/\text{run number} \tag{2.30}$$

$$\text{Average "}H\text{" sequence length} = (H)/\text{run number} \tag{2.31}$$

2.6.2   High-Resolution NMR in the Solid State

While dilute solution ^{1}H and ^{13}C NMR spectra measured under classical Fourier transform conditions tend to be sharp and narrow, similar NMR spectra on solid polymers are usually very broad. Recent advances, however, have made solid-state techniques more valuable to polymer science (23). Improvements include dipolar decoupling, crosspolarization, CP, high-powered decoupling, DD, and magic-angle spinning techniques, MAS, which are often combined (24). Beyond studies of homopolymers and statistical copolymers, solid-state NMR can characterize polymer blends and composites, which frequently have supermolecular organization that disappears in solution. Since most polymers are used in the solid state, studies showing how the polymer is organized, and how organization changes with processing, provide much needed basic and engineering information.

Such studies may combine ^{1}H and ^{13}C spectra to obtain more detailed information. For example, two-dimensional WISE (*WI*deline *SE*paration) experiments allow infor-

---

[†] Note that the number of runs of *both* kinds of mers is twice that calculated here.

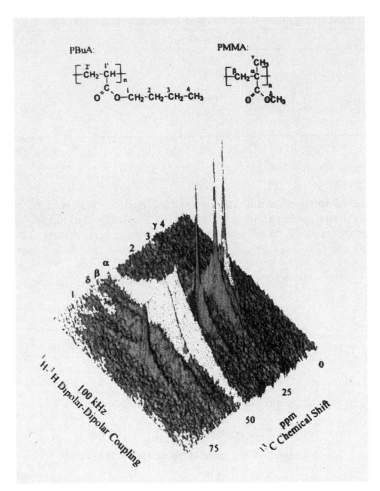

Figure 2.9 A 2D WISE spectrum of a blend of poly(n-butyl acrylate) and poly(methyl methacrylate), with assignment of ^{13}C chemical shifts.

mation obtained from the isotropic chemical shift in the ^{13}C spectra and the proton line shape in the ^{1}H spectra, respectively, to be displayed; see Figure 2.9 (25). In Figure 2.9 an NMR spectra of poly(n-butyl acrylate)–*blend*–poly(methyl methacrylate) is displayed. (See Section 2.7 for the definition of *blend*.) The material was synthesized in latex form (see Section 4.5) using two sequential free-radical polymerizations: first, poly(n-butyl acrylate) was synthesized, and then poly(methyl methacrylate), forming a core-shell structure. This spectrum reveals the existence of a pure poly(n-butyl acrylate) phase, a pure poly(methyl methacrylate) phase, and an *interphase* region where the two components are mixed. The strength of a film formed from such a latex depends on the thickness of the interphase; see Chapter 13.

Two-dimensional NMR studies of polymer solutions (26) can also be used to detect and assign NMR resonances from lesser chain structures in polymers (e.g., chain ends, defects, branches, and block junctions), critical in characterizing many synthetic polymers.

2.7 MULTICOMPONENT POLYMERS

The statistical, random, and alternating copolymers above describe sequence lengths of one, two, three, or at most several mers. This section treats cases where whole polymer chains are linked together to form still larger polymer structures (11). These structures have been variously named "polymer alloys," or "polymer blends," but the term "multi-component polymers" is used here to describe this general class of materials.

2.7.1 Block Copolymers

A block copolymer contains a linear arrangement of blocks, a block being defined as a portion of a polymer molecule in which the monomeric units have at least one constitutional or configurational feature absent from the adjacent portions. A block copolymer of A and B may be written

$$\cdots-A-A-A-A-A-A-B-B-B-B-B-B-B-B-B-B\cdots \qquad (2.32)$$

Note that the blocks are linked end on end. Since the individual blocks are usually long enough to be considered polymers in their own right, the polymer is named (19)

$$\text{poly } A-block-\text{poly } B \qquad (2.33)$$

An especially important block copolymer is the triblock copolymer of styrene and butadiene (11),

$$\text{polystyrene}-block-\text{polybutadiene}-block-\text{polystyrene} \qquad (2.34)$$

In the older literature, $-b-$ was used for $-block-$, and $-g-$ was used for $-graft-$ (below). Only the first poly was indicated. Structure (2.34) was then written

$$\text{poly(styrene}-b-\text{butadiene}-b-\text{styrene)} \qquad (2.35)$$

Table 2.7 defines a number of terms used in block copolymer terminology, as well as other structures described later (20). These structures are also illustrated in Figure 2.10 (20).

2.7.2 Graft Copolymers

A graft copolymer comprises a backbone species, poly A, and a side-chain species, poly B. The side chains comprise units of mer that differ from those comprising the backbone chain. If the two mers are the same, the polymer is said to be branched. The name of a graft copolymer of A and B is written (19) in this order:

$$\text{poly } A-graft-\text{poly } B \qquad (2.36)$$

Table 2.7   Specialized nomenclature terms (20)

Link	Covalent chemical bond between two monomeric units, or between two chains.
Chain	Linear polymer formed by covalent linking of monomeric units.
Backbone	Used in graft copolymer nomenclature to describe the chain onto which the graft is formed.
Side chain	Grafted chain in a graft copolymer.
Cross-link	Structure bonding two or more chains together.
Network	Three-dimensional polymer structure, where (ideally) all the chains are connected through cross-links.
Multicomponent polymer, multipolymer, and multicomponent molecule	General terms describing intimate solutions, blends, or bonded combinations of two or more polymers.
Copolymer	Polymers that are derived from more than one species of monomer.
Block	Portion of a polymer molecule in which the monomeric units have at least one constitutional or configurational feature absent from the adjacent portions.
Block copolymer	Combination of two or more chains of constitutionally or configurationally different features linked in a linear fashion.
Graft copolymer	Combination of two or more chains of constitutionally or configurationally different features, one of which serves as a backbone main chain, and at least one of which is bonded at some point(s) along the backbone and constitutes a side chain.
Polymer blend	Intimate combination of two or more polymer chains of constitutionally or configurationally different features, which are not bonded to each other.
Conterminous	At both ends or at points along the chain.
AB-cross-linked copolymer	Polymer chain that is linked at both ends to the same or to constitutionally or configurationally different chain or chains; a polymer cross-linked by a second species of polymer.
Interpenetrating polymer network	Intimate combination of two polymers both in network form, at least one of which is synthesized and/or cross-linked in the immediate presence of the other.
Semi-interpenetrating polymer network[a]	Combination of two polymers, one cross-linked and one linear, at least one of which was synthesized and/or cross-linked in the immediate presence of the other.
Star polymer	Three or more chains linked at one end through a central moiety.
Star block copolymer	Three or more chains of different constitutional or configurational features linked at one end through a central moiety.

[a] Also called a pseudo-interpenetrating polymer network. See D. Klempner, K. C. Frisch, and H. L. Frisch, *J. Elastoplastics*, **5**, 196 (1973).

Although many of the block copolymers reported in the literature are actually highly blocked, some of the most important "graft copolymers" described in the literature have been shown to be only partly grafted, with much homopolymer being present. To some extent, then, the term graft copolymer may also mean, "polymer *B* synthesized in the immediate presence of polymer *A*." Only by a reading of the context can the two meanings be distinguished.

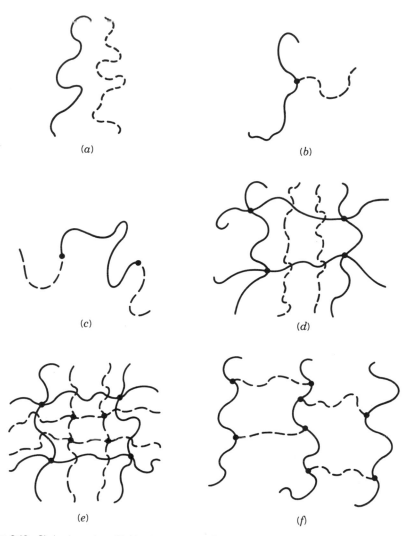

Figure 2.10 Six basic modes of linking two or more polymers are identified (20). (*a*) A polymer blend, constituted by a mixture or mutual solution of two or more polymers, not chemically bonded together. (*b*) A graft copolymer, constituted by a backbone of polymer I with covalently bonded side chains of polymer II. (*c*) A block copolymer, constituted by linking two polymers end on end by covalent bonds. (*d*) A semi-interpenetrating polymer network constituted by an entangled combination of two polymers, one of which is cross-linked, that are not bonded to each other. (*e*) An interpenetrating polymer network, abbreviated IPN, is an entangled combination of two cross-linked polymers that are not bonded to each other. (*f*) *AB*-cross-linked copolymer, constituted by having the polymer II species linked, at both ends, onto polymer I. The ends may be grafted to different chains or the same chain. The total product is a network composed of two different polymers.

2.7.3   AB–Cross-linked Copolymer

The polymers of Section 2.7.2 are soluble, at least in the ideal case. A conterminously grafted copolymer has polymer B grafted at both ends, or at various points along the structure to polymer A, and hence it is a network and not soluble. See structure (f) in Figure 2.10, which is sometimes called a conterminously grafted copolymer.

2.7.4   Interpenetrating Polymer Networks

This is an intimate combination of two polymers in network form. At least one of the polymers is polymerized and/or cross-linked in the immediate presence of the other (27). While ideally the polymers should interpenetrate on the molecular level, actual interpenetration may be limited owing to phase separation. (Phase separation in polymer blends, grafts, blocks, and interpenetrating polymer networks is the more usual case and is discussed in Chapter 4.)

2.7.5   Other Polymer–Polymer Combinations

According to new nomenclature, a polymer blend is accorded the connective *–blend–*. Many of these blends are prepared by highly sophisticated methods and are actually on a parallel with blocks, grafts, and interpenetrating polymer networks.

   Block copolymers may also be arranged in various star arrangements. In this case polymer A radiates from a central point, with a number of arms to be specified. Then polymer B is attached to the end of each arm.

2.7.6   Separation and Identification of Multicomponent Polymers

The methods of separation and identification of multicomponent polymers are far different from the methods described previously for the statistical type of polymer. First, only the blends are separable by extraction techniques. The remainder are bound together by either chemical bonds or interpenetration. The interpenetrating polymer networks and the conterminously grafted polymers are insoluble in all simple solvents and do not flow on heating. The graft and block copolymers, on the other hand, do dissolve and flow on heating above T_f and/or T_g.

   Most, but not all, of the multicomponent polymer combinations exhibit some type of phase separation, which is discussed in Chapter 4. Where the polymers are stainable and observable under the electron microscope, characteristic morphologies are often manifest. The principal polymers that are stainable include the diene types and those containing ester groups. For those combinations exhibiting phase separation, two characteristic glass temperatures are also usually observed.

2.8   CONFORMATIONAL STATES IN POLYMERS

This chapter would not be complete without a further discussion of the various conformational states in polymers (28–34). The rotational potential energy diagram (Figure 2.11) (23) indicates three stable positions or conformations—the *trans*, the *gauche* plus, and the *gauche* minus.

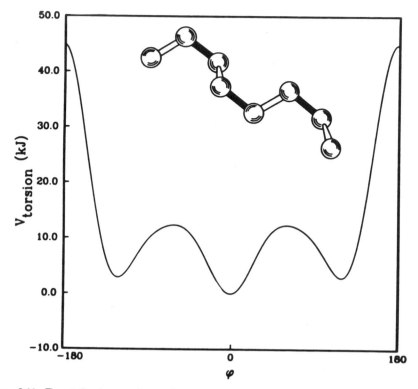

Figure 2.11 The rotational energy diagram for carbon–carbon single bonds in a hydrocarbon polymer such as polyethylene. Illustrated are the energy wells of the *trans, gauche* plus, and the *gauche* minus positions (28).

The barriers separating the three conformational states have heights several times the thermal energy, kT, which means that the lifetime in a given state will be much longer than the vibration periods within the well. The quantity k is Boltzmann's constant. The sequence of bond conformations at a given instant defines the rotational isomeric state of the chain.

Helfand and co-workers investigated the various transitions among the conformational states by means of computer simulations (30,31) and by applications of a kinetic theory (32–34). This analysis yielded the details of the long periods of motion near the bottom of the conformational wells, and the occasional transition to a different well. The activation energy for the transition was found to be approximately equal to the barrier height between the two states, as is required.

One surprising finding was that the transitions frequently occur in pairs, cooperatively. Immediately following the transition of one bond, a strong increase in the transition rate of its second-neighbor bonds was found. The intermediate bond usually remained unchanged. Thus the transitions might be

$$g^{\pm} tt \rightleftharpoons ttg^{\pm} \tag{2.37}$$

$$ttt \rightleftharpoons g^{\pm} tg^{\mp} \tag{2.38}$$

The significance of this observation arises from the geometric properties of the two transitions. In both cases the first two and last two bonds translate relative to each other in opposite directions. Except for the central bond, the final state of each bond is parallel to its initial state. The cooperative pair transitions of equations (2.37) and (2.38) greatly reduce the motion of the long tail chains attached to the rotating segment, and hence the frictional resistance that the tails would present to the transition (28).

The rate of the transitions is given by two factors. First is the Arrhenius factor $\exp(-E_{act}/kT)$, where E_{act} represents the energy of activation. The Arrhenius factor yields the probability of being near the saddle point joining the two energy wells in question. This is multiplied by a factor reflecting the frequency of saddle traversal. The "reaction" coordinate moves along the path of steepest descent from the saddle point. Helfand (28) points out that the two cooperative bond changes must take place in a coherent, sequential fashion to minimize the effect of the activation energy barrier.

The *trans–gauche* transitions underlie the diffusional motions of de Gennes (Section 5.4), the Shatzki transition (Section 6.4.1), and the glass transition itself.

2.9 ANALYSIS OF POLYMERS DURING MECHANICAL STRAIN

Many polymers in the solid (bulk) state undergo strain, either during processing such as extrusion, molding, and spinning or when in service and under load. Studies using solid-state NMR and FTIR showing how polymers respond to strain have contributed greatly to improving their mechanical behavior.

As *st*-polypropylene chains became oriented on stretching, Sozzani et al. (35) found that the *trans–gauche* ratio of bond placements shifted in favor of the *trans*. This was especially noticeable in the necking region; see Figure 2.12 (36), where the switch from *tggt* and *gttg* to nearly all *tttt* placements was observed.

Table 2.8 summarizes some of the effects noted during mechanically induced strain. In each case, of course, it takes work to orient the sample. Part of that work is absorbed by molecular orientation, rupturing hydrogen bonding, increased crystallization, and so on.

An interesting question in the literature relates to the actual stretching of covalent bonds during mechanical strain. It has been postulated (see Section 11.5) that at high strains, the backbone carbon atom bond distances increase, with concomitant excitation of the bond above the ground state. Can this be observed instrumentally?

Bretzlaff and Wool (37) performed stress–strain studies on *it*-polypropylene, finding that the frequency shift followed the relation

$$\Delta v_\sigma = v(\sigma) - v(0) = \alpha_x \sigma \tag{2.39}$$

where Δv_σ represents the mechanically induced peak frequency shift, α_x is the mechanically induced frequency shifting coefficient at constant temperature T, and σ is the applied uniaxial stress. Most of the frequency shifts were to lower frequencies.

However, the interpretation of the data was complicated by the existence of anisotropic crystal field forces, in addition to interchain perturbing forces, and the question remains unresolved.

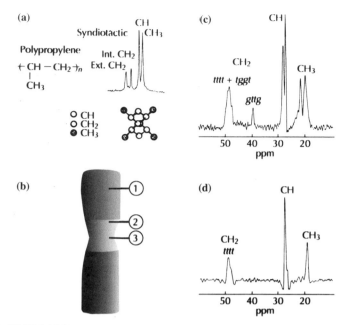

Figure 2.12 CP/MAS NMR spectrum of *syn*-polypropylene: (*a*) Spectrum of a crystalline powder; (*b*) A sample of cold-stretched *st*-polypropylene, showing elongation with necking; (*c*) Sample of cold-stretched *st*-polypropylene, far from the neck region (positions 1 and 2), and (*d*) Spectrum of a cold-stretched *st*-polypropylene, in the necking region (position 3). (Reproduced, courtesy of P. Sozzani).

Table 2.8   Molecular properties of polymers during strain

Instrument	Quantity Measured	Reference
CP MAS ^{13}C NMR	*Trans–gauche* shifting to higher *trans* levels with increasing *st*-polypropylene chain orientation	(a)
CP MAS ^{13}C NMR	Crystallization of poly(tetramethylene oxide) block on orientation of stretched poly(butylene terephthalate)–*block*–poly(tetramethylene oxide) elastomers ($\sim$700% strain)	(b)
FTIR	Interchain hydrogen bonding in polyurethanes decreasing with increasing strain	(c)
FTIR	Chain orientation in glassy epoxy resins increasing with plastic deformation (10% strain); absorbances measured parallel and perpendicular to the stretching direction	(d)

References: (a) P. Sozzani, M. Galimberi, and G. Balbontin, *Makromol. Chem., Rapid Commun.*, **13**, 305 (1992). (b) A. Schmidt, W. S. Veeman, V. M. Litvinov, and W. Gabriëlse, *Macromolecules*, **31**, 1652 (1998). (c) S. L. Huang and J. Y. Lai, *Eur. Polym. J.* **33**, 1563 (1997). (d) T. Scherzer, *J. Polym. Sci., Part B: Polym. Phys. Ed.*, **34**, 459 (1996).

2.10 PHOTOPHYSICS OF POLYMERS

Photophysics is the science of the absorption, transfer, localization, and emission of electromagnetic energy, with no chemical reactions occurring. By contrast, photo-chemistry deals with those processes by which light interacts with matter so as to induce chemical reactions (38). The portion of the electromagnetic spectrum of interest to pho-tophysics includes both the ultraviolet and the visible wavelength ranges. In many of the experiments performed, light is absorbed in the ultraviolet range, and a fluorescence is measured in the visible range. Often two molecular moieties must be in proper juxtapo-sition for the phenomenon of interest to be measured.

The first step, of course, is the absorption of electromagnetic energy, transforming it into excited molecular states,

$$A + h\nu = A^* \tag{2.40}$$

where A is the molecule to be excited, A^* represents the excited state, and $h\nu$ represents the electromagnetic energy absorbed.

Next most important is energy migration, either along the chain or among the chains. This allows the energy to reach the sites of interest. Such energy migration mimics that observed in the ordered chlorophyll regions of green plant chloroplasts, that is, the an-tenna chlorophyll pigments (38–40). These light-gathering antennas are composed of chlorophylls, carotenoids, and special pigment-containing proteins. These large organic molecules, some of them natural polymers, harvest light energy by absorbing a photon of light and storing the absorbed energy temporarily in the form of an electron in an excited singlet energy state. The energy migrates throughout the system of antennas within about 100 ps, being transmitted to the reaction center protein (40). Hence, in polymer photophysics, this phenomenon is termed the "antenna effect."

2.10.1 Quenching Phenomena

In situations where bimolecular encounters dominate, typical for polymers, such en-counters may lead to an electronic relaxation of the system, termed quenching. In gen-eral, such collisions may be written

$$A^* + B = A + B^* \tag{2.41}$$

where the excited molecule A^* encounters another molecule B. Most often, the bimo-lecular interaction is between an excited molecule in the singlet state and a quencher molecule in the ground state.

The possible bimolecular quenching process includes (41) (a) chemical reaction, (b) enhancement of nonradiative decay, (c) electronic energy transfer, or (d) complex formation.

Chemical reactions involve cross-linking, degradation, and rearrangement. Enhance-ment of nonradiative decay is not discussed. Electronic energy transfer involves exo-thermic processes, where part of the energy is absorbed as heat, and part is emitted via fluorescence or phosphorescence from the donor molecule.

Polarized energy is absorbed in fluorescence depolarization. This phenomenon is also known as luminescence anisotropy (39). If the chain portions are moving at about the same rate as the reemission, the energy is partly depolarized. The extent of depolarization is related to the various motions and their relative rates. Important are the inherent degree of anisotropy of the fluorescent chromophore, the degree of energy migration, and rotation of the chromophore during its excited state lifetime. From steady-state and transient emission anisotropic measurements, rotational relaxation times can be deduced.

Complex formation between two species is very important in photophysics. Two terms need definition—exciplex and excimer. An exciplex is an *excited* state *complex* between two different kinds of molecules, one being initially excited and the other in the ground state. A complex between an excited molecule and a ground-state molecule of the same species is called an excimer, being derived from the phrase *excited dimer* (38). Excimer formation is emphasized in this discussion. In excimer formation, excited-state complexes are usually formed between two aromatic structures. Resonance interactions lead to a weak intermolecular force, which binds the two species together, involving π bonds. Such excimers exhibit strong fluoresence characterized by being red-shifted with respect to the uncomplexed fluorescence.

Features of excimer fluorescence important in polymer characterization include intensity of radiation, intensity changes, decay rates, extent of frequency shifting of the fluorescence, and depolarization effects.

2.10.2 Excimer Formation

The formation of an excimer from an excited-state moiety A* and a ground-state moiety A may be illustrated as

$$A^* + A = (AA)^* \tag{2.42}$$

The excimer, (AA)*, decomposes due to a variety of interactions, the most important one being the emission of fluorescence:

$$(AA)^* = 2A + h\nu_E \tag{2.43}$$

where the emitted frequency, ν_E, is lower than the input frequency, the remaining energy being required to separate the two moieties, and h is Planck's constant.

The stability of excimers can be examined with the aid of Figure 2.13 (41). Highly stable excimers lie in the bottom of the energy well. Figure 2.14 illustrates the relationships among the initial light frequency, single mer emission, and excimer emission. The fluoresced light is more red-shifted in the more stable excimers, because it takes more energy to break them apart. There are three parameters determining excimer stability. With reference to the relative positions of the two aromatic groups in Figure 2.15 (38), these are the angle that the two planes make with each other, their distance apart, and their lateral displacement.

2.10.3 Experimental Studies

2.10.3.1 Microstructure of Polystyrene One of the first polymers studied was polystyrene (Figure 2.16) (42). The single mer emission is at about 290 nm, with the band at 335 nm being attributed to excimer emission. In dilute solutions, excimer for-

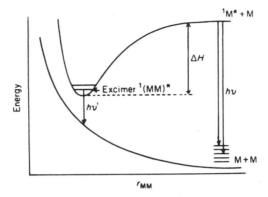

Figure 2.13   An energy-well diagram for excimer formation, illustrating the effects as a function of the distance, r_{MM}, between the two moieties (41).

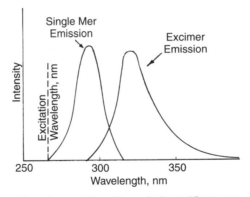

Figure 2.14   Schematic description of the excitation and fluorescence phenomena.

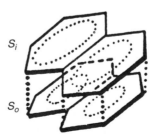

Figure 2.15   Geometry of the naphthalene excimer (38).

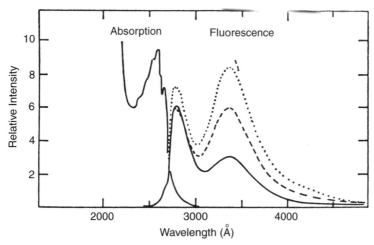

Figure 2.16 Room-temperature absorption and fluorescence spectra of atactic polystyrene in 1,2-dichloro-ethane (1×10^{-3} M). The three fluorescence curves denote nitrogen solution (dotted line), aerated solution (broken line), and oxygenated solution (solid line) (42).

mation is largely intramolecular rather than intermolecular, the excimers arising from adjacent phenyl groups. This is because the chains are far separated from one another. In the bulk state, excimers may involve neighboring chains as well. Excimer formation in atactic polystyrene requires the tt conformation of the meso-isomer, or the tg^- or g^-t isomers of the racemic isomer.

David et al. (43) found that the ratio of excimer emission to single mer emission, I_D/I_M, increased from 10 to 100 times with increasing degree of tacticity; see Table 2.9. Since isotactic polystyrene exists in a 3-1 helix in the crystalline state with ring spacings of the order of 3 Å, this arrangement provides more excimer sites than in the atactic configuration (43).

There is generally an increase in the ratio I_D/I_M as the molecular weight of the polymer increases. This has been taken as a confirmation of the energy migration along polymer chains.

2.10.3.2 Excimer Stability While the adjacent phenyl rings in polystyrene form relatively stable excimers, putting substituent groups on the phenyls may alter the bonding energy. Chakraborty et al. (44) substituted bulky t-butyl groups in the para

Table 2.9 Ratio of excimer to normal fluorescence intensity (I_D/I_M) in polystyrene at 77°C (43)

Polystyrene Type	I_D/I_M
Atactic, unoriented	1.43
Atactic, oriented ($F = 0.07$)	1.67
Atactic, oriented ($F = 0.1$)	2.08
Isotactic (25% crystalline)	10
Isotactic (35% crystalline)	100

Table 2.10 Relative phenyl positions

Quantity	Polystyrene	Poly(p-t-butylstyrene)
Angle from parallel, degrees	6	22
Center of ring distance, Å	3.0	3.5
Lateral displacement, Å	0.2	1.3

position,

$$(2.44)$$

The t-butyl group forces the phenyl groups away from their ideal excimer positions, resulting in a blue shift of the fluorescence to 320 nm. Thus excimers in poly(p-t-butylstyrene) are less stable than those in unsubstituted polystyrene. Analysis of the resulting blue shift, together with molecular modeling, led to a comparison of the relative phenyl positions, shown in Table 2.10.

2.11 CONFIGURATION AND CONFORMATION

Through the development of instrumental techniques such as infrared, Raman spectroscopy, nuclear magnetic resonance, X-ray diffraction, fluorescence, and other methods, the organization of the individual atoms along the chain has gradually become clear. In many cases two or more isomeric forms may be simultaneously present. The configurational properties of a polymer determine if it is crystallizable and, if so, its melting temperature.

If two or more monomeric units are used to make one polymer, a copolymer is formed. The statistical, random, alternating, and periodic copolymers show the relationship of two mers on an individual basis. The block, graft, conterminously grafted, and interpenetrating polymer network copolymers comprise large portions of chain or chains containing only one mer. It must be pointed out that each of these may be subject to being composed of the various tacticities and so forth that describe the configurational properties.

By contrast, the conformational properties of a polymer are determined by rotations about single bonds. The overall shape and size of the chain and many of its motions are determined by its conformation. Both conformation and configuration contribute, albeit in different ways, to the behavior of the polymer.

REFERENCES

1. F. W. Harris, *J. Chem. Ed.*, **58** (11), 836 (1981).

2. M. Malanga and O. Vogl, *Polym. Eng. Sci.*, **23**, 597 (1983).

3. J. L. Koenig, *Chemical Microstructure of Polymer Chains*, Wiley-Interscience, New York, 1980, chs. 6–8.

4. P. C. Painter and M. M. Coleman, in *Static and Dynamic Properties of the Polymeric Solid State*, R. A. Pethrick and R. W. Richards, eds., D. Reidel, Boston, 1982.

5. A. Winston and P. Wichackeewa, *Macromolecules*, **6**, 200 (1973).

6. J. L. Koenig, *Spectroscopy of Polymers*, 2nd ed., Elsevier, Amsterdam, 1999.

7. G. Natta, *Makromol. Chem.*, **35**, 94 (1960).

8. S. W. Cornell and J. L. Koenig, *J. Polym. Sci.*, *A2*, **7**, 1965 (1969).

9. H. W. Siesler and K. Holland-Moritz, *Infrared and Raman Spectroscopy of Polymers*, Dekker, New York, 1980, Chap. 2.

10. F. A. Bovey, G. V. D. Tiers, and G. Filipovitch, *J. Polym. Sci.*, **38**, 73 (1959).

11. J. A. Manson and L. H. Sperling, *Polymer Blends and Composites*, Plenum, New York, 1976, Chap. 1.

12. G. Odian, *Principles of Polymerization*, 3rd ed., Wiley-Interscience, New York, 1991, Chap. 8.

13. F. A. Bovey, Chap. 1 in *NMR and Macromolecules*, J. C. Randall Jr., ed., ACS Symposium Series No. 247, American Chemical Society, Washington, DC, 1984.

14. F. A. Bovey and G. V. D. Tiers, *J. Polym. Sci.*, **44**, 173 (1960).

15. F. A. Bovey, *Polymer Conformation and Configuration*, Academic Press, New York, 1969, Chap. 1.

16. D. O. Hummel, *Atlas of Polymer and Plastics Analysis*, 2nd ed., Hanser, Munich, 1978.

17. W. Klopffer, *Introduction to Polymer Spectroscopy*, Springer, Berlin, 1984.

18. J. R. Fried, *Polymer Science and Technology*, Prentice-Hall, Englewood Cliffs, NJ, 1995.

19. W. Ring, I. Mita, A. D. Jenkins, and N. M. Bikales, *Pure and Appl. Chem.*, **57**, 1427 (1985).

20. J. Kahovec, P. Kratochvil, A. D. Jenkins, I. Mita, I. M. Papisov, L. H. Sperling, and R. F. T. Stepto, *Pure & Appl. Chem.*, **69**, 2511 (1997).

21. J. C. Randall and E. T. Hsieh, in *NMR and Macromolecules*, J. C. Randall Jr., ed., American Chemical Society, Washington, DC, 1984.

22. H. J. Harwood and W. M. Ritchey, *Polym. Lett.*, **2**, 601 (1964).

23. J. L. Koenig, *Spectroscopy of Polymers*, 2nd ed., Elsevier, Amsterdam, 1999.

24. L. Mathias, ed., *Solid State NMR of Polymers*, Plenum Publishers, New York, 1989.

25. K. Landfester, C. Boeffel, M. Lambla, and H. W. Spiess, *Macromolecules*, **29**, 5972 (1996).

26. P. L. Rinaldi, D. G. Ray III, V. E. Litman, and P. A. Keifer, *Polym. International*, **36**, 177 (1995).

27. L. H. Sperling, *Polymeric Multicomponent Materials: An Introduction*, Wiley, New York, 1997.

28. E. Helfand, *Science*, **226** (4675), 647 (1984).

29. W. H. Stockmeyer, *Pure Appl. Chem. Suppl. Macromol. Chem.*, **8**, 379 (1973).

30. E. Helfand, Z. R. Wasserman, and T. A. Weber, *Macromolecules*, **13**, 526 (1980).

31. T. A. Weber and E. Helfand, *J. Phys. Chem.*, **87**, 2881 (1983).

32. E. Helfand, *J. Chem. Phys.* **54**, 4651 (1971).

33. J. Skolnick and E. Helfand, *J. Chem. Phys.*, **72**, 5489 (1980).

34. E. Helfand and J. Skolnick, *J. Chem. Phys.*, **77**, 3275 (1982).

35. P. Sozzani, M. Galimberti, and G. Balbontin, *Makromol. Chem., Rapid Commun.*, **13**, 305 (1992).

36. A. L. Segre and D. Capitani, *TRIP (Trends in Polym. Sci.)*, **1**, 280 (1993).

37. R. S. Bretzlaff and R. P. Wool, *Macromolecules*, **16**, 1907 (1983).

38. J. Guillet, *Polymer Photophysics and Photochemistry*, Cambridge University Press, Cambridge, England, 1985.

39. C. E. Hoyle, in *Photophysics of Polymers*, C. E. Hoyle and J. M. Torkelson, eds., ACS Symposium Series No. 358, ACS Books, Washington, DC, 1987.

40. J. R. Norris and M. Schiffer, *C&E News*, **68**(31), July 30, 22 (1990).

41. D. Phillips, in *Polymer Photophysics*, D. Phillips, ed., Chapman and Hall, London, 1985.

42. M. T. Vala, Jr., J. Haebig, and S. A. Rice, *J. Chem. Phys.*, **43**, 886 (1965).

43. C. David, N. Putman-de Lavarielle, and G. Geuskens, *Eur. Polym. J.*, **10**, 617 (1974).

44. D. K. Chakraborty, K. D. Heitzhaus, F. J. Hamilton, H. J. Harwood, and W. L. Mattice, *Polym. Prepr.*, **31**(2), 590 (1990).

GENERAL READING

H. Duddeck, W. Dietrich, and G. Toth, *Structure Elucidation by Modern NMR: A Workbook*, 3rd ed., Springer, Darmstadt, 1998.

C. E. Hoyle and J. M. Torkelson, eds., *Photophysics of Polymers*, ACS Books, Washington, DC, 1987.

J. L. Koenig, *Spectroscopy of Polymers*, 2nd ed., Elsevier, Amsterdam, 1999.

G. Odian, *Principles of Polymerization*, 3rd ed., Wiley-Interscience, New York, 1991.

P. C. Painter, M. M. Coleman, and J. L. Koenig, *The Theory of Vibrational Spectroscopy and Its Application to Polymeric Materials*, Wiley-Interscience, New York, 1982.

S. R. Sandler and W. Karo, *Sourcebook of Advanced Polymer Laboratory Preparation*, Academic Press, San Diego, 1998.

S. R. Sandler, W. Karo, J. A. Bonesteel, and E. M. Pearce, *Polymer Synthesis and Characterization: A Laboratory Manual*, Academic Press, San Diego, 1998.

STUDY PROBLEMS

1. What are the chemical structures of isotactic, syndiotactic, and atactic polystyrene?

2. (a) What are the chemical structures of *cis*- and *trans*-polybutadiene, and (b) the 1,2- and 3,4-structures of polyisoprene?

3. How do head-to-head and head-to-tail structures of poly(methyl methacrylate) differ?

4. Show the structures of statistical and alternating copolymers of vinyl chloride and ethyl acrylate.

5. *Cis*-polyisoprene has been totally hydrogenated. What is the name of the new polymer formed?

6. What are the two possible triblock copolymer structures of polybutadiene and cellulose?

7. Using Table 2.6, calculate the run numbers and average sequence lengths for the two poly(ethylene–*stat*–1-hexene) copolymers. Do they indeed appear to be statistical copolymers?

8. A graft copolymer is formed with polybutadiene as the backbone and polystyrene as the side chains. What is the name of this material?

9. Compare and contrast infrared and Raman spectra with NMR techniques for their capability of characterizing (a) tacticity and (b) *cis* and *trans* double bonds in polymers.

10. Chemical nomenclature forms the alphabet of polymer science. (a) What is the chemical structure of *it*–poly(vinyl chloride)–*block*–cis-1,4-polyisoprene? (b) Poly(vinyl acetate) is totally hydrolyzed. What new polymer is formed? What polymer is formed if the hydrolysis is only partial?

11. In the accompanying structures, P_1 is poly(vinyl acetate), P_2 is poly(ethyl acrylate), and P_3 is polystyrene. What are the chemical names of these structures?

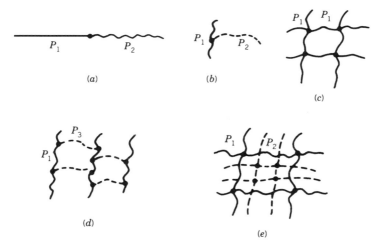

Figure P2.11 Various polymer structures.

12. Your new assistant copolymerized styrene and *n*-butyl acrylate, 50/50 mole-%. "I made it in such a way as to produce an alternating copolymer," he said.

"No," you replied, "I think you really made a statistical copolymer."

At that instant, you boss walks in. "I'll bet this new synthesis really made a block copolymer," she volunteers.

(a) Using the concepts of photophysics, devise an experiment to distinguish the three possibilities. How would the resulting data look as a function of composition? (Prepare appropriate plots.) (b) NMR is also a powerful tool. Use a hydrogen NMR experiment in dilute solution to distinguish the three possibilities, again preparing hypothetical figures or tables of resulting data.

13. There was an old Prof.
Who lived in a lab
He had so many students
He became an old crab.

He gave some an IR
And some an NMR
"Show me a stretch
With an old fashioned *kvetch*,
And earn an *A* who you are!"

Saying so, he handed the students a polyamide-66 sample that had been stretched (while hot) ×2, ×4, and ×6, as well as the unstretched sample, all now at room temperature, crystallized. "What molecular changes do you think took place?" He asked. Plot your anticipated results as a function of strain.

APPENDIX 2.1   ASSORTED ISOMERIC AND COPOLYMER MACROMOLECULES

In addition to the types of isomeric and copolymer structures illustrated in the text of the chapter, there are several other structures of which the student must be aware.

Proteins

Proteins have the general structure

$$\left(\begin{array}{c} \text{H} \quad \text{H} \quad \text{O} \\ \| \quad\;\; | \quad\;\; \| \\ \text{N} - \text{C} - \text{C} \\ | \\ \text{R} \end{array}\right)_n \tag{A2.1.1}$$

and hence are sometimes called nylon 2. There are 20 common types of amino acids (A1), each with a specific type of R group. In proteins these amino acids follow very specific sequences, which frequently differ from species to species of plant or animal. (Note the various kinds of insulin, for example.) However, in the broad sense of the term, they are copolymers.

In addition to the copolymer structure, proteins are also optically active. All amino acids except glycine possess at least one asymmetric carbon atom, as illustrated in equation (A2.1.1). The language commonly used to characterize the structure is different, however. The Fischer convention of designating the active group as *D*- or *L*- is used (A1). All the natural proteins contain the *L* configuration. Woe to the person who feeds an earth-bound creature the *D* configuration! This difference, of course, has generated many a science fiction story.

The proteins also have specific spatial arrangements. This is partly aided by sulfur–sulfur cross-links linking cystine residues. On cooking, these bonds break or rearrange, which is called denaturing. The proteins and the cellulose and starch that follow are also examples of biopolymers, synthesized by Mother Nature.

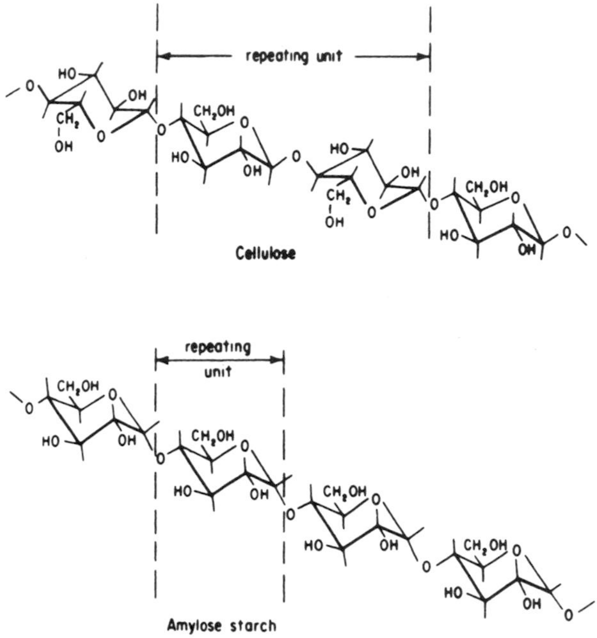

Figure A2.1.1 A comparison of the structures of cellulose and starch. Note that cellulose has two mers in the repeat unit, caused by β-1,4-glucoside linkage (12).

Cellulose and Starch

Both of these natural polymers are composed of glucose, a six-membered ring (see Chapter 1). Both are linked together at the 1,4 position but differ in that cellulose has a β linkage and starch has the α linkage. The β linkage has the effect of alternating the structure of the glucosides in an up-down-up-down configuration, while the α linkage makes them all up-up-up-up (See Figure A2.1.1). The difference in physical properties reflects the altered chemistry. Cellulose is highly crystalline and nondigestible by humans. Starch is much less crystalline but highly digestible, as too many overweight people know. The difference in digestibility is caused by the lack of an enzyme (a protein!) to attack the β linkage. It must be pointed out that there are many structurally different polysaccharides in nature.

Ditactic Polymers

These structures are generated by polymerizing 1,2-disubstituted ethylenes having the general structure

$$\left(\begin{array}{cc} H & H \\ | & | \\ -C - C - \\ | & | \\ R' & R \end{array} \right)_n \tag{A2.1.2}$$

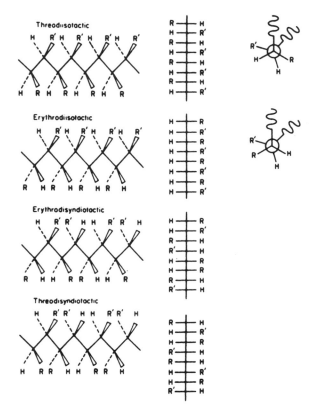

Figure A2.1.2   Ditactic structures from 1,2-disubstituted ethylenes (12).

Ditacticity occurs when the individual carbon atoms possess specific stereoisomerism. Four different stereoregular structures may be identified, as shown in Figure A2.1.2 (12).

REFERENCE

A1. H. K. Salzberg, in *Encyclopedia of Polymer Science and Technology*, Vol. 11, N. M. Bikales, ed., Interscience, New York, 1969, p. 620.

3

MOLECULAR WEIGHTS
AND SIZES

3.1 INTRODUCTION

3.1.1 Polymer Size and Shape

The general problem of the size and shape of polymer molecules stands at the very heart of polymer science and engineering. If the molecular weight[†] and molecular weight distribution (MWD) are known along with a good understanding of the polymer chain conformation, many mechanical and rheological properties can be predicted. While the question of molecular weights pervades the entire area of the chemical arts, polymers have classically presented several special problems:

1. The molecular weights of polymers are very high, ranging from about 25,000 to 1,000,000 g/mol or higher. At lower molecular weights, the term *oligomer* is used for a polymer with only a few repeat units, having degrees of polymerization of not more than 10 to 100 (1). Assuming a molecular weight of 50 g/mol for each mer, a range of molecular weights of up to 500 to 5000 g/mol result. Although the term *telomer* is sometimes used for higher molecular weight materials, it is used primarily for polymers whose molecular weights are in the oligomer range. It refers especially to those materials formed by chain transfer reactions. The term *telechelic* polymers is used for oligomers and telomers with functional groups on both ends (2). Telechelic polymers are usually intended for further chemical reaction, particularly polymer network formation.

   Polymers having a reactive group at only one end are called *macromers* (3) (an abbreviation for *macromonomers*), and they are intended for further reaction. Super high molecular weight polymers with molecular weights greater than 1×10^7 g/mol have been called *pleistomers* (3). This chapter will be concerned primarily with linear polymers having molecular weights above 25,000 g/mol.

[†] Some authors prefer the term "relative molar mass," using units of daltons.

2. The molecular weight of ordinary size molecules is fixed (e.g., benzene has a molecular weight of 78 g/mol regardless of its source). Most polymer molecular weights, on the other hand, vary greatly depending on the method of preparation. In addition most polymers are polydisperse; that is, the sample contains more than one species.

3. The spatial arrangement of the polymer chain is called its "conformation." Conformations can be determined in dilute solutions by light-scattering, and in the bulk state by small-angle neutron scattering (SANS). Conformations can also be estimated theoretically, from the structure and molecular weight of the polymer.

4. Polymer chain conformations are functions of temperature, solvent, structure, crystallization, extension, and the presence of other polymers.

Most of the methods for determining the molecular weight and size of polymers (except for small-angle neutron scattering) depend on dissolving the polymer in an appropriate solvent and measuring the required properties in dilute solution. The molecular weights of gaseous molecules such as oxygen or methane can be determined via an understanding of the gas laws and the fact that the molecules are separated in space. The same is true for polymer solutions. Solution laws must be understood, and the polymer chains must be separated from each other in the solution. The following sections develop the basic polymer solution thermodynamics for the purposes of determining molecular weights and sizes. In Chapter 4, these same concepts are developed further in the study of concentrated solutions, phase separation, and diffusion problems.

3.1.2 How Does a Polymer Dissolve?

Once placed in a solvent, polymers dissolve in several steps. First the solvent must wet the polymer. Then the solvent diffuses into the polymer, swelling it. For polymers of high molecular weight, this process may take several hours or longer, depending on sample size, temperature, and so on. Contrary to many low molecular weight substances, the polymer does not initially diffuse into the solvent.

There are two distinguishable modes of solvent diffusion into a polymer. If the polymer is amorphous and above its glass transition temperature (i.e., a polymer melt), the diffusion of the solvent into the polymer forms a smooth composition curve, with the most highly swollen material at the outer edge. This is referred to as Fickian diffusion, following Fick's laws; see Section 4.4.

If the polymer is significantly below its glass transition temperature, however, the non-Fickian phenomenon known as case II swelling may predominate (4). In this situation the diffusion into the glassy polymer is slow. First the solvent must plasticize the polymer, lowering its glass transition temperature until it is below ambient. Then swelling is rapid. A rather sharp, moving boundary between highly swollen material and that substantially not swollen results. Frequently the stresses at the swelling boundary cause the sample to craze or fracture. A similar phenomenon is sometimes found with semicrystalline polymers.

Finally the polymer diffuses out of the swollen mass into the solvent, completing the solution process. Very dilute solutions are usually required for molecular weight determination.

If the polymer is cross-linked, the polymer only swells, reaching an equilibrium degree of swelling; see Section 9.12. There are also some polymers, particularly with high melting temperatures or strong internal secondary bonds, that cannot be dissolved without degradation. In these last cases the polymer molecular weight cannot be determined directly, and sometimes not at all.

3.2 THE SOLUBILITY PARAMETER

One of the simplest notions in chemistry is that "like dissolves like." Qualitatively, "like" may be defined variously in terms of similar chemical groups or similar polarities.

Quantitatively, solubility of one component in another is governed by the familiar equation of the free energy of mixing,

$$\Delta G_M = \Delta H_M - T\Delta S_M \tag{3.1}$$

where ΔG_M is the change in Gibbs' free energy on mixing, T is the absolute temperature, and ΔS_M is the entropy of mixing. A negative value of ΔG_M indicates that the solution process will occur spontaneously. The term $T\Delta S_M$ is always positive because there is an increase in the entropy on mixing. Therefore the sign of ΔG_M depends on ΔH_M, the enthalpy of mixing.

Surprisingly, the heat of mixing is usually positive, opposing mixing. This is true for big and little molecules alike. Exceptions occur most frequently when the two species in question attract one another in some way, perhaps by having opposite polarities, being acid and base relative to one another, or through hydrogen bonding. However, positive heats of mixing are the more usual case for relatively nonpolar organic compounds. On a quantitative basis, Hildebrand and Scott (5) proposed that, for regular solutions,

$$\Delta H_M = V_M \left[\left(\frac{\Delta E_1}{V_1} \right)^{1/2} - \left(\frac{\Delta E_2}{V_2} \right)^{1/2} \right]^2 v_1 v_2 \tag{3.2}$$

where V_M represents the total volume of the mixture, ΔE represents the energy of vaporization to a gas at zero pressure (i.e., at infinite separation of the molecules), and V is the molar volume of the components, for both species 1 and 2. The quantity v represents the volume fraction of component 1 or 2 in the mixture. The quantity $\Delta E/V$ represents the energy of vaporization per unit volume. This term is sometimes called the cohesive energy density. By convention, component 1 is the solvent, and component 2 is the polymer.

The reader should note that according to equation (3.2), "like dissolves like" means that the two terms $\Delta E_1/V_1$ and $\Delta E_2/V_2$ have nearly the same numerical values. Equation (3.2) also yields only positive values of ΔH_M, a serious fault in the theory. However, since the majority of polymer solutions do have positive heats of mixing, the theory has found very considerable application.

The square root of the cohesive energy density is widely known as the solubility parameter,

$$\delta = \left(\frac{\Delta E}{V} \right)^{1/2} \tag{3.3}$$

Thus the heat of mixing of two substances is dependent on $(\delta_1 - \delta_2)^2$.

Table 3.1 Solubility parameters of some common solvents

Solvent	δ (cal/cm^3)$^{1/2}$	H-bonding[a] Group	Specific Gravity[b] 20°C (g/cm^3)
Acetone	9.9	m	0.7899
Benzene	9.2	p	0.87865
n-Butyl acetate	8.3	m	0.8825
Carbon tetrachloride	8.6	p	1.5940
Cyclohexane	8.2	p	0.7785
n-Decane	6.6	p	—
Dibutyl amine	8.1	s	—
Difluorodichloromethane	5.1	p	—
1,4-Dioxane	10.0	m	1.0337
Low odor mineral spirits	6.9	p	—
Methanol	14.5	s	0.7914
Toluene	8.9	p	0.8669
Turpentine	8.1	p	—
Water	23.4	s	0.99823
Xylene	8.8	p	0.8611

Source: J. Brandrup and E. H. Immergut, eds., *Polymer Handbook*, 2nd ed., Wiley-Interscience, New York, 1975, sec. IV.

[a] Hydrogen bonding is an important secondary parameter in predicting solubility. p, Poorly H-bonded; m, moderately H-bonded; and s, strongly H-bonded.

[b] J. Brandrup and E. H. Immergut, *Polymer Handbook*, 3rd ed., Wiley-Interscience, New York, 1989, sec. III, p. 29.

These relationships are meaningful only for positive heats of mixing; that is, when the heat of mixing term opposes solution. Since $(\delta_1 - \delta_2)^2$ cannot be negative, equations (3.2) and (3.3) break down for negative heats of mixing.

3.2.1 Solubility Parameter Tables

Tables 3.1 (6) and 3.2 (6) present the solubility parameters of common solvents and polymers, respectively. These tables provide a quantitative basis for understanding why methanol or water does not dissolve polybutadiene or polystyrene. However, benzene and toluene are predicted to be good solvents for these polymers, which they are. While solubility of a polymer also depends on its molecular weight, the temperature, and so on, it is frequently found that polymers will dissolve in solvents having solubility parameters within about one unit of their own, in (cal/cm^3)$^{1/2}$.

3.2.2 Experimental Determination

The solubility parameter of a new polymer may be determined by any of several means. If the polymer is cross-linked, the solubility parameter may be determined by swelling experiments (7). The best solvent is defined for the purposes of the experiment as the one with the closest solubility parameter. This solvent also swells the polymer the most. Several solvents of varying solubility parameter are selected, and the cross-linked polymer is swelled to equilibrium in each of them. The swelling coefficient, Q, is plotted against the various solvent's solubility parameter, the maximum defining the solubility

Table 3.2 Solubility parameters and densities of common polymers (6)

Polymer	δ (cal/cm^3)$^{1/2}$	Density (g/cm^3)
Polybutadiene	8.4	1.01
Polyethylene	7.9	0.85 (amorphous)
Poly(methyl methacrylate)	9.45	1.188
Polytetrafluorethylene	6.2	2.00 amorphous, estimated
Polyisobutene	7.85	0.917
Polystyrene	9.10	1.06
Cellulose triacetate	13.60	1.28[a]
Cellulose tributyrate	—	1.16[a]
Polyamide 66	13.6	1.24
Poly(ethylene oxide)	9.9	1.20
Poly(vinyl chloride)	9.6	1.39

Note: 1 (cal/cm^3)$^{1/2}$ = 2.046 × 10^3 (J/m^3)$^{1/2}$.

[a] C. J. Malm, C. R. Fordyce, and H. A. Tanner, *Ind. Eng. Chem.*, **34**, 430 (1942).

parameter of the polymer. The theoretical extent of swelling is predicted by the Flory–Rehner theory on the basis of the cross-link density and the attractive forces between the solvent and the polymer (see Section 9.12).

The swelling coefficient, Q, is defined by

$$Q = \frac{m - m_0}{m_0} \times \frac{1}{\rho_s} \tag{3.4}$$

where m is the weight of the swollen sample, m_0 is the dry weight, and ρ_s is the density of the swelling agent (8,9). Typical results are shown in Figure 3.1 (9). Here, the swelling behavior of a cross-linked polyurethane and a cross-linked polystyrene are shown, together with the 50/50 interpenetrating polymer network made from these two polymers. Both the homopolymers and the interpenetrating polymer network exhibit single peaks, albeit that the IPN peak is somewhat broader and appears in-between its two homopolymers. Polymer networks swollen to equilibrium are discussed in Section 9.12.

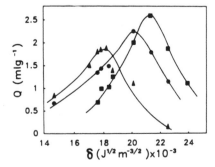

Figure 3.1 The swelling coefficient, Q, reaches a maximum when the solubility parameter of the solvent nearly matches that of the polymer, for several cross-linked systems: polyurethane (■), polystyrene (▲), and a polyurethane–polystyrene interpenetrating polymer networks (●) (9).

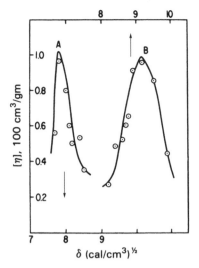

Figure 3.2 Determination of the solubility parameter, using the intrinsic viscosity method (8), for poly-isobutene (A) and polystyrene (B). The intrinsic viscosity, [η], is a measure of the individual chain size. See Section 3.8.

Alternatively, the solubility parameter may be determined by measuring the intrinsic viscosity of the polymer in these solvents, if the polymer is soluble in them. Then the intrinsic viscosity is plotted against the solubility parameter of the several solvents. Since the chain conformation is most expanded in the best solvent [see equation (3.83)], the intrinsic viscosity will be highest for the best match in solubility parameter. Such an experiment is illustrated in Figure 3.2 (10) for polyisobutene and polystyrene. The results of such experiments are collected in Table 3.3.

3.2.3 Theoretical Calculation: An Example

Values of the solubility parameter may be calculated from a knowledge of the chemical structure of any compound, polymer or otherwise. Use is made of the group molar attraction constants, **G**, for each group,

$$\delta = \frac{\rho \sum \mathbf{G}}{M} \tag{3.5}$$

where ρ represents the density and M is the molecular weight. For a polymer, M is the mer molecular weight.

Group molar attraction constants have been calculated by Small (11) and Hoy (12). Table 3.3 (11) presents a wide range of values of **G** for chemical groups.

For example, the solubility parameter of polystyrene may be estimated from Table 3.3. The structure is

$$-CH_2-CH-$$

Table 3.3 Group molar attraction constants at 25°C (according to Small; derived from measurement of heat of evaporation)†

Group		G^a	Group		G	Group		G
—CH₃		214	Ring	5-membered	105–115	Br	single	340
—CH₂—	single-bonded	133	Ring	6-membered	95–105	I	single	425
—CH<		28	Conjugation		20–30	CF₂	} n-fluorocarbons only	150
>C<		−93	H	(variable)	80–100	CF₃		274
CH₂=		190	O	ethers	70	S	sulfides	225
—CH=	double-bonded	111	CO	ketones	275	SH	thiols	315
>C=		19	COO	esters	310	ONO	nitrates	~440
—CH=C<		285	CN		410	NO₂	(aliphatic nitro-compounds)	~440
—C≡C—		222	Cl	(mean)	260	PO₄	(organic phosphates)	~500
Phenyl		735	Cl	single	270	Si	(in silicones)	−38
Phenylene	(o, m, p)	658	Cl	twinned as in >CCl₂	260			
Naphthyl		1146	Cl	triple as in —CCl₃	250			

Source: P. A. Small, *J. Appl. Chem.*, **3**, 71 (1953).

^{a} Units of $G = (\text{cal-cm}^3)^{1/2}/\text{mol}$.

† The solubility parameter can be calculated *via* $\delta = \rho \Sigma G / M$, where M is the mer molecular weight.

which contains $—CH_2—$ with a **G** value of 133, a $—\overset{|}{CH}—$ with **G** equal to 28, and a phenyl group with **G** equal to 735. The density of polystyrene is 1.05 g/cm^3, and the mer molecular weight is 104 g/mol. Then equation (3.5) gives

$$\delta = \frac{1.05}{104}(133 + 28 + 735) \tag{3.6}$$

$$\delta = 9.05 \ (\text{cal/cm}^3)^{1/2} \tag{3.7}$$

Table 3.2 gives a value of $9.1 \ (\text{cal/cm}^3)^{1/2}$ for polystyrene.

3.3   THERMODYNAMICS OF MIXING

3.3.1   Types of Solutions

3.3.1.1   The Ideal Solution   In the previous section the solubility of a polymer in a given solvent was examined on the basis of their respective solubility parameters, which was governed by the heats of mixing. The entropy of mixing was entirely ignored.

In an ideal solution, the circumstances are reversed, and the heat of mixing is zero, by definition. Raoult's law is obeyed.

$$p_1 = p_1^\circ n_1 \tag{3.8}$$

where p_1 is the partial vapor pressure, n_1 is the mole fraction of component 1, and p_1° is the vapor pressure of the pure component.

The free energy of mixing is given as the sum of the free energies of dilution per molecule, incompressibility of the mixture implicitly assumed:

$$\Delta G_M = N_1 \Delta G_1 + N_2 \Delta G_2 \tag{3.9}$$

$$\Delta G_M = kT\left[N_1 \ln\left(\frac{p_1}{p_1^\circ}\right) + N_2 \ln\left(\frac{p_2}{p_2^\circ}\right) \right] \tag{3.10}$$

where N_1 and N_2 are the numbers of molecules of the 1 and 2 species, respectively. Then, from equation (3.8), for small molecules,

$$\Delta G_M = kT(N_1 \ln n_1 + N_2 \ln n_2) \tag{3.11}$$

Since $\Delta H_M = 0$,

$$\Delta S_M = -k(N_1 \ln n_1 + N_2 \ln n_2) \tag{3.12}$$

Since the entropy of mixing is always positive, and the heat of mixing is zero for an ideal solution, mixing in all proportions always occurs spontaneously.

3.3.1.2   Statistical Thermodynamics of Mixing   Equations (3.8) to (3.12) present the classically derived entropy of mixing. More generally, equation (3.12) can be derived through the application of statistical thermodynamics. According to statistical thermodynamics, the entropy of mixing is determined by counting the number of possible arrangements in space that the molecules may assume, Ω. The entropy of mixing is given by Boltzmann's relation

$$\Delta S_M = k \ln \Omega \qquad (3.13)$$

For small molecules of about the same size, this is given by the total number of ways of arranging the N_2 identical molecules of the solute on a lattice comprising $N_0 = N_1 + N_2$ cells (see Figure 3.3). The total number of such arrangements is given by $\Omega = N_0!/N_1!N_2!$.

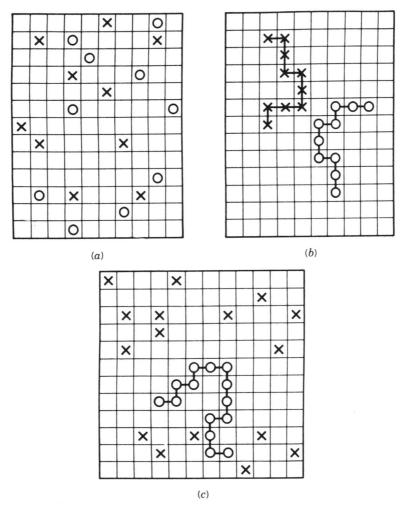

(a)

(b)

(c)

Figure 3.3 Illustration of two types of molecules on quasilattice structures. (a) Two types of small molecules; (b) a blend of two types of polymer molecules; (c) a polymer dissolved in a solvent. The entropy of mixing decreases from (a) to (c) to (b) because the number of different ways of arranging the molecules in space decreases. Note that the mers of the polymer chains are constrained to remain in juxtaposition with their neighbors.

Applying Stirling's approximation,

$$\ln N! = N \ln N - N \tag{3.14}$$

the entropy of mixing is given by

$$\Delta S_M = k[(N_1 + N_2) \ln(N_1 + N_2) - N_1 \ln N_1 - N_2 \ln N_2] \tag{3.15}$$

which rearranges to give equation (3.12).

When the polymer has x chain segments (mers), the entropy of mixing is given by

$$\Delta S_M = -k(N_1 \ln v_1 + N_2 \ln v_2) \tag{3.16}$$

Equation (3.16) yields the total entropy. The entropy per unit volume is given by

$$\frac{\Delta S_M}{kV} = -\frac{\sum_i N_i \ln v_i}{\sum_i V_i} \tag{3.16a}$$

for i components. In equation (3.16), v_1 and v_2 represent the volume fractions of solvent and polymer, respectively, and

$$v_1 = \frac{N_1}{N_1 + xN_2} \tag{3.17}$$

$$v_2 = \frac{xN_2}{N_1 + xN_2} \tag{3.18}$$

For polymer–polymer solutions,

$$v_1 = \frac{x_1 N_1}{x_1 N_1 + x_2 N_2} \tag{3.19}$$

$$v_2 = \frac{x_2 N_2}{x_1 N_1 + x_2 N_2} \tag{3.20}$$

It must be pointed out that ΔS_M is the combinatorial entropy computed by considering the possible arrangements of the molecules on the lattice in Figure 3.3. Furthermore the number of ways that the system can be rearranged in space is reduced when one or both of the species exist as long chains. In the equations above, the subscript 1 usually represents the solvent, and the subscript 2 the polymer.

3.3.1.3 Other Types of Solutions

According to Hildebrand, the *ideal* solution has a zero heat of mixing. Other types of solutions include the *athermal* solutions, in which ΔH_M is still zero, but the entropy of mixing is not given by equation (3.16). A *regular solution* is defined as one in which ΔS_M has zero value but ΔH_M is finite. In an *irregular solution* both ΔH_M and ΔS_M deviate from ideal values. As will be shown later, a principal cause of nonideality arises from changes in volume nonadditivity on mixing. The reader should observe that Hildebrand's use of the term *ideal* as a good solvent differs from the θ-temperature definition (Section 3.5), which involves a thermodynamically poor solvent.

3.3.2 Dilute Solutions

The Flory–Huggins theory (13–17) introduces the unitless quantity χ_1 to represent the heat of mixing,

$$\chi_1 = \frac{\Delta H_M}{kTN_1 v_2} \tag{3.21}$$

which combined with equation (3.16) leads to the free energy of mixing in statistical thermodynamic terms[†]:

$$\Delta G_M = kT(N_1 \ln v_1 + N_2 \ln v_2 + \chi_1 N_1 v_2) \tag{3.22}$$

The first two terms of equation (3.22) are entropic, while the last term on the right is enthalpic. Again, incompressibility is assumed, meaning that the free volume in the system is constant. Equation (3.22) provides a starting point for many equations of interest. The partial molar free energy of mixing may be written, after multiplying by Avogadro's number,

$$\overline{\Delta G_1} = RT\left[\ln(1 - v_2) + \left(1 - \frac{1}{x}\right)v_2 + \chi_1 v_2^2\right] \tag{3.23}$$

Since the osmotic pressure is given by

$$\pi = -\frac{\overline{\Delta G_1}}{v_1} \tag{3.24}$$

then

$$\pi = -\frac{RT}{V_1}\left[\ln(1 - v_2) + \left(1 - \frac{1}{x}\right)v_2 + \chi_1 v_2^2\right] \tag{3.25}$$

where V_1 represents the molar volume of the solvent.

The second virial coefficient [see equation (3.41)] may be given a theoretical interpretation. On expanding equation (3.25), the theoretical value for the second virial coefficient can be separated out,

$$A_2 = \frac{\bar{v}_2^2}{N_A V_1}\left(\frac{1}{2} - \chi_1\right) \tag{3.26}$$

where $\bar{v}_2$ is the specific volume of the polymer. The Flory θ-temperature can now be seen as the point where $\chi_1 = \frac{1}{2}$.

At the temperature where $\chi_1 = \frac{1}{2}$, A_2 is zero. Another form of equation (3.26) is

$$A_2 = \frac{N_A u}{2M^2} \tag{3.27}$$

[†] The convention assumed is that a positive ΔG_M leads to phase separation, and a negative value leads to molecular solutions.

where N_A represents Avogadro's number, u is the intermolecular excluded volume between molecular pairs and M is the polymer molecular weight. When the quantity u is zero, A_2 is zero, and the chains mix without recognizing each other's presence. Also, since the forces are balanced out, the chains cannot distinguish between solvent molecules and segments of the other polymer chains. Thus the simplified conditions of mixing existing at the Flory θ-temperature become apparent.

Another equation interrelating thermodynamic terms may be written (18)

$$\chi_1 = \beta_1 + \frac{V_1}{RT}(\delta_1 - \delta_2)^2 \tag{3.28}$$

Thus the Flory–Huggins heat of mixing term is related to the solubility parameters. While β_1, sometimes called the lattice constant of entropic origin, is often assigned a value of 0.35 in the older literature, a value of zero yields better correlations with other quantities in most cases. The quantity α equals r/r_0, the expansion of the polymer chain over that of the theta condition state. In fact χ_1 can be positive, negative, or zero.

The quantity α may be expressed in terms of the Flory θ-temperature (19),

$$\alpha^5 - \alpha^3 = 2C_m\psi_1\left(1 - \frac{\theta}{T}\right)M^{1/2} \tag{3.29}$$

where ψ_1 is an entropic factor and C_m is a constant. Thus, at $T = \theta$, $\alpha = 1$. For high molecular weights, equation (3.29) shows that the chains expand as $M^{0.1}$ power. Since the Mark–Houwink equation depends on $M^{0.5}$ in the θ condition, the quantity a is seen to vary theoretically from 0.5 to 0.8; see Section 3.8.

While the relations above must be used with caution, they provide a clear theoretical interpretation and interrelationships for the equations used in molecular weight determination. As will be seen later, they can also express quantities useful for phase separation problems. The theory expressed in equations (3.21) to (3.29) has been reviewed many times (20) and derived by Flory (17).

3.3.3   Values for the Flory–Huggins χ_1 Parameter

The Flory–Huggins χ_1 parameter has been one of the most widely used quantities, characterizing a variety of polymer–solvent and polymer–polymer interactions. It is a unitless number. Sometimes the Flory–Huggins parameter is written $\chi_{1,2}$, and sometimes just plain χ. While the original theory proposed that χ_1 be concentration independent, many polymer–solvent systems exhibit increases of χ_1 with polymer concentration (21). In that case the analytical representation of the experimentally found concentration dependence can be written as a power series,

$$\chi_1 = \chi^0 + \chi^1 v_2 + \chi^2 v_2^2 \tag{3.30}$$

where χ^0, χ^1, and χ^2 are determined experimentally.

However, for many simple calculations, it is valuable to have a single number parameter. Typical values of χ_1 are illustrated in Table 3.4, where values were selected at low concentrations of polymer. If the value of χ_1 is below 0.5, the polymer should be soluble if amorphous and linear. When χ_1 equals 0.5, as in the case of the polystyrene–

Table 3.4   Flory–Huggins χ_1 values

Polymer	Solvent	$T, °C$	χ_1
Polystyrene	Toluene	25	0.37
Polystyrene	Cyclohexane	34	0.50
Polyisoprene	Benzene	25	0.40
Cellulose nitrate	Acetone	20	0.14
Cellulose nitrate	n-Propylacetate	20	−0.38
Poly(ethylene oxide)	Benzene	70	0.19
Poly(dimethyl siloxane)	Toluene	20	0.45
Polyethylene	n-Heptane	109	0.29
Poly(butadiene–stat–styrene)	Toluene	25	0.39
Poly(ethylene oxide)	Water	25	0.4[a]

Source: J. Brandrup and E. H. Immergut, eds., *Polymer Handbook*, 3rd ed., Wiley, New York, 1989, sec. VII, pp. 176–178.

[a]CRC *Handbook of Polymer–Liquid Interaction Parameters and Solubility Parameters*, Part II, p. 178 (major entrees).

cyclohexane system at 34°C in Table 3.4, then the Flory θ conditions exist; see Section 4.1. If the polymer is crystalline, as in the case of polyethylene, it must be heated to near its melting temperature, so that the total free energy of melting plus dissolving is negative. For very many nonpolar polymer–solvent systems, χ_1 is in the range of 0.3 to 0.4.

3.3.4   A Worked Example for the Free Energy of Mixing

What is the free energy of mixing polystyrene of molecular weight 10,000 g/mol with cyclohexane at 34°C, to make up a 10% solution by volume?

The starting point is equation (3.22),

$$\Delta G_M = kT(N_1 \ln v_1 + N_2 \ln v_2 + \chi_1 N_1 v_2)$$

The basis usually taken is 1 cm^3. The number of molecules of each species needs to be calculated for $v_1 = 0.90$ and $v_2 = 0.10$. Densities are shown in Tables 3.1 and 3.2. For cyclohexane, C_6H_{12}, the molecular weight is 84 g/mol and the molar volume is (84 g/mol)/(0.7785 g/cm^3) $\cong$ 108 cm^3/mol. For 0.90 cm^3, there are 0.0083 mol, or, using Avogadro's number, 5.02×10^{21} molecules, N_1.

The density of polystyrene is 1.06 g/cm^3, yielding a molar volume of 9.43×10^3 cm^3/mol for a molecular weight of 10,000 g/mol, and for 0.10 cm^3, 1.06×10^{-5} mol of polystyrene, or 6.38×10^{18} molecules, N_2. The value of χ_1 is 0.50, taken from Table 3.4. The free energy of mixing is given by

$$\Delta G_M = 1.38 \times 10^{-23} \text{ J/K} \times 307 \text{ K}$$

$$\times (5.02 \times 10^{21} \ln 0.90 + 6.38 \times 10^{18} \ln 0.10 + 0.50 \times 5.02 \times 10^{21} \times 0.10)$$

$$\Delta G_M = -1.24 \text{ J}$$

for each cm^3 of solution. The value of ΔG_M is small and negative because of the entropy of mixing term.

3.4   MOLECULAR WEIGHT AVERAGES

There are four molecular weight averages in common use; the number-average molecular weight, M_n; the weight-average molecular weight, M_w; the z-average molecular weight, M_z; and the viscosity-average molecular weight, M_v. These are defined below in terms of the numbers of molecules N_i having molecular weights M_i, or in terms of w_i, the weight of species with molecular weights M_i.

$$M_n = \frac{\sum_i N_i M_i}{\sum_i N_i} = \frac{\sum_i w_i}{\sum_i (w_i / M_i)} \tag{3.31}$$

$$M_w = \frac{\sum_i N_i M_i^2}{\sum N_i M_i} = \frac{\sum_i w_i M_i}{\sum w_i} \tag{3.32}$$

$$M_z = \frac{\sum_i N_i M_i^3}{\sum N_i M_i^2} = \frac{\sum_i w_i M_i^2}{\sum w_i M_i} \tag{3.33}$$

$$M_v = \left[\frac{\sum_i N_i M_i^{1+a}}{\sum_i N_i M_i} \right]^{1/a} \tag{3.34}$$

For a random molecular weight distribution, such as produced by many free radical or condensation syntheses, $M_n : M_w : M_z = 1 : 2 : 3$. This is illustrated in Figure 3.4. A convenient measure of the width of the molecular weight distribution is the ratio M_w / M_n, called the polydispersity index (PDI). The quantity a in equation (3.34) varies from 0.5 to 0.8 (see Section 3.8.3).

An absolute method of measuring the molecular weight is one that depends solely on theoretical considerations, counting molecules and their weight directly. The relative methods require calibration based on an absolute method and include intrinsic viscosity and gel permeation chromatography (GPC).

Absolute methods of determining the number-average molecular weight include osmometry and other colligative methods, and end-group analysis. Light-scattering yields

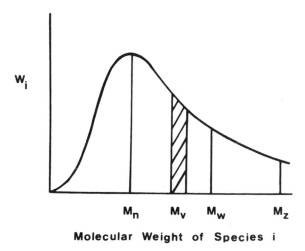

Figure 3.4   Schematic of a simple molecular weight distribution, showing the various averages.

an absolute weight-average molecular weight. The problem is compounded for light-scattering and SANS studies, because these techniques measure not only M_w but also the z-average radius of gyration (R_g^z). A knowledge of the molecular weight distribution is then required to obtain R_g^w. Only then can values of R_g be studied in relation to M_w.

3.5 DETERMINATION OF THE NUMBER-AVERAGE MOLECULAR WEIGHT

The number-average molecular weight, M_n, involves a count of the number of molecules of each species, $N_i M_i$, summed over i, divided by the total number of molecules; see equation (3.31). It is the simple average that most people think about. For many simple single-peaked distributions, M_n is near the peak (see Figure 3.4). There are two important groups of methods for determining M_n.

3.5.1 End-Group Analyses

The first group of methods involves end-group analyses. Many types of syntheses leave a special group on one or both ends of the molecule, such as hydroxyl and carboxyl. These can be titrated or analyzed instrumentally by such methods as infrared. For molecular weights above about 25,000 g/mol, however, the method becomes insensitive because the end groups are present in too low a concentration.

3.5.2 Colligative Properties

The second group of methods makes use of the colligative properties of solutions. Colligative properties depend on the number of molecules in a solution, and not their chemical constitution (22). The colligative properties include boiling point elevation, melting point depression, vapor pressure lowering, and osmotic pressure. The basic equations for the first two may be written (23)

$$\lim_{c \to 0} \frac{\Delta T_b}{c} = \frac{RT^2}{\rho \Delta H_v} \left(\frac{1}{M_n} \right) \tag{3.35}$$

$$\lim_{c \to 0} \frac{\Delta T_f}{c} = -\frac{RT^2}{\rho \Delta H_f} \left(\frac{1}{M_n} \right) \tag{3.36}$$

where ΔT_b and ΔT_f are the boiling point elevation and freezing point depression, respectively; ρ is the solvent density; ΔH_v and ΔH_f are the latent heats of vaporization and fusion per gram of solvent, and c is the solute concentration in grams per cubic centimeter.

If the vapor pressure of the solute is small, and the solvent follows Raoult's vapor pressure law,

$$\frac{P_1^\circ - P_1}{P_1^\circ} = X_2 \tag{3.37}$$

where P_1° is the vapor pressure of the pure solvent, P_1 is that of the solution, and X_2 is the mole fraction of the solute.

Table 3.5   Comparison of the colligative solution properties of a 1% polymer solution with $M = 20,000$ g/mol (23)

Property	Value
Vapor pressure lowering	4×10^{-3} mm Hg
Boiling point elevation	1.3×10^{-3} °C
Freezing point depression	2.5×10^{-3} °C
Osmotic pressure	15 cm solvent

The osmotic pressure π depends on the molecular weight as follows:

$$\lim_{c \to 0} \frac{\pi}{c} = \frac{RT}{M_n} \tag{3.38}$$

Typical values for the colligative properties for a polymer having a molecular weight of 20,000 g/mol are shown in Table 3.5 (23). Only osmotic pressure is large enough for fruitful studies at this molecular weight or higher.

3.5.3   Osmotic Pressure

3.5.3.1   Thermodynamic Basis   Polymer solutions exhibit osmotic pressures because the chemical potentials of the pure solvent and the solvent in the solution are unequal. Because of this inequality, there is a net flow of solvent, through a connecting membrane, from the pure solvent side to the solution side. When sufficient pressure is built up on the solution side of the membrane, so that the two sides have the same activity, equilibrium will be restored (24).

While the discussion above gives an exact thermodynamic interpretation of the phenomenon, a consideration in terms of the number of solute molecules per unit volume is useful for practical calculations. An analogy exists between equation (3.38) and the ideal gas law:

$$PV = nRT \tag{3.39}$$

where n is in moles, as usual. The quantity n/V is equal to c/M, yielding

$$P = \frac{c}{M} RT \tag{3.40}$$

Setting the gas pressure equal to the osmotic pressure, $P = \pi$, and rearranging, we obtain equation (3.38).

3.5.3.2   Instrumentation   A typical static osmometer design includes a membrane that permeates only the solvent, a capillary, and a reference capillary. Typical membranes are made from regenerated cellulose or other microporous materials. Such an instrument usually requires 24 h to reach equilibrium (20).

There are now several types of automatic osmometers that operate with essentially zero flow and that reach equilibrium very rapidly, usually within minutes. Osmotic equilibrium depends on an equal and opposite pressure being developed. The critical part of their design relates to the method of automatic adjustment of the osmotic pressure of the solution side so that the activity of the two sides is equal. Since several concentrations usually need to be run, the time required to determine a molecular weight by osmometry has been reduced from a week to a few hours by these automatic instruments.

3.5.3.3 *The Flory θ-Temperature* The basis for determining the molecular weight by osmometry has been given in equation (3.38). At finite concentrations, interactions between the solvent and the solute result in the virial coefficients A_2, A_3, and so on. The full equation may be written

$$\frac{\pi}{c} = RT\left(\frac{1}{M_n} + A_2c + A_3c^2 + \cdots\right) \tag{3.41}$$

Interactions between one polymer molecule and the solvent result in the second virial coefficient, A_2. Multiple polymer–solvent interactions produce higher virial coefficients, A_3, A_4, and so on. A further interpretation of the virial coefficients is considered along with solution theory (Section 3.3). For medium molecular weights, the slope is substantially linear below about 1% solute concentration. Of course, $A_1 = 1/M_n$.

The quantity A_2 depends on both the temperature and the solvent, for a given polymer. A unique and much desired state arises when A_2 equals zero. The temperature at which this condition holds is called the Flory θ-temperature (17). In this state, π/c is independent of concentration, so that only one concentration need be studied to determine M_n. Since it is also the state where an infinite molecular weight polymer just precipitates (see Section 4.1) and $\chi_1 = 0.5$, considerable care must be taken to keep the polymer in solution.

Figure 3.5 illustrates the determination of the Flory θ-temperature for cellulose tricaproate dissolved in dimethylformamide (25). Table 3.6 presents a selected list of polymers and their theta solvents (26).

It must be emphasized that molecular weights determined by any of the colligative properties, and osmometry in particular, are absolute molecular weights; that is, the values are determined by theory and not by prior calibration. The practical limit of osmometry is about 500,000 g/mol because the pressures become too small.

3.6 WEIGHT-AVERAGE MOLECULAR WEIGHTS AND RADII OF GYRATION

The principal method of determining the weight-average molecular weight is light-scattering, although small-angle neutron scattering is now becoming important, especially in the bulk state, and X-ray scattering is also sometimes employed. For generality, all three methods are considered.

First, a few general terms need to be defined. Radiation from an object is said to be reflected when the object is much larger than the wavelength of the radiation. The radiation is said to be scattered when the object begins to approach the wavelength of the radiation in size, down to atomic dimensions. Common examples of light-scattering include the blue of the sky and the rainbow. The latter, involving scattering from spheres of about 50 μm in diameter, is connected to other types of scattering through the Mie (27) equations.

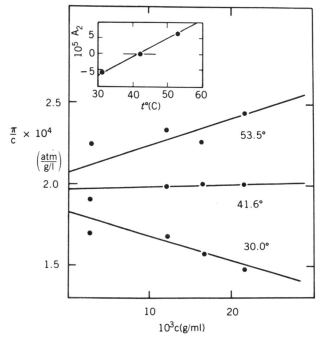

Figure 3.5 The osmotic pressure data for cellulose tricaproate in dimethylformamide at three temperatures. The Flory θ-temperature was determined to be 41 ± 1°C (25).

Scattering from a single electron isolated in space provides a beginning. As the electron is approached by an electromagnetic wave, the electron absorbs the energy and begins to oscillate (see Figure 3.6). Now an accelerating electric charge is itself a radiator, and the energy is re-radiated, but in all directions. This re-radiation of energy is called scattering.

For more complex systems, there are two general cases. First, if the atoms, molecules, or particles are organized into a regular array, the radiation will be diffracted; that is, scattering will be observed only at special angles. This arises because at all other angles there is total destructive interference among the scattered radiation arising from different parts of the array. An example is the diffraction of X rays by a crystal.

Table 3.6   Polymers and their θ-solvents (26)

Polymer	Solvent	Temperature (°C)
cis-Polybutadiene	n-Heptane	−1
Polyethylene	Biphenyl	125
Poly(n-butyl acrylate)	Benzene/methanol 52/48	25
Polystyrene	Cyclohexane	34
Poly(oxytetramethylene)	Chlorobenzene	25
Cellulose tricaproate	Dimethylformamide	41

Source: J. Brandrup and E. H. Immergut, eds., *Polymer Handbook*, 2nd ed., Wiley-Interscience, New York, 1975.

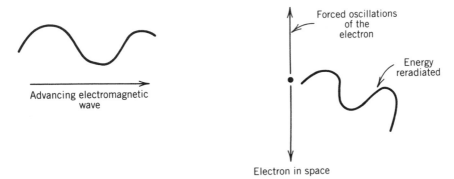

Advancing electromagnetic wave

Forced oscillations of the electron

Energy reradiated

Electron in space

Figure 3.6 The effect of an electromagnetic wave on a free electron. The forced oscillations of the electron involve accelerations and decelerations, which cause the electromagnetic energy to be re-radiated.

If the atoms, molecules, or particles are not organized into a regular array, scattering will be observed at all angles. In general, the angular variation of the scattering intensity provides a measure of the size of the structures.

The phenomenon of light-scattering is caused by fluctuations in the refractive index of the medium on the molecular or supermolecular scale. For example, the blue-of-the-sky scattering mentioned previously is caused by the random presence of gas molecules in what is otherwise a vacuum. The blue of large bodies of water is caused by slight fluctuations in the spacing of the water molecules. In both cases the scattered intensity varies as the wavelength to the inverse fourth power, which causes the characteristic blue color.

The application of light-scattering to polymer molecular weights follows a trail begun in part by Smoluchowski (28,29) and Einstein (30), who considered the fluctuations of refractive index in liquids. The basic equations used in light-scattering of polymer solutions today were developed by Debye (31) and Zimm (32,33). They replaced fluctuations in the refractive index of the solvent itself by the changes caused by the polymer molecules. The final result relates the observed light-scattering intensity to the osmotic pressure, π, of the polymer as follows:

$$\frac{Hc}{R(\theta)} = \frac{1}{RT} \left(\frac{\partial \pi}{\partial c} \right)_T \tag{3.44}$$

where $R(\theta)$ is called Rayleigh's ratio and is equal to $I_\theta w^2 / I_0 V_s$ where I_θ represents the light intensity observed at angle θ scattered from a volume V_s, if the distance from the source is w and the intensity of the incident light is I_0. The optical constant H is given by[†]

$$H = \frac{2\pi_1^2 n_0^2 (dn/dc)^2}{N_A \lambda^4} \tag{3.45}$$

where n_0 represents the refractive index at wavelength λ and N_A is Avogadro's number,

[†]Note: π = osmotic pressure, π_1 = 3.1416.

and $\pi_1 = 3.14$. The quantity dn/dc must be determined for each polymer–solvent pair. Then H is a constant for a particular polymer and solvent, but determined theoretically rather than empirically.

3.6.1 Scattering Theory and Formulations

The basic theory of light-scattering applied to polymer solutions dates from the works of Debye (31), who formulated the absolute molecular weight determination in terms of an optical constant H, as shown in equation (3.44). The corresponding theory for X rays was developed by Guinier and Fournet (34). The theory for small-angle neutron scattering was derived by Kirste, Ballard, and Ibel (35). Several reviews have been written (36–39).

The basic equation for the molecular weight and size for all three modes of scattering can be written

$$\frac{Hc}{R(\theta) - R(\text{solvent})} = \frac{1}{M_w P(\theta)} + 2A_2 c \tag{3.46}$$

This equation corrects $R(\theta)$ for the Einstein–Smoluchowski solvent scattering. The factor of 2 appearing in the second term, RHS of equation (3.46) arises from first multiplying equation (3.41) by c, then differentiating as shown in equation (3.44). If the particles are very small compared to the wavelength of the radiation, $R(\theta)$ reduces to $3\tau/16\pi_1$, where τ is the turbidity in Beer's law[†]:

$$I = I_0 e^{-\tau x} \tag{3.47}$$

where x is the sample thickness. Under these conditions, $P(\theta)$, the scattering form factor, equals unity. The calibration and use of light-scattering instrumentation is discussed in Appendix 3.1 for the case of small molecules, where $P(\theta) = 1$.

Incidentally, the inverse fourth power of the wavelength shown in equation (3.45) can be quantified,

$$\tau = \frac{\tilde{K}}{\lambda^4} \tag{3.48}$$

where $\tilde{K}$ is a constant that can be calculated from the above relationships for any particular system. As a general phenomenon for particles much smaller than the wavelength (e.g., gas molecules), equation (3.48) then quantifies the blue of the sky.

For particles (or molecules) larger than about 0.05 times the wavelength, $P(\theta)$ differs from unity. The quantity $P(\theta)$ is called the single chain form factor, which describes the angular scattering arising from the conformation of an individual chain. This molecular structure factor becomes independent of particle shape as θ approaches zero. The region of very small angles, $K^2 R_g^2 < 1$, is known as the Guinier region. In the Guinier region, $P(\theta)$ becomes a measure of the radius of gyration, R_g.

For a random coil (31), $P(\theta)$ in equation (3.46) is expressed by

$$P(\theta) = \frac{2}{R_g^4 K^4} \{R_g^2 K^2 - [1 - \exp(-R_g^2 K^2)]\} \tag{3.49}$$

[†] Use of τ in equation (3.46) leads to an optical constant of $32\pi_1^3 n_0^2 (dn/dc)^2 / 3N_A \lambda^4$.

Equation (3.49) may be given in the expanded form as

$$P(\theta) = 1 - \frac{K^2 R_g^2}{3} + \cdots \tag{3.50}$$

where

$$K = \frac{4\pi_1}{\lambda} \sin\left(\frac{\theta}{2}\right) \tag{3.51}$$

The quantity λ represents the wavelength of the radiation, and θ is the angle of scatter. The quantity K, sometimes written q or Q, is variously called the wave vector or the range of momentum transfer, especially for neutron scattering.

According to Zimm (32,33), the key equations at the limit of zero angle and zero concentration, respectively, relating the light-scattering intensity to the weight-average molecular weight M_w and the z-average radius of gyration R_g, may be written

$$\left(H\frac{c}{R(\theta)}\right)_{\theta=0} = \frac{1}{M_w} + 2A_2 c + \cdots \tag{3.52}$$

$$\left(H\frac{c}{R(\theta)}\right)_{c=0} = \frac{1}{M_w}\left[1 + \frac{1}{3}\left(\frac{4\pi_1}{\lambda'}\right)^2 R_g^2 \sin^2\frac{\theta}{2} + \cdots\right] \tag{3.53}$$

where λ' is the wavelength of the light in solution (λ_0/n_0). To construct a Zimm plot, equations (3.52) and (3.53) are added together. To make a more aesthetic plot, the concentration term is usually multiplied by an arbitrary factor (see Section 3.6.3).

Thus three quantities of interest can be determined from the same experiment: the weight-average molecular weight, the z-average radius of gyration, and the second virial coefficient. Note that the numeral 2, which appears in front of A_2 in equations (3.46) and (3.52), arises from the partial differentiation given in equation (3.44). A useful practical equation for the determination of R_g from a plot of $H[c/R(\theta)]$ versus $\sin^2\theta/2$ is

$$R_g^2 = \frac{3(\lambda')^2(\text{initial slope})}{16\pi_1^2(\text{intercept})} \tag{3.54}$$

Convenient light-scattering units are shown in Table 3.7.

The quantity H depends on the kind of radiation. For light-scattering, the formulation has been given in equation (3.45).

Table 3.7 Convenient light-scattering units

$H = \dfrac{\text{mol} \cdot \text{cm}^2}{\text{g}^2}$	$A_2 = \dfrac{\text{mol} \cdot \text{cm}^3}{\text{g}^2}$
$c = \dfrac{\text{g}}{\text{cm}^3}$	$R(\theta) = \text{cm}^{-1}$
	$\theta = \text{degrees}$
	$(\dfrac{\sin^2\theta}{2}$ for 90° is 0.500, unitless)
$M_w = \dfrac{\text{g}}{\text{mol}}$	

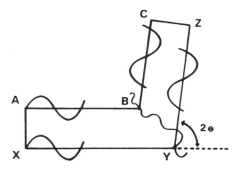

Figure 3.7 Schematic of the scattering phenomenon. When the waves are out of phase, the intensity of scattered light is reduced.

For X-ray scattering,

$$H = N_a i_e \left(\frac{\partial \rho_e}{\partial c} \right)^2 = \frac{N_a i_e}{e^2} (\rho_{e_s} - \rho_{e_p})^2 \tag{3.55}$$

and for neutron scattering,

$$H = \frac{N_a}{M_p^2} \left[a_s \left(\frac{V_p}{V_s} \right) - a_p \right]^2 \tag{3.56}$$

where M_p is the mer[†] molecular weight, and a_p and a_s are the coherent neutron scattering lengths of the polymer mer units and solvent, respectively. The quantity ρ is the density of the polymer (38). The quantity i_e is the Thomson scattering factor for a single electron, and ρ_e is the electron density of the solution; subscripts p and s represent polymer and solvent, respectively. Of course, the solvent can also be polymeric, and frequently the "solvent" differs from the polymer by having hydrogen or deuterium (H or D) atoms where the polymer has D or H atoms, especially for neutron scattering, and V_p and V_s represent the volumes of polymer and solvent, respectively.

3.6.2 The Appropriate Angular Range

To determine the radius of gyration, R_g, the basic mathematical relationship must hold,

$$K^2 R_g^2 < 1 \tag{3.57}$$

The corresponding physical requirement indicates that there must be only partial destructive interference between two waves striking the same particle, so that the waves should not be out of phase by more than 180° (see Figure 3.7). Referring to Figure 3.7, note that the distance A, B, C appears to be somewhat different from the distance X, Y, Z, where the line $A–X$ is perpendicular to the direction of radiation flux before scattering, and the line $C–Z$ is perpendicular to the scattered flux (36–40).

[†] Recall that the term "mer," derived from poly*mer*, means the individual monomeric unit molecular entity. Some texts call this entity the monomeric unit.

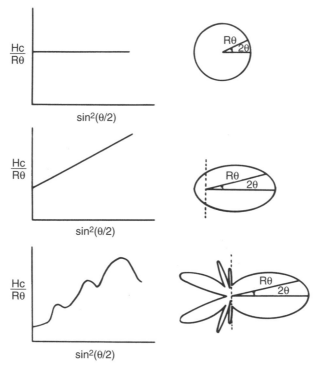

Figure 3.8 Scattering intensity envelopes for small, medium, and large particles. The Guinier region contains both the small and medium ranges, but the medium range is far more useful for scattering experiments.

Before scattering, the radiation has a certain degree of coherency; that is, the waves tend to be in phase. (Ordinary light is coherent over the size of the object; laser light is coherent over larger distances.) On scattering from points B and Y at an angle 2θ, the waves become out of phase. If the angle of scatter is large enough, the waves are out of phase by 180°; that is, one wave lags behind the other by $\frac{1}{2}\lambda$, and the radiation intensity observed at angle 2θ will be at the minimum. At still higher angles, the intensity increases again until a 360° phase difference is attained. The scattering must be observed for angles smaller than $KR_g = 1$ [see equation (3.57)] for radii of gyration to be determined. The partial destructive interference between the waves causes the intensity to vary according to angle (see Figure 3.8). Until 180° phase difference is attained, scattering intensity decreases with increasing angle.

For ordinary-size polymer molecules, R_g is of the order of 100 to 200 Å. For light-scattering θ is usually measured in the range of 45 to 135°, because $\lambda \simeq 5000$ Å. For X rays ($\lambda \simeq 1$–2 Å) and thermal neutrons ($\lambda \simeq 5$ Å), scattering measurements are usually made below 1°.

The requirement of small angles for X-ray and neutron scattering has led to the construction of huge instruments. For example, X-ray scattering instruments may be 10 m long. The giant small-angle neutron scattering instruments are located at the Institute Laue–Langevin, Grenoble, France (80 m), Jülich, West Germany (40 m), Oak Ridge National Laboratory, Tennessee (30 m) (41) and the several at NIST. The work-

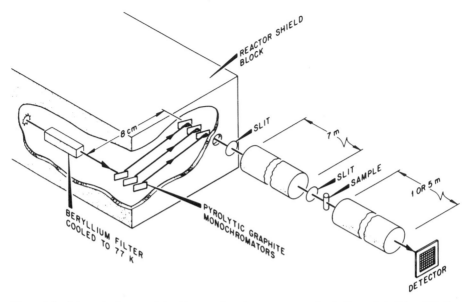

Figure 3.9 Schematic diagram of the 10-m small-angle neutron scattering facility at Oak Ridge, TN. The neutron beam used by the instrument is transported from the beam room to the SANS facility level by Bragg reflection from pyrolitic graphite crystals scattering at 90°. Courtesy of Dr. G. D. Wignall.

ings of the Oak Ridge instruments are illustrated in Figure 3.9 for the 10-m instrument. These instruments employ thermal or cool neutrons to increase their wavelengths.

Before proceeding further, a word should be said about the term "radius of gyration." The quantity R_g^2 is defined as the mean square distance away from the center of gravity, $R_g^2 = (1/N) \sum_{i=1}^{N} r_i^2$, for N scattering points of distance r_i. It is sometimes helpful to view R_g as a mechanical term wherein the radius of gyration of a body is the radius of a thin ring that has the same mass and same moment of inertia as the body when centered at the same axis (39). For a random coil, it is related to the end-to-end distance, r, by

$$R_g^2 = \frac{r^2}{6} \tag{3.58}$$

where r is the distance between the ends of the chain. Different relationships hold for spheres, rods, and coils. These differences are expressed quantitatively in terms of the quantity $P(\theta)$ (40):

$$\text{Sphere} \quad P(\theta) = \left[\frac{3}{x^3} (\sin x - x \cos x) \right]^2 \qquad x = \frac{ksD}{2} \tag{3.59}$$

$$\text{Rod} \quad P(\theta) = \frac{1}{x} \int_0^{2x} \frac{\sin w}{w} dw - \left(\frac{\sin x}{x} \right)^2 \qquad x = \frac{ksL}{2} \tag{3.60}$$

$$\text{Coil} \quad P(\theta) = \frac{2}{x^2} [e^{-x} - (1 - x)] \qquad x = \frac{k^2 s^2 r^2}{6} \tag{3.61}$$

where D = diameter of sphere, L = length of rod, $ks = K$, and r = root mean square of the distance between ends of the random coil. Rearranged, equation (3.61) yields equation (3.49), discussed previously. While the present text is interested primarily in random coils, scattering phenomena are widely used to determine quantities related to many shapes (39).

3.6.3 The Zimm Plot

Equations (3.52) and (3.53) show the function $H[c/R(\theta)]$ in the limit of $\theta = 0$ and $c = 0$, respectively. Three important pieces of information can be extracted from this experiment: the weight-average molecular weight, z-average radius of gyration, and the second virial coefficient. A simple but laborious method of plotting the data is shown in Figure 3.10. Here, (a) shows a plot of $H[c/R(\theta)]$ versus concentration for the several angles, extrapolated to $c = 0$. Part (b) is a replot of the intercepts from (a) versus $\sin^2(\theta/2)$, which yields M_w and R_g. In (c), the same data as (a) are replotted, this time against

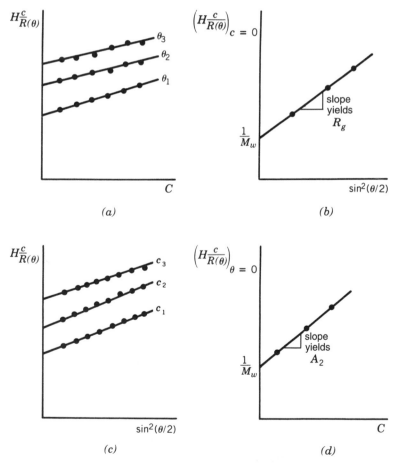

Figure 3.10 Illustration of a light-scattering calculation.

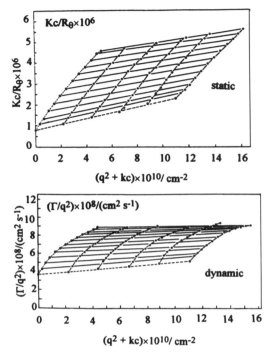

Figure 3.11 Zimm plots of a bacterial polysaccharide in 0.1 M NaCl. Top: Static, bottom: Dynamic light scattering. $\lambda_0 = 488$ nm, $dn/dc = 0.145$ cm³/g. Static light-scattering yields R_g, M_w, and A_2. Dynamic light-scattering yields the diffusion coefficient, D. Results: $M_w = 1.29 \times 10^6$ g/mol, $D_z = 3.58 \times 10^{-8}$ cm²/s. Terms: $q = (4\pi/\lambda) \sin(\theta/2)$, $\Gamma = Dq$.

$\sin^2(\theta/2)$ first. Then, in (d), the intercepts of (c) are plotted against concentration, yielding M_w and A_2. The intercepts from (b) and (d) must be equal, since they both equal $1/M_w$. This last provides an internal test of the data.

A most powerful advance was the introduction of the Zimm plot (32,33), which enabled the radius of gyration, the molecular weight, and the second virial coefficient to be calculated from a single master figure, by plotting $H[c/R(\theta)]$ versus a function of both angle and concentration (see Figure 3.11) (42). In a Zimm plot, the concentration is multiplied by an arbitrary constant, in this case 1×10^3, which is, of course, divided out of the final answer.

3.6.4 Polymer Chain Dimensions

For random coils obeying Gaussian statistics, the end-to-end distance squared depends on the molecular weight,

$$r^2 = CM \qquad (3.62)$$

where C is a function of the chain molecular structure. Equation (3.62) has the same form as the relation equating the total distance traveled for a particle undergoing Brownian motion as a function of time, except that in Brownian motion the distance traveled under each impact is variable where mer bond lengths are fixed. In addition the Brownian particle may turn at any angle, whereas carbon–carbon bonds are fixed. Perhaps a more fundamental difference is that the path of a Brownian particle may cross itself, whereas a polymer chain may not. All three of these effects tend to increase the constant C but do not change the fundamental relationship expressed in equation (3.62).

Another basic relationship is

$$R_g^2 = \tfrac{1}{6} r^2 \tag{3.62a}$$

good for random coils.

Light-scattering studies such as illustrated in Figure 3.11 show that for high enough molecular weight, polymer chains become random coils. During the 1940s and 1950s scientists theorized that the conformations of polymer chains in an amorphous bulk polymer are similar to that which exists under Flory θ-conditions. In fact there was only limited direct experimental evidence to support this theory.

The problem was that nobody could figure out a good way to measure the molecular dimensions in the bulk state, although on theoretical grounds (see Chapter 5) they were thought to be similar to the dilute solution results. After about 1960, polymer scientists turned their attention to other problems, and there the results stood until the development of small-angle neutron scattering (SANS) in the early 1970s.

3.6.5 Scattering Data

One problem that had to be solved, and was, relates to the fact that all these scattering techniques yield the z-average radius of gyration and the weight-average molecular weight. To correct the data for proper comparison, the molecular weight distribution needs to be known, or very sharp molecular weight distributions need to be made. The preferred solution has been to work with nearly monodisperse polymer samples, such as prepared by anionic polymerization. In other cases, the ratio of M_z to M_w is known or estimated from polymerization kinetics (see Section 3.7.2). To properly estimate the molecular weight distribution, the number-average molecular weight is required, bringing into play osmometry or gel permeation chromatography (see Sections 3.5.3 and 3.9). Generally, the z-average R_g values are corrected back to the weight-average, and the weight-averages of both quantities are reported.

When SANS instrumentation became available in the early 1970s, polymer scientists sought to reexamine chain conformation behavior, this time directly in the bulk amorphous state. The earliest studies were on amorphous poly(methyl methacrylate) (43,44) and polystyrene (45–47). These and several subsequent papers (48–56) indeed confirmed that values of $(R_g^2/M_w)^{1/2}$ [see equation (3.62)] were substantially the same in the bulk as in Flory θ-solvents. The results of the SANS experiments are substantially the same as those obtained in θ-solvents (36).

The quantity $(R_g^2/M_w)^{1/2}$ is, of course, a measure of chain stiffness. For example, polycarbonate, with $(R_g^2/M_w)^{1/2} = 0.457$, is stiffer than polystyrene, which has a value of 0.275. The importance of these quantities lies in their relation to physical and mechanical behavior. Both melt and solution viscosities depend directly on the radius of gyration of the polymer and on the chain's capability of being deformed. The theory of the random coil (Section 5.3), strongly supported by these measurements, is used in rubber elasticity theory (Chapter 9) and many mechanical and relaxation calculations.

3.6.6 Dynamic Light-Scattering

Dynamic light-scattering, also known as quasi-elastic light scattering and photon correlation spectroscopy, makes use of scattering from very small volumes at very short time
intervals, and correlating the scattering intensity fluctuations with time. Dynamic light-
scattering can be used to measure molecular motion and diffusion coefficients in polymers and colloids.

Historically there are two possible approaches to dynamic light-scattering. First, one
may consider a modified Doppler experiment. If the molecule or particle is moving
toward the beam, its frequency will be increased; if it is moving away, the frequency
is decreased. However, the broadening of the beam caused by Doppler effects is of the
order of 10 to 1000 Hz, difficultly resolvable with optical tools (57), noting that green
light has a frequency of 6×10^{14} Hz.

Better, one may focus a laser light beam down to a waist of 0.1 mm, yielding a scattering volume of 1×10^{-3} mm^3. The length of time of each measurement is on the order
of 1 μs. Under these conditions there is a time dependence of the scattered light, leading
to the construction of an intensity–time correlation function. This carries the old concept
of the observation of Brownian motion to new heights. The fluctuations are then correlated *via* computer programs in real time, yielding the required information.

In dynamic light-scattering, the scattered intensity is mostly due to concentration
fluctuations. Two different types of motion in polymer solutions mainly contribute to the
fluctuations: (a) the translational motion of the center of mass of the individual molecules, and (b) the internal modes of motion of segments with respect to their centers of
mass (58).

In a simple case the diffusion coefficients of spheres can be determined via the
Stokes–Einstein relationship; see Section 12.6.4. While such a relationship was originally
derived with hard spheres in mind, it can also be used with globular proteins such as
bovine serum albumin (59). Dynamic light-scattering can be used to study aggregation,
adsorption, and structural changes as well as chain dynamics.

For dilute polymer solutions the basic equation for the diffusion coefficient can be
written (60)

$$D_{\text{app}}(K, c) = D_{z0}(1 + CR_g^2 K^2)(1 + k_D c) \tag{3.63}$$

where $D_{\text{app}}(K, c)$ is the diffusion coefficient at finite angles and concentration, D_{z0} represents the corresponding z-average value at zero concentration and angle, C is a coefficient characteristic of the polymer, and k_D is a dynamic interaction parameter.

Equation (3.63) above represents a condensed equivalent of equations (3.52) and
(3.53) for dynamic light scattering, suggesting a Zimm plot-like analysis. Figure 3.11
illustrates both the static light-scattering Zimm plot and the dynamic light-scattering
Zimm plot equivalent for a microbial polysaccharide (61). Here, $\Gamma = DK^2$, and $K \equiv q$.
The key results in this case are $M_w = 1.29 \times 10^6$ g/mol and $D_z = 3.58 \times 10^{-8}$ cm^2/s,
suggestive of a very large macromolecule.

3.7 MOLECULAR WEIGHTS OF POLYMERS

3.7.1 Molecular Weight of Commercial Polymers

The molecular weight of polymers used in commerce varies from about 30,000 to over
1,000,000 g/mol. Sometimes conflicting requirements include the use of high enough

molecular weights to obtain good physical properties, and low enough molecular weights to permit reasonable processing conditions, such as melt viscosity.

Poly (vinyl chloride) Commercial poly(vinyl chloride) "vinyl" polymers range from 60,000 to about 90,000 g/mol. The restrictions above hold in this case.

Poly (methyl methacrylate) Those polymers that are used in such products as Plexiglas® have high molecular weights with broad distributions. A viscous syrup containing low-molecular-weight polymer and monomer is poured into a mold and allowed to polymerize. Late in the polymerization, the phenomenon known as autoacceleration takes place, where the molecular weight increases dramatically owing to a suppression of the termination step. This high molecular weight produced at the end may be over 1×10^6 g/mol, contributing strength and toughness to the final sheet. This material is used, however, without further processing.

Cellulose This natural polymer occurs with extremely high molecular weights, sometimes in the several million range, and with molecular weight distributions of M_w/M_n in the range of 10 to 50. For commercial applications such as rayon, the polymer is deliberately degraded down to the 50,000 to 80,000 g/mol range to increase processibility. The better products often utilize the higher end of this range.

3.7.2 Thermodynamics and Kinetics of Polymerization

The synthesis of polymers, with attendant aspects of the thermodynamics and kinetics of polymerization that occupy entire textbooks (62,63), is to a very significant extent beyond of the coverage of this text. Indeed, organic polymer science is often taught as a mate course to physical polymer science. However, since the thermodynamics and kinetics of polymerization affect both the molecular weights and the polydispersity index obtained, the most salient features of these areas will be explored. Polymer synthesis itself was briefly discussed in Section 1.4.

3.7.2.1 Thermodynamics of Chain Polymerization
Under standard conditions, the Gibbs free energy, ΔG^0, is related to the equilibrium constant of the polymerization, K, by

$$\Delta G^0 = -RT \ln K \tag{3.64}$$

Consider a chain polymerization,

$$M_n \bullet + M \leftrightarrow M_{n+1} \bullet \tag{3.65}$$

where the rate constant of the forward reaction, polymerization, is k_p and the rate constant of the reverse reaction, depolymerization, is given by k_{dp}. Then

$$K = \frac{k_p}{k_{dp}} = \frac{[M_{n+1}\bullet]}{[M_n\bullet][M]} = \frac{1}{[M]} \tag{3.66}$$

When the forward and reverse reactions have equal rates, namely when polymerization-depolymerization propagation rates are equal, the concept of ceiling and floor temperatures arises. Most polymerizations have ceiling temperatures, temperatures above which the monomer cannot be polymerized, but the polymer will spontaneously depolymerize back to the monomer. Commercially this fact leads to an important method of polymer recycling whereby scrapped polymer is heated under anaerobic conditions to allow distilling off the resultant monomers.

3.7.2.2 Kinetics of Chain Polymerization   The kinetics of the chain reactions in Section 1.4 can be written, for initiation, as

$$I \overset{k_i}{\rightarrow} 2R\bullet \tag{3.67}$$

$$R\bullet + M \overset{k_2}{\rightarrow} RM\bullet \tag{3.68}$$

and for propagation, as

$$RM\bullet + M \overset{k_p}{\rightarrow} RM_2\bullet \tag{3.69}$$

or

$$RM_n\bullet + M \overset{k_p}{\rightarrow} RM_{n+1}\bullet \tag{3.70}$$

following the usual assumption that k_p is independent of the degree of polymerization. For termination by combination,

$$RM_n\bullet + RM_m\bullet \overset{k_t}{\rightarrow} RM_{n+m}R \tag{3.71}$$

where k_i, k_p, and k_t stand for the rate constants of initiation, propagation, and termination, respectively, and I, R, and M stand for initiator, initiator radical, and monomer, respectively, and M_n or M_m represent a polymer of degree of polymerization of n or m, respectively.

The preceding kinetics of chain polymerization leads to a first-order rate dependence on monomer concentration and a half-order dependence on initiator:

$$R_p = k_p \left(\frac{k_i}{k_t}\right)^{1/2} [M][I]^{1/2} \tag{3.72}$$

where R_p represents the rate of polymerization. Typical values of k_p and k_t at 60°C are shown in Table 3.8 (71). A typical initiator is benzoyl peroxide, which at 60°C has a value of k_i of 2.76×10^{-6} s^{-1} (64). Brackets indicate concentrations.

The kinetic chain length, v, of a radical chain polymerization is the average number of monomer molecules consumed for each radical initiating a chain. Thus, at steady state,

Table 3.8   Chain polymerization kinetics at 60°C^a

Monomer	k_p (1/mol · s)	$k_t \times 10^{-6}$ (1/mol · s)	Comments
Methyl methacrylate	573	2.0	Termination by combination
Styrene	187.1	29.4	Steady state
Acrylonitrile	1960	782	Solvent: dimethyl formamide

^{a}Based on J. Brandrup and E. H. Immergut, eds., *Polymer Handbook*, 3rd ed., Wiley, 1989, sec. II.

$$v = \frac{R_p}{R_i} = \frac{R_p}{R_t} = \frac{k_p[M]}{2(fk_ik_t[I])^{1/2}} \tag{3.73}$$

where R_i, R_p, and R_t represent the rates of initiation, propagation, and termination, respectively. The quantity f is the initiator efficiency factor, the fraction of initiator molecules that actually initiate a polymerization on decomposition. Frequently f is about 0.8. The number-average degree of polymerization, DP_n in reaction (3.73), is equal to $2v$ for termination by combination. Termination by combination yields a polydispersity index of 1.5, while termination by chain transfer yields a PDI of 2.0, ideally.

3.7.2.3 Thermodynamics of Step Polymerization

Many step polymerizations involve equilibrium reactions. Consider a simple polyesterification where carboxyl groups and hydroxyl groups react to form a polyester and water; see Section 1.4.2. The equilibrium constant, K, is given schematically by

$$-COOH + -OH \overset{K}{\leftrightarrow} -COO- + H_2O \tag{3.74}$$

where

$$K = \frac{[COO][H_2O]}{[COOH][OH]} \tag{3.75}$$

For step-growth polymerizations, the fractional conversion p is given by $[COO] = p[M]_0$, where $[M]_0$ is the concentration of ester groups. Then DP_n is given by

$$DP_n = \frac{1}{1-p} \tag{3.76}$$

and the corresponding weight-average degree of polymerization is given by

$$DP_w = \frac{(1+p)}{(1-p)} \tag{3.77}$$

This leads directly to the polydispersity index, DP_w/DP_n,

$$PDI = 1 + p \tag{3.78}$$

At high conversion, where p approaches one, the molecular weight distribution approaches two.

3.7.2.4 Kinetics of Step Polymerizations

Step polymerizations can be divided into two broad categories, those catalyzed by some externally added chemical such as an acid, and those that are self-catalyzed. For simplicity, the self-catalyzed polyesterification reaction will be considered further. The reaction can be written

$$\frac{-d[COOH]}{dt} = k[COOH]^2[OH] \tag{3.79}$$

where for each carboxyl molecule reacting, another one is required for acid catalysis, and t is time. Assuming equal molar concentrations of carboxyl and hydroxyl groups to start, the extent of reaction is given by

$$\frac{1}{(1-p)^2} = 2[M]_0 kt + 1 \tag{3.80}$$

The key finding is that the thermodynamics and kinetics of polymerization provides information relative to degrees of polymerization and polydispersity index. Most of the expression above is cast in terms of DP or kinetic chain length. The actual molecular weight can be easily obtained by multiplying by the mer molecular weight.

3.7.3   Molecular Weight Distributions

If the termination reaction in chain polymerization is by disproportionation, then the polydispersity index, M_w/M_n, is 2. Termination by combination yields a polydispersity index of 1.5. Stepwise polymerizations, such as polyester formation, yield a value of 2. Anionic polymerizations yield surprisingly narrow distribution, with values sometimes less than 1.05.

Proteins are almost the only source of truly monodisperse polymers. Nature makes all these molecules exactly alike. Polymers like cellulose have very broad distributions, as mentioned previously.

Of course, polymerization need not be ideal in its kinetics. Branching may occur, which broadens the molecular weight distribution. There may even be two or more peaks in the molecular weight distribution. A powerful method for directly observing the shape of the distribution curve is gel permeation chromatography (see Section 3.9).

The various molecular weight distributions have been modeled. Two of the most important are the Schultz (65) distribution and the Poisson (66) distribution. Taking n to be the degree of polymerization and w_n as the differential distribution by weight, the Schultz distribution is

$$w_n = \frac{a}{x_n \Gamma(a+1)} \left(\frac{an}{n_n}\right)^a \exp\left(-\frac{an}{n_n}\right) \tag{3.81}$$

where $n_w/n_n = (a+1)/a$, and $\Gamma(a+1)$ is the gamma function of $a+1$. The subscripts w and n stand for weight and number average, respectively. When $a = 1$ and n is large, the polydispersity index is near 2.

The Poisson distribution assumes that all the chains are initiated simultaneously. Growth continues at approximately the same rate in each chain, until the monomer runs out. An analogy would be a horse race, where all the horses start at the bell and finish at nearly the same time. The result is a narrow molecular weight distribution. (For anionic polymerizations following these statistics, the ends of the chains remain "living," even though the monomer has run out. Addition of a second monomer then yields well-defined block copolymers; see Section 4.3.9.)

The Poisson distribution is given by (66)

$$(P_r)_x = \frac{e^{1-n_n}(n_n - 1)^{n-1}}{(n-1)!} \tag{3.82}$$

Table 3.9 Typical molecular weight distributions

Method	Polydispersity	Stereospecificity
Natural proteins	1.0	Perfect
Anionic polymerization	1.02–1.5	None
Chain polymerization	1.5–3	None
Step polymerization	2.0–4	None
Ziegler-Natta	2–40	High
Cationic	Broad	None
Metallocene	2–2.5	High

where $(P_r)_n$ is the mole fraction of n-mer. Then

$$\frac{n_w}{n_n} = 1 + \frac{n_n - 1}{n_n^2} \tag{3.83}$$

For reasonable values of n_n, the polydispersity index is nearly unity. This distribution is realized for carefully prepared anionic polymerizations.

For stepwise polymerizations, the kinetics are considered in terms of the extent of reaction, p, defined as the fraction of the functional groups reacted at time t. If only bifunctional reactants are present, then the degree of polymerization may be deduced by considering the total number of structural units present, N_0, and the total number of molecules, N:

$$n_n = \frac{N_0}{N} = \frac{1}{1 - p} \tag{3.81}$$

and

$$n_w = \frac{1 + p}{1 - p} \tag{3.82}$$

An important difference between stepwise reactions and chain polymerization is that the former type of molecules can keep on reacting. As p approaches unity, both n_w and n_n increase. Their ratio, however, approaches 2, called the most probable distribution.

The molecular weight distributions ordinarily obtained by the more important polymerization methods are shown in Table 3.9. Natural proteins, nature's best, are both monodisperse and 100% stereospecific. The Ziegler–Natta polymerization route, while highly stereospecific, has a broad molecular weight distribution. In general, the polydispersity index can be determined from an analysis of the kinetics of the reaction; in practice, various phenomena cause the products to be much broader in molecular weight distributions (67).

3.7.4 Gelation and Network Formation

If the functionality of the monomer is 2, as in the case of vinyl groups for chain polymerization,

$$C{=}C \rightarrow -C{-}C-$$

or monomers containing one carboxyl group and one hydroxyl group for stepwise polymerization,

$$HO - R - COOH \rightarrow \; - O - R - COO - \; + H_2O$$

linear polymers are formed. If some trifunctional, tetrafunctional, or higher monomers are incorporated, the polymer will be either branched or cross-linked.

An example of a trifunctional monomer is glycerol, with three hydroxyl groups,

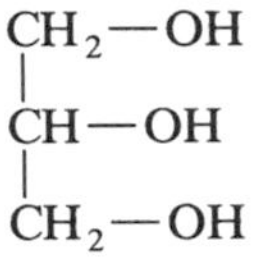

$$CH_2 - OH$$
$$CH - OH$$
$$CH_2 - OH$$

Divinyl benzene, with two vinyl groups, is a common cross-linker for chain polymerizations,

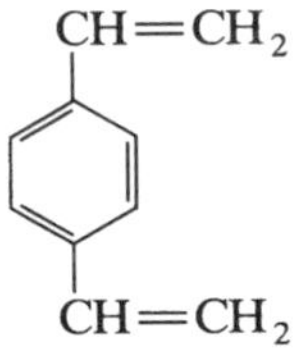

$$CH{=}CH_2$$

$$CH{=}CH_2$$

An important question is: When in a polymerization will gelation occur? Gelation is defined as when a single molecule, connected by ordinary covalent bonds, extends throughout the polymerization vessel. (Most of the material in the vessel need not be part of the molecule, however.) Alternately, gelation may be conceived as the point where a three-dimensional network is formed. From a physical point of view, the viscosity of the reacting mass goes to infinity at the gelation point.

According to Flory and Stockmayer (68), the critical extent of reaction, P_c, at the gel point is given by

$$P_c = \frac{1}{(f-1)^{1/2}} \tag{3.83}$$

where f represents the functionality of the branch units—that is, of the monomer with functionality greater than 2. This simple equation has been modified many times for particular stoichiometries and mixtures of monomers. Gelation is considered further in Section 9.1 and in Section 9.14.

3.8 INTRINSIC VISCOSITY

Both the colligative and the scattering methods result in absolute molecular weights; that is, the molecular weight can be calculated directly from first principles based on theory. Frequently these methods are slow, and sometimes expensive. In order to handle large numbers of samples, especially on a routine basis, rapid, inexpensive methods are required. This need is fulfilled by intrinsic viscosity and by gel permeation chromatography. The latter is discussed in the next section.

Intrinsic viscosity measurements are carried out in dilute solution and result in the

Relative velocity of solvent in capillary

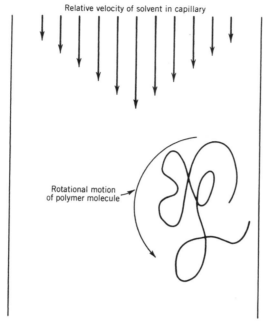

Rotational motion
of polymer molecule

Figure 3.12 The effect of shear rates on polymer chain rotation. Hydrodynamic work is converted into heat, resulting in an increased solution viscosity.

viscosity-average molecular weight; see Figure 3.4 and equation (3.34). Consider such a dilute solution flowing down a capillary tube (Figure 3.12). The flow rate, and hence the shear rate, is different depending on the distance from the edge of the capillary. The polymer molecule, although small, is of finite size and "sees" a different shear rate in different parts of its coil. This change in shear rate results in an increase in the frictional drag and rotational forces on the molecule, yielding the mechanism of viscosity increase by the polymer in the solution.

3.8.1 Definition of Terms

Several terms need defining. The solvent viscosity is η_0, usually expressed in poises, Stokes, or, more recently, Pascal seconds, Pa · s. The viscosity of the polymer solution is η. The relative viscosity is the ratio of the two,

$$\eta_{rel} = \frac{\eta}{\eta_0} \qquad (3.84)$$

where η_0 is the viscosity of the solvent.

Of course, the relative viscosity is a quantity larger than unity. The specific viscosity is the relative viscosity minus one:

$$\eta_{sp} = \eta_{rel} - 1 \qquad (3.85)$$

Usually η_{sp} is a quantity between 0.2 and 0.6 for the best results.

The specific viscosity, divided by the concentration and extrapolated to zero concentration, yields the intrinsic viscosity:

$$\left[\frac{\eta_{sp}}{c}\right]_{c=0} = [\eta] \tag{3.86}$$

For dilute solutions, where the relative viscosity is just over unity, the following algebraic expansion is useful:

$$\ln \eta_{rel} = \ln(\eta_{sp} + 1) \cong \eta_{sp} - \frac{\eta_{sp}^2}{2} + \cdots \tag{3.87}$$

Then, dividing $\ln \eta_{rel}$ by c and extrapolating to zero concentration also yields the intrinsic viscosity:

$$\left[\frac{\ln(\eta_{rel})}{c}\right]_{c=0} = [\eta] \tag{3.88}$$

Note that the natural logarithm of η_{rel} is divided by c in equation (3.73), not η_{rel} itself. The term $(\ln \eta_{rel})/c$ is called the inherent viscosity. Also note that the intrinsic viscosity is written with η enclosed in brackets. This is not to be confused with the plain η, which is used to indicate solution or melt viscosities.

Two sets of units are in use for $[\eta]$. The "American" units are 100 cm^3/g, whereas the "European" units are cm^3/g. Of course, this results in a factor of 100 difference in the numerical result. Lately, the European units are becoming preferred.

3.8.2   The Equivalent Sphere Model

In assuming a dilute dispersion of uniform, rigid, noninteracting spheres, Einstein (69, 70) derived an equation expressing the increase in viscosity of the dispersion,

$$\eta = \eta_0(1 + 2.5v_2) \tag{3.89}$$

where the quantity v_2 represents the volume fraction of spheres. The intrinsic viscosity of a dispersion of Einstein spheres is 2.5 for v_2, or 0.025 for concentration in units of g/100 cm^3.

Now consider a coiled polymer molecule as being impenetrable to solvent in the first approximation. A hydrodynamic sphere of equivalent radius R_e will be used to approximate the coil dimensions (see Figure 3.13). In shear flow, it exhibits a frictional coefficient of f_0. Then according to Stokes law,

$$f_0 = 6\pi_1\eta_0 R_e \tag{3.90}$$

where R_e remains quantitatively undefined.

The Einstein viscosity relationship for spheres may be written

$$\frac{\eta - \eta_0}{\eta_0} = \eta_{sp} = 2.5\left(\frac{n_2}{V}\right)V_e \tag{3.91}$$

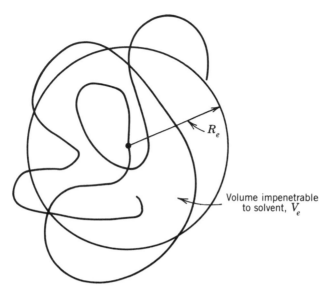

Figure 3.13   The equivalent sphere model.

where n_2/V is the number of molecules per unit volume. Of course, $V_e = (4\pi/3)R_e^3$. The quantity $n_2 V_e/V$ is the volume fraction of equivalent spheres, yielding the familiar result that the viscosity of an assembly of spheres is independent of the size of the spheres, depending only on their volume fraction.

Writing

$$\frac{n_2}{V} = \frac{cN_A}{M}$$ (3.92)

where c is the concentration and N_A is Avogadro's number,

$$\left[\frac{\eta_{sp}}{c}\right]_{c=0} = [\eta] = 2.5\frac{N_A V_e}{M}$$ (3.93)

Note that

$$\frac{V_e}{M} = \frac{4\pi_1}{3}\frac{R_e^3}{M} = \frac{4\pi_1}{3}\left(\frac{R_e^2}{M}\right)^{3/2}M^{1/2}$$ (3.94)

and

$$R_e = R_{e0}\alpha$$ (3.95)

where α is the expansion of the coil in a good solvent over that of a Flory θ-solvent.

The quantity R_{e0}^2/M is roughly constant. The same constant appears in Brownian motion statistics, where time takes the place of the molecular weight. This expresses the

Table 3.10 Selected intrinsic viscosity–molecular weight relationships, $[\eta] = KM_v^a$ (77)

Polymer	Solvent	$T(°C)$	$K \times 10^{3a}$	a^*
cis-Polybutadiene	Benzene	30	33.7	0.715
it-Polypropylene	1-Chloronaphthalene	139	21.5	0.67
Poly(ethyl acrylate)	Acetone	25	51	0.59
Poly(methyl methacrylate)	Acetone	20	5.5	0.73
Poly(vinyl acetate)	Benzene	30	22	0.65
Polystyrene	Butanone	25	39	0.58
Polystyrene	Cyclohexane (θ-solvent)	34.5	84.6	0.50
Polytetrahydrofuran	Toluene	28	25.1	0.78
Polytetrahydrofuran	Ethyl acetate hexane (θ-solvent)	31.8	206	0.49
Cellulose trinitrate	Acetone	25	6.93	0.91

Source: J. Brandrup and E. H. Immergut, eds., *Polymer Handbook*, 2nd ed., Wiley, New York, 1975, sec. IV.

[a] European units, concentrations in g/ml. Units do not vary with a.

[*] The quantity a, last column, is the exponent in equation (3.97).

distance traveled by the chain in a random walk as a function of molecular weight. According to Flory (17), the expansion of the coil increases with molecular weight for high molecular weights as $M^{0.1}$, yielding

$$[\eta] = 2.5\frac{4\pi_1}{3} N_A \left(\frac{R_{e0}^2}{M}\right)^{3/2} M^{1/2}\alpha^3 \tag{3.96}$$

3.8.3 The Mark–Houwink Relationship

In the late 1930s and 1940s Mark, Houwink and Sakurada arrived at an empirical relationship between the molecular weight and the intrinsic viscosity (71):

$$[\eta] = KM_V^a \tag{3.97}$$

where **K** and a are constants for a particular polymer–solvent pair at a particular temperature. Equation (3.97) is known today as the Mark–Houwink–Sakurada equation. This equation is in wide use today, being one of the most important relationships in polymer science and probably the single most important equation in the field. Values of **K** and a for selected polymers are given in Table 3.10 (72). It must be pointed out that since viscosity-average molecular weights are difficult to obtain directly, the weight-average molecular weights of sharp fractions or narrow molecular weight distributions are usually substituted to determine **K** and a.

According to equation (3.96) the value of a is predicted to vary from 0.5 for a Flory θ-solvent to about 0.8 in a thermodynamically good solvent. This corresponds to α increasing from a zero dependence on the molecular weight to a 0.1 power dependence. More generally, it should be pointed out that a varies from 0 to 2; see Table 3.11.

The quantity **K** is often given in terms of the universal constant Φ,

$$\mathbf{K} = \Phi \left(\frac{\overline{r_0^2}}{M}\right)^{3/2} \tag{3.98}$$

Table 3.11   Values of the Mark–Houwink–Sakurada exponent _a_

_a_	Interpretation
0	Spheres
0.5–0.8	Random coils
1.0	Stiff coils
2.0	Rods

where $\overline{r_0^2}$ represents the mean square end-to-end distance of the unperturbed coil. If the number-average molecular weights are used, then Φ equals 2.5×10^{21} dl/mol · cm³. A theoretical value of 3.6×10^{21} dl/mol · cm³ can be calculated from a study of the chain frictional coefficients (17).[†] For many theoretical purposes, it is convenient to express the Mark–Houwink equation in the form

$$[\eta] = \Phi \left(\frac{\overline{r_0^2}}{M} \right)^{3/2} M^{1/2} \alpha^3 = KM^{1/2}\alpha^3 \qquad (3.99)$$

If the intrinsic viscosity is determined in both a Flory θ-solvent and a "good" solvent, the expansion of the coil may be estimated. From equation (3.95),

$$[\eta]/[\eta]_\theta = \alpha^3 \qquad (3.100)$$

Values of α vary from unity in Flory θ-solvents to about 2 or 3, increasing with molecular weight.

3.8.4   Intrinsic Viscosity Experiments

In most experiments, dilute solutions of about 1% polymer are made up. The quantity η_{rel} should be about 1.6 for the highest concentration used. The most frequently used instrument is the Ubbelhode viscometer, which equalizes the pressure above and below the capillary.

Several concentrations are run and plotted according to Figure 3.14. Two practical points must be noted:

1. Both lines must extrapolate to the same intercept at zero concentration.
2. The sum of the slopes of the two curves is related through the Huggins (73) equation,

$$\frac{\eta_{sp}}{c} = [\eta] + k'[\eta]^2 c \qquad (3.101)$$

and the Kraemer (74) equation,

$$\frac{\ln \eta_{rel}}{c} = [\eta] - k''[\eta]^2 c \qquad (3.102)$$

[†] A widely used older value of Φ is 2.1×10^{21} dl/mol · cm³.

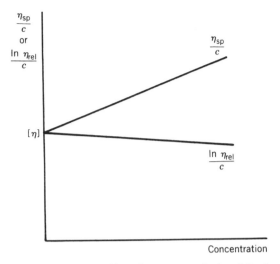

Figure 3.14 Schematic of a plot of η_{sp}/c and $\ln \eta_{rel}/c$ versus c, and extrapolation to zero concentration to determine $[\eta]$.

Algebraically

$$k' + k'' = 0.5 \qquad (3.103)$$

If either of these requirements is not met, molecular aggregation, ionic effects, or other problems may be indicated. For many polymer–solvent systems, k' is about 0.35, and k'' is about 0.15, although significant variation is possible.

Noting the negative sign in equation (3.102), k'' is actually a negative number.

The molecular weight is usually determined through light-scattering, as indicated previously. In order to determine the constants **K** and a in the Mark–Houwink equation, a double logarithmic plot of molecular weight versus intrinsic viscosity is prepared (see Figure 3.15) (75). Results of this type of experiment were used in compiling Table 3.11.

3.8.5 Example Calculation Involving Intrinsic Viscosity

Say we are interested in a fast, approximate molecular weight of a polystyrene sample. We dissolve 0.10 g of the polymer in 100 ml of butanone and measure the flow times at 25°C in an Ubbelhode capillary viscometer. The results are

Pure butanone   110 s

0.10% Polystyrene solution   140 s

Starting with equation (3.69)

$$\eta_{rel} = \frac{\eta}{\eta_0} = \frac{140}{110} = 1.273$$

$$\eta_{sp} = \eta_{rel} - 1 = 0.273$$

$$\frac{\eta_{sp}}{c} = \frac{0.273}{0.001} \text{ ml/g} = 2.73 \times 10^2 \text{ ml/g}$$

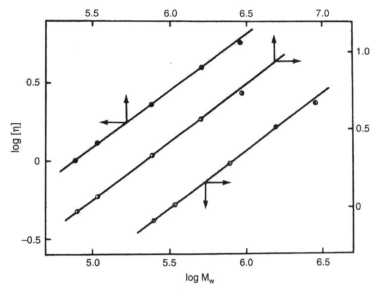

Figure 3.15 Double logarithmic plots of $[\eta]$ versus M_w for anionically synthesized polystyrenes, which were then fractionated leading to values of M_w/M_n of less than 1.06. Filled circles in benzene, half-filled circles in toluene, and open circles in dichloroethylene, all at 30°C (75). The arrows indicate the axes to be used. Units for [7] in 100 ml/g.

As an approximation, assume that the concentration is near zero, and that $[\eta] = 2.73 \times 10^2$ ml/g, equation (3.86). This obviates the extrapolation in Figure 3.14 that is required for more accurate results. Using the Mark–Houwink–Sakurada relation, equation (3.97) and Table 3.11, we have

$$[\eta] = \mathbf{K}M_V^a$$

$$2.73 \times 10^2 = 39 \times 10^{-3} M_V^{0.58}$$

$$M_V = 4.26 \times 10^6 \text{ g/mol}$$

Note that the units of $\mathbf{K}$ are irregular, depending on the value of a. Usually the units of $\mathbf{K}$ are omitted from tables.

3.9 GEL PERMEATION CHROMATOGRAPHY

Gel permeation chromatography (GPC) makes use of the size exclusion principle. The size of the molecule, defined by its hydrodynamic radius, can or cannot enter small pores in a bed of cross-linked polymer particles (76), the most common form of the *stationary phase*. The smaller molecules diffuse into the pores via Brownian motion (see Figure 3.16) and are delayed. The larger molecules pass by and continue in the *mobile* phase.

The instrumentation most commonly used in GPC work is illustrated in Figure 3.17 (77). The stationary phase consists of small, porous particles. While the mobile phase flows at a specified rate controlled by the solvent delivery system, the sample is injected

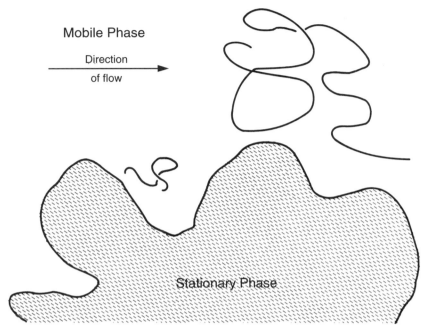

Figure 3.16 The size exclusion effect. The short chain can enter the pore, whereas the long chain will pass by.

into the mobile phase and enters the columns. The length of time that a particular fraction remains in the columns is called the *retention time* (78). As the mobile phase passes the porous particles, the separation between the smaller and the larger molecules becomes greater (see Figure 3.18) (76). While separation of polymer chains according to size remains the most important experiment, there are many other aspects, as described below.

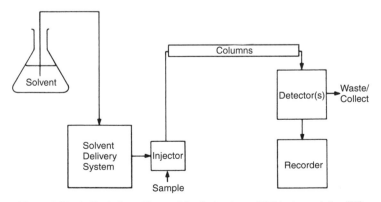

Figure 3.17 An illustration of the modules that make up GPC instrumentation (77).

Gel Permeation Chromatography

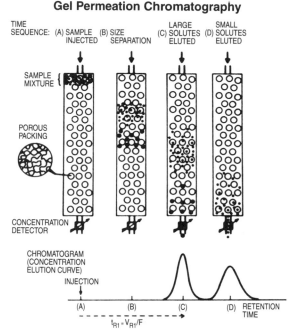

Figure 3.18 Illustration of the GPC experiment (81). The sample is injected into a solvent, which flows into a porous packed bed. The larger molecules flow straight through, whereas the small ones are temporarily held up.

3.9.1 Theory of Gel Permeation Chromatography

With the exception of proteins and a very few other macromolecules, most polymers exhibit some form(s) of heterogeneity (79). The most important is the molecular weight distribution (MWD), sometimes called a molar mass distribution, or polydispersity index (PDI), equal to M_w/M_n. Another type of heterogeneity involves a distribution of chemical composition, including statistical, alternating, block, and graft copolymers, as described in Chapter 2. Still another type of heterogeneity relates to functionality, particularly end groups. Macromolecules with terminal functional groups are usually called telechelics or macromonomers. Molecular architecture provides yet another type of heterogeneity, dictated by the shape of the chain. While most synthetic polymers form random coils, an increasing number of polymers are rod shaped, or form rings. Very many polymers are branched. Each of these types of heterogeneity must be taken into account when measuring molecular weights by relative methods.

There are two very popular relative methods of characterizing polymers with one or more of the above heterogeneities. Gel permeation chromatography (GPC), also known as size exclusion chromatography (SEC) or gel filtration chromatography, is one of several chromatographic methods available for molecular weight (molar mass) and molecular weight distributions. While GPC has its greatest value for measuring the molecular weight and polydispersity of synthetic polymers, a closely related method, high-performance liquid chromatography (HPLC) is more useful for separating and characterizing polymers containing functional groups, such as proteins and pharmaceutical polymers containing special active groups. Both of these methods depend on distribution coefficients.

3.9.2   Utilization of Distribution Coefficients in GPC and HPLC

Both GPC and HPLC processes relate to the selective distribution of an analyte between a mobile phase and a stationary phase. In general, the distribution coefficient, K_d, is given by (79)

$$K_d = \frac{V_R - V_i}{V} \tag{3.104}$$

where V_R represents the retention volume of the solute, V_i represents the interstitial volume of the column, and V is the volume of the stationary phase. Expressed qualitatively, K_d is defined as the ratio of the analyte concentration in or attached to the stationary phase to that in the mobile phase, namely the partitioning between the mobile and stationary phases. The volume V can be comprised of the pore volume of porous particles, the active surface area, or volume of chemically bonded stationary phase, depending on the separation mode.

As with all such distribution coefficients, the quantity K_d is related to the Gibbs free energy, ΔG,

$$-RT \ln K_d = \Delta H - T\Delta S = \Delta G \tag{3.105}$$

After rearranging, we have

$$K_d = \exp\left(\frac{\Delta S}{R} - \frac{\Delta H}{RT}\right) \tag{3.106}$$

Now the physical picture determines the sign and magnitude of the quantities above. The limited dimensions of the pores relative to the size of the polymer chains causes ΔS of the polymer chains to decrease. Interactions between the pore walls and the polymer chains are expressed in changes in ΔH, and are negative if the polymer and the wall are attracted to each other.

In the general case, K_d may be expressed

$$K_d = K_{GPC} K_{HPLC} \tag{3.107}$$

where the subscripts GPC and HPLC indicate quantities involving only entropic or enthalpic interactions, respectively. In the ideal GPC case, K_{HPLC} equals unity, and $K_d = K_{GPC}$.

In the ideal GPC case, $K_{HPLC} = 1$, and

$$K_d = \exp\left(\frac{\Delta S}{R}\right) \tag{3.108}$$

Of course, for the ideal HPLC case, the reverse is true. Equation (3.106) expresses the general case, where both entropy and enthalpy are involved. Usually one or the other must be reduced to substantially zero to use either GPC or HPLC.

3.9.3 Types of Chromatography

All types of chromatography utilize columns containing a finely divided stationary phase and a solution of a mixture that passes through the columns, called the mobile phase. Analyte mixtures separate as they travel through the columns due to the differences in their partitioning between the mobile phase and the stationary phase.

The present interest primarily involves GPC, which uses porous particles to separate molecules of different sizes (80,81). Its most important use has been to determine molecular weights and molecular weight distributions of polymers. Polymer molecules that are smaller than the pore sizes in the particles can enter the pores, and therefore they have a longer path and longer transit time than larger molecules that cannot enter the pores. Motion in and out of the pores is governed by Brownian motion. Thus the larger molecules elute earlier in the chromatogram, while molecules that entered more and more pores elute later and later.

HPLC, by contrast, utilizes interactions between the polymers and the surface of the particles composing the stationary phase (82). Important HPLC methods include reverse-phase partition chromatography, normal-phase partition chromatography, adsorption chromatography, chiral stationary phases, partition chromatography, and ion chromatography. In reverse-phase chromatography, the groups being analyzed for are more polar than the stationary phase. In normal-phase chromatography, the groups are less polar. There are also various hybrid methods such as HPLC size exclusion chromatography, particularly for proteins and functional group macromolecules. The best instruments today are sometimes called universal HPLCs, containing both GPC and HPLC columns and measurement capability; see Table 3.12 for specifications.

3.9.4 GPC Instrumentation

The most important parts of the instruments are the pumps for maintaining constant, pulseless rates of flow, the column types for the molecular weight range of analysis, and the detector system, see Table 3.12. Some of the popular types of solvent delivery pumps include syringe pumps, the plunger of which is actuated by a screw-fed stepper motor, and the dual piston reciprocating pumps, which may have the pistons in either series or parallel, see Figure 3.19 (79).

The type of column packing depends on whether the polymer is water soluble or organic soluble. For water soluble polymers, column packings consist of a range of materials, including silicas, hydroxyethyl methacrylate copolymers, chitosan, and highly cross-linked poly(vinyl alcohol). Organic soluble polymer-based columns most often

Table 3.12 GPC/HPLC universal type instruments available

Company	Features
Waters	Alliance Systems HPLC w/RE + UV detectors; HPLC oriented
Polymer Laboratories	LC Accord w/UV-vis detectors, integrated HPLC/GPC RI, LS evaporation detectors added
Hewlett Packard	HP 1100 Quaternary pump w/degasser, thermostatted column compartment, w/fluorescence and refractive index detectors

Note: Each of these instruments is equipped with full sets of columns, a computer, and printer.

Abbreviations: GPC, gel permeation chromatography; HPLC, high-performance liquid chromatography; RI, refractive index; LS, light-scattering; UV, ultraviolet, usually for measuring fluorescence; vis, visible light.

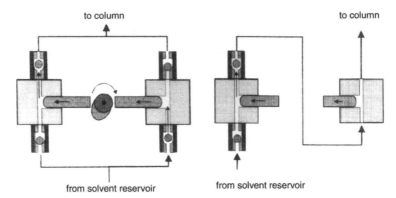

to column   to column

from solvent reservoir   from solvent reservoir

Figure 3.19 Dual piston pumps with parallel pistons deliver a smooth flow, while those in series are easier to maintain, since they have two check valves instead of four.

contain porous, densely crosslinked polystyrene, but porous silicas and highly cross-linked poly(vinyl alcohol) are also used. The size of the pores determines the size of the molecule that can diffuse in and out by Brownian motion, the larger molecules being restricted to entering only the larger pores. Since the motion in and out of the pores is random, the residence time of the short chains is longer. Hence the larger, high molecular weight polymer chains elute first. There are two major types of such columns. First, there are a series of columns each containing a specific pore size range. A series of three or even four such columns are required for the general case of any molecular weight between, say, 10,000 and 2,000,000 g/mol. The newer types are called mixed bed columns, because a range of pore sizes is included in each. Two such columns may be sufficient for molecular weight determination, speeding up the flow time. Usually one adds a guard column up front, which absorbs the gels and other unwanted material and thus prolongs the calibration and life of the columns.

There are several types of detectors; see Figure 3.17 (79). These are classified as either concentration-sensitive detectors, or molar mass (molecular weight) sensitive detectors. The refractive index detector is most popular concentration-sensitive detector, measuring the change in refractive index as the concentration of polymer in the solution changes. These usually operate on some type of differential refractive index method or Fresnel refraction. Of course, most of the time substantially pure solvent flows. When the polymer chains arrive at the detector, then the refractive index of the solution changes, providing a measure of the polymer concentration. While most polymers have a different refractive index than the solvent (usually higher), if the refractive indices of both the polymer and solvent are substantially the same, the method cannot be used.

Another detector group of methods involves the input of ultraviolet light, with the output being fluorescence or absorption by the polymer. There are two different methods of detecting the ultraviolet light interaction with the polymer. If the polymer fluoresces, then the detector can be placed at an angle to the light beam path and the fluorescence intensity level measured, usually in the visible range. Another method measures the attenuation of the beam directly, from whatever cause. Polymers that neither absorb nor fluoresce at the incoming wavelength cannot be detected by this method. However, instruments that measure the attenuation of the beam directly are more popular than the fluorescence units in practice, apparently because they are more useful. Some models come with diode array detectors to give greater control over the wavelengths of light being utilized. Other types have variable wavelength inputs, or a dual wavelength detector system, such that two ultraviolet wavelengths can be utilized simultaneously, enhancing copolymer analysis, for example. For aromatic polymers such as polystyrene, this provides an alternate and very powerful detector system, since it absorbs strongly in the ultraviolet.

Other methods include a density detector utilizing a mechanical oscillator, and also an evaporative light-scattering method. In this latter method the sample is nebulized (evaporated). Each droplet that contained nonvolatile material will form a particle. This aerosol causes light to be scattered, resulting in a method to determine the concentration of solute.

Molecular-weight-sensitive methods include light-scattering, viscometry, and the like. Since light-scattering and viscometry measure different averages (see Section 3.6 and Section 3.8, respectively), the results will be somewhat different. If two angles are used in the light-scattering detectors, the radius of gyration of the polymer chains can be determined. Thus not only can the molecular weight distribution be determined, but also the radii of gyration distribution.

Analysis of the literature shows a surprising tendency to "mix and match" parts of instrumentation from different supply sources, including software (83–85).

3.9.5 Calibration

Noting GPC is a relative molecular weight method, such instrumentation needs to be calibrated. Narrow molecular weight distribution, anionically synthesized polystyrenes are used most often for the purpose. Other polymers used for calibration include poly(methyl methacrylate), polyisoprene, polybutadiene, poly(ethylene oxide), and the sodium salt of poly(methacrylic acid). Molecular weight ranges available start at low oligomers of only a few hundred g/mol, up to 20,000,000 g/mol. In all cases, use of narrow molecular weight distribution standards is preferred.

Since the polymer chains are separated on the basis of their size, rather than their molecular weight per se, calibration via polymers other than the one of actual interest carries an absolute error. Clearly, selecting a polymer with an R_g^2/M ratio similar to that of interest is preferred. Today, except for special purposes, most polymer scientists are willing to accept the absolute error to be able to determine the approximate molecular weight and the molecular weight distribution of a polymer rapidly, usually in about half an hour. For most random coil synthetic polymers, the error is less than about 30%. If the actual molecular weight needed is important, then the instrument must be calibrated with the polymer in question. Section 3.9.7 describes a universal method to obtain more realistic results in the general case.

Calibration usually involves the determination of the elution volume for a series of narrow molecular weight standards. If the molecular weight is very low, as in oligomers, peaks from individual species may be used. For higher molecular weights, the peak average is used. Usually an assumption is made of a linear relation between log M and a function of the elution volume, such as a polynomial. As with all such fits, care must be exercised to use standards that cover the entire molecular weight range to be studied, and perhaps a bit more. For completion, other types of chromatography include thin layer (86), gas (87), and supercritcal chromatography (88), among others.

Today, data acquisition and processing are usually computer controlled. There are four transformations required of the raw chromatographic data to provide results as

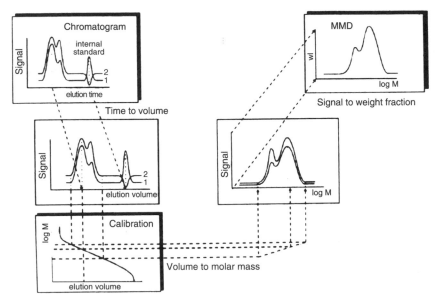

Figure 3.20 The molecular weight and molecular weight distribution are determined with standards precalibrated *via* an absolute method such as light-scattering.

usually reported; see Figure 3.20 (79): (a) conversion of elution time to elution volume, (b) conversion of elution volume to molecular weight, (c) conversion of detector response to polymer concentration, and (d) conversion of polymer concentration to weight fraction. Quantification of plate count and resolution, as well as calibration are discussed further in ASTM D5296-97 (89).

3.9.6 Selected Current Research Problems

Pasch and Trathnigg (79) describe a basic determination of the molecular weight distribution of a suspension polymerized polystyrene, using dibenzoyl peroxide as an initiator. Calibration of the GPC instrument utilized anionically polymerized polystyrene standards. Five µm cross-linked polystyrene particles were used in the columns, the mobile phase solvent was chloroform, and the detectors were refractive index and density based. Ethanol was added as an internal standard for flow rate correction. The results for the refractive index detector are shown in Figure 3.21 (79). The weight-average molecular weight of this sample is approximately 63,000 g/mol.

In the case of copolymers, any single detection method will have variable sensitivity for each type of mer. If the copolymer composition is itself a variable, then the use of dual or even multiple detectors will be required for accurate results. Calibrations for both homopolymers should be followed, if possible, by an analysis of homopolymers of similar molecular weights, before attempting an analysis of the copolymers themselves.

The molecular weight of each component in a polymer blend may also be determined. In a model experiment, poly(methyl methacrylate), PMMA, molecular weights were estimated in the presence of polystyrene, PS (90). Anionically polymerized polystyrene and free radically polymerized poly(methyl methacrylate) were dissolved in

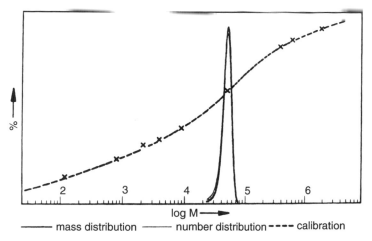

Figure 3.21 The molecular weight distribution of polystyrene PS 50000 with stationary phase Phenogel M, mobile phase, butanone.

tetrahydrofuran in a 50/50 w/w mix. A dual detector GPC was used, equipped with refractive index (RI) and ultraviolet (UV) detectors.

The refractive index detector measures both polymers simultaneously; see Figure 3.22 (90). However, the sensitivity to each polymer differs, being a function of the difference between each polymer refractive index and the solvent. The values are summarized in Table 3.13 (90). This particular system is nearly three times more sensitive to the polystyrene.

The ultraviolet detector only observes the phenyl groups on the polystyrene, which fluoresce. After subtracting out the polystyrene component, the poly(methyl methacrylate) component remains (bottom curve, Figure 3.22). The results in Table 3.13

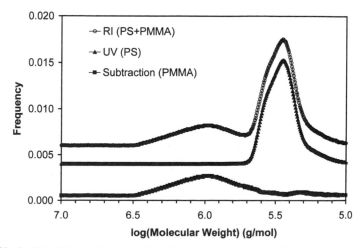

Figure 3.22 Analysis of the molecular weights of the polymers in a polymer blend of polystyrene and poly (methyl methacrylate). This method requires a dual detector GPC system.

Table 3.13 Molecular weight determination of poly(methyl methacrylate) in blend

Component	Refractive Index	Theoretical Area under Curve, %	Molecular Weight, nominal, g/mol	Molecular Weight, determined, g/mol
Polystyrene	1.592	69.4	$M_n = 2.66 \times 10^5$ $M_w = 2.90 \times 10^5$	—
Poly(methyl methacrylate)	1.4893	30.6	$M_n = 4.01 \times 10^5$ $M_w = 8.82 \times 10^5$	$M_n = 6.3 \times 10^5$ $M_w = 1.06 \times 10^6$
Tetrahydrofuran	1.408	—	—	—

show that the result is some 30% in error, but not so bad considering that poly(methyl methacrylate) is the less sensitive polymer in the measurements.

While much of the work done today involves more or less straightforward characterization of molecular weights and molecular weight distributions of polymers, a great deal of research involving much greater complexity is in progress. Current research problems include chain geometry and solution aggregation (83), ABC triblock copolymers as topological isomers (84), and hyperbranched polymers (85); see Section 14.5.

3.9.7 The Universal Calibration

Beginning with the Mark–Houwink relationship, equation (3.97), it is easy to show that the average molecular size is given by

$$[\eta]M = \Phi(\overline{r_0^2})^{3/2} \alpha^3 \tag{3.109}$$

where $\overline{r_0^2}$ represents the root-mean-square end-to-end distance of the polymer chain.

The right-hand side is proportional to the polymer's hydrodynamic volume (91). A new aspect of GPC calibration arises from the recognition that a polymer's hydrodynamic volume might form the basis for molecular weight determination. Since GPC depends on the hydrodynamic volume per se rather than its molecular weight per se, a new calibration method is suggested. This is the "universal calibration," which calls for a plot of $[\eta]M$ versus elution volume.

Figure 3.23 (92) illustrates the universal calibration procedure for poly(vinyl acetate) and polystyrene. Note that the two sets of data lie on the same straight line. The universal calibration is valid for a range of topologies and chemical compositions. However, it cannot be used for highly branched materials or polyelectrolytes, which have different or varying hydrodynamic volume relationships. The universal calibration procedure is especially useful for estimating the molecular weight of new polymers, since the intrinsic viscosity is usually easy to obtain. The procedure also tends to correct for differences in the hydrodynamic relationships when several polymers are compared, and only one of them (e.g., polystyrene) is used as the calibration material.

3.10 MASS SPECTROMETRY

Mass spectrometry is the study of the mass, or molecular weight, of ions created via ionization or fragmentation and determined electrically in the gas phase. In the study of polymers, mass spectrometry has two broad applications:

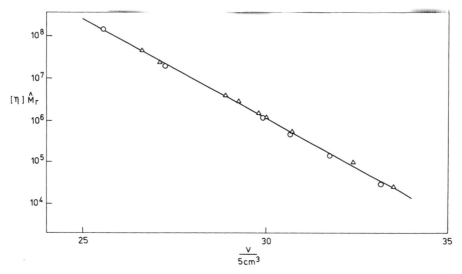

Figure 3.23 The universal calibration curves for polystyrene and poly(vinyl acetate) (92). The number 5 in the x-axis units means that the scale is in siphon "counts" of 5 cm^3, so that the x-ordinate 30 corresponds to an elution volume of 150 cm^3. (R. Dietz, private communication, November 1984.) $\hat{M}_r$ is the "peak" GPC molecular weight, usually the unknown, $\hat{M}_r$ values are close to the geometric mean of M_n and M_w.

1. To characterize functionality. Unknown polymers, residual volatile chemicals, and additives can be identified. This application depends on the fragmentation or degradation of the polymer chain or chemicals during the ionization; see Chapter 2.

2. To provide a new basis for the determination of absolute molecular weights. Novel techniques, developed below, now allow for the determination of absolute molecular weights and molecular weight distributions for polymers. In some cases individual molecular species can be identified.

The mass spectrum obtained is a plot of the ion abundance against mass-to-charge ratio, m/z, where m represents the mass and z the charge (93). The most important peak is the molecular ion, M^+.

The main problem with mass spectrometry involves the low volatility of polymers. The classical limit was about 10,000 g/mol without encountering significant degradation. The need, of course, is for stable, charged macromolecules in the vapor phase.

Mass spectroscopy instrumentation has four basic components, sample input, ion source, a mass analyzer, and an ion detection and recording device. Once the data are gathered, the interpretive problem requires decisions as to whether the detected species results from vaporization of the polymer chain, decomposition, double charges, and so on.

Significantly higher molecular weights can be characterized through newer methods. Two such improvements are critical for polymeric applications:

1. Matrix-assisted laser desorption ionization (MALDI), a *soft* ionization technique for transferring large molecular ions into a mass spectrometer with minimum fragmentation, and

2. Time of flight (TOF), techniques which measure the flight time of ions over fixed distances. The ions drift to the detector, the heaviest ions arriving last.

These two methods can be combined to allow improved analysis at higher molecular weights, around 100,000 g/mol being the upper limit at present.

Data obtained by Danis et al. (94) on narrow molecular weight distribution poly(butyl methacrylate), Figure 3.24, illustrate both the power and the problems of the method. The spectrum shows three different charge states centered at about m/z of 17,000, 32,000, and 65,000 g/mol. The m/z 17,000 g/mol distribution represent the doubly charged ions, while the distribution centered at m/z of 65,000 g/mol corresponds to dimers formed by the aggregation of two polymer chains. The peak centered at $M_w = 32,000$ g/mol represents the single-charged species of interest. The extraneous distributions can be removed mathematically to allow a better estimate of the molecular weight distribution. Mass spectrometry of lower molecular weight samples suffer less from dimers and doubly charged species illustrated in Figure 3.24, and also suffer less degradation.

The poly(butyl methacrylate) examined in Figure 3.24 was also subjected to different GPC studies, calibrated with different narrow molecular weight polystyrenes. A weight-average molecular weight of 32,200 g/mol was obtained. While the result agrees with the mass spectrometry data, problems exist with GPC calibration also. As pointed out

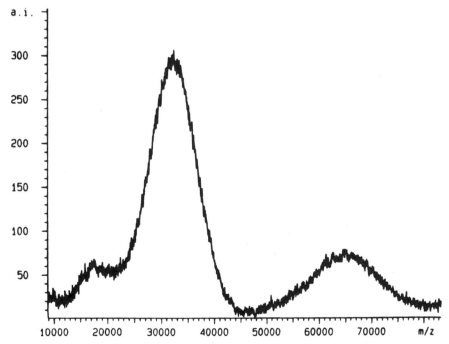

Figure 3.24 Positive ion MALDI/TOF mass spectrum of poly(butyl methacrylate). The data were acquired in the linear mode and 11-point smoothed.

above, mass spectrometry is an absolute molecular weight method, while GPC is a relative method. An important advantage is that MALDI-TOF data can be obtained from submilligram quantities of sample in less than 15 minutes. At the present time, mass spectrometry of polymers appears to be evolving rapidly (95,96).

3.11 SOLUTION THERMODYNAMICS AND MOLECULAR WEIGHTS

A knowledge of solution thermodynamics is critically important in determining the suitability of a solvent for a molecular weight determination. Once having decided on a suitable solvent, solution thermodynamics provides a basis for determining how an extrapolation to zero concentration is to be carried out.

Below the Flory θ-temperature, polymer solutions may phase-separate. The higher the molecular weight is, the higher the upper critical solution temperature. At infinite molecular weight, the Flory θ-temperature is reached. Thus the Flory θ-temperature is defined by several different criteria:

1. It is the temperature where A_2 is zero for dilute solutions, and $\chi_1 = \frac{1}{2}$.
2. It is the temperature where the radius of gyration approximates that of the bulk polymer (see Chapter 5).
3. It is the temperature at which an infinite molecular weight fraction would just precipitate (see Chapter 4).

The molecular weight and polydispersity of polymers remain among the most important properties that are measured. The methods are divided into absolute methods, which determine the molecular weight from first principles, and relative methods, which depend on prior calibration. The latter are usually selected because they are fast and inexpensive. Values obtained from the several methods are summarized in Table 3.14.

While polymer molecular weights vary from about 20,000 to over 1,000,000 g/mol for linear polymers, many polymers used in commerce have molecular weights around 10^5 g/mol and polydispersity indexes of about 2. This is governed by polymerization kinetics and by the balance between good physical properties and processibility.

It must be noted that sometimes the molecular weight distribution can be important in ways that are not obvious. For example, the low-molecular-weight component behaves

Table 3.14 Summary of molecular weight methods and results

	Absolute	Relative	M_n	M_w	Other
Gel permeation chromatography		X	X	X	Molecular-weight Distribution
Scattering—light, neutrons	X			X	R_g^z, A_2
Osmometry	X		X		A_2
Intrinsic viscosity [η]		X			M_v
Mass spectrometry	X		X	X	Best for lower molecular weights

Source: D. A. Thomas, unpublished observations.

substantially as a plasticizer, weakening the material rather than strengthening it. The high-molecular-weight tail adds much to the melt viscosity, since the melt viscosity of high-molecular-weight polymers depends on M_w to the 3.4 power (see Chapter 10). Thus most industries try to minimize the polydispersity index of their polymers.

REFERENCES

1. C. V. Uglea, *Oligomer Technology and Applications*, Dekker, New York, 1998.

2. R. Faust, A. Fehervari, and J. P. Kennedy, in *Reactive Oligomers*, F. W. Harris and H. J. Spinelli, eds., American Chemical Society, Washington, DC, 1985.

3. C. V. Uglea and I. I. Negulescu, *Synthesis and Characterization of Oligomers*, CRC Press, Boca Raton, FL, 1991.

4. P. Neogi, *Diffusion in Polymers*, P. Neogi, ed., Dekker, New York, 1996.

5. J. Hildebrand and R. Scott, *The Solubility of Nonelectrolytes*, 3rd ed., Reinhold, New York, 1949.

6. J. Brandrup and E. H. Immergut, eds., *Polymer Handbook*, 2nd ed., Wiley-Interscience, New York, 1975, Sec. IV.

7. G. M. Bristow and W. F. Watson, *Trans. Faraday Soc.*, **54**, 1731, 1742 (1958).

8. H. Mark and A. V. Tobolsky, *Physical Chemistry of High Polymer Systems*, 2nd ed., Interscience, New York, 1950.

9. S. K. Kim and S. C. Kim, *Polym. Bull.*, **23**, 141 (1990).

10. D. Mangaraj, S. K. Bhatnagar, and S. B. Rath, *Makromol. Chem.* **67**, 75 (1963).

11. P. A. Small, *J. Appl. Chem.*, **3**, 71 (1953).

12. K. L. Hoy, *J. Paint Technol.*, **46**, 76 (1970).

13. P. J. Flory, *J. Chem. Phys.*, **10**, 51 (1942).

14. M. L. Huggins, *Ann. N. Y. Acad. Sci.*, **42**, 1 (1942).

15. M. L. Huggins, *J. Phys. Chem.*, **46**, 151 (1942)

16. M. L. Huggins, *J. Am. Chem. Soc.*, **64**, 1712 (1942).

17. P. J. Flory, *Principles of Polymer Chemistry*, Cornell University Press, Ithaca, NY, 1953.

18. G. M. Bristow and W. F. Watson, *Trans. Faraday Soc.*, **54**, 1731 (1958).

19. P. J. Flory and W. R. Krigbaum, *J. Chem. Phys.*, **18**, 1086 (1950).

20. H.-G. Elias, *An Introduction to Polymer Science*, VCH, New York, 1997.

21. F. Gundert and B. A. Wolf, in J. Brandrup and E. H. Immergut, eds., *Polymer Handbook*, 3rd ed., Wiley, New York, 1989, Sec. VII, p. 173.

22. C. A. Glover, Chap. 4 in *Polymer Molecular Weights*, Part I, P. E. Slade Jr., ed., Dekker, New York, 1975.

23. F. W. Billmeyer Jr., *Textbook of Polymer Sciences*, Interscience, New York, 1962.

24. R. D. Ulrich, Chap. 2 in *Polymer Molecular Weights*, Part I, P. E. Slade Jr., ed., Dekker, New York, 1975.

25. W. R. Krigbaum and L. H. Sperling, *J. Phys. Chem.*, **64**, 99 (1960).

26. J. Brandrup and E. H. Immergut, eds., *Polymer Handbook*, 2nd ed., Wiley-Interscience, New York, 1975.

27. G. Mie, *Ann. Phys.*, **25**, 377 (1908).

28. M. Smoluchowski, *Ann. Phys.*, **25**, 205 (1908).

29. M. Smoluchowski, *Philos. Mag.*, **23**, 165 (1912).

30. A. Einstein, *Ann. Phys.*, **33**, 1275 (1910).

31. P. Debye, *J. Phys. Coll. Chem.*, **51**, 18 (1947).

32. B. H. Zimm, *J. Chem. Phys.*, **16**, 1093 (1948).

33. B. H. Zimm, *J. Chem. Phys.*, **16**, 1099 (1948).

34. A. Guinier and G. Fournet, *Small-Angle Scattering of X-Rays*, trans. by C. B. Walker, Wiley, New York, 1955.

35. R. G. Kirste, W. A. Kruse, and K. Ibel, *Polymer*, **16**, 120 (1975).

36. L. H. Sperling, *Polym. Eng. Sci.*, **24**, 1 (1984).

37. K. A. Stacy, *Light-Scattering in Physical Chemistry*, Academic Press, Orlando, FL, 1956.

38. G. Oster, *Chem. Rev.*, **43**, 319 (1948).

39. M. Bender, *J. Chem. Ed.*, **29**, 15 (1952).

40. P. Doty and J. T. Edsall, *Advances in Protein Chemistry VI*, Academic Press, Orlando, 1951, pp. 35–121.

41. J. J. Rush, *Current Status of Neutron-Scattering Research and Facilities in the United States*, National Academy Press, Washington, DC, 1984.

42. M. Dentini, T. Coviello, W. Burchard, and V. Crescenzi, *Macromolecules*, **21**, 3312 (1988).

43. R. G. Kirste, W. A. Kruse, and J. Schelten, *Makromol. Chem.*, **162**, 299 (1973).

44. J. Schelten, W. A. Kruse, and R. G. Kirste, *Kolloid Z. Z. Polym.*, **251**, 919 (1973).

45. J. P. Cotton, B. Farnoux, G. Jannink, J. Mons, and C. Picot, *C. R. Acad. Sci. (Paris)*, **C275**, 175 (1972).

46. H. Benoit, D. Decker, J. S. Higgins, C. Picot, J. P. Cotton, B. Farnoux, G. Jannink, and R. Ober, *Nature*, **245**, 13 (1973).

47. D. G. H. Ballard, J. Schelten, and G. D. Wignall, *Eur. Polym. J.*, **9**, 965 (1973).

48. J. P. Cotton, D. Decker, H. Benoit, B. Farnoux, J. S. Higgins, G. Jannink, R. Ober, C. Picot, and J. des Cloiseaux, *Macromolecules*, **7**, 863 (1974).

49. G. D. Wignall, D. G. Ballard, and J. Schelten, *Eur. Polym. J.*, **10**, 861 (1974).

50. J. Schelten, D. G. H. Ballard, G. Wignall, G. Longman, and W. Schmatz, *Polymer*, **17**, 751 (1976).

51. G. Lieser, E. W. Fischer, and K. Ibel, *J. Polym. Sci. Polym. Lett. Ed.*, **13**, 39 (1975).

52. R. G. Kirste and B. R. Lehnen, *Makromol. Chem.*, **177**, 1137 (1976).

53. G. Allen, *Proc. R. Soc. London Ser. A*, **351**, 381 (1976).

54. P. Herchenroeder, M. Dettenmaier, E. W. Fischer, M. Stamm, J. Hass, H. Reimann, B. Tieke, G. Wegner, and E. L. Zichny, *Europhys. Conf. Abstr.*, C.A. **89**, 198133Z (1978).

55. R. G. Kirste, W. A. Kruse, and K. Ibel, *Polymer*, **16**, 120 (1975).

56. J. Schelten, G. D. Wignall, and D. G. H. Ballard, *Polymer*, **15**, 682 (1974).

57. M. Helmstedt, *Makromol. Chem., Macromol. Symp.*, **18**, 37 (1988).

58. W. Burchard, *Light Scattering Principles and Development*, W. Brown, ed., Clarendon Press, Oxford, 1996.

59. G. Chirico, M. Placidi, and S. Cannistraro, *J. Phys. Chem. B*, **103**, 1746 (1999).

60. W. Burchard, *Makromol. Chem., Macromol. Symp.*, **18**, 1 (1988).

61. M. Dentini, T. Coriello, W. Burchard, and V. Crescenzi, *Macromolecules*, **21**, 3312 (1988).

62. A. Ravve, *Principles of Polymer Chemistry*, Plenum Press, New York, 1995.

63. G. Odian, *Principles of Polymerization*, 3rd ed., Wiley, New York, 1991.

64. J. Brandrup, E. H. Immergut, and E. Grulke, eds., *Polymer Handbook*, 4th ed., Wiley, New York, 1999.

65. G. V. Schultz, *Z. Phys. Chem.*, **B43**, 25 (1939).

66. See M. Swarc, *Polymerization and Polycondensation Processes*, Adv. Chem. Ser. 34, American Chemical Society, Washington, DC, 1962, p. 96.

67. G. C. Odian, *Principles of Polymerization*, 3rd ed., Wiley, New York, 1991.

68. W. H. Stockmayer, *J. Chem. Phys.*, **11**, 45 (1943).

69. A. Einstein, *Ann. Phys.*, **19**, 289 (1906).

70. A. Einstein, *Ann. Phys.*, **34**, 591 (1911).

71. H. Mark, *Der feste Korper*, Hirzel, Leipzig, 1938, p. 103.

72. J. Brandrup and E. H. Immergut, eds., *Polymer Handbook*, 3rd ed., Wiley, New York, 1989, Sec. IV.

73. M. L. Huggins, *J. Am. Chem. Soc.*, **64**, 2716 (1942).

74. E. O. Kraemer, *Ind. Eng. Chem.*, **30**, 1200 (1938).

75. A. Yamamoto, M. Fujii, G. Tanaka, and H. Yamakowa, *Polym. J.*, **2**, 799 (1971).

76. W. W. Yau, J. J. Kirkland, and D. D. Bly, *Modern Size-Exclusion Liquid Chromatography*, Wiley-Interscience, New York, 1979.

77. Waters Associates Liquid Chromatography School, Manual, *LC Short Course*, Waters Associates, Morristown, NJ, 1983.

78. F. M. Rabel, *J. Chromatogr. Sci.*, **18**, 394 (1980).

79. H. Pasch and B. Trathnigg, *HPLC of Polymers*, Springer, Berlin, 1998.

80. T. Provder, ed., *Chromatography of Polymers: Characterization by SEC and FFF*, American Chemical Society, Washington, DC, 1993.

81. M. Potschka and P. L. Dublin, eds., *Strategies in Size Exclusion Chromatography*, American Chemical Society, Washington, DC, 1996.

82. E. J. Swadesh, *HPLC Practical and Industrial Applications*, CRC Press, Boca Raton, FL, 1997.

83. M. Grell, D. D. C. Bradley, X. Long, T. Chamberlain, M. Inbaselain, E. P. Woo, and M. Soliman, *Acta Polymerica*, **49**, 439 (1998).

84. C. S. Patrickios, A. B. Lowe, S. P. Armes, and N. C. Billingham, *J. Polym. Sci., Part A, Polym. Chem.*, **36**, 617 (1998).

85. J. F. Miravet and J. M. J. Frechet, *Macromolecules*, **31**, 3461 (1998).

86. B. Fried and J. Sherma, *Thin-Layer Chromatography: Techniques and Applications*, 2nd ed., Dekker, New York, 1986.

87. R. P. W. Scott, *Introduction to Analytical Gas Chromatography*, 2nd ed., Dekker, New York, 1998.

88. K. Anton and C. Berger, eds., *Supercritical Fluid Chromatography with Packed Columns: Techniques and Applications*, Dekker, New York, 1998.

89. Annon., *Ann. ASTM Stds.*, **08.03**, 433 (1998).

90. S. D. Kim, A. Klein, and L. H. Sperling, accepted, *J. Mater. Sci.*, 2000.

91. T. C. Ward Jr., *Chem. Ed.*, **58**, 867 (1981).

92. C. M. L. Atkinson and R. Dietz, *Eur. Polym. J.*, **15**, 21 (1979).

93. J. L. Koenig, *Spectroscopy of Polymers*, 2nd ed., Elsevier, Amsterdam, 1999.

94. P. O. Danis, D. E. Karr, J. W. J. Simonsick Jr., and D. T. Wu, *Macromolecules*, **28**, 1229 (1995).

95. M. W. F. Nielen, *Mass Spectrometry Rev.*, **18**, 309 (1999).

96. H. S. Creel, *Trends in Polym. Sci. (TRIP)*, **1**, 336 (1993).

GENERAL READING

S. T. Balke, *Quantitative Column Liquid Chromatography*, Elsevier, New York, 1984.

W. Brown, ed., *Light Scattering Principles and Development*, Clarendon Press, Oxford, 1996.

E. Katz, R. Eksteen, P. Schoenmakers, and N. Miller, eds., *Handbook of HPLC*, Dekker, New York, 1999.

H. Pasch and B. Trathnigg, *HPLC of Polymers*, Springer, Berlin, 1998.

P. E. Slade Jr., ed., *Polymer Molecular Weights*, Dekker, New York, 1975, Parts I and II.

STUDY PROBLEMS

1. Calculate the solubility parameter of polyisobutene from Table 3.3. How does this value compare with that shown in Figure 3.2?

2. What is the free energy of mixing of one mole of polystyrene, $M = 2 \times 10^5$ g/mol, with 1×10^4 liters of toluene, at 298 K?

3. What is the molecular weight of a 10 g rubber band?

4. What are the values of K and a in the Mark–Houwink–Sakurada equation for polystyrene in benzene from Figure 3.15?

5. A trifunctional carboxylic acid monomer is being reacted to form a polyanhydride. What is the critical extent of reaction at gelation?

6. A 5-g sample of a polyester having one carboxyl group per molecule is to be titrated by sodium hydroxide solutions to determine its number-average molecular weight. How much 0.01 molar solution is required if the polymer has a molecular weight of approximately 1000 g/mol? 10,000 g/mol? 100,000 g/mol? Discuss the practicality of the experiment, with special reference to the upper molecular weight limit that can be analyzed.

7. What are ΔH_v and ΔH_f based on Table 3.5, assuming that $T_b = 150$ C, $T_f = 10$ C, and $\rho = 1.0$ g/cm^3?

8. What is the molecular weight of the cellulose tricaproate sample in Figure 3.5? Note that $R = 0.08205$ atm $\cdot$ l/mol $\cdot$ K for this problem.

9. What are the units of A_2 in both cgs and SI unit systems?

10. What is the estimated radius of gyration of a 5×10^5 g/mol polystyrene in the bulkstate? Hint: Note Table 3.10.

11. A sample of polystyrene has a radius of gyration of approximately 150 Å. Calculate the maximum usable angle to determine R_g for (a) light-scattering, (b) small-angle neutron scattering, and (c) small-angle X-ray scattering.

12. Values of **K** and a are listed in Table 3.10 for polystyrene in cyclohexane, a θ-solvent at 34.5°C. What is the calculated value of $\overline{r_0^2}/M$ [see equation (3.98)]?

13. The intrinsic viscosity of a sample of poly(methyl methacrylate) in acetone at 20°C was found to be 6.7 ml/g. What is its viscosity-average molecular weight?

14. Prove the relation $k' + k'' = 0.5$, equation (3.103).

15. Estimate the Mark–Houwink–Sakurada constants from the data in Figure 3.15.

16. The chain expansion quantity α varies with both R_g (by definition) and $[\eta]$. Using Table 3.10, show the relationships between α, R_g, and $[\eta]$. Can α and $[\eta]$ be related theoretically?

17. What is the z-average molecular weight of the poly(methyl methacrylate) shown in Table 3.13?

18. Given the following data, what is the number-average molecular weight and the second virial coefficient? ($T = 25°C$; density $= 1.0$ g/cm^3.)

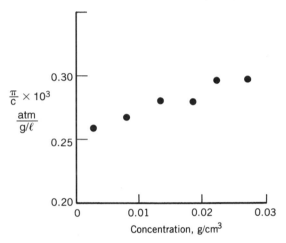

Figure P18. 1   Osmotic pressure data.

19. "But I know its a block copolymer!" your technician exclaimed. "Poly$A$ is a living polymer, which adds poly$B$ at its end!" You are not so sure. The synthesis might have taken another route, ending in a blend of the two polymers. The starting molecular weight of poly$A$ is known to be 100,000 g/mol, and poly$B$ is estimated to be around 15,000 g/mol. Devise a GPC experiment for your technician, which will distinguish (a) the 100% block copolymer from (b) a 50% block and 50% blend, from (c) a simple blend. Illustrate the probable resulting GPC chromatograms.

20. A new polymer has $[\eta] = 5.5$ cm^3/g, and an elution volume of 160 cm^3. Based on the method of Figure 3.23, what is its molecular weight?

21. What are the solubility parameters of the polyurethane and polystyrene in Figure 3.1?

22. We tend to think of molecules as being of finite size. The polymer networks used in Figure 3.1 are clearly the size of the sample, while the molecules used in Figure 3.2 are clearly finite and soluble. In what philosophical sense are (or are not) the materials making up Figure 3.1 molecules?

23. Two syntheses of the same polymer are made, but with different molecular weights, M_A and M_B, with intrinsic viscosities $[\eta]_A$ and $[\eta]_B$. Derive a general equation to express the intrinsic viscosity of various blends of the two syntheses as a function of composition.

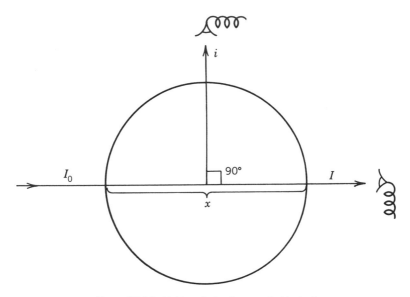

Figure A3.1.1   Light-scattering from a cylindrical cell.

APPENDIX 3.1   CALIBRATION AND APPLICATION OF LIGHT-SCATTERING INSTRUMENTATION FOR THE CASE WHERE $P(\theta) = 1$

A cylindrical cell with flat portions where the beam enters and leaves is presumed (Figure A3.1.1). A turbid calibrating liquid such as a soap solution or a Ludox® dispersion is made up. Ideally, the light intensity after passing through the solution, I, should be about $0.7I_0$. Beer's law gives the turbidity, τ,

$$I = I_0 e^{-\tau x} \tag{A3.1.1}$$

where x is the path length of the light. The receiving photomultiplier tube is then turned to 90°, and the scattered light intensity i recorded. The quantity i is related to the turbidity through the relation

$$\tau = ki \tag{A3.1.2}$$

where k is a proportionality constant.

The polymer solution of interest is then placed in the cell, and the scattering intensity i determined at 90°. The turbidity of the solution is determined through equation (A3.1.2). The turbidity is determined for several concentrations and extrapolated to zero concentration. For this simple experiment,

$$H\frac{c}{\tau} = \frac{1}{M_w} + 2A_2 c \tag{A3.1.3}$$

In reality, a secondary standard is also calibrated, so that the electronic sensitivity of the instrument may be varied widely and τ still calculated easily.

4

CONCENTRATED SOLUTIONS AND PHASE SEPARATION BEHAVIOR

Polymer solutions are important for a variety of purposes. In Chapter 3 dilute polymer solutions were required for the determination of molecular weights and sizes.

A wide range of modern research as well as a variety of engineering applications exist for polymers in solution (see Table 4.1). These range from reducing turbulent flow in heat exchangers to the production of paints, varnishes, and glues.

4.1 PHASE SEPARATION AND FRACTIONATION

4.1.1 Motor Oil Viscosity Example

As an example, the maintenance of near constant viscosity in motor oils over wide operating temperatures will be considered. In the old days, people changed their oil for summer and winter use; otherwise, at low temperatures the summer oil would become too "thick," the reverse being true for using winter oil in summer. Of course, "thickness" in this case refers to viscosity. The viscosity of today's motor oils bears designations such as SAE 5W-30. According to crankcase oil viscosity specification SAE J300a, the first number refers to the viscosity at $-18°C$, and the second number at $99°C$. The closer the two numbers are, the less the temperature variation of the viscosity.

One of the earlier solutions to the problem of wide temperature variations of oil viscosity involved the use of tapered diblock copolymers dissolved in the motor oil; see Figure 4.1. Tapered block copolymers of A and B mers have substantially all A mers in one block, and substantially all B mers in the other block, and a tapered transition region in between. The tapered region provides for greater mutual miscibility. At low temperatures, one block phase separates, creating a colloidal structure that contributes little to the viscosity of the oil. At higher temperatures, the polymer dissolves, raising the viscosity. Since the viscosity of the oil itself decreases rapidly with increasing temperature, the effect of the diblock copolymer is to keep the viscosity of the solution nearly constant.

Table 4.1 Selected industrial applications of polymer solutions and precipitates

Polymer	Solvent	Effect	Application
Sodium carboxymethyl cellulose	Soapy water	Selective precipitation onto clothing fibers	Prevents oils from redepositing on clothing during detergent washing: antiredeposition agent
Diblock copolymers	Motor oil	Colloidal suspensions dissolve at high temperatures, raising viscosity	Multiviscosity (constant viscosity) motor oil. Example: 10W-40
Poly(ethylene oxide) $M = 10^6$ g/mol	Water	Reduces turbulent flow	Heat exchange systems, reduces pumping costs
Proteins	Wine	Gels on reacting with tannin	Clarification of wines, removes colloidal matter
Polystyene, various	Triglyceride oils	Viscosity control, phase-separates during oil polymerization	Oil-based house paints, makes coatings harder, tougher
Polyurethanes, various	Esters, alcohols, various	Solvent vehicle evaporates, leaving polymer film for glues, solvent enters mating surfaces	Varnishes, shellac, and glues (adhesives)
Poly(vinyl chloride)	Dibutyl phthalate	Plasticizes polymer	Lower polymer T_g, soften polymer, makes "vinyl"
Polystyrene	Poly(2,6-dimethyl-1,4-phenyleneoxide)	Mutual solution; toughens polystyrene	Impact-resistant objects, such as appliances
Poly(methyl methacrylate)	Poly(vinylidine fluoride)	Increases PMMA oil and solvent resistance	Automotive applications, parts that might contact gasoline

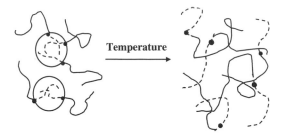

Figure 4.1 Viscosity control of motor oils with diblock copolymers. At low temperatures, the styrenic block (dashed) forms micelles. Hydrogenated polyisoprene (solid lines) remains in solution. At higher temperatures, the polymers dissolve, raising the viscosity. Note the chain overlap and entanglements. This suggests that the polymer is in the semidilute region (see Section 4.2).

The basic composition of the tapered block copolymers involves hydrogenated polyisoprene and polystyrene, with polystyrene being the block that is insoluble at low temperatures (1,2). Since then, a number of improvements have been made, including the use of star polymers and star block copolymers (3–5). Star polymers contain many arms emanating from a central source. The arms may be homopolymers or block copolymers themselves, but not all the arms need to be identical in the chemical structure. Seven to 15 arms are common in these materials, at a concentration of several percent. However, all of these commercial materials are based on combinations of hydrogenated polyisoprene and polystyrene. Besides phase separation and dissolution with temperature, polymer chain entanglement is cited as contributing to the viscosity at high temperatures (6). The end result for the consumer is a near constant-viscosity motor oil for all seasons.

Not every polymer will dissolve in every solvent, however. When one attempts to dissolve a polymer in solvents selected at random, many, perhaps most, will not work. The experimenter rapidly discovers that the higher the molecular weight of the polymer, the more difficult it is to select a good solvent. Polymer–polymer mutual solutions are even more difficult to attain.

Attempts to understand polymer–solvent and polymer–polymer mutual solution behavior led to many new theoretical developments, particularly in thermodynamics. To an increasing extent, the polymer scientist is now able to predict phase relationships involving polymers. The development of polymer solution and phase separation behavior is the subject of this chapter.

4.1.2 Polymer–Solvent Systems

According to thermodynamic principles, the condition for equilibrium between two phases requires that the partial molar free energy of each component be equal in each phase. This condition requires that the first and second derivatives of $\overline{\Delta G_1}$ in equation (3.23) with respect to v_2 be zero, where v_2 represents the volume fraction of polymer. The critical concentration at which phase separation occurs may be written

$$v_{2c} = \frac{1}{1 + n^{0.5}} \tag{4.1}$$

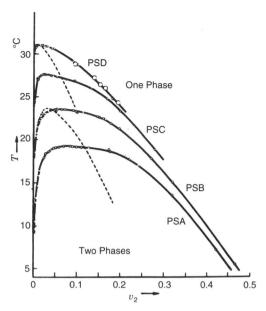

Figure 4.2   Phase diagrams for polystyrene fractions in cyclohexane. Circles and solid lines, experimental. Theoretical curves are shown for two of the fractions. The viscosity-average molecular weights are PSA, 43,600; PSB, 89,000; PSC, 250,000; PSD, 1,270,000 g/mol (7).

For large n, the right-hand side of equation (4.1) reduces to $1/n^{0.5}$. For $n = 10^4$, v_{2c} equals about 0.01, a very dilute solution. The critical value of the Flory–Huggins polymer-solvent interaction parameter, χ_1, is given by

$$\chi_{1c} = \frac{(1 + n^{1/2})^2}{2n} \cong \frac{1}{2} + \frac{1}{n^{1/2}} \tag{4.2}$$

which suggests further that as n approaches infinity, χ_{1c} approaches $\frac{1}{2}$.

Experimental data are given in Figure 4.2 (7) for the system of polystyrene in cyclohexane. The Flory θ-temperature for this system was already given as 34.5°C.

The critical temperature is the highest temperature of phase separation. The equation for the critical temperature is given by

$$\frac{1}{T_c} = \frac{1}{\theta}\left[1 + \frac{1}{\psi_1}\left(\frac{1}{n^{1/2}} + \frac{1}{2n}\right)\right] \tag{4.3}$$

where ψ_1 is a constant. Thus a plot of $1/T_c$ versus $1/n^{0.5} + 1/2n$ should yield the Flory θ-temperature at $n = $ infinity (see Figure 4.3) (7).

If a range of molecular species exists, which is true for all synthetic polymers, then the lower molecular weight species will tend to remain in solution at a temperature where the higher molecular weights phase-separate. Actually there is always a partition of molecular weights between the more concentrated and more dilute phases. The fractionation becomes more efficient at very low concentrations.

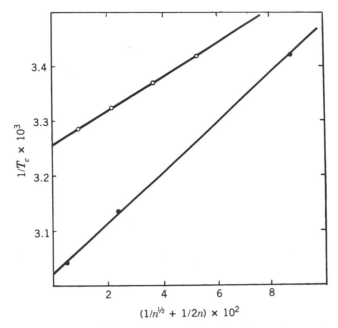

Figure 4.3 Dependence of the critical temperature on the number of segments per polymer chain. Polystyrene in cylcohexane, open circles; polyisobutene in diisobutyl ketone, solid circles (7).

4.1.3 Vitrification Effects

Figure 4.2 shows that the phase separation curve, called the *binodal line*, may cover a range of polymer volume fractions. The effect of the solvent at high polymer volume fraction is to plasticize the polymer. However, if the polymer is below its glass transition temperature, the concentrated polymer solution may *vitrify*, or become glassy.

The vitrification line generally curves down to lower temperatures from pure polymer as it becomes more highly plasticized; see Figure 4.4 (8). Frequently the curvature is well represented by the Fox equation (see Section 8.8).

The interception of these two curves is known as Berghmans' point (BP) and defined as the point where the liquid–liquid phase separation binodal line is intercepted by the vitrification curve (9,10); see Figure 4.4. When the vitrification curve intercepts the binodal line, the development of ordinary tie lines is inhibited because molecular diffusion in the glassy state is far slower than in the liquid state; see Figure 4.4.

4.2 REGIONS OF THE POLYMER–SOLVENT PHASE DIAGRAM

A polymer dissolves in two stages. First, solvent molecules diffuse into the polymer, swelling it to a gel state. Then the gel gradually disintegrates, the molecules diffusing into the solvent-rich regions. In this discussion, linear amorphous polymers are assumed. Cross-linked polymers may reach the gel state, but they do not dissolve.

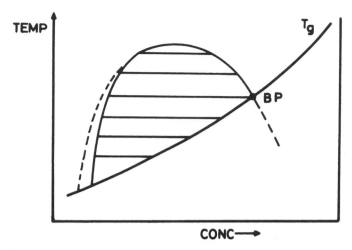

Figure 4.4 Hypothetical effective phase diagram, showing the effects of vitrification. --- lines represent un-realized thermodynamic equilibrium boundary lines. Horizontal lines are classical tie lines, with BP indicated. [Original copyright: F. C. Frank and A. Keller, *Polym. Comm.*, **29**, 186 (1988)].

The concentration of the final solution, of course, depends on the relative proportions of polymer and solvent. In Chapter 3 the solutions were assumed to be dilute, generally below 1% concentration, because this is required to obtain molecular weights. However, many solutions are used in the 10% to 50% concentration range. More concentrated systems are better described as plasticized polymers. Daoud and Jannink (11) and others divided polymer–solvent space into several regions, plotting the volume fraction of polymer, φ, vs the excluded volume parameter, v (see Figure 4.5). Each of these regions exhibits distinct molecular characteristics, as derived by scaling concepts.

Scaling concepts in polymer physics were brought to their modern state of importance by de Gennes (12). They consist of a series of proportionalities, showing the algebraic relationships between many quantities, particularly between macroscopic and microscopic variables, molecular dimensions, and thermodynamics; see Appendix 4.1.

The dilute solution regime in Figure 4.5 is shown in the upper left. According to the scaling laws, the dilute regime is delimited by the crossover volume fraction φ_{ov} on the right, and v_c on the bottom. Next is the semidilute regime, where the chains begin to overlap; see Figure 4.6. The curve dividing the dilute and semidilute regimes, as well as the behavior of the system within the semidilute regimes, is treated via the scaling concepts in Appendix 4.1. Still more concentrated are the marginal and concentrated regimes.

The ideal regime covers the behavior of the polymer in θ-solvent conditions, which is somewhat broader than defined in Chapter 3. In the lower left portion of Figure 4.5 the system phase-separates. This models the result shown in Figure 4.2, where below a certain temperature (χ significantly larger than $\frac{1}{2}$), there is phase separation.

An important quantity in Figure 4.5 is the screening length, ξ, first introduced by Edwards (13). This quantity takes slightly different meanings in different regimes. In the dilute solution regime, $\xi = R_g$. In the semidilute regime, ξ measures the distance between chain contacts (14). If the polymer is cross-linked, ξ provides a measure of the net size.

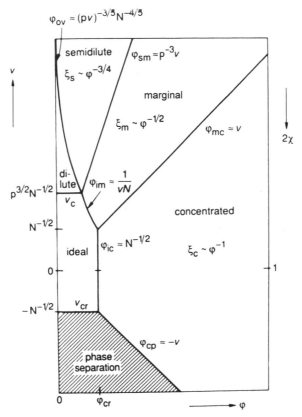

Figure 4.5 Illustration of the positions of the five stable solution regimes: ideal, dilute, semidilute, marginal, and concentrated, and the phase separated region, hatched, for polymer solutions. The dependencies of the various crossovers on N, p, and v and the dependence of the correlation lengths on the concentration φ are indicated. (G. J. Fleer, M. A. Cohen Stuart, J. M. H. M. Scheutjens, T. Cosgrove, and B. Vincent, *Polymers at Interfaces*, Chapman and Hall, London, 1993).
Notation: φ = volume fraction, ξ = correlation length, $p = C_\infty/6$, v = excluded volume parameter, $1 - 2\chi$, N = number of bonds per chain, χ = Flory-Huggins polymer-solvent interaction parameter, v_c = cross-over from swollen to ideal. Subscripts: cr = critical, i = ideal, d = dilute, s = semi-dilute, m = marginal regime, c = concentrated regime, ov = cross-over from dilute to semi-dilute, cp = cross-over from concentrated to phase separated, sm = cross-over from semi-dilute to marginal, mc = cross-over from marginal to concentrated.

For semidilute solutions, the dependence of ξ on φ follows the scaling law

$$\xi_s \sim \varphi^{-3/4} \tag{4.4}$$

Figure 4.5 also shows how ξ varies in the marginal and concentrated regimes. In all cases, through the use of scaling concepts, only the exponents of the various quantities are determined. The coefficients, in general, depend on the individual polymer.

Another quantity of interest in semidilute solutions is called the *blob*. It contains a number of mers on the same chain defined by the mesh volume $\xi_s{}^3$, inside of which excluded volume effects are operative. Some texts define the blob as the number of mers

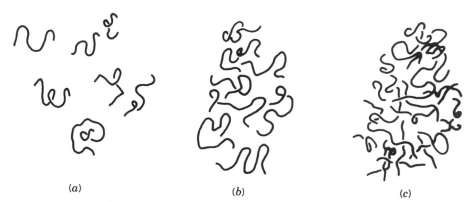

(a)   (b)   (c)

Figure 4.6 Relationships of polymer chains in solution at different concentration regions. (a) Dilute solution regime, where $\varphi < \varphi_{ov}$. (b) The transition regions, where $\varphi = \varphi_{ov}$. (c) Semi-dilute regime, $\varphi > \varphi_{ov}$. Note overlap of chain portions in space.

between adjacent entanglements, distance ξ_s apart. These blobs are large enough to be self-similar to the whole polymer chain coiling characteristics; they are coils within a coil. The concept of the blob has been important in theoretical calculations. With increasing concentration, the blobs become smaller, becoming equal to that of an individual mer in polymer melts (no solvent).

4.3 POLYMER–POLYMER PHASE SEPARATION

When two polymers are mixed, the most frequent result is a system that exhibits almost total phase separation. Qualitatively, this can be explained in terms of the reduced combinatorial entropy of mixing of two types of polymer chains, as illustrated in Figure 3.3b. In this case neither type of chain can interchange its segments, because of covalent bonding.

Prior to the 1970s the scientific literature on polymer blends was dominated by the idea that polymer–polymer miscibility would always be the rare exception (15–18). This was based on numerous experiments, and the theoretical work of Scott (18). Thus the usual endothermic heat of mixing and the very small combinatorial entropy of mixing made it seem unlikely to realize the necessary negative free energy of mixing.

When two polymers do mutually dissolve, they are generally found to phase-separate at some higher temperature rather than at some lower temperature. This is called a lower critical solution temperature (LCST) (see Figure 4.7, upper portion). This unusual result can be interpreted by considering the unusual features of the mixing process. At the critical point, the heat of mixing must balance the entropy of mixing times the absolute temperature. The latter is known to be unusually small (see above), so that the free energy of mixing, according to the Flory–Huggins theory, is the difference between two small quantities.

The Flory–Huggins theory does not permit volume change on mixing, and it ignores the equation of state properties of the pure components. It also does not properly consider the enormous size differences between polymer and solvent.

In response to the developing field of polymer blends, two new theories of polymer mixing were developed. The first was the Flory equation of state theory (19,20), and the

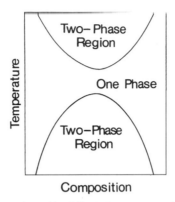

Two–Phase
Region

One Phase

Two–Phase
Region

Temperature

Composition

Figure 4.7 Phase diagram for a polymer blend illustrating an upper critical solution temperature, UCST (apex of lower curve), and a lower critical solution temperature, LCST (apex of upper curve). Because of the low entropy of mixing, high-molecular-weight polymer blends exhibit the LCST phenomenon. If the chains are short, an UCST may be seen.

second was Sanchez's lattice fluid theory (21–24). These theories were expressed in terms of the reduced temperature, $\tilde{T} = T/T^*$, reduced pressure, $\tilde{P} = P/P^*$, reduced volume, $\tilde{V} = V/V^*$, and reduced density, $\tilde{\rho} = 1/\tilde{V}$. The starred quantities represent characteristic values for particular polymers, often referred to as the "hard core" or "close-packed" values, that is, with no free volume.

The new theories pointed out that LCST behavior is characteristic of exothermic mixing and negative excess entropy (25). This last is caused by densification of the polymers on mixing. The entropy of volume change, which ordinarily is relatively small compared to other quantities, comes to the fore in polymer–polymer blends (24). While an introduction to phase separation in polymer blends is presented here, a more detailed development of polymer blends in general is given in Chapter 13.

4.3.1 Phase Diagrams

Phase separation and dissolution are controlled by three variables: temperature, pressure, and concentration. While simultaneous variation of all three yields a three-dimensional figure of significant complexity, the effect of systematic variation of two of these quantities, holding the third constant, leads to the main stream of polymer research. The schematic phase diagrams in Figure 4.8, while drawn specifically for polymer–polymer miscibility analysis, are quite general. The solid line is called the binodal and the dashed line the spinodal (26). As defined below, these two lines demarcate regions of different kinetics of phase separation. All the phase diagrams in Figure 4.8 are for liquid–liquid phase separation. If one of the components crystallizes, quite different phase diagrams emerge (27). While the figures are drawn symmetrically, the shapes and positions of the various curves depend on the ratio of the two polymer molecular weights.

Figure 4.8*a* shows the LCST behavior of a typical polymer blend. Again, as in Figure 4.7, as the temperature is raised through the binodal line, phase separation will occur. Figure 4.8*b* shows the corresponding behavior as a function of pressure. While not widely investigated, early results show that increasing pressure increases miscibility (28). Thus the pressure coefficient of an LCST is positive. This result agrees with the finding that miscible polymer pairs undergo densification on mixing.

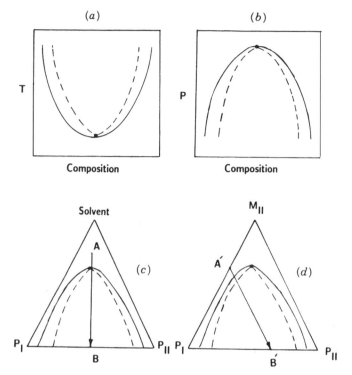

Figure 4.8   Methods of causing phase separation in multiconstituent polymer materials.

Figure 4.8c illustrates the phase diagram of a solvent and two polymers, P_I and P_{II}. As the solvent is removed, the mixture becomes less miscible and undergoes phase separation on going from point A to point B.

The case of a mutual solution of a polymer, P_I, and a monomer M_{II} is illustrated in Figure 4.8d. As monomer II polymerizes to form polymer II, the composition follows the trajectory A' to B'. Here, phase separation occurs during polymerization. While this case is the most important industrially, encompassing such materials as high-impact polystyrene and ABS resins, it is the least understood. The reasons lie in the complexity of the thermodynamics of a system undergoing two simultaneous changes: (a) removal of the monomer solvent (as in Figure 4.8c) and (b) replacement of the solvent by polymer II. Grafting during the polymerization, if present, constitutes another complication. Generally, grafted polymers are more miscible than the corresponding blends.

4.3.2   Thermodynamics of Phase Separation

The basic thermodynamic relationships developed in Section 3.2 hold for polymer blends. The basic equation for mixing of blends reads

$$\frac{\Delta G_M}{kT} = \frac{V}{V_r} v_1 v_2 \chi_{12} \left(1 - \frac{2}{z} \right) + N_c [v_1 \ln v_1 + v_2 \ln v_2] \tag{4.5}$$

where V is the volume of the sample (usually taken as 1 cm³), V_r is the volume of one cell (see Figure 3.3), z is the lattice coordination number (z is usually between 6 and 12), and N_c is the number of molecules in 1 cm³. Then, the term V/V_r is a count of the number of cells in 1 cm³, the first term on the right being the heat of mixing term, ΔH_M. This may be positive or negative, depending on the sign of χ_{12}. The second term on the right is the statistical entropy of mixing term, ΔS_M. Since v_1 and v_2 are always fractions, this second term is always negative. Of course, the classical representation of equation (4.5) is

$$\Delta G_M = \Delta H_M - T\Delta S_M \tag{4.6}$$

The statistical thermodynamic approach provides a molecular basis for calculating the properties of interest.

4.3.3 An Example Calculation: Molecular Weight Miscibility Limit

What is the molecular weight limit for mixing polystyrene and polybutadiene at 25°C and obtaining one phase? For simplicity, we assume $v_1 = v_2 = 0.5$, unit density, and a mer molecular weight of 100 g/mol for both polymers. This last relieves the requirement somewhat of having exactly one mer in each cell; see Figure 3.3. The quantity $\Delta G_M/kT$ equals zero at the critical point. The basic unknown is reached through N_c.

The quantity χ_{12} is found through Table 3.2 and equation (3.28):

$$\chi_{12} = \frac{V_1}{RT}(\delta_1 - \delta_2)^2 = \frac{(100\ \text{cm}^3/\text{mol})(9.1 - 8.4)^2\ \text{cal/cm}^3}{(1.987\ \text{cal/mol} \cdot \text{K}) \times 298\ \text{K}}$$

$$\chi_{12} = +0.083$$

Assuming that $z = 6$, the first term in equation (4.5) may be evaluated as

$$\frac{V}{V_r} = \frac{1\ \text{cm}^3 \times 6.023 \times 10^{23}\ \text{molecules/mol}}{(100\ \text{g/mol}) \times 1\ \text{cm}^3/\text{g}}$$

$$\frac{V}{V_r} = 6.023 \times 10^{21}\ \text{mers(or cells)/cm}^3$$

leading to

$$\frac{V}{V_r} v_1 v_2 \chi_{12}\left(1 - \frac{2}{z}\right) = 8.3 \times 10^{19}$$

The second term on the right of equation (4.5) is

$$N_c[v_1 \ln v_1 + v_2 \ln v_2] = -0.69 N_c$$

Then from equation (4.5),

$$0 = 8.3 \times 10^{19} - 0.69 N_c$$

$N_c = 1.2 \times 10^{20}$ chains/cm^3, which occupy 6.023×10^{21} cells:

$$\frac{6.023 \times 10^{21} \text{ cells(mers)/cm}^3}{1.2 \times 10^{20} \text{ chains/cm}^3} = 50 \text{ mers/chain}$$

Since each "mer" weighs 100 g/mol, each chain is 5000 g/mol.

4.3.4 Equation of State Theories

However, specific new theories, called the equation of state theories, have been worked out in an effort to understand the special conditions of mixing long chains (29–34). At equilibrium, an equation of state is a constitutive equation that relates the thermodynamic variables of pressure, volume, and temperature. A simple example is the ideal gas equation, $PV = nRT$. The van der Waals equation provides a fundamental correction for molecular volume and attractive forces. Equations of state may also include mechanical terms. Thus rubber elasticity phenomena are also described by equations of state; see Chapter 9.

A serious deficiency in the classical theory, as exemplified by equation (4.5), is the assumption of incompressibility. This deficiency can easily be remedied by the addition of free volume in the form of "holes" to the system. These holes will be about the size of a mer and occupy one lattice site. In materials science and engineering, "holes" are frequently called "vacancies." Imagine that a multicomponent mixture is mixed with N_0 holes of volume fraction v_0. Then the entropy of mixing is

$$\frac{\Delta S_M}{k} = -N_0 \ln v_0 - \sum_i N_i \ln v_i \tag{4.7}$$

which follows from equation (3.15). This allows for compressibility, since the number of holes may be varied. Noting that the fractional free volume is given by $1 - \tilde{\rho}$, the entropy of mixing vacant sites with the molecules in equation of state terminology is given by

$$\Delta S_M = -\frac{k}{\tilde{\rho}}[(1 - \tilde{\rho}) \ln(1 - \tilde{\rho}) + \tilde{\rho} \ln \tilde{\rho}] \tag{4.8}$$

where $\tilde{\rho}$ is less than unity. When all the sites are occupied, $\tilde{\rho} = 1$, and the right hand side is zero. In reality the cell size also increases with temperature, to accommodate increased vibrational amplitudes.

The Gibbs free energy of mixing is given by

$$\Delta G_M = \tilde{\rho}\varepsilon^* - T\Delta S_M \tag{4.9}$$

where the quantity ε^* is a van der Waals type of energy of interaction.

Note that $\tilde{\rho} = \tilde{\rho}(P, T)$; $\tilde{\rho} \to 1$ as $T \to 0$; $\tilde{\rho} \to 1$ as $P \to \infty$.

By taking $\partial \Delta G_M / \partial \tilde{\rho} = 0$, the equation of state via the lattice fluid theory (31) is obtained

$$\tilde{\rho}^2 + \tilde{P} + \tilde{T}\left[\ln(1 - \tilde{\rho}) + \left(1 - \frac{1}{r}\right)\tilde{\rho}\right] = 0 \tag{4.10}$$

where r is the number of sites (mers, if each occupies a site) in the chains, and

$$\frac{1}{r} = \frac{v_1}{r_1} + \frac{v_2}{r_2} \tag{4.11}$$

For high polymers, r goes substantially to infinity, yielding a general equation of state for both homopolymers and miscible polymer blends,

$$\tilde{\rho}^2 + \tilde{P} + \tilde{T}[\ln(1 - \tilde{\rho}) + \tilde{\rho}] = 0 \tag{4.12}$$

The corresponding equation of state derived by Flory (29) is

$$\tilde{\rho}^2 + \tilde{P} - \tilde{T}\tilde{\rho}(1 - \tilde{\rho}^{1/3})^{-1} = 0 \tag{4.13}$$

Workers in the field prefer to state the equations in terms of density relations, because for condensed systems, density is easier to measure than volume. Again, $\tilde{\rho} = \rho/\rho^* = V^*/V$, where V^* represents the close-packed volume per unit mass.

Data for several polymers of interest are shown in Table 4.2 (31). This information may be used to determine miscibility criteria. Basically, T^* values must be similar for miscibility. If $T_1^* > T_2^*$, then it is desirable to have $P_1^* > P_2^*$. Polymers must have similar coefficients of expansion for miscibility; that is, the ratio of the densities must remain similar as the temperature is changed. The quantities T^* and P^* represent theoretical values at close packing.

From such information, it may be found that the system poly(2,6-dimethyl phenylene oxide)–blend–polystyrene is miscible but that poly(dimethyl siloxane), which is so different from the others, should be immiscible with all of them.

The basis for phase separation in polymer blends must be considered further here. For the Flory–Huggins theory of incompressible polymer mixtures (see Section 3.3), the small entropy of mixing is dominated by the heat of mixing, leading to the conclusion that the heat of mixing must be zero or negative to induce miscibility. When the system is compressible, the reverse may be true. "Compressible" in the current sense means that the theory includes terms for volume changes.

Table 4.2   Equation of state parameters for some common polymers

Polymer	T^* (K)	P^* (bars)	ρ^* (g/cm^3)
Poly(dimethyl siloxane), PDMS	6050	3020	1.104
Poly(vinyl acetate), PVAc	6720	5090	1.283
Poly(n-butyl methacrylate), PnBMA	7000	4310	1.125
Polyisobutylene, PIB	8030	3540	0.974
Polyethylene (low density), LDPE	7010	4250	0.887
Poly(methyl methacrylate), PMMA	8440	5030	1.269
Polystyrene (atactic), PS	7950	3570	1.105
Poly(2,6-dimethylphenylene oxide), PPO	8260	5170	1.161
Polyethylene (high density)[a]	6812	4720	0.995
Polypropylene[a]	6802	4080	0.992
Polybutadiene[a]	4665	5400	0.932

[a] D. J. Walsh, W. W. Graessley, S. Datta, D. J. Lohse, and L. J. Fetters, *Macromolecules*, **25**, 5236 (1992).

When two compressible polymers are mixed together, negative heats of mixing cause a negative volume change. Since reducing the volume of the system reduces the number of holes, the entropy of mixing change is negative. At high enough temperatures, the unfavorable entropy change associated with the densification of the mixture becomes prohibitive, that is, $T\Delta S > \Delta H$, and the mixture phase-separates. The two-phase system has a larger volume as the holes are reintroduced, and hence a larger positive contribution to the entropy. This scenario leads to lower critical solution temperatures. Significantly differing coefficients of expansion contribute to phase separation. In virtually all polymer–polymer systems exhibiting critical phenomena, both ΔH and $T\Delta S$ are relatively small quantities. Thus relatively modest changes in either the enthalpy or the entropy alter the phase diagram significantly.

4.3.5   Kinetics of Phase Separation

There are two major mechanisms by which two components of a mutual solution can phase-separate: nucleation and growth, and spinodal decomposition; see Figure 4.9 (34). Nucleation and growth are associated with metastability, implying the existence of an energy barrier and the occurrence of large composition fluctuations. Domains of a minimum size, the so-called critical nuclei, are a necessary condition. Nucleation and growth (NG) are the usual mechanisms of phase separation of salts from supersaturated aqueous solutions, for example. Spinodal decomposition (SD), on the other hand, refers to

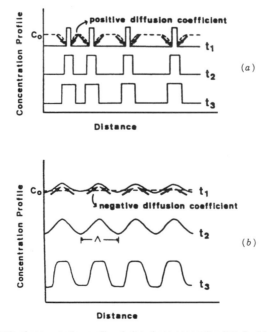

Figure 4.9   A schematic of concentration profiles during phase separation (34), for (a) nucleation and growth and (b) spinodal decomposition.

Table 4.3   Mechanisms of phase separation

1. Nucleation and growth
   (a) Initial fragment of more stable phase forms.
   (b) Two contributions to free energy: (i) work spent in forming the surface and (ii) work gained in forming the interior.
   (c) Concentration in immediate vicinity of nucleus is reduced; hence diffusion into this region is *downhill*. (Diffusion coefficient is positive.)
   (d) Droplet size increases by growth initially.
   (e) Requires activation energy.
2. Spinodal decomposition
   (a) Initial small-amplitude composition fluctuations.
   (b) Amplitude of wavelike composition fluctuations increases with time.
   (c) Diffusion is *uphill* from the low concentration region into the domain. (Diffusion coefficient is negative.)
   (d) Unstable process: no activation energy required.
   (e) Phases tend to be interconnected.

phase separation under conditions in which the energy barrier is negligible, so even small fluctuations in composition grow.

The two kinetic modes are compared in Table 4.3. The most interesting differences are the growth mechanisms. As illustrated in Figure 4.9, the nucleation and growth mechanism results in domain size increasing with time. The domains tend to be spheroidal in nature. For spinodal decomposition, interconnected cylinders tend to form. Their initial growth mechanism involves an exchange of mass across the boundary, purer phases forming with time. Thus the amplitude of the waves increases, but not necessarily the wavelength. The domains are of about the same size as the original wavelength of the wavelike fluctuation during the early stages of phase separation. On annealing, it must be pointed out, both kinds of domains (NG and SD) may coalesce, forming larger and larger spheroidal structures.

The best characterized system has been polystyrene blended with poly(vinyl methyl ether). Several studies showed that this system is miscible in all proportions below about 80°C or so, depending on the molecular weight. Nishi et al. (35) prepared a number of mutual solutions at low temperatures. After equilibration, they rapidly raised the temperature, then held the temperature constant again, making observations in the microscope. Nucleation and growth could be seen as tiny spheres, while spinodal decomposition looked like tiny overlapping worms. The more unusual of these two results, that of spinodal decomposition, has been modeled by Reich (36), see Figure 4.10. The results are shown in Figure 4.11 (35). Experiments like these delineate the regions of nucleation and growth from those of spinodal decomposition in phase-separating polymer blends.

The kinetics of phase separation proper have been described by Cahn and Hilliard (37,38). For the early stages of decomposition, the diffusion equation is of the following form:

$$\frac{\partial \phi}{\partial t} = M\left[\left(\frac{\partial^2 G}{\partial \phi^2}\right)\nabla^2 \phi - 2K\nabla^4 \phi\right] \tag{4.14}$$

where G is the free energy, ϕ is the composition, M is the mobility coefficient, and K is the energy gradient coefficient arising from contributions of composition gradients to the

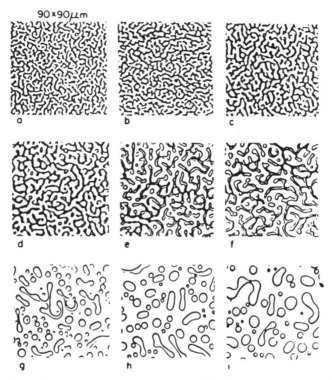

90 x 90 μm

Figure 4.10   Coarsening during the latter stages of spinodal phase separation (36), polystyrene-*blend*-poly(vinyl methyl ether).

free energy. Solution of this equation leads to theories of the wavelength of the system and the rate at which the domains will form. The spinodal line is defined by the locus of temperatures in the phase diagram, where $\partial^2 G/\partial \phi^2 = 0$. The critical point is given where the third derivative is equal to zero. It is interesting to note that the Cahn-Hilliard theory was originally derived to obtain an understanding of the phase-separation kinetics in certain inorganic glasses.

4.3.6   Miscibility in Statistical Copolymer Blends

As stated previously, most homopolymer blends are immiscible due to the negative entropy of mixing and negative heats of mixing. Sometimes, however, miscibility can be achieved with the introduction of comonomers. These may be, in general,

$$\text{poly}(\text{A}-stat\ \text{B})\ blend-\text{poly}(\text{C}-stat-\text{D}) \tag{4.15}$$

or more frequently

$$\text{poly}(\text{A}-stat-\text{B})-blend-\text{poly}(\text{A}-stat-\text{C}) \tag{4.16}$$

The phenomenon makes use of the fact that miscibility in copolymer–copolymer blends

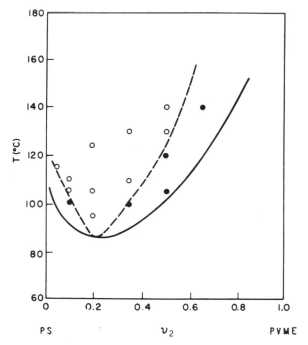

Figure 4.11 Phase separation by spinodal decomposition (○) and nucleation and growth (●), as observed under a microscope (35).

may be enhanced by the repulsion of dissimilar mers on the *same* chain: A hates B more than it hates C. An example is poly(vinyl chloride)–*blend*–poly(ethylene–*stat*–vinyl acetate), where poly(vinyl chloride) is immiscible with either polyethylene or poly(vinyl acetate).

Karasz and MacKnight (39) approached the problem through mean field thermodynamic considerations, arguing that negative net interactions are necessary to induce miscibility. The general equation for the free energy was adopted from equation (4.5):

$$\frac{\Delta G_M}{RT} = \frac{v_1}{n_1} \ln v_1 + \frac{v_2}{n_2} \ln v_2 + v_1 v_2 \chi_{\text{blend}} \tag{4.17}$$

where v_1 and v_2 are volume fractions, n_1 and n_2 are degrees of polymerization, and χ_{blend} is a dimensionless interaction parameter defined as

$$\chi_{\text{blend}} = \sum_{i,j} c_{ij} \chi_{ij} \tag{4.17a}$$

where the coefficients c_{ij} are functions of the copolymer compositions, with $0 \leq c_{ij} \leq 1$. For $A_n/(B_x C_{1-x})_{n'}$ blends,

$$\chi_{\text{blend}} = x\chi_{\text{AB}} + (1 - x)\chi_{\text{AC}} - x(1 - x)\chi_{\text{BC}} \tag{4.18}$$

where x represents the mole fraction.

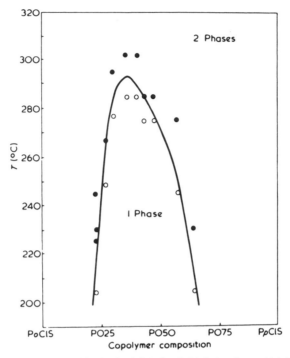

Figure 4.12 Window of miscibility for blends of poly(2.6-dimethyl-1,4-phenylene oxide), PPO, and statistical copolymers of *ortho*-chlorostyrene and *para*-chlorostyrene, PO, 40/60 PO/PPO. For midrange compositions and lower temperatures, the polymers mix. For higher temperatures and other compositions, there is phase separation (40).

"Windows of miscibility" result when $\Delta G_M < 0$. This is illustrated for the system poly(2,6-phenylene oxide)–*blend*–poly(*o*-chlorostyrene–*stat*–*p*-chlorostyrene) in Figure 4.12 (40). The experimental determination of miscibility involved a DSC characterization of each sample for one or two glass transition temperatures. As mentioned previously, poly(2,6-phenylene oxide) is miscible with polystyrene in all proportions; indeed, this forms the basis for an important impact-resistant plastic, Noryl®. However, miscibility is limited with the halogenated styrenes.

Miscibility for $(A_xB_{1-x})_n/C_yD_{1-y})_{n'}$ may even be brought about if all the χ_{ij} values are positive, as illustrated in Figure 4.13 (39). The ellipse shown cannot extend to the corners of the diagram because these areas correspond to immiscible blends of the respective homopolymers.

The history of polymer blending has taken several interesting turns. Before about 1974, virtually all polymer blends known were immiscible. Polymer scientists thought that thermodynamics was against such mixing, particularly that the very small entropies of mixing would always dominate the circumstances. Kinetically, diffusion was thought to be so slow that even if a miscible polymer pair could be found, inordinate amounts of time would elapse before mixing would occur.

With the development of the equation of state theories, however, phase diagrams began to be constructed. To everyone's surprise, virtually all these systems exhibited LCST behavior. (Earlier, many scientists were trying to *raise* the temperature to bring

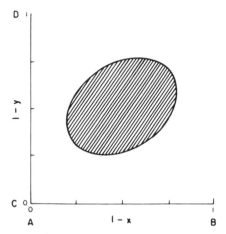

Figure 4.13 Theoretical miscibility domain (shaded region) in copolymer blends, assuming infinite degree of polymerization. The Flory interaction parameters are χ_{AB}, χ_{AC}, χ_{AD}, χ_{BC}, χ_{BD}, and χ_{CD} which are 0.3, 0.1, 0.2, 0.2, 0.1, and 0.4, respectively (39).

about miscibility.) It was discovered that for very small domains, equilibrium mixing could be achieved in a matter of minutes. The concept of specific interactions was developed, where hydrogen bonds, polar attractions, and so on, between the polymers each played a role in developing miscibility. Today, very many miscible polymer pairs are known. While the science of polymer blending was qualitative in nature in the 1950s and 1960s, and primarily concerned with toughness and impact resistance (still important!), major advances in the field have now turned toward thermodynamics and kinetics.

From an engineering point of view, the need is to control the miscibility and the phase domain size. Very small sizes (200 Å) serve to make good damping compositions, while domains of the order of 1000 Å make better impact-resistant materials. The extent of mixing, particularly at the interface between the domains, has now been found to be critical in determining many properties. All these are subjects of current research (26,30,41–46).

4.3.7 Polymer Blend Characterization

There has been increasing interest in mutually miscible polymer blends (38–43). Rather than trying to find polymer–polymer systems that are 'alike," the emphasis has been on finding systems that are complementary, attracting one another by hydrogen bonds or polarity. The first important polymer–polymer phase diagram was worked out by McMaster (47) on poly(styrene–*stat*–acrylonitrile)–*blend*–poly(methyl methacrylate), who determined the phase diagram by cloud point determination. He found a LCST behavior. Other miscible systems include poly(methyl methacrylate)–*blend*–poly(vinylidene chloride) (48) and polystyrene–*blend*–poly(2,6-dimethyl–1,4-phenylene oxide) (49). These materials are characterized by a negative heat of mixing.

Polymer–polymer interactions in the bulk state have been studied by a number of experimental tools; see Table 4.4. For phase-separated compositions, dynamic mechanical spectroscopy (Chapter 8) measures shifts in the glass transition temperature, which

Table 4.4   Methods of characterizing polymer blends

Method	Parameter Determined	Reference
Inverse gas chromatography	χ_{23}	(a)
Differential scanning calorimetry	ΔC_p	(b)
Dynamic mechanical spectroscopy	ΔT_g	(c)
NMR	τ	(d)
ESR	τ	(e)
Solvent vapor sorption	χ_{23}	(f)
Heat of mixing	ΔH_M	(g)
Light-, X-ray, and neutron scattering	A_2	(h)
Photophysics	I_D/I_M	(i)
FT-IR	Δv	(j)
Electron microscopy	Domain size and shape	(k)

References: (a) M. Galin and L. Maslinka, *Eur. Polym. J.*, **23**, 923 (1987). (b) A. Natansohn, *J. Polym. Sci. Polym. Lett. Ed.*, **23**, 305 (1985). (c) L. H. Sperling, J. J. Fay, C. J. Murphy, and D. A. Thomas, *Macromol. Chem. Macromol. Symp.*, **38**, 99 (1990). (d) T. K. Kwei, T. Nishi, and R. F. Roberts, *Macromolecules*, **7**, 667 (1974). (e) Y. Shimada and H. Kashiwabara, *Macromolecules*, **21**, 3454 (1988). (f) S. Saeki, S. Tsubotani, H. Kominami, M. Tsubokawa, and T. Yamaguchi, *J. Polym. Sci. Polym. Phys. Ed.*, **24**, 325 (1986). (g) R. E. Bernstein, D. C. Wahrmund, J. W. Barlow, and D. R. Paul, *Polym. Eng. Sci.*, **18**, 1220 (1978). (h) R. J. Roe and W. C. Sin, *Macromolecules*, **13**, 1221 (1980). (i) C. W. Frank and M. A. Gashgari, *Macromolecules*, **12**, 163 (1979). (j) P. Cousin and R. E. Prud'homme, in *Multicomponent Polymer Materials*, D. R. Paul and L. H. Sperling, eds., American Chemical Society, Washington, DC, 1986. (k) A. A. Donatelli, L. H. Sperling, and D. A. Thomas, *Macromolecules*, **9**, 671 (1976).

provides a measure of the extent of molecular mixing in partially mixed systems. Electron microscopy is used to determine domain size and shape; see Section 4.5.6 (50–52).

For measurements on miscible systems, a variety of instruments are used. These include small-angle neutron scattering (see Section 5.2.2), which provides a measure of the second virial coefficient thermodynamic interaction between the two polymers in miscible blends, inverse gas chromatography, and solvent vapor sorption; the last two measure the Flory interaction parameter between the two polymers. (*Note*: In the following text subscript 1 stands for the solvent; subscripts 2 and 3 stand for the two polymers when addressing volume fractions, thermodynamic quantities, and so on.) NMR and ESR instrumentation each measure different relaxation times for segments of one component in contact with segments of the other component. FT-IR provides a measure of the specific interactions occurring between polymer segments. For example, the carbonyl vibration band of polycaprolactone shifts to lower frequencies on blending with poly(vinyl chloride), providing a measure of the strength of interaction between the carbonyl and the chlorine. Heats of mixing of two polymers, of course, provide a sum of all such energies of interaction. Differential scanning calorimetry measures the changes in heat capacity between the individual components and the blend.

Another method involves excimer fluorescence as a molecular probe; see Section 2.9. The question may be raised as to whether polymer blends will become more miscible if the differences in their solubility parameters are reduced. Excimer fluorescence provides some evidence; see Figure 4.14 (52). Here, 0.2 wt.% of poly(2-vinyl naphthalene), P2VN, is dispersed in a series of poly(alkyl methacrylates). These include the following, which are identified in Figure 4.14 by acronym: methyl, PMMA; ethyl, PEMA; *n*-propyl, PnPMA; isopropyl, PiPMA; *n*-butyl, PnBMA; isobutyl, PiBMA; *sec*-butyl, PsBMA; *tert*-butyl, PtBMA; phenyl, PPhMA, isobornyl, PiBoMA; benzyl, PBzMA; and cyclo-

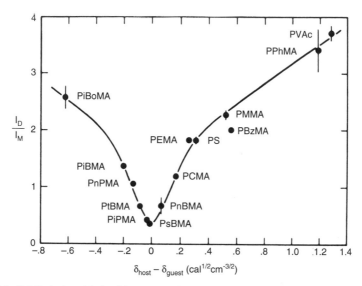

Figure 4.14 Poly(2-vinyl naphthalene) becomes quite miscible with acrylic polymers when the differences between their solubility parameters become small (52).

hexyl, PCMA. Two other host polymers were polystyrene, PS, and poly(vinyl acetate), PVAc.

Figure 4.14 shows the average values of the excimer to monomer ratio, I_D/I_M, plotted against the difference in host and guest solubility parameters. There are two phenomena involved. The intramolecular site concentration depends on the population of suitable local conformation states of the chain, that is, rotational dyads. However, the concentration of intermolecular sites is sensitive solely on clustering of the guest P2VN, or the chains bending back on themselves. In a thermodynamically good host medium, there will be extensive mixing, causing the local concentration of aromatic rings to drop. In addition, chain expansion reduces the likelihood of chain back-bending. Both effects lead to a minimum in I_D/I_M for $\Delta\delta = 0$. It must be noted that in all cases $\delta_{host} - \delta_{guest}$ is less than unity, and at the extremes studied I_D/I_M has not yet leveled out, which would be a sign of more or less complete phase separation.

4.3.8 Graft Copolymers and IPNs

Most polymer blends, grafts, blocks, and IPNs exhibit phase separation, as discussed previously. It must be emphasized that their wide application in commerce arises largely because of the synergistic properties exhibited by these materials. Applications have included impact-resistant plastics, thermoplastic elastomers, coatings, and adhesives.

The morphology of some graft copolymers and IPNs based on SBR[†] and polystyrene is illustrated in Figure 4.15 (50). Here the various morphologies, and hence their subsequent physical and mechanical properties, are controlled by cross-linking and/or mixing.

[†] Styrene–butadiene rubber, poly(butadiene–*stat*–styrene).

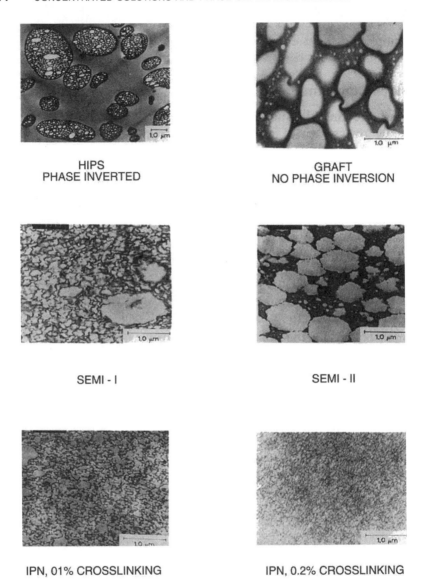

HIPS
PHASE INVERTED

GRAFT
NO PHASE INVERSION

SEMI - I

SEMI - II

IPN, 01% CROSSLINKING

IPN, 0.2% CROSSLINKING

Figure 4.15   Phase morphology of graft copolymers and IPNs of SBR rubber and polystyrene by transmission electron microscopy. Upper two figures, graft copolymers; middle two, semi-IPNs; bottom two, full IPNs. Diene portion stained dark with OsO_4 (50).

The graft copolymer in the upper left of Figure 4.15 is called high-impact polystyrene (HIPS). The SBR phase, stained dark, toughens the otherwise brittle polystyrene. Important features of the rubber phase include its low glass transition temperature, small domain size, and extent of grafting to the polystyrene phase. These features will be discussed further in Chapter 11. It must be pointed out, however, that actual extraction studies show that most of the polystyrene is not grafted. Only that portion in contact with polybutadiene is significantly bonded. The upper right structure is the graft copolymer made without proper stirring. Note that the rubber phase is continuous. As a consequence the material is much softer than the HIPS composition.

The bottom four structures are IPNs or semi-IPNs, as indicated. Note that all six compositions have essentially the same chemical composition, except for cross-linking. Higher cross-linking makes the domains smaller.

4.3.9 Block Copolymers

In the discussion above, the limited solubility of one polymer in another was ascribed to the unusually low entropy of mixing. In a block copolymer, one chain portion is attached to another, end on end. It is interesting to note the miscibility characteristics and morphology of these materials.

First of all, the presence of the chemical bond between the blocks definitely improves the mutual miscibility[†] of the two polymers (53–55). Thus block copolymers are miscible to higher block molecular weight than their corresponding blends.

If block copolymers phase-separate, how big are the domains?[‡] If the bonds holding the blocks together are maintained, then the domains must be small enough that one block resides in one phase while the other block is in the neighboring phase. The junction, or bond between the blocks, tends to be in the interface between the two blocks (57, 58).

The morphology of the phases changes from spheres to cylinders to alternating lamellae depending on the relative length of the two blocks. Spheres containing the short blocks are formed within the continuous phase of the longer block (see Figure 4.16). Alternating lamellae form when the blocks are about the same length. Cylinders are formed for intermediate cases as illustrated in Figure 4.17 (59).

The actual size of the three types of domains was calculated by Meier (49,50):

$$\text{Spheres} \qquad R = 1.33\alpha\bar{K}M^{1/2} \tag{4.19}$$

$$\text{Cylinders} \qquad R = 1.0\alpha\bar{K}M^{1/2} \tag{4.20}$$

$$\text{Lamellae} \qquad R = 1.4\alpha\bar{K}M^{1/2} \tag{4.21}$$

where $\bar{K}$ is the experimental constant relating the unperturbed root-mean-square end-to-end distance to the molecular weight, r_0^2/M, and R is a characteristic dimension, the radius of the spheres and cylinders, and the half-thickness of the corresponding lamellae. Typically the minor component forms spheres up to 25% concentration, cylinders between 25 and 40%, and alternating lamellae in midrange. For some common polymers, values of $\bar{K}$ for R in Ångstroms are

---

[†] The term "miscibility" is preferred by Olabisi et al. (51,53–55) and Paul (56) and others rather than "solubility" because the very large size of the polymer molecules has sometimes created the appearance of slight demixing, when they were in fact mixed on a molecular level according to all thermodynamic criteria. An older term for miscibility is "compatibility." Use of that term is now recommended in terms of satisfactory engineering properties.

[‡] A domain is a discrete region of space occupied by one phase and surrounded by another.

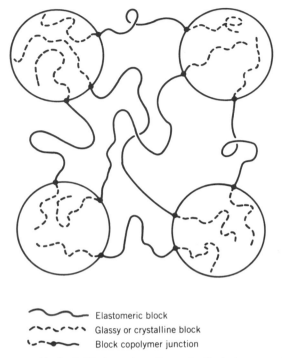

Elastomeric block
Glassy or crystalline block
Block copolymer junction

Figure 4.16 Idealized triblock copolymer thermophastic elastomer morphology.

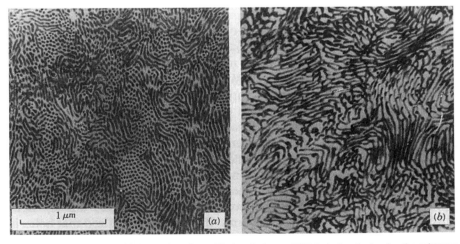

Figure 4.17 Polystyrene–*block*–polybutadiene–*block*–polystyrene (40% butadiene), showing the existence of the cylindrical structures. Polymer cast from toluene solution, and the polybutadiene portion stained with OsO$_4$. Two cuts through the film are (*a*) normal and (*b*) parallel. Apparent spheres in (*a*) are end-on cuts (59).

Polystyrene	670×10^{-3}
Polyisobutylene	700×10^{-3}
Poly(dimethyl siloxane)	880×10^{-3}
Poly(methyl methacrylate)	565×10^{-3}

Values of α are slightly larger than unity because the chains are slightly strained at the common junction points between the two phases. For practical calculations, a value of 1.2 might be assumed, since α varies between 1.0 and 1.5 for most cases of interest.

It must be pointed out that many block copolymers are synthesized in the form of triblock or multiblock copolymers. As an example, the triblock copolymer polystyrene–*block*–polybutadiene–*block*–polystyrene is widely used as a rubbery shoe sole material. In this case the long, rubbery center block forms the continuous phase whereas the short, glassy polystyrene blocks form submicroscopic spheres (see Figure 4.16). The hard domains constitute a type of physical cross-link, holding the whole mass together. On heating above the glass transition of the hard phase, the material softens and flows. Other examples are the so-called segmented polyurethanes, which form the elastic thread in clothing, particularly undergarments, a multiblock material.

4.3.10 Example Calculations with Block Copolymers

Suppose that a polystyrene–*block*–polybutadiene–*block*–polystyrene triblock copolymer is synthesized with block molecular weights of 10,000–80,000–10,000 g/mol. With a total molecular weight of 100,000 g/mol, it is 20% polystyrene, in the spherical domain range. First, what is the volume of one domain(60)? Equation (4.19) can be used:

$$R = 1.33 \times 1.25 \times 670 \times 10^{-3}(10{,}000)^{1/2} \text{ Å}$$

Then $R = 111$ Å, yielding a volume of 5.72×10^6 Å^3 for one spherical domain.

Second, how many chains are in a domain? Polystyrene has a density of 1.06 g/cm^3 (Table 3.2), with a mer molecular weight of 104 g/mol. The number of mers per cm^3 is $\{(1.06 \text{ g/cm}^3)/(104 \text{ g/mol})\} \times 6.02 \times 10^{23}$ mers/mol $= 6.14 \times 10^{21}$ mers/cm^3, yielding 1.63×10^2 Å^3/mer. Then 5.72×10^6 Å$^3/1.63 \times 10^2$ Å^3/mer $= 3.51 \times 10^4$ mers per domain. Since a 10,000 g/mol polystyrene chain contains 96.1 mers, this yields 365 polystyrene blocks per domain. This result comes from having, for many commercial materials of interest, 200 to 500 blocks per domain. It is unusual in that people are used to thinking about astronomical numbers of molecules in most locations except high vacuums.

4.3.11 Ionomers

Ionomers are polymers that contain 5% to 15% ionic groups (61). While these materials are statistical copolymers, the ionic groups usually phase separate from their hydrocarbon-like surroundings thus providing properties resembling multiblock copolymers. The ionic groups, phase separated, form a type of physical cross-linking. The level of bonding increases with the valence of the ions; thus $Na^+ < Zn^{+2} < Al^{+3}$.

An important ionomer is polyethylene containing about 5 mole-% of sodium acrylate mers. An application of this material involves its capability of melt bonding in the presence of aqueous fluids, such as are encountered in the meat-packing industry.

4.4   DIFFUSION AND PERMEABILITY IN POLYMERS

Permeation is the rate at which a gas or vapor passes through a polymer. The mechanism by which permeation takes place involves three steps: (a) absorption of the permeating species into the polymer; (b) diffusion of the permeating species through the polymer, traveling, on average, along the concentration gradient; and (c) desorption of the permeating species from the polymer surface and evaporation or removal by other mechanisms (62).

Factors affecting permeability include the solubility and diffusivity of the penetrant into the polymer, polymer packing and side-group complexity, polarity, crystallinity, orientation, fillers, humidity, and plasticization. For example, polymers with high crystallinity usually are less permeable because their ordered structure has fewer holes through which gases may pass.

The concept of holes, or free volume, in polymers has already been introduced in relation to phase diagrams. It must be emphasized that holes in materials are required for all types of molecular motion beyond simple vibrational and rotational states. One must ask the question: When a molecule moves from position A to position B, into what does it move, and what does it leave behind? The answer is that it moves into a hole. The hole and the molecule are transposed, so the hole is where the molecule was before the action started. The general concept of free volume is developed in Chapter 8, in relation to the glass transition.

4.4.1   Swelling Phenomena

Consider a polymer in contact with a solvent. Diffusion takes place in both directions, the polymer into the solvent, and vice versa. However, the rate of diffusion of the solvent, being a small molecule, is much faster. Hence, for a time, the polymer really acts as the solvent.

If the polymer is glassy, the solvent lowers the T_g by a plasticizing action. Polymer molecular motion increases. Diffusion rates above T_g are far higher than below T_g. Thus diffusion may depend on the concentration of the diffusing species (63,64).

4.4.2   Fick's Laws

Perhaps one of the most interesting case of the polymer behaving as the solvent has to do with permeability of water and gases. Often polymers, in the form of films, are used as barriers to keep out water and air. In the case of food wrappers, it is often desired to keep in water but keep out oxygen.

The general case of diffusion in materials is given by Fick's laws (65). His first law governs the steady-state diffusion circumstance:

$$J = -D\frac{\partial c}{\partial x} \tag{4.22}$$

Simple steady-state diffusion through a film is modeled in Figure 4.18. The flux J gives the quantity of permeant passing through a unit cross section of membrane per unit time. Thus equation (4.22) leads to first-order transport kinetics, where the quantity transported at any instant depends on the concentration on the high concentration side to the first power.

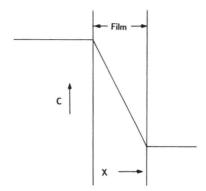

Figure 4.18 Illustration of permeability through a film under steady-state conditions, showing the decrease in concentration, c.

Fick's second law controls the unsteady state:

$$\frac{\partial c}{\partial t} = \frac{\partial}{\partial x}\left[D\frac{\partial c}{\partial x}\right] \qquad (4.23)$$

The quantity J represents the net flux of diffusing material across unit area of a reference plane and has the units of $mol \cdot cm^{-2} \cdot s^{-1}$; c is the vapor concentration, x represents the distance diffused in time t, and D is the diffusion coefficient.

The permeability coefficient, P, is defined as the volume of vapor passing per unit time through unit area of polymer having unit thickness, with a unit pressure difference across the sample. The solubility coefficient, S, determines the concentration. For the simplest case

$$P = D \cdot S \qquad (4.24)$$

which expresses the permeability in terms of solubility and diffusivity.

The temperature dependence of the solubility obeys the Clausius-Clapeyron equation written in the form (66)

$$\Delta H_s = -R\frac{d \ln S}{d(1/T)} \qquad (4.25)$$

A study of vapor solubility as a function of temperature allows the heat of solution ΔH_s to be evaluated. Since the size of the diffusing molecule per se is important, it turns out that the logarithm of the diffusion coefficient depends inversely on the molar volume (67,68).

Of course, the permeability coefficients depend on the temperature according to the Arrhenius equation,

$$P = P_0 e^{-\Delta E/RT} \qquad (4.26)$$

where ΔE is the activation energy for permeation (69).

Similar relations hold for solubility and diffusivity.

4.4.3 Permeability Units

While there are many dimensions and units found in the literature for permeability, the preferred current dimensions are (70)

$$P = \frac{\text{(quantity of permeant)} \times \text{(film thickness)}}{\text{(area)} \times \text{(time)} \times \text{(pressure drop across the film)}} \tag{4.27}$$

The preferred SI unit of the permeability coefficient is (70)

$$\text{Units of } P = \frac{cm^3(273.15 \text{ K}, 1.013 \times 10^5 \text{ Pa}) \times cm}{cm^2 \times s \times Pa} \tag{4.28}$$

Thus permeability coefficients range from 10^{-11} to 10^{-16} $cm^3 \times cm/cm^2 \times s \times Pa$ for many polymers and permeants. The units of activation energy are kJ/mol.

Similarly the units of D and S are

$$\text{Units of } D = cm^2/s \tag{4.29}$$

$$\text{Units of } S = \frac{cm^3(273.15 \text{ K}; 1.013 \times 10^5 \text{ Pa})}{cm^3 \times Pa} \tag{4.30}$$

4.4.4 Permeability Data

Factors controlling permeability of small molecules through polymers include: (a) solubility and diffusivity, (b) polymer packing and side groups, (c) polarity, (d) crystallinity, (e) orientation, (f) fillers, (g) humidity, and (h) plasticization.

Typical permeability data are shown in Table 4.5 (70) for a range of elastomers,

Table 4.5 Permeability in polymers at 25°C (70)

Polymer	Permeant	$P \times 10^{13}$	$P_0 \times 10^7$	ΔE
Polyethylene (LDPE)	O_2	2.2	66.5	42.7
	N_2	0.73	329	49.9
	CO_2	9.5	62	38.9
	H_2O	68	48.8	33.5
cis-1,4-Polybutadiene	N_2	14.4	0.078	21.3
	He	24.5	0.0855	20.3
Poly(ethyl methacrylate)	O_2	0.889	2.1	36.4
	CO_2	3.79	0.435	28.9
Poly(ethylene terephthalate)[a]	O_2	0.0444	0.227	37.7
(amorphous)	CO_2	0.227	0.00021	27.0
	CH_4	0.0070	0.9232	24.7
Poly(vinylidene chloride)[b]	N_2	0.00070	900	70.3
	O_2	0.00383	825	66.6
	CO_2	0.0218	24.8	51.5
	H_2O	7.0	863	46.1
Cellulose	H_2O	18900	—	—

[a] Major component of plastic soft drink bottles.
[b] Major component of Saran®.
Units: $P, P_0 = cm^3$ (273 K; 1×10^5 Pa) $\times (cm/cm^2) \times s^{-1} \times Pa^{-1}$.
     ΔE = kJ/mol.

Table 4.6 Kinetic diameters[a] of various penetrants (72), nm

Molecule	Diameter	Molecule	Diameter
He	0.26	C_2H_4	0.39
H_2	0.289	Xe	0.396
NO	0.317	C_3H_8	0.43
CO_2	0.33	$n\text{-}C_4H_{10}$	0.43
Ar	0.34	CF_2Cl_2	0.44
O_2	0.346	C_3H_6	0.45
N_2	0.364	CF_4	0.47
CO	0.376	$i\text{-}C_4H_{10}$	0.50
CH_4	0.38		

[a] Zeolite sieving diameters.

amorphous plastics, and semicrystalline plastics. In general, permeability decreases from elastomers to amorphous plastics to semicrystalline plastics.

One of the more recent applications is the use of amorphous poly(ethylene terephthalate)[†] for soft drink bottles. The major requirements are to keep carbon dioxide and water in and to keep oxygen out. One must realize that these gases are continuously being transported across the plastic even if at a low rate, causing the soft drink eventually to go "flat." Thus these drinks have a "shelf life," after which they must be discarded if not sold.

4.4.5 Effect of Permeant Size

For elastomers, concerted movements of several adjacent chain segments take place to provide rapid transport. Such motions are restricted in glassy polymers (71). Also free volume is much less in glasses than in plastics. The size of the permeant (Table 4.6) (72) is critical in determining its diffusion rate in polymers. Sizes range from 2 to 5 Å for many molecules. The larger the molecule, the smaller the diffusion rate; see Figure 4.19 (73). In Figure 4.19 diffusion is seen to be significantly lower in poly(vinyl chloride), a glassy polymer, than in natural rubber. Of course, polymers have a distribution of hole sizes (74). Note that the permeability of a given molecule depends on both its diffusion coefficient and its solubility; see equation (4.24). Solubility depends on the solubility parameter, as described in Section 3.2.

An important application relates to oxygen diffusion through soft contact lenses (75). Soft contact lenses are made of poly(2-hydroxyethyl methacrylate) and its copolymers, in the form of cross-linked networks. These are swollen to thermodynamic equilibrium in water or saline solution. The hydroxyl group provides the hydrophilic characteristic and is also important for oxygen permeability. Oxygen permeability is important because of the physiological requirements of the eye. Thus the polymer is highly swollen with water and also serves as a semipermeable material.

On the other hand, ion exchange membranes require large permeabilities for ions. This is obtained through the use of polymers with large bonded ion concentrations such as $-SO_4Na$ and $-NH_4Cl$, which allow the polymer to swell in water. In both contact lenses and ion exchange membranes, large permeation rates take place because the polymer is so swollen with water that diffusion substantially takes place in the aqueous medium.

[†] Small amounts of comonomer destroy crystallinity.

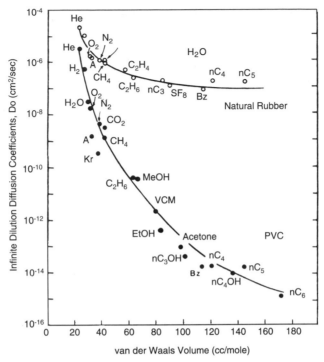

Figure 4.19 Rigid poly(vinyl chloride) has lower diffusion coefficients than natural rubber for a variety of gases and liquids (73).

4.4.6 Permselectivity of Polymeric Membranes and Separations

Today polymeric membranes are widely used to produce potable water from seawater, treat industrial effluents, for controlled drug delivery systems, separate common gases, pesticide release systems, and in prosthetic devices for humans, among others (76). Most of these methods require the separation of two or more components. Membrane-based separation processes are environmentally green, economic, and frequently more efficient than conventional methods.

For any membrane-based separation process to be successful, the membrane must possess two key attributes: high flux and good selectivity. Flux, which depends directly on permeability, was treated in Sections 4.4.2 and 4.4.4. Selectivity depends in part on differences in permeant size and solubility in the membrane (Section 4.4.5). Separation will then occur because of differences in the transport rates of molecules within the membrane. This rate of transport is determined by the mobility and concentration of the individual components as well as the driving force, which is the chemical potential gradient across the membrane.

4.4.6.1 Types of Membranes
A membrane is best described by its mode of transport. Three different modes of transport can be distinguished (76):

1. Passive transport, where there is a simple chemical potential gradient.

2. Facilitated transport, where different components are coupled to specific carriers in the membrane phase.

3. Active transport, where components of a mixture are transported *against* the chemical potential gradient.

This last usually the driving force involves a chemical reaction or a sequence of chemical reactions. Active transport is found mainly in biological membranes. From a macroscopic point of view, there are several types of fabrication methods, producing different membrane systems. These include:

1. Microporous membranes, consisting of a solid matrix having well-defined pores with diameters ranging from 5 nm to 50 μm. Separation is achieved via a molecular sieving mechanisms.

2. Homogeneous membranes, consisting of a dense film, leading to normal diffusion processes as discussed above. Two configurations are in wide use: hollow fibers and spiral wound sheets.

3. Ion exchange membranes, consisting of a highly swollen gel. The polymer may contain either positive or negative charges.

4. Asymmetric membranes, usually consisting of a very thin polymer layer on a highly porous thick support. The thin skin representing the actual membrane may be too weak to be used independently, thus requiring a backing.

A simple example of a use of homogeneous membranes involves reverse osmosis in the purification of seawater or brackish waters for drinking and agriculture. The transmembrane pressure must be greater than the osmotic pressure of the salt water so that the solvent flux is reversed. Some 50 to 80 bars are required. Cellulose membranes are used because of their high permeability coefficient to water; see Table 4.5. Applications include potable water aboard ships and agricultural uses where arid lands border seas.

4.4.6.2 Gas Separations   Membrane separation of gases has emerged into an important unit operations technique offering specific advantages over such conventional separation procedures such as cryogenic distillation and adsorption (77,78). One example is the separation of hydrogen from synthesis gas, another is the removal of carbon dioxide from natural gas. Gas selectivity is the ratio of permeability coefficients of two gases,

$$\alpha_{A/B} = \frac{P_A}{P_B} \tag{4.31}$$

where P_A and P_B represent the permeability of the more permeable and less permeable gases, respectively (79). While both high permeability and high selectivity are desirable, a general trade-off relationship has been recognized: Polymers that are more permeable are generally less selective, and vice versa.

On the basis of an exhaustive literature survey, Robeson quantified this notion by graphing the available data; see Figure 4.20 (77). The most desirable performance would be in the upper right-hand corner of this figure. However, there is an upper bound line above which permeability/selectivity combinations are exceptionally rare.

While Freeman (79) has developed a theory about the upper bound line depending on gas size, condensability, and an adjustable parameter, another direction was taken

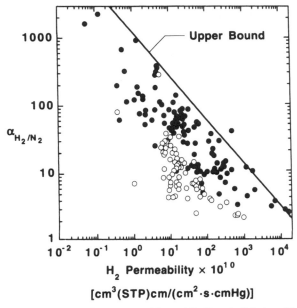

Figure 4.20   Relationship between hydrogen permeability and H_2/N_2 selectivity for rubbery ($\bigcirc$) and glassy ($\bullet$) polymers. Note the empirical upper bound relationship, above which high selectivity and high permeability have proved difficult to pass.

by Robeson et al. (78), who developed a group contribution (80) approach to predict permselectivity. The intent of the latter is to maximize the proximity to the upper bound line based on a knowledge of the permeability contribution of each specific group in a wide variety of polymers. Ideally polymers synthesized on the basis of the most permselective groups would yield the best possibility to increase selectivity.

4.5   LATEXES AND SUSPENSIONS

Polymers in water dispersion serve as coatings, adhesives, and the basis for many plastics and elastomers. The two most important classes are latexes (or latices) and suspensions. These differ by size and method of synthesis; see Table 4.7. While latex particles are normally small enough to be kept in dispersion by Brownian motion, most suspensions will "cream" on standing and hence must be stirred continuously.

Table 4.7   Latexes and suspensions

Property	Latex	Suspension
Size	0.04–5 µm	5–1000 µm
Shape	Spheres	Spheres to potato
Emulsifier	Surfactant	Polar polymer
Initiator	Water soluble	Oil soluble

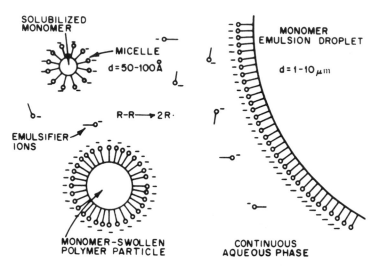

Figure 4.21 Illustration of the emulsion polymerization process during interval I (early stage). The monomer emulsion droplet is large and hence has a very small surface area compared to the soap micelles and monomer swollen particles (81).

Basically a latex is synthesized by mixing monomer, surface active agent (surfactant), and an initiator (free radical source) into the water; see Figure 4.21 (81). The initiator is usually activated by heat. A widely used initiator is potassium persulfate. Surfactants are molecules that have a hydrophilic end and a hydrophobic end. These form micelles in the water, with the hydrophilic end facing outward, shielding the hydrophobic portion. The surfactant can be a soap or detergent, sodium lauryl sulfate being a typical example. The turbid or hazy appearance of soap in water is due to the presence of micelles. The surfactant dissolves the oil-soluble monomer in much the same way as the soap micelles dissolve skin oil in ordinary toiletry. Some of the surfactant remains water soluble, as does some of the monomer. Monomer diffuses from the large droplets into the water, and to the micelles. Some of it begins to polymerize on contact with the free radicals. After adding a few mers, the much less water-soluble free radical diffuses (randomly) into a micelle, where it begins to polymerize the monomer present there. It is terminated when a second free radical enters the micelle. The fully polymerized latex then is a sub-microscopic sphere of polymer emulsified by a layer of surfactant (82). The kinetics of emulsion polymerization are treated in many organic polymer science textbooks.

The reasons why emulsion polymerization is so popular include the following: (a) heat removal through the aqueous phase during the polymerization is very efficient; (b) the final latex has a very low viscosity, approaching that of water; and (c) the polymer is easily recovered via coagulation and/or film formation.

By contrast, a suspension polymerization behaves kinetically much more like a bulk polymerization. Heat removal during polymerization is also important. Special advantages include ease of handling the filterable powder and dry blending of the powder with plasticizer. This last step involves diffusion of the plasticizer into the particles, aided by a large surface area. For such cases the particles are made with a porous structure, which easily adsorbs plasticizer. The plasticizer then diffuses in when the polymer is heated above its glass transition temperature.

4.5.1 Natural Rubber Latex

Natural rubber latexes occur in many plants. The white sap of the common milkweed and dandelion is a rubber latex. The most important source of natural rubber latex is the Hevea brasiliensis tree, now grown in plantations in many tropical parts of the world.

The polymer is a highly linear *cis*-1,4-polyisoprene of about 5×10^5 g/mol. It owes its colloid stability to the presence of adsorbed proteins at the surface of the rubber particles. These adsorbed proteins are in the anionic state so that the rubber particles carry negative charges at their surfaces (83).

4.5.2 Colloidal Stability and Film Formation

Latexes constitute a subgroup of colloid systems known as lyophobic sol. Sometimes they are called polymer colloids. The stability of these colloids is determined by the balance between attractive and repulsive forces affecting two particles as they approach one another. Stability is conferred on these latexes by electrostatic forces, which arise because of the counterion clouds surrounding the particles. Other forces of an enthalpic or entropic nature arise when the lyophilic molecules on the surfaces of the latexes interact on close approach. These can be overcome by evaporation of the water, heating, freezing, or by chemically modifying the surfactant, such as by acidification.

If the polymer in the latex is above its glass transition temperature,* it may form a film on evaporation of the water. A simple example is the drying of a latex paint film on a wall. As the water evaporates, coalescence surface tension forces proceed from the presence of water menisci of very small radii of curvature. These menisci develop between the particles as the last traces of water are removed. The forces that these menisci generate drive the particles together. Interdiffusion of the polymer chains takes place, forming coherent films.

4.6 MULTICOMPONENT AND MULTIPHASED MATERIALS

Solutions may be dilute or concentrated up to substantially 100% solute. A question arises in considering concentrated polymer solutions: When do we stop considering the polymer dissolved in the solvent, and start considering the solvent dissolved in the polymer? It must be pointed out that many polymer solutions containing more than about 25% polymer may no longer be pourable but behave like plasticized solid materials. A key issue for the "semidilute" systems is the overlap of the polymer chains, at which point the properties shift to much higher viscosities and/or solid behavior profiles. Another critical problem is diffusion of the polymer through the solvent and the diffusion of the solvent through the polymer. Of course, a multicomponent solution is one that contains more than one component, while a multiphase system is one in which the various components have phase separated. Polymer blends usually exhibit phase separation, principally because the entropy of mixing and the enthalpy of mixing are both negative. Phase-separated systems exhibit quite complex morphologies.

Many polymer blends, especially those containing a dispersed phase of a low glass transition rubber possess great toughness and/or impact resistance (explored in Chapters 11 and 13). Grafts between the two polymers in the *interface* or *interphase*, as it is

---

*Actually, the threshold point for film formation is called the *minimum film formation temperature*, which is slightly lower than T_g.

sometimes called, tend to bond the phases together, making them stronger (the subject of Chapter 12). In this chapter the size, shape, and interfacial characteristics of the domains are explored along with interphase mixing. Block copolymers also serve a variety of purposes, many as novel types of elastomer called thermoplastic elastomers (Chapter 13) forming the basis for many shoe soles now in use.

REFERENCES

1. Anon., Shellvis 40 Viscosity Improver, SC:1064-89, 1989.

2. R. J. A. Eckert, U. S. 3,775,329, 1973.

3. R. J. A. Eckert, U. S. 4,156,673, 1979.

4. R. B. Rhodes and A. R. Bean, U. S. 4,156,673, 1989.

5. Anon., Shellvis 200 Viscosity Improver, SC:1066-89, 1989.

6. G. Holden, Private communication, March, 2000.

7. A. Shultz and P. J. Flory, *J. Am. Chem. Soc.*, **74**, 4760 (1952).

8. F. C. Frank and A. Keller, *Polym. Commun.*, **29**, 186 (1988).

9. J. Arnauts and H. Berghmans, *Polym. Commun.*, **28**, 66 (1987).

10. S. Callister, A. Keller and R. M. Hikmet, *Makromol. Symp. Chem.*, **39**, 19 (1990).

11. M. Daoud and G. Jannink, *J. Phys. Paris*, **37**, 973 (1976).

12. P. G. de Gennes, *Scaling Concepts in Polymer Physics*, Cornell University Press, Ithaca, NY, 1979.

13. S. F. Edwards, *Proc. Phys. Soc.*, **88**, 265 (1966).

14. S. Candau, J. Bastide, and M. Delsanti, Polymer networks, *Adv. Polym. Sci.*, **44**, 27 (1982).

15. G. J. Fleer, M. A. Cohen Stuart, J. M. H. M. Scheutjens, T. Cosgrove, and B. Vincent, *Polymers at Interfaces*, Chapman and Hall, London, 1993.

16. A. Dobry and F. Boyer-Kawenoki, *J. Polym. Sci.*, **2**, 90 (1947).

17. L. Bohn, *Rubber Chem. Tech.*, **41**, 495 (1968).

18. S. Krause, *J. Macromol. Sci. Rev. Macromol. Chem.*, **C7**, 251 (1972).

19. R. L. Scott, *J. Chem. Phys.*, **17**, 279 (1949).

20. P. J. Flory, R. A. Orwall, and A. Vrij, *J. Am. Chem. Soc.*, **86**, 3515 (1964).

21. P. J. Flory, *Discuss. Faraday Soc.*, **49**, 7 (1970).

22. I. C. Sanchez and R. H. Lacombe, *J. Phys. Chem.*, **80**, 2352 (1976).

23. I. C. Sanchez and R. H. Lacombe, *J. Polym. Sci. Polym. Lett. Ed.*, **15**, 71 (1977).

24. R. H. Lacombe and I. C. Sanchez, *J. Phys. Chem.*, **80**, 2568 (1976).

25. I. C. Sanchez, Chap. 3 in *Polymer Blends*, Vol. I, D. R. Paul and S. Newman, eds., Academic Press, Orlando, 1978.

26. L. D. Taylor and L. D. Cerankowski, *J. Polym. Sci. Polym. Chem. Ed.*, **13**, 2551 (1975).

27. O. Olabisi, L. M. Robeson, and M. T. Shaw, *Polymer–Polymer Miscibility*, Academic Press, Orlando, 1979.

28. D. R. Paul, in *Multicomponent Polymer Materials*, Advances in Chemistry Series No. 211, ACS Books, Washington, DC, 1986.

29. D. S. Lee and S. C. Kim, *Macromolecules*, **17**, 268 (1984).

30. P. J. Flory, *J. Am. Chem. Soc.*, **87**, 1833 (1965).

31. L. A. Utracki, *Polymer Alloys and Blends*, Hanser, New York, 1990.

32. (a) I. C. Sanchez, Chap. 3 in *Polymer Blends*, Vol. I, D. R. Paul and S. Newman, eds., Academic Press, Orlando, 1978; (b) I. C. Sanchez, *Encyclopedia of Physical Science and Technology*, R. A. Meyers, ed., Vol. 11, Academic Press, Orlando, 1987, p. 1.

33. P. J. Flory, R. A. Orwoll, and A. Vrij, *J. Am. Chem. Soc.*, **86**, 3515 (1964).

34. D. Patterson, *Polym. Eng. Sci.*, **22**, 64 (1982).

35. J. H. An and L. H. Sperling, in *Cross-Linked Polymers: Chemistry, Properties and Applications*, R. A. Dickie S. S. Labana and R. A. Baues, eds., American Chemical Society, Washington, DC, 1988.

36. T. Nishi, T. T. Wang, and T. K. Kwei, *Macromolecules*, **8**, 227 (1975).

37. S. Reich, *Phys. Lett.*, **114A**, 90 (1986).

38. J. W. Cahn, *J. Chem. Phys.*, **422**, 93 (1963).

39. J. W. Cahn and J. E. Hilliard, *J. Chem. Phys.*, **28**, 258 (1958).

40. F. E. Karasz and W. J. MacKnight, Chap. 4 in *Multicomponent Polymer Materials*, D. R. Paul and L. H. Sperling, eds., Advances in Chemistry Series No. 211, ACS Books, Washington, DC, 1986.

41. P. Alexandrovich, F. E. Karasz, and W. J. MacKnight, *Polymer*, **17**, 1023 (1977).

42. L. A. Utracki and R. A. Weiss, eds., *Multiphase Polymers: Blends and Ionomers*, ACS Symposium Series No. 395, ACS Books, Washington, DC, 1989.

43. D. R. Paul and L. H. Sperling, eds., *Multicomponent Polymer Materials*, Advances in Chemistry Series No. 211, ACS Books, Washington, DC, 1986.

44. B. M. Culbertson, eds., *Multiphase Macromolecular Systems*, Plenum, New York, 1989.

45. C. R. Riew, ed., *Rubber-Toughened Plastics*, Advances in Chemistry Series No. 222, ACS Books, Washington, DC, 1989.

46. D. Klempner and K. C. Frisch, eds., *Advances in Interpenetrating Polymer Networks*, Vol. II, Technomic, Lancaster, PA, 1990.

47. T. Saegusa, T. Higashimura, and A. Abe, eds., *Frontiers of Macromolecular Science*, Blackwell, Oxford, England, 1989.

48. L. P. McMaster, in *Copolymers, Polyblends, and Composites*, N. A. J. Platzer, ed., Advances in Chemistry Series No. 142, American Chemical Society, Washington, DC, 1975.

49. J. S. Noland, N. N. C. Hsu, R. Saxon, and J. M. Schmitt, Chap. 2 in *Multicomponent Polymer Systems*, N. A. J. Platzer, ed., Advances in Chemistry Series No. 99, American Chemical Society, Washington, DC, 1971.

50. J. Stoelting, F. E. Karasz, and W. J. MacKnight, *Polym. Eng. Sci.*, **10**, 133 (1970).

51. A. A. Donatelli, L. H. Sperling, and D. A. Thomas, *Macromolecules*, **9**, 671 (1976).

52. O. Olabisi, L. M. Robeson, and M. T. Shaw, *Polymer–Polymer Miscibility*, Academic, Orlando, 1979.

53. C. W. Frank and M. A. Gashgari, *Macromolecules*, **12**, 163 (1979).

54. S. Krause, *J. Polym. Sci.*, *A-2*, **7**, 249 (1969).

55. S. Krause, *Macromolecules*, **3**, 84 (1970).

56. S. Krause, in *Colloidal and Morphological Behavior of Block and Graft Copolymers*, G. E. Molair, ed., Plenum, New York, 1971.

57. D. R. Paul, *Polym. Mater. Sci. Eng. Prepr.*, **50**, 1 (1984).

58. D. J. Meier, *J. Polym. Sci.*, **26C** 81 (1969).

59. D. J. Meier, *Polym. Preprints*, **11**, 400 (1970).

60. M. Matsuo, *Jpn. Plastics*, **2**, 6 (1968).

61. L. H. Sperling, *Polymeric Multicomponent Materials: An Introduction*, Wiley, New York, 1997.

62. A. Eisenberg and J.-S. Kim, *Introduction to Ionomers*, Wiley, New York, 1998.

63. W. A. Combellick, in *Encyclopedia of Polymer Science and Engineering*, Vol. 2, Wiley, New York, 1985, p. 176.

64. J. Crank and G. S. Park, *Diffusion in Polymers*, Academic Press, Orlando, 1968.

65. L. F. Rogers and D. Machin, *Crit. Rev. Macromol. Sci.*, **1**, 245 (1972).

66. A. Fick, *Ann. Phys. (Leipzig)*, **170**, 59 (1855).

67. P. Meares, *Polymers: Structure and Bulk Properties*, Van Nostrand, New York, 1965, Chap. 12.

68. G. van Amerongen, *J. Appl. Phys.*, **17**, 972 (1946).

69. G. J. van Amerongen, *J. Polym. Sci.*, **5**, 307 (1950).

70. V. T. Stannett, W. J. Koros, D. R. Paul, H. K. Lonsdale, and R. W. Baker, *Adv. Polym. Sci.*, **32**, 69 (1979).

71. S. Pauly, in *Polymer Handbook*, 3rd ed., J. Brandrup and E. H. Immergut, eds., Wiley-Interscience, New York, 1989, sec. VI, p. 435.

72. W. J. Koros and M. W. Hellums, in *Encyclopedia of Polymer Science and Engineering*, 2nd ed., Supplement Volume, Wiley, New York, 1989, p. 724.

73. D. W. Breck, *Zeolite Molecular Sieves*, Wiley, New York, 1974, Chap. 8.

74. R. T. Chern, W. J. Koros, H. B. Hopfenberg, and V. T. Stannett, in *Materials Science of Synthetic Membranes*, D. R. Lloyd, ed., ACS Symposium Series No. 269, ACS Books, Washington, DC, 1985.

75. S. Arizzi and U. W. Suter, *Polym. Mater. Sci. Eng. Prepr.*, **61**, 481 (1989).

76. N. A. Peppas and W. H. Yang, *Contact Introcular Lens Med. J.*, **7**, 300 (1981).

77. T. M. Aminabhavi, H. G. Niak, U. S. Toti, R. H. Balundgi, A. M. Dave, and M. H. Mehta, *Polymer News*, **24**, 294 (1999).

78. L. M. Robeson, *J. Membr. Sci.*, **62**, 165 (1991).

79. L. M. Robeson, C. D. Smith and M. Langsam, *J. Membr. Sci.*, **132**, 33 (1997).

80. B. D. Freeman, *Macromolecules*, **32**, 375 (1999).

81. D. W. Van Krevelen, *Properties of Polymers*, 3rd ed., Elsevier, Amsterdam, 1990.

82. J. W. Vanderhoff, Chap. 1 in *Vinyl Polymerization*, Vol. 1, Part II, G. E. Ham, ed., Dekker, New York, 1969.

83. G. W. Poehlein, in *Encylopedia of Polymer Science and Engineering*, 2nd ed., Vol. 6, Wiley, New York, 1986.

84. D. C. Blackley, in *Encylopedia of Polymer Science and Engineering*, 2nd ed., Vol. 8, J. I. Kroschwitz, ed., Wiley, New York, 1987.

GENERAL READING

A. W. Birley, B. Haworth, and J. Batchelor, *Physics of Plastics: Processing, Properties, and Materials Engineering*, Hanser, Munich, 1992, Chap. 10.

D. C. Blackley, *Polymer Latices: Science and Technology*, 2nd ed., Chapman and Hall, London, 1997.

J. Comyn, ed., *Polymer Permeability*, Elsevier, London, 1985.

G. J. Fleer, M. A. Cohen Stuart, J. M. H. M. Scheutjens, T. Cosgrove, and B. Vincent, *Polymers at Interfaces*, Chapman and Hall, London, 1993.

W. J. Koros, ed., *Barrier Polymers and Structures*, ACS Symposium Series 423, ACS Books, Washington, DC, 1990.

P. A. Lovell and M. S. El-Aasser, *Emulsion Polymerization and Emulsion Polymers*, Wiley, Chichester, 1997.

J. A. Manson and L. H. Sperling, *Polymer Blends and Composites*, Plenum, New York, 1976.

D. H. Napper, *Polymeric Stabilization of Colloidal Dispersions*, Academic Press, London, 1983.

L. C. Sawyer and D. T. Grubb, *Polymer Microscopy*, 2nd ed., Chapman and Hall, London, 1996.

L. H. Sperling, *Polymeric Multicomponent Materials: An Introduction*, Wiley, New York, 1997.

STUDY PROBLEMS

1. Why do most polymer blends exhibit lower critical solution temperatures, rather than upper critical solution temperatures?

2. A sample of polystyrene–*block*–polybutadiene has block molecular weights of 20,000 g/mol and 80,000 g/mol, respectively. What is the diameter of the polystyrene domains, assuming spheres? How many blocks are in one domain? How many of these domains are there per cm^3 of the whole polymer?

3. A child's polybutadiene balloon, filled with helium, is accidently released into the air. What factors control its rate of rise into the atmosphere and its eventual fall back to earth? Assuming ground level is at 25°C and 760 mm Hg pressure, plot the projected rise and fall of the balloon, making any realistic assumptions required. How long will it stay in the air?

4. What is the analytical expression for χ_{blend} for the general system of two statistical copolymers, $(A_x B_{1-x})_n / (C_y D_{1-y})_{n'}$?

5. Based on equation-of-state concepts, how does the density of polystyrene vary (at 125°C) with hydrostatic pressure?

6. Polystyrene and polybutadiene are glassy and rubbery, respectively. They are substantially immiscible over their entire composition range. Show how both soft (low modulus) and stiff (high modulus) materials may be made from a 50/50 composition.

7. Obtain a two-liter plastic soft drink bottle and measure the thickness of its walls. Assuming it is made of poly(ethylene terephthalate), what is the estimated length of time for the level of carbonation to decrease by a factor of 2? (Afterward, make sure the bottle is recycled!) Why isn't low-density polyethylene used for this purpose? (Pressure in fresh soda bottles is around 4 Atm.)

8. Obtain a sample of latex paint. What is its composition? Coat it on a piece of wood or cardboard and observe as it dries. Touch or rub it periodically with a paper towel. Does the film become strong immediately? Later? Why?

9. Figure 4.14 illustrates the effects of miscibility on excimer fluorescence. (a) If we assume that I_D/I_M arises solely from intermolecular interactions, what value is predicted for total phase separation? (b) If the phase-separated value of I_D/I_M is assumed to be 10, and the system is completely mixed at $\Delta\delta = 0$, what fraction of the excimer fluorescence arises from intramolecular contacts?

10. What is the calculated heat of mixing of 1000 g of polystyrene with 1000 g of polybutadiene? If the molecular weight of both polymers is on the order of 1×10^5 g/mol, would you anticipate that the two polymers are mutually miscible at 150°C? What are the experimental results in this case?

11. What is the entropy of mixing of the red and black checkers on an ordinary checkerboard? Assuming an ideal solution, what is the free energy of mixing? After "polymerizing" the checkers, what is the new entropy and free energy of mixing of the blend?

12. Starting with Table 3.10, what is the critical volume fraction for cis-polybntadiene of 2×10^5 g/mol in benzene entering the semidilute region?

13. At $2x\varphi_{ov}$ in Problem 12, what is the value of ξ?

APPENDIX 4.1   SCALING LAW THEORIES AND APPLICATIONS

Scaling law theories were developed by de Gennes (A1) and others to provide novel solutions to a series of physical problems, including some in polymer science. Scaling law is concerned with *exponents*, sometimes called universal properties, rather than *coefficients*, sometimes called local properties (A2). Scaling law starts with known relationships, with special concern with transitions between regimes of physical behavior.

Fundamental Postulate

The theory supposes that there exists a transition at some critical value x^* of a parameter x such that the variable S changes its form. The fundamental postulate of scaling law theory can be written,

$$S = S_0 f\left(\frac{x}{x^*}\right) \tag{A4.1}$$

where the two regimes of x in function f are specified by

$$f\left(\frac{x}{x^*}\right) = 1 \quad \text{if} \quad \frac{x}{x^*} < 1 \tag{A4.2}$$

and

$$f\left(\frac{x}{x^*}\right) = \left(\frac{x}{x^*}\right)^m \quad \text{if} \quad \frac{x}{x^*} > 1 \tag{A4.3}$$

where m represents the universal scaling exponent.

The Dilute to Semidilute Transition

Flory has shown that

$$R_g = K n^\gamma \tag{A4.4}$$

where K is a constant (which the theory ignores), n is the degree of polymerization (the number of mers in the chain), and γ equals $\frac{1}{2}$ for a θ-solvent (see Section 3.5) and $\frac{3}{5}$ for a thermodynamically good solvent, in the limit of high molecular weight.

At the critical crossover volume fraction from dilute to semidilute, φ_{ov}, the entire volume is just taken up by the molecules. Therefore

$$\varphi_{ov} = \frac{K' M}{N_A R_g^3} \tag{A4.5}$$

where M, the molecular weight, of course is proportional to $n^{1.0}$. Then

$$\varphi_{ov} = \frac{K' n^1}{N_A (K n^\gamma)^3} = K'' n^{1-3\gamma} \tag{A4.6}$$

Taking $\gamma = 3/5$ leads to

$$1 - 3\gamma = 1 - 3\left(\tfrac{3}{5}\right) = 1 - \tfrac{9}{5} = -\tfrac{4}{5} \tag{A4.7}$$

and hence

$$\varphi_{ov} = K'' n^{-4/5} \tag{A4.8}$$

A more comprehensive treatment (A3) yields $\varphi_{ov} \sim (pv)^{-3/5} n^{-4/5}$, where the persistence length, $p = C_\infty/6$, and $v = 1 - 2\chi$, the excluded volume parameter, and χ represents the Flory–Huggins polymer–solvent interaction parameter. This relation for φ_{ov} defines the line dividing the dilute and semidilute regions in Figure 4.5.

Semidilute Regime Scaling Laws

As noted above, in a dilute good solvent $R_{g,d} \sim n^{3/5}$, where R_g is independent of polymer concentration within the dilute concentration range. Using the subscript s for semidilute and d for dilute, the scaling law can be written

$$R_{g,s} = R_{g,d} f\left(\frac{\varphi}{\varphi_{ov}}\right) = R_{g,d}\left(\frac{\varphi}{\varphi_{ov}}\right)^m \tag{A4.9}$$

From equation (A4.8), $\varphi_{ov} \sim n^{-4/5}$; thus

$$R_{g,s} \sim R_{g,d} \varphi^m n^{(4/5)m} \tag{A4.10}$$

showing that the radius of gyration in the semidilute region is a function of both volume fraction polymer and molecular weight.

The assumption that $R_{g,d} \sim n^{3/5}$ follows from both theory and experiment. In the bulk case the radius of gyration goes as $n^{1/2}$, identical to that of θ-solvents—a point first recognized by Flory (A4) in 1953, and experimentally verified only after small-angle neutron scattering studies were done around 1980. (How to win a Nobel Prize!) It is now known experimentally that $R_{g,s} \sim n^{1/2}$ as well. Then

$$n^{1/2} \sim n^{3/5} \varphi^m n^{(4/5)m} \tag{A4.11}$$

Equating the exponents for n,

$$\tfrac{1}{2} = \tfrac{3}{5} + \left(\tfrac{4}{5}\right)m \tag{A4.12}$$

leads to $m = -1/8$. Therefore

$$R_{g,s} \sim n^{1/2} \varphi^{-1/8} \tag{A4.13}$$

Thus, for a good solvent, $R_{g,s}$ decreases slowly with volume fraction of polymer.

The Correlation Length, ξ, in the Semidilute Solution

Using the scaling law approach, we have

$$\xi_s = \xi_d \left(\frac{\varphi}{\varphi_{ov}}\right)^m \tag{A4.14}$$

In dilute solutions, $\xi_d = R_{g,d}$, and $\varphi_{ov} \sim n^{-4/5}$, equation (A4.8); then

$$\xi_s \sim R_{g,d}\varphi^m N^{(4/5)m} \tag{A4.15}$$

Noting that $R_{g,d} \sim n^{3/5}$,

$$\xi_s \sim \varphi^m n^{\{3/5+(4/5)m\}} \tag{A4.16}$$

The assumption is made that ξ_s depends only on the volume fraction of polymer for high enough molecular weight. Then

$$\xi_s \sim n^0 \varphi^m \tag{A4.17}$$

and

$$0 = \tfrac{3}{5} + \left(\tfrac{4}{5}\right)m \tag{A4.18}$$

Thus $m = -3/4$, and $\xi_s \sim \varphi^{-3/4}$, as shown in Figure 4.4. Similar derivations result in the relations shown for the marginal and concentrated regimes, Figure 4.4.

REFERENCES

A1. P. G. de Gennes, *Scaling Concepts in Polymer Physics*, Cornell University Press, Ithaca, NY, 1979.

A2. D. H. Napper, *Polymeric Stabilization of Colloidal Dispersions*, Academic Press, San Diego, CA, 1983.

A3. G. J. Fleer, M. A. C. Stuart, J. M. H. M. Scheutjens, T. Cosgrove, and B. Vincent, *Polymers at Interfaces*, Chapman and Hall, London, 1993.

A4. P. J. Flory, *Principles of Polymer Chemistry*, Cornell University Press, Ithaca, NY, 1953.

5

THE AMORPHOUS STATE

The bulk state, sometimes called the condensed or solid state, includes both amorphous and crystalline polymers. As opposed to polymer solutions, generally there is no solvent present. This state comprises polymers as ordinarily observed, such as plastics, elastomers, fibers, adhesives, and coatings.

While amorphous polymers do not contain any crystalline regions, "crystalline" polymers generally are only semicrystalline, containing appreciable amounts of amorphous material. When a crystalline polymer is melted, the melt is amorphous. In treating the kinetics and thermodynamics of crystallization, the transformation from the amorphous state to the crystalline state and back again is constantly being considered. The subjects of amorphous and crystalline polymers are treated in the next two chapters. This will be followed by a discussion of liquid crystalline polymers, Chapter 7. Although polymers in the bulk state may contain plasticizers, fillers, and other components, this chapter emphasizes the polymer molecular organization itself.

A few definitions are in order. Depending on temperature and structure, amorphous polymers exhibit widely different physical and mechanical behavior patterns. At low temperatures, amorphous polymers are glassy, hard, and brittle. As the temperature is raised, they go through the glass–rubber transition. The glass transition temperature (T_g) is defined as the temperature at which the polymer softens because of the onset of long-range coordinated molecular motion. This is the subject of Chapter 8.

Above T_g, cross-linked amorphous polymers exhibit rubber elasticity. An example is styrene–butadiene rubber (SBR), widely used in materials ranging from rubber bands to automotive tires. Rubber elasticity is treated in Chapter 9. Linear amorphous polymers flow above T_g.

Polymers that cannot crystallize usually have some irregularity in their structure. Examples include the atactic vinyl polymers and statistical copolymers.

5.1   THE AMORPHOUS POLYMER STATE

5.1.1   Solids and Liquids

An amorphous polymer does not exhibit a crystalline X-ray diffraction pattern, and it does not have a first-order melting transition. If the structure of crystalline polymers is taken to be regular or ordered, then by difference, the structure of amorphous polymers contains greater or lesser amounts of disorder.

The older literature often referred to the amorphous state as a liquid state. Water is a noncrystalline (amorphous) condensed substance and is surely a liquid. However, polymers such as polystyrene or poly(methyl methacrylate) at room temperature are glassy, taking months or years for significant creep or flow. By contrast, skyscrapers are also undergoing creep (or flow), becoming measurably shorter as the years pass, as the steel girders creep (or flow). Today, amorphous polymers in the glassy state are better called amorphous solids.

Above the glass transition temperature, if the polymer is amorphous and linear, it will flow, albeit the viscosity may be very high. Such materials are liquids in the modern sense of the term. It should be noted that the glass transition itself is named after the softening of ordinary glass, an amorphous inorganic polymer. If the polymer is crystalline, the melting temperature is always above the glass transition temperature.

5.1.2   Possible Residual Order in Amorphous Polymers?

As a point of focus, the evidence for and against partial order in amorphous polymers is presented. On the simplest level, the structure of bulk amorphous polymers has been likened to a pot of spaghetti, where the spaghetti strands weave randomly in and out among each other. The model would be better if the strands of spaghetti were much longer, because by ratio of length to diameter, spaghetti more resembles wax chain lengths than it does high polymers.

The spaghetti model provides an entry into the question of residual order in amorphous polymers. An examination of relative positions of adjacent strands shows that they have short regions where they appear to lie more or less parallel. One group of experiments finds that oligomeric polymers also exhibit similar parallel regions (1,2). Accordingly, the chains appear to lie parallel for short runs because of space-filling requirements, permitting a higher density (3). This point, the subject of much debate, is discussed further later.

Questions of interest to amorphous state studies include the design of critical experiments concerning the shape of the polymer chain, the estimation of type and extent of order or disorder, and the development of models suitable for physical and mechanical applications. It must be emphasized that our knowledge of the amorphous state remains very incomplete, and that this and other areas of polymer science are the subjects of intensive research at this time. Pechhold and Grossmann (4) capture the spirit of the times exactly:

> Our current knowledge about the level of order in amorphous polymers should stimulate further development of competing molecular models, by making their suppositions more precise in order to provide a bridge between their microscopic structure description and the understanding of macroscopic properties, thereby predicting effects which might be proved experimentally.

The subject of structure in amorphous polymers has been entensively reviewed (5–11) and has been the subject of two published symposia (12,13).

5.2 EXPERIMENTAL EVIDENCE REGARDING AMORPHOUS POLYMERS

The experimental methods used to characterize amorphous polymers may be divided into those that measure relatively short-range interactions (nonrandom versus random chain positions) (14), below about 20 Å, and those that measure longer-range interactions. In the following paragraphs the role of these several techniques will be explored. The information obtainable from these methods is summarized in Table 5.1.

5.2.1 Short-Range Interactions in Amorphous Polymers

Methods that measure short-range interactions can be divided into two groups: those that measure the orientation or correlation of the mers along the *axial* direction of a chain, and those that measure the order between chains, in the *radial* direction. Figure 5.1 illustrates the two types of measurements.

There are several measures of the axial direction in the literature. Two of the more frequently used are the *Kuhn segment length* (see Section 5.3.1.2) and the *persistence length* (15). The latter is defined qualitatively as the length down a chain from a given point where the polymer's direction is random with respect to starting point. The IUPAC definition, with recommended symbol a, is the average projection of the end-to-end vector on the tangent to the chain contour at a chain end in the limit of infinite chain length. For example, the value of the persistence length for polyethylene is 5.75 Å, comprising only a few mers.

One of the most powerful experimental methods of determining short-range order in polymers utilizes birefringence (6). Birefringence measures orientation in the axial direction. The birefringence of a sample is defined by

$$\Delta n = n_1 - n_2 \tag{5.1}$$

where n_1 and n_2 are the refractive indexes for light polarized in two directions 90° apart. If a polymer sample is stretched, n_1 and n_2 are taken as the refractive indexes for light polarized parallel and perpendicular to the stretching direction.

The anisotropy of refractive index of the stretched polymer can be demonstrated by placing a thin film between crossed polaroids. The field of view is dark before stretching, but vivid colors develop as orientation is imposed. For stretching at 45° to the polarization directions, the fraction of light transmitted is given by (6)

$$\mathbf{T} = \sin^2\left(\frac{\pi' d \Delta n}{\lambda_0}\right) \tag{5.2}$$

where d represents the thickness and λ_0 represents the wavelength of light in vacuum.

By measuring the transmitted light quantitatively, the birefringence is obtained. The birefringence is related to the orientation of molecular units such as mers, crystals, or even chemical bonds by

$$\Delta n = \frac{2}{9}\pi \frac{(\bar{n}^2 + 2)^2}{\bar{n}} \sum_i (b_1 - b_2)_i f_i \tag{5.3}$$

Table 5.1 Selected studies of the amorphous state

Method	Information Obtainable	Principal Findings	Reference
A. Short-Range Interactions			
Stress–optical coefficient	Orientation of segments in isolated chain	Orientation limited to 5–10 Å	(a)
Depolarized light-scattering	Segmental orientation correlation	2–3 — CH_2 — units along chain correlated	(b)
Magnetic birefringence	Segmental orientation correlation	Orientation correlations very small	(b)
Raman scattering	Trans and gauche populations	Little or no modification in chain conformation initiated by intermolecular forces	(b)
NMR relaxation	Relaxation times	Small fluctuating bundles in the melt	(c)
Small-angle X-ray scattering, SAXS	Density variations	Amorphous polymers highly homogeneous; thermal fluctuations predominate	(d)
Birefringence	$n_1 - n_2$	Orientation	
B. Long-Range Interactions			
Small-angle neutron scattering	Conformation of single chains	Radius of gyration the same in melt as in θ-solvents	(e, f)
Electron microscopy	Surface inhomogeneities	Nodular structures of 50–200 Å in diameter	(g, h)
Electron diffraction and wide-angle X-ray diffraction	Amorphous halos	Bundles of radial dimension = 25 Å and axial dimension = 50 Å, but order may extend to only one or two adjacent chains	(i, j)
C. General			
Enthalpy relaxation	Deviations from equilibrium state	Changes not related to formation of structure	(k)
Density	Packing of chains	Density in the amorphous state is about 0.9 times the density in the crystalline state	(l, m)

References: (a) R. S. Stein and S. D. Hong, *J. Macromol. Sci. Phys.*, **B12** (11), 125 (1976). (b) E. W. Fischer, G. R. Strobl, M. Dettenmaier, M. Stamm, and N. Steidle, *Faraday Discuss. Chem. Soc.*, **68**, 26 (1979). (c) W. L. F. Golz and H. G. Zachmann, *Makromol. Chem.*, **176**, 2721 (1975). (d) D. R. Uhlmann, *Faraday Discuss. Chem. Soc.*, **68**, 87 (1979). (e) H. Benoit, *J. Macromol. Sci. Phys.*, **B12** (1), 27 (1976). (f) G. D. Wignall, D. G. H. Ballard, and J. Schelten, *J. Macromol. Sci. Phys.*, **B12** (1), 75 (1976). (g) G. S. Y. Yeh, *Crit. Rev. Macromol. Sci.*, **1**, 173 (1972). (h) R. Lam and P. H. Gell, *J. Macromol. Sci. Phys.*, **B20** (1), 37 (1981). (i) Yu. K. Ovchinnikov, G. S. Markova, and V. A. Kargin, *Vysokomol. Soedin.* **AII** (2), 329 (1969). (j) R. Lovell, G. R. Mitchell, and A. H. Windle, *Faraday Discuss. Chem. Soc.*, **68**, 46 (1979). (k) S. E. B. Petrie, *J. Macromol. Sci. Phys.*, **B12** (2), 225 (1976). (l) R. E. Robertson, *J. Phys. Chem.*, **69**, 1575 (1965). (m) R. F. Boyer, *J. Macromol. Sci. Phys.*, **B12**, 253 (1976).

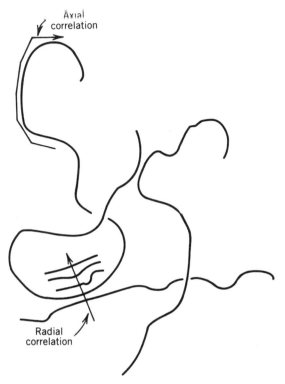

Figure 5.1 Schematic diagram illustrating the axial and radial correlation directions.

where f_i is an orientation function of such units given by

$$f_i = \frac{3 \cos^2 \theta_i - 1}{2} \tag{5.4}$$

where θ_i is the angle that the symmetry axis of the unit makes with respect to the stretching direction, $\bar{n}$ is the average refractive index, and b_1 and b_2 are the polarizabilities along and perpendicular to the axes of such units.

Equation (5.4) contains two important solutions for fibers and films:

$$\theta = 0°, \text{ perfect orientation}$$

$$\theta = 54°, \text{ zero orientation}$$

Many commercial fibers such as nylon or rayon will have θ equal to about 5°.

The stress–optical coefficient (SOC) is a measure of the change in birefringence on stretching a sample under a stress σ (16)

$$\text{SOC} = \frac{\Delta n}{\sigma} \tag{5.5}$$

If the polymer is assumed to obey rubbery elasticity relations (see Chapter 9), then

$$\text{SOC} = \frac{\Delta n}{\sigma} = \frac{2\pi_1}{45kT} \frac{(\bar{n}^2 + 2)^2}{\bar{n}} (b_1 - b_2) \tag{5.6}$$

where $\pi_1 = 3.14$, and $\bar{n}$ represents the average refractive index. The change in birefringence that occurs when an amorphous polymer is deformed yields important information concerning the state of order in the amorphous solid. It should be emphasized that the theory expressed in equation (5.6) involves the orientation of segments within a single isolated chain. From an experimental point of view, it has been found that the strain–optical coefficient (STOC) is independent of the extension (17) but that the SOC is not.

The anisotropy of a segment is given by $b_1 - b_2$. Experiments carried out by Stein and Hong (16) on this quantity as a function of swelling and extension show no appreciable changes, leading to the conclusion that the order within a chain (axial correlation) does not change beyond a range of 5 to 10 Å, comparable with the range of ordering found for low-molecular-weight liquids.

Depolarized light-scattering (DPS) is a related technique whereby the intensity of scattered light is measured when the sample is irradiated by visible light. During this experiment, the sample is held between crossed Nicols. Studies on DPS on n-alkane liquids (13) reveal that there is a critical chain length of 8 to 9 carbons, below which there is no order in the melt. For longer chains, only 2 to 3 $-CH_2-$ units in one chain are correlated with regard to their orientation, indicating an extremely weak orientational correlation.

Other electromagnetic radiation interactions with polymers useful for the study of short-range interactions in polymers include:

1. *Rayleigh scattering*: elastically scattered light, usually measured as a function of scattering angle.
2. *Brillouin scattering*: in essence a Doppler effect, which yields small frequency shifts.
3. *Raman scattering*: an inelastic process with a shift in wavelength due to chemical absorption or emission.

Results of measurements utilizing SOC, DPS, and other short-range experimental methods such as magnetic birefringence, Raman scattering, Brillouin scattering, NMR relaxation, and small-angle X-ray scattering are summarized in Table 5.1. The basic conclusion is that intramolecular orientation is little affected by the presence of other chains in the bulk amorphous state. The extent of order indicated by these techniques is limited to at most a few tens of angstroms, approximately that which was found in ordinary low-molecular-weight liquids.

5.2.2 Long-Range Interactions in Amorphous Polymers

5.2.2.1 *Small-Angle Neutron Scattering* The long-range interactions are more interesting from a polymer conformation and structure point of view. The most powerful of the methods now available is small-angle neutron scattering (SANS). For these experiments, the de Broglie wave nature of neutrons is utilized. Applied to polymers, SANS techniques can be used to determine the actual chain radius of gyration in the bulk state (18–24).

The Theory Strategy of SANS follows the development of light-scattering (see Section 3.6.1). For small-angle neutron scattering, the weight-average molecular weight, M_w, and the z-average radius of gyration, R_g, may be determined (18–19):

$$\frac{Hc}{R(\theta) - R(\text{solvent})} = \frac{1}{M_w P(\theta)} + 2A_2 c \tag{5.7}$$

where $R(\theta)$ is the scattering intensity known as the "Rayleigh ratio,"

$$R(\theta) = \frac{I_\theta \omega^2}{I_0 V_s} \tag{5.8}$$

where ω represents the sample-detector distance, V_s is the scattering volume, and I_θ/I_0 is the ratio of scattered radiation intensity to the initial intensity (20–22). The quantity $P(\theta)$ is the scattering form factor, identical to the form factor used in light-scattering formulations [see equation (3.49)]. The formulation for $P(\theta)$, originally derived by Peter Debye (25), forms one of the mainspring relationships between physical measurements in both the dilute solution and solid states and in the interpretation of the data. For very small particles or molecules, $P(\theta)$ equals unity. In both equation (5.7) (explicit) and equation (5.8) (implicit), the scattering intensity of the solvent or background must be subtracted.

In SANS experiments of the type of interest here, a deuterated polymer is dissolved in an ordinary hydrogen-bearing polymer of the same type (or vice versa). The calculations are simplified if the two polymers have the same molecular weight. The background to be subtracted originates from the scattering of the protonated species, and the coherent scattering of interest originates from the dissolved deuterated species. The quantity H in equation (5.7) was already defined in equation (3.56) for neutron scattering. SANS instrumentation has evolved through several generations, as delineated in Table 5.2.

As currently used (26–27), the coherent intensity in a SANS experiment is described by the cross section, $d\Sigma/d\Omega$, which is the probability that a neutron will be scattered in a solid angle, Ω, per unit volume of the sample. This cross section, which is normally used to express the neutron scattering power of a sample, is identical with the quantity R defined in equation (5.8).

Then it is convenient to express equation (5.7) as

$$\frac{C_N}{d\Sigma/d\Omega} = \frac{1}{M_w P(\theta)} \tag{5.9}$$

Table 5.2 Evolution of SANS instrumentation

Method	Location	Comments
Long flight path	(a) ILL, Grenoble (b) Oak Ridge (c) Jülich	Inverse square distance law means long experimental times
Long wavelength	NIST	Neutrons cooled via liquid He or H_2
Time-of-flight[a] (TOF)	Los Alamos Nat. Lab	"White" neutrons, liquid H_2, pulsed source

[a]Pulsed neutrons are separated via TOF according to wavelength. TOF is to long flight path as FT-IR is to IR.

Table 5.3 Scattering lengths of elements (20,21)

Element	Coherent Scattering Length[a] $b \times 10^{12}$ cm
Carbon, ^{12}C	0.665
Oxygen, ^{16}O	0.580
Hydrogen, ^{1}H	−0.374
Deuterium, ^{2}H	0.667
Fluorine, ^{19}F	0.560
Sulfur, ^{32}S	0.280

[a] Here $a = \Sigma_i b_i$.

where C_N, the analogue of H, may be expressed

$$C_N = \frac{(a_H - a_D)^2 N_a \rho (1-n) n}{M_p^2} \tag{5.10}$$

The quantities a_H and a_D are the scattering length of a normal protonated and deuterated (labeled) structural unit (mer), and n is the mole fraction of labeled chains. Thus C_N contains the concentration term as well as the "optical" constants. The quantities a_H and a_D are calculated by adding up the scattering lengths of each atom in the mer (see Table 5.3). In the case of high dilution, the quantity $(1-n)n$ reduces to the concentration c, as in equation (5.7).

After rearranging, equation (5.7) becomes

$$\left[\frac{d\Sigma}{d\Omega}\right]^{-1} = \frac{1}{C_N M_w}\left(1 + \frac{K^2 R_g^2}{3} + \cdots\right) \tag{5.11}$$

where K is the wave vector. Thus the mean square radius of gyration, R_g^2, and the polymer molecular weight, M_w, may be obtained from the ratio of the slope to the intercept and the intercept, respectively, of a plot of $[d\Sigma/d\Omega]^{-1}$ versus K^2. If $A_2 = 0$, this result is satisfactory. For finite A_2 values, a second extrapolation to zero concentration is required (see below).

A problem in neutron scattering and light-scattering alike stems from the fact that R_g is a z-average quantity, whereas the molecular weight is a weight-average quantity. The preferred solution has been to work with nearly monodisperse polymer samples, such as prepared by anionic polymerization. If the molecular weight distribution is known, an approximate correction can be made.

Typical data for polyprotostyrene dissolved in polydeuterostyrene are shown in Figure 5.2 (28). Use is made of the Zimm plot, which allows simultaneous plotting of both concentration and angular functions for a more compact representation of the data (29).

From data such as presented in Figure 5.2, both R_g and M_w may be calculated. The results are tabulated in Table 5.3. Also shown in Table 5.3 are corresponding data obtained by light-scattering in Flory θ-solvents, where the conformation of the chain is unperturbed because the free energies of solvent–polymer and polymer–polymer interactions are all the same (see Section 3.3).

Values of $(R_g^2/M)^{1/2}$ are shown in Table 5.4 because this quantity is independent of the molecular weight when the chain is unperturbed (30), being a constant characteristic

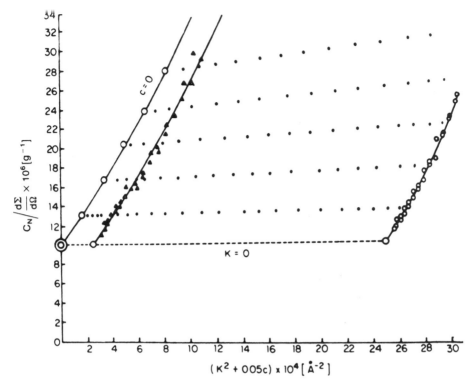

Figure 5.2 Small-angle neutron scattering of polyprotostyrene dissolved in polydeuterostyrene. A Zimm plot with extrapolations to both zero angle and zero concentration. Note that the second virial coefficient is zero, because polystyrene is essentially dissolved in polystyrene (28).

of each polymer. An examination of Table 5.4 reveals that the values in θ-solvents and in the bulk state are identical within experimental error. This important finding confirms earlier theories (30) that these two quantities ought to be equal, since under these conditions the polymer chain theoretically is unable to distinguish between a solvent molecular and a polymer segment with which it may be in contact.[†] Since it was believed that polymer chains in dilute solution were random coils, this finding provided powerful evidence that random coils also existed in the bulk amorphous state.

The reader should note the similarities between $R_g = KM^{1/2}$ and the Brownian motion relationship, $X = k't^{1/2}$, where X is the average distance traversed. For random coils, the end-to-end distance $r = 6^{1/2}R_g$.

5.2.2.2 Electron and X-Ray Diffraction

Under various conditions, crystalline substances diffract X rays and electrons to give spots or rings. According to Bragg's law,[‡] these can be interpreted as interplanar spacings. Amorphous materials, including ordi-

---

[†] This finding was predicted in 1953 by P. J. Flory, 20 years before it was confirmed experimentally.

[‡] See Section 6.2.2. Bragg's law: $n\lambda = 2d \sin \theta$, where $n = 1$ here, d is the distance between chains, and θ is the angle of diffraction.

Table 5.4   Molecular dimensions in bulk polymer samples (20)

| | | | $(R_g^2/M_w)^{1/2} \dfrac{\text{Å} \cdot \text{mol}^{1/2}}{\text{g}^{1/2}}$ | | |
Polymer	State of Bulk	SANS Bulk	Light-Scattering θ-Solvent	SAXS	Reference
Polystyrene	Glass	0.275	0.275	0.27 (i)	(a)
Polystyrene	Glass	0.28	0.275	—	(b)
Polyethylene	Melt	0.46	0.45	—	(c)
Polyethylene	Melt	0.45	0.45	—	(d)
Poly(methyl methacrylate)	Glass	0.31	0.30	—	(e)
Poly(ethylene oxide)	Melt	0.343	—	—	(f)
Poly(vinyl chloride)	Glass	0.30	0.37	—	(g)
Polycarbonate	Glass	0.457	—	—	(h)

References: (a) J. P. Cotton, D. Decker, H. Benoit, B. Farnoux, J. Higgins, G. Jannink, R. Ober, C. Picot, and J. desCloizeaux, *Macromolecules,* **7** 863 (1974). (b) G. D. Wignall, D. G. Ballard, and J. Schelten, *Eur. Polym. J.,* **10**, 861 (1974). (c) J. Schelten, D. G. H. Ballard, G. Wignall, G. Longman, and W. Schmatz, *Polymer,* **17**, 751 (1976). (d) G. Lieser, E. W. Fischer, and K. Ibel, *J. Polym. Sci. Polym. Lett. Ed.,* **13**, 39 (1975). (e) R. G. Kirste, W. A. Kruse, and K. Ibel, *Polymer,* **16**, 120 (1975). (f) G. Allen, *Proc. R. Soc. Lond., Ser. A,* **351**, 381 (1976). (g) P. Herchenroeder and M. Dettenmaier, Unpublished manuscript (1977). (h) D. G. H. Ballard, A. N. Burgess, P. Cheshire, E. W. Janke, A. Nevin, and J. Schelten, *Polymer,* **22**, 1353 (1981). (i) H. Hayashi, F. Hamada, and A. Nakajima, *Macromolecules,* **9**, 543 (1976). (i) G. J. Fleer, M. A. C. Stuart, J. M. H. M. Scheutjens, T. Cosgrove, and B. Vincent, *Polymers at Interfaces,* Chapman and Hall, London, 1993.

nary liquids, also diffract X-rays and electrons, but the diffraction is much more diffuse, sometimes called halos. For low-molecular-weight liquids, the diffuse halos have long been interpreted to mean that the nearest-neighbor spacings are slightly irregular and that after two or three molecular spacings all sense of order is lost. The situation is complicated in the case of polymers because of the presence of long chains. Questions to be resolved center about whether or not chains lie parallel for some distance, and if so, to what extent (31–34).

X-ray diffraction studies are frequently called wide-angle X-ray scattering, or WAXS. Typical data are illustrated in Figure 5.3 for polytetrafluorethylene (33). The first scattering maximum indicates the chain spacing distance. Maxima at larger values of s indicate other, shorter spacings. The (33) reduced intensity data are plotted as a function of angle,[†] $s = 4\pi' \sin\theta/\lambda$, which is sometimes called inverse space because the dimensions are Å^{-1}. The diffracted intensity is plotted in the y axis multiplied by the quantity s to permit the features to be more evenly weighted. Lovell et al. (33) fitted the experimental data with various theoretical models, also illustrated in Figure 5.3.

In analyzing WAXS data, the two different molecular directions must be borne in mind: (a) conformational orientation in the axial direction, which is a measure of how ordered or straight a given chain might be, and (b) organization in the radial direction, which is a direct measure of intermolecular order. WAXS measures both parameters. Lovell et al. (33) concluded from their study that the axial direction of molten polyethylene could be described by a chain with three rotational states, $0°$ and $\pm 120°$, with an average *trans* sequence length of three to four backbone bonds. The best radial packing model consisted of flexible chains arranged in a random manner (see Figure 5.3*b*).

[†] Variously, K is used by SANS experimenters for the angular function. The quantity s is called the scattering vector.

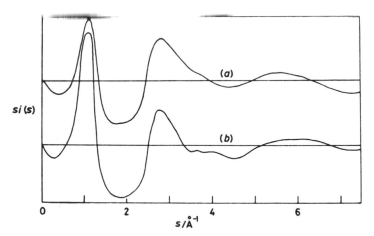

Figure 5.3 WAXS data on polytetrafluorethylene: (*a*) experimental data and (*b*) theory. Model is based on a disordered helix arranged with fivefold packing in a 24-Å diameter cylinder.

Polytetrafluoroethylene, on the other hand, was found to have more or less straight chains in the axial direction for distances of at least 24 Å (30). Many other studies have also shown that this polymer has extraordinarily stiff chains, because of its extensive substitution. In the radial direction, a model of parallel straight-chain segments was strongly supported by the WAXS data, although the exact nature of the packing and the extent of chain disorder are still the subject of current research. Poly(methyl methacrylate) and polystyrene were found to have a level of order intermediate between polyethylene and polytetrafluoroethylene.

The first interchain spacing of typical amorphous polymers is shown in Table 5.5. The greater interchain spacing of polystyrene and silicone rubbers is in part caused by bulky side groups compared with polyethylene.

The interpretation of diffraction data on amorphous polymers is currently a subject of debate. Ovchinnikov et al. (31,34) interpreted their electron diffraction data to show considerable order in the bulk amorphous state, even for polyethylene. Miller and co-workers (35,36) found that spacings increase with the size of the side groups, supporting the idea of local order in amorphous polymers. Fischer et al. (32), on the other hand,

Table 5.5   Interchain spacing in selected amorphous polymers

Polymer	Spacing, Å	Reference
Polyethylene	5.5	(a)
Silicone rubber	9.0	(a)
Polystyrene	10.0	(b)
Polycarbonate	4.8	(c)

References: (a) Y. K. Ovchinnikov, G. S. Markova, and V. A. Kargin, *Vysokomol. Soyed.*, **A11**, 329 (1969). (b) A. Bjornhaug, O. Ellefsen, and B. A. Tonnesen, *J. Polym. Sci.*, **12** 621 (1954). (c) A. Siegmann and P. H. Geil, *J. Macromol. Sci. (Phys.)*, **4** (2), 239 (1970).

found that little or no order fits their data best. Schubach et al. (37) take an intermediate position, finding that they were able to characterize first- and second-neighbor spacings for polystyrene and polycarbonate, but no further.

5.2.2.3 General Properties Two of the most important general properties of the amorphous polymers are the density and the excess free energy due to nonattainment of equilibrium. The latter shows mostly smooth changes on relaxation and annealing (38) and is not suggestive of any particular order. Changes in enthalpy on relaxation and annealing are touched on in Chapter 8.

However, many polymer scientists have been highly concerned with the density of polymers (3,32). For many common polymers the density of the amorphous phase is approximately 0.85 to 0.95 that of the crystalline phase (3,10). Returning to the spaghetti model, some scientists think that the polymer chains have to be organized more or less parallel over short distances, or the experimental densities cannot be attained. Others (32) have pointed out that different statistical methods of calculation lead, in fact, to satisfactory agreement between the experimental densities and a more random arrangement of the chains.

Using computer simulation of polymer molecular packing, Weber and Helfand (39) studied the relative alignment of polyethylene chains expected from certain models. They calculated the angle between pairs of chords of chains (from the center of one bond to the center of the next), which showed a small but clear tendency toward alignment between closely situated molecular segments, and registered well-developed first and second density peaks. However, no long-range order was observed.

Table 5.1 summarized the several experiments designed to obtain information about the organization of polymer chains in the bulk amorphous state, both for short- and long-range order and for the general properties. Table 5.6 outlines some of the major order–disorder arguments. Some of these are discussed below. The next section is concerned with the development of molecular models that best fit the data and understanding obtained to date.

5.3 CONFORMATION OF THE POLYMER CHAIN

One of the great classic problems in polymer science has been the determination of the conformation of the polymer chain in space. The data in Table 5.4 show that the radius of gyration divided by the square root of the molecular weight is a constant for any given polymer in the Flory θ-state, or in the bulk state. However, the detailed arrangement in space must be determined by other experiments and, in particular, by modeling. The resulting models are important in deriving equations for viscosity, diffusion, rubbery elasticity, and mechanical behavior.

5.3.1 Models and Ideas

5.3.1.1 The Freely Jointed Chain The simplest mathematical model of a polymer chain in space is the freely jointed chain. It has n links, each of length l, joined in a linear sequence with no restrictions on the angles between successive bonds (see Figure 5.4). By analogy with Brownian motion statistics, the root-mean-square end-to-end distance is

Table 5.6   Major order–disorder arguments in amorphous polymers

Order	Disorder
Conceptual difficulties in dense packing without order (a, b)	Rubber elasticity of polymer networks (e, f)
Appearance of nodules (b, c)	Absence of anomalous thermodynamic dilution effects (g)
Amorphous halos intensifying on equatorial plane during extension (d)	Radii of gyration the same in bulk as in θ-solvents (h)
Nonzero Mooney–Rivlin C_2 constants (a)	Fit of $P(\theta)$ for random coil model to scattering data (i)
Electron diffraction (l) lateral order to 15–20 Å	Rayleigh–Brillouin scattering, x-ray diffraction (j, k), stress–optical coefficient, etc. studies showing only modest (if any) short-range order (j, k, m, n)

References: (a) R. F. Boyer, *J. Macromol. Sci. Phys.*, **B12** (2), 253 (1976). (b) G. S. Y. Yeh, *Crit. Rev. Macromol. Sci.*, **1**, 173 (1972). (c) P. H. Geil, *Faraday Discuss. Chem. Soc.*, **68**, 141 (1979); but see S. W. Lee, H. Miyaji, and P. H. Geil, *J. Macromol. Sci., Phys.*, **B22** (3), 489 (1983); and D. R. Uhlmann, *Faraday Discuss. Chem. Soc.*, **68**, 87 (1979). (d) S. Krimm and A. V. Tobolsky, *Text. Res. J.*, **21**, 805 (1951). (e) P. J. Flory, *J. Macromol. Sci. Phys.*, **B12** (1), 1 (1976). (f) P. J. Flory, *Faraday Discuss. Chem. Soc.*, **68**, 14 (1979). (g) P. J. Flory, *Principles of Polymer Chemistry*, Cornell University Press, Ithaca, NY, 1953. (h) J. S. Higgins and R. S. Stein, *J. Appl. Crystallog.* **11**, 346 (1978). (i) H. Hayashi, F. Hamada, and A. Nakajima, *Macromolecules*, **9**, 543 (1976). (j) E. W. Fischer, J. H. Wendorff, M. Dettenmaier, G. Leiser, and I. Voigt-Martin, *J. Macromol. Sci. Phys.*, **B12** (1), 41 (1976). (k) D. R. Uhlmann, *Faraday Discuss. Chem. Soc.*, **68**, 87 (1979). (l) Yu. K. Ovchinnikov, G. S. Markova, and V. A. Kargin, *Vysokomol. Soyed*, **A11** (2), 329 (1969); *Polym. Sci. USSR*, **11**, 369 (1969). (m) R. S. Stein and S. O. Hong, *J. Macromol. Sci. Phys.*, **B12** (1), 125 (1976). (n) R. E. Robertson, *J. Phys. Chem.*, **69**, 1575 (1965).

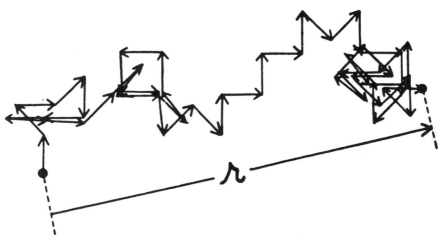

Figure 5.4   A vectorial representation of a freely jointed chain in two dimensions. A random walk of 50 steps (42).

given by (38–42)

$$(\overline{r_f^2})^{1/2} = ln^{1/2} \tag{5.12}$$

where the subscript f indicates free rotation.

A more general equation yielding the average end-to-end distance of a random coil, r_0, is given by

$$r_0^2 = l^2 n \frac{(1 - \cos \theta)(1 + \cos \phi)}{(1 + \cos \theta)(1 - \cos \phi)} \tag{5.13}$$

where θ is the bond angle between atoms, and ϕ is the conformation angle. This latter is as the angle of rotation for each bond, defined as the angle each bond makes from the plane delineated by the previous two bonds.

However, equation (5.13) still underestimates the end-to-end distance of the polymer chain, omitting such factors as excluded volume. This last arises from the fact that a chain cannot cross itself in space.

The placement of regular bond angles (109° between carbon atoms) expands the chain by a factor of $[(1 - \cos \theta)/(1 + \cos \theta)]^{1/2} = \sqrt{2}$, and the three major positions of successive placement obtained by rotation about the previous bond (two gauche and one trans positions) results in the chain being extended still further. Other short-range interactions include steric hindrances. Long-range interactions include excluded volume, which eliminates conformations in which two widely separated segments would occupy the same space. The total expansion is represented by a constant, C_∞, after squaring both sides of equation (5.12):

$$r^2 = nl^2 C_\infty \tag{5.14}$$

The characteristic ratio $C_\infty = r^2/l^2 n$ varies from about 5 to about 10, depending on the foliage present on the individual chains (see Table 5.7). The values of $l^2 n$ can be calculated by a direct consideration of the bond angles and energies of the various states and a consideration of longer-range interactions between portions of the chain (40).

Table 5.7   Typical values of the characteristic ratio C_∞

Polymer	C_∞
Polyethylene	6.8
Polystyrene	9.85
it-Polyproplyene	5.5
Poly(ethylene oxide)	4.1
Polyamide 66	5.9
Polybutadiene, 98% *cis*	4.75
cis-polyisoprene	4.75
trans-polyisoprene	7.4
polycarbonate	2.4
poly(methyl methacrylate)	8.2
poly(vinyl acetate)	9.4

Source: J. Brandrup and E. H. Immergut, eds., *Polymer Handbook*, 2nd ed., Wiley-Interscience, New York, 1975, and R. Wool, *Polymer Interfaces*, Hanser, Munich, 1995.

5.3.1.2 Kuhn Segments There are several approaches for dividing the polymer chain into specified lengths for conceptual or analytic purposes. For example, Section 4.2 introduced the blob, useful for semidilute solution calculations. In the bulk state, the *Kuhn segment* serves a similar purpose. The Kuhn segment length, b, depends on the chain's end-to-end distance under Flory θ-conditions, or its equivalent in the unoriented, amorphous bulk state, r_θ,

$$b = \frac{r_\theta^2}{L} \tag{5.15}$$

where L represents the chain contour length (43). The Kuhn segment length is the basic scale for specifying the size of a chain segment, providing a quantitative basis for evaluating the axial correlation length of Section 5.2.1. For flexible polymers, the Kuhn segment size varies between six and 12 mers (44), having a value of eight mers for polystyrene, and six for poly(methyl methacrylate). The Kuhn segment also expresses the idea of how far one must travel along a chain until all memory of the starting direction is lost, similar to the axial correlation distance, Figure 5.1.

5.3.2 The Random Coil

The term "random coil" is often used to describe the unperturbed shape of the polymer chains in both dilute solutions and in the bulk amorphous state. In dilute solutions the random coil dimensions are present under Flory θ-solvent conditions, where the polymer–solvent interactions and the excluded volume terms just cancel each other. In the bulk amorphous state the mers are surrounded entirely by identical mers, and the sum of all the interactions is zero. Considering mer–mer contacts, the interaction between two distant mers on the same chain is the same as the interaction between two mers on different chains. The same is true for longer chain segments.

In the limit of high molecular weight, the end-to-end distance of the random coil divided by the square root of 6 yields the radius of gyration (Section 3.6.2). Since the n links are proportional to the molecular weight, these relations lead directly to the result that $R_g/M^{1/2}$ is constant.

Of course, there is a distribution in end-to-end distances for random coils, even of the same molecular weight. The distribution of end-to-end distances can be treated by Gaussian distribution functions (see Chapter 9). The most important result is that, for relaxed random coils, there is a well-defined maximum in the frequency of the end-to-end distances, this distance is designated as r_0.

Appendix 5.1 describes the historical development of the random coil.

5.3.3 Models of Polymer Chains in the Bulk Amorphous State

Ever since Hermann Staudinger developed the macromolecular hypothesis in the 1920s (41), polymer scientists have wondered about the spatial arrangement of polymer chains, both in dilute solution and in the bulk. The earliest models included both rods and bedspring-like coils. X-ray and mechanical studies led to the development of the random coil model. In this model the polymer chains are permitted to wander about in a space-filling way as long as they do not pass through themselves or another chain (excluded-volume theory).

Table 5.8   Major models of the amorphous polymer state

Principals	Description of Model	Reference
H. Mark and P. J. Flory	Random coil model; chains mutually penetrable and of the same dimension as in θ-solvents	(a, b)
B. Vollmert	Individual cell structure model, close-packed structure of individual chains	(c, d)
P. H. Lindenmeyer	Highly coiled or irregularly folded conformational model, limited chain interpenetration	(e)
T. G. F. Schoon	Pearl necklace model of spherical structural units	(f)
V. A. Kargin	Bundle model, aggregates of molecules exist in parallel alignment	(g)
W. Pechhold	Meander model, with defective bundle structure, with meander-like folds	(h, i)
G. S. Y. Yeh	Folded-chain fringed-micellar grain model. Contains two elements: grain (ordered) domain of quasi-parallel chains, and intergrain region of randomly packed chains	(j)
V. P. Privalko and Y. S. Lipatov	Conformation having folded structures with R_g equaling the unperturbed dimension	(k)
R. Hosemann	Paracrystalline model with disorder within the lamellae (see Figure 6.38)	(l, m, n)
S. A. Arzhakov	Folded fibril model, with folded chains perpendicular to fibrillar axis	(o)

References: (a) P. J. Flory, *Principles of Polymer Chemistry*, Cornell University Press, Ithaca, NY, 1953. (b) P. J. Flory, *Faraday Discuss. Chem. Soc.*, **68**, 14 (1979). (c) B. Vollmert, *Polymer Chemistry*, Springer-Verlag, Berlin, 1973, p. 552. (d) B. Vollmert and H. Stuty, in *Colloidal and Morphological Behavior of Block and Graft Copolymers*, G. E. Molau, ed., Plenum, New York, 1970. (e) P. H. Lindenmeyer, *J. Macromol. Sci. Phys.*, **8**, 361 (1973). (f) T. G. F. Schoon and G. Rieber, *Angew. Makromol. Chem.*, **15**, 263 (1971). (g) Y. K. Ovchinnikov, G. S. Markova, and V. A. Kargin, *Polym. Sci. USSR* (Eng. Transl.), **11**, 369 (1969); V. A. Kargin, A. I. Kitajgorodskij, and G. L. Slonimskii, *Kolloid-Zh.*, **19**, 131 (1957). (h) W. Pechhold, M. E. T. Hauber, and E. Liska, *Kolloid Z. Z. Polym.*, **251**, 818 (1973). (i) W. R. Pechhold and H. P. Grossmann, *Faraday Discuss. Chem. Soc.*, **68**, 58 (1979). (j) G. S. Y. Yeh, *J. Macromol. Sci. Phys.*, **6**, 451 (1972). (k) V. P. Privalko and Yu. S. Lipatov, *Makromol. Chem.*, **175**, 641 (1974). (l) R. Hosemann, *J. Polym. Sci.*, **C20**, 1 (1967). (m) R. Hosemann, *Colloid Polym. Sci.*, **260**, 864 (1982). (n) R. Hosemann, *CRC Crit. Rev. Macromol. Sci.*, **1**, 351 (1972). (o) S. A. Arzhakov, N. F. Bakeyev, and V. A. Kabanov, *Vysokomol. Soyed.*, **A15** (5), 1154 (1973).

The development of the random coil model by Mark, and the many further developments by Flory (5–8,42), led to a description of the conformation of chains in the bulk amorphous state. Neutron-scattering studies found the conformation in the bulk to be close to that found in the θ-solvents, strengthening the random coil model. On the other hand, some workers suggested that the chains have various degrees of either local or long-range order (45–48).

Some of the better-developed models are described in Table 5.8. They range from the random coil model of Mark and Flory (37) to the highly organized meander model of Pechhold et al. (48). Several of the models have taken an intermediate position of suggesting some type of tighter than random coiling, or various extents of chain folding in the amorphous state (46–48). A collage of the most different models is illustrated in Figure 5.5.

The most important reasons why some polymer scientists are suggesting nonrandom chain conformations in the bulk state include the high amorphous/crystalline density ratio, and electron and X-ray diffraction studies, which suggest lateral order (see Table

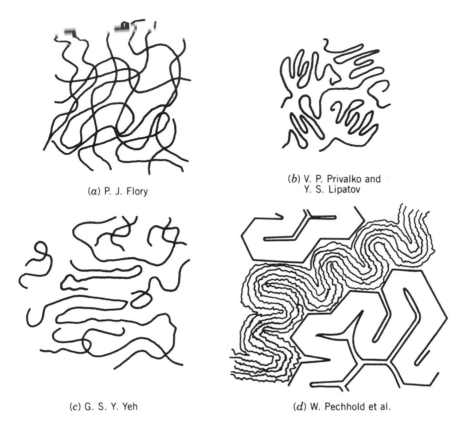

(a) P. J. Flory

(b) V. P. Privalko and
Y. S. Lipatov

(c) G. S. Y. Yeh

(d) W. Pechhold et al.

Figure 5.5 Models of the amorphous state in pictorial form. (a) Flory's random coil model; the (b) Privalko and Lipatov randomly folded chain conformations; (c) Yeh's folded-chain fringed-micellar model; and (d) Pechhold's meander model. Models increase in degree of order from (a) to (d). *References:* (a) P. J. Flory, *Principles of Polymer Chemistry*, Cornell University Press, Ithaca, NY, 1953. (b) V. P. Privalko and Y. S. Lipatov, *Makromol. Chem.*, **175**, 641 (1972). (c) G. S. Y. Yeh, *J. Makoromol. Sci. Phys.*, **6**, 451 (1972). (d) W. Pechhold, M. E. T. Hauber, and E. Liska, *Kolloid Z. Z. Polym.*, **251**, 818 (1973). W. Pechhold, IUPAC Preprints, 789 (1971).

5.6). Experiments that most favor the random coil model include small-angle neutron scattering and a host of short-range interaction experiments that suggest little or no order at the local level. Both the random coil proponents and the order-favoring proponents claim points in the area of rubber elasticity, which is examined further in Chapter 9.

The SANS experiments bear further development. As shown above, the radius of gyration (R_g) of the chains is the same in the bulk amorphous state as it is in θ-solvents. However, virtually the same values of R_g are also obtained in rapidly crystallized polymers (49–53), where significant order is known to exist. This finding at first appeared to support the possibility of short-range order of the type suggested by the appearance of X-ray halos. Two points need to be mentioned. (a) A more sensitive indication of random chains is the Debye scattering form factor for random coils [see equations (3.49) and (5.11)]. Plots of $P(\theta)$ versus $\sin^2(\theta/2)$ follow the experimental data over surprisingly long ranges of θ, including regions where the Guinier approximation, implicit in equation (5.11), no longer holds (54,55). (b) The cases where R_g is the same in the melt as in the crystallized polymer appear to be in crystallization regime III, where chain folding is significantly reduced. (See Section 6.6.2.5.)

A major advantage of the random coil model, interestingly, is its simplicity. By not assuming any particular order, the random coil has become amenable to extensive mathematical development. Thus detailed theories have been developed including rubber elasticity (Chapter 9) and viscosity behavior (Section 3.8), which predict polymer behavior quite well. By difference, little or no analytical development of the other models has taken place, so few properties can be quantitatively predicted. Until such developments have taken place, their absence alone is a strong driving force for the use of the random coil model.

Some of the models may not be quite as far apart as first imagined, however. Privalko and Lipatov have pointed out some of the possible relationships between the random, Gaussian coil, and their own folded-chain model (48) (see Figure 5.5b). As a result of thermal motion, they suggest that both the size and location of regions of short-range order in amorphous polymers depend on the time of observation, assuming that the polymers are above T_g and in rapid motion. The instantaneous conformation of the polymer corresponds to a loosely folded chain. However, when the time of observation is long relative to the time required for molecular motion (see Section 5.4), the various chain conformations will be averaged out in time, yielding radii of gyration more like the unperturbed random coil. For polymers in the glassy state, a similar argument holds, because the very many different chains and their respective conformations replace the argument of a single chain varying its conformation with time.

Clearly, the issue of the conformation of polymer chains in the bulk amorphous state is not yet settled; indeed it remains an area of current research. The vast bulk of research to date strongly suggests that the random coil must be at least close to the truth for many polymers of interest. Points such as the extent of local order await further developments. Thus this book will expound the Mark–Flory theory of the random coil, except where specifically mentioned to the contrary.

5.4 MACROMOLECULAR DYNAMICS

Since the basic notions of chain motion in the bulk state are required to understand much of physical polymer science, a brief introduction is given here. Applications include chain crystallization (to be considered beginning in Chapter 6), the onset of motions in the glass transition region (Chapter 8), and the extension and relaxation of elastomers (Chapters 9 and 10).

Small molecules move primarily by translation. A simple case is of a gas molecule moving in space, following a straight line until hitting another molecular or a wall. In the liquid state, small molecules also move primarily by translation, although the path length is usually only of the order of molecular dimensions.

Polymer motion can take two forms: (a) the chain can change its overall conformation, as in relaxation after strain, or (b) it can move relative to its neighbors. Both motions can be considered in terms of self-diffusion. All such diffusion is a subcase of Brownian motion, being induced by random thermal processes. For center-of-mass diffusion, the center-of-mass distance diffused depends on the square root of time. For high enough temperatures, an Arrhenius temperature dependence is found.

Polymer chains find it almost impossible to move "sideways" by simple translation, for such motion is exceedingly slow for long, entangled chains. This is because the sur-

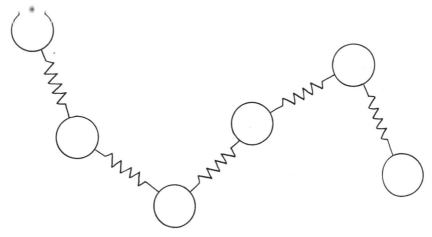

Figure 5.6 Rouse–Bueche bead and spring model of a polymer chain.

rounding chains that block sideways diffusion are also long and entangled, and sideways diffusion can only occur by many cooperative motions. Thus polymer chain diffusion demands separate theoretical treatment.

5.4.1 The Rouse–Bueche Theory

The first molecular theories concerned with polymer chain motion were developed by Rouse (56) and Bueche (57), and modified by Peticolas (58). This theory begins with the notion that a polymer chain may be considered as a succession of equal submolecules, each long enough to obey the Gaussian distribution function; that is, they are random coils in their own right. These submolecules are replaced by a series of beads of mass M connected by springs with the proper Hooke's force constant; see Figure 5.6. The beads also act as universal joints. This model resembles a one-dimensional crystal.

In the development of rubber elasticity theory (Section 9.7.1), it will be shown that the restoring force, f, on a chain or chain portion large enough to be Gaussian, is given by

$$f = \frac{3kT\Delta x}{\overline{r^2}} \tag{5.16}$$

where Δx is the displacement and r is the end-to-end distance of the chain or Gaussian segment. Thus the springs (but not the whole mass) have an equivalent modulus.

If the beads in Figure 5.6 are numbered $1, 2, 3 \ldots, z$, so that there are z springs and $z + 1$ beads, the restoring force on the ith bead may be written

$$f_i = \frac{-3kT}{a^2}(-x_{i-1} + 2x_i - x_{i+1}), \qquad 1 \le i \le z - 1 \tag{5.17}$$

where f_i represents the force on the ith bead in the x direction, x_i represents the amount by which the bead i has been displaced from its equilibrium position, and a is the end-to-end distance of the segment.

The segments move through a viscous medium (other polymer chains and segments) in which they are immersed. This viscous medium exerts a drag force on the system, damping out the motions. It is assumed that the force is proportional to the velocity of the beads, which is equivalent to assuming that the bead behaves exactly as if it were a macroscopic bead in a continuous viscous medium. The viscous force on the ith bead is given by

$$f_i = \rho \left(\frac{dx_i}{dt} \right) \tag{5.18}$$

where ρ is the friction factor.

Zimm (59) advanced the theory by introducing the concepts of Brownian motion and hydrodynamic shielding into the system. One advantage is that the friction factor is replaced by the macroscopic viscosity of the medium. This leads to a matrix algebra solution with relaxation times of

$$\tau_{p,i} = \frac{6\eta_0 M_i^2}{\pi^2 cRTM_w p^2} \tag{5.19}$$

where η_0 is the bulk-melt viscosity, p is a running index, and c is the polymer concentration.

The Rouse–Bueche theory is useful especially below 1% concentration. However, only poor agreement is obtained on studies of the bulk melt. The theory describes the relaxation of deformed polymer chains, leading to advances in creep and stress relaxation. While it does not speak about the center-of-mass diffusional motions of the polymer chains, the theory is important because it serves as a precursor to the de Gennes reptation theory, described next.

5.4.2 Reptation and Chain Motion

5.4.2.1 The de Gennes Reptation Theory   While the Rouse–Bueche theory was highly successful in establishing the idea that chain motion was responsible for creep, relaxation, and viscosity, quantitative agreement with experiment was generally unsatisfactory. More recently, de Gennes (60) introduced his theory of reptation of polymer chains. His model consisted of a single polymeric chain, P, trapped inside a three-dimensional network, G, such as a polymeric gel. The gel itself may be reduced to a set of fixed obstacles—$O_1, O_2, \ldots, O_n \ldots$. His model is illustrated in Figure 5.7 (60). The chain P is not allowed to cross any of the obstacles; however, it may move in a snakelike fashion among them (61).

The snakelike motion is called reptation. The chain is assumed to have certain "defects," each with stored length, b (see Figure 5.8) (60). These defects migrate along the chain in a type of defect current. When the defects move, the chain progresses, as shown in Figure 5.9 (60). The velocity of the nth mer is related to the defect current J_n by

$$\frac{d\vec{r}_n}{dt} = bJ_n \tag{5.20}$$

where $\vec{r}_n$ represents the position vector of the nth mer.

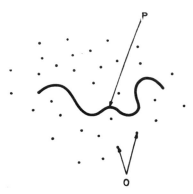

Figure 5.7 A model for reptation. The chain *P* moves among the fixed obstacles, *O*, but cannot cross any of them (60).

The reptation motion yields forward motion when a defect leaves the chain at the extremity. The end of the chain may assume various new orientations. In zoological terms, the head of the snake must decide which direction it will go through the bushes. De Gennes assumes that this choice is at random.

Using scaling concepts, see Appendix 4.1, de Gennes (60) found that the self-diffusion coefficient, *D*, of a chain in the gel depends on the molecular weight *M* as

$$D \propto M^{-2} \tag{5.21}$$

Numerical values of the diffusion coefficient in bulk systems range from 10^{-12} to 10^{-6} cm^2/s. In data reviewed by Tirrell (62), polyethylene of 1×10^4 g/mol at 176°C has a value of *D* near 1×10^{-8} cm^2/s. Polystyrene of 1×10^5 g/mol has a diffusion coefficient of about 1×10^{-12} cm^2/s at 175°C. The inverse second-power molecular weight relationship holds. The temperature dependence can be determined either through acti-

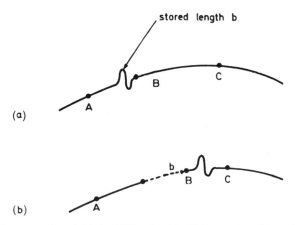

Figure 5.8 Reptation as a motion of defects. (*a*) The stored length *b* moves from A toward C along the chain. (*b*) When the defect crosses mer B, it is displaced by an amount *b* (60).

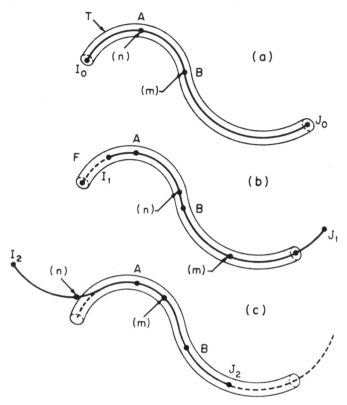

Figure 5.9 The chain is considered within a tube. (*a*) Initial position. (*b*) The chain has moved to the right by reptation. (*c*) The chain has moved to the left, the extremity *I* choosing another path, I_2J_2. A certain fraction of the chain, I_1J_2, remains trapped within the tube at this stage (60).

vation energies ($E_a = 90$ kJ/mol for polystyrene and 23 kJ/mol for polyethylene) or through the WLF equation (Chapter 8). Calculations using the diffusion coefficient are illustrated in Appendix 5.2.

The reptation time, T_r, depends on the molecular weight as

$$T_r \propto M^3 \tag{5.22}$$

Continuing these theoretical developments, Doi and Edwards (63) developed the relationship of the dynamics of reptating chains to mechanical properties. In brief, expressions for the rubbery plateau shear modulus, G_N^0, steady-state viscosity, η_0, and the steady-state recoverable compliance, J_e^0, were found to be related to the molecular weight as follows:

$$G_N^0 \propto M^0 \tag{5.23}$$

$$\eta_0 \propto M^3 \tag{5.24}$$

$$J_e^0 \propto M^0 \tag{5.25}$$

An important reason why the modulus and the compliance are independent of the molecular weight (above about 8 M_c) is that the number of entanglements (contacts between the reptating chain and the gel) are large for each chain and occur at roughly constant intervals. Experimentally, the viscosity is found to depend on the molecular weight to the 3.4 power (Chapter 10), more rapidly than predicted. The de Gennes theory of reptation as a mechanism for diffusion has also seen applications in the dissolution of polymers, termination by combination in free radical polymerizations, and polymer–polymer welding (62). Graessley (64) reviewed the theory of reptation recently.

Dr. Pierre de Gennes was awarded the 1991 Nobel Prize in Physics for his work in polymers and liquid crystals, as detailed in part above. Several sections following also discuss his contributions. The field of polymer science and engineering has, in fact, several Nobel Prize winners, as delineated in Appendix 5.3.

5.4.2.2 *Fickian and Non-Fickian Diffusion* The three-dimensional self-diffusion coefficient, D, of a polymer chain in a melt is given by

$$X = (6Dt)^{1/2} \tag{5.26}$$

where X is the center-of-mass distance traversed in three dimensions, and t represents the time (65). For one-dimensional diffusion in a particular direction, the six is replaced by a two. This is a simple case of Fickian diffusion, noting the time to the one-half dependence. The reptation model of de Gennes supports the $t^{1/2}$ dependence, also supplying the molecular weight dependence and leading to other important analytical features, as discussed above.

The initial diffusion rate as the chain leaves the tube goes as $t^{1/4}$ (Section 11.5) representing a case of non-Fickian diffusion.

5.4.3 Nonlinear Chains

The discussion above models the motion of *linear* chains in a tube. Physical entanglements define a tube of some 50 Å diameter. This permits the easy passage of defects but effectively prevents sideways motion of the chain. According to the reptation theory, the chains wiggle onward randomly from their ends. How do branched, star, and cyclic polymers diffuse?

Two possibilities exist for translational motion in branched polymers. First, one end may move forward, pulling the other end and the branch into the same tube. This process is strongly resisted by the chains as it requires a considerable decrease in entropy to cause a substantial portion of a branch to lie parallel to the main chain in an adjacent tube (66).

Instead, it is energetically cheaper for an entangled branched-chain polymer to renew its conformation by retracting a branch so that it retraces its path along the confining tube to the position of the center mer. Then it may extend outward again, adopting a new conformation at random; see Figure 5.10 (67). The basic requirement is that the branch not loop around another chain in the process, or it must drag it along also.

De Gennes (64) calculated the probability P_1 of an arm of n-mers folding back on itself as

$$P_1 = \exp\left(\frac{-\alpha n}{n_c}\right) \tag{5.27}$$

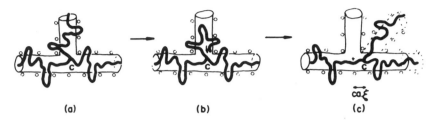

(a)   **(b)**   **(c)**

Figure 5.10   The basic diffusion steps for a branched polymer. Note motion of mer C, which requires a fully retracted branch before it can take a step into a new topological environment (67).

where n_c is the critical number of mers between physical entanglements and α is a constant.

The result is that diffusion in branched-chain polymers is much slower than in linear chains. For rings, diffusion is even more sluggish, because the ring is forced to collapse into a quasilinear conformation in order to have center-of-mass motion. Since many commercial polymers are branched or star-shaped, the self-diffusion of the polymer is correspondingly decreased, and the melt viscosity increased.

5.4.4   Experimental Methods of Determining Diffusion Coefficients

Two general methods exist for determining the translational diffusion coefficient, D, in polymer melts: (a) by measuring the broadening of concentration gradients as a function of time (in such cases, two portions of polymer, which differ in some identifiable mode, are placed in juxtaposition and the two polymer portions are allowed to interdiffuse) and (b) by measuring the translation of molecules directly using local probes such as NMR.

Diffusion broadening has been the more widely reported. Small-angle neutron scattering (69), forward recoil spectrometry (70), radioactive labeling (71), infrared absorption methods (72), secondary mass spectroscopy (SIMS) (73) and dynamic mechanial spectroscopy (74) have been employed, among others. A generalized scheme of sample preparation and analysis is shown in Figure 5.11 (72). Of course, pure labeled polymers may be used as well as the blend. The objective is to measure some characteristic change at the interface.

Today, the self-diffusion coefficients, D, of many polymers are known (75), having been measured by a number of investigators using a variety of techniques. Of course, the diffusion coefficient depends on the inverse square of the molecular weight. Above the glass transition temperature, the temperature dependence can be estimated either through activation energies and the Arrhenius equation (Section 8.5) or through the WLF equation (see Section 8.6). Below the glass transition temperature, Fickian diffusion of polymer chains is substantially absent and vibrational modes of motion dominate.

On normalizing the data to 150,000 g/mol and 135°C, the average values of the diffusion coefficients for a few common polymers are

Polystyrene	1.2×10^{-15} cm^2/s
Poly(methyl methacrylate)	6.9×10^{-17} cm^2/s
Poly(n-butyl methacrylate)	8.0×10^{-11} cm^2/s

The much larger value of D for poly(n-butyl methacrylate) is due to its much lower glass transition. For polystyrenes and poly(methyl methacrylate)s as used in many applica-

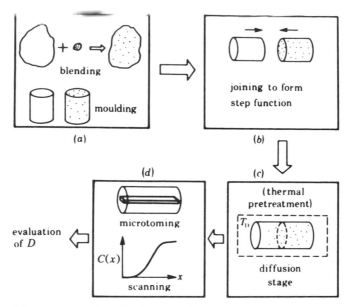

Figure 5.11 Schematic flowchart showing the stages involved in an interdiffusion experiment (67). The slices are scanned in an IR microdensitometer to obtain the broadened concentration profile, from which D is evaluated.

tions, the molecules would diffuse at the rate of a few Ångstroms a minute at 135°C. This is convenient for scientific research. However, commercial molding usually takes place at significantly higher temperatures, as delineated in Appendix 5.2.

Bartels et al. (76) reacted H_2 or D_2 with polybutadiene to saturate the double bond, making the equivalent of polyethylene. Films with up to 25 alternating layers of HPB and DPB for the diffusion studies were prepared with layer thicknesses ranging from 3 to 15 μm, chosen to keep homogenization times in the order of a few hours. SANS measurements were made with the incident beam perpendicular to the film surface. The resulting diffusion coefficient on the linear polymer is compared to values obtained on three-armed stars of the same molecular weight in Table 5.9. The stars diffuse almost

Table 5.9 Diffusion coefficients of hydrogenated polybutadienes[†]

Method	Shape	T, °C	M_w, g/mol	D, cm²/s	Reference
SANS	Linear	125	7.3×10^4	4.8×10^{-11}	(a)
Forward recoil Spectrometry	3-arm	125	7.5×10^4	2.4×10^{-14}	(b)
SANS	3-arm	165	7.5×10^4	1.4×10^{-13}	(c)

References: (a) C. R. Bartels, B. Crist, and W. W. Graessley, *Macromolecules*, **17**, 2702 (1984). (b) B. Crist, P. F. Green, R. A. L. Jones, and E. J. Kramer, *Macromolecules*, **22**, 2857 (1989). (c) C. R. Bartels, B. Crist Jr., L. J. Fetters, and W. W. Graessley, *Macromolecules*, **19**, 785 (1986).

[†] Hydrogenated polybutadiene makes polyethylene. In this case, a narrow polydispersity polymer not available otherwise.

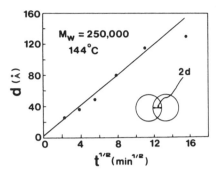

Figure 5.12 Interdiffusion depth, d, of polystyrene latex chains during film formation. The latex particle diameter was 544 Å (78).

three orders of magnitude slower than linear polymer, illustrating the detriment that long side chains present to reptation.

Mixtures of deuterated and protonated polystyrene latexes have also been studied by SANS (77,78). The main advantage is the relatively large surface area presented for interdiffusion, especially if the latex particles are small. Figure 5.12 (78) shows that the interdiffusion distance increases as the square root of time, theoretically predicted.

Of course, there are other methods for measuring interdiffusion in latexbased films. More recently Winnik et al. (79–82) employed a direct nonradiative energy transfer, DET, fluorescence technique to measure diffusion. In this method, latexes are prepared in two different batches. In one batch, the chains contain a "donor" group, while in the other, an "acceptor" group is attached. When the two groups are close to one another, the excitation energy of the donor molecules may be transferred by the resonance dipole–dipole interaction mechanism known as DET, if the emission spectrum of the "donor" overlaps the absorption spectrum of the "acceptor." The ratio of emission intensities to the acceptor intensities changes with interdiffusion, permitting the calculation of self-diffusion coefficients, and hence the interdiffusion depth. For poly(methyl methacrylate)-based latexes, self-diffusion coefficients of approximately 10^{-15} cm^2/s were calculated, similar to the results obtained by Yoo et al. (78). The field has been reviewed by Morawetz (83), who pointed out that changes in the emission spectrum in nonradiative energy transfer between fluorescent labels has been used to characterize polymer miscibility, interpenetration of chain molecules in solution, micelle formation in graft copolymers, and other important effects.

5.5 CONCLUDING REMARKS

The amorphous state is defined as a condensed, noncrystalline state of matter. Many polymers are amorphous under ordinary use conditions, including polystyrene, poly(methyl methacrylate), and poly(vinyl acetate). Crystalline polymers such as polyethylene, polypropylene, and nylon become amorphous above their melting temperatures.

In the amorphous state the position of one chain segment relative to its neighbors is relatively disordered. In the relaxed condition, the polymer chains making up the amorphous state form random coils. The chains are highly entangled with one another, with physical cross-links appearing at about every 600 backbone atoms.

While the amorphous polymer state is "liquid-like" in the classical sense, if the polymer is glassy, a better term would be "amorphous solid," since measurable flow takes years or centuries. Ordinary glass, an inorganic polymer, is such a glassy polymer. The chains are rigidly interlocked in glassy polymers, motion being restricted to vibrational modes.

Above the glass transition, the polymer may flow if it is not in network form. On a submicroscopic scale, the chains interdiffuse with one another with a reptating motion. Common values of the diffusion coefficient vary from 10^{-10} to 10^{-17} cm^2/s, depending inversely on the square of the molecular weight.

REFERENCES

1. E. W. Fischer, G. R. Strobl, M. Dettenmaier, M. Stamm, and N. Steidle, *Faraday Discuss. Chem. Soc.*, **68**, 26 (1979).

2. F. J. Balta-Calleja, K. D. Berling, H. Cackovic, R. Hosemann, and J. Loboda-Cackovic, *J. Macromol. Sci. Phys.*, **B12**, 383 (1976).

3. R. E. Robertson, *J. Phys. Chem.*, **69**, 1575 (1965).

4. W. R. Pechhold and H. P. Grossmann, *Faraday Discuss. Chem. Soc.*, **68**, 58 (1979).

5. P. J. Flory, *J. Macromol. Sci. Phys.*, **B12** (1), 1 (1976).

6. R. S. Stein, *J. Chem. Ed.*, **50**, 748 (1973).

7. P. J. Flory, *Faraday Discuss. Chem. Soc.*, **68**, 15 (1979).

8. P. J. Flory, *Pure Appl. Chem. Macromol. Chem.*, **8**, 1 (1972); reprinted in *Rubber Chem. Tech.*, **48**, 513 (1975).

9. G. S. Y. Yeh, *Crit. Rev. Macromol. Sci.*, **1**, 173 (1972).

10. R. F. Boyer, *J. Macromol. Sci. Phys.*, **B12**, 253 (1976).

11. V. P. Privalko, Yu. S. Lipatov, and A. P. Lobodina, *J. Macromol. Sci. Phys.*, **B11** (4), 441 (1975).

12. Symposium on "Physical Structure of the Amorphous State," *J. Macromol. Sci. Phys.*, **B12** (1976).

13. "Organization of Macromolecules in the Condensed Phase," *Faraday Discuss. Chem. Soc.*, **68** (1979).

14. E. W. Fischer, G. R. Strobl, M. Dettenmaier, M. Stamm, and N. Steidle, *Faraday Discuss. Chem. Soc.*, **68**, 26 (1979).

15. B. Erman, P. J. Flory and J. P. Hummel, *Macromolecules*, **13**, 484 (1980).

16. R. S. Stein and S. D. Hong, *J. Macromol. Sci. Phys.*, **B12** (1), 125 (1976).

17. G. M. Estes, R. W. Seymour, D. S. Huh, and S. L. Cooper, *Polym. Eng. Sci.*, **9**, 383 (1969).

18. R. G. Kirste, W. A. Kruse, and K. Ibel, *Polymer*, **16**, 120 (1975).

19. D. G. H. Ballard, A. N. Burgess, P. Cheshire, E. W. Janke, A. Nevin, and J. Schelten, *Polymer*, **22**, 1353 (1981).

20. J. S. Higgins and R. S. Stein, *J. Appl. Cryst.*, **11**, 346 (1978).

21. A. Maconnachie and R. W. Richards, *Polymer*, **19**, 739 (1978).

22. L. H. Sperling, *Polym. Eng. Sci.*, **24**, 1 (1984).

23. H. Benoit, *J. Macromol. Sci. Phys.*, **B12** (1), 27 (1976).

24. G. D. Wignall, D. G. H. Ballard, and J. Schelten, *J. Macromol. Sci. Phys.*, **B12** (1), 75 (1976).

25. P. Debye, *J. Phys. Coll. Chem.*, **51**, 18 (1947). See also B. H. Zimm, R. S. Stein, and P. Duty, *Polym. Bull.*, **1**, 90 (1945).

26. *National Center for Small-Angle Neutron Scattering Research User's Guide*, Oak Ridge National Laboratory, 1980. Solid State Divisions, Oak Ridge National Laboratory, Oak Ridge, TN 37830.

27. W. C. Koehler, R. W. Hendricks, H. R. Child, S. P. King, J. S. Lin, and G. D. Wignall, in *Proceedings of NATO Advanced Study Institute on Scattering Techniques Applied to Supramolecular and Nonequilibrium Systems*, Vol. 73, S. H. Chen, B. Chu, and R. Nossal, eds., Plenum, New York, 1981.

28. G. D. Wignall, D. G. H. Ballard, and J. Schelten, *Eur. Polym. J.*, **10**, 861 (1974); reprinted in *J. Macromol. Sci. Phys.*, **B12** (1), 75 (1976).

29. B. H. Zimm, *J. Chem. Phys.* **16**, 1098 (1948).

30. P. J. Flory, *Principles of Polymer Chemistry*, Cornell University Press, Ithaca, NY, 1953.

31. K. C. Honnell, J. D. McCoy, J. G. Curro, K. C. Schweizer, A. H. Narten, and A. Habenshuss, *J. Chem. Phys.*, **94**, 4659 (1991).

32. E. W. Fischer, J. H. Wendorff, M. Dettenmaier, G. Lieser, and I. Voigt-Martin, *J. Macromol. Sci. Phys.*, **B12** (1), 41 (1976).

33. R. Lovell, G. R. Mitchell, and A. H. Windle, *Faraday Discuss. Chem. Soc.*, **68**, 46 (1979).

34. Yu. K. Ovchinnikov, Ye. M. Antipov, and G. S. Markova, *Polymer Sci. USSR*, **17**, 2081 (1975).

35. R. L. Miller, R. F. Boyer, and J. Heijboer, *J. Polym. Sci. Polym. Phys. Ed.*, **22**, 2021 (1984).

36. R. L. Miller and R. F. Boyer, *J. Polym. Sci. Polym. Phys. Ed.*, **22**, 2043 (1984).

37. H. R. Schubach, E. Nagy, and B. Heise, *Coll. Polym. Sci.* (*Koll. Z.z. Polym.*), **259** (8), 789 (1981).

38. S. E. B. Petrie, *J. Macromol. Sci. Phys.*, **B12**, 225 (1976).

39. T. A. Weber and E. Helfand, *J. Chem. Phys.*, **71**, 4760 (1979).

40. P. J. Flory, *Statistical Mechanics of Chain Molecules*, Interscience, New York, 1969.

41. H. Staudinger, *Die Hochmolekularen Organischen Verbindung*, Springer, Berlin, 1932.

42. P. J. Flory, *Principles of Polymer Chemistry*, Cornell University Press, Ithaca, NY, 1953.

43. H. Fujita, *Polymer Solutions*, Elsevier, Amsterdam, 1990.

44. V. A. Bershtein, V. M. Egorov, L. M. Egorova, and V. A. Ryzhov, *Thermochim. Acta*, **238** (1994).

45. W. Pechhold, M. E. T. Hauber, and E. Liska, *Kolloid Z.z. Polym.*, **251**, 818 (1973).

46. B. Vollmert, *Polymer Chemistry*, Springer, Berlin, 1973, p. 552.

47. P. H. Lindenmeyer, *J. Macromol. Sci. Phys.*, **8**, 361 (1973).

48. V. P. Privalko and Yu. S. Lipatov, *Makromol. Chem.*, **175**, 641 (1974).

49. J. Schelten, D. G. H. Ballard, G. Wignall, G. Longman, and W. Schmatz, *Polymer*, **17**, 751 (1976).

50. J. Schelten, G. D. Wignall, D. G. H. Ballard, and G. W. Longman, *Polymer*, **18**, 1111 (1977).

51. D. G. H. Ballard, P. Cheshire, G. W. Longman, and J. Schelten, *Polymer*, **19**, 379 (1978).

52. E. W. Fischer, M. Stamm, M. Dettenmaier, and P. Herschenraeder, *Polym. Prepr. Am. Chem. Soc. Div. Polym. Chem.*, **20** (1), 219 (1979).

53. J. M. Guenet, *Polymer*, **22**, 313 (1981).

54. F. S. Bates, C. V. Berney, R. E. Cohen, and G. D. Wignall, *Polymer*, **24**, 519 (1983).

55. A. M. Fernandez, J. M. Widmaier, G. D. Wignall, and L. H. Sperling, *Polymer*, **25**, 1718 (1984).

56. P. E. Rouse, *J. Chem. Phys.*, **21**, 1272 (1953).

57. F. Bueche, *J. Chem. Phys.*, **22**, 1570 (1954).

58. W. L. Peticolas, *Rubber Chem. Tech.* **36**, 1422 (1963).

59. B. H. Zimm, *J. Chem. Phys.*, **24**, 269 (1956).

60. P. G. de Gennes, *J. Chem. Phys.*, **55**, 572 (1971).

61. P. G. de Gennes, *Phys. Today*, **36** (6), 33 (1983).

62. M. Tirrell, *Rubber Chem. Tech.*, **57**, 523 (1984).

63. M. Doi and S. F. Edwards, *J. Chem. Soc. Faraday Trans. 2*, **74**, 1789, 1802, 1818 (1978); **75**, 38 (1979).

64. W. W. Graessley, *Adv. Polym. Sci.*, **47**, 67 (1982).

65. K. Binder and H. Sillescu, *Encyclopedia of Polymer Science and Engineering, Supplementary Volume*, J. I. Kroschwitz, ed., Wiley, New York, 1989.

66. J. Klein, in *Encyclopedia of Polymer Science and Engineering*, 2nd ed., Vol. 9, J. I. Kroschwitz, ed., Wiley, New York, 1987.

67. J. Klein, *Macromolecules*, **19**, 105 (1986).

68. P. G. de Gennes, *J. Phys. (Les Ulis, Fr.)*, **36**, 1199 (1975).

69. C. R. Bartels, B. Crist, Jr., L. J. Fetters, and W. W. Graessley, *Macromolecules*, **19**, 785 (1986).

70. H. Yokoyama, E. J. Kramer, D. A. Hajduk, and F. S. Bates, *Macromolecules*, **32**, 3353 (1999).

71. F. Bueche, W. M. Cashin, and P. Debye, *J. Chem. Phys.*, **20**, 1956 (1952).

72. J. Klein and B. J. Briscoe, *Proc. R. Soc. Lond. A*, **365**, 53 (1979).

73. S. J. Whitlow and R. P. Wool, *Macromolecules*, **24**, 5926 (1991).

74. H. Qiu and M. Bousmina, *J. Rheology*, **43**, 551 (1999).

75. L. H. Sperling, A. Klein, M. Sambasivam, and K. D. Kim, *Polym. Adv. Technol.*, **5**, 453 (1994).

76. C. R. Bartels, B. Crist, and W. W. Graessley, *Macromolecules*, **20**, 2702 (1984).

77. J. H. Jou and J. E. Anderson, *Macromolecules*, **20**, 1544 (1987).

78. J. N. Yoo, L. H. Sperling, C. J. Glinka, and A. Klein, *Macromolecules*, **23**, 3962 (1990).

79. O. Pekan, M. A. Winnik, and M. D. Croucher, *Macromolecules*, **23**, 2673 (1990).

80. C. L. Zhao, W. C. Wang, Z. Hruska, and M. A. Winnik, *Macromolecules*, **23**, 4082 (1990).

81. Y. C. Wang and M. A. Winnik, *Macromolecules*, **23**, 4731 (1990).

82. Y. Wang and M. A. Winnik, *J. Phys. Chem.*, **97**, 2507 (1993).

83. H. Morawetz, *Science*, **240**, 172 (1988).

GENERAL READING

M. Doi and S. F. Edwards, *The Theory of Polymer Dynamics*, Oxford University Press, New York, 1986.

P. G. de Gennes, *Scaling Concepts in Polymer Physics*, Cornell University Press, Ithaca, NY, 1979.

J. S. Higgins and H. C. Benoit, *Polymers and Neutron Scattering*, Oxford University Press, Oxford, 1994.

S. E. Keinath, R. E. Miller, and J. K. Reike, eds., *Order in the Amorphous State of Polymers*, Plenum, New York, 1987.

V. P. Privalko, *Molecular Structure and Properties of Polymers*, Khimia, Leningrad, 1986.

R.-J. Roe, *Methods of X-Ray and Neutron Scattering in Polymer Science*, Oxford University Press, New York, 2000.

STUDY PROBLEMS

1. Why is the radius of gyration of a polymer in the bulk state essentially the same as measured in a θ-solvent but not the same as in other solvents?

2. Estimate the radius of gyration and end-to-end of a polystyrene sample having $M_w = 1 \times 10^5$ g/mol, in the bulk state.

3. In an actual kitchen experiment, one quart of cooked spaghetti was measured out level with cold water so that the spaghetti strands just break the water surface. Nine ounces of water were drained.

   (a) Assuming the spaghetti strands were polymer chains, what is the ratio of the specific volumes of the perfectly packed state to the actual disordered state? Assume a hexagonal close pack array. [*Hint:* Allow for water between perfectly aligned spaghetti strands.]

   (b) Calculate the average angle between strands, $\theta/2$, given by the ratio of the specific volumes (specific volume is the reciprocal of the density),[†]

$$\frac{v_c}{v_a} = \left(\frac{3}{2}\right)^3 \left\{ \left[\frac{1 - \cos^3(\theta/2)}{\sin^3(\theta/2)} + 1 \right]^2 \left(1 - \cos^3\left(\frac{\theta}{2}\right) \right) \right\}^{-1}$$

   (c) Interpret $\theta/2$ in terms of intermolecular orientation and the randomness of the "amorphous state."

4. Compare the Rouse–Bueche theory with the de Gennes theory. How do they model molecular motion?

5. What is the Kuhn segment length of 1×10^5 g/mol polystyrene? What does the result suggest about the chain conformation?

6. With the advent of small-angle neutron scattering, molecular dimensions can now be determined in the bulk state. A polymer scientist determined the following data on a new deuterated polymer dissolved in a sample of (protonated) polymer:

$\left[\dfrac{d\Sigma}{d\Omega}\right]^{-1}$ (cm)	0.50	0.72	1.20
$K^2 \times 10^4$ (Å^{-2})	1.00	3.70	10.1

   The constant C_N for this system was determined to be 10.0×10^{-5} mol/g·cm. What is the weight-average molecular weight and the z-average radius of gyration of the deuterated polymer? What third quantity is implicit in this experiment, and what is its probable numerical value?

7. What is the activation energy for the three-armed star's diffusion coefficient in Table 5.9, assuming an Arrhenius relationship? How do you interpret this result?

8. Calculate the first interchain radial spacing for polytetrafluoroethylene from the data given in Figure 5.3a. How are these data best interpreted? [*Hint:* Use Bragg's law.]

[†] R. E. Robertson, *J. Phys. Chem.*, **69**, 1575 (1965).

APPENDIX 5.1   HISTORY OF THE RANDOM COIL MODEL FOR POLYMER CHAINS[†]

Introduction

Advances in science and engineering have never been completely uniform nor followed an orderly pattern. The truth is that science advances by fits and jerks, with ideas propounded by individuals who see the world in a different light. Frequently they face adversity when putting their ideas forward.

Before polymer science came to be, people had the concept of colloids. There were both inorganic and organic colloids, but they shared certain facts. They both were large compared to ordinary molecules, and both were of irregular sizes and shapes. While this concept "explained" certain simple experimental results, it left much to be desired in the way of understanding the properties of rubber and plastics, which were then considered to be colloids.

In 1920 Herman Staudinger formulated the macromolecular hypothesis: there was a special class of organic colloids of high viscosity that were composed of long chains (A1,A2). This revolutionary idea was argued throughout important areas of chemistry (A3). One of the most important experiments was provided by Herman Mark, who showed that crystalline polymers that had cells of ordinary sizes had only a few mers in each cell, but that the mers were connected to those in the next cell. Eventually the idea of long-chained molecules formed one of the most important cornerstones in the development of modern polymer science.

Early Ideas of Polymer Chain Shape

If one accepts the idea of long-chain macromolecules, the next obvious question relates to their conformation or shape in space. This was especially important since it was early thought that the physical and mechanical properties of the material were determined by the spatial arrangement of the long chains. Staudinger himself thought that most amorphous high polymers such as polystyrene were rod-shaped, and when in solution, the rods lay parallel to each other (A2).

Rubbery materials were different, however. According to early scientists (A4), elastomers were coils or spirals resembling bedsprings (see Figure A5.1.1). Staudinger himself described the idea as follows (A2):

> In order to clarify the elasticity of rubber, several investigators have stated that long molecules form spirals, and to be sure the spiral form of the molecules is promoted through the double bonds. By this arrangement, the secondary valences of the double bonds can be satisfied. The elasticity of rubber depends upon the extensibility of such spirals.

The Random Coil

According to H. Mark (A5), the story of the development of the random coil began with the X-ray work of Katz on natural rubber in 1925 (A6–A9). Katz studied the X-ray patterns of rubber both in the relaxed state and the extended or stretched state. In the

[†] L. H. Sperling, in *Pioneers in Polymer Science*, R. B. Seymour, ed., Kluwer Academic Publishers, Dordrecht, Germany, 1989.

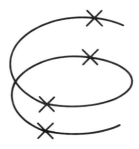

Figure A5.1.1 The spiral structure of natural rubber proposed to explain long-range elasticity. X indicates double bond locations.

stretched state, Katz found a characteristic fiber diagram, with many strong and clear diffraction spots, indicating a crystalline material. This contrasted with the diffuse halo found in the relaxed state, indicating that the chains were amorphous under that condition. The fiber periodicity of the elementary cell was found to be about 9 Å, which could only accommodate a few isoprene units. Since the question of how a long chain could fit into a small elementary cell is fundamental to the macromolecular hypothesis, Hauser and Mark repeated the "Katz effect" experiment and, on the basis of improved diagrams and X-ray techniques, established the exact size of the elementary cell (A10). Of course, the answer to the question of the cell size is that the cell actually accommodates the mer, or repeat unit, rather than the whole chain.

The "Katz effect" was particularly important because it established the first relationship between mechanical deformation and concomitant molecular events in polymers (A5). This led Mark and Valko (A11) to carry out stress–strain studies over a wide temperature range together with X-ray studies in order to analyze the phenomenon of rubber reinforcement. This paper contains the first clear statement that the contraction of rubber is not caused by an increase in energy but by the decrease in entropy on elongation.

This finding can be explained by assuming that the rubber chains are in the form of flexible coils (see Figure A5.1.2) (A12). These flexible coils have a high conformational entropy, but they lose their conformational entropy on being straightened out. The fully extended chain, which is rod-shaped, can have only one conformation, and its entropy is

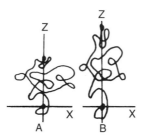

Figure A5.1.2 Early drawing of the random coil. (A) Relaxed state. (B) Effect of deformation in the Z direction. After W. Kuhn (A12).

zero. This concept was extended to all elastic polymers by Meyer et al. (A13) in 1932. These ideas are developed quantitatively in Chapter 9. Although thermal motion and free chain rotation are required for rubber elasticity, the idea of the random coil was later adopted for glassy polymers such as polystyrene as well.

The main quantitative developments began in 1934 with the work of Guth and Mark (A14) and Kuhn (A15). Guth and Mark chose to study the entropic origin of the rubber elastic forces, whereas Kuhn was more interested in explaining the high viscosity of polymeric solutions. Using the concept of free rotation of the carbon–carbon bond, Guth and Mark developed the idea of the "random walk" or "random flight" of the polymer chain. This led to the familiar Gaussian statistics of today, and eventually to the famous relationship between the end-to-end distance of the chain and the square root of the molecular weight. Three stages in the development of the random coil model have been described by H. Mark (see Table A5.1.1) (A16). Like many great ideas, it apparently occurred to several people nearly simultaneously. It is also clear from Table A5.1.1 that Mark played a central role in the development of the random coil.

The random coil model has remained essentially the same until today (A17,A18), although many mathematical treatments have refined its exact definition. Its main values are twofold: by all experiments, it appears to be the best model for amorphous polymers, and it is the only model that has been extensively treated mathematically. It is interesting to note that by its very randomness, the random coil model is easier to understand quantitatively and analytically than models introducing modest amounts of order (A19).

Table A5.1.1   Early references on randomly linked flexible macromolecules (16)

Phase 1	Hypothetic and accidental remarks:
	E. Wachlisch, *Z. S. Biol.*, **85**, 206 (1927)
	H. Mark and E. Valko, *Kautschuk*, **6**, 210 (1930)
	W. Haller, *Koll. Z. S.*, **56**, 257 (1931)
Phase 2	Elaborate *qualitative* interpretation:
	K. H. Meyer, G. von Susich, and E. Valko, *Koll. Z. S.*, **59**, 208 (1932)
Phase 3	*Quantitative* mathematical treatment:
	E. Guth and H. Mark, *Monatschefte Chemie*, **65**, 93 (1934)
	H. Mark, IX Congress for Chemistry, Madrid 1934, Vol. **4**, p. 197 (1934)
	W. Kuhn, *Koll. Z. S.*, **68**, 2 (1934)
	H. Mark, *Der Feste Koerper*, 1937, p. 65

Source: H. Mark, Private communication, May 2, 1983.

The random coil model has been supported by many experiments over the years. The most important of these has been light-scattering from dilute solutions and, more recently, small-angle neutron scattering from the bulk state (see earlier sections of Chapter 5). Both of these experiments support the famous relationships between the square root of the molecular weight and the end-to-end distance. The random coil model has been used to explain not only rubber elasticity and dilute solution viscosities but a host of other physical and mechanical phenomena, such as melt rheology, diffusion, and the equilibrium swelling of cross-linked polymers. Some important reviews include the works of Flory (A17), Treloar (A18), Staverman (A21), and Guth and Mark (A22).

REFERENCES

A1. H. Staudinger, *Ber. Dtsch. Chem. Ges.*, **53**, 1074 (1920).

A2. H. Staudinger, *Die Hochmolekularen Organischen Verbindung*, Springer, Berlin, 1932, at pp. 116–123. Reprinted, 1960.

A3. G. A. Stahl, ed., "*Polymer Science: Overview A Tribute to Herman F. Mark*," ACS Symposium No. 175, American Chemical Society, Washington, DC, 1981.

A4. F. Kirchhof, *Kautschuk*, **6**, 31 (1930).

A5. H. Mark, unpublished, 1982.

A6. J. R. Katz, *Naturwissenschaften*, **13**, 410 (1925).

A7. J. R. Katz, *Chem. Ztg.*, **19**, 353 (1925).

A8. J. R. Katz, *Kolloid Z.*, **36**, 300 (1925).

A9. J. R. Katz, *Kolloid Z.*, **37**, 19 (1925).

A10. E. A. Hauser and H. Mark, *Koll. Chem. Beih.*, **22**, 63; **23**, 64 (1929).

A11. H. Mark and E. Valko, *Kautschuk*, **6**, 210 (1930).

A12. W. Kuhn, *Angew. Chem.*, **49**, 858 (1936).

A13. K. H. Meyer, G. V. Susich, and E. Valko, *Koll. Z.*, **41**, 208 (1932).

A14. E. Guth and H. Mark, *Monatsh. Chem.*, **65**, 93 (1934).

A15. W. Kuhn, *Kolloid Z.*, **68**, 2 (1934).

A16. H. Mark, private communication, May 2, 1983.

A17. P. J. Flory, *Principles of Polymer Chemistry*, Cornell University Press, Ithaca, NY, 1953.

A18. L. R. G. Treloar, *The Physics of Rubber Elasticity*, 3rd ed., Clarendon Press, Oxford, 1975.

A19. R. F. Boyer, *J. Macromol. Sci. Phys.*, **B12**, 253 (1976).

A20. P. J. Flory, *Statistical Mechanics of Chain Molecules*, Wiley, New York, 1969.

A21. A. J. Staverman, *J. Polym. Sci.*, Symposium No. 51, 45 (1975).

A22. E. Guth and H. F. Mark, *J. Polym. Sci. B Polym. Phys.*, **29**, 627 (1991).

APPENDIX 5.2   CALCULATIONS USING THE DIFFUSION COEFFICIENT

The diffusion of polystyrene of various molecular weights in high-molecular-weight polystyrene $(M_w = 2 \times 10^7$ g/mol) was measured by Mills et al. (B1) using forward recoil spectrometry. At 170°C, the diffusion coefficient was found to depend on the weight-average molecular weight as

$$D = 8 \times 10^{-3} \, M_w^{-2} \, \text{cm}^2/\text{s} \qquad (\text{A5.2.1})$$

If the concentration gradient of the diffusing species is given in concentration per unit distance, the unit

$$\frac{\text{mol/cm}^3}{\text{cm}} \times \frac{\text{cm}^2}{\text{s}}$$

yields the flux in mol/(cm$^2 \cdot$ s)—that is, the number of moles of polystyrene crossing a square centimeter of area per second.

Consider a polymer having a weight-average molecular weight of 1×10^6 g/mol. With a density of 1.05 g/cm^3, a bulk concentration of about 1×10^{-6} mol/cm^3 is obtained. Consider a diffusion over a 100-Å distance, or 1×10^{-6} cm from the bulk concentration to a zero concentration. In 1 sec, $8 \times 10^{-15} \cong 1 \times 10^{-14}$ mol will diffuse through a 1-cm^2 area. This is a significant part of the polymer that is lying on the surface of the hypothetical 1 cm^3 under consideration.

Bulk polymeric materials being pressed together under molding conditions require diffusion of the order of 100 Å to produce a significant number of entanglements, thereby fusing the interfacial boundary. In the commercial molding of polystyrene at 170°C, times of the order of 2 minutes might be employed. The largest portion of this time is actually required for heat transfer to be complete and for a uniform temperature to be achieved. Since the weight-average molecular weight of many polystyrenes is about 1×10^5 g/mol, the original boundary will be obliterated in a few seconds under these conditions. Thus the values obtained via polymer physics research confirm the values used in practice.

REFERENCE

B1. P. J. Mills, P. F. Green, C. J. Palmstrom, J. W. Mayer, and E. J. Kramer, *Appl. Phys. Lett.*, **45** (9), 957 (1984).

APPENDIX 5.3 NOBEL PRIZE WINNERS IN POLYMER SCIENCE AND ENGINEERING

From the time of Hermann Staudinger's enunciation of the macromolecular hypothesis in 1920, polymer science and engineering has had many fundamental advances leading

Table C5.3.1   Nobel Prize winners for advances in polymer science and engineering

Scientist	Year	Field	Research and Discovery
Hermann Staudinger	1953	Chemistry	Macromolecular Hypothesis
Karl Ziegler and Giulio Natta	1963	Chemistry	Ziegler–Natta catalysts and resulting stereospecific polymers like isotactic polypropylene
Paul J. Flory	1974	Chemistry	Random coil and organization of polymer chains
Pierre G. de Gennes	1991	Physics	Reptation in polymers and polymer structures at interfaces
A. J. Heeger, A. G. MacDiarmid and H. Shirakawa	2000	Chemistry	Discovery and development of conductive polymers

to the understanding of plastics, rubber, adhesives, coatings, and fibers of today. The discoveries that these Nobel Prize winners made are summarized in Table C5.3.1 (C1,C2). These people have revolutionized life in the modern world.

REFERENCES

C1. P. Canning, ed., *Who's Who in Science and Engineering*, 4th ed., Marquis Who's Who, New Providence, NJ, 1998–9.

C2. F. N. Magill, ed., *The Nobel Prize Winners in Chemistry*, Salem Press, Pasadena, CA, 1990.

6

THE CRYSTALLINE STATE

6.1 GENERAL CONSIDERATIONS

In the previous chapter the structure of amorphous polymers was examined. In this chapter the study of crystalline polymers is undertaken. The crystalline state is defined as one that diffracts X rays and exhibits the *first-order* transition known as melting.

A first-order transition normally has a discontinuity in the volume–temperature dependence, as well as a heat of transition, ΔH_f, also called the enthalpy of fusion or melting. The most important *second-order* transition is the glass transition, Chapter 8, in which the volume–temperature dependence undergoes a change in slope, and only the derivative expansion coefficient, dV/dT, undergoes a discontinuity. There is no heat of transition at T_g, but rather a change in the heat capacity, ΔC_p.

Polymers crystallized in the bulk, however, are never totally crystalline, a consequence of their long-chain nature and subsequent entanglements. The melting (fusion) temperature of the polymer, T_f, is always higher than the glass transition temperature, T_g. Thus the polymer may be either hard and rigid or flexible. An example of the latter is ordinary polyethylene, which has a T_g of about $-80°C$ and a melting temperature of about $+139°C$. At room temperature it forms a leathery product as a result.

The development of crystallinity in polymers depends on the regularity of structure in the polymer (see Chapter 2). Thus isotactic and syndiotactic polymers usually crystallize, whereas atactic polymers, with a few exceptions (where the side groups are small or highly polar), do not. Regular structures also appear in the polyamides (nylons), polyesters, and so on, and these polymers make excellent fibers.

Nonregularity of structure first decreases the melting temperature and finally prevents crystallinity. Mers of incorrect tacticity (see Chapter 2) tend to destroy crystallinity, as does copolymerization. Thus statistical copolymers are generally amorphous. Blends of isotactic and atactic polymers show reduced crystallinity, with only the isotactic portion crystallizing. Under some circumstances block copolymers containing a crystallizable block will crystallize; again, only the crystallizable block crystallizes.

Table 6.1   Properties of selected crystalline polymers (1)

Polymer	T_f, °C	$\Delta H_f, \dfrac{\text{kJ}}{\text{mol}}$
Polyethylene	139	7.87[a]
Poly(ethylene oxide)	66	8.29
it-Polystyrene	240	8.37
Poly(vinyl chloride)	212	3.28
Poly(ethylene terephthalate)[b]	265	24.1
Poly(hexamethylene adipamide)[c]	265	46.5
Cellulose tributyrate	207	12.6
cis-Polyisoprene[d]	28	4.40
Polytetrafluoroethylene[e]	330	5.74
it-polypropylene	171	8.79
Poly(oxymethylene)[f]	182	10.6

[a] Per $-CH_2-CH_2-$. Note that values for $-CH_2-$ alone are sometimes reported.
[b] Dacron.
[c] Nylon 66 or Polyamide 66.
[d] Natural rubber.
[e] Teflon.
[f] Delrin.

Factors that control the melting temperature include polarity and hydrogen bonding as well as packing capability. Table 6.1 lists some important crystalline polymers and their melting temperatures (1).

6.1.1   Historical Aspects

Historically the study of crystallinity of polymers was important in the proof of the Macromolecular Hypothesis, developed originally by Staudinger. In the early 1900s when X-ray studies were first applied to crystal structures, scientists found that the cell size of crystalline polymers was of normal size (about several Ångstoms on a side). This was long before they developed an understanding of the chain nature of polymers required to completely characterize the cell contents. In the case of ordinary sized organic molecules, each cell was found to contain only a few molecules. If polymers were composed of long chains, they asked Staudinger, how could they fit into the small unit cells? The density of such a material would have to be 50 times lead! The answer, developed by Mark and co-workers (2–7), and others, was that the unit cell contains only a few mers that are repeated in adjacent unit cells. The molecule continues in the adjacent axial directions. Mark showed that the bond distances both within a cell and between cells were consistent with covalent bond distances (1.54 Å for carbon–carbon bonds) and inconsistent with the formation of discrete small molecules. For these and many other advances (see Section 3.8), and a life-long leadership in polymers (he died at age 96 in 1992), Herman Mark was called the *Father of Polymer Science*.

One of the first structures to be determined was the natural polysaccharide cellulose. In this case the repeat unit is cellobiose, composed of two glucoside rings. In the 1980s, ^{13}C NMR experiments established that native cellulose is actually a composite of a triclinic parallel-packed unit cell called cellulose I_α, and a monoclinic parallel-packed unit uun uunied cellulose I_β. Experimentally, the structures are only difficultly distinguishable *via* X-ray analysis (7a,7b). Figure 6.1 (3) illustrates the general form of the cellulose unit cell.

When vinyl, acrylic, and polyolefin polymers were first synthesized, the only microstructure then known was atactic. Most of these materials were amorphous. Scientific advancement waited until Ziegler's work on novel catalysts (8), together with Natta's work on X-ray characterization of the stereospecific polymers subsequently synthesized (9), when isotactic and syndiotactic crystalline polymers became known. For this great pioneering work, Ziegler and Natta were jointly awarded the 1963 Nobel Prize in Chemistry (see Appendix 5.3). The general class of these catalysts are known today as Ziegler-Natta catalysts.

Of course, crystalline polymers constitute many of the plastics and fibers of commerce. Polyethylene is used in films to cover dry-cleaned clothes, and as water and solvent containers (e.g., wash bottles). Polypropylene makes a highly extensible rope, finding particularly important applications in the marine industry. Polyamides (nylons) and polyesters are used as both plastics and fibers. Their use in clothing is world famous. Cellulose, mentioned above, is used in clothing in both its native state (cotton) and its regenerated state (rayon). The film is called cellophane.

6.1.2 Melting Phenomena

The melting of polymers may be observed by any of several experiments. For linear or branched polymers, the sample becomes liquid and flows. However, there are several possible complications to this experiment, which may make interpretation difficult. First of all, simple liquid behavior may not be immediately apparent because of the polymer's high viscosity. If the polymer is cross-linked, it may not flow at all. It must also be noted that amorphous polymers soften at their glass transition temperature, T_g, which is emphatically not a melting temperature but may resemble one, especially to the novice (see Chapter 8). If the sample does not contain colorants, it is usually hazy in the crystalline state because of the difference in refractive index between the amorphous and crystalline portions. On melting, the sample becomes clear, or more transparent.

The disappearance of crystallinity may also be observed in a microscope, for example, between crossed Nicols. The sharp X-ray pattern characteristic of crystalline materials gives way to amorphous halos at the melting temperature, providing one of the best experiments.

Another important way of observing the melting point is to observe the changes in specific volume with temperature. Since melting constitutes a first-order phase change, a discontinuity in the volume is expected. Ideally, the melting temperature should give a discontinuity in the volume, with a concomitant sharp melting point. In fact, because of the very small size of the crystallites in bulk crystallized polymers (or alternatively, their imperfections), most polymers melt over a range of several degrees (see Figure 6.2) (12). The melting temperature is usually taken as the temperature at which the last trace of crystallinity disappears. This is the temperature at which the largest and/or most perfect crystals are melting. The volumetric coefficients of expansion in Figure 6.2 can be calculated from $\alpha = (1/V)(dV/dT)_p$.

Alternatively, the melting temperature can be determined thermally. Today, the differential scanning calorimeter (DSC) is popular, since it gives the heat of fusion as well as the melting temperature. Such an experiment is illustrated in Figure 6.3 for *it*-polypropylene (13). The heat of fusion, ΔH_f, is given by the area under the peak.

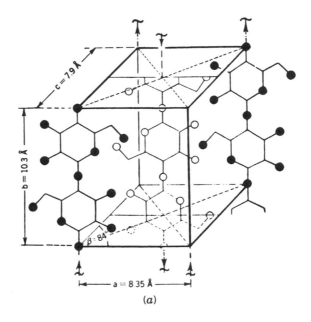

(a)

Parameter	Cellulose I[a]	Cellulose II[a]	Cellulose III[b]	Cellulose IV[a] Cellulose x
a-axis (Å)	8.35	8.02	7.74	8.12
b-axis (Å)	10.30	10.30	10.30	10.30
c-axis (Å)	7.90	9.03	9.96	7.99
β (degrees)	83.3	62.8	58	90
Density (g/cm^3)	1.625	1.62	1.61	1.61

[a]Ø. Ellefsen, J. G. Ønnes, and N. Norman, *Acta Chem. Scand.*, **13**, 853 (1959).
[b]C. Legrand, *J. Polym. Sci.*, **7**, 333 (1951).

(b)

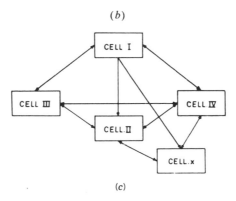

(c)

Figure 6.1 The crystalline structure of cellulose. (*a*) The unit cell of native cellulose, or cellulose I, as determined by x-ray analysis (2,10,11). Cellobiose units are not shown on all diagonals for clarity. The volume of the cell is given by $V = abc \sin \beta$. (*b*) Unit cell dimensions of the four forms of cellulose. (*c*) Known pathways to change the crystalline structure of cellulose. The lesser known form of cellulose x is also included (10).

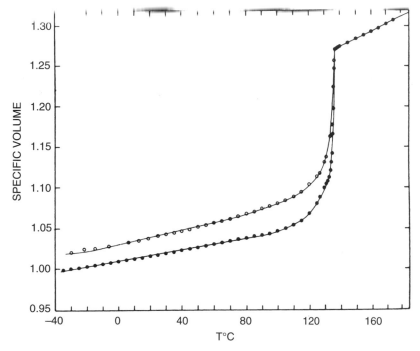

Figure 6.2 Specific volume–temperature relations for linear polyethylene. Open circles, specimen cooled relatively rapidly from the melt to room temperature before fusion experiments; solid circles, specimen crystallized at 130°C for 40 days, then cooled to room temperature prior to fusion (12).

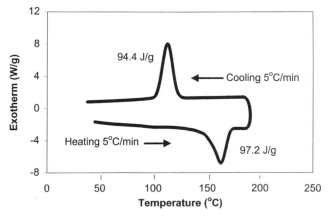

Figure 6.3 Differential scanning calorimetry of a commercial isotactic polypropylene sample, generously provided by Dr. S. J. Han of the Exxon Research and Engineering Company. Note the supercooling effect on crystallization, but the equal and opposite heats of melting and crystallization (13). Experiment by S. D. Kim.

Further general studies of polymer fusion are presented in Sections 6.8 and 6.9, after the introduction of crystallographic concepts and the kinetics and thermodynamics of crystallization.

6.1.3 Example Calculation of Percent Crystallinity

Exactly how crystalline is the *it*-polypropylene in Figure 6.3? The heat of fusion, ΔH_f, of the whole sample, amorphous plus crystalline parts, is 97.2 J/g, determined by the area under the melting curve. Table 6.1 gives the heat of fusion for *it*-polypropylene as 8.79 kJ/mol. This latter value is for the crystalline component only. Noting that *it*-polypropylene has a mer molecular weight of 42 g/mol,

$$\frac{8790 \text{ J/mol}}{42 \text{ g/mol}} = 209 \text{ J/g}$$

Then the percent crystallinity is given by

$$\frac{97.2 \text{ J/g}}{209 \text{ J/g}} \times 100 = 46\% \text{ crystallinity}$$

Many semicrystalline polymers are between 40% and 75% crystalline. (See Section 6.5.4 for further information.)

6.2 METHODS OF DETERMINING CRYSTAL STRUCTURE

6.2.1 A Review of Crystal Structure

Before beginning the study of the structure of crystalline polymers, the subject of crystallography and molecular order in crystalline substances is reviewed. Long before X-ray analysis was available, scientists had already deduced a great deal about the atomic order within crystals.

The science of geometric crystallography was concerned with the outward spatial arrangement of crystal planes and the geometric shape of crystals. Workers of that day arrived at three fundamental laws: (a) the law of constancy of interfacial angles, (b) the law of rationality of indexes, and (c) the law of symmetry (14).

Briefly, the law of constancy of interfacial angles states that for a given substance, corresponding faces or planes that form the external surface of a crystal always intersect at a definite angle. This angle remains constant, independent of the sizes of the individual faces.

The law of rationality of indexes states that for any crystal a set of three coordinate axes can be chosen such that all the faces of the crystal will either intercept these axes at definite distances from the origin or be parallel to some of the axes. In 1784 Hauy showed that it was possible to choose among the three coordinate axes unit distances (a, b, c) of not necessarily the same length. Furthermore, Hauy showed that it was possible to choose three coefficients for these three axes—*m*, *n*, and *p*—that are either integral whole numbers, infinity, or fractions of whole numbers such that the ratio of the three intercepts of any plane in the crystal is given by (*ma*: *nb*: *pc*). The numbers *m*, *n*, and *p* are known as the Weiss indexes of the plane in question. The Weiss indexes have from which are the Miller indexes, which are obtained by taking the reciprocals of the Weiss coefficients and multiplying through by the smallest number that will express the reciprocals as integers. For example, if a plane in the Weiss notation is given by $a : \infty b : \frac{1}{4} c$, the Miller indexes become $a : 0b : 4c$, or more simply written (104), which is the modern way of expressing the indexes in the Miller system of crystal face notation.

The third law of crystallography states that all crystals of the same compound possess the same elements of symmetry. There are three types of symmetry: a plane of symmetry, a line of symmetry, and a center of symmetry (14). A plane of symmetry passes through the center of the crystal and divides it into two equal portions, each of which is the mirror image of the other. If it is possible to draw an imaginary line through the center of the crystal and then revolve the crystal about this line in such a way as to cause the crystal to appear unchanged two, three, four, or six times in 360° of revolution, then the crystal is said to possess a line of symmetry. Similarly, a crystal possesses a center of symmetry if every face has an identical atom at an equal distance on the opposite side of this center. On the basis of the total number of plane, line, and center symmetries, it is possible to classify the crystal types into six crystal systems, which may in turn be grouped into 32 classes and finally into 230 crystal forms.

The scientists of the pre–X-ray period postulated that any macroscopic crystal was built up by repetition of a fundamental structural unit composed of atoms, molecules, or ions, called the unit crystal lattice or space group. This unit crystal lattice has the same geometric shape as the macroscopic crystal. This line of reasoning led to the 14 basic arrangements of atoms in space, called space lattices. Among these are the familiar simple cubic, hexagonal, and triclinic lattices.

There are four basic methods in wide use for the study of polymer crystallinity: X-ray diffraction, electron diffraction, infrared absorption, and Raman spectra. The first two methods constitute the fundamental basis for crystal cell size and form, and the latter two methods provide a wealth of supporting data such as bond distances and inter-molecular attractive forces. These several methods are now briefly described.

6.2.2 X-Ray Methods

In 1895 X-rays were discovered by Roentgen. The new X-rays were first applied to crystalline substances in 1912 and 1913, following the suggestion by Von Laue that crystalline substances ought to act as a three-dimensional diffraction grating for X-rays.

By considering crystals as reflection gratings for X-rays, Bragg (15) derived his now famous equation for the distance d between successive identical planes of atoms in the crystal:

$$d = \frac{n\lambda}{2 \sin \theta} \tag{6.1}$$

where λ is the X-ray wavelength, θ is the angle between the X-ray beam and these atomic planes, and n represents the order of diffraction, a whole number. It turns out that both the X-ray wavelength and the distance between crystal planes, d, are of the order of 1 Å. Such an analysis from a single crystal produces a series of spots.

However, not every crystalline substance can be obtained in the form of macroscopic crystals. This led to the Debye–Scherrer (16) method of analysis for powdered crystalline solids or polycrystalline specimens. The crystals are oriented at random so the spots be-come cones of diffracted beams that can be recorded either as circles on a flat photo-

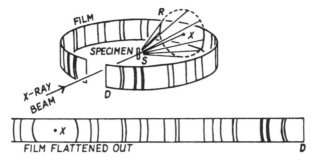

Figure 6.4 The Debye–Scherrer method for taking powder photographs. The angle RSX is 2θ, where θ is the angle of incidence on a set of crystal planes (17).

graphic plate or as arcs on a strip of film encircling the specimen (see Figure 6.4) (17). The latter method permits the study of back reflections as well as forward reflections.

Basically the intensity of the diffraction spot or line depends on the scattering power of the individual atoms, which in turn depends on the number of electrons in the atom. Other quantities of importance include the arrangement of the atoms with regard to the crystal planes, the angle of reflection, the number of crystallographically equivalent sets of planes contributing, and the amplitude of the thermal vibrations of the atoms. Both the intensities of the spots or arcs and their positions are required to calculate the crystal lattice, plus lots of imagination and hard work. The subject of X-ray analysis of crystalline materials has been widely reviewed (14,17).

6.2.3 Electron Diffraction of Single Crystals

Electron microscopy provides a wealth of information about the very small, including a view of the actual crystal cell size and shape. In another mode of use, the electrons can be made to diffract, using their wavelike properties. In this regard they are made to behave like the neutrons considered earlier.

In the case of X-ray studies, the polymer samples are usually uniaxially oriented and yield fiber diagrams that correspond to single-crystal rotation photographs. Electron diffraction studies utilize single crystals.

Since the polymer chains in single crystals are most often oriented perpendicular to their large flat surface, diffraction patterns perpendicular to the 001 plane are common. Tilting of the sample yields diffraction from other planes. The interpretation of the spots obtained utilizes Bragg's law in a manner identical to that of X-rays.

6.2.4 Infrared Absorption

Tadokoro (18) summarized some of the specialized information that infrared absorption spectra yield about crystallinity:

1. Infrared spectra of semicrystalline polymers include "crystallization-sensitive bands." The intensities of these bands vary with the degree of crystallinity and have been used as a measure of the crystallinity.

2. By measuring the polarized infrared spectra of oriented semicrystalline polymers, information about both the molecular and crystal structure can be obtained. Both uniaxially and biaxially oriented samples can be studied.

3. The regular arrangement of polymer molecules in a crystalline region can be treated theoretically, utilizing the symmetry properties of the chain or crystal. With the advent of modern computers, the normal modes of vibrations of crystalline polymers can be calculated and compared with experiment.

4. Deuteration of specific groups yields information about the extent of the contribution of a given group to specific spectral bands. This aids in the assignment of the bands as well as the identification of bands owing to the crystalline and amorphous regions.

6.2.5 Raman Spectra

Although Raman spectra have been known since 1928, studies on high polymers and other materials became popular only after the development of efficient laser sources. According to Tadokoro (18), some of the advantages of Raman spectra are the following:

1. Since the selection rules for Raman and infrared spectra are different, Raman spectra yield information complementary to the infrared spectra. For example, the S—S linkages in vulcanized rubber and the C=C bonds yield strong Raman spectra but are very weak or unobservable in infrared spectra.

2. Since the Raman spectrum is a scattering phenomenon, whereas the infrared methods depend on transmission, small bulk, powdered, or turbid samples can be employed.

3. On analysis, the Raman spectra provide information equivalent to very low-frequency measurements, even lower than 10 cm^{-1}. Such low-frequency studies provide information on lattice vibrations.

4. Polarization measurements can be made on oriented samples.

Of course, much of the above is widely practiced by spectroscopists on small molecules as well as big ones. Again, it must be emphasized that polymer chains are ordinary molecules that have been grown long in one direction.

6.3 THE UNIT CELL OF CRYSTALLINE POLYMERS

When polymers are crystallized in the bulk state, the individual crystallites are microscopic or even submicroscopic in size. They are an integral part of the solids and cannot be isolated. Hence studies on crystalline polymers in the bulk were limited to powder diagrams of the Debye-Scherrer type, or fiber diagrams of oriented materials.

It was only in 1957 that Keller (19) and others discovered a method of preparing single crystals from very dilute solutions by slow precipitation. These too were microscopic in size (see Section 6.4). However, X-ray studies could now be carried out on single crystals, with concomitant increases in detail obtainable.

Of course a major difference between polymers and low-molecular-weight compounds relates to the very existence of the macromolecule's long chains. These long chains traverse many unit cells. Their initial entangled nature impedes their motion, however, and leaves regions that are amorphous. Even the crystalline portions may be less than perfectly ordered.

This section describes the structure of the unit cell in polymers, principally as determined by X-ray analysis. The following sections describe the structure and morphology of single crystals, bulk crystallized crystallites, and spherulites and develop the kinetics and thermodynamics of crystallization.

6.3.1 Polyethylene

One of the most important polymers to be studied is polyethylene. It is the simplest of the polyolefins, those polymers consisting only of carbon and hydrogen, and polymerized through a double bond. Because of its simple structure, it has served as a model polymer in many laboratories. Also polyethylene's great commercial importance as a crystalline plastic has made the results immediately usable. It has been investigated both in the bulk and in the single-crystal state.

The unit cell structure of polyethylene was first investigated by Bunn (20). A number of experiments were reviewed by Natta and Corradini (21). The unit cell is orthorhombic, with cell dimensions of $a = 7.40$, $b = 4.93$, and $c = 2.534$ Å. The unit cell contains two mers (see Figure 6.5) (22). Not unexpectedly, the unit cell dimensions are substantially the same as those found for the normal paraffins of molecular weights in the range 300 to

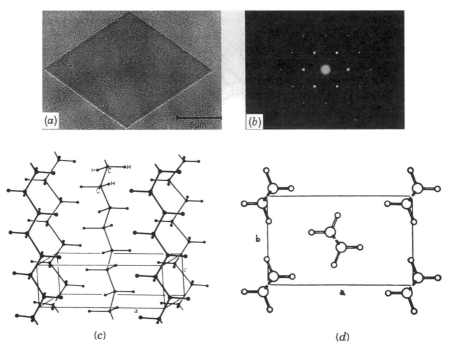

Figure 6.5 A study of polyethylene single-crystal structure. (*a*) A single crystal of polyethylene, precipitated from xylene, as seen by electron microscopy. (*b*) Electron diffraction of the same crystal, with identical orientation. (*c*) Perspective view of the unit cell of polyethylene, after Bunn. (*d*) View along chain axis. This latter corresponds to the crystal and diffraction orientation in (*a*) and (*b*) (22). Courtesy of A. Keller and Sally Argon.

600 g/mol. The chains are in the extended zigzag form, that is, the carbon–carbon bonds are *trans* rather than *gauche*. The zigzag form may also be viewed as a twofold screw axis.

The single-crystal electron diffraction pattern shown in Figure 6.5 was obtained by viewing the crystal along the *c* axis. Also shown is the single-crystal structure of polyethylene, which is typically diamond-shaped (see below). The unit cell is viewed from the *c*-axis direction, perpendicular to the diamonds.

6.3.2 Other Polyolefin Polymers

Because of the need for regularity along the chain, only vinyl polymers that are either isotactic or syndiotactic will crystallize. Thus isotactic polypropylene crystallizes well and is a good fiber former, whereas atactic polypropylene is essentially amorphous.

The idea of a screw axis along the individual extended chains needs developing. Such chains may be viewed as having a n/p-fold helix, where n is the number of mer units and p is the number of pitches within the identity period. Of course, n/p will be a rational number. Some of the possible types of helices for isotactic polymers are illustrated in Figure 6.6 (23). Group I of Figure 6.6 has a helix that makes one complete turn for every

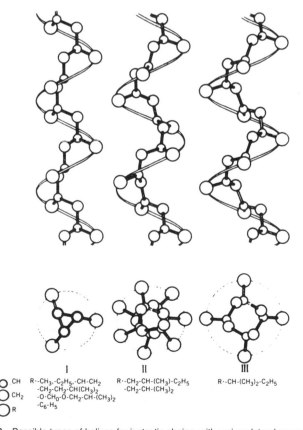

		I	II	III
○	CH	R·-CH₃,-C₂H₅,-CH-CH₂	R·-CH₂-CH-(CH₃)-C₂H₅	R·-CH-(CH₃)₂-C₂H₅

Figure 6.6 Possible types of helices for isotactic chains, with various lateral groups (23).

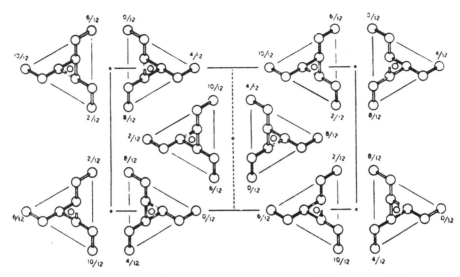

Figure 6.7 Enantiomorphous mode of packing polybutene-1 chains in a crystal (21).

three mer units, so $n = 3$ and $p = 1$. Group II shows seven mer units in two turns, so $n = 7$ and $p = 2$. Group III shows four mer units per turn.

Of course, both left- and right-handed helices are possible. The isotactic hydrocarbon polymers in question occur as enantiomorphic pairs that face each other, a closer packing being realized through the operation of a glide plane with translation parallel to the fiber axis. The enantiomorphic crystal structure of polybutene-1 is illustrated in Figure 6.7 (21). Note that better packing is achieved through the chains' having the opposite sense of helical twist.

6.3.3 Polar Polymers and Hydrogen Bonding

The hydrocarbon polymers illustrated above are nonpolar, being bonded together only by van der Waals–type attractive forces. When the polymers possess polar groups or hydrogen bonding capability, the most energetically favored crystal structures will tend to capitalize on these features. Figure 6.8 (24) illustrates the molecular organization within crystallites of a polyamide, known as polyamide 66 or nylon 66. The chains in the crystallites are found to occur as fully extended, planar zigzag structures.

X-ray analysis reveals that poly(ethylene terephthalate) (Dacron®) belongs to the triclinic system (25). The cell dimensions are $a = 4.56$, $b = 5.94$, $c = 10.75$ Å, with the angles being $\alpha = 98.5°$, $\beta = 118°$, $\gamma = 112°$. Both the polyamides and the aromatic polyesters are high melting polymers because of hydrogen bonding in the former case and chain stiffness in the latter case (see Table 6.1). As is well known, both of these polymers make excellent fibers and plastics.

The polyethers are a less polar group of polymers. Poly(ethylene oxide) will be taken as an example (26). Four (7/2) helical molecules pass through a unit cell with parameters $a = 8.05$, $b = 13.04$, $c = 19.48$ Å, and $\beta = 125.4°$ with the space group $P2_1/a - C_{2h}$. [Space groups are discussed by Tadokoro (27,28). These symbols represent the particular

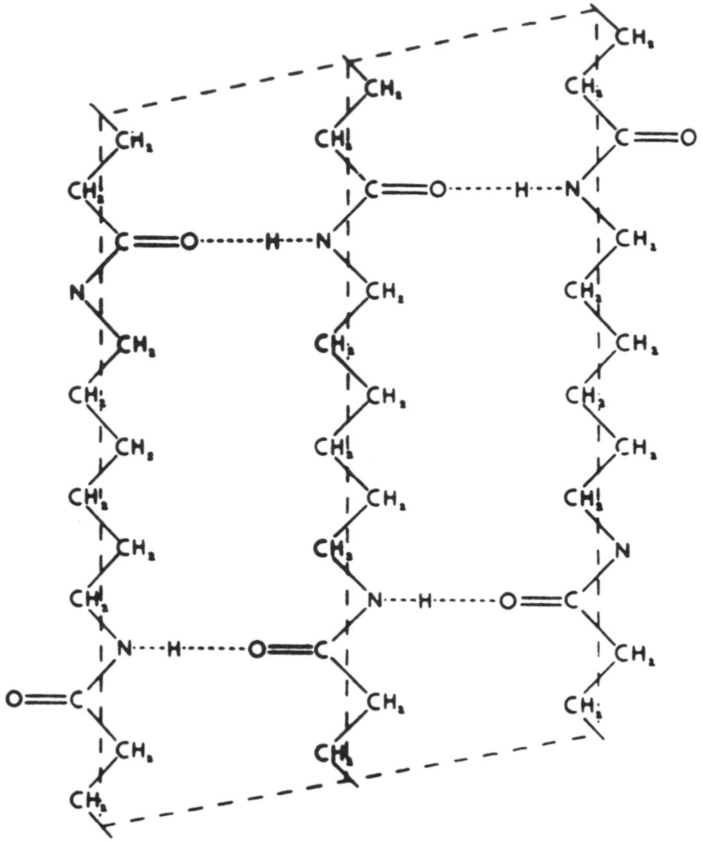

Figure 6.8 The hydrogen-bonded structure of polyamide 66. The unit cell face is shown dotted (24).

one of the 230 possible space groups to which poly(ethylene oxide) belongs; see Section 6.2.]

Table 6.2 (27) summarizes the crystallographic data of some important polymers. Several polymers have more than one crystallographic form. Polytetrafluoroethylene, for example, undergoes a first-order crystal–crystal transition at 19°C.

6.3.4 Polymorphic Forms of Cellulose

A few words have already been said about the crystalline structure of cellulose (see Figure 6.1). The monoclinic unit cell structure illustrated for cellulose I was postulated many years ago by Meyer et al. (3–6) and has been confirmed many times. The b axis is the fiber direction, and the cell belongs to the space group P2, containing four glucose residues.

The triclinic structure is also known. Note that the structure shown in Figure 6.1a illustrates the chains running in alternating directions, up and down, now not thought true for most celluloses, the assumption of all one way or of alternating opposite directions is very important in the development of crystalline models (see below).

Table 6.2 Selected crystallographic data (27)

Polymer	Crystal System, Space Group, Lattice Constants, and Number of Chains per Unit Cell[a]	Molecular Conformation	Crystal Density (g/cm³)
Polyethylene $-(CH_2-CH_2-)_n$	Stable form, orthorhombic, $Pnam\text{-}D_{2h}^{16}$, $a = 7.417\,\text{Å}$, $b = 4.945\,\text{Å}$, $c = 2.547\,\text{Å}$, $N = 2$	Planar zigzag (2/1)	1.00
	Metastable form, monoclinic, $C2/m-C_{2h}^3$, $a = 8.09\,\text{Å}$, b (f.a.) $= 2.53\,\text{Å}$, $c = 4.79\,\text{Å}$, $\beta = 107.9°$, $N = 2$	Planar zigzag (2/1)	0.998
	High-pressure form, orthohexagonal (assumed), $a = 8.42\,\text{Å}$, $b = 4.56\,\text{Å}$, c (f.a.) has not been determined		
Polytetrafluoroethylene $-(CF_2-CF_2-)_n$	Below 19°C, pseudohexagonal (triclinic), $a' = b' = 5.59\,\text{Å}$, $c = 16.88\,\text{Å}$, $\gamma' = 119.3°$, $N = 1$	Helix (13/6) 163.5°	2.35
	Above 19°C, trigonal, $a = 5.66\,\text{Å}$, $c = 19.50\,\text{Å}$, $N = 1$	Helix (15/7) 165.8°	2.30
	High-pressure form I[12], orthorhombic, $Pnam\text{-}D_{2h}^{16}$, $a = 8.73\,\text{Å}$, $b = 5.69\,\text{Å}$, $c = 2.62\,\text{Å}$, $N = 2$	Planar zigzag (2/1)	2.55
	High-pressure form II,[193] monoclinic, $B2/m\text{-}C_{2h}^3$, $a = 9.50\,\text{Å}$, $b = 5.05\,\text{Å}$, $c = 2.62\,\text{Å}$, $\gamma = 105.5°$, $N = 2$	Planar zigzag (2/1)	2.74
it-Polypropylene $-(CH_2-CH-)_n$ $\quad\;\; CH_3$	α-Form, monoclinic, $C2/c\text{-}C_{2h}^6$ of $Cc\text{-}C_s^4$, $a = 6.65\,\text{Å}$, $b = 20.96\,\text{Å}$, $c = 6.50\,\text{Å}$, $\beta = 99°20'$. $N = 4$	Helix (3/1) $(TG)_3$	0.936
	β-Form, hexagonal, $a = 19.08\,\text{Å}$, $c = 6.49\,\text{Å}$, $N = 9$	Helix (3/1) $(TG)_3$	0.922
	γ-Form, trigonal, $P3_121\text{-}D_3^4$ or $P3_221\text{-}D_3^6$, $a = 6.38\,\text{Å}$, $c = 6.33\,\text{Å}$, $N = 1$	Helix (3/1) $(TG)_3$	0.939
it-Polystyrene $-(CH_2-CH-)_n$ $\quad\;\; C_6H_5$	Trigonal, $R3c\text{-}C_{3v}^6$ or $R\bar{3}c\text{-}D_{3d}^6$, $a = 21.90\,\text{Å}$, $c = 6.65\,\text{Å}$, $N = 6$	Helix (3/1) $(TG)_3$	1.13

Polymer	Crystal data	Conformation	Density
cis-1,4-Polyisoprene $+CH_2-C(CH_3)=CH-CH_2+_n$	Monoclinic, $P2_1/a\text{-}C_{2h}^5$, $a = 12.46$ Å, $b = 8.89$ Å, $c = 8.10$ Å, $\beta = 92°$, $N = 4$	$cis\text{-}ST\bar{S}\text{-}cis\text{-}\bar{S}TS$, (2/0)	1.02
Poly(vinyl chloride) $+CH_2-CHCl+_n$	Orthorhombic, $Pcam\text{-}D_{2h}^{11}$, $a = 10.6$ Å, $b = 5.4$ Å, $c = 5.1$ Å, $N = 2$	Planar zigzag	1.42
Polytetrahydrofuran $+(CH_2)_4-O+_n$	Monoclinic, $C2/c\text{-}C_{2h}^6$, $a = 5.59$ Å, $b = 8.90$ Å, $c = 12.07$ Å, $\beta = 134.2°$, $N = 2$	Planar zigzag (2/1)	1.11
polyamide 6 $+(CH_2)_5-CONH+_n$	α-Form, monoclinic, $P2_1\text{-}C_2^2$, $a = 9.56$ Å, b (f.a.) $= 17.2$ Å, $c = 8.01$ Å, $\beta = 67.5°$, $N = 4$	Planar zigzag (2/1)	1.23
	γ-Form, monoclinic, $P2_1/a\text{-}C_{2h}^5$, $a = 9.33$ Å, b (f.a.) $= 16.88$ Å, $c = 4.78$ Å, $\beta = 121°$, $N = 2$	Helix (2/1) $(T_4 ST\bar{S})^2$	1.17
polyamide 66 $+NH-(CH_2)_6NHCO-(CH_2)_4-CO+_n$	α-Form, triclinic, $P\bar{1}\text{-}C_i^1$, $a = 4.9$ Å, $b = 5.4$ Å, $c = 17.2$ Å, $a = 48.5°$, $\beta = 77°$, $\gamma = 63.5°$, $N = 1$	Planar zigzag (1/0)	1.24
	β-Form, triclinic, $P\bar{1}\text{-}C_i^1$, $a = 4.9$ Å, $b = 8.0$ Å, $c = 17.2$ Å, $\alpha = 90°$, $\beta = 77°$, $\gamma = 67°$, $N = 2$	Planar zigzag (1/0)	1.248
Poly(ethylene oxide) $+CH_2-CH_2-O+_n$	Form I, monoclinic, $P2_1/a\text{-}C_{2h}^5$, $a = 8.05$ Å, $b = 13.04$ Å, $c = 19.48$ Å, $\beta = 125.4°$, $N = 4$	Helix (7/2)	1.228
	Form II, triclinic, $P\bar{1}\text{-}C_i^1$, $a = 4.17$ Å, $b = 4.44$ Å, $c = 7.12$ Å, $\alpha = 62.8°$, $\beta = 93.2°$, $\gamma = 111.4°$, $N = 1$	Planar zigzag (2/1)	1.197
	β-Form, triclinic, $P\bar{1}\text{-}C_i^1$, $a = 4.9$ Å, $b = 8.0$ Å, $c = 22.4$ Å, $\alpha = 90°$, $\beta = 77°$, $\gamma = 67°$, $N = 2$	Planar zigzag (1/0)	1.196
Poly(ethylene terephthalate) $+O-(CH_2)_2-O-CO-\!\!\bigcirc\!\!-CO+_n$	Triclinic, $P\bar{1}\text{-}C_i^1$, $a = 4.56$ Å, $b = 5.94$ Å, $c = 10.75$ Å, $\alpha = 98.5°$, $\beta = 112°$, $N = 1$	Nearly planar	1.455

As shown further in Figure 6.1*b, c,* four different polymorphic forms of crystalline cellulose exist. Cellulose I is native cellulose, the kind found in wood and cotton. Cellulose II is made either by soaking cellulose I in strong alkali solutions (e.g., making mercerized cotton) or by dissolving it in the viscose process, which makes the labile but soluble cellulose xanthate. The regenerated cellulose II products are known as rayon for the fiber form and cellophane for the film form. Cellulose III can be made by treating cellulose with ethylamine. Cellulose IV may be obtained by treatment with glycerol or alkali at high temperatures (29,30).

Going from the polymorphic forms of cellulose back to cellulose I is difficult but can be accomplished by partial hydrolysis. The subject of the polymorphic forms of cellulose has been reviewed (10,11).

6.3.5   Principles of Crystal Structure Determination

Through the use of X-ray analysis, electron diffraction, and the supporting experiments of infrared absorption and Raman spectroscopy, much information has been collected on crystalline polymers. Today, the data are analyzed through the use of computer techniques. Somewhere along the line, however, the investigator is required to use intuition to propose models of crystal structure. The proposed models are then compared to experiment. The models are gradually refined to produce the structures given above. It must be emphasized, that the experiments do not yield the crystal structure; only researchers' imagination and hard work yield that.

However, it is possible to simplify the task. Natta and Corradini (23) postulated three principles for the determination of crystal structures, which introduce considerable order into the procedure. These are:

1. *The Equivalence Postulate.* It is possible to assume that all mer units in a crystal occupy geometrically equivalent positions with respect to the chain axis.
2. *The Minimum Energy Postulate.* The conformation of the chain in a crystal may be assumed to approach the conformation of minimum potential energy for an isolated chain oriented along an axis.
3. *The Packing Postulate.* As many elements of symmetry of isolated chain as possible are maintained in the lattice, so equivalent atoms of different mer units along an axis tend to assume equivalent positions with respect to atoms of neighboring chains.

The equivalence postulate is seen in the structures given in Figure 6.6. Here, the chain mers repeat their structure in the next unit cell.

Energy calculations made for both single molecules and their unit cells serve three purposes: (a) they clarify the factors governing the crystal and molecular structure already tentatively arrived at experimentally, (b) they suggest the most stable molecular conformation and its crystal packing starting from the individual mer chemical structure, and (c) they provide a collection of reliable potential functions and parameters for both intra- and intermolecular interactions based on well-defined crystal structures (27). An example of intermolecular interactions is hydrogen bonding in the polyamide structures described in Figure 6.8.

The packing postulate is seen at work in Figure 6.7, where enantiomorphic structures pack closer together in space than if the chains had the same sense of helical twist.

T̶h̶i̶s̶ ̶m̶o̶d̶e̶l̶ ̶t̶e̶n̶d̶s̶ ̶t̶o̶ ̶n̶e̶g̶l̶e̶c̶t̶ ̶t̶h̶e̶ ̶v̶e̶r̶y̶ simple but all important density. The crystalline cell is usually about 10% more dense than the bulk amorphous polymer. Significant deviations from this density must mean an incorrect model.

Because of the importance of polymer crystallinity generally, and the unit cell in particular, the subject has been reviewed many times (17,21,27,31–41).

6.4 STRUCTURE OF CRYSTALLINE POLYMERS

6.4.1 The Fringed Micelle Model

Very early studies on bulk materials showed that some polymers were partly crystalline. X-ray line broadening indicated that the crystals were either very imperfect or very small (42). Assuming the latter, in 1928 Hengstenberg and Mark (43) estimated that the crystallites of ramie, a form of native cellulose, were about 55 Å wide and over 600 Å long by this method. It had already been established that the polymer chain passed through many unit cells. Because of the known high molecular weight, the polymer chain was calculated to be even longer than the crystallites. Hence it was reasoned that they passed in and out of many crystallites (32,44–45). These findings led to the fringed micelle model.

According to the fringed micelle model, the crystallites are about 100 Å long (Figure 6.9). The disordered regions separating the crystallites are amorphous. The chains wander from the amorphous region through a crystallite, and back out into the amorphous region. The chains are long enough to pass through several crystallites, binding them together.

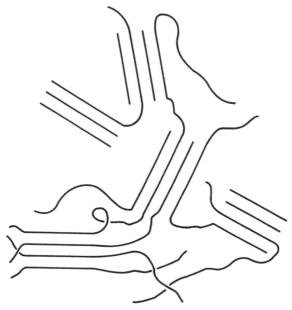

Figure 6.9 The fringed micelle model. Each chain meanders from crystallite to crystallite, binding the whole mass together.

The fringe micelle model was used with great success to explain a wide range of behavior in semicrystalline plastics, and also in fibers. The amorphous regions, if glassy, yielded a stiff plastic. However, if they were above T_g, then they were rubbery and were held together by the hard crystallites. This model explains the leathery behavior of ordinary polyethylene plastics, for example. The greater tensile strength of polyethylene over that of low-molecular-weight hydrocarbon waxes was attributed to amorphous chains wandering from crystallite to crystallite, holding them together by primary bonds. The flexible nature of fibers was explained similarly; however, the chains were oriented along the fiber axis (see Section 6.3). The exact stiffness of the plastic or fiber was related to the degree of crystallinity, or fraction of the polymer that was crystallized.

6.4.2 Polymer Single Crystals

Ideas about polymer crystallinity underwent an abrupt change in 1957 when Keller (19) succeeded in preparing single crystals of polyethylene. These were made by precipitation from extremely dilute solutions of hot xylene. These crystals tended to be diamond-shaped and of the order of 100 to 200 Å thick (see Figure 6.5) (21). Amazingly electron diffraction analysis showed that the polymer chain axes in the crystal body were essentially perpendicular to the large, flat faces of the crystal. Since the chains were known to have contour lengths of about 2000 Å and the thickness of the single crystals was in the vicinity of 110 to 140 Å, Keller concluded that the polymer molecules in the crystals had to be folded upon themselves. These observations were immediately confirmed by Fischer (47) and Till (48).[†]

6.4.2.1 The Folded-Chain Model This led to the folded-chain model, illustrated in Figure 6.10 (41). Ideally the molecules fold back and forth with hairpin turns. While adjacent reentry has been generally confirmed by small-angle neutron scattering and infrared studies for single crystals, the present understanding of bulk crystallized polymers indicates a much more complex situation (see below).

Figure 6.10 uses polyethylene as the model material. The orthorhombic cell structure and the a and b axes are illustrated. The c axis runs parallel to the chains. The dimension ℓ is the thickness of the crystal. The predominant fold plane in polyethylene solution-grown crystals is along the (110) plane. Chain folding is also supported by NMR studies (see Section 6.7) (49–51).

For many polymers, the single crystals are not simple flat structures. The crystals often occur in the form of hollow pyramids, which collapse on drying. If the polymer solution is slightly more concentrated, or if the crystallization rate is increased, the polymers will crystallize in the form of various twins, spirals, and dendritic structures, which are multilayered (see Figure 6.11) (34). These latter form a preliminary basis for understanding polymer crystallization from bulk systems.

Simple homopolymers are not the only polymeric materials capable of forming single crystals. Block copolymers of poly(ethylene oxide) crystallize in the presence of considerable weight fractions of amorphous polystyrene (see Figure 6.12) (52). In this case

---

[†] The early literature reveals significant premonitions of this discovery. K. H. Storks, *J. Am. Chem. Soc.*, **60**, 1753 (1938) suggested that the macromolecules in crystalline gutta percha are folded back and forth upon themselves in such a way that adjacent sections remain parallel. R. Jaccodine, *Nature (London)*, **176**, 305 (1955) showed the spiral growth of polyethylene single crystals by a dislocation mechanism.

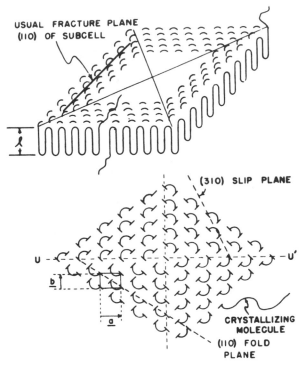

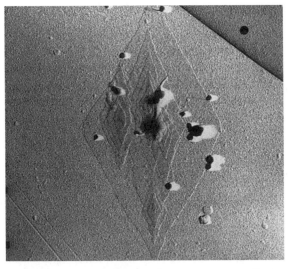

Figure 6.10 Schematic view of a polyethylene single crystal exhibiting adjacent reentry. The orthorhombic subcell with dimensions *a* and *b*, typical of many *n*-paraffins, is illustrated below (41).

Figure 6.11 Single crystal of polyamide 6 precipitated from a glycerol solution. The lamellae are about 60 Å thick. Black marks indicate 1 μm (34).

Figure 6.12 Single crystals of poly(ethylene oxide)–*block*–polystyrene diblock copolymers. (*a*) Optical micrograph. (*b*) Electron micrograph. M_n (PS) $= 7.3 \times 10^3$ g/mol; M_n (PEO) $= 10.9 \times 10^3$; weight fraction polystyrene is 0.34 (52).

square-shaped crystals with some spirals are seen. The crystals reject the amorphous portion (polystyrene), which appears on the surfaces of the crystals.

Amorphous material also appears on the surfaces of homopolymer single crystals. As will be developed below, causes of this amorphous material range from chain-end cilia to irregular folding.

6.4.2.2 The Switchboard Model In the switchboard model the chains do not have a reentry into the lamellae by regular folding; they rather reenter more or less randomly (53). The model more or less resembles an old-time telephone switchboard. Of course, both the perfectly folded chain and switchboard models represent limiting cases. Real systems may combine elements of both. For bulk systems, this aspect is discussed in Section 6.7.

6.5 CRYSTALLIZATION FROM THE MELT

6.5.1 Spherulitic Morphology

In the previous sections it was observed that when polymers are crystallized from dilute solutions, they form lamellar-shaped single crystals. These crystals exhibit a folded-chain habit and are of the order of 100 to 200 Å thick. From somewhat more concentrated solutions, various multilayered dendritic structures are observed.

When polymers crystallize from the melt, they usually *supercool* to greater or lesser extents; see Figure 6.3. Thus, the crystallization temperature may be 10 to 20°C lower than the melting temperature. Supercooling arises from the extra free energy required to

Figure 6.13 Spherulites of low-density polyethylene, observed through crossed polarizers. Note characteristic Maltese cross pattern (34).

align chain segments, common in the crystallization of many complex organic compounds as well as polymers.

When polymer samples are crystallized from the melt, the most obvious of the observed structures are the spherulites (33). As the name implies, spherulites are sphere-shaped crystalline structures that form in the bulk (see Figure 6.13) (34). One of the more important problems to be addressed concerns the form of the lamellae within the spherulite.

Spherulites are remarkably easy to grow and observe in the laboratory (36). Simple cooling of a thin section between crossed polarizers is sufficient, although controlled experiments are obviously more demanding. It is observed that each spherulite exhibits an extinction cross, sometimes called a Maltese cross. This extinction is centered at the origin of the spherulite, and the arms of the cross are oriented parallel to the vibration directions of the microscope polarizer and analyzer.

Usually the spherulites are really spherical in shape only during the initial stages of crystallization. During the latter stages of crystallization, the spherulites impinge on their neighbors. When the spherulites are nucleated simultaneously, the boundaries between them are straight. However, when the spherulites have been nucleated at different times, so that they are different in size when impinging on one another, their boundaries form hyperbolas. Finally, the spherulites form structures that pervade the entire mass of the material. The kinetics of spherulite crystallization are considered in Section 6.7.

Figure 6.14 Surface replica of polyoxymethylene fractured at liquid nitrogen temperatures. Lamellae at lower left are oriented at an angle to the fracture surface. Lamellae elsewhere are nearly parallel to the fracture surface, being stacked up like cards or dishes in the bulk state. These structures closely resemble stacks of single crystals, and they have led to ideas about chain folding in bulk materials (34).

Electron microscopy examination of the spherulitic structure shows that the spherulites are composed of individual lamellar crystalline plates (see Figure 6.14) (34). The lamellar structures sometimes resemble staircases, being composed of nearly parallel (but slightly diverging) lamellae of equal thickness. Amorphous material usually exists between the staircase lamellae.

X-ray microdiffraction (37) and electron diffraction (38) examination of the spherulites indicates that the c axis of the crystals is normal to the radial (growth) direction of the spherulites. Thus the c axis is perpendicular to the lamellae flat surfaces, showing the resemblance to single-crystal structures.

For some polymers, such as polyethylene (37), it was shown that their lamellae have a screwlike twist along their unit cell b axis, on the spherulite radius. The distance corresponding to one-half of the pitch of the lamellar screw is just in accordance with the extinction ring interval visible on some photographs.

The growth and structure of spherulites may also be studied by small-angle light scattering (39,40). The sample is placed between polarizers, a monochromatic or laser light beam is passed through, and the resultant scattered beam is photographed. Two types of scattering patterns are obtained, depending on polarization conditions (54). When the polarization of the incident beam and that of the analyzer are both vertical, it is called a V_v type of pattern. When the incident radiation is vertical in polarization but the analyzer is horizontal (polarizers crossed), an H_v pattern is obtained.

Figure 6.15 Different types of light-scattering patterns are obtained from spherulitic polyethylene using (a) V_v and (b) H_v polarization (54). Note the twofold symmetry of the V_v pattern, and the fourfold symmetry of the H_v pattern. This provided direct experimental evidence that spherulites were anisotropic.

The two types of scattering patterns are illustrated in Figure 6.15 (54). These patterns arise from the spherulitic structure of the polymer, which is optically anisotropic, with the radial and tangential refractive indexes being different.

The scattering pattern can be used to calculate the size of the spherulites (40) (see Figure 6.16). The maximum that occurs in the radial direction, U, is related to R, the radius of the spherulite by

$$U_{max} = \left(\frac{4\pi R}{\lambda}\right) \sin\left(\frac{\theta_{max}}{2}\right) = 4.1 \tag{6.2}$$

where θ_{max} is the angle at which the intensity maximum occurs and λ is the wavelength. As the spherulites get larger, the maximum in intensity occurs at smaller angles.

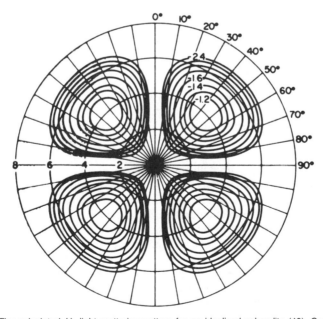

Figure 6.16 The calculated H_v light-scattering pattern for an idealized spherulite (40). Compare with the actual result, Figure 6.15b.

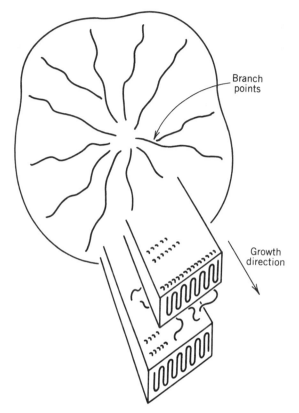

Figure 6.17   Model of spherulitic structure. Note the growth directions and lamellar branch points that fill the space uniformly with crystalline material. After J. D. Hoffman et al. (41).

Conversely, Stein (54) points out that in very rapidly crystallized polymers, spherulites are often not observed. The smaller amount of scattering observed results entirely from local structure. These structures are highly disordered.

Mandelkern recently drew a morphological map for polyethylene (55). He showed that the supermolecular structures become less ordered as the molecular weight is increased or the temperature of crystallization is decreased.

A model of the spherulite structure is illustrated in Figure 6.17 (41). The chain direction in the bulk crystallized lamellae is perpendicular to the broad plane of the structure, just like the dilute solution crystallized material.

The spherulite lamellae also contain low-angle branch points, where new lamellar structures are initiated. The new lamellae tend to keep the spacing between the crystallites constant.

While the lamellar structures in the spherulites are the analogue of the single crystals, the folding of the chains is much more irregular, as will be developed further in Section 6.6.2.3. In between the lamellar structures lies amorphous material. This portion is rich in components such as atactic polymers, low-molecular-weight material, or impurities of various kinds.

Figure 6.18 Electron micrographs of replicas of hedrites formed in the same melt-crystallized thin film of poly(4-methylpentene-1). (*a*) An edge-on view of a hedrite. Note the distinctly lamellar character and the "sheaflike" arrangement of the lamellae. (*b*) A flat-on view of a hedrite. Note the degenerate overall square outline of the object, whose lamellar texture is evident (33).

The individual lamellae in the spherulites are bonded together by tie molecules, which lie partly in one crystallite and partly in another. Sometimes these tie molecules are actually in the form of what are called intercrystalline links (56–59), which are long, threadlike crystalline structures with the *c* axis along their long dimension. These intercrystalline links are thought to be important in the development of the great toughness characteristic of semicrystalline polymers. They serve to tie the entire structure together by crystalline regions and/or primary chain bonds.

6.5.2 Mechanism of Spherulite Formation

On cooling from the melt, the first structure that forms is the single crystal. These rapidly degenerate into sheaflike structures during the early stages of the growth of polymer spherulites (see Figure 6.18) (33). These sheaflike structures have been variously called axialites (60) or hedrites (61). These transitional, multilayered structures represent an intermediate stage in the formation of spherulites (62). It is evident from Figure 6.18 that as growth proceeds, the lamellae develop on either side of a central reference plane. The lamellae fan out progressively and grow away from the plane as the structure begins to mature.

The sheaflike structures illustrated in Figure 6.18 are modeled in Figure 6.19 (33). As in Figure 6.18, both edge-on and flat-on views are illustrated. Figure 6.18*a* is modeled by Figure 6.19, row *a*, column III, and Figure 6.18*b* is modeled by row *b*, column III. Gradually the lamellae in the hedrites diverge or fan outward in a splaying motion. Repeated splaying, perhaps aided by lamellae that are intrinsically curved, eventually leads to the spherical shape characteristic of the spherulite.

6.5.3 Spherulites in Polymer Blends and Block Copolymers

There are two cases to be considered. Either the two polymers composing the blend may be miscible and form one phase in the melt, or they are immiscible and form two phases. Martuscelli (63) pointed out that if the glass transition of the miscible noncrystallizing component is lower than that of the crystallizing component (i.e., its melt viscosity will be lower, other things being equal), then the spherulites will actually grow faster, although

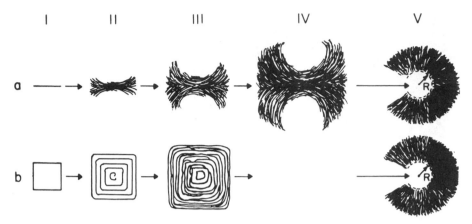

Figure 6.19 Schematic development of a spherulite from a chain-folded precursor crystal. Rows (*a*) and (*b*) represent, respectively, edge-on and flat-on views of the evolution of the spherulite (33).

the system is diluted. This is in general agreement with Section 6.6.2.2, which shows why crystallizable polymers containing low-molecular-weight fractions (which are not incorporated in the spherulite) also crystallize faster. Martuscelli also pointed out that the inverse was also true, especially if the noncrystallizing polymer was glassy at the temperature of crystallization.

The crystallization behavior is quite different if the two polymers are immiscible in the melt. Figure 6.20 (64) shows droplets of polyisobutylene dispersed in isotactic polypropylene. On spherulite formation, the droplets, which are noncrystallizing, become ordered within the growing arms of the crystallizing component.

Block copolymers also form spherulites (65–67). The morphology develops on a finer scale, however, because the domains are constrained to be of the order of the size of the individual blocks. In the case of a triblock copolymer, for example, the chain may be modeled as wandering from one lamella to another through an amorphous phase consisting of the center block (see Figure 6.21) (67).

Figure 6.22 (66) illustrates the spherulite morphology for poly(ethylene oxide)–*block*–polystyrene. Two points should be made. First, the glass transition temperature of the polystyrene component is higher than the temperature of crystallization (65); this particular sample was made by casting from chloroform. Second, the amount of polystyrene is small, only 19.6%. When the polystyrene component is increased, it disturbs the ordering process (see Figure 6.23) (66). This figure shows spheres rich in poly(ethylene oxide) lamellae but containing some polystyrene segments (dark spots) embedded in the poly(ethylene oxide) spherulites, as well as forming the more continuous phase. The size of the fine structure is of the order of a few hundred angstroms, because the two blocks must remain attached, even though they are in different phases.

In the case where the polymer forms a triblock copolymer of the structure $A—B—A$, or a multiblock copolymer $(A—B)_m$, where A is crystalline at use temperature and B is rubbery (above T_g), then a thermoplastic elastomer is formed (see Figure 6.21). The material exhibits some degree of rubber elasticity at use temperature, the crystallites serving as cross-links. Above the melting temperature of the crystalline

PiB_MM (80/20)   Tc=133°C

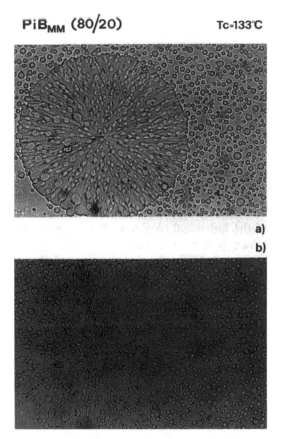

Figure 6.20 Optical micrograph of a thin film of isotactic polypropylene/polyisobutylene blend. (*a*) *it*-PP spherulite (T_c = 133°C), surrounded by melt blend at the early stage of crystallization. (*b*) The same region of film after melting of spherulite. Note the multiphase morphology common to many blends (63).

phase, the material is capable of flowing: that is, it is thermoplastic. It must be pointed out that a very important kind of thermoplastic elastomer is when the A polymer is glassy rather than crystalline (see Section 9.16).

Two important kinds of $+A—B+_n$ block copolymers, where A is amorphous and above T_g, and B is crystalline are the segmented polyurethanes and poly(ester–ether) materials. Both are fiber formers; see Chapter 13. These fibers tend to be soft and elastic.

6.5.4 Percent Crystallinity in Polymers

As suggested above, most polymers are semicrystalline; that is, a certain fraction of the material is amorphous, while the remainder is crystalline. The reason why polymers fail to attain 100% crystallinity is kinetic, resulting from the inability of the polymer chains to completely disentangle and line up properly in a finite period of cooling or annealing.

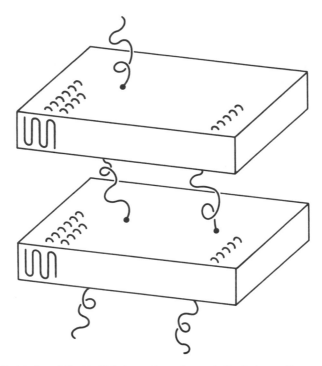

Figure 6.21 Model of crystallizable triblock copolymer thermoplastic elastomer. The center block, amorphous, is rubbery, whereas the end blocks are crystalline (64).

Figure 6.22 An optical micrograph of a poly(ethylene oxide)–*block*–polystyrene copolymer containing 19.6% polystyrene (66). Crystals cast from chloroform and observed through crossed polarizers. White markers are 250 μm apart.

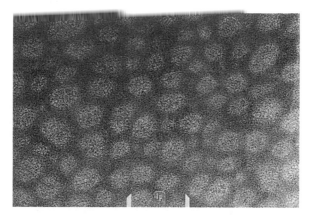

Figure 6.23 A transmission electron micrograph of poly(ethylene oxide)–*block*–polystyrene, containing 70% polystyrene. The polystyrene phase is stained with OsO_4 (66).

There are several methods for determining the percent crystallinity in such polymers. The first involves the determination of the heat of fusion of the whole sample by calorimetric methods such as DSC; see Figure 6.3. The heat of fusion per mole of crystalline material can be estimated independently by melting point depression experiments; see Section 6.8.

A second method involves the determination of the density of the crystalline portion via X-ray analysis of the crystal structure, and determining the theoretical density of a 100% crystalline material. The density of the amorphous material can be determined from an extrapolation of the density from the melt to the temperature of interest; see Figure 6.24. Then the percent crystallinity is given by

$$\% \text{ Crystallinity} = \left[\frac{\rho_{\text{exptl}} - \rho_{\text{amorph}}}{\rho_{100\% \text{ cryst}} - \rho_{\text{amorph}}} \right] \times 100 \qquad (6.3)$$

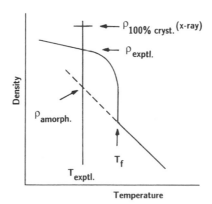

Figure 6.24 The experimental determination of the extent of polymer crystallinity using the density method.

where ρ_{exptl} represents the experimental density, and ρ_{amorph} and $\rho_{100\% \; cryst}$ are the densities of the amorphous and crystalline portions, respectively.

A third method stems from the fact that the intensity of X-ray diffraction depends on the number of electrons involved and is thus proportional to the density. Besides Bragg diffraction lines for the crystalline portion, there is an amorphous halo caused by the amorphous portion of the polymer. This last occurs at a slightly smaller angle than the corresponding crystalline peak, because the atomic spacings are larger. The amorphous halo is broader than the corresponding crystalline peak, because of the molecular disorder.

This third method, sometimes called wide-angle X-ray scattering (WAXS), can be quantified by the crystallinity index (68), CI,

$$CI = \frac{A_c}{A_a + A_c} \qquad (6.4)$$

where A_c and A_a represent the area under the Bragg diffraction line (or equivalent crystalline Debye-Scherrer diffraction line; see Figure 6.4) and corresponding amorphous halo, respectively.

Naturally these methods will not yield the same answer for a given sample, but surprisingly good agreement is obtained. For many semicrystalline polymers the crystallinity is in the range of 40% to 75%. Polymers such as polytetrafluoroethylene achieve 90% crystallinity, while poly(vinyl chloride) is often down around 15% crystallinity. The latter polymer is largely atactic, but short syndiotactic segments contribute greatly to its crystallinity. Of course, annealing usually increases crystallinity, as does orienting the polymer in fiber or film formation.

6.6 KINETICS OF CRYSTALLIZATION

During crystallization from the bulk, polymers form lamellae, which in turn are organized into spherulites or their predecessor structures, hedrites. This section is concerned with the rates of crystallization under various conditions of temperature, molecular weight, structure, and so on, and the theories that provide not only an insight into the molecular mechanisms but considerable predictive power.

6.6.1 Experimental Observations of Crystallization Kinetics

It has already been pointed out that the volume changes on melting, usually increasing (see Figure 6.2). This phenomenon may be used to study the kinetics of crystallization. Figure 6.25 (69) illustrates the isothermal crystallization of poly(ethylene oxide) as determined dilatometrically. From Table 6.1 the melting temperature of poly(ethylene oxide) is 66°C, where the rate of crystallization is zero. The rate of crystallization increases as the temperature is decreased. This follows from the fact that the driving force increases as the sample is supercooled.

Crystallization rates may also be observed microscopically, by measuring the growth of the spherulites as a function of time. This may be done by optical microscopy, as has been done by Keith and Padden (70,71), or by transmission electron microscopy of thin sections (72). The isothermal radial growth of the spherulites is usually observed to be linear (see Figure 6.26) (71). This implies that the concentration of impurity at the growing tips of the lamellae remains constant through the growth process. The more

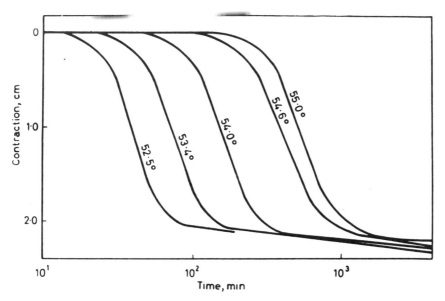

Figure 6.25 Dilatometric crystallization isotherms for poly(ethylene oxide), $M = 20,000$ g/mol. The Avrami exponent n falls from 4.0 to 2.0 as crystallization proceeds (69).

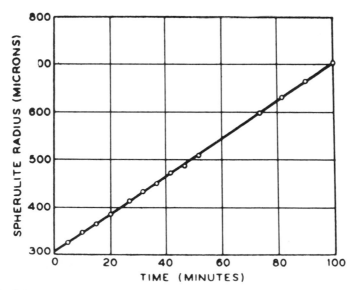

Figure 6.26 Spherulite radius as a function of time, grown isothermally (at 125°C) in a blend of 20% isotactic and 80% atactic ($M = 2600$) polypropylene. Note the linear behavior (71).

Table 6.3   Blends of unextracted isotactic polypropylene with atactic polypropylene (71)

Temperature (°C)	Radial Growth Rates (µm/min) for Various Compositions				
	100% Isotactic	90% Isotactic	80% Isotactic	60% Isotactic	40% Isotactic
120	29.4	29.4	26.4	22.8	21.2
125	13.0	12.0	11.0	8.90	8.57
131	3.88	3.60	3.03	2.37	2.40
135	1.63	1.57	1.35	1.18	1.12
Melting point (°C)	171	169	167	165	162

impurity, the slower is the growth rate (Table 6.3) (71). However, linearity of growth rate is maintained. In such a steady state, the radial diffusion of rejected impurities is outstripped by the more rapidly growing lamellae so that impurities diffuse aside and are trapped in interlamellar channels. In the case illustrated in Figure 6.26, the main "impurity" is low-molecular-weight atactic polypropylene.

When the radial growth rate is plotted as a function of crystallization temperature, a maximum is observed (see Figure 6.27) (72). As mentioned earlier, the increase in rate of crystallization as the temperature is lowered is controlled by the increase in the driving

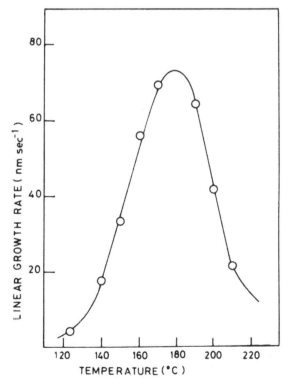

Figure 6.27   Plot of linear growth rate versus crystallization temperature for poly(ethylene terephthalate) (72). $T_f = 265°C$, and $T_g = 67°C$, at which points the rates of crystallization are theoretically zero.

ñ⋯⋯. As the temperature is lowered still further, molecular motion becomes sluggish as the glass transition is approached, and the crystallization rate decreases again. Below T_g, molecular motion is so sluggish that the rate of crystallization effectively becomes zero. This is another example of supercooling, as illustrated earlier in Figure 6.3.

There is an interesting rule-of-thumb for determining a good temperature to crystallize a polymer, if the melting temperature is known. It is called the *eight-ninths temperature-of-fusion* rule, where T_f is in absolute temperature. At $(8/9)T_f$ the polymer is supposed to crystallize readily. Starting with $T_f = 265°C$, this yields $205°C$ for the poly(ethylene terephthalate) shown in Figure 6.27. While this result is somewhat higher than the maximum rate of crystallization temperature, it remains a handy rule if no other data are available.

6.6.2 Theories of Crystallization Kinetics

Section 6.4.2 described Keller's early preparations of single crystals from dilute solutions. Since the crystals were only about $100\,\text{Å}$ thick and the chains were oriented perpendicular to the flat faces, Keller postulated that the chains had to be folded back and forth.

Similar structures, called lamellae, exist in the bulk state. While their folding is now thought to be much less regular, their proposed molecular organization remains similar. In the bulk state, however, these crystals are organized into the larger structures known as spherulites. Section 6.6.1 showed that the rate of radial growth of the spherulites was linear in time and that the rate of growth goes through a maximum as the temperature of crystallization is lowered. These several experimental findings form the basis for three theories of polymer crystallization kinetics.

The first of these theories is based on the work of Avrami (73–74), which adapts formulations intended for metallurgy to the needs of polymer science. The second theory was developed by Keith and Padden (70,71), providing a qualitative understanding of the rates of spherulitic growth. More recently Hoffman and co-workers (41,77–80) developed the kinetic nucleation theory of chain folding, which provides an understanding of how lamellar structures form from the melt. This theory continues to be developed even as this material is being written. Together, these theories provide insight into the kinetics not only of crystallization but also of the several molecular mechanisms taking part.

6.6.2.1 The Avrami Equation The original derivations by Avrami (73–75) have been simplified by Evans (81) and put into polymer context by Meares (82) and Hay (83). In the following, it is helpful to imagine raindrops falling in a puddle. These drops produce expanding circles of waves that intersect and cover the whole surface. The drops may fall sporadically or all at once. In either case they must strike the puddle surface at random points. The expanding circles of waves, of course, are the growth fronts of the spherulites, and the points of impact are the crystallite nuclei.

The probability p_x that a point P is crossed by x fronts of growing spherulites is given by an equation originally derived by Poisson (84):

$$p_x = \frac{e^{-E}E^x}{x!} \tag{6.5}$$

where E represents the average number of fronts of all such points in the system. The probability that P will not have been crossed by any of the fronts, and is still amorphous, is given by

$$p_0 = e^{-E} \tag{6.6}$$

since E^0 and 0! are both unity. Of course, p_0 is equal to $1 - X_t$, where X_t is the volume fraction of crystalline material, known widely as the degree of crystallinity. Equation (6.6) may be written

$$1 - X_t = e^{-E} \tag{6.7}$$

which for low degrees of crystallinity yield the useful approximation

$$X_t \cong E \tag{6.8}$$

For the bulk crystallization of polymers, X_t (in the exponent) may be considered related to the volume of crystallization material, V_t:

$$1 - X_t = e^{-V_t}. \tag{6.9}$$

The problem now resides on the evaluation of V_t. There are two cases to be considered: (a) the nuclei are predetermined—that is, they all develop at once on cooling the polymer to the temperature of crystallization—and (b) there is sporadic nucleation of the spheres.

For case (a), L spherical nuclei, randomly placed, are considered to be growing at a constant rate, g. The volume increase in crystallinity in the time period t to $t + dt$ is

$$dV_t = 4\pi r^2 L\, dr \tag{6.10}$$

where r represents the radius of the spheres at time t; that is,

$$r = gt \tag{6.11}$$

and

$$V_t = \int_0^1 4\pi g^2 t^2 Lg\, dt \tag{6.12}$$

Upon integration

$$V_t = \tfrac{4}{3}\pi g^3 L t^3 \tag{6.13}$$

For sporadic nucleation the argument above is followed, but the number of spherical nuclei is allowed to increase linearly with time at a rate 1. Then spheres nucleated at time t_i will produce a volume increase of

$$dV_t = 4\pi g^2 (t - t_i)^2 ltg\, dt \tag{6.14}$$

Upon integration

$$V_t = \tfrac{2}{3}\pi g^3 l t^4 \tag{6.15}$$

The quantities on the right of equations (6.12) and (6.14) can be substituted into equation (6.8) to produce the familiar form of the Avrami equation:

$$1 - X_t = e^{-Zt^n} \tag{6.16}$$

which is often written in the logarithmic form:

$$\ln(1 - X_t) = -Zt^n \tag{6.17}$$

The quantity Z is replaced by K in some books (82).

The above derivation suggests that the quantity n in equations (6.16) or (6.17) should be either 3 or 4. [If rates of crystallization are diffusion controlled, which occurs in the presence of high concentrations of noncrystallizable impurities (70,71) $r = gt^{1/2}$, leading to half-order values of n.]

Both Z and n are diagnostic of the crystallization mechanism. The equation has been derived for spheres, discs, and rods, representing three-, two-, and one-dimensional forms of growth. The constants are summarized in Table 6.4 (83). It must be emphasized that the approximation given in equation (6.8) limits the equations to low degrees of crystallinity. In practice, the quantity n frequently decreases as the crystallization proceeds. Values for typical polymers are summarized in Table 6.5 (83).

The Avrami equation represents only the initial portions of polymer crystallization correctly. The spherulites grow outward with a constant radial growth rate until impingement takes place when they stop growth at the intersection, as illustrated in Figures 6.13 and 6.22. Then a secondary crystallization process is often observed after the initial spherulite growth in the amorphous interstices (85).

As an example, miscible blends of high-density (and higher melting temperature) polyethylene (HDPE) and low density polyethylene (LDPE), which contains both short and long branches, and a lower melting temperature (see Chapter 14), are often utilized commercially. On cooling from the melt, the HDPE portion crystallizes first, forming the spherulites, while the LDPE tends to be preferentially located in the amorphous interlamellar regions (86,87), and partly crystallize later in the remaining space.

Table 6.4 The Avrami parameters for crystallization of polymers (83)

	Crystallization Mechanism	Avrami Constants		Restrictions
		Z	n	
Spheres	Sporadic	$2/3\pi g^3 l$	$4\cdot0$	3 dimensions
	Predetermined	$4/3\pi g^3 L$	$3\cdot0$	3 dimensions
Discs[a]	Sporadic	$\pi/3g^2 ld$	$3\cdot0$	2 dimensions
	Predetermined	$\pi g^3 Ld$	$2\cdot0$	2 dimensions
Rods[b]	Sporadic	$\pi/4gld^2$	$2\cdot0$	1 dimension
	Predetermined	$\tfrac{1}{2}\pi gLd^2$	$1\cdot0$	1 dimension

[a] Constant thickness d.

[b] Constant radius d.

Table 6.5   Range of the Avrami constant for typical polymers (83)

Polymer	Range of n	Reference
Polyethylene	2.6–4.0	(a)
Poly(ethylene oxide)	2.0–4.0	(b, c)
Polypropylene	2.8–4.1	(d)
Poly(decamethylene terephthalate)	2.7–4.0	(e)
it-Polystyrene	2.0–4.0	(f, g)

References: (a) W. Banks, M. Gordon, and A. Sharples, *Polymer*, **4**, 61, 289 (1963). (b) J. N. Hay, M. Sabin, and R. L. T. Stevens, *Polymer*, **10**, 187 (1969). (c) W. Banks and A. Sharples, *Makromol. Chem.*, **59**, 283 (1963). (d) P. Parrini and G. Corrieri, *Makromol. Chem.*, **62**, 83 (1963). (e) A. Sharples and F. L. Swinton, *Polymer*, **4**, 119 (1963). (f) I. H. Hillier, *J. Polym. Sci.*, **A-2** (4), 1 (1966). (g) J. N. Hay, *J. Polym. Sci.*, **A-3**, 433 (1965).

If the system is considered as two-phased, then the volume of the amorphous phase is V_a and the volume of the crystalline phase is V_c. The total volume, V, is given by

$$V = X_t V_c + (1 - X_t) V_a \tag{6.18}$$

Then

$$1 - X_t = \frac{V - V_c}{V_a - V_c} \tag{6.19}$$

or for dilatometric experiments,

$$1 - X_t = \frac{h_0 - h_t}{h_0 - h_\infty} \tag{6.20}$$

where h_0, h_t, and h_∞ represent capillary dilatometric heights at time zero, time t, and the final dilatometric reading. Substitution of equation (6.20) into (6.17) yields a method of determining the constants Z and n experimentally (e.g., see Figure 6.25).

6.6.2.2 Keith–Padden Kinetics of Spherulitic Crystallization

Although the Avrami equation provides useful data on the overall kinetics of crystallization, it provides little insight as to the molecular organization of the crystalline regions, structure of the spherulites, and so on.

Section 6.5 described how the spherulites are composed of lamellar structures that grow outward radially. The individual chains are folded back and forth tangentially to the growing spherical surface of the spherulite (see Figure 6.28) (71). Normally, the rate of growth in the radial direction is constant until the spherulites meet (see Figure 6.26). As the spherulites grow, the individual lamellae branch. Impurities, atactic components, and so on, become trapped in the interlamellar regions.

The first theory to address the kinetics of spherulitic growth in crystallizing polymers directly was developed by Keith and Padden (70,71,88). According to Keith and Padden (70), a parameter of major significance is the quantity

$$\delta = \frac{D}{G} \tag{6.21}$$

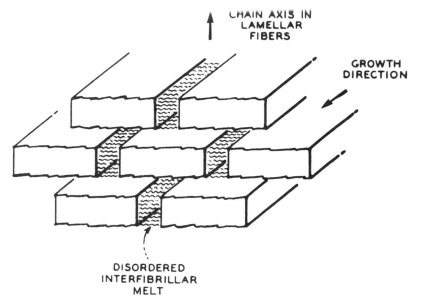

CHAIN AXIS IN
LAMELLAR
FIBERS

GROWTH
DIRECTION

DISORDERED
INTERFIBRILLAR
MELT

Figure 6.28 Schematic representation of the distribution of residual melt and disordered material among lamellae in a spherulite (71).

where D is the diffusion coefficient for impurity in the melt and G represents the radial growth rate of a spherulite. The quantity δ, whose dimension is that of length, determines the lateral dimensions of the lamellae, and that noncrystallographic branching should be observed when δ becomes small enough to be commensurate with the dimensions of the disordered regions on their surfaces. Thus δ is a measure of the internal structure of the spherulite, or its coarseness.

By logarithmic differentiation of equation (6.21),

$$\frac{1}{\delta}\left(\frac{d\delta}{dT}\right) = \frac{1}{D}\left(\frac{dD}{dT}\right) - \frac{1}{G}\frac{dG}{dT} \qquad (6.22)$$

The derivative dD/dT always has a positive value. However, dG/dT may be positive or negative (see Figure 6.27). The coarseness of the spherulites depends on which of the two terms on the right of equation (6.22) is the larger. If the quantity on the right-hand side of the equation is positive, an increase in coarseness is expected as the temperature is increased.

The radial growth rate, G, may be described by the equation

$$G = G_0 e^{\Delta E/RT} e^{-\Delta F^*/RT} \qquad (6.23)$$

where ΔF^* is the free energy of formation of a surface nucleus of critical size, and ΔE is the free energy of activation for a chain crossing the barrier to the crystal. Equation (6.23) allows the temperature dependence of spherulite growth rates to be understood in terms of two competing processes. Opposing one another are the rate of molecular transport in the melt, which increases with increasing temperature, and the rate of

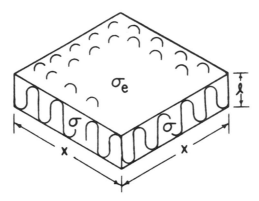

Figure 6.29   Thin chain-folded crystal showing σ and σ_e (schematic) (41).

nucleation, which decreases with increasing temperature (see Figure 6.27). According to Keith and Padden, diffusion is the controlling factor at low temperatures, whereas at higher temperatures the rate of nucleation dominates. Between these two extremes the growth rate passes through a maximum where the two factors are approximately equal in magnitude.

6.6.2.3 Hoffman's Nucleation Theory

The major shortcoming of the Keith–Padden theory resides in its qualitative nature. Although great insight into the morphology of spherulites was attained, little detail was given concerning growth mechanisms, particularly the thermodynamics and kinetics of the phenomenon.

More recently Hoffman and co-workers attacked the kinetics of polymer crystallization anew (41,76–80). Hoffman began with the assumption that chain folding and lamellar formation are kinetically controlled, the resulting crystals being metastable. The thermodynamically stable form is the extended chain crystal, obtainable by crystallizing under pressure (89).

The basic model is illustrated in Figure 6.29 (41), where ℓ is the thin dimension of the crystal, x the large dimension, σ_e the fold surface interfacial free energy, and σ the lateral surface interfacial free energy, a single chain making up the entire crystal. The free energy of formation of a single chain-folded crystal may be set down in the manner of Gibbs as

$$\Delta G_{\text{crystal}} = 4x\ell\sigma + 2x^2\sigma_e - x^2\ell(\Delta f) \tag{6.24}$$

where the quantity Δf represents the bulk free energy of fusion, which can be approximated from the entropy of fusion, ΔS_f, by assuming that the heat of fusion, Δh_f, is independent of the temperature:

$$\Delta f = \Delta h_f - T\,\Delta S_f = \Delta h_f - \frac{T\,\Delta h_f}{T_f^0} = \frac{\Delta h_f(\Delta T)}{T_f^0} \tag{6.25}$$

At the melting temperature of the crystal, the free energy of formation is zero. For $x \gg \ell$,

$$T_f = T_f^0 \left| 1 - \frac{2\sigma_e}{\Delta h_f \ell} \right| \tag{6.26}$$

Equation (6.26) yields the melting point depression in terms of fundamental parameters. The quantity σ_e may be interpreted in terms of the fold structure. For the actual crystallization, chains are added from the melt or solution to the surface of the crystal defined by the area $x\ell$, and T_f^0 represents the large crystal limit.

6.6.2.4 *Example Calculation of the Fold Surface Free Energy*   Find the fold surface interfacial free energy of a new polymer with the following characteristics: the actual melting temperature is 350°C, while after extended annealing, it melts at 360°C. X-ray analysis shows its lamellae are 150 Å thick. The heat of fusion is 12.0 kJ/mol. (See Table 6.1.)

Equation (6.26) can be rewritten

$$\sigma_e = \frac{\Delta h_f \ell}{2} \left(1 - \frac{T_f}{T_f^0} \right)$$

Substituting the given values obtains

$$\sigma_e = \frac{12.0 \text{ kJ/mol} \times 1.50 \times 10^{-6} \text{ cm}}{2} \left(1 - \frac{623 \text{ K}}{633 \text{ K}} \right)$$

$$= 1.42 \times 10^{-7} \text{ kJ} \cdot \text{cm/mol}$$

This polymer has a mer molecular weight of 100 g/mol and a density of 1.0 g/cm^3. Then

$$100 \text{ g/mol} \times 1 \text{ cm}^3/\text{g} = 100 \text{ cm}^3/\text{mol}$$

and

$$\sigma_e = \frac{1.42 \times 10^{-7} \text{ kJ} \cdot \text{cm/mol}}{100 \text{ cm}^3/\text{mol}} = 1.42 \times 10^{-9} \text{ kJ/cm}^2$$

6.6.2.5 *Three Regimes of Crystallization Kinetics*   Hoffman defined three regimes of crystallization kinetics from the melt, which differ according to the rate that the chains are deposited on the crystal surface.

Regime I. One surface nucleus causes the completion of the entire substrate of length L (see Figure 6.30) (77); that is, one chain is crystallizing at a time. Many molecules may be required to complete L. The term "surface nucleus" refers to a segment of a chain sitting down on a preexisting crystalline lamellar structure, as opposed to the nucleus, which initiates the lamellae from the melt in the first place.

These nuclei are deposited sporadically in time on the substrate at a rate i per unit length in a manner that is highly dependent on the temperature. Substrate completion at a rate g begins at the energetically favorable niche that occurs on either side of the surface nucleus, chain folding assumed.

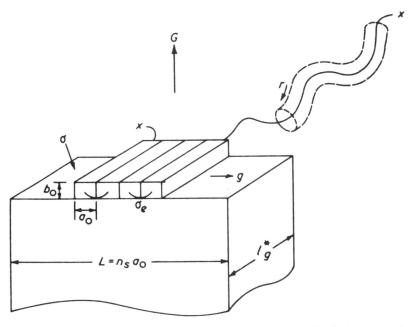

Figure 6.30 Surface nucleation and substrate completion with reptation in regime I, where one surface nucleus deposited at rate i causes completion of substrate of length L, giving overall growth rate $G_I = b_0 i L$. Multiple surface nuclei occur in regime II (not shown) and lead to $G_{II} = b_0(2ig)^{1/2}$, where g is the substrate completion rate. The substrate completion rate, g, is associated with a "reeling in" rate $r = (\ell_g^*/a_0)g$ for the case of adjacent reentry (77).

As illustrated in Figure 6.30, the overall growth rate is given by G, and g is the substrate completion rate. The quantities a_0 and b_0 refer to the molecular width and layer thickness, respectively. The quantity l_g^* refers to the initial fold thickness of the lamellae. The portion of chain occupying the length l_g^* is called a stem.

Again noting Figure 6.30, the "reeling in" or reptation rate r (see Section 5.4) is given by

$$r = \left(\frac{\ell_g^*}{a_0}\right)g \tag{6.27}$$

A significant question, still being debated in the literature, is whether or not reptation type diffusion is sufficiently rapid to supply chain portions as required (90). An alternate theory, proposed by Yoon and Flory (90) suggests that disengagement of the macromolecule from its entanglements with other chains in the melt is necessary for regular folding. In this theory, the 100 to 200 skeletal bonds corresponding to one traversal of the crystal lamella readily undergo the conformational rearrangements required for their deposition in the growth layer. On the other hand, Hoffman (77) concludes that the reptation rate characteristic of the melt is fast enough to allow a significant degree of adjacent reentry "regular" folding during crystallization.

According to Hoffman, the overall growth rate is given by

$$G_I = b_0 iL = b_0 a_0 n_s i \tag{6.28}$$

where G_I is the growth rate in regime I, n_s represents the number of stems of width a_0 that make up this length, and i is the rate of deposition of surface nuclei.

The free energy of crystallizing v stems and $v_f = v - 1$ folds can be generalized from equation (6.24), using Figure 6.30:

$$\Delta G_v = 2b_0 \ell \sigma + 2v_f a_0 b_0 \sigma_e - v a_0 b_0 \ell\, \Delta f \tag{6.29}$$

which for v large becomes

$$\Delta G_v = 2b \ell_g^* \sigma + v a_0 b_0 (2\sigma_e - \ell_g^*\, \Delta f) \tag{6.30}$$

Regime II. In this regime multiple surface nuclei occur on the same crystallizing surface, because the rate of nucleation is larger than the crystallization rate of each molecule. This, in turn, results from the larger undercooling necessary to reach regime II. As in regime I each molecule is assumed to fold back and forth to give adjacent reentry (see Figure 6.31) (76). An important parameter in this regime is the niche separation distance, S_n.

The rate of growth of the lamellae is

$$G_{II} = b_0 (2ig)^{1/2} \tag{6.31}$$

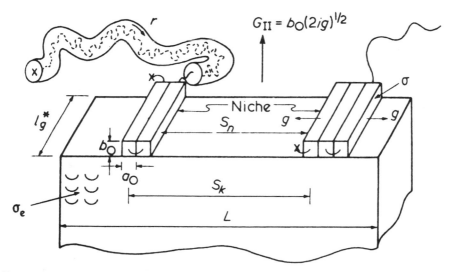

Figure 6.31 Model for regime II growth showing multiple nucleation. The quantity S_k represents the mean separation between the primary nuclei, and S_n denotes the mean distance between the associated niches. The primary nucleation rate is i, and the substrate completion rate is g. The overall observable growth rate is G_{II}. Reptation tube contains molecule being reeled at rate r onto substrate (75).

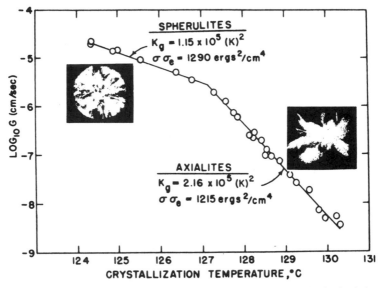

Figure 6.32 Change in growth kinetics from regime I to regime II in melt-crystallized polyethylene. Growth rate in melt-crystallized polyethylene is shown as a function of the crystallization temperature for fraction with molecular weight of 30,600 g/mol. The pronounced change in slope is accompanied by a change in morphology, as shown by optical micrographs of the type from which the growth rate data were obtained (41).

The number of nucleation sites per centimeter in regime II is given by

$$N_k = \left(\frac{i}{2g}\right)^{1/2} \tag{6.32}$$

Then the mean separation distance between the sites is

$$S_k = \frac{1}{N_k} = \left(\frac{2g}{i}\right)^{1/2} \tag{6.33}$$

Regimes I and II differ morphologically as well as kinetically. In regime I, polyethylene usually forms axialites. In regime II, normal spherulites are formed. The rates of growth show different slopes, as illustrated in Figure 6.32 (41). For very high or very low molecular weights, Hoffman et al. (91) found that the rates of growth of polyethylenes gave curves, rather than a sharp break at the dividing point between the two regimes.

Regime III. This regime becomes important when the niche separation characteristic of the substrate in regime II approaches the width of a stem, producing a snow storm effect. Regime III is very important industially, where rapid cooling is employed. In this regime, the crystallization rate is very rapid. The growth rate is given by

$$G_{III} = b_0 iL = b_0 i n_s a_0 \tag{6.34}$$

where n_ν is the mean number of atoms laid down in the niche adjacent to the newly nucleated stem.

In regime III the chains do not undergo repeated adjacent reentry into the lamellae but have only a few folds before entering the amorphous phase. Then they are free to reenter the same lamella via a type of switchboard model, or go on to the next lamella.

As the temperature is lowered through regimes I, II, and III, substrate completion rates per chain decrease. However, more chains are crystallizing simultaneously. At temperatures approaching T_g, de Gennes reptation is severely limited, as illustrated in Figure 6.27.

6.7 THE REENTRY PROBLEM IN LAMELLAE

In the discussion above the lamellae were assumed to the formed through regular adjacent reentry, although it was recognized that this was an oversimplification. The concept of the switchboard and folded-chain models were briefly developed in Section 6.4. The question of the molecular organization within polymer single crystals as well as the bulk state has dogged polymer science since the discovery of lamellar-shaped single crystals in 1957 (19). Again, X-ray and other studies show that the chains are perpendicular to the lamellar surface (see Section 6.4.2).

Since the chain length far exceeds the thickness of the crystal, the chains must either reenter the crystal or go elsewhere. However, the relative merits of the switchboard versus folded-chain models remained substantially unresolved for several years for lack of appropriate instrumentation.

6.7.1 Infrared Spectroscopy

Beginning in 1968, Tasumi and Krimm (92) undertook a series of experiments using a mixed crystal infrared spectroscopy technique. Mixed single crystals of protonated and deuterated polymer were made by precipitation from dilute solution. The characteristic crystal field splitting in the infrared spectrum was measured and analyzed to determine the relative locations of the chain stems of one molecule, usually the deuterated portion, in the crystal lattice. The main experiments involved blending protonated and deuterated polyethylenes (93–95).

The main findings were that folding takes place with adjacent reentry along (110) planes for dilute solution-grown crystals. In addition, it was also concluded that there is a high probability for a molecule to fold back along itself on the next adjacent (110) plane.

Melt-crystallized polyethylene was shown to be organized differently, with a much lower (if any) extent of adjacent reentry (95). However, significant undercooling was required to prevent segregation of the deuterated species from the ordinary, hydrogen-bearing species.[†] Since the experimental rate of cooling was estimated to be 1°C per minute down to room temperature, it may be that some crystallization occurred in all three regimes (see Section 6.6.2.3).

[†] In fact, one early problem that continues to plague many studies is that deuterated polyethylene tends to phase-separate from ordinary, protonated polymer, even though they are chemically identical. The cause has been related to slightly different crystallization rates owing to polydeuteroethylene melting 6°C lower than ordinary polyethylene.

6.7.2 Carbon-13 NMR

Additional evidence for chain folding in solution-grown crystals comes from carbon-13 NMR studies of partially epoxidized 1,4-trans-polybutadiene crystals (49,50). This polymer was crystallized from dilute heptane solution and oxidized with *m*-chloroperbenzoic acid. This reaction is thought to epoxidize the amorphous portions present in the folds, while leaving the crystalline stem portions intact.

The result was a type of block copolymer with alternating epoxy and double-bonded segments. NMR analyses showed that for the two samples studied the chain-folded portion was about 2.4 and 3 mers thick, whereas the stems were 15.2 and 40.8 mers thick, respectively. Since the number of mer units to complete the tightest fold in this polymer has been calculated to be about three (51), the NMR study strongly favors a tight adjacent reentry fold model for single crystals.

Broad line proton NMR (96) and Raman analyses (97) of polyethylenes indicated three major regions for crystalline polymers: the crystalline region, the interfacial or interzonal region, and the amorphous or liquidlike region. For molecular weight of 250,000 g/mol, the three regions were 75, 10, and 15% of the total, respectively. The presence of the interfacial regions reduces the requirements for chain folding (98).

6.7.3 Small-Angle Neutron Scattering

6.7.3.1 Single-Crystal Studies With the advent of small-angle neutron scattering (SANS), the several possible modes of chain reentry could be put to the test anew (79,98–110). Sadler and Keller (108–110) prepared blends of deuterated and normal (protonated) polyethylene and crystallized them from dilute solution. The radius of gyration, R_g, was determined as a function of molecular weight.

The several models possible are illustrated in Figure 6.33 (111). For adjacent reentry (Figure 6.33c), R_g should vary as $M^{1.0}$ for high enough molecular weights, since a type of rod would be generated. For the switchboard model (Figure 6.33d), R_g is expected to vary close to $M^{0.5}$, since the chains would be expected to be nearly Gaussian in conformation. The several possible relationships between R_g and M are set out in Table 6.6 (112).

For solution-grown crystals, Sadler and Keller (109) found that R_g depended on M only to the 0.1 power. Such a situation could arise only if the stems folded up on themselves beyond a certain number of entries (see Figure 6.33), called superfolding. However, the 0.1 power dependence appears to hold only for intermediate molecular weight ranges. For low enough M, there should not be superfolding. For high enough molecular weight, a square plate with a 0.5 power dependence would be generated.

Recent quantitative calculations of the absolute scattering intensities expected from various crystallite models for single crystals by Keller (110) and Yoon and Flory (114–116) (Figure 6.34) on polyethylene suggested that the model for adjacent reentry does not correlate with experiment. Rather, Yoon and Flory put forward a model requiring a stem dilution by a factor of 2–3. The calculated scattering functions are shown in Figure 6.34; this leads to the model in Figure 6.33b. This last suggests a type of skip mechanism, with two or three chains participating.

6.7.3.2 Melt-Crystallized Polymers Upon crystallization from the melt, an entirely different result emerges. Experiments by Sadler and Keller (109–111) showed that nearly random stem reentry was most likely; that is, some type of switchboard model was cor-

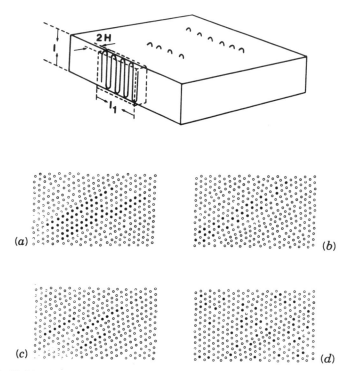

Figure 6.33 Models of stem reentry for chain sequences in a lamellar-shaped crystal. (*a*) Regular reentry with superfolding; (*b*) partial nonadjacency (stem dilution) as required by closer matching of the experimental data in accord with Yoon and Flory; (*c*) adjacent stem positions without superfolding; (*d*) the switchboard model. All reentry is along the (110) plane; superfolding is along adjacent (110) planes. View is from the (001) plane, indicated by dots (111).

rect. Quantitative calculations by Yoon and Flory (114–116) and by Dettenmaier et al. (117,118) on melt-crystallized polyethylene (119) and isotactic polypropylene (120) also showed that adjacent reentry should occur only infrequently on cooling from the melt.

Three regions of space were defined by Yoon (115): a crystalline lamellar region about 100 Å thick, an interfacial region about 5 to 15 Å thick, and an amorphous region about 50 Å thick. A nearby reentry model constrains the chain within the interfacial layer during the irregular folding process. The calculated reentry dimensions for solution and melt-crystallized polyethylene are summarized in Table 6.7 (115). A major problem,

Table 6.6 Relationships among geometric shape, R_g, and M (112)

Geometric Shape	R_g Equals	Molecular Weight Dependence
Sphere	$D/\sqrt{20/3}$	$M^{1/3}$
Rod	$L/\sqrt{12}$	M^{1}
Random coil	$r/\sqrt{6}$	$M^{0.5}$
Rectangular plate	$(b^2 + l^2)^{1/2}/\sqrt{12}$	M^{variable}
Square plate	$A^{1/2}/\sqrt{6}$	$M^{0.5}$

Symbols: D, diameter; L, length of rod; r, end-to-end distance; A, area; b, width of plate; l, length of plate.

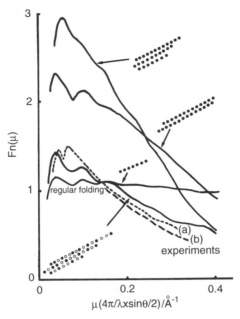

Figure 6.34 Scattering functions $F_n(\mu)$ calculated for various arrangements of $n_s = 40$ PED stems, each stem containing 80 bonds, in a PEH single-crystal matrix (110).

of course, was that the samples had to be severely undercooled to prevent segregation of the two species during crystallization. Undoubtedly, large portions of the crystallization took place in regime III.

Crist et al. (121) deuterated or protonated a slightly branched polybutadiene to produce a type of polyethylene. Blends of these two materials were used in SANS studies. The scattering curves indicated identical dimensions (R_g values) for both the melt and melt-crystallized materials; these were the same as expected for Flory θ-solvent values for polyethylene. Using wide-angle neutron scattering, Wignall et al. (102) concluded that the number of stems that could be regularly folded had an upper limit of about four.

Of course, other crystallizable polymers have been studied (103–108). These include polypropylene (103–106), poly(ethylene oxide) (107), and isotactic polystyrene (108). Wignall et al. (102) have summarized the values of $R_g/M_w^{1/2}$ in both the melt and crystalline states (see Table 6.8). Most interestingly, the dimensions in the crystalline state and in the melt state are virtually identical for all these polymers. None of these data

Table 6.7 Polyethylene lamellar reentry dimensions (114)

	Reentry Statistics	
Case	Probability of Reentering Same Crystal	Average Displacement on Reentry
Solution-crystallized	1.0	10–15 Å
Melt-crystallized	0.7	25–30 Å

Table 6.6   Comparison of molecular dimensions in molten and crystallized polymers (102)

Polymer	Method of Crystallization	$R_g/M_w^{1/2}$ Å/(g/mol)$^{1/2}$ Melt	Crystallized
Polyethylene	Rapidly quenched from melt	0.46	0.46
it-Polypropylene	Rapidly quenched	0.35	0.34
	Isothermally crystallized at 139°C	0.35	0.38
	Rapidly quenched from melt and subsequently annealed at 137°C	0.35	0.36
Poly(ethylene oxide)	Slowly cooled	0.42	0.52
it-polystyrene	Crystallized at 140°C (5 h)	0.26–0.28[a]	0.24–0.27
	Crystallized at 140°C (5 h) then at 180°C (50 min)		0.26
	Crystallized at 200° (1 h)	0.22[a]	0.24–0.29

[a]Dimensions in the melt were not available. The values quoted are for atactic polystyrene annealed in the same way as the crystalline material.

show a decrease in R_g on crystallization. That the R_g values in the melt and in the crystallized material are the same all but rules out regular folding. In fact, the data for poly(ethylene oxide) (107) shows a slight increase, if anything. By way of summarizing the above studies on melt-crystallized polymers, it seems that adjacent reentry occurs much less than in solution-crystallized polymers. Some experiments suggest very little adjacent reentry.

Hoffman (76) and Frank (122) point out, however, that some folding is required. Alternatively, the crystals must have the chains at an oblique angle to the crystal surface. If neither condition is met, a serious density anomaly at the crystalline–amorphous interface is predicted: the density is too high. These workers point out that the difficulty can be mitigated by interspersing some tight folds between (or among) longer loops in the amorphous phase (see Figure 6.35) (122). Frank (122) derived a general equation that combines the probability of back folding, p, and the obliquity angle, θ, with the crystalline stem length, l, and the contour length of the chain, L, to yield the minimum

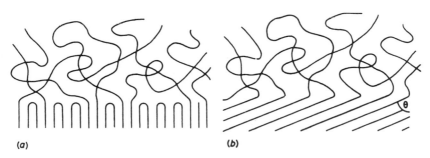

(a)   (b)

Figure 6.35 Alternative resolutions of the density paradox $(1 - p - 2\ell/L)\cos\theta < 3/10$ (116): (a) increased chain folding beyond critical value and (b) oblique angle crystalline stems to reduce amorphous chain density at interface.

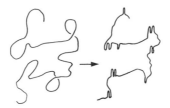

Figure 6.36 The variable cluster model, showing how a chain can crystallize from the melt with some folding and some amorphous portions and retain, substantially, its original dimensions and its melt radius of gyration (112).

conditions to prevent an anomalous density in the amorphous region:

$$\left(1 - p - \frac{2l}{L}\right) \cos \theta \leqq \frac{3}{10} \tag{6.35}$$

The findings above led to two different models. In 1980 Dettenmaier et al. (117) proposed their solidification model, whereby it was assumed that crystallization occurred by a straightening out of short coil sequences without a long-range diffusion process. Thus these sequences of chains crystallized where they stood, following a modified type of switchboard model (53). This was the first model to illustrate how R_g values could remain virtually unchanged during crystallization.

On the other hand, Hoffman (76) showed that the density of the amorphous phase is better accounted for by having at least about 2/3 adjacent reentries, which he calls the variable cluster model. An illustration of how a chain can crystallize with a few folds in one lamella, then move on through an amorphous region to another lamella, where it folds a few more times and so on, is illustrated in Figure 6.36 (113). Thus a regime III crystallization according to the variable cluster model will substantially retain its melt value of R_g.

The general conclusion from these studies is that the fringed micelle model fails, because it predicts that the density of the amorphous polymer at both ends of the crystal will be higher than that of the crystal itself. While the chains could be laid over at a sharp angle (Figure 6.35b), the most viable alternative is to introduce a significant amount of chain folding.

It is of interest to compare the results of this modern research with Hosemann's paracrystalline model, first published in 1962. As illustrated in Figure 6.37 (123), this model emphasizes lattice imperfections and disorder, as might be expected from regime III crystallization. This model also serves as a bridge between the concepts of crystalline and amorphous polymers (see Figure 5.3). More recent research by Hosemann has continued to examine the partially ordered state (124,125).

By way of summary, for dilute solution-grown crystals a modified regular reentry model fits best, with the same molecule forming a new stem either after immediate reentry or after skipping over one or two nearest-neighbor sites. For melt-formed crystals the concept of folded chains is considerably modified. Since active research in this area is now in progress, perhaps a more definitive set of conclusions will be forthcoming.

It must be remembered that the formation of lamellae, whether with adjacently folded chains or with a switchboardlike structure, is kinetically controlled by the degree of

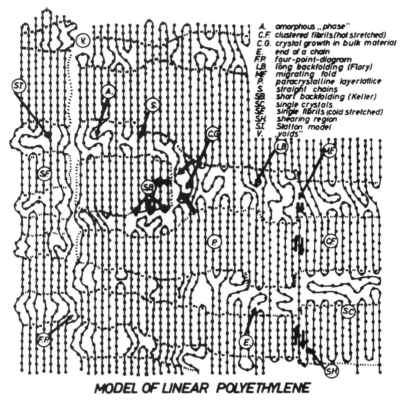

A.	amorphous „phase"
C.F.	clustered fibrils (hot stretched)
C.G.	crystal growth in bulk material
E.	end of a chain
F.P.	four-point-diagram
LB.	long backfolding (Flory)
MF.	migrating fold
P.	paracrystalline layerlattice
S.	straight chains
SB.	short backfolding (Keller)
SC.	single crystals
SF.	single fibrils (cold stretched)
SH.	shearing region
ST.	Statton model
V.	voids"

MODEL OF LINEAR POLYETHYLENE

Figure 6.37 The paracrystalline model of Hosemann (123). Amorphous structures are illustrated in terms of defects. A radius of gyration approaching amorphous materials might be expected.

undercooling and finite rates of molecular motion. The thermodynamically most stable crystal form is thought to have extended chains.

6.7.4 Extended Chain Crystals

Wunderlich (126,127) pointed out that the thermodynamic equilibrium crystalline state has an extended chain macroconformation when the crystallization is carried out under great hydrostatic pressure. Thus polyethylene forms extended chain single crystals at pressures approaching 5 kbar (128,129). These crystals form long needlelike structures, which may be several μm in length. In the discussion above it was pointed out that polymer chains fold during crystallization at atmospheric pressure, in significant measure because of kinetic circumstances. It is now thought that the appearance of folded chain instead of extended chain alternative phase variants depends on a complex interaction between thermodynamics and kinetics.

The development of a pressure–temperature phase diagram (130) for polyethylene showed that *orthorhombic* (folded chain), *o*, and *hexagonal* (extended chain), *h*, crystal domains were placed in such a way that on cooling from the melt above about 4 kbar, first the hexagonal crystal structure was encountered and then the orthorhombic. The

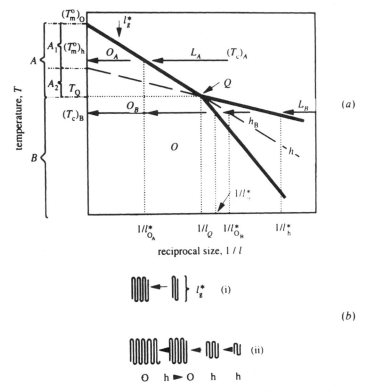

Figure 6.38 Variation of lamellar crystal thickness *vs.* temperature. (*a*) Phase growth in terms of a *phase stability* diagram. Notation *h* and *o* refer to crystal forms in polyethylene, and L stands for the liquid melt. The two sets of horizontal arrows pointing towards $1/l = 0$ denote isothermal growth pathways at the two selected crystallization temperatures, T_c, chosen to be in the two temperature regimes A and B. (*b*) Schematic representation of chain folded polymer crystal grown: (*i*) Region A leading to lamellae of a specific thickness l_g, in which they continue to grow laterally through direct growth in phase *o*. (*ii*) Region B where crystals arise in the *h* phase and develop by simultaneous lateral and thickening growth with the latter stopping (or slowing down) on the *h* to *o* transformation and/or impingement. T_m: melting temperature.

surprising conclusion was that at room temperature and one atmosphere, the hexagonal structure was metastable.

The thermodynamic stability of a crystal depends on its dimensions, and in particular, on its thickness (Section 6.6.2.3). Extended crystals can be considered the limit of thick crystals, with molecules adding end-on-end as well as side-to-side. From a study of a phase stability diagram (Figure 6.38), it was shown that the relative stability of the hexagonal and orthorhombic phases can invert with size; that is, above a certain thickness the folded chain is more stable (131).

Figure 6.38 shows the relation between reciprocal lamellar crystal thickness, $1/l$, and temperature under isobaric conditions (131). There are several regions of isothermal growth possible, indicated by horizontal arrows pointing toward $1/l = 0$ (i.e., infinite thickness) chosen to lie in the two principal temperature regions, above and below T_Q, denoted by A and B, respectively. While in the liquid, L, stability region, any crystalli-

..utiOn is transient until the size corresponding to a phase line at a specified temperature is reached. At this point the new crystal phase will become stable and capable of continued growth. In region A the first crystal to appear is thought to be in the o phase, but growth will proceed only up to a limited value of l, $l_g{}^*$, after which there is an $h \rightarrow o$ transformation, and further growth will be in the lateral direction with constant l. Thus it substantially passes straight into the region of ultimate stability, the orthorhombic phase structure, o.

At a lower temperature, region B, first the hexagonal crystal structure forms, h. The lamellar thickness increases in the course of continuing crystal growth. This hexagonal form remains only stable within a limited size range at modest pressures, being metastable for larger dimensions, with $1/l_{tr}{}^*$ representing the boundary between the two phase regimes in $T - 1/l$ space.

For very high pressures, $P > P_Q$ (132), the h phase remains stable even for infinite l. However, from a kinetic point of view, lateral growth in folded-chain crystals is thought to be faster than thickening growth, favoring the appearance of folded-chain morphologies under most circumstances. The o phase is much less mobile than the h phase, slowing down growth of the h phase in the presence of the o phase.

There is a continuing connection with kinetics of crystallization in this argument. When region A is small, the polymer must be close to $(T_m{}^o)_o$ throughout the region, hence at small supercooolings. Consequently crystallization in this region will be slow, and may be unrealizable. In this case, on cooling, crystallization will take place in region B, resulting in hexagonal crystallization. If region A is wide, however, the crystallization may start at the ultimately more stable orthorhombic state before it reaches T_Q.

6.8 THERMODYNAMICS OF FUSION

In the previous sections it was shown that the formation of lamellae with folded chains was essentially a kinetically controlled phenomenon. This section treats the free energy of polymer crystallization and melting point depression.

Melting is a first-order transition, ordinarily accompanied by discontinuities of such functions as the volume and the enthalpy. Ideal and real melting in polymers is illustrated in Figure 6.39. Ideally polymers should exhibit the behavior shown in Figure 6.39a, where the volume increases a finite amount exactly at the melting (fusion) temperature, T_f. (The subscript M, for melting, is also in wide use. In this text, M represents

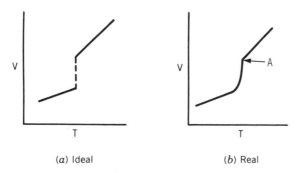

(a) Ideal   (b) Real

Figure 6.39 Dilatometric behavior of polymer melting.

mixing.) Note that the coefficient of thermal expansion also increases above T_f. Owing to the range of crystallite sizes and degrees of perfection in the real case, a range of melting temperatures is usually encountered, as shown experimentally in Figure 6.2. The classic melting temperature is usually taken where the last trace of crystallinity disappears, point A in Figure 6.39b.

The free energy of fusion, ΔG_f, is given by the usual equation,

$$\Delta G_f = \Delta H_f - T\,\Delta S_f \tag{6.36}$$

where ΔH_f and ΔS_f represent the molar enthalpy and entropy of fusion. At the melting temperature, ΔG_f equals zero, and

$$T_f = \frac{\Delta H_f}{\Delta S_f} \tag{6.37}$$

Thus a smaller entropy or a larger enthalpy term raises T_f. Since highly polar molecules such as polyamides tend to have large values of ΔH_f (see Table 6.1), they have high melting temperatures.

6.8.1 Theory of Melting Point Depression

The melting point depression in crystalline substances from the pure state T_f^0 is given by the general equation (132)

$$\frac{1}{T_f} - \frac{1}{T_f^0} = -\frac{R}{\Delta H_f}\ln a \tag{6.38}$$

where a represents the activity of the crystal in the presence of the impurity.

The thermodynamics of melting in polymers was developed by Flory and his co-workers (134–136). To a first approximation, the melting point depression depends on the mole fraction of impurity, X_B, the mole fraction of crystallizable polymer being X_A. Substituting X_A for a in equation (6.38),

$$\frac{1}{T_f} - \frac{1}{T_f^0} = -\frac{R}{\Delta H_f}\ln X_A \tag{6.39}$$

For small values of X_B,

$$-\ln X_A = -\ln(1 - X_B) \cong X_B \tag{6.40}$$

In the following discussion, ΔH_f is the heat of fusion per mole of crystalline mers. There are three important cases in which the melting temperature may be depressed. If X_B represents the mole fraction of noncrystallizable comonomer incorporated in the chain,

$$\frac{1}{T_f} - \frac{1}{T_f^0} = \frac{R}{\Delta H_f}X_B \tag{6.41}$$

The mmm unit at the end of the chain must always have a different chemical structure from those of the mers along the chain. Thus end mers constitute a special type of impurity, and the melting point depends on the molecular weight. If M_0 is the molecular weight of the end mer (and assuming that both ends are identical), the mole fraction of the chain ends is given approximately by $2M_0/M_n$. Thus

$$\frac{1}{T_f} - \frac{1}{T_f^0} = \frac{R}{\Delta H_f} \frac{2M_0}{M_n} \tag{6.42}$$

which predicts that the highest possible melting temperature will occur at infinite molecular weight.

If a solvent or plasticizer is added, the case is slightly more complicated. Here, the molar volume of the solvent, V_1, and the molar volume of the polymer repeat unit, V_u, cannot be assumed to be equal. Also, the interaction between the polymer and the solvent needs to be taken into account. The result may be written (136)

$$\frac{1}{T_f} - \frac{1}{T_f^0} = \frac{R}{\Delta H_f} \frac{V_u}{V_1} (v_1) \tag{6.43}$$

where v_1 represents the volume fraction of diluent.

The quantity χ_1 has been interpreted in several ways (134,135). Principally it is a function of the energy of mixing per unit volume. For calculations involving plasticizers, the form using the solubility parameters δ_1 and δ_2 is particularly easy to use (137) (see Section 3.3.2):

$$\chi_1 = \frac{(\delta_1 - \delta_2)^2 V_1}{RT} \tag{6.44}$$

Corresponding relations for the depression of the glass transition temperature, T_g, by plasticizer are given in Section 8.8.1.

Corresponding equations for the dependence of the melting point on pressure were derived by Karasz and Jones (138). For a pressure P_f,

$$P_f - P_f^0 = \frac{RT_f}{\Delta V_f} \frac{V_u}{V_1} (v_1) \tag{6.45}$$

where ΔV_f is the volume change on fusion.

6.8.2 Example Calculation of Melting Point Depression

Suppose that we swell poly(ethylene oxide) with 10% of benzene. What will be the new melting temperature, if it melted at 66°C dry?

Equation (6.43) can be solved, with the aid of Tables 3.1, 3.2, 3.4, and 6.1. First, the molar volumes of the polymer and the solvent must be computed: for benzene, with six carbons and six hydrogens, $M = 78$ g/mol. Its density is 0.878 g/cm^3. Division yields 88 cm^3/mol. For poly(ethylene oxide), the mer molecular weight is 44 g/mol, and its density is 1.20 g/cm^3, yielding a molar volume of 36.6 cm^3/mol. Equation (6.43), with appropriate numerical values, is

$$\frac{1}{T_f} = -\frac{R}{\Delta H_f}\left(\frac{V_u}{V_1}\right)(v_1) + \frac{1}{T_f^0}$$

$$\frac{1}{T_f} = +\frac{0.0083 \text{ kJ/mol} \cdot \text{K}}{8.29 \text{ kJ/mol}}\left(\frac{36.6 \text{ cm}^3/\text{mol}}{88 \text{ cm}^3/\text{mol}}\right)(0.1) + \frac{1}{339 \text{ K}}$$

$$T_f = 334.3 \text{ K or } 61.4°\text{C}$$

If room temperature is about 25°C, then the crystallinity will not be destroyed by the benzene plasticizer.

6.8.3 Experimental Thermodynamic Parameters

The quantity T_f^0 may be determined either directly on the pure polymer or by a plot of $1/T_f$ versus v_1. The latter is very useful in the case where the polymer decomposes below its melting temperature.

Once T_f^0 is determined, equations (6.43) and (6.44) permit the calculation of both the Flory interaction parameter and the heat of fusion of the polymer from the slope and intercept of a plot of $(1/T_f) - (1/T_f^0)$ versus $(1/T_f)$ (135). The heat of fusion determined in this way measures only the crystalline portion. If heat of fusion data are compared with corresponding data obtained by DSC (see Figure 6.3), which measures the heat of fusion for the whole polymer, the percent crystallinity may be obtained.

6.8.4 Entropy of Melting

In classical thermodynamics, the change in Gibbs' free energy is zero at the melting point,

$$\Delta G_f = \Delta H_f - T\,\Delta S_f = 0 \tag{6.46}$$

where $T = T_f$. For polyethylene, per $-\text{CH}_2-$ group, with $\Delta H_f = 3.94$ kJ/mol; see Table 6.1. Then

$$\Delta S_f = \Delta H_f/T_f = 9.61 \text{ J/mol} \cdot \text{K} \tag{6.47}$$

at a T_f of 410 K.

Statistical thermodynamics asks how many conformational changes are involved in the melting process. If the polyethylene is crystallized in the all *trans* conformation, and two *gauche* plus the *trans* are possible in the melt (see Section 2.1.2), then the polymer goes from one possible conformation to three on melting (139,140). From equation (3.13),

$$\Delta S = R \ln \Omega = R \ln 3 = 9.13 \text{ J/mol} \cdot \text{K} \tag{6.48}$$

The agreement between the classical and statistical results, equations (6.47) and (6.48), is seen to be excellent, noting the approximations involved.

A more general statistical thermodynamic theory can be obtained with the quasi-lattice models; see Figure 3.3. If a coordination number z of the lattice is assumed, then there are $z - 1$ choices of where to put the next bond in the chain. (This is a little smaller

Table 6.9   Entropies of fusion for various polymers (135)

Polymer	Repeating Unit	Entropy of fusion, J/mol·K	
		E.u./mol of Repeating Unit	E.u./No. Bonds Permitting Rotation
Polyethylene	$-CH_2-$	8.37	8.37
Cellulose tributyrate	$-C_{18}H_{28}O_{18}-$	25.9	13.0
Poly(decamethylene sebacate)		145	6.3
	$-O-(CH_2)_{10}-O-CO-(CH_2)_8CO-$		
Poly(N,N'-sebacoylpiperazine)		57.3	5.0

$$-N\begin{matrix} CH_2CH_2 \\ \diagup \qquad \diagdown \\ \diagdown \qquad \diagup \\ CH_2CH_2 \end{matrix}N-CO(CH_2)_8CO-$$

with excluded volume considered.) Then the entropy varies with dimensionality of the quasi-lattice according to

$$3\text{-D:} \quad \Delta S_f = R \ln 5 = 13.4 \text{ J/mol} \cdot \text{K} \tag{6.49}$$

$$2\text{-D:} \quad \Delta S_f = R \ln 3 = 9.13 \text{ J/mol} \cdot \text{K} \tag{6.50}$$

$$1\text{-D:} \quad \Delta S_f = R \ln 1 = 0 \tag{6.51}$$

If more than one group needs to be considered, the values are multiplicative, rather than additive, Taking two $-CH_2-$ groups, for example, yields nine conformations, rather than six. The general equation can be written

$$\Delta S = R \ln(A^a B^b C^c \cdots) \tag{6.52}$$

where $A, B, C, \ldots$ are the number of ways the various moieties can be arranged in space, and $a, b, c, \ldots$ are the number of appearances of the moiety in each mer.

Values of entropy of fusion are shown in Table 6.9 (135). While the entropies per mer varied widely, as might be expected from the enormous differences in the sizes and structures of the units, values divided by the number of chain bonds about which free rotation is permitted gave more nearly uniform values. According to Flory (141), the configurational entropy of fusion per segment should be $R \ln(Z' - 1)$, where Z' is the coordination number of the lattice. Values of E.u./No. bonds permitting rotation in Table 6.9 are in rough agreement with Flory's calculation. Based on equation (6.37), it is easy to understand why large heats of fusion produce high melting polymers (see Table 6.1). The quantity E.u. represents entropy units per mer.

In experiments such as the above, heating and cooling are usually done very slowly. Therefore regime I structures may predominate. The melting temperature is higher under these conditions than when cooling or heating is rapid, in which case regime II and III kinetics apply.

6.8.5   The Hoffman–Weeks Equilibrium Melting Temperature

There are several definitions of the equilibrium melting temperature currently in use. According to equation (6.42) the highest melting temperature (and presumably the equilibrium melting temperature) is reached at infinite molecular weight. Another definition assumes infinitely thick crystalline lamellae (142); see Section 6.7.4.

According to Hoffman and Weeks (143) the equilibrium melting temperature of a polymer, T_f^*, is defined as the melting point of an assembly of crystals, each of which is so large that surface effects are negligible and that each such large crystal is in equilibrium with the normal polymer liquid. Furthermore the crystals at the melting temperature must have the equilibrium degree of perfection consistent with the minimum free energy at T_f^*.

While this definition holds for most pure compounds, polymers as ordinarily crystallized tend to melt below T_f^* because the crystals are small and all too imperfect. Thus the temperature of crystallization, usually still lower because of supercooling (see Figure 6.3), has an important influence on the experimentally observed melting point. Hoffman and Weeks (143) found the following relation to hold:

$$T_f^* - T_f = \phi'(T_f^* - T_c) \tag{6.53}$$

where ϕ' represents a stability parameter that depends on crystal size and perfection. The quantity ϕ' may assume all values between 0 and 1, where $\phi' = 0$ implies that $T_f = T_f^*$, whereas $\phi' = 1$ implies that $T_f = T_c$. Therefore crystals are most stable at $\phi' = 0$ and inherently unstable at $\phi' = 1$. Values of ϕ' near $\frac{1}{2}$ are common.

The experiment generally involves rapidly cooling the polymer from the melt to some lower temperature, T_c, where it is then crystallized isothermally. For higher crystallization temperatures the polymer forms more perfect crystals; that is, it is better annealed. Hence its melting temperature increases.

To determine T_f^*, a plot of T_c versus T_f is prepared (143) (Figure 6.40). A line is drawn where $T_c = T_f$. The experimental data are extrapolated to the intersection with the line. The temperature of intersection is T_f^*.

6.9   EFFECT OF CHEMICAL STRUCTURE ON THE MELTING TEMPERATURE

The actual values of the enthalpy and entropy of fusion are, of course, controlled by the chemical structure of the polymer. The most important inter- and intramolecular structural characteristics include structural regularity, bond flexibility, close packing ability, and interchain attraction (144,145). In general, high melting points are associated with highly regular structures, rigid molecules, close packing capability, strong interchain attraction, or several of these factors combined.

The effect of structural irregularities can be illustrated by a study of polyesters (145) having the general structure

$$\left\{ \begin{matrix} \text{O} \\ \| \\ \text{C} \end{matrix} - \langle \ \rangle - \begin{matrix} \text{O} \\ \| \\ \text{C} \end{matrix} - \text{O} - \text{R} - \text{O} \right\}_n \tag{6.54}$$

when $-\text{R}-$ is $-\text{CH}_2-\text{CH}_2-$, the structure is Dacron®. The melting temperature depends on the regularity of the group R. For aliphatic groups, the size and regularity of R

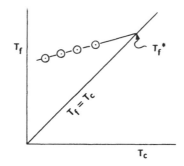

Figure 6.40 Idealized Hoffman-Weeks plot, showing the extrapolation to T_f^*.

are both important:

R	T_f, °C
$-CH_2-CH_2-$	265
$+CH_2 +_3$	220
$-CH_2-CH-$ $\quad\quad\;\;\mid$ $\quad\quad\;CH_3$	Noncrystalline

The irregularity of the atactic propylene unit destroys crystallinity entirely.

The effect of bond flexibility may also be examined utilizing the polyester structure (145). In this case substitutions in the rigid aromatic group are made:

$$+O-\overset{\overset{\displaystyle O}{\|}}{C}-R'-\overset{\overset{\displaystyle O}{\|}}{C}-O-CH_2-CH_2+_n$$

R'	T_f, °C	
⬡	265	(6.55)
⬡⬡	355	
$+CH_2+_4$	50	

The flexible aliphatic group, although having dimensions similar to those of the phenyl group, has a much lower melting temperature. It should be pointed out that the aliphatic polyesters, first synthesized by Carothers (146), failed as clothing fibers because they melted during washing or ironing. Aromatic polyesters (147) as well as aliphatic nylons achieved the necessary high melting temperatures; see Chapter 7.

Interchain forces can be illustrated by the following substitutions (145) of increasingly polar groups:

$$
\left[O - \overset{\displaystyle O}{\underset{\displaystyle \|}{C}} - \text{⟨benzene⟩} - R'' - \text{⟨benzene⟩} - \overset{\displaystyle O}{\underset{\displaystyle \|}{C}} - O - CH_2 - CH_2 \right]_n
\tag{6.56}
$$

R''	T_f, °C
$-(CH_2)_4-$	170
$-O-CH_2-CH_2-O-$	240
$-NH-CH_2-CH_2-NH-$	273

Similar effects are caused by bulky substituents and by odd or even numbers of carbon atoms in hydrocarbon segments of the chain. Generally, bulky groups lower T_f, because they separate the chains. The odd or even number of carbon atoms affects the regularity of packing. Of course, the frequency of occurrence of polar groups is very important. As the length of the aliphatic group R in equation (6.54) is increased, the melting point gradually approaches that of polyethylene, or about 139°C.

In each of the cases above, of course, the crystal structure is governed by the principles laid down in Section 6.3.5 and elsewhere. Generally, those structures that are most tightly bonded, fit the most closely together, and are held in place the most rigidly will have the highest melting temperatures.

6.10 FIBER FORMATION AND STRUCTURE

The synthetic fibers of today, the polyamides, polyesters, rayons, and so on, are manufactured by a process called spinning. Spinning involves extrusion through fine holes known as spinnerets. Immediately after the spinning process, the polymer is oriented by stretching or drawing. This both increases polymer chain orientation and degree of crystallinity. As a result the modulus and tensile strength of the fiber are increased.

Fiber manufacture is subdivided into three basic methods, melt spinning, dry spinning, and wet spinning; see Table 6.10 (148). Melt spinning is the simplest but requires that the polymer be stable above its melting temperature. Polyamide 66 is a typical example. Basically, the polymer is melted and forced through spinnerets, which may have from 50 to 500 holes. The fiber diameter immediately after the hole and before attenuation begins

Table 6.10 Spinning processes (148)

		Solution Spinning	
		Wet Spinning	
Melt Spinning	Dry Spinning	Coagulation	Regeneration
Polyamide	Cellulose acetate		Viscose rayon
Polyester	Cellulose triacetate		Cupro
Polyethylene	Acrylic	Acrylic	
	Modacrylic	Modacrylic	
Polypropylene	Aramid	Aramid	
PVDC	Elastane	Elastane	
	PVC	PVC	
	Vinylal		

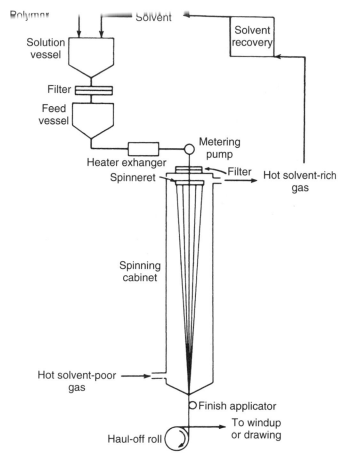

Figure 6.41 In typical dry-spinning operations, hot gas is used to evaporate the solvent in the spinning cabinet. The fibers are simultaneously oriented (148).

is larger than the hole diameter. This is called die swell, which is due to a relaxation of the viscoelastic stress-induced orientation in the hole; see Section 5.4 and Figure 10.20.

During the cooling process the fiber is subjected to a draw-down force, which introduces the orientation. Additional orientation may be introduced later by stretching the fiber to a higher draw ratio.

In dry spinning, the polymer is dissolved in a solvent. A typical example is polyacrylonitrile dissolved in dimethylformamide to 30% concentration. The polymer solution is extruded through the spinnerets, after which the solvent is rapidly evaporated (Figure 6.41) (148). After the solvent is evaporated, the fiber is drawn as before.

In wet spinning, the polymer solution is spun into a coagulant bath. An example is a 7% aqueous solution of sodium cellulose xanthate (viscose), which is spun into a dilute sulfuric acid bath, also containing sodium sulfate and zinc sulfate (149). The zinc ions form temporary ionic cross-links between the xanthate groups, holding the chains together while the sulfuric acid, in turn, removes the xanthate groups, thus precipitating the polymer. After orientation, and so on, the final product is known as rayon.

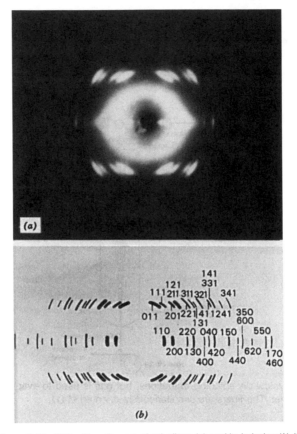

Figure 6.42 X-Ray fiber diagram of polyallene (a), and its indexing (b) (27).

6.10.1 X-Ray Fiber Diagrams

For the purpose of X-ray analyses, the samples should be as highly oriented and crystalline as possible. Since these are also the conditions required for strong, high-modulus fibers, basic characterization and engineering requirements are almost identical.

Figure 6.42 (27,150,151) illustrates a typical X-ray fiber diagram, for polyallene, $+CH_2-C(=CH_2)+_n$. The actual fiber orientation is vertical. The most intense diffractions are on the equitorial plane; note the 110 and 200 reflections. Note the rather intense amorphous halo, appearing inside the 011 reflection.

Because of the imperfect orientation of the polymer in the fiber, arcs are seen, rather than spots. The variation in the intensity over the arcs can be used, however, to calculate the average orientation.

A further complication in the interpretation of the fiber diagram arises because it actually is a "full rotation photograph." In these ways fiber diagrams differ from those of single crystals. Vibrational analyses via infrared and Raman spectroscopy studies play an important role in the selection of the molecular model, as described above.

Table 6.11 Chemical nature of natural fibers

Cellulose	Protein
Cotton	Wool
Tracheid (wood)	Hair
Flax	Silk
Hemp	Spider webs
Coir	
Ramie	
Jute	

6.10.2 Natural Fibers

Natural fibers were used long before the discovery of the synthetics in the twentieth century. Natural fibers are usually composed of either cellulose or protein, as shown in Table 6.11. Animal hair fibers belong to a class of proteins known as keratin, which serve as the protective outer layer of the higher vertebrates. The silks are partly crystalline protein fibers. The crystalline portions of these macromolecules are arranged in antiparallel pleated sheets, a form of the folded-chain lamellae (152).

The morphology of the natural fibers is often quite complex; see Figure 6.43 (153). The cellulose making up these trachieds is a polysaccharide; see Table 1.4. The crystalline portion of the cellulose making up the trachieds is highly oriented, following the various patterns indicated in Figure 6.43. The winding angles of the cellulose form the basis for a natural composite of great strength and resilience. A similar morphology exists in cotton cellulose.

The fibrous proteins (keratin) are likewise highly organized; see Figure 6.44 (154). Proteins are actually polyamide derivatives, a copolymeric form of polyamide 2, where the mers are amino acids. For example, the structure of the amino acid phenylalanine in a protein may be written

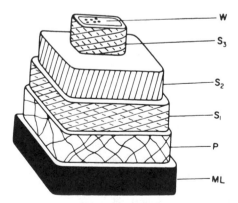

Figure 6.43 The cell walls of a tracheid or wood fiber have several layers, each with a different orientation of microfibrils (153). *ML*, middle lamella, composed of lignin; *P*, primary wall; S_1, S_2, S_3, layers of the secondary wall; *W*, warty layer. The lumen in the interior of the warty layer is used to transport water.

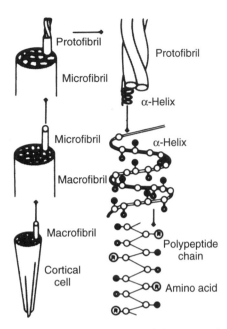

Figure 6.44   A wool-fiber cortical cell is a complex structure, being composed ultimately of proteins (154).

$$
\begin{array}{ccc}
\text{H} & \text{H} & \text{O} \\
| & | & \| \\
-\text{N}-\text{C}-\text{C}- \\
& | \\
& \text{CH}_2\!-\!\bigcirc
\end{array}
\tag{6.57}
$$

There are some 20 amino acids in nature. These are organized into an α-helix in the fibrous proteins, which in turn are combined to form protofibrils as shown. In addition to being crystalline, the fibrous proteins are cross-linked though disulfide bonds contained in the cystine amino acid mer, which is especially high in keratin. Animal tendons, composed of collagen, another fibrous protein, have also been shown to have a surprisingly complex hierarchical structure (155).

6.11   THE HIERARCHICAL STRUCTURE OF POLYMERIC MATERIALS

While all polymers are composed of long-chain molecules, it is the organization of such materials at higher and higher levels that progressively determines their properties and ultimately determines their applications. Table 6.12 summarizes the hierarchical structure of the several classes of polymers according to size range. The largest synthetic crystalline polymer structures are the spherulites, while the largest structures of the natural polymers are the walls of the whole (living) cell.

Of course, there are many kinds of natural polymers. Starch and bread are discussed in Section 14.3, and silk fibers in Section 14.4; both are semicrystalline materials. The

Table 6.12 Hierarchical structure of polymers

| Size Range | Synthetic | | Natural | |
	Amorphous	Crystalline	Cellulosic	Protein
Ångstroms	Mer	Mer	Glucose	Amino acid
Nanometers	{ Chain coil Physical entanglement Cross-link	Stem	Stem	α-Helix
Microns	Multicomponent morphologies	Lamellae spherulite	Lamellae microfibril	Protofibril microfibril
100 Microns	—	—	Fiber (cell)	Microfibril cell

hierarchical structure of polymers has been reviewed by Baer and co-workers (156,157). They emphasize that the occurrence of crystallinity in synthetic polymers requires a sufficiently stereoregular chemical structure, so that the chain molecules can pack closely in parallel orientation. Structure at the nanometer level is generally determined by the pronounced tendency of these chains to crystallize by folding back and forth within crystals of thin lamellar habit. Chain folding itself is a kinetic phenomenon. Given sufficient annealing time, the lamellar thickness increases until the chains are all straight at thermodynamic equilibrium.

When crystallization occurs in flowing solutions or melts, fine fibrous crystals may be produced that consist of highly extended chains oriented axially. This will be recognized as the basis for the supermolecular structure in fibers. Even after prolonged crystallization from the melt, there always remains an appreciable fraction of a disordered phase known as amorphous polymer. If the polymer can crystallize, the appearance of amorphous material is also a kinetic phenomenon, which at equilibrium should disappear. However, the physical entanglements and already existing crystalline structure may slow further crystallization down to substantially zero.

While amorphous materials are significantly less organized than their crystalline counterparts, entanglements, branches, and cross-links significantly control their properties. When two polymers are blended, they generally phase-separate to create a supermolecular morphology, which may be of the order of tens of micrometers, while graft and block copolymers exhibit morphologies of the order of hundreds of angstroms.

6.12 HOW DO YOU KNOW IT'S A POLYMER?

Suppose that a polymer scientist watches a demonstration. The contents of two bottles are mixed, and a white precipitate appears. "Here, I have created a new polymer for you," the demonstrator announces. The polymer scientist examines the two bottles, which are unlabeled. The demonstrator mysteriously leaves the building. What experiments should be performed to determine if the white precipitate is polymeric? (A separate, but equally important, question is: What is its structure?)

In 1920 such questions were far from trivial; note Appendix 5.1. Even today, questions of this nature need to be addressed. For example, the structure of ordinary (window) glass is considered polymeric by some, and not by others. It surely exhibits a glass transition (see Chapter 8) and has a high viscosity, but also apparently has a time variable structure.

There are several types of experiments that should be performed in order to ascertain that the material is (or is not) polymeric. A key experiment is the determination of the molecular weight. If the molecular weight is above about 25,000 g/mol, most scientists will consider that evidence in favor of a polymeric structure. However, it may be colloidal, as a sulfur or silver sol, and not polymeric. If the material is semicrystalline, the unit cell should be established. Hence it will become known if a chain running from cell to cell is likely to occur.

A next series of experiments involves ordinary chemistry, as outlined in Section 2.2. Elemental analysis, characterization of degradation products, end-group analysis, and determination of the probable mer structure are all important in solving the puzzle.

Perhaps the material can be reacted or degraded to form a soluble compound, which can then be characterized. An example is cellulose, a natural polymer that was part of the earliest discussions. Insoluble itself, it can be acetylated to form soluble cellulose acetate for molecular weight determination, by osmometry at the time. Then it was subjected to degradation to glucose and cellobiose, which were determined to be the monomer and dimer, respectively.

The history of polymer science (158,159), takes the reader back to before the macromolecular hypothesis. Natural rubber, *cis*-polyisoprene, was investigated around the turn of the twentieth century. Morawetz (158) writes of the destructive distillation of rubber to isoprene, and of the ozonolysis of rubber to levulinic aldehyde. These studies led to the dimer ring structure of dimethylcyclooctadiene for *cis*-polyisoprene:

$$
\begin{array}{ccc}
& CH_3 & CH_3 \\
& | & | \\
\ldots C-CH_2-CH_2-CH & \ldots C-CH_2-CH_2-C\ldots \\
\| & & \| & & \| \\
\ldots CH-CH_2-CH_2-C & \ldots CH-CH_2-CH_2-C\ldots \\
& | & | \\
& CH_3 & CH_3
\end{array}
\tag{6.58}
$$

The strength of the affinity of the carbon–carbon double bonds was supposed to hold the material together with partial valences. Only much later was it determined that the structure was actually a long chain of isoprene mers, connected in the 1,4-positions, with all the double bonds in the *cis*-configuration.

Such structures imply the confusion then reigning between colloids and polymers. Both were large structures. What was the difference between a sulfur colloidal dispersion and a rubber solution? The answer was found after many years of research.

Another class of experiments that is important includes FT-IR, Raman spectroscopy, and NMR, which provide physical evidence as to where each atom is located and what moieties are present; see Section 2.2. In combination with a known high molecular weight and chemical analyses, large parts of the puzzle can be put together.

If the material is crystalline, X-ray and electron diffraction experiments can help decide the relative location of large sections of the chain. Are the presumed "mers" lined up in such a fashion that they can be reasonably bonded together? What is the relation between one cell and its neighbors?

Returning to our mysterious demonstration, the material could have been polyamide 610, synthesized via interfacial polymerization. Then elemental analysis would show the percent nitrogen, carbon, oxygen, and hydrogen, and each of the methods described above would be brought to play their roles. If, however, it was concentrated hydrochloric acid and concentrated sodium hydroxide that were mixed, the result would eventually be shown to be sodium chloride, ordinary salt! Philosophically, of course, conclusions are the product of reasoning power. It must be remembered that experiments provide evidence, and nothing else.

REFERENCES

1. J. Brandrup and E. H. Immergut, eds., *Polymer Handbook*, 2nd ed., Wiley-Interscience, New York, 1975.

2. K. H. Meyer and H. Mark, *Ber. Deutsch. Chem. Ges.*, **61**, 593 (1928).

3. H. Mark and K. H. Meyer, *Z. Phys. Chem.* **B2**, 115 (1929).

4. K. H. Meyer and H. Mark, *Z. Phys. Chem.*, **B2**, 115 (1929).

5. K. H. Meyer and L. Misch, *Ber. Deutsch. Chem. Ges.*, *B.*, **70B**, 266 (1937).

6. K. H. Meyer and L. Misch, *Helv. Chim. Acta*, **20**, 232 (1937).

7. H. Mark, presented at the American Chemical Society meeting, Seattle, Washington, March 1983.

7a. O. L. van der Hart and R. H. Atalla, *Macromolecules*, **17**, 1465 (1984).

7b. S. Neyertz, A. Pizzi, A. Merlin, B. Maigret, D. Brown, and X. Deglise, *J. Appl. Polym. Sci.*, **78**, 1939 (2000).

8. K. Ziegler, E. Holtzkamp, H. Briel, and H. Martin, *Angew. Chem.*, **67**, 426, 541 (1955).

9. G. Natta, P. Pino and G. Mazzanti, *Gazz. Chim. Ital.*, **87**, 528 (1957).

10. O. Ellefsen and B. A. Tonnesen, in *Cellulose and Cellulose Derivatives*, Part IV, N. M. Bikales and L. Segal, eds., Wiley-Interscience, New York, 1971, p. 151.

11. J. A. Hawsman and W. A. Sisson, in *Cellulose and Cellulose Derivatives*, Part I, E. Ott, H. M. Spurlin, and M. W. Grafflin, eds., Wiley-Interscience, New York, 1954, p. 231.

12. L. Mandelkern, *Rubber Chem. Tech.*, **32**, 1392 (1959).

13. S. D. Kim and L. H. Sperling, Unpublished.

14. P. W. Atkins, *Physical Chemistry*, 6th ed., Oxford University Press, Oxford, 1998.

15. W. L. Bragg, *Proc. Camb. Philos. Soc.*, **17**, 43 (1913).

16. P. Debye and P. Scherrer, *Phys. Z.*, **17**, 277 (1916).

17. C. W. Bunn, *Chemical Crystallography*, Oxford University Press, London, 1945, p. 109.

18. H. Tadokoro, *Structure of Crystalline Polymers*, Wiley-Interscience, New York, 1979, Chap. 5.

19. A. Keller, *Philos. Mag.*, **2**, 1171 (1957).

20. C. W. Bunn, *Trans. Faraday Soc.*, **35**, 482 (1939).

21. G. Natta and P. Corradini, *Rubber Chem. Tech.*, **33**, 703 (1960).

22. Courtesy of Dr. A. Keller and Sally Argon.

23. G. Natta and P. Corradini, *J. Polym. Sci.*, **39**, 29 (1959).

24. D. R. Holmes, C. W. Bunn, and D. J. Smith, *J. Polym. Sci.*, **17**, 159 (1955).

25. R. de P. Daubeny, C. W. Bunn, and C. J. Brown, *Proc. R. Soc. (Lond.)*, **A226**, 531 (1954).

26. Y. Takahashi and H. Tadokoro, *Macromolecules*, **6**, 672 (1973).

27. H. Tadokoro, *Structure of Crystalline Polymers*, Wiley-Interscience, New York, 1979.

28. H. Tadokoro, Y. Chatani, T. Yoshihara, S. Tahara, and S. Murahashi, *Makromol. Chem.*, **73**, 109 (1964).

29. B. C. Rånby, *Acta Chem. Scand.*, **6**, 101, 116 (1952).

30. L. Loeb and L. Segal, *J. Polym. Sci.*, **14**, 121 (1954).

31. J. W. S. Hearle, *Polymers and Their Properties*, Vol. I, *Fundamentals of Structure and Mechanics*, Ellis Horwood Ltd., Chichester, England, 1982.

32. J. W. S. Hearle and R. H. Peters, *Fiber Structure*, Textile Institute, Manchester, England, 1963.

33. F. Khoury and E. Passaglia, in *Treatise on Solid State Chemistry*, Vol. 3, *Crystalline and Noncrystalline Solids*, N. B. Hannay, ed., Plenum, New York, 1976, Ch. 6.

34. P. H. Geil, *Polymer Single Crystals*, Interscience, New York, 1963.

35. D. W. Van Krevelen, *Properties of Polymers*, 3rd ed., Elsevier, Amsterdam, 1990, Chap. 19.

36. J. E. Mark, A. Eisenberg, W. W. Graessley, L. Mandelkern, E. T. Samulski, J. L. Koenig, and G. D. Wignall, *Physical Properties of Polymers*, 2nd ed., American Chemical Society, Washington, DC, 1993, Chap. 4.

37. A. Keller, M. Warner, A. H. Windle, eds., *Self-order and Form in Polymers*, Chapman and Hall, London, 1995.

38. E. J. Roche, R. S. Stein, and E. L. Thomas, *J. Polym. Sci. Polym. Phys. Ed.*, **18**, 1145 (1980).

39. R. S. Stein and A. Misra, *J. Polym Sci. Polym. Phys. Ed.*, **18** 327 (1980).

40. R. S. Stein, in *Rheology, Theory and Applications*, Vol. 5, F. R. Eirich, ed., Academic New York 1969, Chap. 6.

41. J. D. Hoffman, G. T. Davis, and J. I. Lauritzen Jr., in *Treatise on Solid State Chemistry*, Vol. 3, *Crystalline and Noncrystalline Solids*, N. B. Hannay, ed., Plenum, New York, 1976, Chap. 7.

42. W. A. Sisson, in *Cellulose and Cellulose Derivatives*, Interscience, New York, 1943, pp. 203–285.

43. J. Hengstenberg and J. Mark, *Z. Kristallogr.*, **69**, 271 (1928).

44. W. O. Statton, *J. Polym. Sci.*, **20C**, 117 (1967).

45. K. Herrmann and O. Gerngross, *Kautschuk*, **8**, 181 (1932).

46. K. Herrmann, O. Gerngross, and W. Abitz, *Z. Phys. Chem.*, **10**, 371 (1930).

47. E. W. Fischer, *Z. Naturforsch.*, **12a**, 753 (1957).

48. P. H. Till, Jr., *J. Polym. Sci.*, **24**, 301 (1957).

49. F. A. Bovey, *Org. Coat. Appl. Polym. Sci Prepr.*, **48** (1), 76 (1983).

50. F. C. Schilling, F. A. Bovey, S. Tseng, and A. E. Woodward, *Macromolecules*, **16**, 808 (1983).

51. T. Oyama, K. Shiokawa, and Y. Murata, *Polym. J.*, **6**, 549 (1974).

52. A. J. Kovacs, J. A. Manson, and D. Levy, *Kolloid Z.*, **214**, 1 (1966).

53. P. J. Flory, *J. Am. Chem. Soc.*, **84**, 2857 (1962).

54. R. S. Stein, *J. Chem. Ed.*, **50**, 748 (1973).

55. L. Mandelkern, in *Physical Properties of Polymers*, J. E. Mark, A. Eisenberg, W. W. Graessley, L. Mandelkern, and J. L. Koenig, eds., American Chemical Society, Washington, DC, 1984.

56. H. D. Keith, F. J. Padden, Jr., and R. G. Vadimsky, *J. Polym. Sci.*, *A-2*, **4**, 267 (1966).

57. H. D. Keith, F. J. Padden, and R. G. Vadimsky, *J. Appl. Phys.*, **42**, 4585 (1971).

58. Y. Hase and P. H. Geil, *Polym. J.* (*Jpn.*), **2**, 560, 581 (1971).

59. F. Rybnikar and P. H. Geil, *J. Macromol. Sci. Phys.*, **B7**, 1 (1973).

60. D. C. Bassett, A. Keller, and S. Mitsuhashi, *J. Polym. Sci.*, **A1**, 763 (1963).

61. P. H. Geil, in *Growth and Perfection of Crystals*, R. H. Doremus, B. W. Roberts, and D. Turnbull, eds., Wiley, New York, 1958, pp. 579–585.

62. H. D. Keith, *J. Polym. Sci.*, **A2**, 4339 (1964).

63. E. Martuscelli, Multicomponent Polymer Blends Symposium, Capri, Italy, May 1983.

64. E. Martuscelli, *Polym. Eng. Sci.*, **24**, 563 (1984).

65. A. J. Kovacs, *Chim. Ind. Genie Chim.*, **97**, 315 (1967).

66. R. G. Crystal, P. F. Erhardt, and I. L. O'Malloy, in *Block Copolymers*, S. L. Aggarwal, ed., Plenum, New York, 1970.

67. K. E. Hardenstine, C. J. Murphy, R. B. Jones, L. H. Sperling, and G. E. Manser, *J. Appl. Polym. Sci.*, **30**, 2051 (1985).

68. H. Cornélis, R. G. Kander, and J. P. Martin, *Polymer*, **37**, 4573 (1996).

69. J. N. Hay and M. Sabin, *Polymer*, **10** (3), 203 (1969).

70. H. D. Keith and F. J. Padden Jr., *J. Appl. Phys.*, **35**, 1270 (1964).

71. H. D. Keith and F. J. Padden Jr., *J. Appl. Phys.*, **35**, 1286 (1964).

72. L. H. Palys and P. J. Phillips., *J. Polym. Sci. Polym. Phys. Ed.*, **18**, 829 (1980).

73. M. Avrami, *J. Chem. Phys.*, **7**, 1103 (1939).

74. M. Avrami, *J. Chem. Phys.*, **8**, 212 (1940).

75. M. Avrami, *J. Chem. Phys.*, **9**, 177 (1941).

76. J. D. Hoffman, *Polymer*, **24**, 3 (1983).

77. J. D. Hoffman, *Polymer*, **23**, 656 (1982).

78. E. A. DiMarzio, C. M. Guttman, and J. D. Hoffman, *Faraday Discuss. Chem. Soc.*, **68**, 210 (1979).

79. C. M. Guttman, J. D. Hoffman, and E. A. DiMarzio, *Faraday Discuss. Chem. Soc.*, **68**, 297 (1979).

80. J. D. Hoffman, C. M. Guttman, and E. A. DiMarzio, *Faraday Discuss. Chem. Soc.*, **68**, 177 (1979).

81. U. R. Evans, *Trans. Faraday Soc.*, **41**, 365 (1945).

82. P. Meares, *Polymers: Structure and Bulk Properties*, Van Nostrand, New York, 1965, Chap. 5.

83. J. N. Hay, *Br. Polym. J.*, **3**, 74 (1971).

84. S. D. Poisson, *Recherches sur la Probabilite des Judgements en Matiere Criminelle et en Matiere Civile*, Bachelier, Paris, 1837, p. 206.

85. W. Liu, B. S. Hsiao, and R. S. Stein, *Polym. Mater. Sci. Eng. (Prepr.)*, **81**, 363 (1999).

86. R. G. Alamo, J. D. Londono, L. Mandelkern, F. C. Stehling, and G. D. Wignall, *Macromolecules*, **27**, 411 (1994).

87. G. D. Wignall, J. D. Londono, J. S. Lin, R. G. Alamo, M. J. Galante, and L. Mandelkern, *Macromolecules*, **28**, 3156 (1995).

88. H. D. Keith and F. J. Padden Jr., *J. Appl. Phys.*, **34**, 2409 (1963).

89. B. Wunderlich and L. Melillo, *Makromol. Chem.*, **118**, 250 (1968).

90. D. O. Yoon and P. J. Flory, *Faraday Discuss. Chem. Soc.*, **68**, 288 (1979).

91. J. D. Hoffman, L. J. Frolen, G. S. Ross, and J. I. Lauritzen Jr., *J. Res. NBS*, **79A** (6), 671 (1975).

92. M. Tasumi and S. Krimm, *J. Polym. Sci.*, *A-2*, **6**, 995 (1968).

93. M. I. Bank and S. Krimm, *J. Polym. Sci.*, *A-2*, **7**, 1785 (1969).

94. S. Krimm and T. C. Cheam, *Faraday Discuss. Chem. Soc.*, **68**, 244 (1979).

95. X. Jing and S. Krimm, *Polym. Lett.*, **21**, 123 (1983).

96. R. Kitamarn, F. Horii, and S. H. Hyon, *J. Polym. Sci. Polym. Phys. Ed.*, **15**, 821 (1977).

97. M. Glotin and L. Mandelkern, *Colloid Polym. Sci.*, **260**, 182 (1982).

98. L. Mandelkern, in *Physical Properties of Polymers*, J. E. Mark, A. Eisenberg, W. W. Graessley, L. Mandelkern, and J. L. Koenig, eds. American Chemical Society, Washington, DC, 1984.

99. M. Stamm, E. W. Fischer, M. Dettenmaier, and P. Convert, *Faraday Discuss. Chem. Soc.*, **68**, 263 (1979).

100. D. G. H. Ballard, A. N. Burgess, T. L. Crawley, G. W. Longman, and J. Schelten, *Faraday Discuss. Chem. Soc.*, **68**, 279 (1979).

101. M. Stamm, *J. Polym. Sci. Polym. Phys. Ed.*, **20**, 235 (1982).

102. G. D. Wignall, L. Mandelkern, C. Edwards, and M. Glotin, *J. Polym. Sci. Polym. Phys. Ed.*, **20**, 245 (1982).

103. D. M. Sadler and R. Harris, *J. Polym. Sci. Polym. Phys. Ed.*, **20**, 561 (1982).

104. J. Schelten, G. D. Wignall, D. G. H. Ballard, and G. W. Longman, *Polymer*, **18**, 1111 (1977).

105. J. Schelten, A. Zinken, and D. G. H. Ballard, *Colloid Polym. Sci.*, **259**, 260 (1981).

106. D. G. H. Ballard, P. Cheshire, G. W. Longman, and J. Schelten, *Polymer*, **19**, 379 (1978).

107. E. W. Fischer, M. Stamm, M. Dettenmaier, and P. Herschenraeder, *Polym. Prepr. Am. Chem. Soc. Div. Polym. Chem.*, **20** (1), 219 (1979).

108. J. M. Guenet, *Polymer*, **22**, 313 (1981).

109. D. M. Sadler and A. Keller, *Macromolecules*, **10**, 1128 (1977).

110. D. M. Sadler and A. Keller, *Science*, **203**, 263 (1979).

111. A. Keller, *Faraday Discuss. Chem. Soc.*, **68**, 145 (1979).

112. D. M. Sadler and A. Keller, *Polymer*, **17**, 37 (1976).

113. L. H. Sperling, *Polym. Eng. Sci.*, **24**, 1 (1984).

114. D. Y. Yoon and P. J. Flory, *Polymer*, **18**, 509 (1977).

115. D. Y. Yoon, *J. Appl. Cryst.*, **11**, 531 (1978).

116. D. Y. Yoon and P. J. Flory, *Faraday Discuss. Chem. Soc.*, **68**, 289 (1979).

117. M. Dettenmaier, E. W. Fischer, and M. Stamm, *Colloid Polym. Sci.*, **258**, 343 (1979).

118. M. Stamm, E. W. Fischer, and M. Dettenmaier, *Faraday Discuss. Chem. Soc.*, **68**, 263 (1979).

119. J. Schelten, D. G. H. Ballard, G. D. Wignall, G. W. Longman, and W. Schmatz, *Polymer*, **17**, 751 (1976).

120. D. G. H. Ballard, P. Cheshire, G. W. Longman, and J. Schelten, *Polymer*, **19**, 379 (1978).

121. B. Crist, W. W. Graessley, and G. D. Wignall, *Polymer*, **23**, 1561 (1982).

122. F. C. Frank, *Faraday Discuss. Chem. Soc.*, **68**, 7 (1979).

123. R. Hosemann, *Polymer*, **3**, 349 (1962).

124. F. J. Balta-Calleja and R. Hosemann, *J. Appl. Crystallogr.*, **13**, 521 (1980).

125. R. Hosemann, *Colloid Polym. Sci.*, **260**, 864 (1982).

126. B. Wunderlich, in *Macromolecular Physics*, Vol. I, Academic Press, Orlando, FL, 1973.

127. B. Wunderlich and L. Melillo, *Makromol. Chem.*, **118**, 250 (1968).

128. E. Hellmuth 2nd, and B. Wunderlich, *J. Appl. Phys.*, **36**, 3039 (1965).

129. B. Wunderlich, *J. Polym. Sci. Symp.*, **43**, 29 (1973).

130. D. C. Bassett and B. Turner, *Phil. Mag.*, **29**, 925 (1974).

131. A. Keller, M. Hikosaka, S. Rastogi, A. Toda, P. J. Barham, and G. Goldbeck-Wood, in *Self-order and Form in Polymeric Materials*, A. Keller, M. Warner, and A. H. Windle, eds., Chapman and Hall, London, 1995.

132. A. Keller, M. Hikosaka, S. Rastogi, A. Toda, P. J. Barham, and G. Goldbeck-Wood, *J. Mater. Sci.*, **29**, 2579 (1994).

133. W. J. Moore, *Physical Chemistry*, 4th ed., Prentice-Hall, Englewood Cliffs, NJ, 1972, p. 134.

134. P. J. Flory, *J. Chem. Phys.*, **17**, 223 (1949).

135. L. Mandelkern and P. J. Flory, *J. Am. Chem. Soc.*, **73**, 3206 (1951).

136. L. Mandelkern, R. R. Garrett, and P. J. Flory, *J. Am. Chem. Soc.*, **74**, 3949 (1952).

137. G. M. Bristow and W. F. Watson, *Trans. Faraday Soc.*, **54**, 1731 (1958).

138. F. E. Karasz and L. D. Jones, *J. Phys. Chem.*, **71**, 2234 (1967).

139. A. V. Tobolsky, *Properties and Structure of Polymers*, Wiley, New York, 1960.

140. M. Warner, in *Side-Chain Liquid Crystals*, C. B. McArdle, ed., Blackie, Glasgow, 1989.

141. P. J. Flory, *J. Chem. Phys.*, **10**, 51 (1942).

142. B. Wunderlich, *Macromolecular Physics*, Vol. 1, Academic Press, Orlando, FL, 1973.

143. J. D. Hoffman and J. J. Weeks, *J. Res. Natl. Bur. Stand.*, **66A**, 13 (1962).

144. R. E. Wilfong, *J. Polym. Sci.*, **54**, 385 (1961).

145. R. W. Lenz, *Organic Chemistry of Synthetic High Polymers*, Interscience, New York, 1967, pp. 91–95.

146. W. H. Carothers, *J. Am. Chem. Soc.*, **51**, 2548, 2560 (1929).

147. J. R. Whinfield, *Nature*, **158**, 930 (1946).

148. J. E. McIntire and M. J. Denton, in *Encyclopedia of Polymer Science and Engineering*, Vol. 6, J. I. Kroschwitz, ed., Wiley, New York, 1986.

149. J. W. Schappel and G. C. Bockno, in *Cellulose and Cellulose Derivatives*, N. M. Bikales and L. Segal, eds., Vol. 5, Part 5, High Polymer Series, Wiley-Interscience, New York, 1971.

150. H. Tadokoro, Y. Takahasi, S. Otsuka, K. Mori, and F. Imaizumi, *J. Polym. Sci.*, *3B*, **3B**, 697 (1965).

151. H. Tadokoro, M. Kobayaski, K. Mori, Y. Takahashi, and S. Taniyama, *J. Polym. Sci. C*, **22**, 1031 (1969).

152. L. H. Sperling and C. E. Carraher, in *Encyclopedia of Polymer Science and Engineering*, 2nd ed., Vol. 12, J. I. Kroschwitz, ed., Wiley, New York, 1988.

153. G. Tsoumis, *Wood as a Raw Material*, Pergamon, New York, 1968.

154. W. S. Boston, in *Encylopedia of Textiles, Fibers, and Nonwoven Fabrics*, M. Grayson, ed., Wiley-Interscience, New York, 1984.

155. K. Kastelic, A. Galeski, and E. Baer, *Conn. Tiss. Res.*, **6**, 11 (1978).

156. E. Baer, A. Hiltner, and H. D. Keith, *Science*, **235**, 1015 (1987).

157. E. Baer, *Sci. Am.* **254** (10), 179 (1986).

158. H. Morawetz, *Polymers: The Origins and Growth of a Science*, Wiley, New York, 1985.

159. Y. Furukawa, *Inventing Polymer Science*, University of Pennsylvania Press, Philadelphia, 1998.

GENERAL READING

M. Dosiere, ed., *Crystallization of Polymers*, Kluwer, Dordrecht, 1993.

A. Keller, M. Warner, and A. H. Windle, eds., *Self-order and Form in Polymeric Materials*, Chapman and Hall, London, 1995.

R. S. Porter and L. H. Wang, *Rev. Macromol. Chem. Phys.*, **C35** (1), 63 (1995).

H. Tadokoro, *Structure of Crystalline Polymers*, Wiley-Interscience, New York, 1979.

STUDY PROBLEMS

1. Based on the unit cell structure for cellulose I, calculate its theoretical crystal density. (See Figure 6.1.)

2. If polyethylene of z-average molecular weight 30,000 g/mol is cooled from the melt at 1°C/min and 100°C/min, estimate the fractions of polymer crystallized in regimes I, II, and III. [*Hint:* Assume an instantaneous nucleation density of $10^4/cm^3$.]

3. A difficultly crystallizable high-molecular-weight polymer was finally crystallized in regime I. Compare and contrast the properties of the crystallized polymer and the amorphous polymer at the same temperature and pressure. Specifically, how do the densities, radii of gyration, and morphology via optical microscopy differ?

4. Wood and wool are based on renewable-resources, called "green" materials today. What is happening in either one of these fields today, bearing on polymer science? Write a brief summary of your findings.

5. Polymers are supposed to consist of long chains, yet the unit cell, by X-ray studies, is about the same size as those of ordinary molecules, containing only relatively few atoms. How can this be?

6. Compare and contrast the Avrami, Keith–Padden, and Hoffman theories of crystallization.

7. Note the volume–temperature data for polyethylene in Figure 6.2. What are the volume coefficients of expansion of the melt and two crystalline samples? Why are they different?

8. Given the unit cell structure of polyethylene (Figure 6.5), compute the theoretical density of the 100% crystalline product. [*Hint:* see Table 6.2.]

9. Equation (6.43) shows corrections to the melting point depression due to mismatch of molar volumes of solvent and mer. Should the corresponding equations for copolymers and finite molecular weight be corrected similarly? See equations (6.41) and (6.42). If so, derive suitable relations.

10. Poly(decamethylene adipate), $+CO-(CH_2)_4-CO-(CH_2)_{10}-O-]_n$, density = 0.99 g/cm^3, was mixed with various quantities of dimethylformamide, $(CH_3)_2NCHO$, $d = 0.9445$ g/cm^3 and the melting temperatures observed:[†]

v_1	T_f, °C
0.078	72.5
0.202	66.5
0.422	61.5
0.603	57.5

(a) What is the melting temperature of the pure polymer? (b) What is the heat of fusion of poly(decamethylene adipate)?

11. What spherulite radius can be calculated from Figure 6.16?

12. Devise an NMR experiment to study chain folding in (a) cellulose triacetate, (b) isotactic polystyrene, and (c) transpolyisoprene (Gutta percha). [*Hint:* What chemical modifications, if any, are required?]

13. Compare infrared, NMR, and SANS results on chain folding in single crystals. Can you devise a new experiment to investigate the problem?

[†] L. Mandelkern, R. R. Garrett, and P. J. Flory, *J. Am. Chem. Soc.*, **74**, 3949 (1952).

14. Read an original paper published in the last 12 months on crystalline polymer behavior or theory, and write a brief report on it in your own words. Cite the authors and exact reference. Does it support the present text? Add new ideas or data? Contradict present theories or ideas?

15. Single crystals of polyethylene are grown from different molecular weight materials from $M = 2000$ to 5×10^7 g/mol. The crystals are all 150 Å thick, with adjacent reentry and superfolding after each 20 stems. How does R_g depend on M in this region? Plot the results. What dependence of R_g on M is predicted as M goes to infinity?

16. The lattice constants of orthorhombic polyethylene have been determined as a function of temperature:[†]

	Lattice Constants, Å		
T, K	a	b	c
4	7.121	4.851	2.548
77	7.155	4.899	2.5473
293	7.399	4.946	2.543
303	7.414	4.942	2.5473

(a) What is the theoretical volume coefficient of expansion of 100% crystalline polyethylene? (The volume coefficient of expansion is given by

$$\alpha = \frac{1}{V} \left(\frac{\partial V}{\partial T} \right)_P$$

where the change in volume V is measured as a function of temperature T at constant pressure P.) (b) Why are the c-axis lattice constants substantially independent of the temperature?

17. You were handed bottles of dimethylcyclooctadiene and *cis*-polyisoprene, but they became mixed up. What experiments would you perform to identify them?

18. Toothbrushes are available with soft, medium, and hard bristles, generally made of polyamides. (a) Propose at least two distinctly different ways by which the bristles could be controlled to provide soft, medium, or hard performance. (b) If you had bristles from different toothbrushes, what tests or experiments would you perform to determine the ways that were actually used to control the hardness? (c) Why polyamide?

19. You are handed a glass fiber reinforced sample. "Do not damage this very valuable material," your boss asks, "but we need to know what is in it!" What nondestructive experiments would you perform in situ to determine the chemical structure, crystallinity (if any), orientation, and so on, in the material?

[†] H. Tadokoro, *Structure of Crystalline Polymers*, Wiley-Interscience, New York, 1979, p. 375.

7

POLYMERS IN THE LIQUID
CRYSTALLINE STATE

7.1 DEFINITION OF A LIQUID CRYSTAL

Liquid crystals, LCs, are substances that exhibit long-range order in one or two dimensions, but not all three. Both small molecules and polymers may exist in the liquid crystalline state, but generally special chemical structures are required. Thus, while amorphous substances are more or less entirely disordered (see Chapter 5) and crystalline materials are ordered in all three dimensions (see Chapter 6), the LCs lie in-between in properties. Liquid crystals are ordered in one or two dimensions only. Liquid crystals all exhibit some degree of fluidity. LC polymers are a relatively new discovery dating from about 1950 (1), with the field growing extremely rapidly. Engineering advances utilizing polymers that go through an LC stage include new classes of high-modulus fibers, high-temperature plastics, and a host of new electronic and data storage materials.

The formation of liquid crystals is a direct consequence of molecular asymmetry. It arises because two molecules cannot occupy the same space at the same time and is largely entropically derived.

A simple example of a "liquid crystal" may be considered using a sink filled with water. Onto the water surface, toothpicks are slowly added without overlap. At first, the toothpicks are random in order. However, as the surface becomes more concentrated, the toothpicks begin to align into small groups more or less side by side. Now there are two phases present: the dilute, disordered phase, and the LC phase. The reason why the toothpicks order themselves is to prevent one toothpick from having to go through (or over) another. Much of what follows can use this ordering behavior as a model.

7.2 ROD-SHAPED CHEMICAL STRUCTURES

In the preceding chapters the concept of the random coil was developed, where axial correlation of mers is only of very short distance; see Figure 5.1. Now a number of polymers have been synthesized which are substantially rod-shaped. An example is poly(p-phenylene terephthalamide), PPD-T, with the chemical structure

$$\left(\mathrm{NH}-\!\!\!\left\langle\bigcirc\right\rangle\!\!\!-\mathrm{NH}-\!\!\!\underset{\underset{\displaystyle O}{\overset{\displaystyle O}{\overset{\|}{C}}}}{}\!\!\!-\!\!\!\left\langle\bigcirc\right\rangle\!\!\!-\!\!\!\underset{}{\overset{\displaystyle O}{\overset{\|}{C}}}\right)_{n} \tag{7.1}$$

Because the amide groups are located in the trans positions on the phenylene rings, there is minimal bond rotation out of the chain axis.

Another example is a copolymer of p-hydroxybenzoic acid, p,p'-biphenol, and terephthalic acid, creating the polyester structure

$$\left(\mathrm{O}-\!\!\!\left\langle\bigcirc\right\rangle\!\!\!-\!\!\!\underset{}{\overset{\displaystyle O}{\overset{\|}{C}}}\!\!\!-\mathrm{O}-\!\!\!\left\langle\bigcirc\right\rangle\!\!\!\left\langle\bigcirc\right\rangle\!\!\!-\mathrm{O}-\!\!\!\underset{}{\overset{\displaystyle O}{\overset{\|}{C}}}\!\!\!-\!\!\!\left\langle\bigcirc\right\rangle\!\!\!-\!\!\!\underset{}{\overset{\displaystyle O}{\overset{\|}{C}}}\right)_{n} \tag{7.2}$$

Again, the ester groups, together with the para-bonded phenylene rings, form strictly rodlike structures.

Another type of liquid crystalline polymers are those having rod-shaped side chains. Thus, the backbone may be a random coil, but the side chains are organized into one- or two-dimensional liquid crystals. The stiff backbone types make very high modulus fibers and high temperature plastics. The side chain types are useful for their action in magnetic and electrical fields. (Note the behavior of battery-driven liquid crystalline watches, for example. The liquid crystalline material is held between crossed Nicols; the orientation of the molecules is controlled by an electric field.)

7.3   LIQUID CRYSTALLINE MESOPHASES

7.3.1   Mesophase Topologies

Liquid crystalline structures can be organized into several classes, much the same as crystalline materials are organized into body-centered, triclinic, and so on lattices. The major LC mesophase topologies are shown in Figure 7.1 (2). The nematic LCs are or-

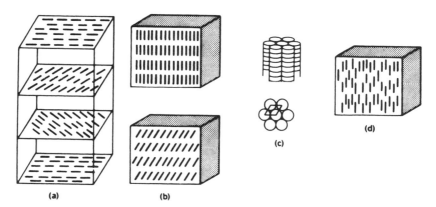

Figure 7.1   Liquid crystal structures: (*a*) cholesteric (*b*) smectic A (top) and smectic C (bottom), (*c*) discotic, and (*d*) nematic (2).

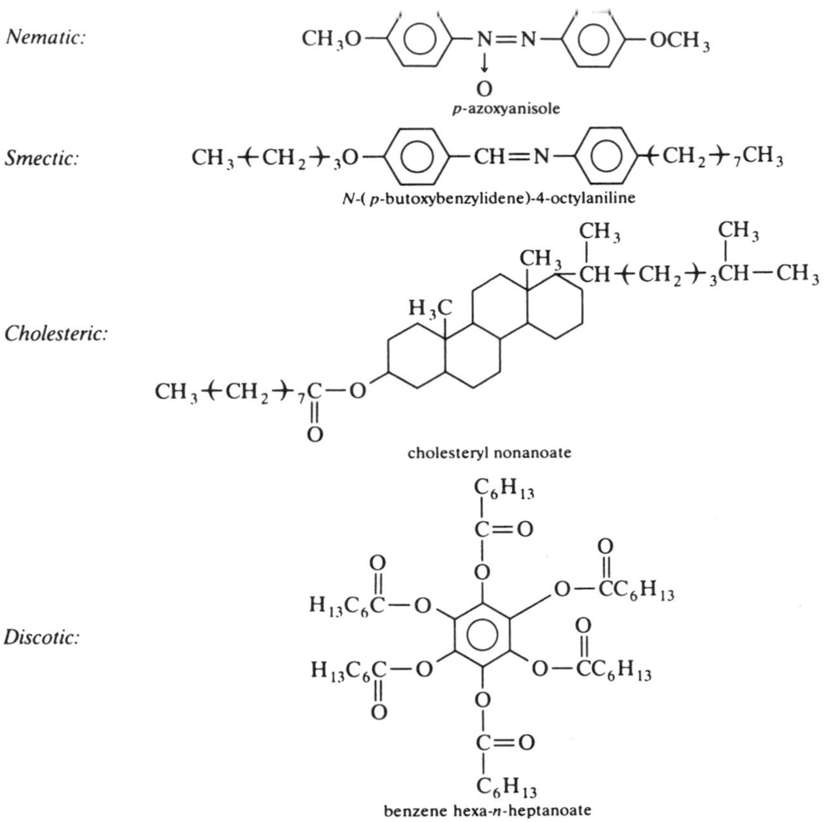

Nematic:

$$CH_3O-\langle\bigcirc\rangle-N=N-\langle\bigcirc\rangle-OCH_3$$
$$\downarrow$$
$$O$$

p-azoxyanisole

Smectic:

$$CH_3(CH_2)_3O-\langle\bigcirc\rangle-CH=N-\langle\bigcirc\rangle(CH_2)_7CH_3$$

N-(p-butoxybenzylidene)-4-octylaniline

Cholesteric:

cholesteryl nonanoate

Discotic:

benzene hexa-n-heptanoate

Figure 7.2 Asymmetric organic molecules in the form of rods or plates may form liquid crystal structures (3).

ganized in one dimension only, with their chains lying parallel to each other at equilibrium. (The toothpicks described earlier also formed a nematic mesophase.) The smectic LCs are ordered in two dimensions. Polymer scientists now recognize a large number of smectic mesophases. Figure 7.1b illustrates the smectic A and C structures, two of the more important mesophases. The cholesteric mesophase, Figure 7.1a, is like a two-dimensional twisted nematic mesophase. The discotic mesophase resembles stacks of dishes or coins.

The common feature of the structures shown in Figure 7.1 is asymmetry of shape, manifested either as rods of axial ratio greater than about 3, or by thin platelets of biaxial order. Of course, actual materials are made up of molecules. Some LC chemical structures of low molecular weight are shown in Figure 7.2 (3). The reader will imagine how the various chemical structures fit into the indicated mesophases of Figure 7.1. Thus the old concept of all matter being composed only of crystalline, liquid, and gaseous phases is obsolete. Besides amorphous solids [e.g., polystyrene or poly(methyl methacrylate)], there are the liquid crystalline structures.

A LC-forming polymer may exhibit multiple mesophases at different temperatures or pressures. As the temperature is raised, the polymer then goes through multiple first-

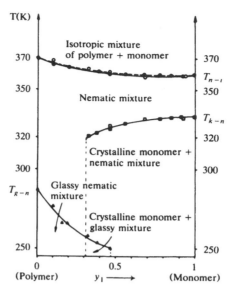

Figure 7.3   Phase diagrams illustrate the expected behavior of a material as a function of temperature and composition. Here, a mixture of the polymer

and the monomer

are shown. The value y_1 is the weight fraction. The full symbols represent DSC measurements; open symbols represent data obtained by polarizing microscope (4,5).

order transitions from a more ordered to a less ordered state. Scientists speak of a "clearing temperature," where the last (or only) LC phase gives way to the isotropic melt or solution.

7.3.2   Phase Diagrams

A phase diagram for a mesogenic side chain polymer and a related monomer is illustrated in Figure 7.3 (4,5). Note that the monomer by itself crystallizes. The nematic mixture clears in the temperature range of 360 to 370 K.

7.3.3   First-Order Transitions

The various mesophasic transitions are first-order transitions, because volume and heat capacity go through discontinuities when plotted against temperature or pressure; see Section 6.8. (The glass transition is ideally a second-order transition; see Sections 8.2 and 8.6.3.) The various transitions that LC polymers undergo as the temperature increases are illustrated schematically in Figure 7.4 (4). Here, a semicrystalline polymer melts first

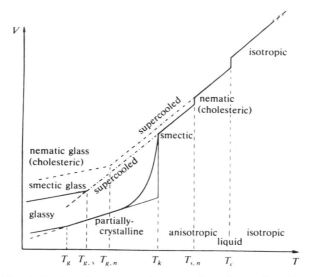

Figure 7.4 Liquid crystal-forming polymers may undergo many first-order transitions. Here, as the temperature is raised, the polymer first melts to a smectic structure, then to a nematic structure, and then to an isotropic melt (4).

to a smectic mesophase, then to a nematic mesophase, and finally to the isotropic liquid state as the temperature increases. Both the volume and the enthalpic jumps through the several transitions are additive; that is, the sum of the changes exactly equals that of going directly from the crystalline state to the isotropic melt.

The nature of the first-order transitions is best illustrated by observing the behavior of these materials in a differential scanning calorimeter (DSC). Figure 7.5 shows the heat-

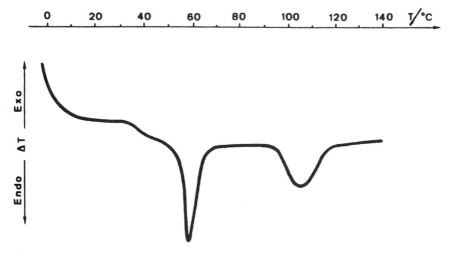

Figure 7.5 DSC heating trace of acrylic, liquid crystal-forming polymer. Note two first order transitions.

ing trace of a side-chain liquid crystalline acrylate with the structure

$$+CH_2-CH+_n$$
$$O=C$$
$$O+CH_2+_{11}O-\bigcirc-\overset{O}{\overset{\|}{C}}-O-\bigcirc-O-CH_2-\overset{CH_3}{\overset{|}{CH}}-CH_2-CH_3$$

where the tail has a chiral carbon (6). The presence of multiple endotherms provides direct evidence of the multiplicity of first-order transitions in this polymer.

This acrylate exhibits a melting point at 59°C, becoming a birefringent melt that exhibits the characteristic features of a smectic A phase via microscopy. Near 100°C there is some indication that the polymer transforms to a cholesteric phase; however, the isotropic melt phase appears at nearly the same temperature, with the polymer clearing at 108°C. These overlapping transitions result in the broadening of the endotherm, which appears between 97 and 120°C in Figure 7.5.

The kinetics of forming many LC mesophases is fairly slow. In the case of the overlapping cholesteric/isotropic transitions above, it is possible that heating the polymer at a slower rate may separate the transitions. Generally, if transitions are overlapping or skipped due to rapid heating, the volume and enthalpy changes observed will exactly equal the sum of the changes of going through each of the mesophases.

Again, it must be emphasized that not all polymers go through LC mesophases. Those polymers that form random coils melt directly to the isotropic liquid state. Some portion of the chain or side chain must be rod- or disk-shaped to form a LC mesophase.

7.4 LIQUID CRYSTAL CLASSIFICATION

Section 7.3 classified LCs according to their mesophase structure (2). Another important classification method divides LC polymers into lyotropic, thermotropic, and mesogenic side group compositions. The lyotropic LCs form ordered states in concentrated solutions, similar again to the toothpick analogy previously described. Frequently rod-shaped polymer chains are dissolved to concentrations of about 30% for processing purposes, particularly for fiber formation. The main reason for using solutions rather than melts is that the polymers in question decompose below their very high melting temperatures. Thermotropic LCs exist as ordered melts, being able to remain in the LC state without decomposition. The mesogenic side-chain compositions usually have random coil backbone chains, with rod-shaped side groups. It is these latter that form the LC structures in this case. The mesogenic side-chain polymers are actually a subclass of the thermotropic LCs, since they are usually processed in the melt state.

7.4.1 Lyotropic Liquid Crystalline Chemical Structures

Most often the lyotropic LC polymers form nematic mesophases. Most of the polymers in this class are aromatic polyamides with aromatic ring structures, as shown in Table 7.1 (7). Several of the polymers in Table 7.1 form very high-modulus fibers; see Chapter 11. The fibers are crystalline after formation.

Table 7.1 Important polyamides yielding liquid crystalline mesophases (7)

_a_ The basis for the fiber "Kevlar®."

Heterocyclic polymers yield materials with outstanding high-temperature performance; see Table 7.2 (2). Many of these polymers have ladder or semiladder chemical structures. If one bond in a ladder structure is broken by heat or oxidation, the chain may retain its original molecular weight. If a single carbon atom chain is oxidized, usually the chain is degraded, with concomitant loss of properties.

Table 7.2 Lyotropic solutions of polyheterocyclic compounds (2)

Compound	Structure	Lyotropic Solution	Reference
Poly(1,4-phenylene-2,6-benzobisimidazole)		Methanesulfonic acid	(a)
Poly(1,4-phenylene-2,6-benzobisoxazole) (PBO)		Methanesulfonic acid Chlorosulfonic acid 100% sulfuric acid	(b, c) (d)
Poly(1,4-phenylene-2,6-benzobisthiazole) (PBT)		5–10% in polyphosphoric acid Methanesulfonic acid	(e) (c, e, f)
Poly[2,6-(1,4-phenylene)-4-phenylquinoline]		1.0–1.5% in m-cresol-di-m-cresyl phosphate	(g)
Poly[1,1'-(4,4'-biphenylene)-6,6'-bis(4-phenylquinoline)]		>9% in m-cresol-di-m-cresyl phosphate	(g)
Poly[2,5(6)-benzimidazole] (AB–PBI)		Methanesulfonic acid	(h)

References: (a) T. E. Helminiak, F. E. Arnold, and C. L. Benner, *Polym. Prepr. Am. Chem. Soc. Div. Polym. Chem.*, **16** (2), 659 (1975). (b) G. C. Berry, P. M. Cotts, and S.-G. Chu, *Br. Polym. J.*, **13**, 47 (1981); C. P. Wong, H. Ohnuma, and G. C. Berry, *J. Polym. Sci. Polym. Symp.*, **65**, 173 (1978). (c) J. F. Wolfe, B. H. Loo, and F. E. Arnold, *Macromolecules*, **14**, 915 (1981). (d) E. W. Choe and S. N. Kim, *Macromolecules*, **14**, 920 (1981). (e) J. F. Wolfe, B. H. Loo, and E. R. Sevilla, *Polym. Prepr. Am. Chem. Soc. Div. Polym. Chem.*, **22** (1), 60 (1981). (f) S.-G. Chu, S. Venkatraman, G. C. Berry, and Y. Einaga, *Macromolecules*, **14**, 939 (1981). (g) P. D. Sybert, W. H. Beever, and J. K. Stille, *Macromolecules*, **14**, 493 (1981). (h) A. Wereta Jr. and M. T. Gehatia, *Polym. Eng. Sci.*, **18**, 204 (1978).

Some natural polymers also form lyotropic systems. These include cellulose deriva tives, cellulose being a stiff molecule but not entirely rod-shaped. Another polymer is the polypeptide poly(γ-benzyl L-glutamate) (8,9),

$$\left(\begin{array}{c} \text{H} \qquad \text{O} \\ | \qquad\quad \| \\ \text{N}-\text{CH}-\text{C} \\ | \\ (\text{CH}_2)_2 \\ | \\ \text{O}=\text{C}-\text{O}-\text{CH}_2-\bigcirc \end{array} \right)_n$$

(7.3)

which forms birefringent, anisotropic phases in organic solutions. The shape of these polypeptides is in the form of α-helices, with regular intramolecular hydrogen bonding and association of the solvent with the substituents to stabilize the structure in the form of a rod.

7.4.2 Thermotropic Liquid Crystalline Chemical Structures

Most of the thermotropic LC polymers are aromatic copolyesters, some being terpoly-mers (10). Simple homopolymer aromatic polyesters generally have too high a melting temperature to form thermotropic mesophases without decomposition, but copoly-merization reduces their melting temperatures; see Section 6.8.1. In order to reduce the melting temperature of these materials, several techniques are available; see Table 7.3 (11). These include (11)

1. Copolymerization of several mesogenic monomers, which produces random co-polymers with depressed melting temperatures.
2. Use of monomers with bulky side groups, which prevents close packing in the LC mesophases, sometimes referred to as "frustrated chain packing."
3. Use of bent comonomers to interrupt the order in the system.
4. Incorporation of flexible spacers, to decrease rigidity. This permits the develop-ment of bends or elbows in the chain.

Flexible spacers placed in the main chain achieve both improved solubility and the lowering of transition temperatures (12,13). The flexible spacer may be hydrocarbon based, as in Table 7.3, or a short series of dimethyl siloxane groups. The use of spacers is illustrated schematically for both main-chain and side-chain mesogenic units in Figure 7.6 (12).

There are several commercial thermotropic polyesters that exhibit outstanding high-temperature capabilities. These include (14,15) an increasing number of fibers and high temperature plastics. Similar to the lyotropic liquid crystalline polymers, the thermo-tropics exhibit unusually low viscosities because of orientation and lack of entanglement. Of course, the orientation serves to improve their mechanical properties. The chemical structure can be varied significantly.

Table 7.3 Melting point versus structure for selected thermotropic LC polyesters (10)

Bulky side group	*T (melting)* °C

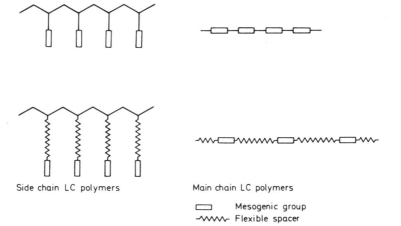

	>600
	≈340
Bent monomer	*T (melting)* °C
	≈400
	≈350
Flexible group	*T (melting)* °C
	>400
	≈210

Side chain LC polymers   **Main chain LC polymers**

▭ Mesogenic group
~~~ Flexible spacer

Figure 7.6 One way of classifying thermotropic LC polymers is to examine whether the mesogenic unit is in the main chain or in the side chain (12).

For example, the chemical structure of one copolyester is:

$$+ (O - \bigcirc - \bigcirc - O - \underset{\underset{O}{\parallel}}{C} - \bigcirc - \underset{\underset{O}{\parallel}}{C})_{0.66} (O - \bigcirc - \underset{\underset{O}{\parallel}}{C})_{0.33}]_{n}$$

where the use of two different mers allows a more precise control of properties.

It must be emphasized that neither the lyotropic nor the thermotropic polymers are ordinarily used in the liquid crystalline state. The LC state is highly convenient for processing, yielding highly ordered structures with low viscosities. However, both types are highly crystalline in actual usage, going from the LC to the crystalline state by either removing the solvent or cooling the system. Still, the side-chain mesogenic materials usually remain in the LC state for their intended uses.

7.4.3 Side-Chain Liquid Crystalline Chemical Structures

Polymers here have rod- or disk-shaped side groups placed on ordinary random coil polymers, frequently acrylics or siloxanes; see Table 7.4. The first attempts to form side-chain LC structures involved attachment of short rod-shaped moieties directly to the main chain (Figure 7.7) (16). These efforts were disappointing. Two reasons were identified (3):

1. The glass transition of the polymers is always at much higher temperatures than for the corresponding polymers without mesogenic side chains.
2. During the polymerization process, the LC packing of the monomers was destroyed by steric requirements.

It was reasoned by Plate and Shibaev (17) that if the mesogenic moieties were to be linked to the backbone through alkyl spacer groups, they should develop a LC structure similar to that attained before polymerization. The effects of backbone type, spacer group, and length of the "free" substituent at the end of the mesogenic moiety all must be considered. For example, as the length of the spacer group is increased, the low enthalpies of phase transformation of the nematic polymers become larger and the better ordered smectic phases form in their place. The structure of these LC phases was determined by X-ray analysis, as shown in Figure 7.8 (17). Similar considerations apply to the "free" substituent.

The packing arrangement of these materials is illustrated in Figure 7.9 (16). The side chains can be arranged in either single-layer or double-layer packing arrangements, Figure 7.9a and b, respectively. Packing with partial overlap is also possible; see Figure 7.9c and d. The selection of the probable packing mode was based on the d spacing shown in Figure 7.9, which corresponds to the thickness of the smectic layer.

In contrast to the backbone types of LC, the side-chain types exhibit neither high modulus nor high strength. However, they are proving highly interesting for their structural and optical properties, especially as transformed by electric and magnetic fields.

Table 7.4 Examples of smectic polymers by lengthening the segments *A* and *B*

Number	Polymer	*A*	Phase Transitions (K)	$\Delta \bar{H}_{LC-i}$ (J/g)
1	···—CH$_2$—C(CH$_3$)—··· COO—(CH$_2$)$_2$—O—⟨C$_6$H$_4$⟩—COO—⟨C$_6$H$_4$⟩—OCH$_3$	—OCH$_3$	g 369, n 394, i	2.3
2		—OC$_6$H$_{13}$	g 410, s 451, i	11.5
3	···—Si(CH$_3$)—O—··· (CH$_2$)$_3$—O—⟨C$_6$H$_4$⟩—OCH$_3$	—OCH$_3$	g 288, n 334, i	2.2
4		—OC$_6$H$_{13}$	g 288, s 385, i	11.6

Number	Polymer *B*		Phase Transitions (K)	$\Delta \bar{H}_{LC-i}$ (J/g)
1	···—CH$_2$—C(CH$_3$)—··· COO—(CH$_2$)$_2$—O—⟨C$_6$H$_4$⟩—COO—⟨C$_6$H$_4$⟩—CH=N—⟨C$_6$H$_4$⟩—CN		g 361, n 580, i	—
2		—(CH$_2$)$_6$	g 324, s 607, i	—
3	···—CH$_2$—C(CH$_3$)—··· COO—(CH$_2$)$_2$—O—⟨C$_6$H$_4$⟩—⟨C$_6$H$_4$⟩—OCH$_3$		g 393, n, 425, i	2.8
4		—(CH$_2$)$_6$	g, 119, s 136, i	7.1

Note: g, glassy; n, nematic; s, smectic; i, isotropic.

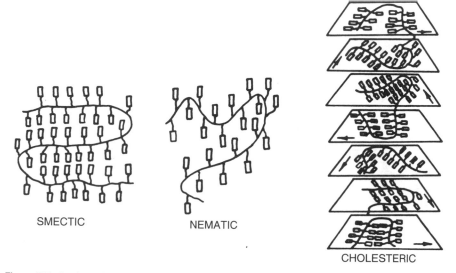

Figure 7.7 A schematic showing how mesogenic LC side-chain polymers can be organized into different mesophases (16).

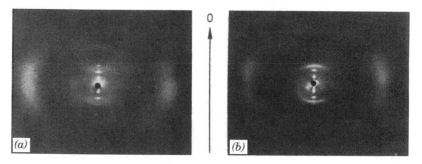

Figure 7.8 Oriented smectic LC polymers yield very distinctive X-ray patterns. Here, the smectic A mesophase (S_A) is shown for polymers having the structures (17)

(a) polymer

$$[-CH_2-C(CH_3)-]$$
$$O\overset{|}{C}-O-(CH_2)_{11}-O-\langle O \rangle-CH=N-\langle O \rangle-CN$$

and (b) polymer

$$[-CH_2-C(CH_3)-]$$
$$O\overset{|}{C}-O(CH_2)_{10}-\overset{O}{\overset{\|}{C}}-O-\langle O \rangle-\langle O \rangle-CN$$

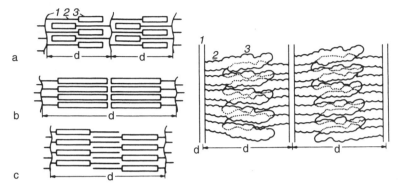

Figure 7.9 LC side chains can be packed in several arrangements, even considering only the smectic A mesophase: (*a*) single-layer packing, (*b*) two-layer packing, (*c*) packing with overlapping alkyl "tails," and (*d*) packing with partial overlapping of the mesogenic side chains. 1, Main chain; 2, spacer; 3, mesogenic group; *d*, repeat distance, as revealed by X rays (17).

7.5 THERMODYNAMICS AND PHASE DIAGRAMS

7.5.1 Historical Aspects

Liquid crystals were first noted for organic molecules about 100 years ago, when Reinitzer (18) observed a peculiar melting behavior of a number of cholesterol esters. He found that the crystals of the substances melted sharply to form an opaque melt instead of the usual clear melt. The conclusion Reinitzer drew from this observation was that some type of order still existed in the molten state. Furthermore Reinitzer observed that the opacity vanished at a higher temperature, called the clearing temperature (19). Shortly thereafter, Lehmann (20) reported that ammonium oleate and *p*-azoxyphenetole exhibited turbid states between the truly crystalline and the truly isotropic fluid state. To describe the strange behavior of the new state, Lehmann introduced the term "Flussige Kristalle," or liquid crystals. As mentioned previously, the first polymeric liquid crystal discovered was poly(γ-benzyl L-glutamate) (1), in 1950.

In 1956 Paul Flory contributed a pair of papers based on statistical thermodynamics, which proved to have enormous predictive power (21,22). The model was a rod of x isodiametric segments, each of a size that occupied one lattice cell. The quantity x also constitutes the axial ratio of the molecule. The molecules were assumed to lie at some arbitrary angle from the preferred axis. The positioning probabilities were formulated as the product of the number of lattice sites available to the initial segment of the chain times the probabilities that the sites required for each successive segment of the chain would be unoccupied, and hence accessible (23). This was entirely an entropic calculation; heats of mixing were ignored. Most important, a transition from complete disorder to partial order was predicted to occur abruptly and discontinuously beyond a critical volume concentration, v_2^*, of polymer,

$$v_2^* = \frac{8}{x}\left(1 - \frac{2}{x}\right) \tag{7.4}$$

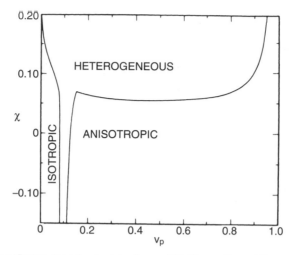

Figure 7.10   Phase diagram for rods of axial ratio $x = 100$. The positions of the coexisting phases are determined by the value of the parameter χ. The minimum of the shallow concave portion of the diagram is a critical point marking the emergence of two anisotropic phases, in addition to the isotropic phase appearing on the left of the figure (23).

which is approximately equal to $8/x$ for high values of x. Equation (7.4) yields the threshold volume fraction for appearance of a stable anisotropic phase.

The first *synthetic* LC polymers, poly(*p*-benzamide) and poly(*p*-phenylene terephthalamide), followed in 1966 and 1968, respectively (24,25); the latter is known as Kevlar[TM]. Thus theoretical papers based only on the appearance of scant data on protein materials led the way to major advances in the synthetic arcas.[†]

7.5.2   Importance of the χ_1 Parameter

The statistical thermodynamic theory described above pertains to rods devoid of interactions other than the short-range repulsions, which preclude intrusion of one rod on the space occupied by another. Thus theoretical deductions stem solely from the geometrical aspects of the molecules. Flory then introduced the polymer–solvent interaction parameter, χ_1, for lyotropic systems. The resulting phase diagram is illustrated in Figure 7.10 (23). The narrow biphasic gap is little affected by the interactions for negative values of χ_1, because the rods are attracted to the solvent. If, however, χ_1 is positive, a critical point emerges at $\chi_1 = 0.055$. For values of χ_1 immediately above this critical limit, the shallow concave curve delineates the loci of coexisting anisotropic phases, these being in addition to the isotropic and nematic phases of lower concentration within the narrow biphasic gap on the left. At $\chi_1 = 0.070$, a triple point is reached at the cusp, where three phases coexist, dilute isotropic, dilute anisotropic, and concentrated anisotropic. The heterogeneous regions contain both isotropic and anisotropic phase domains.

[†] It must be pointed out that Flory made a large number of theoretical predictions, which, together with his outstanding research in the field of polymers, led to his winning the Nobel prize in chemistry in 1974. See Table C5.1.

Figure 7.11 Optical microscopy reveals a Schlieren texture for a copolyester formed by the transesterification of poly(ethylene-1,2-diphenoxyethane-p,p'-dicarboxylate) with p-acetoxybenzoic acid. Recognizable singularities $S = \pm\frac{1}{2}$ are marked with arrows. Crossed polarizers, 260°C, ×200 (26).

7.6 MESOPHASE IDENTIFICATION IN THERMOTROPIC POLYMERS

A major question in the liquid crystal field relates to the identification of the various mesophases. Several methods may be used (26):

1. Optical pattern or texture observations with a polarizing microscope. Isotropic liquids have no texture, while both regular crystals and liquid crystals do. However, the texture of each type of phase is different.

2. Nematic and smectic phases can be distinguished by DSC on the basis of the magnitude of the enthalpy change accompanying the transition to the isotropic phase.

3. Miscibility with known liquid crystals, to form isomorphous mesophases; see Figure 7.3.

4. Possibilities of inducing significant molecular orientation by either supporting surface treatments, or external electrical or magnetic fields.

5. Small-angle X-ray studies can be used to establish molecular long-range order; see Figure 7.8.

6. Small-angle light scattering, particularly between crossed nicols, yields different patterns for different mesophases.

The most widely used method is optical microscopy, especially between crossed polarizers. It is used to identify the appearance of mesophases and transitions between the various mesophases and isotropic materials.

Many liquid crystalline polymers exhibit Schlieren textures, which display dark brushes. These correspond to extinction positions of the mesophase. At certain points, two or more dark brushes meet at points called disclinations; see Figure 7.11 (27). The disclination strength is calculated from the number of dark brushes meeting at one point:

$$|S| = \frac{\text{number of brushes meeting}}{4} \tag{7.5}$$

The sign of S is positive when the brushes turn in the same direction as the rotated polarizers, and negative when they turn in the opposite direction. Mesophase identification by this procedure requires that

$$S = \pm\tfrac{1}{2}; \pm 1 \qquad \text{Nematic}$$

$$S = \pm 1 \qquad \text{Smectic}$$

Thus a nematic phase is indicated by a mixture of two and four point disclinations, while a smectic phase exhibits only four point disclinations.

Disclinations are like dislocations in crystalline solids, where domains of differing orientations meet. The disclinations cause distortion of the director field of the polymer chains, giving rise to an excess free energy of the liquid crystal material (28).

7.7 FIBER FORMATION

7.7.1 Viscosity of Lyotropic Solutions

Rod-shaped polymer chains make unusually strong, high modulus fibers, primarily because it is easy to orient the chains in the fiber direction, without folding; see Chapter 11. Important examples are the aromatic polyamides, which are soluble in high-strength sulfuric acid or in mixed solvents containing N,N-dimethylacetamide. As the concentration of the polymer is increased, the viscosity first increases, then shows an abrupt decrease; see Figure 7.12 (29,30). This has been interpreted (30) as the formation of the nematic mesophase, v_2^* having been exceeded; see equation (7.1). Of course, the mesophase is optically anisotropic, providing a quantitative measure of the orientation. Interestingly, the viscosity of lyotropic solutions in shear flow is *lower* than that of random coil solutions of the same molecular weight and concentration. The reason is the lack of molecular entanglements in the nematic mesophase. Thus, from a strict engineering point of view, it cost less energy to pump the solutions around, and the resulting fibers are stronger.

7.7.2 Molecular Orientation

The general orientation of the nematic morphology is increased by the shearing action of flow accompanying fiber spinning. Thus, at the instant the concentrated polymer solution exits the spinnerette, ideally all the molecules are highly oriented in the direction of flow. Referring to the toothpick model introduced earlier, all the toothpicks are now aligned in the flow direction, rather than having domains of differing orientation, as when the system is at rest.

It must be emphasized that after the solvent is removed, the polymer crystallizes. Thus the aromatic polyamide fibers, while existing in liquid crystal mesophases during preparation, are highly crystalline when fully formed. Again, because of the lack of chain folds and other imperfections, the fibers have higher moduli and higher strength.

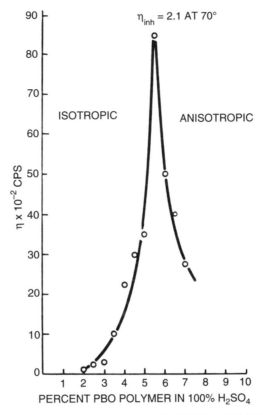

Figure 7.12   The viscosity of rod-shaped polymers usually goes through a maximum as the chains organize from the isotropic state to a mesostate. Data for poly(p-phenylene benzobisoxazole) is concentrated sulfuric acid (30).

7.8   COMPARISON OF MAJOR POLYMER TYPES

7.8.1   Molecular Conformation

The dilute solution, conformation, and bulk behaviors of rigid rods are compared with random coils in Table 7.5 (31,32). Also included are the extended chains, both those chains that are stiffer than usual and those rod-shaped chains that contain flexible spacers. For example, v_2^* is higher for the extended chain than the rigid rod, as might be expected by the decreased order, while this feature is missing entirely from the random coil solutions. The Mark-Houwink value of a [see equation (3.97) and Table 3.11] reaches a theoretical maximum value of 2.0 for rods, while stiff chains such as cellulose trinitrate have $a = 1.0$. Random coils, sometimes called flexible chains in this context, have values in the range of 0.5 to 0.8, depending on how thermodynamically "good" the solvent is. Spheres that follow the Einstein viscosity relationship have $a = 0$, the viscosity of the solution ideally depending only on the concentration of the solution and not on the size of the spheres.

The relative orientation of rigid-rod polymers is schematically compared with that of

Table 7.5   Comparison of major polymer types (32)

Property	Rigid Rod	Extended Chain	Random Coils
LC v_2^* critical concentration, %	4–5	14–15	None
Dilute solution			
conformation	Rod	Worm	Coil
T_g	No[a]	No[a]	Yes
T_f	No[b]	No[b]	Yes
Persistence length, Å	>500	90–130	10
Mark-Houwink a value	1.8	1.0	0.5–0.8
Catenation angle, degrees	180	150–162	109–120
Max $[\eta]$, dl/g	48–60	15–25	0.1–3

[a] May be difficult to observe if highly crystalline.

[b] May decompose before melting.

the semiflexible polymers in Figure 7.13 (33). Here v_2^* is depicted as being between 5 and 10% concentration, with nearly total alignment occurring above 15% concentration.

Typically rigid-rod systems such as the aromatic polyamides or the polyheterocyclic polymers do not exhibit a glass transition, because their amorphous fraction is negligibly small, and they decompose before their melting temperature. By comparison, many of the crystalline random coil polymers, such as the aliphatic nylons, exhibit normal melting phenomena, as well as glass transitions. Thus the properties of the new rigid-rod polymers are quite different from those of random coil polymers.

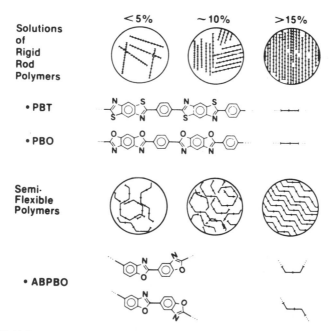

Figure 7.13   Rigid-chain and semiflexible-chain ordered assemblies as the concentration of the polymer is increased (33).

7.8.2 Properties in Bulk

These last three chapters delineate the three bulk forms of polymers: amorphous, crystalline, and liquid crystalline. How are they best distinguished in the laboratory? Key experiments involve X-ray diffraction, optical microscopy, and methods of observing transitions. Typical results are as follows:

Experiment	Amorphous	Crystalline	Liquid Crystalline
X-ray	Amorphous halo	3-D order	1- or 2-D order
Polarizing microscope	no texture	spherulites	Schlieren texture
DSC, dilatometric, etc.	only T_g	T_g and T_f	T_g and two or more first-order transitions

7.9 BASIC REQUIREMENTS FOR LIQUID CRYSTAL FORMATION

A liquid crystal must satisfy three basic requirements, regardless of molecular size or shape:

1. There must be a first-order transition between the true crystalline state at the lower temperature bound leading to the liquid crystalline state, and another first-order transition leading to the isotropic liquid state (or another liquid crystal state) at the upper temperature bound of the liquid crystalline state.

2. A liquid crystal must exhibit one- or two-dimensional order only; true crystals have three-dimensional order, and the isotropic liquid is completely disordered.

3. A liquid crystalline material must display some degree of fluidity, although for polymers the viscosity may be high.

Usually the conditions for liquid crystal formation are best met when the molecules have at least some portion of their structure in the form of rods or disks. However, this is not an absolute requirement, and a surprisingly large number of polymers are now being found to exhibit liquid crystalline phases. While truly rigid-rod polymers ordinarily do not have a glass transition, if some degree of flexibility is built into the polymer chain, there may be one.

The backbone types of polymeric liquid crystals may be lyotropic, requiring a solvent for entrance into the liquid crystalline phase, or thermotropic, requiring heat for entrance into the liquid crystalline phase. However, when dry and at service temperature, both types ordinarily exhibit three-dimensional crystalline order.

REFERENCES

1. A. Elliott and E. J. Ambrose, *Discuss. Faraday Soc.*, **9**, 246 (1950).

2. S. L. Kwolek, P. W. Morgan, and J. R. Schaefgen, in *Encyclopedia of Polymer Science and Engineering*, 2nd ed., Vol. 9, J. I. Kroschwitz, ed., Wiley, New York, 1987.

3. J. L. White, *J. Appl. Polym. Sci. Polym. Symp.*, **41**, 3 (1985).

4. H. Finkelmann and G. Rehage, *Adv. Polym. Sci.*, **60/61**, 99 (1984).

5. H. Finkelmann, J. J. Kock, and G. Rehage, *Mol. Cryst. Liq. Cryst.*, **89**, 23 (1982).

6. G. Decorbert, J. C. Dubois, S. Esselin, and C. Noël, *Liquid Crystals*, **1**, 307 (1986).

7. J. R. Schaefgen, T. I. Bair, J. W. Ballou, S. L. Kwolek, P. W. Morgan, M. Panar, and J. Zimmerman, in *Ultra-High Modulus Polymers*, A. Ciferri and I. M. Ward, eds., Applied Science, London, 1979.

8. W. G. Miller, C. C. Wu, G. L. Antee, J. H. Rai, and K. G. Goebel, *Pure Appl. Chem.*, **38**, 37 (1974).

9. R. Sakamoto, *Colloid Polym. Sci.*, **262**, 788 (1984).

10. A. J. East, L. F. Charbonneau, and G. W. Calundann, U.S. Pat. No. 4,330,457 (1982).

11. G. Huynh-Ba and E. F. Cluff, in *Polymeric Liquid Crystals*, A. Blumstein, ed., Plenum, New York, 1985.

12. C. K. Ober, J. I. Jin, and R. W. Lenz, in *Liquid Crystal Polymers I*, N. A. Plate, ed., Advances in Polymer Science Vol. 59, Springer, Berlin, 1984.

13. W. J. Jackson and F. H. Kuhfuss, *J. Polym. Sci. Polym. Chem. Ed.*, **14**, 2043 (1976).

14. P. K. Bhowmik, E. O. T. Atkins, and R. W. Lenz, *Macromolecules*, **26**, 447 (1993).

15. A. K. Rath and S. Ponrathnam, *J. Appl. Polym. Sci.*, **49**, 391 (1993).

16. N. A. Plate, R. V. Talroze, and V. P. Shibaev, in *Polymer Yearbook 3*, R. A. Pethrick, ed., Harwood, Chur, United Kingdom, 1986.

17. N. A. Plate and V. P. Shibaev, *Comb-Shaped Polymers and Liquid Crystals*, Plenum Press, New York, 1987, translated from the Russian by S. L. Schnur.

18. F. Reinitzer, *Monatsh. Chem.*, **9**, 421 (1888).

19. J. H. Wendorff, in *Liquid Crtystalline Order in Polymers*, A. Blumstein, ed., Academic Press, Orlando, 1978.

20. O. Lehmann, *Z. Kristallogr. Kristallgeom. Kristallphys. Kristallchem.*, **18**, 464 (1890).

21. P. J. Flory, *Proc. R. Soc. London Ser. A.*, **234**, 60 (1956).

22. P. J. Flory, *Proc. R. Soc. London Ser. A.*, **234**, 73 (1956).

23. P. J. Flory, in *Liquid Crystal Polymers I*, M. Gordon, ed., Advances in Polymer Science, Vol. 59, Springer-Verlag, Berlin, 1984.

24. S. L. Kwolek, B.P. 1,198,081 (priority June 13, 1966).

25. S. L. Kwolek, B.P. 1,283,064 (priority June 12, 1968).

26. C. Noel, in *Polymeric Liquid Crystals*, A. Blumstein, ed., Plenum, New York, 1985.

27. C. Noel, F. Laupretre, C. Friedrich, B. Fayolle, and L. Bosio, *Polymer*, **25**, 808 (1984).

28. T. Hashimoto, A. Nakai, T. Shiwaku, H. Hasegawa, S. Rojstaczer, and R. S. Stein, *Macromolecules*, **22**, 422 (1989).

29. D. B. DuPre, "Liquid Crystals," in *Encyclopedia of Chemical Technology*, 3rd ed., Vol. 14, M. Grayson, ed., Wiley, New York, 1981, p. 417.

30. E. W. Choe and S. N. Kim, *Macromolecules*, **14**, 920 (1981).

31. J. F. Wolfe, B. H. Loo, and F. E. Arnold, *Macromolecules*, **14**, 915 (1981).

32. J. F. Wolfe, Presented at Pacifichem, Honolulu, December 1989.

33. D. R. Ulrich, *Polymer*, **28**, 533 (1987).

GENERAL READING

W. Brostow, ed., *Mechanical and Thermophysical Properties of Polymeric Liquid Crystals*, Chapman and Hall, London, 1998.

E. Chiellini, M. Girodano, and D. Leporini, eds., *Structure and Transport in Organized Polymeric Materials*, World Scientific, Signapore, 1997.

A. Ciferri, ed., *Liquid Crystallinity in Polymers: Principles and Fundamental Properties*, VCH Publishers, New York, 1991.

P. G. de Gennes, *The Physics of Liquid Crystals*, Oxford Press, New York, 1975.

A. Keller, M. Warner, and A. H. Windle, eds., *Self-Order and Form in Polymeric Materials*, Chapman and Hall, London, 1995.

C. B. McArdle, ed., *Side Chain Liquid Crystal Polymers*, Chapman and Hall, New York, 1989.

N. A. Plate, ed., *Liquid Crystal Polymers I*, Advances in Polymer Science, Vol. 59, Springer, Berlin, 1984.

N. A. Plate, ed., *Liquid Crystal Polymers II/III*, Advances in Polymer Science, Vol. 60/61, Springer, Berlin, 1984.

N. A. Plate and V. P. Shibaev, *Comb-Shaped Polymers and Liquid Crystals*, Plenum, New York, 1987.

N. A. Plate, R. V. Talroze, and V. P. Shibaev, in *Polymer Yearbook 3*, R. A. Pethrick, ed., Harwood, Chur, United Kingdom, 1986.

V. V. Shilov, *Structure of Polymeric Liquid Crystals*, Naukova Dumka, Kiev, 1990.

D. R. Ulrich, *Polymer*, **28**, 533 (1987).

G. Vertogen and W. H. de Jeu, *Thermotropic Liquid Crystals, Fundamentals*, Springer-Verlag, Berlin, 1988.

R. A. Weiss and C. K. Ober, eds., *Liquid-Crystalline Polymers*, ACS Books, Washington, DC, 1990.

STUDY PROBLEMS

1. As the temperature is raised, some polymers melt from a regular three-dimensional crystal to a smectic phase, then to a nematic phase, and then finally to an isotropic melt. How would the appropriate DSC curve look for such a polymer?

2. Based on Figure 7.10, discuss whether lyotropic polymer systems exhibit two or more phases simultaneously and, if so, under what conditions?

3. There are three major types of polymers: semicrystalline, liquid crystalline (or goes through an L. C. phase), and amorphous. (a) Give an example of each, naming the polymer. (b) Given an unknown polymer, what one experiment would you perform to identify which type you had? Give an illustration of the data that you might obtain. (c) What major behavioral differences would you expect among the three types?

4. (a) Compare and contrast the intrinsic viscosity Mark-Houwink-Sakarada values of *a* for rigid rods and random coils. (b) Why is the viscosity higher for rigid rods in the case of dilute solutions, but lower in the case of concentrated solutions?

5. Read any scientific or engineering paper written in the last three years on the topic of liquid crystalline polymers, and briefly report on its findings. How does it update this textbook? (Give the complete reference.)

6. In an actual experiment, 5.9 cm × 1.5 mm × 1 mm Forster™ flat toothpicks were added to a water surface one at a time with no overlapping under mild agitation. A distinct nematic phase formed at about a total surface concentration of 3%. Thereafter the remaining disordered phase maintained a 0.8% surface concentration up to a total surface concentration of 14%. How would these toothpicks behave if they were first cut in half? In fourths? (Try the experiment in a sink!)

8

GLASS–RUBBER
TRANSITION BEHAVIOR

The state of a polymer depends on the temperature and on the time allotted to the experiment. While this is equally true for semicrystalline and amorphous polymers, although in different ways, the discussion in this chapter centers on amorphous materials.

At low enough temperatures, all amorphous polymers are stiff and glassy. This is the glassy state, sometimes called the vitreous state, especially for inorganic materials. On warming, the polymers soften in a characteristic temperature range known as the glass–rubber transition region. Here, the polymers behave in a leathery manner. The importance of the glass transition in polymer science was stated by Eisenberg:[†] "The glass transition is perhaps the most important single parameter that determines the application of many noncrystalline polymers now available."

The glass transition is named after the softening of ordinary glass. On a molecular basis, the glass transition involves the onset of long-range coordinated molecular motion, the beginning of reptation. The glass transition is a second-order transition. Rather than dicontinuities in enthalpy and volume, their temperature derivatives, heat capacity and coefficients of expansion shift. By difference, melting and boiling are first-order transitions, exhibiting discontinuities in enthalpy and volume, with heats of transition.

For amorphous polymers, the glass transition temperature, T_g, constitutes their most important mechanical property. In fact, upon synthesis of a new polymer, the glass transition temperature is among the first properties measured. This chapter describes the behavior of amorphous polymers in the glass transition range, emphasizing the onset of molecular motions associated with the transition. Before beginning the main topic, two introductory sections are presented. The first defines a number of mechanical terms that will be needed, and the second describes the mechanical spectrum encountered as a polymer's temperature is raised.

[†] In J. E. Mark, A. Eisenberg, W. W. Graessley, L. Mandelkern, E. T. Samulski, J. L. Koenig, and G. D. Wignall, *Physical Properties of Polymers* 2nd ed., American Chemical Society, Washington, DC, 1993.

8.1 SIMPLE MECHANICAL RELATIONSHIPS

Terms such as "glassy," "rubbery," and "viscous" imply a knowledge of simple material mechanical relationships. Although such information is usually obtained by the student in elementary courses in physics or mechanics, the basic relationships are reviewed here because they are used throughout the text. More detailed treatments are available (1–4).

8.1.1 Modulus

8.1.1.1 Young's Modulus   Hook's law assumes perfect elasticity in a material body. Young's modulus, E, may be written

$$E = \sigma/\varepsilon \tag{8.1}$$

where σ and ε represent the tensile stress and strain, respectively. Young's modulus is a fundamental measure of the stiffness of the material. The higher its value, the more resistant the material is to being stretched.

The tensile stress is defined in terms of force per unit area. If the sample's initial length is L_0 and its final length is L, then the strain is $\varepsilon = (L - L_0)/L_0$† (see Figure 8.1).

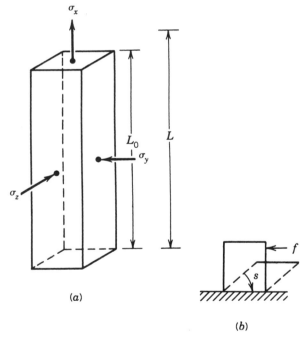

(a)

(b)

Figure 8.1 Mechanical deformation of solid bodies. (a) Triaxial stresses on a material body undergoing elongation. (b) Simple shear deformation.

---

† In more graphic language, the amount of stress applied is measured by the amount of grunting the investigator does, and the strain is measured by the sample's groaning.

Table 8.1 Some mechanical terms

Term	Definition
σ	Stress
ε	Strain
E	Young's modulus
G	Shear modulus
B	Bulk modulus
v	Poisson's ratio
η	Viscosity
J	Tensile compliance
s	Shear strain
f	Shear stress
R	Gas constant, 8.31×10^7 (dynes $\cdot$ cm)/(mol $\cdot$ K)
t	Time

The forces and subsequent work terms have some good simple examples. Consider a postage stamp, which weighs about 1 g and is about 1 cm × 1 cm in size. It requires about 1 dyne of force to lift it from the horizontal position to the vertical position, as in turning it over. One dyne of force through 1 cm gives 1 erg, the amount of work done. Modulus is usually reported in dynes/cm^2, in terms of force per unit area. Frequently the pascal unit of modulus is used, 10 dynes/cm^2 = 1 pascal.

8.1.1.2 Shear Modulus   Instead of elongating (or compressing!) a sample, it may be subjected to various shearing or twisting motions (see Figure 8.1). The ratio of the shear stress, f, to the shear strain, s, defines the shear modulus, G:

$$G = \frac{f}{s} \tag{8.2}$$

These and other mechanical terms are summarized in Table 8.1.

8.1.2 Newton's Law

The equation for a perfect liquid exhibiting a viscosity, η, may be written

$$\eta = f/(ds/dt) \tag{8.3}$$

where f and s represent the shear stress and strain, respectively, and t is the time. For simple liquids such as water or toluene, equation (8.3) reasonably describes their viscosity, especially at low shear rates. For larger values of η, flow is slower at constant shear stress. While neither equation (8.1) nor equation (8.3) accurately describes polymer behavior, they represent two important limiting cases.

The basic definition of viscosity should be considered in terms of equation (8.3). Consider two 1-cm^2 planes 1 cm apart imbedded in a liquid. If it takes 1 dyne of force to move one of the planes 1 cm/s relative to the other in a shearing motion, the liquid has a viscosity of 1 poise. Viscosity is also expressed in pascal-seconds, with 1 Pa $\cdot$ s = 10 poise.

Table 8.2   Values of Poisson's ratio

Value	Interpretation
0.5	No volume change during stretch
0.0	No lateral contraction
0.49–0.499	Typical values for elastomers
0.20–0.40	Typical values for plastics

8.1.3   Poisson's Ratio

When a material body is elongated (or undergoes other modes of deformation), in general, the volume changes, usually increasing as elongational strains are applied. Poisson's ratio, v, is defined as

$$-v\varepsilon_x = \varepsilon_y = \varepsilon_z \tag{8.4}$$

for very small strains, where the strain ε_x is applied in the x direction and the strains ε_y and ε_z are responses in the y and z directions, respectively (see Figure 8.1). Table 8.2 summarizes the behavior of v under several circumstances.

For analytical purposes, Poisson's ratio is defined on the differential scale. If V represents the volume,

$$V = xyz \tag{8.5}$$

Then

$$\frac{d \ln V}{d \ln x} = \frac{d \ln x}{d \ln x} + \frac{d \ln y}{d \ln x} + \frac{d \ln z}{d \ln x} \tag{8.6}$$

and

$$-\frac{d \ln y}{d \ln x} = -\frac{d \ln z}{d \ln x} = v = -\frac{dy}{y_0} \Big/ \frac{dx}{x_0} \tag{8.7}$$

Since $d \ln x / d \ln x = 1$, for no volume change $v = 0.5$ (see Table 8.2).

On extension, plastics exhibit considerable volume increases, as illustrated by the values of v in Table 8.2. The physical separation of atoms provides a major mechanism for energy storage and short-range elasticity.

Poisson's ratio is only useful for very small strains. Poisson's ratio was originally developed for calculations involving metals, concrete, and other materials with limited extensibility. The approximations built into the theory make rubber elasticity results unrealistic. Glazebrook (5) presents a more general treatment of Poisson's ratio.

8.1.4   The Bulk Modulus and Compressibility

The bulk modulus, B, is defined as

$$B = -V \left(\frac{\partial P}{\partial V} \right)_T \tag{8.8}$$

where P is the hydrostatic pressure. Normally a body shrinks in volume on being exposed to increasing external pressures, so the term $(\partial P/\partial V)_T$ is negative.

The inverse of the bulk modulus is the compressibility, β,

$$\beta = \frac{1}{B} \tag{8.9}$$

which is strictly true only for a solid or liquid in which there is no time-dependent response. Bulk compression usually does not involve long-range conformational changes but rather a forcing together of the chain atoms. Of course, materials ordinarily exist under a hydrostatic pressure of 1 atm (1 bar) at sea level.

8.1.5 Relationships among *E*, *G*, *B*, and *v*

A three-way equation may be written relating the four basic mechanical properties:

$$E = 3B(1 - 2v) = 2(1 + v)G \tag{8.10}$$

Any two of these properties may be varied independently, and conversely, knowledge of any two defines the other two. As an especially important relationship, when $v \cong 0.5$,

$$E \cong 3G \tag{8.11}$$

which defines the relationship between E and G to a good approximation for elastomers.

Equation (8.10) can also be used to evaluate Poisson's ratio for elastomers. Rearranging the two left-hand portions, we have

$$1 - 2v = \frac{E}{3B} = \frac{\beta E}{3} \tag{8.12}$$

Because the quantity $1 - 2v$ is close to zero for elastomers (but cannot be exactly so), exact evaluation of v depends on the evaluation of the right-hand side of equation (8.12). Values in the literature for elastomers vary from 0.49 to 0.49996 (2).

Thus, in contrast to plastics, separation of the atoms plays only a small role in the internal storage of energy. Instead, conformational changes in the chains come to the fore, the main subject of Chapter 9.

8.1.6 Compliance versus Modulus

If the modulus is a measure of the stiffness or hardness of an object, its compliance is a measure of softness. In regions far from transitions, the elongational compliance, J, is defined as

$$J \simeq \frac{1}{E} \tag{8.13}$$

For regions in or near transitions, the relationship is more complex. Shear and other compliances can also be defined. Ferry has reviewed this topic (2).

Table 8.3 Numerical values of Young's modulus

Material	E (dyne/cm^2)	E (Pa)
Copper	1.2×10^{12}	1.2×10^{11}
Polystyrene	3×10^{10}	3×10^9
Soft rubber	2×10^7	2×10^6

8.1.7 Numerical Values for *E*

Before proceeding with the description of the temperature behavior of polymers, it is of interest to establish some numerical values for Young's modulus (see Table 8.3). Polystyrene represents a typical glassy polymer at room temperature. It is about 40 times as soft as elemental copper, however. Soft rubber, exemplified by such materials as rubber bands, is nearly 1000 times softer still. Perhaps the most important observation from Table 8.3 is that the modulus varies over wide ranges, leading to the wide use of logarithmic plots to describe the variation of modulus with temperature or time.

8.1.8 Storage and Loss Moduli

The quantities E and G refer to quasistatic measurements. When cyclical or repetitive motions of stress and strain are involved, it is more convenient to talk about dynamic mechanical moduli. The complex Young's modulus has the formal definition

$$E^* = E' + iE'' \tag{8.14}$$

where E' is the storage modulus and E'' is the loss modulus. Note that $E = |E^*|$. The quantity i represents the square root of minus one. The storage modulus is a measure of the energy stored elastically during deformation, and the loss modulus is a measure of the energy converted to heat. Similar definitions hold for G^*, J^*, and other mechanical quantities.

8.2 FIVE REGIONS OF VISCOELASTIC BEHAVIOR

Viscoelastic materials simultaneously exhibit a combination of elastic and viscous behavior. While all substances are viscoelastic to some degree, this behavior is especially prominent in polymers. Generally, viscoelasticity refers to both the time and temperature dependence of mechanical behavior.

The states of matter of low-molecular-weight compounds are well known: crystalline, liquid, and gaseous. The first-order transitions that separate these states are equally well known: melting and boiling. Another well-known first-order transition is the crystalline–crystalline transition, in which a compound changes from one crystalline form to another.

By contrast, no high-molecular-weight polymer vaporizes to a gaseous state; all decompose before the boiling point. In addition no high-molecular-weight polymer attains a totally crystalline structure, except in the single-crystal state (see Section 6.4.2).

In fact many important polymers do not crystallize at all but form glasses at low temperatures. At higher temperatures they form viscous liquids. The transition that separates the glassy state from the viscous state is known as the glass–rubber transition.

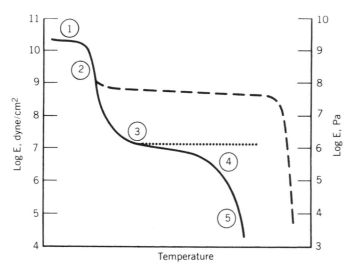

Figure 8.2 Five regions of viscoelastic behavior for a linear, amorphous polymer. Also illustrated are effects of crystallinity (dashed line) and cross-linking (dotted line).

According to theories to be developed later, this transition attains the properties of a second-order transition at very slow rates of heating or cooling.

Before entering into a detailed discussion of the glass transition, the five regions of viscoelastic behavior are briefly discussed to provide a broader picture of the temperature dependence of polymer properties. In the following, quasi-static measurements of the modulus at constant time, perhaps 10 or 100 s, and the temperature being raised 1°C/min will be assumed.

8.2.1 The Glassy Region

The five regions of viscoelastic behavior for linear amorphous polymers (3, 6–8) are shown in Figure 8.2. In region 1 the polymer is glassy and frequently brittle. Typical examples at room temperature include polystyrene (plastic) drinking cups and poly(methyl methacrylate) (Plexiglas® sheets).

Young's modulus for glassy polymers just below the glass transition temperature is surprisingly constant over a wide range of polymers, having the value of approximately 3×10^{10} dynes/cm^2 (3×10^9 Pa). In the glassy state, molecular motions are largely restricted to vibrations and short-range rotational motions.

A classical explanation for the constant modulus values in the glassy state starts with the Lennard-Jones potential, describing the energy of interaction between a pair of isolated molecules. From this, the molar lattice energy between polymer segments was calculated. The bulk modulus, B, was calculated in terms of the cohesive energy density, CED, which represents the energy theoretically required to move a detached segment into the vapor phase. This, in turn, is related to the square of the solubility parameter:

$$B = 8.04(\text{CED}) = 8.04\delta^2 \tag{8.15}$$

The factor 8.04 arises from Lennard-Jones considerations (2,3).

Using polystyrene as an example, its CED is 83 cal/cm^3 (3.47 × 10^9 ergs/cm^3). Its Poisson ratio is approximately 0.30. From equation (8.10), $E \simeq 1.2B$, and hence

$$E \simeq 9.6(\text{CED}) \tag{8.16}$$

For polystyrene, a value of $E = 3.3 \times 10^{10}$ dynes/cm^2 (3.3 × 10^9 Pa) is calculated, which is surprisingly close to the value obtained via modulus measurements, 3 × 10^9 Pa. It should be noted that many hydrocarbon and not-too-polar polymers have CED values within a factor of 2 of the value for polystyrene.

A more modern explanation of the glassy modulus of polymers starts with a consideration of the carbon–carbon bonding force fields adopted for the explanation of vibrational frequencies (9). For perfectly oriented polyethylene, a value of $E = 3.4 \times 10^{11}$ Pa was calculated, with a slightly lower value, 1.6×10^{11} Pa for polytetrafluoroethylene.

Using scanning force microscopy, Du et al. (4) determined Young's modulus of polyethylene single crystals in the chain direction to be 1.7×10^{11} Pa. Ultradrawn polyethylene fibers also have an experimental Young's modulus of 1.7×10^{11} Pa; see Table 11.2. These slightly lower values may be attributed to less than perfect orientation of the chains.

For unoriented, amorphous glassy polymers, the theoretical values should be divided by three. Further consideration is required for the somewhat greater mer cross-sectional area of polymers such as polystyrene or poly(methyl methacrylate), since the side groups should not contribute to the modulus according to this theory. Thus values of approximately 3 × 10^9 Pa may again be estimated.

Interestingly, both the CED and the vibrational force field theories of glassy polymer modulus arrive at very similar values, although their approaches are entirely different. Further reductions in the experimental modulus may be imagined as caused by free volume, end groups, and so on. It may be that the two theoretical values should better be considered as *additive*, since intermolecular attractive forces and carbon–carbon covalent bonding forces both play roles. This is yet to be determined experimentally.

8.2.2 The Glass Transition Region

Region 2 in Figure 8.2 is the glass transition region. Typically the modulus drops a factor of about 1000 in a 20 to 30°C range. The behavior of polymers in this region is best described as leathery, although a few degrees of temperature change will obviously affect the stiffness of the leather.

For quasi-static measurements such as illustrated in Figure 8.2, the glass transition temperature, T_g, is often taken at the maximum rate of turndown of the modulus at the elbow, where $E \cong 10^9$ Pa. Often the glass transition temperature is defined as the temperature where the thermal expansion coefficient (Section 8.3) undergoes a discontinuity. (Enthalpic and dynamic definitions are given in Section 8.2.9. Other, more precise definitions are given in Section 8.5.)

Qualitatively, the glass transition region can be interpreted as the onset of long-range, coordinated molecular motion. While only 1 to 4 chain atoms are involved in motions below the glass transition temperature, some 10 to 50 chain atoms attain sufficient thermal energy to move in a coordinated manner in the glass transition region (8,10–13) (see Table 8.4) (8,14). The number of chain atoms (10–50) involved in the coordinated motions was deduced by observing the dependence of T_g on the molecular

Table 8.4   Glass transition parameters (0, 14)

Polymer	T_g, °C	Number of Chain Atoms Involved
Poly(dimethyl siloxane)	−127	40
Poly(ethylene glycol)	−41	30
Polystyrene	+100	40–100
Polyisoprene	−73	30–40

weight between cross-links, M_c. When T_g became relatively independent of M_c in a plot of T_g versus M_c, the number of chain atoms was counted. It should be emphasized that these results are tenuous at best.

The glass transition temperature itself varies widely with structure and other parameters, as will be discussed later. A few glass transition temperatures are shown in Table 8.4. Interestingly, the idealized map of polymer behavior shown in Figure 8.2 can be made to fit any of these polymers merely by moving the curve to the right or left, so that the glass transition temperature appears in the right place.

8.2.3 The Rubbery Plateau Region

Region 3 in Figure 8.2 is the rubbery plateau region. After the sharp drop that the modulus takes in the glass transition region, it becomes almost constant again in the rubbery plateau region, with typical values of 2×10^7 dynes/cm² (2×10^6 Pa). In the rubbery plateau region, polymers exhibit long-range rubber elasticity, which means that the elastomer can be stretched, perhaps several hundred percent, and snap back to substantially its original length on being released.

Two cases in region 3 need to be distinguished:

1. The polymer is linear. In this case the modulus will drop off slowly, as indicated in Figure 8.2. The width of the plateau is governed primarily by the molecular weight of the polymer; the higher the molecular weight, the longer is the plateau (see Figure 8.3) (15).

   An interesting example of such a material is unvulcanized natural rubber. When Columbus came to America (16), he found the American Indians playing ball with natural rubber. This product, a linear polymer of very high molecular weight, retains its shape for short durations of time. However, on standing overnight, it creeps, first forming a flat spot on the bottom, and eventually flattening out like a pancake. (See Section 9.2 and Chapter 10.)

2. The polymer is cross-linked. In this case the dotted line in Figure 8.2 is followed, and improved rubber elasticity is observed, with the creep portion suppressed. The dotted line follows the equation $E = 3nRT$, where n is the number of active chain segments in the network and RT represents the gas constant times the temperature; see equation (9.36). An example of a cross-linked polymer above its glass transition temperature obeying this relationship is the ordinary rubber band. Cross-linked elastomers and rubber elasticity relationships are the primary subjects of Chapter 9.

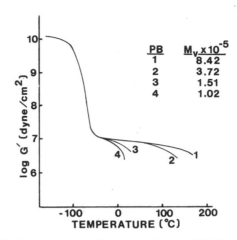

Figure 8.3   Effect of molecular weight on length of plateau (15). PB = polybutadiene.

The rapid, coordinated molecular motion in this region is governed by the principles of reptation and diffusion laid down in Section 5.4. Thus, when the elastomer is stretched, the chains deform with a series of rapid motions of the de Gennes type. The model must be altered slightly for cross-linked systems, for then the chain ends are bound at the cross-links. The motion is thought to become a more complex affair involving the several chain segments that are bound together.

So far the discussion has been limited to amorphous polymers. If a polymer is semi-crystalline, the dashed line in Figure 8.2 is followed. The height of the plateau is governed by the degree of crystallinity. This is so because of two reasons: first, the crystalline regions tend to behave as a filler phase, and second, because the crystalline regions also behave as a type of physical cross-link, tying the chains together.

The crystalline plateau extends until the melting point of the polymer. The melting temperature, T_f, is always higher than T_g, T_g being from one-half to two-thirds of T_f on the absolute temperature scale (see Section 8.9.3 for further details).

8.2.4   The Rubbery Flow Region

As the temperature is raised past the rubbery plateau region for linear amorphous polymers, the rubbery flow region is reached—region 4. In this region the polymer is marked by both rubber elasticity and flow properties, depending on the time scale of the experiment. For short time scale experiments, the physical entanglements are not able to relax, and the material still behaves rubbery. For longer times, the increased molecular motion imparted by the increased temperature permits assemblies of chains to move in a coordinated manner (depending on the molecular weight), and hence to flow (see Figure 8.3) (15). An example of a material in the rubbery flow region is Silly Putty®, which can be bounced like a ball (short-time experiment) or pulled out like taffy (a much slower experiment).

It must be emphasized that region 4 does not occur for cross-linked polymers. In that case, region 3 remains in effect up to the decomposition temperature of the polymer (Figure 8.2).

8.2.5   The Liquid Flow Region

At still higher temperatures, the liquid flow region is reached—region 5. The polymer flows readily, often behaving like molasses. In this region, as an idealized limit, equation (8.3) is obeyed. The increased energy allotted to the chains permits them to reptate out through entanglements rapidly and flow as individual molecules.

For semicrystalline polymers the modulus depends on the degree of crystallinity. The amorphous portions go through the glass transition, but the crystalline portion remains hard. Thus a composite modulus is found. The melting temperature is always above the glass transition temperature (see below). At the melting temperature the modulus drops rapidly to that of the corresponding amorphous material, now in the liquid flow region. It must be mentioned that modulus and viscosity are related through the molecular relaxation time, also discussed below.

8.2.6   Effect of Plasticizers

Polymers are frequently plasticized to "soften" them. These plasticizers are usually small, relatively nonvolatile molecules that dissolve in the polymer, separating the chains from each other and hence making reptation easier. In the context of Figure 8.2, the glass transition temperature is lowered, and the rubbery plateau modulus is lowered. If the polymer is semicrystalline, the plasticizer reduces the melting temperature and/or reduces the extent of crystallinity.

An example is poly(vinyl chloride), which has a T_g of $+80\,°C$. Properly plasticized, it has a T_g of about $+20\,°C$ or lower, forming the familiar "vinyl." A typical plasticizer is dioctyl phthalate, with a solubility parameter of 8.7 $(cal/cm^3)^{1/2}$, fairly close to that of poly(vinyl chloride), 9.6 $(cal/cm^3)^{1/2}$; see Table 3.2. A significant parameter in this case is thought to be hydrogen bonding between the hydrogen on the same carbon as the chlorine and the ester group on the dioctyl phthalate. Other factors influencing the glass transition temperature are considered in Sections 8.6.3.2, 8.7, and 8.8.

8.2.7   Definitions of the Terms "Transition," "Relaxation," and "Dispersion"

The term "transition" refers to a change of state induced by changing the temperature or pressure.

The term "relaxation" refers to the time required to respond to a change in temperature or pressure. It also implies some measure of the molecular motion, especially near a transition condition. Frequently an external stress is present, permitting the relaxation to be measured. For example, one could state that $1/e$ (0.367) of the polymer chains respond to an applied stress in 10 s at the glass transition temperature, providing a simple molecular definition.

The term "dispersion" refers to the emission or absorption of energy—that is, a loss peak—at a transition. In practice, the literature sometimes uses these terms somewhat interchangeably.

8.2.8   Melt Viscosity Relationships near T_g

The discussion above emphasizes changes in the modulus with temperature. Equally large changes also take place in the viscosity of the polymer. In fact the term "glass transition" refers to the temperature in which ordinary glass softens and flows (17).

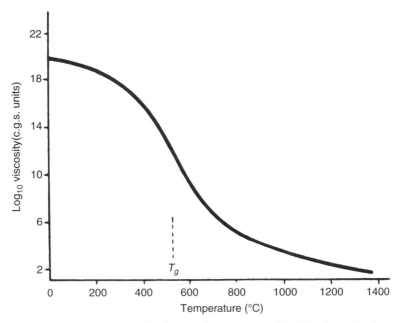

Figure 8.4   Viscosity–temperature relation of a soda–lime–silica glass (18). Soda–lime–silica glass is one of the commonly used glasses for windows and other items.

(Glass is an inorganic polymer, held together with both covalent —Si—O—Si— bonds and ionic bonds.) In this case viscoelastic region 3 is virtually absent.

The viscosity–temperature relationship of glass is shown in Figure 8.4 (18). A criterion sometimes used for T_g for both inorganic and organic polymers is the temperature at which the melt viscosity reaches a value of 1×10^{13} poises (1×10^{12} Pa·s) (19) on cooling.

8.2.9 Dynamic Mechanical Behavior through the Five Regions

The change in the modulus with temperature has already been introduced (see Section 8.2.2). More detail about the transitions is available through dynamic mechanical measurements, sometimes called dynamic mechanical spectroscopy (DMS) (see Section 8.1.8).

The shear storage modulus, G', and the shear loss modulus, G'', are the shear counter parts of E' and E''. While the temperature dependence of G' is similar to that of G, the quantity G'' behaves quite differently (see Figure 8.5) (8). The loss quantities behave somewhat like the absorption spectra in infrared spectroscopy, where the energy of the electromagnetic radiation is just sufficient to cause a portion of a molecule to go to a higher energy state. (Infrared spectrometry is usually carried out by varying the frequency of the radiation at constant temperature.) This exact analogue is frequently carried out for polymers also (see below). Of course, in DMS, the energy is imparted by mechanical waves. The subject has been reviewed (17).

The reader will note that in Figure 8.5 there is no discontinuity in the V–T or H–T plots, only a change in slope. This is characteristic of a *second-order transition*. In metallurgy a second-order transition of note is the Curie temperature of iron at 1043 K,

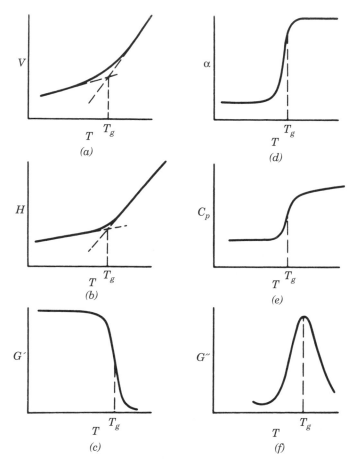

Figure 8.5 Idealized variations in volume, V, enthalpy, H, and storage shear modulus, G' as a function of temperature. Also shown are α, the volume coefficient of expansion, and C_p, the heat capacity, which are, respectively, the first derivatives of V and H with respect to temperature, and the loss shear modulus, G'' (8).

where it looses its ferromagnetic capability (20). *First-order transitions* include melting, boiling, and changes in crystalline structure with temperature. Ordinary melting and boiling of water are common examples. These are characterized by discontinuities in V–T and H–T plots. (Note that Figure 6.2 illustrates such melting.)

Measurements by DMS refer to any one of several methods where the sample undergoes repeated small-amplitude strains in a cyclic manner. Molecules perturbed in this way store a portion of the imparted energy elastically and dissipate a portion in the form of heat (1–4). The quantity E', Young's storage modulus, is a measure of the energy stored elastically, whereas E'', Young's loss modulus, is a measure of the energy lost as heat (see Figure 8.6) (21).

Another equation in wide use is

$$\frac{E''}{E'} = \tan \delta \qquad (8.17)$$

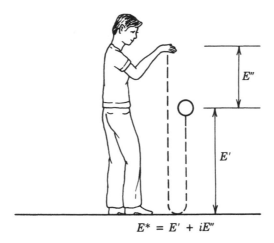

$$E^* = E' + iE''$$

Figure 8.6 Simplified definition of E' and E'' (21). When a viscoelastic ball is dropped onto a perfectly elastic floor, it bounces back to a height E', a measure of the energy stored elastically during the collision between the ball and the floor. The quantity E'' represents the energy lost as heat during the collision.

where tan δ is called the loss tangent, δ being the angle between the in-phase and out-of-phase components in the cyclic motion. Tan δ also goes through a series of maxima. The maxima in E'' and tan δ are sometimes used as the definition of T_g. For the glass transition, the portion of the molecule excited may be from 10 to 50 atoms or more (see Table 8.4).

The dynamic mechanical behavior of an ideal polymer is illustrated in Figure 8.7. The storage modulus generally follows the behavior of Young's modulus as shown in Figure 8.2. In detail, it is subject to equation (8.14), so the storage modulus is slightly smaller, depending on the value of E''.

The quantities E'' and tan δ display decided maxima at T_g, the tan δ maximum appearing several degrees centigrade higher than the E'' peak. Also shown in Figure 8.7 is the β peak, generally involving a smaller number of atoms. The area under the peaks, especially when plotted with a linear y axis, is related to the chemical structure of the polymer (22). The width of the transition and shifts in the peak temperatures of E'' or tan δ are sensitive guides to the exact state of the material, molecular mixing in blends, and so on. The smaller transitions, such as the β peak, usually appear at other temperatures, such as the Schatzki crankshaft transition (see Section 8.4.1), or at higher temperatures (see Section 8.4.2).

Also included in the x axis of Figure 8.7 are the log time and the minus log frequency axes. As discussed in Sections 8.5.3 and 8.6.1.2, the minus log frequency dependence of the mechanical behavior takes the same form as temperature. As the imposed frequency on the sample is raised, it goes through the glass transition in much the same way as when the temperature is lowered. The theory of the glass transition will be discussed in Section 8.6.

For small stresses and strains, if both the elastic deformation at equilibrium and the rate of viscous flow are simple functions of the stress, the polymer is said to exhibit linear viscoelasticity. While most of the material in this chapter refers to linear viscoelasticity, much of the material in Chapters 9 and 10 concerns nonlinear behavior.

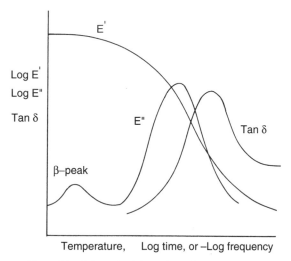

Figure 8.7 Schematic of dynamic mechanical behavior.

A principal value of the loss quantities stems from their sensitivity to many other types of transitions besides the glass transition temperature. The maxima in E'', G'', tan δ, and so on, provide a convenient and reproducible measure of each transition's behavior from dynamic experiments. From an engineering point of view, the intensity of the loss quantities can be utilized in mechanical damping problems such as vibration control.

In the following section several types of instrumentation used to measure the glass transition and other transitions are discussed.

8.3 METHODS OF MEASURING TRANSITIONS IN POLYMERS

The glass transition and other transitions in polymers can be observed experimentally by measuring any one of several basic thermodynamic, physical, mechanical, or electrical properties as a function of temperature. Recall that in first-order transitions such as melting and boiling, there is a discontinuity in the volume–temperature plot. For second-order transitions such as the glass transition, a change in slope occurs, as illustrated in Figure 8.5 (8).

The volumetric coefficient of expansion, α, is defined as

$$\alpha = \frac{1}{V}\left(\frac{\partial V}{\partial T}\right)_p \tag{8.18}$$

where V is the volume of the material, and α has the units K^{-1}. While this quantity increases as the temperature increases beyond T_g, it usually changes over a range of 10–30°C. The elbow shown in Figure 8.5a is not sharp. Similar changes occur in the enthalpy, H, and the heat capacity at constant pressure, C_p (Figure 8.5b and e, respectively).

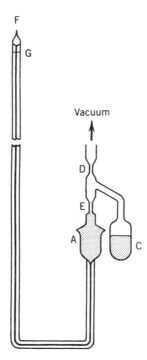

Figure 8.8 A mercury-based dilatometer (23). Bulb *A* contains the polymer (about 1 g), capillary *B* is for recording volume changes (Hg + polymer), *G* is a capillary for calibration, sealed at point *F*. After packing bulb *A*, the inlet is constricted at *E*, *C* contains weighed mercury to fill all dead space, and *D* is a second constriction.

8.3.1 Dilatometry Studies

There are two ways of characterizing polymers via dilatometry. The most obvious is volume–temperature measurements (23), where the polymer is confined by a liquid and the change in volume is recorded as the temperature is raised (see Figure 8.8). The usual confining liquid is mercury, since it does not swell organic polymers and has no transition of its own through most of the temperature range of interest. In an apparatus such as is shown in Figure 8.8, outgassing is required.

The results may be plotted as specific volume versus temperature (see Figure 8.9) (23). Since the elbow in volume–temperature studies is not sharp (all measurements of T_g show a dispersion of some 20–30°C), the two straight lines below and above the transition are extrapolated until they meet; that point is usually taken as T_g. Dilatometric and other methods of measuring T_g are summarized in Table 8.5.

The straight lines above and below T_g in Figure 8.9, of course, yield the volumetric coefficient of expansion, α. The linear coefficient of expansion, β, can also be employed (24). Note that $\alpha = 3\beta$.

Dilatometric data agree well with modulus–temperature studies, especially if the heating rates and/or length of times between measurements are controlled. (Raising the temperature 1°C/min roughly corresponds to a 10 s mechanical measurement.) Besides

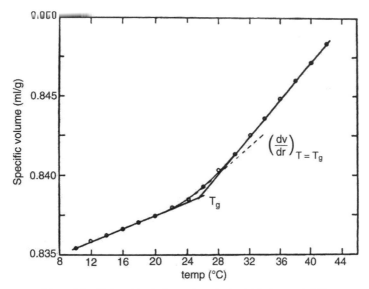

Figure 8.9 Dilatometric studies on branched poly(vinyl acetate) (23).

being a direct measure of T_g, dilatometric studies provide free volume information, of use in theoretical studies of the glass transition phenomenon (see Section 8.6.1).

8.3.2 Thermal Methods

Two closely related methods dominate the field—the older method, differential thermal analysis (DTA), and the newer method, differential scanning calorimetry (DSC). Both methods yield peaks relating to endothermic and exothermic transitions and show

Table 8.5 Methods of measuring the glass transition and thermal properties

Method	Representative Instrumentation
Dilatometry	
Volume-temperature	Polymer confined by mercury (home made)
Linear expansivity	TMS + computer
Thermal	
DSC	Modulated DSC 2920 (TA Instruments)
DTA	DuPont 900
Calorimetric, C_p	Perkin-Elmer modulated DSC, Pyris 1
Mechanical	
Static	Gehman, Clash-Berg
Dynamic	Rheometrics RDA2 (Rheometrics Scientific)
Torsional Braid Analysis	TBA (Plastics Analysis Instruments)
Dielectric and magnetic	
Dielectric loss	DuPont Dielectric Analyzer, DEA 2920
Broad-line NMR	Joel JNH 3H60 Spectrometer
Melt viscosity	Weissenberg Rheogoniometer
TGA	HiRes TGA 2950 (TA Instruments)

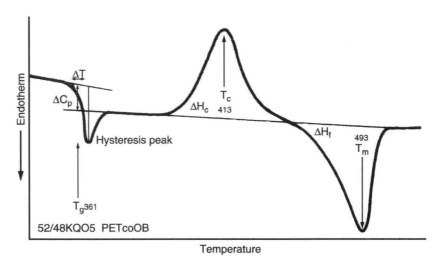

Figure 8.10 Example of a differential scanning calorimetry trace of poly(ethylene terephthalate-*stat-p*-oxbenzoate), quenched, reheated, and cooled at 0.5 K/min through the glass transition, and reheated for measurement at 10 K/min (28). T_g is taken at the temperature at which half the increase in heat capacity has occurred. The width of the glass transition is indicated by ΔT. Note that ΔH_c and ΔH_f are equal in magnitude but opposite in sign.

changes in heat capacity. The DSC method also yields quantitative information relating to the enthalpic changes in the polymer (25–27) (see Table 8.5).

The DSC method uses a servo system to supply energy at a varying rate to the sample and the reference, so that the temperatures of the two stay equal. The DSC output plots energy supplied against average temperature. By this improved method, the areas under the peaks can be directly related to the enthalpic changes quantitatively.

As illustrated by Figure 8.10 (28), T_g can be taken as the temperature at which one-half of the increase in the heat capacity, ΔC_p, has occurred. The increase in ΔC_p is associated with the increased molecular motion of the polymer.

Figure 8.10 shows a hysteresis peak associated with the glass transition. Such hysteresis peaks appear frequently, but not all the time. They are most often associated with some physical relaxation, such as residual orientation. On repeated measurement, the hysteresis peak usually disappears.

An improved method of separating a transient phenomenon such as the hysteresis peak from the reproducible result of the change in heat capacity is obtained *via* the use of modulated DSC (24,25); see Table 8.5. Here, a sine wave is imposed on the temperature ramp. A real-time computer analysis allows a plot of not only the whole data but also its transient and reproducible components.

Figure 8.10 also illustrates the crystalline and melting behavior of this material. Due to the thermal treatment of the sample, it is amorphous before the last heating. Above the glass transition temperature, the increased molecular motion allows the sample to crystallize. At some higher temperature, the polymer melts. The heats of crystallization, ΔH_c, and the heat of fusion, ΔH_f, determined by the areas under the curves must be equal, of course.

This type of difficultly crystallizable copolymer of poly(ethylene terephthalate), is widely used in today's two-liter soda pop bottles. At room temperature the polymer is in

Its glassy state, with some order but little actual crystallinity. Besides being transparent and mechanically tough, such copolymers have very low permeability to both oxygen and carbon dioxide; see Table 4.5.

A related technique, thermogravimetric analysis (TGA), must be introduced at this point. In using TGA, the weight of the sample is recorded continuously as the temperature is raised. Volatilization, dehydration, oxidation, and other chemical reactions can easily be recorded, but the simple transitions are missed, as no weight changes occur.

8.3.3 Mechanical Methods

Since the very notion of the glass–rubber transition stems from a softening behavior, the mechanical methods provide the most direct determination of the transition temperature. Two fundamental types of measurement prevail—the static or quasistatic methods, and the dynamic methods.

Results of the static type of measurement have already been shown in Figure 8.2. For amorphous polymers and many types of semicrystalline polymers in which the crystallinity is not too high, stress relaxation, Gehman, and/or Glash–Berg instrumentation provide rapid and inexpensive scans of the temperature behavior of new polymers before going on to more complex methods.

Several instruments are employed to measure the dynamic mechanical spectroscopy (DMS) behavior (See Table 8.5). The Rheovibron (29) requires a sample that is self-supporting and that yields absolute values of the storage modulus and tan δ. The value of E'' is calculated by equation (8.17). Typical data are shown in Figure 8.11 (30). Although the instrument operates at several fixed frequencies, 110 Hz is most often employed. The sample size is about that of a paper match stick. This method provides excellent results with thermoplastics (29) and preformed polymer networks (30).

An increasingly popular method for studying the mechanical spectra of all types of polymers, especially those that are not self-supporting, is torsional braid analysis (TBA). In this case the monomer, prepolymer, polymer solution, or melt is dipped onto a glass braid, which supports the sample. The braid is set into a torsional motion. The sinusoidal decay of the twisting action is recorded as a function of time as the temperature is

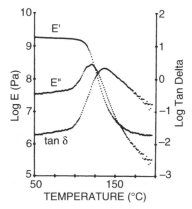

Figure 8.11 Dynamic mechanical spectroscopy on polystyrene cross-linked with 2% divinyl benzene. Data taken with a Rheovibron at 110 Hz. Note that the tan δ peak occurs at a slightly higher temperature than the E'' peak (30). Experiment run by J. J. Fay, Lehigh University.

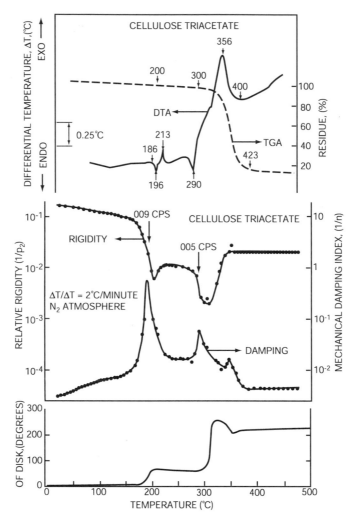

Figure 8.12 Comparison of torsional braid analysis, differential thermal analysis, and thermogravimetric analysis data for cellulose triacetate. The bottom figure shows the twisting of the sample in the absence of oscillations as a result of expansion or contraction of the sample at T_g and T_f (34).

changed (31–34). Because the braid acts as a support medium, the absolute magnitudes of the transitions are not obtained; only their temperatures and relative intensities are recorded. The TBA method appears to have largely replaced the torsional pendulum (32).

Figure 8.12 shows typical TBA data for cellulose triacetate. Also shown by way of comparison are DTA and TGA results. In Figure 8.12, p represents the period of oscillation—that is, the inverse of the natural frequency of the sample, braid, and attachments. The quantity $1/p^2$ is a measure of the stiffness of the system, proportional to the modulus. Since the modulus of the glass braid stays nearly constant through the range of measurement, all changes are representative of the polymer. The quantity n represents the number of oscillations required to reduce the angular amplitude, A, by a fixed ratio. In this case, $A_i/A_{i+n} = 20$ (33). By way of comparison, earlier measurements via volume–temperature had placed T_g at 172°C (35,36) and T_f at 307°C (35), in good agreement with the TBA results in Figure 8.12.

The DTA results in Figure 8.12 show T_f at 290°C while the TBA shows first an exothermic, then an endothermic decomposition at 356°C and 400°C, respectively, corresponding to the weight loss shown by the TGA study. Cellulose triacetate is known to have three second-order transitions (36); it is not clear whether the lower temperature transitions associated with the DTA plot represent these or other motions (37).

8.3.4 Dielectric and Magnetic Methods

As stated previously, part of the work performed on a sample will be converted irreversibly into random thermal motion by excitation of the appropriate molecular segments. In Section 8.3.3 the loss maxima so produced through mechanical means were used to characterize the glass transition. The two important electromagnetic methods for the characterization of transitions in polymers are dielectric loss (3) and broad-line nuclear magnetic resonance (NMR) (38–40).

The dielectric loss constant, ε'', or its associated $\tan\delta$ can be measured by placing the sample between parallel plate capacitors and alternating the electric field. Polar groups on the polymer chain respond to the alternating field. When the average frequency of molecular motion equals the electric field frequency, absorption maxima will occur.

If the dielectric measurements are carried out at the same frequency as the DMS measurements, the transitions will occur at the same temperatures (see Figure 8.13) (40). The glass transition for the polytrifluorochloroethylene at 52°C shown is due to static measurements. The values at 100°C shown are close to those reported for dynamic measurements (14).

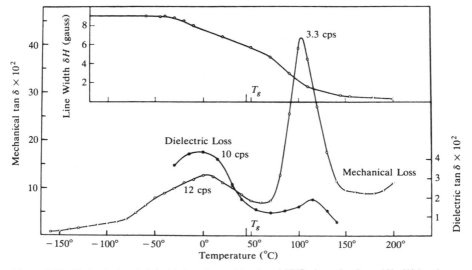

Figure 8.13 Mechanical and dielectric loss tangent $\tan\delta$ and NMR absorption line width δH (maximum slope, in gauss) of polytrifluorochloroethylene (Kel-F) (40).

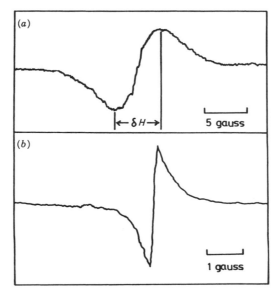

Figure 8.14   Broad-line NMR spectra of a cured epoxy resin. (*a*) Broad line at 291 K; (*b*) motionally narrowed line at 449 K (T_g + 39 K) (38).

Broad-line NMR measurements depend on the fact that hydrogen nuclei, being simply protons, possess a magnetic moment and therefore precess about an imposed alternating magnetic field, especially at radio frequencies. Stronger interactions exist between the magnetic dipoles of different hydrogen nuclei in polymers below the glass transition temperature, resulting in a broad signal. As the chain segments become more mobile with increasing temperature through T_g, the distribution of proton orientations around a given nucleus becomes increasingly random, and the signal sharpens. The behavior of NMR spectra below and above T_g is illustrated in Figure 8.14 (38). The narrowing of the line width through the glass transition for polytrifluorochloroethylene is shown in the inset of Figure 8.13. A number of other methods of observing T_g are discussed by Boyer (8).

8.3.5   A Comparison of the Methods

All the methods of measuring T_g depend on either a basic property or some derived property. The principal ones have already been discussed.

Basic Property	Derived Property
Volume	Refractive index
Modulus	Penetrometry
Dielectric loss	Resistivity

From a practical point of view, since the derived property frequently represents the quantity of interest, it is measured rather than the basic property.

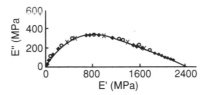

Figure 8.15 Cole-Cole plot for an epoxy network through the glass transition region. Usually, these plots illustrate dynamic mechanical behavior, where the storage quantity is plotted against the loss quantity. Inverted, nearly semi-circular curves as shown are common.

The methods most commonly used at the present time (8) include direct-recording DSC units, the Rheovibron, and the torsional braid. The special value of the mechanical units lies in the fact that loss and storage moduli are frequently of prime engineering value. Thus the instrument supplies basic scientific information about the transitions while giving information about the damping and stiffness characteristics.

On the other hand, DSC supplies thermodynamic information about T_g. Of particular interest is the change in the heat capacity, which reflects fundamental changes in molecular motion. Thus values of C_p are of broad theoretical significance, as described in Section 8.6.

8.3.6 The Cole–Cole Plot

The Cole–Cole plot usually presents a loss term *vs.* a storage term as a function of frequency or time element (41). Figure 8.15 (42) illustrates a plot of Young's loss modulus against. Young's storage modulus of an epoxy network as a function of frequency through the glass transition region. The epoxy was based on diglycidyl ether of bisphenol A and diaminodiphenyl methane, the latter serving as the cross-linker. Note the characteristic near-semicircular appearance of this plot.

Although the Cole–Cole plot was first introduced in the context of a dielectric relaxation spectrum, it helped discover that the molecular mechanism underlying both dielectric relaxation and stress relaxation are substantially identical (43). Figure 8.13 provides an illustration, with temperature instead of frequency. Specifically, the same molecular motions that generate a frequency dependence for the dielectric spectrum are also responsible for the relaxation of orientation in polymers above T_g. Subsequently the Cole–Cole type of plot has been applied to the linear viscoelastic mechanical properties of polymers, especially in the vicinity of the glass transition, including the dynamic compliance and dynamic viscosity functions.

8.4 OTHER TRANSITIONS AND RELAXATIONS

As the temperature of a polymer is lowered continuously, the sample may exhibit several second-order transitions. By custom, the glass transition is designated the α transition, and successively lower temperature transitions are called the $\beta, \gamma, \ldots$ transitions. One important second-order transition appears above T_g, designated the T_{ll} (liquid–liquid) transition. Of course, if the polymer is semicrystalline, it will also melt at a temperature above T_g.

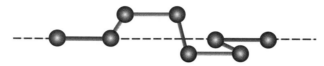

Figure 8.16 Schatzki's crankshaft motion (40) requires at least four —CH$_2$— groups in succession. As illustrated, for eight —CH$_2$— groups, bonds 1 and 7 are colinear and intervening —CH$_2$— units can rotate in the manner of a crankshaft (43).

8.4.1 The Schatzki Crankshaft Mechanism

8.4.1.1 *Main-Chain Motions* There appear to be two major mechanisms for transitions in the glassy state (43a). For main-chain motions in hydrocarbon-based polymers such as polyethylene, the Schatzki crankshaft mechanism (43b), Figure 8.16 (43c), is thought to play an important role. Schatzki showed that eight —CH$_2$— units could be lined up so that the 1–2 bonds and the 7–8 bonds form a colinear axis. Then, given sufficient free volume, the intervening four —CH$_2$— units rotate more or less independently in the manner of an old-time automobile crankshaft. It is thought that at least four —CH$_2$— units in succession are required for this motion. The transition of polyethylene occurring near −120°C is thought to involve the Schatzki mechanism.

It is interesting to consider the basic motions possible for small hydrocarbon molecules by way of comparison. At very low temperatures, the CH$_3$— groups in ethane can only vibrate relative to the other. At about 90 K ethane undergoes a second-order transition as detected by NMR absorption (44), and the two CH$_3$— units begin to rotate freely, relative to one another. For propane and larger molecules, the number of motions becomes more complex (45), as now three-dimensional rotations come into play. One might imagine that *n*-octane itself might have the motion illustrated in Figure 8.16 as one of its basic energy absorbing modes.

8.4.1.2 *Side-Chain Motions* The above considers main-chain motions. Many polymers have considerable side-chain "foliage," and these groups can, of course, have their own motions.

A major difference between main-chain and side-chain motions is the toughness imparted to the polymer. Low-temperature main-chain motions act to absorb energy much better than the equivalent side-chain motions, in the face of impact blows. When the main-chain motions absorb energy under these conditions, they tend to prevent main-chain rupture. (The temperature of the transition actually appears at or below ambient temperature, noting the equivalent "frequency" of the growing crack. The frequency dependence is discussed in Section 8.5.) Toughness and fracture in polymers are discussed in Chapter 11.

8.4.2 The T_{ll} Transition

As illustrated in Figure 8.17 (46), the T_{ll} transition occurs above the glass transition and is thought to represent the onset of the ability of the entire polymer molecule to move as a unit (8,47,48). Above T_{ll}, physical entanglements play a much smaller role, as the molecule becomes able to translate as a whole unit.

Although there is much evidence supporting the existence of a T_{ll} (47–49), it is surrounded by much controversy (50–53). Reasons include the strong dependence of T_{ll} on

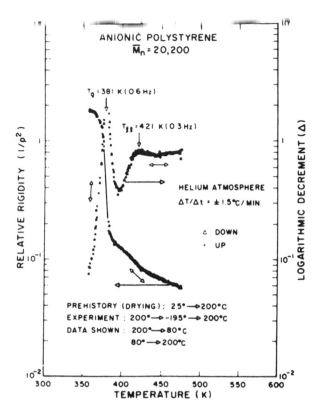

Figure 8.17 Thermomechanical spectra (relative rigidity and logarithmic decrement versus temperature (K) of anionic polystyrene, M_n = 20,200 (46).

molecular weight and an analysis of the equivalent behavior of spring and dashpot models (see Section 10.1). The critics contend that T_{ll} is an instrumental artifact produced by the composite nature of the specimen in torsional braid analysis (TBA), since TBA instrumentation is the principal method of studying this phenomenon (see Figure 8.17). The T_{ll} transition may be related to reptation.

Many polymers show evidence of several transitions besides T_g. Table 8.6 summarizes the data for polystyrene, including the proposed molecular mechanisms for the several transitions. The General Mechanisms column in Table 8.6 follows the results described by Bershtein and Ergos (54) on a number of amorphous polymers. Clearly, different polymers may have somewhat different mechanistic details for the various transitions, especially the lower temperature ones. However, the participating moieties become smaller in size at lower temperatures. The onset of de Gennes reptation is probably associated with T_g, the motions being experimentally identified at $T_g + 20°C$.

8.5 TIME AND FREQUENCY EFFECTS ON RELAXATION PROCESSES

So far the discussion has implicitly assumed that the time (for static) or frequency (for dynamic) measurements of T_g were constant. In fact the observed glass transition tem-

Table 8.6   Multiple transitions in polystyrene and other amorphous polymers

Temperature	Transitions	Polystyrene Mechanism	General Mechanism
433 K (160°C)	T_{ll}	Liquid$_1$ to liquid$_2$	Boundary between rubber elasticity and rubbery flow states
373 K (100°C)	T_g	Long-range chain motions, onset of reptation	Cooperative motion of several Kuhn segments, onset of reptation
325 K (50°C)	β	Torsional vibrations of phenyl groups	Single Kuhn segment motion
130 K	γ	Motion due to four carbon backbone moieties	Small-angle torsional vibrations, 2–3 mers
38–48 K	δ	Oscillation or wagging of phenyl groups	Small-angle vibrations, single mer

perature depends very much on the time allotted to the experiment, becoming lower as the experiment is carried out slower.

For static or quasi-static experiments, the effect of time can be judged in two ways: (a) by speeding up the heating or cooling rate, as in dilatometric experiments, or (b) by allowing more time for the actual observation. For example, in measuring the shear modulus by Gehman instrumentation, the sample may be stressed for 100 s rather than 10 s before recording the angle of twist.

In the case of dynamic experiments, especially where the sample is exposed to a sinusoidal motion, the frequency of the experiment can be varied over wide ranges. For dynamic mechanical spectroscopy, the frequency range can be broadened further by changing instrumentation. For example, DMS measurements in the 20,000-Hz range can best be carried out by employing sound waves. In dielectric studies, the frequency of the alternating electric field can be varied.

The inverse of changing the frequency of the experiment, making measurements as a function of time at constant temperature, is called stress relaxation or creep and is discussed in Section 8.5.2 and Chapter 10. In the following paragraphs a few examples of time and frequency effects are given.

8.5.1   Time Dependence in Dilatometric Studies

It was pointed out in Section 8.3.1 that the elbow in volume–temperature studies constitutes a fundamental measure of the glass transition temperature, since the coefficient of expansion increases at T_g. The heating or cooling rate is important in determining exactly where the transition will be observed, however. As illustrated in Figure 8.18 (55), measuring the result twice after differing by a factor of 500 reduces the glass transition temperature by about 8°C. Similarly T_g varies with heating rate in DSC studies.

8.5.2   Time Dependence in Mechanical Relaxation Studies

If a polymer sample is held at constant strain and measurements of stress are recorded as a function of time, stress relaxation of the type shown in Figure 8.19 (56) will be observed. The shape of the curve shown in Figure 8.19 bears comparison with those in Figure 8.2. In the present case, log time has replaced temperature in the x axis, but the

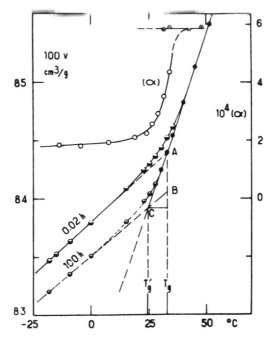

Figure 8.18 Isochronous volumes of poly(vinyl acetate) at two times, 0.02 h and 100 h, after quenching to various isothermal temperatures (55). Also shown is the cubic coefficient of expansion, α, measured at the 0.02 h cooling rate.

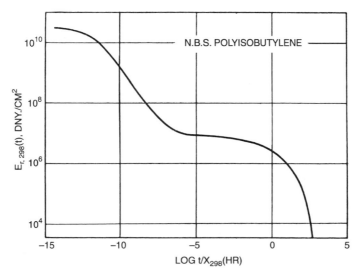

Figure 8.19 Master curve for polyisobutylene (56). The shift factor X, selected here so that log $X_{298} = 0$, is equivalent to the experimentally determined WLF A_t (see Section 8.6.1.2).

phenomenon is otherwise similar. As time is increased, more molecular motions occur, and the sample softens.

It must be emphasized that the sample softens only after the time allowed for relaxation. For example, on a given curve of the type shown in Figure 8.19, the modulus might be 1×10^7 dynes/cm^2 after 10 years, showing a rubbery behavior. Someone coming up and pressing his thumb into the material after 10 years will report the material as being much harder; however, it must be remembered that pressing one's thumb into a material is a short-time experiment, of the order of a few seconds. (This assumes physical relaxation phenomena only, not true chemical degradation.)

The slope corresponding to the glass transition has been quantitatively treated by Aklonis and co-workers (57–59) for relaxation phenomena. Aklonis defined a steepness index (SI) as the maximum of the negative slope of a stress relaxation curve in the glass transition region. They found that while polyisobutylene has a SI of about 0.5, poly(methyl methacrylate) has a value of about 1.0, and polystyrene is close to 1.5. Aklonis treated the data theoretically, using the Rouse–Bueche–Zimm bead and spring model (60–62), based in turn on the Debye damped torsional oscillator model (63,64). Aklonis concluded that values of SI equal to 0.5 represented a predominance of intramolecular forces and that a SI of 1.5 represented a predominance of intermolecular forces. A SI of 1.0 was an intermediate case. Stress relaxation is treated in greater detail in Chapter 10.

8.5.3 Frequency Effects in Dynamic Experiments

The loss peaks such as those illustrated in Figure 8.7 and Figure 8.12 can be determined as a function of frequency. The peak frequency can then be plotted against $1/T$ to obtain apparent activation energies.

Figure 8.20 (41) shows the $\alpha^\dagger$ transition, T_g, of polystyrene increasing steadily in temperature as the frequency of measurement is increased. Both DMS and dielectric measurements are included. Since T_g is usually reported at 10 sec (or 1×10^{-1} Hz), a glass transition temperature of 100°C may be deduced from Figure 8.20, which is, in fact, the usually reported T_g.

The straight line for the α relaxation process, as drawn, corresponds to an apparent activation energy of 84 kcal/mol (65). The β relaxation possesses a corresponding apparent energy of activation of 35 kcal/mol. Section 8.6.2.3 discusses methods of calculating these values. As a first approximation for many polymers, $T_\beta \cong 0.75T_g$ (66) at low frequencies. Bershtein et al. (67) point out that T_g and T_β frequently merge at a frequency of 10^6 to 10^8 Hz.

The WLF equation (Section 8.6.1.2) says that T_g will change 6 to 7°C per decade of frequency. Figure 8.19 yields about 5°C change in T_g for a factor of 10 increase in the time scale, and Figure 8.20 yields about 7.5°C per decade. Obviously this depends on the apparent energy of activation of the individual polymer, but many of the common carbon-backbone polymers have similar energies of activation. The effect of frequency on mechanical behavior is discussed further in Section 10.3.

While each of these second-order transitions has a frequency dependence, the corresponding first-order melting transition for semicrystalline polymers does not.

---

† The peaks in the loss spectrum are sometimes labeled $\alpha, \beta, \gamma, \dots$, with α being the highest temperature peak (T_g).

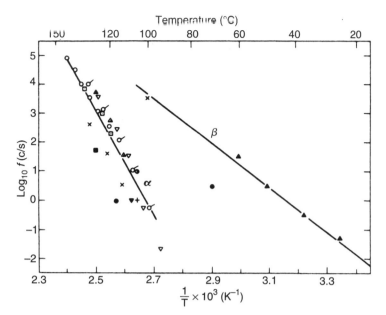

Figure 8.20 The frequency dependence of the α and β transitions of polystyrene (41). Composite of both dynamic mechanical and dielectric studies of several researchers.

8.6 THEORIES OF THE GLASS TRANSITION

The basic experimental behavior of polymers near their glass transition temperatures was explored in the preceding phenomenological description. In Section 8.5, T_g was shown to decrease steadily as the time allotted to the experiment was increased. One may raise the not so hypothetical question, is there an end to the decrease in T_g as the experiment is slowed? How can the transition be explained on a molecular level? These are the questions to which the theories of the glass transition are addressed.

The following paragraphs describe three main groups of theories of the glass transition (1–4,68): free-volume theory, kinetic theory, and thermodynamic theory. Although these three theories may at first appear to be as different as the proverbial three blind men's description of an elephant, they really examine three aspects of the same phenomenon and can be successfully unified, if only in a qualitative way.

8.6.1 The Free-Volume Theory

As first developed by Eyring (69) and others, molecular motion in the bulk state depends on the presence of holes, or places where there are vacancies or voids (see Figure 8.21). When a molecule moves into a hole, the hole, of course, exchanges places with the molecule, as illustrated by the motion indicated in Figure 8.21. (This model is also exemplified in the children's game involving a square with 15 movable numbers and one empty place; the object of the game is to rearrange the numbers in an orderly fashion.) With real materials, Figure 8.21 must be imagined in three dimensions.

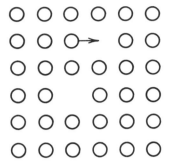

Figure 8.21 A quasi-crystalline lattice exhibiting vacancies, or holes. Circles represent molecules; arrow indicates molecular motion.

Although Figure 8.21 suggests small molecules, a similar model can be constructed for the motion of polymer chains, the main difference being that more than one "hole" may be required to be in the same locality, as cooperative motions are required (see reptation theory, Section 5.4). Thus, for a polymeric segment to move from its present position to an adjacent site, a critical void volume must first exist before the segment can jump.

The important point is that molecular motion cannot take place without the presence of holes. These holes, collectively, are called free volume. One of the most important considerations of the theory discussed below involves the quantitative development of the exact free-volume fraction in a polymeric system.

8.6.1.1 T_g as an Iso–Free-Volume State In 1950 Fox and Flory (70) studied the glass transition and free volume of polystyrene as a function of molecular weight and relaxation time. For infinite molecular weight, they found that the specific free volume, v_f, could be expressed above T_g as

$$v_f = K + (\alpha_R - \alpha_G)T \tag{8.19}$$

where K was related to the free volume at 0 K, and α_R and α_G represented the cubic (volume) expansion coefficients in the rubbery and glassy states, respectively (see Section 8.3.1). (The linear coefficients of expansion are 1/3 of the volumetric values.) Fox and Flory found that below T_g the same specific volume–temperature relationships held for all the polystyrenes, independent of molecular weight. They concluded that (a) below T_g the local conformational arrangement of the polymer segments was independent of both molecular weight and temperature and (b) the glass transition temperature was an iso–free-volume state. Simha and Boyer (71) then postulated that the free volume at $T = T_g$ should be defined as

$$v - (v_{0,R} + \alpha_G T) = v_f \tag{8.20}$$

Figure 8.22 illustrates these quantities.

Substitution of the quantity

$$v = v_{0,R} + \alpha_R T \tag{8.21}$$

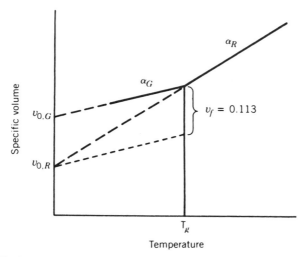

Figure 8.22 A schematic diagram illustrating free volume as calculated by Simha and Boyer.

leads to the relation

$$(\alpha_R - \alpha_G)T_g = K_1 \tag{8.22}$$

In the expressions above, v is the specific volume, and $v_{0,G}$ and $v_{0,R}$ are the volumes extrapolated to 0 K using α_G and α_R as the coefficients of expansion, respectively. Based on the data in Table 8.7 (71), Simha and Boyer concluded that

$$K_1 = (\alpha_R - \alpha_G)T_g = 0.113 \tag{8.23}$$

Equation (8.23) leads directly to the finding that the free volume at the glass transition temperature is indeed a constant, 11.3%. This is the largest of the theoretical values derived, but the first. (It should be pointed out that many simple organic compounds have a 10% volume increase on melting.) Other early estimates placed the free volume at about 2% (68,70).

The use of α_G in equation (8.20) results from the conclusion that expansion in the glassy state occurs at nearly constant free volume; hence $\alpha_G T$ is proportional to the occupied volume. The use of $\alpha_R T_g$ in Table 8.7 arises from the less exact, but simpler relationship

$$\alpha_R T_g = K_2 = 0.164 \tag{8.24}$$

The quantities K_1 and K_2 provide a criterion for the glass temperature, especially for new polymers or when the value is in doubt. This latter arises in systems with multiple transitions, for example, and in semicrystalline polymers, where T_g may be lessened or obscured.

The relation expressed in equation (8.23), though approximate, has been a subject of more recent research. Simha and co-workers (72,73) found equation (8.23) still acceptable, while Sharma et al. (74) found $\alpha_R - \alpha_G$ roughly constant at 3.2×10^{-4} deg^{-1} (see below).

Table 8.7   Glass transition temperature as an iso–free-volume state (71)

Polymer	T_g, K	$\alpha_G \times 10^4$ K^{-1}	$\alpha_R \times 10^4$ K^{-1}	$\alpha_R T_g$	$(\alpha_R - \alpha_G)$	$(\alpha_R - \alpha_G)T_g$
Polyethylene	143	7.1	13.5	0.192	7.4	0.105
Poly(dimethyl siloxane)	150	—	12	0.180	9.3	0.140
Polytetrafluoro-ethylene	160	3.0	8.3	0.133	7.0	0.112
Polybutadiene	188	—	7.8	0.147	5.8	0.109
Polyisobutylene	199.4	—	6.18	0.123	4.70	0.094
Hevea rubber	201	—	6.16	0.124	4.1	0.082
Polyurethane	213	—	8.02	0.171	6.04	0.129
Poly(vinylidene chloride)	256	—	5.7	0.146	4.5	0.115
Poly(methyl acrylate)	282	—	5.6	0.158	2.9	0.082
Poly(vinyl acetate)	302	2.1	5.98	0.18	3.9	0.118
Poly(4-methyl pentene-1)	302	3.4	7.61	0.23	3.78	0.114
Poly(vinyl chloride)	355	2.2	5.2	0.185	3.1	0.110
Polystyrene	373	2.0	5.5	0.205	3.0	0.112
Poly(methyl methacrylate)	378	2.6	5.1	0.182	2.80	0.113

Source: See also J. Brandrup and E. H. Immergut, eds., *Polymer Handbook*, 3rd ed., Wiley, New York, 1989.
Note: The quantity α represents the volume coefficient of expansion. The linear coefficients of expansion are approximately 1/3 of the volume coefficients of expansion. The quantities α_G (determined at 20°C) and α_R represent the glassy and rubbery states, respectively, the former being a typical value for the polymer (semicrystalline for polyethylene and polytetrafluoroethylene) while the latter is calculated for the 100% amorphous polymer. However, the quantities $\alpha_R - \alpha_G$ and $(\alpha_R - \alpha_G)T_g$ are calculated on the basis of 100% amorphous polymer.

8.6.1.2   The WLF Equation   Section 8.2.7 illustrates how polymers soften and flow at temperatures near and above T_g. Flow, a form of molecular motion, requires a critical amount of free volume. This section considers the analytical relationships between polymer melt viscosity and free volume, particularly the WLF (Williams–Landel–Ferry) equation. The WLF equation is derived here because the free volume at T_g arises as a fundamental constant. The application of the WLF equation to viscosity and other polymer problems is considered in Chapter 10.

Early work of Doolittle (75) on the viscosity, η, of nonassociated pure liquids such as n-alkanes led to an equation of the form

$$\ln \eta = B\left(\frac{v_0}{v_f}\right) + \ln A \tag{8.25}$$

where A and B are constants and v_0 is the occupied volume, and as before v_f is the specific free volume. The Doolittle equation can be derived by considering the molecular transport of a liquid consisting of hard spheres (76–80).

An important consequence of the Doolittle equation is that it provides a theoretical basis for the WLF equation (81). One derivation of the WLF equation begins with a consideration of the need of free volume to permit rotation of chain segments, and the hindrance to such rotation caused by neighboring molecules.

The quantity P is defined as the probability of the barriers to rotation per unit time, cooperative motion, or reptation being surmounted (82). An Arrhenius-type relationship is assumed, where ΔE_{act} is the free energy of activation of the process:

$$P = \exp\left(-\frac{\Delta E_{act}}{kT}\right) \qquad (8.26)$$

Of course, P increases with temperature.

Next the time ("time scale") of the experiment is considered. Long times, t, allow for greater probability of the required motion, and P also increases. The theory assumes that tP must reach a certain value for the onset of the motion, and for the associated transition to be recorded:

$$\ln tP = \text{constant} = -\frac{\Delta E_{act}}{kT} + \ln t \qquad (8.27)$$

hence

$$\ln t = \text{constant} + \frac{\Delta E_{act}}{kT} \qquad (8.28)$$

Equation (8.28) equates the logarithm of time with an inverse function of the temperature. Taking the differential obtains

$$\Delta \ln t = -\frac{\Delta E_{act}}{kT^2}\Delta T \qquad (8.29)$$

and the relationships become clearer: an increase in the logarithm of time is equivalent to a decrease in the absolute temperature. This must be understood in the context of the time–temperature relationship for the onset of a particular cooperative motion.

The quantity ΔE_{act} is associated with free volume and qualitatively would be expected to decrease as the fractional free volume increases. It is assumed that

$$\frac{\Delta E_{act}}{kT} = \frac{B'}{f} \qquad (8.30)$$

where B' is a constant and f is the fractional free volume. Noting the similarity of form between equations (8.30) and (8.25), we take B' as equal to B. Then, instead of the Arrhenius relation, we have

$$P = \exp\left(-\frac{B}{f}\right) \qquad (8.31)$$

The quantity tP still remains constant for the particular set of properties to be observed (not necessarily T_g):

$$\ln tP = \text{constant} = -\frac{B}{f} + \ln t \qquad (8.32)$$

Taking the differential,

$$\Delta \ln t = B\Delta\left(\frac{1}{f}\right) \tag{8.33}$$

which states that a change in the fractional free volume is equivalent to a change in the logarithm of the time scale of the event to be observed.

In Section 8.6.1.1, it was concluded (70) that the expansion in the glassy state occurs at constant free volume. (Actually, free volume must increase slowly with temperature, even in the glassy state.) As illustrated in Figures 8.9 and 8.22, the coefficient of expansion increases at T_g, allowing for a steady increase in free volume above T_g. Setting α_f equal to the expansion coefficient of the free volume, and f_0 as the fractional free volume at T_g or other point of interest, the dependence of the fractional free volume on temperature may be written

$$f = f_0 + \alpha_f(T - T_0) \tag{8.34}$$

where T_0 is a generalized transition temperature. Equation (8.33) may be differentiated as

$$\Delta \ln t = B\left(\frac{1}{f} - \frac{1}{f_0}\right) \tag{8.35}$$

Substituting equation (8.34) into (8.35),

$$\Delta \ln t = B\left[\frac{1}{f_0 + \alpha_f(T - T_0)} - \frac{1}{f_0}\right] \tag{8.36}$$

Cross-multiplying yields

$$\Delta \ln t = B\left\{\frac{f_0 - [f_0 + \alpha_f(T - T_0)]}{f_0[f_0 + \alpha_f(T - T_0)]}\right\} \tag{8.37}$$

$$\Delta \ln t = -\frac{B\alpha_f(T - T_0)/f_0}{f_0 + \alpha_f(T - T_0)} \tag{8.38}$$

Dividing by α_f yields

$$\Delta \ln t = -\frac{(B/f_0)(T - T_0)}{f_0/\alpha_f + (T - T_0)} \tag{8.39}$$

Consider the meaning of $\Delta \ln t$:

$$\Delta \ln t = \ln t - \ln t_0 = \ln\left(\frac{t}{t_0}\right) = \ln A_T \tag{8.40}$$

where A_T is called the reduced variables shift factor (1–3). The quantity A_T will be shown to relate not only to the time for a transition with another time but also to many other time-dependent quantities at the transition temperature and another temperature.

The most important of these quantities is the melt viscosity, described below and in Section 10.4.

The theoretical form of the WLF equation can now be written:

$$\ln A_T = -\frac{(B/f_0)(T - T_0)}{f_0/\alpha_f + (T - T_0)} \tag{8.41}$$

Or in log base 10 form, it is

$$\log A_T = -\frac{B}{2.303 f_0} \left[\frac{(T - T_0)}{f_0/\alpha_f + (T - T_0)}\right] \tag{8.42}$$

Equations (8.41) and (8.42) show that a shift in the log time scale will produce the same change in molecular motion as will the indicated nonlinear change in temperature.

The derivation leading to equations (8.41) and (8.42) suggests a generalized time dependence. Before proceeding with an interpretation of the constants in these equations, it is useful to consider the derivation originally presented by Williams, Landel, and Ferry (81).

Beginning with the Doolittle equation, equation (8.25), they note that for small v_f,

$$\frac{v_f}{v_0} \simeq \frac{v_f}{v_0 + v_f} = f \tag{8.43}$$

where $v_0 + v_f$ is the specific volume, and equation (8.43) provides a quantitative definition for f. Equation (8.25) may now be written in terms of the melt viscosity,

$$\ln \eta = \ln A + \frac{B}{f} \tag{8.44}$$

Subtracting conditions at T_0 (or T_g),

$$\ln \eta - \ln \eta_0 = \ln A - \ln A + \frac{B}{f} - \frac{B}{f_0} \tag{8.45}$$

$$\ln\left(\frac{\eta}{\eta_0}\right) = B\left(\frac{1}{f} - \frac{1}{f_0}\right) \tag{8.46}$$

The viscosity is a time (shear rate)-dependent quantity,

$$\ln\left(\frac{\eta}{\eta_0}\right) = \ln A_T = \ln\left(\frac{t}{t_0}\right) \tag{8.47}$$

Note that by equation (8.40), this leads directly back to equation (8.35). Thus equations (8.41) and (8.42) follow directly from the original Doolittle equation, although in a somewhat more limited form.

Now the constants in equation (8.42) may be evaluated. Experimentally, for many linear amorphous polymers above T_g, independent of chemical structure,

$$\log\left(\frac{\eta}{\eta_g}\right) = -\frac{17.44(T - T_g)}{51.6 + T - T_g} \tag{8.48}$$

where T_0 has been set as T_g. (For T_0 equal to an arbitrary temperature, T_s, about 50°C above T_g, the constants in the WLF equation read

$$\log\left(\frac{\eta}{\eta_s}\right) = -\frac{8.86(T - T_s)}{101.6 + T - T_s} \tag{8.49}$$

in an alternately phrased mode of expression.) Comparing equation (8.48) with (8.42), we have

$$\frac{B}{2.303 f_0} = 17.44 \tag{8.50}$$

$$\frac{f_0}{\alpha_f} = 51.6 \tag{8.51}$$

Here, three unknowns and two equations are shown, which can be solved by assigning the constant B a value of unity (81), consistent with the viscosity data of Doolittle. Then $f_0 = 0.025$, and $\alpha_f = 4.8 \times 10^{-4}$ deg^{-1}.

The value of α_f may be verified in a rough way through equation (8.23). Here, if the free volume is constant in the α_G region, then $\alpha_R - \alpha_G \simeq \alpha_f$. The value of $\alpha_f = 4.8 \times 10^{-4}$ deg^{-1} leads to a temperature of -38°C, a temperature at least in the range of the T_g's observed for many polymers. Sharma et al. (74) found $\alpha_f = 3.2 \times 10^{-4}$ deg^{-1}.

The finding of $f_0 = 0.025$ is more significant. It assigns the value of the free volume at the T_g of any polymer at 2.5%. This approximate value has stood the test of time. Wrasidlo (68) suggested a value of 2.35%, based on thermodynamic data, in relatively good agreement with the WLF value.

For numerical results, it must be emphasized that the WLF equation is good for the range T_g to $T_g + 100$. In equations (8.48) and (8.49), T must be larger than T_g or T_0. Its power lies in its generality: no particular chemical structure is assumed other than a linear amorphous polymer above T_g. For a generation of polymer scientists and rheologists, the WLF equation has provided a mainstay both in utility and theory.

8.6.1.3 *An Example of WLF Calculations*

The WLF equation, equation (8.48), can be used to calculate melt viscosity changes with temperature. Suppose a polymer has a glass transition temperature of 0°C. At 40°C, it has a melt viscosity of 2.5×10^5 poises (P) (2.5×10^6 Pa · s). What will its viscosity be at 50°C?

First calculate η_g:

$$\log\left(\frac{\eta}{\eta_g}\right) = -\frac{17.44(T - T_g)}{51.6 + (T - T_g)}$$

$$\log \eta_g = \log 2.5 \times 10^5 + \frac{17.44(313 - 273)}{51.6 + (313 - 273)}$$

$$\log \eta_g = 13.013$$

Polymers often have melt viscosities near 10^{13} P at their glass transition temperature, 1.03×10^{13} P in this case.

Now calculate the new viscosity:

$$\log \eta = 13.013 - \frac{17.44(323 - 273)}{51.6 + (323 - 273)}$$

$$\eta = 2.69 \times 10^4 \ \text{P} \quad \text{or} \quad 2.69 \times 10^5 \ \text{Pa} \cdot \text{s}$$

Thus a 10°C increase in temperature has decreased the melt viscosity by approximately a factor of 10 in this range. The WLF equation can be used to calculate shift factors (Section 8.6.1.2), failure envelopes (Section 11.2.5.2), and much more.

8.6.2 The Kinetic Theory of the Glass Transition

The free-volume theory of the glass transition, as developed in Section 8.6.1, is concerned with the introduction of free volume as a requirement for coordinated molecular motion, leading to reptation. The WLF equation also serves to introduce some kinetic aspects. For example, if the time frame of an experiment is decreased by a factor of 10 near T_g, equations (8.47) and (8.48) indicate that the glass transition temperature should be raised by about 3°C:

$$\lim_{T \to T_g} \left(\frac{\log A_T}{T - T_g} \right) = -0.338 \tag{8.52}$$

$$T - T_g = \frac{-1.0}{-0.338} = +3.0 \tag{8.53}$$

For larger changes in the time or frequency frame, values of 6–7°C are obtained from equation (8.48), in agreement with experiment. For example, if $A_T = 1 \times 10^{-10}$, an average value of 6.9°C per decade change in T_g is obtained. The kinetic theory of the glass transition, to be developed in this section, considers the molecular and macroscopic response within a varying time frame.

8.6.2.1 *Estimations of the Free-Volume Hole Size in Polymers*   Free volume has long been proposed to explain both the molecular motion and physical behavior of polymers. In general, an expression of the free volume, V_f, can be written as the total volume, V_t, minus the occupied volume, V_o. The quantity V_t is usually defined as the specific volume (83). However, V_o has at least three different definitions:

1. Calculated via the van der Walls excluded volume,
2. The crystalline volume at 0 K,
3. The fluctuation volume swept out by the center of gravity of the molecules as the result of thermal motion.

Because of the different ways of defining the free volume, values for it may vary by an order of magnitude!

Free-volume concepts have a long standing in the literature. It goes back to the times of van der Waals and the ideas of molecular mobility in the description of transport phenomena. Next came the Doolittle ideas and the WLF equation (Section 8.6.1). Simha

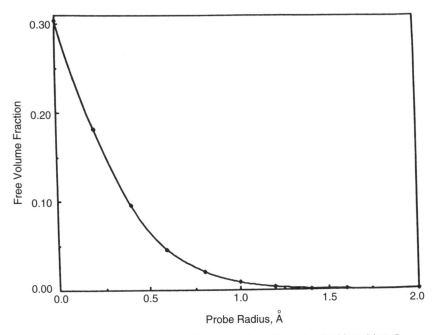

Figure 8.23 Theoretical free volume fraction as a function of the probe sized for polybutadiene.

and Carri (84) examined free volume from an equation of state point of view (see Section 4.3.4), arriving at a general equation between V_f and the hole fraction, h,

$$V_f = a + bh(T/T^*) \tag{8.54}$$

where T^* is given by $T/\tilde{T}$, and $\tilde{T}$ is the reduced temperature.

Misra and Mattice (85) utilized atomistic modeling via computer analysis, assuming hard spheres for the atoms. Central to the analysis was the use of mathematical *probes*, which analyzed the properties of the holes. Figure 8.23 (85) shows that the maximum probe radius is around 1.5 Å for polybutadiene at 300 K.

8.6.2.2 Positron Annihilation Lifetime Spectroscopy
The principal experiment utilized in examination of the free-volume hole size has been positron annihilation lifetime spectroscopy (PALS), first developed by Kobayashi and co-workers (86,87). Positrons from a ^{22}Na source are allowed to penetrate the polymer, and the lifetime of single positrons is registered. The positron can annihilate as a free positron with an electron or form a metastable state, called positronium, together with an electron. If the spins of the positron and the electron are antiparallel, the species is called *para*-positronium, and if they are parallel, *ortho*-positronium, o-PS. Of particular interest is the o-PS lifetime, which is sensitive to the free-volume hole sizes in polymer (and other) materials.

Using PALS, Dammert et al. (88), Yu et al. (89), examined the hole volume of a series of polystyrenes of different tacticity. They arrived at a free-volume hole size distribution maximum of around 110 Å^3 at room temperature; see Figure 8.24 (88). This corresponds

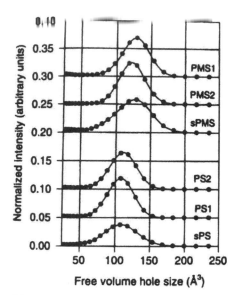

Figure 8.24 PALS study of free volume hole size distribution in polystyrenes and poly(*p*-methylstyrenes) calculated from lifetime distributions of *o*-positronium.

to an effective spherical hole radius of approximately 3 Å. While this radius is somewhat larger than the theoretical value of 1.5 Å found above, if the holes are actually irregular in shape the values are seen to agree quite well.

8.6.3 Thermodynamic Theory of T_g

All noncrystalline polymers display what appears to be a second-order transition in the Ehrenfest sense (90): the temperature and pressure derivatives of both volume and entropy are discontinuous when plotted against T or P, although the volume and the entropy themselves remain continuous (see Figure 8.5).

In Section 8.6.2 it was argued that the transition is primarily a kinetic phenomenon because (a) the temperature of the transition can be changed by changing the time scale of experiment, slower measurements resulting in lower T_g's, and (b) the measured relaxation time near the transition approach the time scale of the experiment.

One can ask what equilibrium properties these glass-forming materials have, even if it is necessary to postulate infinite time scale experiments. A thermodynamically based answer was provided by Gibbs and DiMarzio (91–95), based on a lattice model.

8.6.3.1 *The Gibbs and DiMarzio Theory* Gibbs and DiMarzio (91–95) argued that although the observed glass transitions are indeed a kinetic phenomenon, the underlying true transitions can possess equilibrium properties that may be difficult to realize. At infinitely long times, Gibbs and DiMarzio predict a true second-order transition, when the material finally reaches equilibrium. In infinitely slow experiments, a glassy phase will eventually emerge whose entropy is negligibly higher than that of the crystal. The temperature dependence of the entropy at the approach of T_g is shown in Figure 8.25.

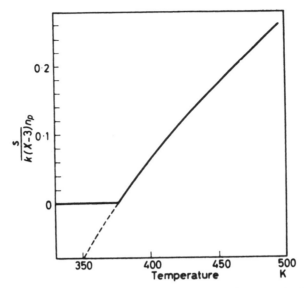

Figure 8.25   Schematic diagram of the conformational entropy of a polymer as a function of temperature according to the Gibbs–DiMarzio theory (93).

This true T_g, designated T_2, as will be shown later, lies some 50°C below the T_g observed at ordinary times.

The central problem of the Gibbs–DiMarzio theory is to find the configurational partition function, Q, from which the expression for the configurational entropy can be calculated. In a manner similar to the kinetic theory described in Section 8.6.2, hindered rotation in the polymer chain is assumed to arise from two energy states: ε_1 is associated with one possible orientation, and ε_2 is associated with all the remaining orientations; $\varepsilon_1 - \varepsilon_2 = \Delta E$, a coordination number, Z, of 4 being assumed. The intermolecular energy is given by the hole energy, α.

The application of the partition function assumes an equilibrium state:

$$Q = \sum_{f, n_x, n_0} W(f_1 n_x \ldots, f_i n_x \ldots, n_0) \exp\left[-\frac{E(f_1 n_x \ldots, f_i n_x \ldots, n_0)}{kT} \right] \qquad (8.55)$$

where $f_i n_x$ is the number of molecules packed in conformation i, and W is the total number of ways that the n_x (x degree of polymerization) molecules can be packed into $x n_x + n_0$ sites on the quasi-lattice, with n_0 being the number of holes. An expression for W was derived earlier by Flory (96) for n_x polymer chains and n_0 solvent molecules, which was used by Gibbs and DiMarzio in their calculations (see Section 3.3).

Once Q is formulated [see equation (8.55)], statistical thermodynamics provides the entropy:

$$S = kT \left(\frac{\partial \ln Q}{\partial T} \right)_{V, n} + k \ln Q \qquad (8.56)$$

Table 8.8 Constants for cross-link effect on T_g (88,93)

Polymer	γ	M/γ	$K \times 10^{23}$	$Z \times 10^4$
Natural rubber	3	22.7	1.30	3.2
Polystyrene	2	52	1.20	4.6
Poly(methyl methacrylate)	4	25	1.38	1.8

from which all necessary calculations can be made (92). This theory has been applied to the variation of the glass transition temperature with the molecular weight (85) (see Section 8.7), random copolymer composition (97,98) (see Section 8.8), plasticization (99), extension (88), and cross-linking (88). This last is briefly explored in Section 8.6.3.2.

8.6.3.2 Effect of Cross-link Density on T_g

The criterion of the second-order transition temperature is that the temperature-dependent conformational entropy, S_c, becomes zero. If S_0 is the conformational entropy for the un-cross-linked system, and ΔS_R is the change in conformational entropy due to adding cross-links (95),

$$S_c = S_0 + \Delta S_R = 0 \tag{8.57}$$

Since cross-linking decreases the conformational entropy, qualitatively it may be concluded that the transition temperature is raised. The final relation may be written

$$\frac{T(\chi') - T(0)}{T(0)} = \frac{KM\chi'/\gamma}{1 - KM\chi'/\gamma} \tag{8.58}$$

where χ' is the number of cross-links per gram, M is the mer molecular weight, and γ is the number of flexible bonds per mer, backbone, and side chain. The quantity K is found by experiment and, interestingly enough, appears independent of the polymer (see Table 8.8).

An alternate relation dates back to Ueberreiter and Kanig (12)

$$\Delta T_{g,c} = Z\chi' \tag{8.59}$$

where the change in the glass temperature with increasing cross-linking, $\Delta T_{g,c}$, is equal to the cross-link density, χ', times a constant, Z. Recently, Glans and Turner (100) compared equations (8.58) and (8.59), using cross-linked polystyrene. The glass transition elevation was observed via DSC analysis (an endothermal peak was reported at T_g; see Section 8.6.2). Plots of straight lines were obtained for $\Delta T_{g,c}$ versus cross-link density, verifying equation (8.59). Some values for Z, with χ' in units of moles per gram and ΔT_c in K, are also shown in Table 8.8. For tetra functional cross-links, $\chi' = 2n$, where n is the number of network chains per unit volume (usually one cm^3).

8.6.3.3 A Summary of the Glass Transition Theories

In the preceding section, three apparently disparate theories of the glass transition were presented. The basic thrust of each is summarized conveniently here and in Table 8.9.

1. The free-volume theory introduces free volume in the form of segment-size voids as a requirement for the onset of coordinated molecular motion. This theory provides

Table 8.9 Glass transition theory box scores

Theory	Advantages	Disadvantages
Free-volume theory	1. Time and temperature of viscoelastic events related to T_g 2. Coefficients of expansion above and below T_g related	1. Actual molecular motions poorly defined
Kinetic theory	1. Shifts in T_g with time frame quantitatively determined 2. Heat capacities determined	1. No T_g predicted at infinite time scales
Thermodynamic theory	1. Variation of T_g with molecular weight, diluent, and cross-link density predicted 2. Predicts true second-order transition temperature	1. Infinite time scale required for measurements 2. True second-order transition temperature poorly defined

relationships between coefficients of expansion below and above T_g and yields equations relating viscoelastic motion to the variables of time and temperature.

2. The kinetic theory defines T_g as the temperature at which the relaxation time for the segmental motions in the main polymer chain is of the same order of magnitude as the time scale of the experiment. The kinetic theory is concerned with the rate of approach to equilibrium of the system, taking the respective motions of the holes and molecules into account. The kinetic theory provides quantitative information about the heat capacities below and above the glass transition temperature and explains the 6 to 7°C shift in the glass transition per decade of time scale of the experiment.

3. The thermodynamic theory introduces the notion of equilibrium and the requirements for a true second-order transition, albeit at infinitely long time scales. The theory postulates the existence of a true second-order transition, which the glass transition approaches as a limit when measurements are carried out more and more slowly. It successfully predicts the variation of T_g with molecular weight and cross-link density (see Section 8.7), diluent content, and other variables.

A summary of the free-volume numbers of the various theories can be made:

Theory	Free-Volume Fraction
WLF	0.025
Hirai and Eyring	0.08
Miller[†]	0.12
Simha-Boyer	0.113

The analytical development of these theories illustrates the power of statistical thermodynamics in providing solutions to important polymer problems. However, much remains to be done. It has been said that less than 5% of all fundamental knowledge has been wrested from nature. This is certainly true in the study of polymer glass transitions.

[†] A. A. Miller, *J. Chem. Phys.*, **49**, 1393 (1968); *J. Polym. Sci.*, **A-2** (6), 249, 1161 (1968).

Insofar as research in this area remains highly active, it is highly probable that new insight will provide an integrated theory in the near future. Attempts to do so up until now are summarized in the following section.

8.6.3.4 A Unifying Treatment Adam and Gibbs (99) attempted to unify the theories relating the rate effect of the observed glass transition and the equilibrium behavior of the hypothetical second-order transition. They proposed the concept of a "cooperatively rearranging region," defined as the smallest region capable of conformational change without a concomitant change outside the region. At T_2 this region becomes equal to the size of the sample, since only one conformation is available.

Adam and Gibbs rederived the WLF equation, putting it in terms of the potential energy hindering the cooperative rearrangement per mer, the molar conformational entropy, and the change in the heat capacity at T_g. By choosing the temperature T in the WLF equation to be T_s [see equation (8.49)] and suitable rearrangements of the WLF formulation to isolate T_2, they found that

$$\frac{T_g}{T_2} = 1.30 \pm 8.4\% \tag{8.60}$$

for a wide range of glass-forming systems, both polymeric and low molecular weight.

For low-temperature elastomers such as the polybutadiene family, $T_g \cong 200$ K. According to equation (8.60), $T_2 \cong 154$ K, or about 50 K below T_g. According to the WLF equation, equation (8.48), the viscosity becomes infinite at $T - T_g = -51.6°$C, which is about the same number. Although this simplified approach yields less quantitative agreement at higher temperatures, the ideas still are interesting.

8.7 EFFECT OF MOLECULAR WEIGHT ON T_g

8.7.1 Linear Polymers

Studies of the increase in T_g with increasing polymer molecular weight date back to the works of Ueberreiter in the 1930s (101). The theoretical analysis of Fox and Flory (70) (see Section 8.6.1.1) indicated that the general relationship between T_g at a molecular weight M was related to the glass temperature at infinite molecular weight, $T_{g\infty}$, by

$$T_g = T_{g\infty} - \frac{K}{(\alpha_R - \alpha_G)M} \tag{8.61}$$

with K being a constant depending on the polymer. Equation (8.61) follows from the decrease in free volume with increasing molecular weight, caused in turn by the increasing number of connected mers in the system, and decreased number of end groups.

The ubiquitous polystyrene seems to have been investigated more than any other polymer (70,101,102). DSC data, first extrapolated to low heating rate, are shown in Figure 8.26 (95). (These data also show an endothermic peak at T_g; see earlier discussions.) The equation for slow heating rates may be expressed

$$T_g = 106°C - \frac{2.1 \times 10^5}{M_n} \tag{8.62}$$

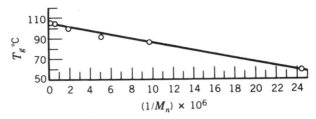

Figure 8.26 The glass transition temperature of polystyrene as a function of $1/M_n$ (102).

For heating rates normally encountered (70)

$$T_g = 100°C - \frac{1.8 \times 10^5}{M_n} \tag{8.63}$$

The molecular weight in equation (8.63) is for fractionated polystrene. For slow experiments, these equations suggest a 6°C increase in T_g at infinite molecular weight.

8.7.2   Effect of T_g on Polymerization

According to equation (8.61) the glass transition depends on the molecular weight. What happens during an isothermal polymerization? When the polymerization begins, the monomers are always in the liquid state. Sometimes, however, the system may go through T_g and the polymer may vitrify as the reaction proceeds. Since molecular motion is much reduced when the system is below T_g, the reaction substantially stops.

Two conditions can be distinguished. First, during a chain polymerization, the monomer effectively acts like a plasticizer for the nascent polymer. An example relates to the emulsion polymerization of polystyrene, often carried out at about 80°C. The reaction will not proceed quite to 100% conversion, because the system vitrifies.

Second, during stepwise polymerization, the molecular weight is continually increasing. An especially interesting case involves gelation. Taking epoxy polymerization as an example, the resin[†] is simultaneously polymerizing and cross-linking (see Section 3.7.3).

Gillham (103–108) pointed out the need to postcure the polymer above $T_{g\infty}$, the glass transition temperature of the fully cured[‡] system. He developed a time–temperature–transformation (TTT) reaction diagram that may be used to provide an intellectual framework for understanding and comparing the cure and glass transition properties of thermosetting systems. Figure 8.27 illustrates the TTT diagram. Besides $T_{g\infty}$, the diagram also displays $_{gel}T_g$, the temperature at which gelation and vitrification occur simultaneously, and T_{g0}, the glass transition temperature of the reactants. The particular S-shaped curve between T_{g0} and $T_{g\infty}$ results because the reaction rate is increased with increasing temperature. At a temperature intermediate between $_{gel}T_g$ and $T_{g\infty}$, the reacting mass first gels, forming a network. Then it vitrifies, and the reaction stops, incomplete. To the novice, the reaction products may appear complete. This last may result in material failure if the temperature is suddenly raised.

[†] Resin is an early term for polymer, often used with epoxies.

[‡] Cure is an early term for cross-linking, also frequently used with epoxies.

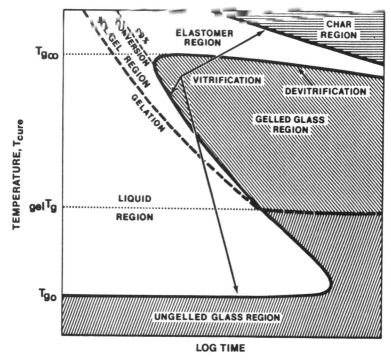

Figure 8.27 The thermosetting process as illustrating by the time–temperature–transformation reaction diagram (108).

The TTT diagram explains why epoxy and similar reactions are carried out in steps, each at a higher temperature. The last step, the postcure, must be done above $T_{g\infty}$. Other points shown in Figure 8.27 include the devitrification region, caused by degradation, and the char region, at still higher temperatures.

8.8   EFFECT OF COPOLYMERIZATION ON T_g

The discussion above relates to simple homopolymers. Addition of a second component may take the form of copolymerization or polymer blending. Addition of low-molecular-weight compounds results in plasticization. Experimentally, two general cases may be distinguished: where one phase is retained and where two or more phases result.

8.8.1   One-Phase Systems

Based on the thermodynamic theory of the glass transition, Couchman derived relations to predict the T_g composition dependence of binary mixtures of miscible high polymers (109) and other systems (110–112). The treatment that follows is easily generalized to the case for random copolymers (109).

Consider two polymers (or two kinds of mers, or one mer and one plasticizer) having pure-component molar entropies denoted as S_1 and S_2, and their respective mole fractions (moles of mers for the polymers) as X_1 and X_2. The mixed system molar entropy may be written

$$S = X_1 S_1 + X_2 S_2 + \Delta S_m \tag{8.64}$$

where ΔS_m represents the excess entropy of mixing. For later convenience, S_1 and S_2 are referred to their respective pure-component glass transition temperatures of T_{g1} and T_{g2}, when their values are denoted as S_1^0 and S_2^0.

Heat capacities are of fundamental importance in glass transition theories, because the measure of the heat absorbed provides a direct measure of the increase in molecular motion. The use of classical thermodynamics leads to an easy introduction of the pure-component heat capacities at constant pressure, C_{p1} and C_{p2}:

$$S = X_1 \left\{ S_1^0 + \int_{T_{g1}}^{T} C_{p1} \, d \ln T \right\} + X_2 \left\{ S_2^0 + \int_{T_{g2}}^{T} C_{p2} \, d \ln T \right\} + \Delta S_m \tag{8.65}$$

The mixed-system glass transition temperature, T_g, is defined by the requirement that S for the glassy state be identical to that for the rubbery state, at T_g. This condition and the use of appropriate superscripts G and R lead to the equation

$$X_1^G \left\{ S_1^{0,G} + \int_{T_{g1}}^{T_g} C_{p1}^G \, d \ln T \right\} + X_2^G \left\{ S_2^{0,G} + \int_{T_{g2}}^{T_g} C_{p2}^G \, d \ln T \right\} + \Delta S_m^G$$

$$= X_1^R \left\{ S_1^{0,R} + \int_{T_{g1}}^{T_g} C_{p1}^R \, d \ln T \right\} + X_2^R \left\{ S_2^{0,R} + \int_{T_{g2}}^{T_g} C_{p2}^R \, d \ln T \right\} + \Delta S_m^R \tag{8.66}$$

Since $S_i^{0,G} = S_i^{0,R}$ $(i = 1, 2)$ and $X_i^G = X_i^R = X_i$, equation (8.66) may be simplified:

$$X_1 \left\{ \int_{T_{g1}}^{T_g} (C_{p1}^G - C_{p1}^R) \, d \ln T \right\} + X_2 \left\{ \int_{T_{g2}}^{T_g} (C_{p2}^G - C_{p2}^R) \, d \ln T \right\} + \Delta S_m^G - \Delta S_m^R = 0 \tag{8.67}$$

In regular small-molecule mixtures, ΔS_m is proportional to $X \ln X + (1 - X) \ln(1 - X)$, where X denotes X_1 and X_2. Similar relations hold for polymer–solvent (plasticizer) and polymer–polymer combinations. Combined with the continuity relation, $\Delta S_m^G = \Delta S_m^R$. For random copolymers these quantities are also equal. Then

$$X_1 \int_{T_{g1}}^{T_g} \Delta C_{p1} \, d \ln T + X_2 \int_{T_{g2}}^{T_g} \Delta C_{p2} \, d \ln T = 0 \tag{8.68}$$

where Δ denotes transition increments. Again, the increase in the heat capacity at T_g reflects the increase in the molecular motion and the increased temperature rate of these motions.

After integration the general relationship emerges,

$$X_1 \Delta C_{p1} \ln \left(\frac{T_g}{T_{g1}} \right) + X_2 \Delta C_{p2} \ln \left(\frac{T_g}{T_{g2}} \right) = 0 \tag{8.69}$$

For later convenience the V_i are exchanged for mass (weight) fractions, M_i (recall that the ΔC_{pi} are then per unit mass), and equation (8.69) becomes

$$\ln T_g = \frac{M_1 \Delta C_{p1} \ln T_{g1} + M_2 \Delta C_{p2} \ln T_{g2}}{M_1 \Delta C_{p1} + M_2 \Delta C_{p2}}$$

(8.70)

or equivalently

$$\ln\left(\frac{T_g}{T_{g1}}\right) = \frac{M_2 \Delta C_{p2} \ln(T_{g2}/T_{g1})}{M_1 \Delta C_{p1} + M_2 \Delta C_{p2}}$$

(8.71)

Equation (8.71) is shown to fit T_g data of thermodynamically miscible blends (see Figure 8.28). Four particular nontrivial cases of the general mixing relation may be derived.

Making use of the expansions of the form $\ln(1 + x) = x$, for small x, and noting that T_{g1}/T_{g2} usually is not greatly different from unity yield

$$T_g \simeq \frac{M_1 \Delta C_{p1} T_{g1} + M_2 \Delta C_{p2} T_{g2}}{M_1 \Delta C_{p1} + M_2 \Delta C_{p2}}$$

(8.72)

which has the same form as the Wood equation (113), originally derived for random copolymers.

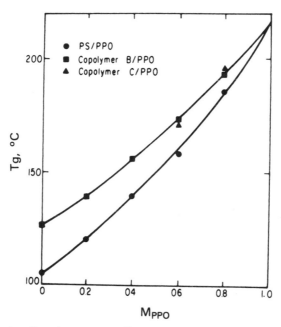

Figure 8.28 Glass-transition temperatures, T_g, of poly(2,6-dimethyl-1,4-phenylene oxide)–*blend*–polystyrene (PPO/PS) blends versus mass fraction of PPO, M_{PPO}. The full curve was calculated from equation (8.71) as circles. $\Delta C_{p1} = 0.0671$ cal $K^{-1} \cdot g^{-1}$, $\Delta C_{p2} = 0.0528$ cal $K^{-1} \cdot g^{-1}$; $T_{g1} = 378$ K, $T_{g2} = 489$ K. PPO was designated as component 2 (110,116).

If $\Delta C_{pi} T_{gi}$ = constant (72–74,114), the familiar Fox equation (115) appears after suitable crossmultiplying:

$$\frac{1}{T_g} = \frac{M_1}{T_{g1}} + \frac{M_2}{T_{g2}} \qquad (8.73)$$

The Fox equation (115) was also originally derived for statistical copolymers (116). This equation predicts the typically convex relationship obtained when T is plotted against M_2 (see Figure 8.28). If $\Delta C_{p1} \simeq \Delta C_{p2}$, the equation of Pochan et al. (117) follows from equation (8.70):

$$\ln T_g = M_1 \ln T_{g1} + M_2 \ln T_{g2} \qquad (8.74)$$

Finally, if both pure-component heat capacity increments have the same value and the log functions are expanded,

$$T_g = M_1 T_{g1} + M_2 T_{g2} \qquad (8.75)$$

which predicts a linear relation for the T_g of the blend, random copolymer, or plasticized system. This equation usually predicts T_g too high. Equations (8.73) and (8.75) are widely used in the literature. Couchman's work (109–112) shows the relationship between them. Previously they were used on a semiempirical basis.

These equations also apply to plasticizers, a low-molecular-weight compound dissolved in the polymer. In this case the plasticizer behaves as a compound with a low T_g. The effect is to lower the glass transition temperature. A secondary effect is to lower the modulus, softening it through much of the temperature range of interest. An example is the plasticization of poly(vinyl chloride) by dioctyl phthalate to make compositions known as "vinyl."

8.8.2 Two-Phase Systems

Most polymer blends, as well as their related graft and block copolymers and interpenetrating polymer networks, are phase-separated (118) (see Section 4.3). In this case each phase will exhibit its own T_g. Figure 8.29 (119,120) illustrates two glass transitions appearing in a series of triblock copolymers of different overall compositions. The intensity of the transition, especially in the loss spectra (E''), is indicative of the mass fraction of that phase.

The storage modulus in the plateau between the two transitions depends both on the overall composition and on which phase is continuous. Electron microscopy shows that the polystyrene phase is continuous in the present case. As the elastomer component increases (small spheres, then cylinders, then alternating lamellae), the material gradually softens. When the rubbery phase becomes the only continuous-phase, the storage modulus will decrease to about 1×10^8 dynes/cm^2.

If appreciable mixing between the component polymers occurs, the inward shift in the T_g of the two phases can each be expressed by the equations of Section 8.8.1 (121). Using equation (8.73), the extent of mixing within each phase in a simultaneous interpenetrating network of an epoxy resin and poly(n-butyl acrylate) was calculated (see Table 8.10). The overall composition was 80/20 epoxy/acrylic, and glycidyl methacrylate

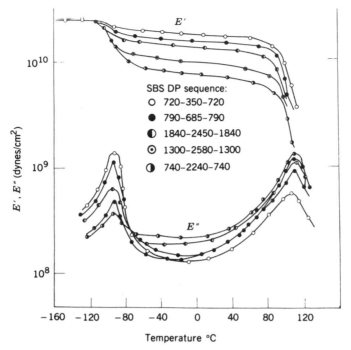

Figure 8.29 Dynamic mechanical behavior of polystyrene–*block*–polybutadiene–*block*–polystyrene, a function of the styrene–butadiene mole ratio (112,113).

is shown to enhance molecular mixing between the chains. Chapter 13 provides additional material on the glass transition behavior of multicomponent materials.

8.9 EFFECT OF CRYSTALLINITY ON T_g

The previous discussion centered on amorphous polymers, with atactic polystyrene being the most frequently studied polymer. Semicrystalline polymers such as polyethylene or polypropylene or of the polyamide and polyester types also exhibit glass transitions,

Table 8.10 Phase composition of epoxy/acrylic simultaneous interpenetrating networks (121)

Glycidyl Methacrylate[a] (%)	Dispersed Phase Weight Fraction		Matrix Phase Weight Fraction	
	PnBA[b]	Epoxy	PnBA	Epoxy
0	0.97	0.03	0.09	0.91
0.3	0.82	0.18	0.12	0.88
3.0	—	—	0.30	0.70

[a] Grafting mer, increases mixing.

[b] Poly(*n*-butyl acrylate).

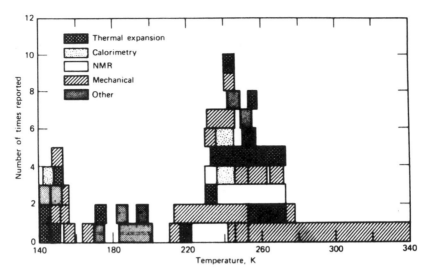

Figure 8.30   Histogram showing number of times a given value of T_g for linear polyethylene has been reported in the literature by various standard methods indicated (115).

though only in the amorphous portions of these polymers. The T_g is often increased in temperature by the molecular-motion restricting crystallites. Sometimes T_g appears to be masked, especially for highly crystalline polymers.

Boyer (8) points out that many semicrystalline polymers appear to possess two glass temperatures: (a) a lower one, $T_g(L)$, that refers to the completely amorphous state and that should be used in all correlations with chemical structure (this transition correlates with the molecular phenomena discussed in previous sections), and (b) an upper value, $T_g(U)$, that occurs in the semicrystalline material and varies with extent of crystallinity and morphology.

8.9.1   The Glass Transition of Polyethylene

Linear polyethylene, frequently referred to as polymethylene, offers a complete contrast with polystyrene in that it has no side groups and has a high degree of crystallinity, usually in excess of 80%. Because of the high degree of crystallinity, molecular motions associated with T_g are partly masked, leading to a confusion with other secondary transitions (see Figure 8.30) (122). Thus various investigators consider the T_g of polyethylene to be in three different regions: $-30°C$, $-80°C$, or $-128°C$.

Davis and Eby support the $-30°C$ value on the basis of volume–time measurements; Stehling and Mandelkern (123) favor the $-128°C$ value based on mechanical measurements. Illers (124) and Boyer (8) support the value of $-80°C$ based on extrapolations of completely amorphous ethylene–vinyl acetate copolymer data with copolymer–T_g relationships. Boyer (8) supports the position that $-80°$ is $T_g(L)$ and $-30°C$ represents $T_g(U)$. The transition at $-128°C$ is thought to be related to the Schatzki crankshaft motion (Section 8.4.1), although the situation apparently is more complicated (124).

Tobolsky (125) obtained $-81°c$ for amorphous polyethylene based on a Fox plot [see equation (8.73)] of statistical copolymers of ethylene and propylene, *it*-polypropylene having a T_g of $-18°C$.

8.9.2 The Nylon Family Glass Transition

Two subfamilies of aliphatic nylons (polyamides) exist:

$$\left[-NH \left(CH_2 \right)_x NH - \overset{\overset{\displaystyle O}{\|}}{C} \left(CH_2 \right)_y \overset{\overset{\displaystyle O}{\|}}{C} - \right]_n \tag{8.76}$$

from diacids and dibases, and

$$\left[-NH \left(CH_2 \right)_x \overset{\overset{\displaystyle O}{\|}}{C} - \right]_n \tag{8.77}$$

originating from ω-amino acids. Both subfamilies are semicrystalline; of course, they form commercially important fibers.

The usually stated T_g range is $T_g \simeq +40°C$ for polyamide 612 to $T_g \simeq 60°C$ for polyamide 6 (8); however, T_g depends on the crystallinity of the particular sample. N-methylated polyamides, with a lower hydrogen bonding, have lower T_g's (126). As x and y increase in equations (8.76) and (8.77), the structure becomes more polyethylene-like, and T_g gradually decreases. Interestingly, when $x > 4$, there is a characteristic mechanical loss peak at about $-130°C$, again suggestive of the Schatzki motion (Section 8.4.1).

8.9.3 Relationships between T_g and T_f

The older literature (127) suggested two relationships between T_g and T_f: $T_g/T_f \simeq \frac{1}{2}$ for symmetrical polymers, and $T_g/T_f \simeq \frac{2}{3}$ for nonsymmetrical polymers. Definitions of symmetry differ, however. One method uses the appearance of atoms down the chain: if a central portion of the chain appears the same when viewed from both ends, it is symmetrical. However, even from the beginning, there were many exceptions to the above. The only rule obeyed in this regard is that T_g is always lower than T_f for homopolymers. This is because (a) the same kinds of molecular motion should occur at T_g and T_f, and (b) short-range order exists at T_g, but long-range order exists at T_f.

Boyer (8) has prepared a cumulative plot of T_g/T_f (see Figure 8.31). Region A (the old $T_g/T_f \simeq \frac{1}{2}$) contains most of the polymers which are free from side groups other than H and F (and hence symmetrical) and contain such polymers as polyethylene, poly(oxymethylene), and poly(vinylidene fluoride). Region B contains most of the common vinyl, vinylidene, and condensation polymers such as the nylons. About 55% of all measured polymers lie in the band $T_g/T_f = 0.667 \pm 0.05$ (8). Region C contains poly(α-olefins) with long alkyl side groups as well as other nontypical polymers such as poly(2,6-dimethylphenylene oxide), which has T_g/T_f approximately equal to 0.93. For an unknown polymer, then, the relationship $T_g/T_f = \frac{2}{3}$ is a good way of providing an estimate of one transition if the temperature of the other is known.

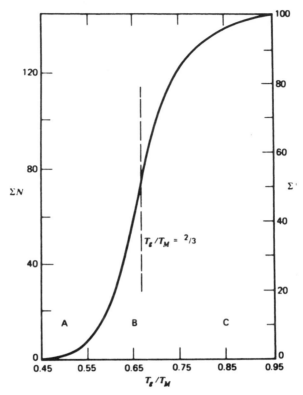

Figure 8.31   Range in T_g/T_f values found in the literature. Region A contains unsubstituted polymers. Region C includes poly(α-olefins) with long side chains. Region B contains the majority of vinyl, vinylidene, and condensation polymers. The left ordinate is cumulative number, N, and the right ordinate is cumulative percentage of all examples reported as having the indicated T_g/T_f values (8).

8.9.4   Heat Distortion Temperature

While the glass transition and melting temperatures define the behavior of polymers from a scientific point of view, the engineers frequently depend on more practical tests. These tests work well for plasticized polymers, blends and composites of various types, and thermosets. These tests originated from the old idea of a *softening temperature*, sometimes defined as the temperature in which a specimen could be easily penetrated with a needle. One such quantitative test is called the Vicat test, where a needle under 1000 g load penetrates the specimen 1 mm (128).

One of the more important of the practical tests is the *heat distortion temperature* (HDT). The HDT is defined as the temperature at which a 100 mm length, 3 mm thick specimen bar at 1.82 MPa in a three-point bending mode deflects 0.25 mm. Young's modulus at the HDT is 7.5 GPa (129,130). For unfilled polymers, both the Vicat and the HDT tests usually record a temperature just above the glass transition temperature, or for melting conditions, just below the temperature of final disappearance of crystallinity. For polymer blends, both the Vicat and the HDT will tend to reflect the properties of the

Table 8.11 Factors affecting T_g (131)

Increase T_g	Decrease T_g
Intermolecular forces	In-chain groups promoting flexibility
High CED	(double-bonds and ether linkages)
Intrachain steric hindrance	Flexible side groups
Bulky, stiff side groups	Symmetrical substitution

continuous phase. If the polymer contains filler which raises the modulus, the HDT will be somewhat increased.

8.10 DEPENDENCE OF T_g ON CHEMICAL STRUCTURE

In Section 8.9 some effects of crystallinity and hydrogen bonding on T_g were considered. The effect of molecular weight was discussed in Section 8.7, and the effect of copolymerization was discussed in Section 8.8. This section discusses the effect of chemical structure in homopolymers.

Boyer (131) suggested a number of general factors that affect T_g (see Table 8.11). In general, factors that increase the energy required for the onset of molecular motion increase T_g; those that decrease the energy requirements lower T_g.

8.10.1 Effect of Aliphatic Side Groups on T_g

In monosubstituted vinyl polymers and at least some other classes of polymers, flexible pendant groups reduce the glass transition of the polymer by acting as "internal diluents," lowering the frictional interaction between chains. The total effect is to reduce the rotational energy requirements of the backbone.

The aliphatic esters of poly(acrylic acid) (132), poly(methacrylic acid) (133), and other polymers (134) (see Figure 8.32) (135) show a decline in T_g as the number of —CH_2— units in the side group increases. At still longer aliphatic side groups, T_g increases as side-chain crystallization sets in, impeding chain motion. In this latter composition range the materials feel waxy. In the ultimate case, of course, the polymer would behave like slightly diluted polyethylene. For cellulose triesters (36) the minimum in T_g is observed at the triheptanoate, probably because of the increased basic backbone stiffness.

8.10.2 Effect of Tacticity on T_g

The discussion so far in this chapter has assumed atactic polymers, which with a few exceptions are amorphous. Other stereo isomers include isotactic and syndiotactic polymers (see Section 2.3).

The effect of tacticity on T_g may be significant, as illustrated in Table 8.12 (136,137). Karasz and MacKnight (137) noted that the effect of tacticity on T_g is expected in view of the Gibbs–DiMarzio theory (Section 8.6.3.1). In disubstituted vinyl polymers, the energy difference between the two predominant rotational isomers is greater for the syndiotactic configuration than for the isotactic configuration. In monosubstituted vinyl

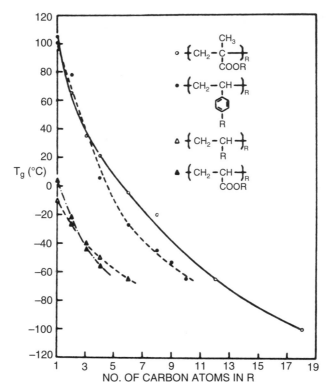

Figure 8.32 Effect of side-chain lengths on the glass transition temperatures of poly-methacrylates [○ (S. S. Rogers and L. Mandelkern, *J. Phys. Chem.*, **61**, 985, 1957)]; poly-*p*-alkyl styrenes [● (W. G. Barb, *J. Polym. Sci.*, **37**, 515, 1959)]; poly-α-olefins [△ (M. L. Dannis, *J. Appl. Polym. Sci.*, **1**, 121, 1959; K. R. Dunham, J. Vandenbergh, J. W. H. Farber, and L. E. Contois., *J. Polym. Sci.*, **1A**, 751, 1963)]; and polyacrylates [▲ (J. A. Shetter, *Polym. Lett.*, **1**, 209, 1963)] (135).

Table 8.12 Effect of tacticity on the glass transition temperatures of polyacrylates and polymethacrylates (137)

	T_g (°C)				
	Polyacrylates		Polymethacrylates		
Side Chain	Isotactic	Dominantly Syndiotactic	Isotactic	Dominantly Syndiotactic	100% Syndiotactic
Methyl	10	8	43	105	160
Ethyl	−25	−24	8	65	120
n-Propyl	—	−44	—	35	—
Iso-Propyl	−11	−6	27	81	139
n-Butyl	—	−49	−24	20	88
Iso-Butyl	—	−24	8	53	120
Sec-Butyl	−23	−22	—	60	—
Cyclo-Hexyl	12	19	51	104	163

polymers, when the other substituent is hydrogen, the energy difference between the rotational states of the two pairs of isomers is the same. Thus the acrylates in Table 8.12 have the same T_g for the two isomers, whereas the methacrylates show distinctly different T_g's, with the isotactic form always having a lower T_g than the syndiotactic form.

8.11 EFFECT OF PRESSURE ON T_g

The discussion above has assumed constant pressure at 1 atm (1 bar). Since an increased pressure causes a decrease in the total volume [see equation (8.8)], an increase in T_g is expected based on the prediction of decreased free volume.

Tamman and Jellinghaus (138) showed that a plot of volume versus pressure at a temperature near the transition shows an elbow reminiscent of the volume–temperature plot (see Figure 8.9). If the temperature is raised at elevated pressures, T_g will in fact show a corresponding increase (see Figure 8.33) (139).

The results in Figure 8.33 can easily be interpreted in terms of the free-volume theory of T_g. In developing the WLF equation (Section 8.6.1.2), it was shown that the free-

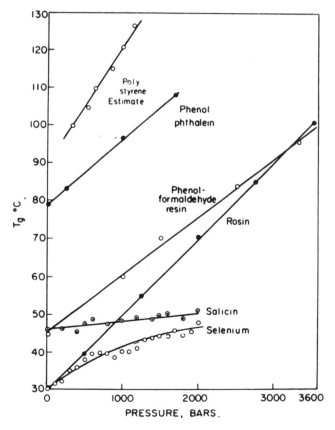

Figure 8.33 Glass transition versus pressure for various substances (139).

volume fraction at any temperature above T_g could be expressed $f = f_0 + \alpha_f(T - T_g)$. If the free-volume compressibility is β_f, then

$$f_{t,p} = f_0 + \alpha_f[T - T_g(0)] - \beta_f P \tag{8.78}$$

where $T_g(0)$ refers to the glass transition at zero pressure. Under particular glass transition temperature and pressure conditions, $f_{t,p} = f_0$ and equation (8.78) becomes

$$\alpha_f[T_g - T_g(0)] = \beta_f P \tag{8.79}$$

By differentiating with respect to pressure (140–143)

$$\left(\frac{\partial T_g}{\partial P}\right)_f = \frac{\Delta\beta_f}{\Delta\alpha_f} \tag{8.80}$$

a relation strongly reminiscent of Ehrenfest's (90) relation for the change of a second-order transition temperature with pressure,

$$\frac{TV\Delta\alpha}{\Delta C_p} = \frac{\Delta\beta}{\Delta\alpha} = \frac{dT_g}{dP} \tag{8.81}$$

where the Δ sign refers to changes from below to above T_g [see also equation (8.69)]. Several representative values of $\partial T_g/\partial P$ are shown in Table 8.13 (135). Since $\Delta\alpha \simeq \alpha_f \simeq 4.8 \times 10^{-4}$ deg^{-1}, $\beta_f \simeq \Delta\beta$ may be estimated. For polystyrene, Table 8.13 predicts a T_g rise of 31°C for a rise in pressure per 1000 atm, in agreement with Figure 8.33.

For polyurethanes, Quested et al. (144) found that $\Delta\beta_f/\Delta\alpha_f$ was greater than dT_g/dP, except at pressures close to 1 bar. At high pressures, dT_g/dP reached a limiting value of 10.4°C/kbar. The effect of pressure has been studied for ultrasonic frequencies (145) and fracture stress differences (146).

In the above, it was demonstrated that an increase in pressure can bring about vitrification. This result is important in engineering operations such as molding or extrusion, where operation too close to T_g (1 bar) can result in a stiffening of the material.

Table 8.13   Pressure coefficients of the glass transition temperatures for selected materials (135)

Material	$T_g(°C)$	dT_g/dP (K/atm)
Natural rubber	−72	0.024
Polyisobutylene	−70	0.024
Poly(vinyl acetate)	25	0.022
Rosin	30	0.019
Selenium	30	0.015–0.004[a]
Salicin	46	0.005
Phenolphthalein	78	0.019
Poly(vinyl chloride)	87	0.016
Polystyrene	100	0.031
Poly(methyl methacrylate)	105	0.020–0.023
Boron trioxide	260	0.020

[a]The variation is probably due to the different compressibilities of ring and chain material.

Thus we may refer to a glass transition pressure. In a broader sense, the glass transition is multidimensional. We could also refer to the glass transition molecular weight (Section 8.7), the glass transition concentration (for diluted or plasticized species), and so forth.

The solubility parameter can also be estimated with the aid of measurements as a function of hydrostatic pressure. Thus

$$\delta = \left\{ T \left(\frac{\alpha}{\beta} \right) \right\}^{1/2} \tag{8.82}$$

where α represents the isobaric volume thermal expansion coefficient and β the isothermal compressibility; see Section 8.1.4. Since β is the most difficult of the three terms in equation (8.82), it is the most likely to be the unknown.

8.12 DAMPING AND DYNAMIC MECHANICAL BEHAVIOR

When the loss modulus or loss tangent is high, as in the glass transition region, the polymers are capable of damping out noise and vibrations, which, after all, are a particular form of dynamic mechanical motion. This section describes some of the aspects of behavior of a polymer under sinusoidal stresses at constant amplitude.

If an applied stress varies with time in a sinusoidal manner, the sinusoidal stress may be written

$$\sigma = \sigma_0 \sin \omega t \tag{8.83}$$

where ω is the angular frequency in radians, equal to $2\pi' \times$ frequency. For Hookian solids, with no energy dissipated, the strain is given by

$$\varepsilon = \varepsilon_0 \sin \omega t \tag{8.84}$$

For real materials, the stress and strain are not in phase, the strain lagging behind the stress by the phase angle δ. The relationships among these parameters are illustrated in Figure 8.34. Of course, the phase angle defines an in-phase and out-of-phase component of the stress, σ' and σ'', as defined in Section 8.1.

Then the relationships between the in-phase and out-of-phase components and δ are given by

$$\sigma' = \sigma_0 \cos \delta \tag{8.85}$$

$$\sigma'' = \sigma_0 \sin \delta \tag{8.86}$$

The dynamic moduli may now be written

$$E' = \frac{\sigma'}{\varepsilon_0} = E^* \cos \delta \tag{8.87}$$

$$E'' = \frac{\sigma''}{\varepsilon_0} = E^* \sin \delta \tag{8.88}$$

$$E^* = \frac{\sigma_0}{\varepsilon_0} = (E'^2 + E''^2)^{1/2} \tag{8.89}$$

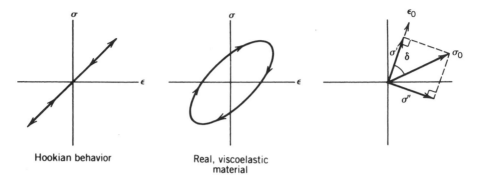

Hookian behavior

Real, viscoelastic
material

Figure 8.34   Simple dynamic relationships between stress and strain, illustrating the role of the phase angle.

In terms of complex notation,

$$E^* = E' + iE'' \tag{8.90}$$

(see Section 8.1) and

$$E = |E^*| = (E'^2 + E''^2)^{1/2} \tag{8.91}$$

where, of course, $E''/E' = \tan \delta$.

Again, the logarithmic decrement, Δ, may be defined as the natural logarithm of the amplitude ratio between successive vibrations (see Section 8.3.4). This is a measure of damping. The quantity $\tan \delta$ is related to the logarithmic decrement,

$$\Delta \cong \pi' \tan \delta$$

$$\Delta \cong \frac{\pi' E''}{E'} \tag{8.92}$$

where $\pi' = 3.14$. The heat energy per unit volume per cycle is $H = \pi' E'' A_0^2$, where A_0 is the maximum amplitude, and Hooke's law is valid. Further relationships are given in Section 10.2.4, where damping is related to models. Hence both the loss tangent and the logarithmic decrement are proportional to the ratio of energy dissipated per cycle to the maximum potential energy stored during a cycle.

These loss terms are at a maximum near the glass transition or a secondary transition. This phenomenon is widely used in engineering for the construction of objects subject to making noise or vibrations. Properly protected by high damping polymers, car doors close more quietly, motors make less noise, and mechanical damage to bridges through vibrations is reduced.

It must be mentioned that the glass transition can be broadened or shifted through various chemical and physical means. These include the use of plasticizers, fillers, or fibers, or through the formation of interpenetrating polymer networks. In this last case the very small phases are formed with variable composition and molecular chains trapped by cross-links promote juxtaposition of the two molecular species. A very broad glass transition may result, spanning the range of the two homopolymer transitions.

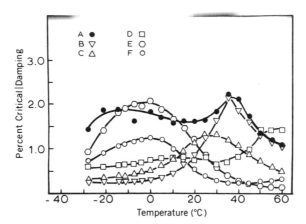

Figure 8.35 Damping as a function of temperature for several polymers. Composition A is an IPN. Note that it damps nearly evenly over a broad temperature range. Compositions B through F are various homopolymers and copolymers that damp over narrow temperature ranges (140).

Thus with *cross*–poly(ethyl acrylate)–*inter*–*cross*–poly(methyl methacrylate), which is composed of chemically isomeric polymers, glass transition behavior 100°C wide is obtained (30,147,148) (see Figure 8.35) (147). Conversely, objects such as rubber tires must be built of low damping elastomers, lest they overheat in service and blow out, see Section 9.16.2.

8.13 DEFINITIONS OF ELASTOMERS, PLASTICS, ADHESIVES, AND FIBERS

This chapter began with an overall approach to the behavior of polymers as a function of temperature. Cast in another way, at ambient temperatures a polymer will be above, below, or at its glass transition temperature, with concomitant properties. Other ways of dividing polymers are according to the presence or absence of crystallinity (Chapter 6) or according to the presence or absence of cross-linking (Chapter 9).

In a certain simplified sense, definitions are given below that will help identify polymers found in the everyday world with their scientific properties.

1. An elastomer is a cross-linked, amorphous polymer above its T_g. An example is the common rubber band.
2. An adhesive is a linear or branched amorphous polymer above its T_g. It must be able to flow on a molecular scale to "grip" surfaces. (This definition is not to be confused with polymerizable adhesive materials, present in monomeric form. These are "tacky" or "sticky" only in the partly polymerized state. Frequently they are cross-linked "thermoset," finally. Contact with the surface to be adhered must be made before gelation, in order to work.) An example is the postage stamp adhesive, composed of linear poly(vinyl alcohol), which is plasticized by water (or saliva) from below its T_g to above its T_g. On migration of the water away from the adhesive surface, it "sticks."
3. A plastic is usually below its T_g if it is amorphous. Crystalline plastics may be either above or below their T_g's.

4. Fibers are always composed of crystalline polymers. Apparel fiber polymers are usually close to T_g at ambient temperatures to allow flexibility in their amorphous portions.

5. House coatings and paints based on either oils or latexes are usually close to T_g at ambient temperatures. A chip of such material is usually flexible but not rubbery, being in the leathery region (see Section 8.2.2).

REFERENCES

1. A. Kaye, R. F. T. Stepto, W. J. Work, J. V. Aleman, and A. Ya. Malkin, *Pure and Applied Chem.*, **70**, 701 (1998).

2. J. D. Ferry, *Viscoelastic Properties of Polymers*, 3rd, ed., Wiley, New York, 1980, Chap. 1.

3. L. E. Nielsen and R. F. Landel, *Mechanical Properties of Polymers and Composites*, 2nd ed., Dekker, New York, 1994.

4. B. Du, J. Liu, Q. Zhang, and T. He, *Polymer*, in press (2000).

5. R. Glazebrook, ed., *Dictionary of Applied Physics*, Vol. 1, Macmillion, London, 1922, pp. 175–176, 244–245.

6. A. V. Tobolsky and J. R. McLoughlin, *J. Polym. Sci.*, **8**, 543 (1952).

7. J. J. Aklonis, *J. Chem. Ed.*, **58** (11), 892 (1981).

8. R. F. Boyer, in *Encyclopedia of Polymer Science and Technology*, Suppl. Vol. 2, N. M. Bikales, ed., Interscience, New York, 1977, p. 745; (a) pp. 822–823.

9. T. J. Shimanouchi, M. Asahina, and S. J. Enomoto, *J. Polym. Sci.*, **59**, 93 (1962).

10. D. Katz and I. G. Zervi, *J. Polym. Sci.*, **46C**, 139 (1974).

11. D. Katz and G. Salee, *J. Polym. Sci.*, **A-2** (6), 801 (1968).

12. K. Ueberreiter and G. Kanig, *J. Chem. Phys.*, **18**, 399 (1950).

13. G. M. Martin and L. Mandelkern, *J. Res. Natl. Bur. Stand.*, **62**, 141 (1959).

14. J. Brandruys and E. H. Immergut, eds., *Polymer Handbook*, 2nd ed., Wiley, New York, 1975; III-139; (a) III-150.

15. A. V. Tobolsky and H. Yu, unpublished.

16. H. J. Stern, *Rubber: Natural and Synthetic*, 2nd ed., Palmerton Publishing, New York, 1967, Chap. 1.

17. A. F. Yee and M. T. Takemori, *J. Polym. Sci. Polym. Phys. Ed.*, **20**, 205 (1982).

18. G. O. Jones, *Glass*, Methuen, London, 1956.

19. R. F. Boyer and R. S. Spencer, *J. Appl. Phys.*, **16**, 594 (1945).

20. D. R. Lide, ed., *CRC Handbook of Chemistry and Physics*, CRC Press, Boca Raton, FL, 1999, sec. 12, p. 119.

21. L. H. Sperling, *J. Polym. Sci. Polym. Symp.*, **60**, 175 (1977).

22. M. C. O. Chang, D. A. Thomas, and L. H. Sperling, *J. Polym. Sci. Part B Polym. Phys.*, **26**, 1627 (1988).

23. P. Meares, *Trans. Faraday Soc.*, **53**, 31 (1957).

24. M. Salmerón, C. Torregrosa, A. Vidaurre, J. M. Meseguer Dueñas, M. Monleón Pradas, and J. L. Gómez Ribelles, *Colloid Polym. Sci.*, **277**, 1033 (1999).

25. J. A. Victor, S. D. Kim, A. Klein, and L. H. Sperling, *J. Appl. Polym. Sci.*, **73**, 1763 (1999).

26. C. Kow, M. Morton, and L. J. Fetters, *Rubber Chem. Technol.*, **55** (1), 245 (1982).

27. K. C. Frisch, D. Klempner, S. Migdal, H. L. Frisch, and H. Ghiradella, *Polym. Eng. Sci.*, **14**, 76 (1974).

28. W. Meesiri, J. Menczel, U. Guar, and B. Wunderlich, *J. Polym. Sci. Polym. Phys. Ed.*, **20**, 719 (1982).

29. (a) M. Takayanagi, *Proc. Polym. Phys.* (*Jpn.*), 1962–1965. (b) Toya Baldwin Co., Ltd. Rheovibron Instruction Manual, 1969.

30. L. H. Sperling, in *Sound and Vibration Damping with Polymers*, R. D. Corsaro and L. H. Sperling, eds., ACS Books Symp. Ser. 424, American Chemical Society, Washington, DC, 1990.

31. Bordon Award Symposium honoring J. K. Gillham, all the papers in *Polym. Eng. Sci.*, **19** (10) (1979).

32. J. K. Gillham, *Polym. Eng. Sci.*, **19**, 749 (1979).

33. R. A. Venditti and J. K. Gillham, *J. Appl. Polym. Sci.*, **64**, 3 (1997).

34. J. K. Gillham, *AICHE J.*, **20**, 1066 (1974).

35. J. Russell and R. G. Van Kerpel, *J. Polym. Sci.*, **25**, 77 (1957).

36. A. F. Klarman, A. V. Galanti, and L. H. Sperling, *J. Polym. Sci.*, **A-2** (7), 1513 (1969).

37. C. J. Malm, J. W. Mench, D. L. Kendall, and G. D. Hiatt, *Ind. Eng. Chem.*, **43**, 688 (1951).

38. L. Banks and B. Ellis, *J. Polym. Sci. Polym. Phys. Ed.*, **20**, 1055 (1982).

39. H. G. Elias, *Macromolecules: Structure and Properties*, Vol. 1, Plenum, New York, 1977, Chap. 10.

40. N. Saito, K. Okano, S. Iwayanagi, and T. Hideshima, in *Solid State Physics*, Vol. 14, F. Seitz and D. Turnbull, eds., Academic Press, Orlando, 1963, p. 344.

41. K. S. Cole and S. Cole, *J. Chem. Phys.*, **9**, 341 (1941).

42. A. Tcharkhtchi, A. S. Lucas, J. P. Trotignon, and J. Verdu, *Polymer*, **39**, 1233 (1998).

43. C. A. Garcia-Franco and D. W. Mead, *Rheological Acta*, **38**, 34 (1999).

43a. N. G. McCrum, B. E. Read, and G. Williams, *Anelastic and Dielectric Effects in Polymeric Solids*, Wiley, New York, 1967.

43b. (a) T. F. Schatzki, *J. Polym. Sci.*, **57**, 496 (1962); (b) J. J. Aklonis and W. J. MacKnight, *Introduction to Polymer Viscoelasticity*, 2nd ed., Wiley-Interscience, New York, 1983, p. 81.

43c. H. A. Flocke, *Kolloid Z.*, **180** 118 (1962).

44. H. S. Gutawsky, G. B. Kistiakowsky, G. E. Pake, and E. M. Purcell, *J. Chem. Phys.*, **17** (10), 972 (1949); see also J. G. Powles and H. S. Gutowsky, *J. Chem. Phys.*, **21**, 1695 (1953), and W. P. Slichter and E. R. Mandell, *J. Appl. Phys.*, **29**, 1438 (1958).

45. J. V. Koleske and J. A. Faucher, *Polym. Eng. Sci.*, **19** (10), 716 (1979).

46. S. J. Stadnicki, J. K. Gillham, and R. F. Boyer, *J. Appl. Polym. Sci.*, **20**, 1245 (1976).

47. J. K. Gillham, J. A. Benci, and R. F. Boyer, *Polym. Eng. Sci.*, **16**, 357 (1976).

48. R. F. Boyer, *Polym. Eng. Sci.*, **19** (10), 732 (1979).

49. S. Hedvat, *Polymer*, **22**, 774 (1981).

50. G. D. Patterson, H. E. Bair, and A. Tonelli, *J. Polym. Sci. Polym. Symp.*, **54**, 249 (1976).

51. L. E. Nielsen, *Polym. Eng. Sci.*, **17**, 713 (1977).

52. R. M. Neumann, G. A. Senich, and W. J. MacKnight, *Polym. Sci. Eng.*, **18**, 624 (1978).

53. J. Heijboer, *Polym. Eng. Sci.*, **19** (10), 664 (1979).

54. V. A. Bershtein and V. M. Ergos, *Differential Scanning Calorimetry*, Ellis Horwood, Chichester, England, 1994.

55. A. J. Kovacs, *J. Polym. Sci.*, **30**, 131 (1958).

56. E. Catsiff and A. V. Tobolsky, *J. Polym. Sci.*, **19**, 111 (1956).

57. V. B. Rele and J. J. Aklonis, *J. Polym. Sci.*, **46C**, 127 (1974).

58. K. C. Lin and J. J. Aklonis, *Polym. Sci. Eng.*, **21**, 703 (1981).

59. J. J. Aklonis, *IUPAC Proceedings*, University of Massachusetts, Amherst, July 12–16, 1982, p. 834.

60. P. E. Rouse, *J. Chem. Phys.*, **21**, 1272 (1953).

61. F. Bueche, *J. Chem. Phys.*, **22**, 603 (1954).

62. B. H. Zimm, *J. Chem. Phys.*, **24**, 269 (1956).

63. A. V. Tobolsky and D. B. DuPre, *Adv. Polym. Sci.*, **6**, 103 (1969).

64. A. V. Tobolsky and J. J. Aklonis, *J. Phys. Chem.*, **68**, 1970 (1964).

65. R. F. Boyer, in *Encyclopedia of Polymer Science and Technology*, Vol. 13, N. M. Bikales, ed., Interscience, New York, 1970, p. 277.

66. R. Boyer, in *Encyclopedia of Polymer Science Technology* Suppl. Vol. II, p. 765, Wiley, New York, 1977.

67. V. A. Bershtein, V. M. Egorov, L. M. Egorova, and V. A. Ryzhov, *Thermochim. Acta*, **238**, 41 (1994).

68. W. Wrasidlo, *Thermal Analysis of Polymers, Advances in Polymer Science*, Vol. 13, Springer-Verlag, New York, 1974. p. 3.

69. H. Eyring, *J. Chem. Phys.*, **4**, 283 (1936).

70. T. G. Fox and P. J. Flory, *J. Appl. Phys.*, **21**, 581 (1950); T. G. Fox and P. J. Flory, *J. Polym. Sci.*, **14**, 315 (1954).

71. R. Simha and R. F. Boyer, *J. Chem. Phys.*, **37**, 1003 (1962).

72. R. Simha and C. E. Weil, *J. Macromol. Sci. Phys.*, **B4**, 215 (1970).

73. R. F. Boyer and R. Simha, *J. Polym. Sci.*, **B11**, 33 (1973).

74. S. C. Sharma, L. Mandelkern, and F. C. Stehling, *J. Polym. Sci.*, **B10**, 345 (1972).

75. A. K. Doolittle, *J. Appl. Phys.*, 1471 (1951).

76. D. Turnbull and M. H. Cohen, *J. Chem. Phys.*, **31**, 1164 (1959).

77. D. Turnbull and M. H. Cohen, *J. Chem. Phys.*, **34**, 120 (1961).

78. F. Bueche, *J. Chem. Phys.*, **21**, 1850 (1953).

79. F. Bueche, *J. Chem. Phys.*, **24**, 418 (1956).

80. F. Bueche, *J. Chem. Phys.*, **30**, 748 (1959).

81. M. L. Williams, R. F. Landel, and J. D. Ferry, *J. Am. Chem. Soc.*, **77**, 3701 (1955).

82. E. H. Andrews, *Fracture in Polymers*, American Elsevier, New York, 1968, pp. 9–16.

83. J. Liu, Q. Deng, and Y. C. Jean, *Macromolecules*, **26**, 7149 (1993).

84. R. Simha and G. Carri, *J. Polym. Sci.: Part B: Polym. Phys.*, **32**, 2645 (1994).

85. S. Misra and W. L. Mattice, *Macromolecules*, **26**, 7274 (1993).

86. Y. Kobayashi, K. Haraya, Y. Kamiya, and S. Hattori, *Bull. Chem. Soc. Jpn.*, **65**, 160 (1992).

87. Y. Kobayashi, *J. Chem. Soc. Faraday Trans.*, **87**, 3641 (1991).

88. R. M. Dammert, S. L. Maunu, F. H. J. Maurer, I. M. Neelow, S. Nievela, F. Sundholm, and C. Wastlund, *Macromolecules*, **32**, 1930 (1999).

89. Z. Yu, Y. Yahsi, J. D. McGervey, A. M. Jamieson, and R. Simha, *J. Polym. Sci.: Part B: Polym. Phys.*, **32**, 2637 (1994).

90. P. Ehrenfest, *Leiden Comm. Suppl.*, 756 (1933).

91. J. H. Gibbs, *J. Chem. Phys.*, **25**, 185 (1956).

92. J. H. Gibbs and E. A. DiMarzio, *J. Chem. Phys.*, **28**, 373 (1958).

93. J. H. Gibbs, in *Modern Aspects of the Vitreous State*, J. D. Mackenzie, ed., Butterworth, London, 1960.

94. E. A. DiMarzio and J. H. Gibbs, *J. Polym. Sci.*, **A1**, 1417 (1963).

95. E. A. DiMarzio, *J. Res. Natl. Bur. Stnds.*, **68A**, 611 (1964).
96. P. J. Flory, *Proc. R. Soc. (Lond.)*, **A234**, 60 (1956).
97. E. A. DiMarzio and J. H. Gibbs, *J. Polym. Sci.*, **40**, 121 (1959).
98. E. A. DiMarzio and J. H. Gibbs, *J. Polym. Sci.*, **1A**, 1417 (1963).
99. G. Adam and J. H. Gibbs, *J. Chem. Phys.*, **43**, 139 (1965).
100. J. H. Glans and D. T. Turner, *Polymer*, **22**, 1540 (1981).
101. E. Jenckel and K. Ueberreiter, *Z. Phys. Chem.*, **A182**, 361 (1938).
102. L. P. Blanchard, J. Hess, and S. L. Malhorta, *Can. J. Chem.*, **52**, 3170 (1974).
103. J. K. Gillham, *Polym. Eng. Sci.*, **19**, 676 (1979).
104. M. T. DeMuse, J. K. Gillham, and F. Parodi, *J. App. Polym. Sci.*, **64**, 15 (1997).
105. J. K. Gillham, in *The Role of the Polymer Matrix in the Processing and Structural Properties of Composite Materials*, J. C. Seferis and L. Nicolais, eds., Plenum, New York, 1983, pp. 127–145.
106. J. B. Enns and J. K. Gillham, in *Polymer Characterization: Spectroscopic, Chromatographic, and Physical Instrumental Methods*, C. D. Craver, ed., Advances in Chemistry Series No. 203, American Chemical Society, Washington, DC, 1983, pp. 27–63.
107. J. B. Enns and J. K. Gillham, *J. Appl. Polym. Sci.*, **28**, 2567 (1983).
108. J. K. Gillham, *Encyclopedia of Polym. Sci. Tech.*, **4**, 519 (1986).
109. P. R. Couchman, *Macromolecules*, **11**, 1156 (1978).
110. P. R. Couchman, *Polym. Eng. Sci.*, **21**, 377 (1981).
111. P. R. Couchman, *J. Mater. Sci.*, **15**, 1680 (1980).
112. P. R. Couchman and F. E. Karasz, *Macromolecules*, **11**, 117 (1978).
113. J. M. Bardin and D. Patterson, *Polymer*, **10**, 247 (1969); L. A. Wood, *J. Polym. Sci.*, **28**, 319 (1958).
114. R. F. Boyer, *J. Macromol. Sci. Phys.*, **7**, 487 (1973).
115. T. G. Fox, *Bull. Am. Phys. Soc.*, **1**, 123 (1956).
116. J. R. Fried, F. E. Karasz, and W. J. MacKnight, *Macromolecules*, **11**, 150 (1978).
117. J. M. Pochan, C. L. Beatty, and D. F. Hinman, *Macromolecules*, **11**, 1156 (1977).
118. L. H. Sperling, *Polymeric Multicomponent Materials: An Introduction*, Wiley, New York, 1997.
119. M. Matsuo, *Jpn. Plastics*, **2**, 6 (1958).
120. M. Matsuo, T. Ueno, H. Horino, S. Chujya, and H. Asai, *Polymer*, **9**, 425 (1968).
121. P. R. Scarito and L. H. Sperling, *Polym. Eng. Sci.*, **19**, 297 (1979).
122. G. T. Davis and R. K. Eby, *J. Appl. Phys.*, **44**, 4274 (1973).
123. F. C. Stehling and L. Mandelkern, *Macromolecules*, **3**, 242 (1970).
124. K. H. Illers, *Kolloid Z. Z. Polym.*, **190**, 16 (1963); **231**, 622 (1969); **250**, 426 (1972).
125. A. V. Tobolsky, *Properties and Structure of Polymers*, Wiley, New York, 1960, App. K.
126. G. Champetier and J. P. Pied, *Makromol. Chem.*, **44**, 64 (1961).
127. R. F. Boyer, *J. Appl. Phys.*, **25**, 825 (1954).
128. ASTM, 08.01, D 1525, American Society for Testing and Materials, 1998.
129. ASTM, *8.01–8.03*, D 648, American Society for Testing and Materials, 1996.
130. M. T. Takemori, *Polym. Eng. Sci.*, **19**, 1104 (1979).
131. R. F. Boyer, *Rubber Chem. Technol.*, **36**, 1303 (1963).
132. J. A. Shetter, *Polym. Lett.*, **1**, 209 (1963).
133. S. S. Rogers and L. Mandelkern, *J. Phys. Chem.*, **61**, 985 (1957).
134. W. G. Barb, *J. Polym. Sci.*, **37**, 515 (1957).

135. M. C. Shen and A. Eisenberg, *Prog. Solid State Chem.*, **3**, 407 (1966); *Rubber Chem. Technol.*, **43**, 95, 156 (1970).

136. S. Bywater and P. M. Toporawski, *Polymer*, **13**, 94 (1972).

137. F. E. Karasz and W. T. MacKnight, *Macromolecules*, **1**, 537 (1968).

138. G. Tamman and W. Jellinghaus, *Ann. Phys.*, [5] **2**, 264 (1929).

139. A. Eisenberg, *J. Phys. Chem.*, **67**, 1333 (1963).

140. J. D. Ferry and R. A. Stratton, *Kolloid Z.*, **171**, 107 (1960).

141. J. M. O'Reilly, *J. Polym. Sci.*, **57**, 429 (1962).

142. J. E. McKinney, H. V. Belcher, and R. S. Marvin, *Trans. Soc. Rheol.*, **4**, 347 (1960).

143. M. Goldstein, *J. Chem. Phys.*, **39**, 3369 (1963).

144. D. L. Quested, K. P. Pae, B. A. Newman, and J. I. Scheinbaum, *J. Appl. Phys.*, **51** (10), 5100 (1980).

145. D. L. Quested and K. D. Pae, *Ind. Eng. Prod. Res. Dev.*, **22**, 138 (1983).

146. Y. Kaieda and K. D. Pae, *J. Mater. Sci.*, **17**, 369 (1982).

147. L. H. Sperling, T. W. Chiu, R. G. Gramlich, and D. A. Thomas, *J. Paint Technol.*, **46**, 47 (1974).

148. J. A. Grates, D. A. Thomas, E. C. Hickey, and L. H. Sperling, *J. Appl. Polym. Sci.*, **19**, 1731 (1975).

GENERAL READING

J. J. Aklonis and W. J. MacKnight, *Introduction to Polymer Viscoelasticity*, 2nd ed., Wiley-Interscience, New York, 1983.

V. A. Bershtein and V. M. Egorov, *Differential Scanning Calorimetry of Polymers: Physics, Chemistry, Analysis*, Ellis-Horwood, New York, 1994.

R. D. Corsaro and L. H. Sperling, eds., *Sound and Vibration Damping with Polymers*, ACS Books, Symp. Ser. No. 424, American Chemical Society, Washington, DC, 1990.

J. D. Ferry, *Viscoelastic Properties of Polymers*, 3rd ed., Wiley, New York, 1980.

L. E. Nielsen and R. F. Landel, *Mechanical Properties of Polymers and Composites*, 2nd ed., Dekker, New York, 1994.

J. P. Silbia, ed., *A Guide to Materials Characterization and Chemical Analysis*, VCH, New York, 1988.

A. K. Sircar, M. L. Galaska, S. Rodrigues, and R. P. Chartoff, *Rubber Chem. Tech.*, **72**, 513 (1999).

STUDY PROBLEMS

1. Name the five regions of viscoelastic behavior, and give an example of a commercial polymer commonly used in each region.

2. Name and give a one-sentence definition of each of the three theories of the glass transition.

3. Polystyrene homopolymer has a $T_g = 100°C$, and polybutadiene has a $T_g = -90°C$. Estimate the T_g of a 50/50 w/w statistical copolymer, poly(styrene–*stat*–butadiene).

4. A new linear amorphous polymer has a T_g of $+10°C$. At $+25°C$, it has a melt viscosity of 6×10^8 poises. Estimate its viscosity at 40°C.

5. Define the following terms: bulk modulus, tan δ; stress relaxation; plasticizer; Schatzki crankshaft motions; A_t; WLF equation; compressibility; Young's modulus.

6. As the newest employee of Polymeric Industries, Inc., you are attending your first staff meeting. One of the company's most respected chemists is speaking: "Yesterday, we completed the preliminary evaluation of the newly synthesized thermoplastic, poly(wantsa cracker). The polymer has a melt viscosity of 1×10^5 Pa·s at 140°C. Our characterization laboratory reported a glass transition temperature of 110°C."
   "You know our extruder works best at 2×10^2 Pa·s," broke in the mechanical engineer, "and you know poly(wantsa cracker) degrades at 160°C. Therefore we won't be able to use poly(wantsa cracker)!"
   As you reach for your trusty calculator to estimate the melt viscosity of poly(wantsa cracker) at 160°C to make a reasoned decision of your own, you realize suddenly that all eyes are on you.
   (a) What is the melt viscosity of poly(wantsa cracker) at 160°C?
   (b) Can Polymeric Industries use the polymer? If not, what can they do to the polymer to increase usability?
   (c) What is the structure of poly(wantsa cracker), anyway?

7. Draw a log E–temperature plot for a linear, amorphous polymer.
   (a) Indicate the position and name the five regions of viscoelastic behavior.
   (b) How is the curve changed if the polymer is semicrystalline?
   (c) How is it changed if the polymer is cross-linked?
   (d) How is it changed if the experiment is run faster—that is, if measurements are made after 1 s rather than 10 s?
   In parts (b), (c), and (d), separate plots are required, each change properly labeled. E stands for Young's modulus.

8. During a coffee break, two chemists, three chemical engineers, and an executive vice president began discussing plastics. "Now everyone knows that such materials as plastics, rubber, fibers, paints, and adhesives have very little in common," began the executive vice president. "For example, nobody manufactures plastics and paints with the same equipment...." Even though you are the most junior member of the group, you interrupt; "That last may be so, but all those materials are closely related because ..."
   Complete the statement in 100 words or less. If you think that the above materials are in fact *not* related, you have 100 words to prove the executive vice president correct.

9. Briefly discuss the salient points in the derivation of the WLF equation.

10. Prepare a "box score" table, laying out the more important advantages and disadvantages of the three theories of the glass transition.

11. A new polymer has been synthesized in your laboratory, and you are proudly discussing the first property studies when your boss walks in. "We need a polymer with a cubic coefficient of thermal expansion of less than 4×10^{-4} deg^{-1} at 50°C. Can we consider your new stuff?"

Your technician hands you the sheet of paper with available data:

$$T_g = 100°C$$

$$\alpha_R = 5.5 \times 10^{-4} \text{ deg}^{-1} \text{ at } 150°C$$

Your boss adds, "By the way, the Board meets in 30 minutes. Any answer by then would surely be valuable." You begin to tear your hair out by its roots, wondering how you can solve this one so fast without going back into the lab, since there really isn't much time.

12. The T_g of poly(vinyl acetate) is listed as 29°C. If 5 mol% of divinyl benzene is co-polymerized in the polymer during polymerization, what is the new glass transition temperature?

13. Noting the instruments mentioned in Section 8.3, what instrument would you most like to have in your laboratory if you were testing (a) each of the three theories of T_g, (b) the molecular weight dependence of T_g, (c) the effect of cross-linking on T_g. Defend your choice.

14. A new atactic polymer has a T_g of 0°C. Your boss asks, "If we made the isotactic form, what is its melting temperature likely to be?" Suddenly, you remember that back in college you took physical polymer science. . . .

15. Rephrase the definitions in Section 8.13, and use other examples.

16. Your assistant rushes in with a new polymer. "It softens at 50°C," she says. "Is it a glass transition or a melting temperature?" you ask. "How would I know?" she answers. "I never took physical polymer science!" Describe two simple but foolproof experiments to distinguish between the two possibilities.

17. A piece of polystyrene is placed under 100 atm pressure at room temperature. What is the fractional volume decrease?

18. What is the free volume of polystyrene at 100°C? What experimental or theoretical evidence supports your conclusion?

19. A rubber ball is dropped from a height of 1 yard and bounces back 18 in. Assuming a perfectly elastic floor, approximately how much did the ball heat up? The heat capacity, C_p, of SBR rubber is about 1.83 kJ kg$^{-1} \cdot$ K^{-1}.

20. Write a 100 to 125-word essay on the importance of free volume in polymer science. This essay is to be accompanied by at least one figure, construction, or equation illustrating your thought train.

21. A new polymer was found to soften at 50°C. Several experiments were performed to determine if the softening was a glass transition or a melting point.
    (a) In interpreting the results for each experiment separately, was it a glass transition? a melting transition? cannot be determined for sure? or was there some mistake in the experiment?
    (b) What is your reasoning for each decision?

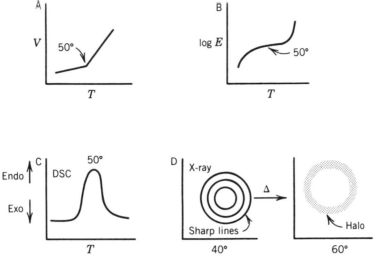

Figure P8.21   Laboratory studies of transitions.

APPENDIX 8.1   MOLECULAR MOTION NEAR THE GLASS TRANSITION[†]

The following experiment is intended for use as a classroom demonstration and takes about 30 minutes. Place the Superball® in the liquid nitrogen 30 minutes before class begins. Have a second such ball for comparison of properties.

Experiment

Time: Actual laboratory time about 1 hour
Level: Physical Chemistry
Principles Illustrated:

1. The onset of molecular motion in polymers.
2. The influence of molecular motion on mechanical behavior.

Equipment and Supplies:

One solid rubber ball (a small Superball® is excellent)
One hollow rubber ball (optional)
One Dewar flask of liquid nitrogen, large enough to hold above rubber balls
One ladle or spoon with long handle, to remove frozen balls (alternately, tie balls with long string)
One yardstick (or meter stick)

[†] Reprinted in part from L. H. Sperling, *J. Chem. Ed.*, **59**, 942 (1982).

One clock (a watch is fine)

One hard surface, suitable for ball bouncing (most floors or desks are suitable)

First, place the yardstick vertical to the surface and drop Superball® from the top height; record percent recovery (bounce). Remove Superball® from liquid nitrogen, and record percent bounce immediately and each succeeding minute (or more often) for the first 15 minutes, then after 20 and 25 minutes and each 5 minutes thereafter until recovery equals that first obtained.

For the hollow rubber ball, first observe toughness at room temperature, then cool in liquid nitrogen, then throw hard against the floor or wall. Observe the glassy behavior of the pieces and their behavior as they warm up. The experimenter should wear safety glasses.

For extra credit, obtain three small dinner bells; coat two of them with any latex paint. (Most latex paints have T_g near room temperature.) After drying, ring the bells. Place one coated bell in a freezer, and compare its behavior cold to the room-temperature coated bell. How can you explain the difference observed?

Another related experiment involves dipping adhesive tape into liquid nitrogen. Outdoor gutter drain tapes are excellent because of their size. Compare the stickiness of the tape before and after freezing. (Note again the definition of an adhesive.)

On warming the frozen solid rubber ball, the percent recovery (bounce) versus time will go through a minimum at T_g, as shown in Figure A8.1.1.

Below T_g, the ball is glassy and bounces much like a marble. At T_g, the bounce is at a

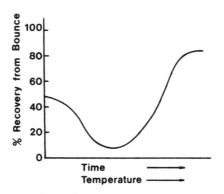

Figure A8.1.1   Ball bounce recovery.

minimum owing to conversion of kinetic energy to heat. (The ball actually warms up slightly.) Above T_g, normal rubber elasticity and bounce characteristics are observed.

For an experiment emphasizing rolling friction near T_g, see G. B. Kauffman, S. W. Mason, and R. B. Seymour, *J. Chem. Ed.*, **67**, 199 (1990).

9

CROSS-LINKED POLYMERS AND RUBBER ELASTICITY

An elastomer is defined as a cross-linked amorphous polymer above its glass transition temperature. Elastomers may be stretched substantially reversibly to several hundred percent. While most of this chapter explores the behavior of elastomers, the study of crosslinking is more general. If the cross-linked polymer is glassy, it is often called a thermoset. Below, the terms elastomer and rubber are often used interchangably.

9.1 CROSS-LINKS AND NETWORKS

During reaction, polymers may be cross-linked to several distinguishable levels. At the lowest level, branched polymers are formed. At this stage the polymers remain soluble, sometimes known as the sol stage. As cross-links are added, clusters form, and cluster size increases. Eventually the structure becomes infinite in size; that is, the composition gels. At this stage a Maxwellian demon could, in principle, traverse the entire macroscopic system stepping on one covalent bond after another. Continued cross-linking produces compositions where, eventually, all the polymer chains are linked to other chains at multiple points, producing, in principle, one giant covalently bonded molecule. This is commonly called a polymer network.

9.1.1 The Sol–Gel Transition

The reaction stage referred to as the sol–gel transition (1–3) is called the gel point. At the gel point the viscosity of the system becomes infinite, and the equilibrium modulus climbs from zero to finite values. In simple terms the polymer goes from being a liquid to being a solid. There are three different routes for producing cross-linked polymers:

1. Step polymerization reactions, where little molecules such as epoxies (oxiranes) react with amines, or isocyanates react with polyols with functionality greater than two to form short, branched chains, eventually condensing it into epoxies or poly-

urethanes, respectively. Schematically

$$
nA \sim A + mB \overset{\displaystyle B}{\underset{\displaystyle}{\overset{\displaystyle \wr}{B}}} \rightarrow
$$

$$
\begin{array}{c}
BA \sim AB \sim \\
\wr \\
\sim AB \overset{\wr}{} BA \sim AB \sim
\end{array}
\tag{9.1}
$$

2. Chain polymerization, with multifunctional molecules present. An example is styrene polymerized with divinyl benzene.
3. Postpolymerization reactions, where a linear (or branched) polymer is cross-linked after synthesis is complete. An example is the vulcanization of rubber with sulfur, which will be considered further below.

The general features of structural evolution during gelation are described by percolation (or connectivity) theory, where one simply connects bonds (or fills sites) on a lattice of arbitrary dimension and coordination number (4–6). Figure 9.1 (6) illustrates a two-dimensional system at the gel point. It must be noted that gels at and just beyond the gel point usually coexist with sol clusters. These can also be seen in Figure 9.1. It is common to speak of the conversion factor, p, which is the fraction of bonds that have been formed between the mers of the system; see Section 3.7. For the two-dimensional schematic illustrated in Figure 9.1, $p = \frac{1}{2}$ yields the gel point.

9.1.2 Micronetworks

A special type of network exists that involves only one or a few polymer chains. Thus microgels may form during specialized reaction conditions where only a few chains are interconnected.

Globular proteins constitute excellent examples of one-molecule micronetworks (7), where a single polymer chain is intramolecularly cross-linked; see Figure 9.2 (8). Here,

Figure 9.1 A square lattice example of percolation, at the gel point (6). Note structures that span the whole "sample."

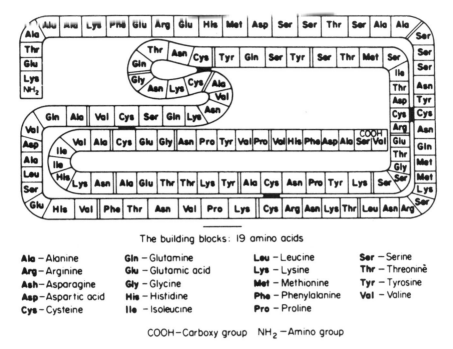

The building blocks: 19 amino acids

Ala – Alanine	**Gln** – Glutamine	**Leu** – Leucine	**Ser** – Serine
Arg – Arginine	**Glu** – Glutamic acid	**Lys** – Lysine	**Thr** – Threonine
Asn – Asparagine	**Gly** – Glycine	**Met** – Methionine	**Tyr** – Tyrosine
Asp – Aspartic acid	**His** – Histidine	**Phe** – Phenylalanine	**Val** – Valine
Cys – Cysteine	**Ile** – Isoleucine	**Pro** – Proline	

COOH–Carboxy group   NH$_2$ –Amino group

Figure 9.2   The structure of ribonuclease A, a micronetwork of one chain. Note cysteine–cysteine bonds (8).

disulfide bonds help keep the three-dimensional structure required for protein bio-polymer activity.

Such proteins may be denatured by any of several mechanisms, especially heat. Thus, when globular proteins are cooked, the intramolecular cross-links become delocalized, forming intermolecular bonds instead. This is the major difference between raw egg white and hard-boiled egg white, for example.

Fully cross-linked on a macroscopic scale, polymer networks fall into different cate-gories. It is convenient to call such polymers, which are used below their glass transition temperature, thermosets, since they are usually plastics incapable of further flow. On the other hand, those networks that are above their glass transition temperature are rubbery unless very heavily cross-linked. While all these kinds of cross-linked polymers are im-portant, the remainder of this chapter is devoted primarily to amorphous, continuous polymer networks above their glass transition temperature, the regime of rubber elasticity.

While the rubber elasticity theory to be described below presumes a randomly cross-linked polymer, it must be noted that each method of network formation described above has distinctive nonuniformities, which can lead to significant deviation of experi-ment from theory. For example, chain polymerization leads first to microgel formation (9,10), where several chains bonded together remain dissolved in the monomer. On continued polymerization, the microgels grow in number and size, eventually forming a macroscopic gel. However, excluded volume effects, slight differences in reactivity be-tween the monomer and cross-linker, and so on lead to systematic variations in cross-link densities at the 100- to 500-Å level.

9.2 HISTORICAL DEVELOPMENT OF RUBBER

9.2.1 Early Developments

A simple rubber band may be stretched several hundred percent; yet on being released, it snaps back substantially to its original dimensions. By contrast, a steel wire can be stretched reversibly for only about a 1% extension. Above that level, it undergoes an irreversible deformation and then breaks. This long-range, reversible elasticity constitutes the most striking property of rubbery materials. Rubber elasticity takes place in the third region of polymer viscoelasticity (see Section 8.2) and is especially concerned with cross-linked amorphous polymers in that region.

Columbus, on his second trip to America, found the American Indians playing a game with rubber balls (11,12) made of natural rubber. These crude materials were un-cross-linked but of high molecular weight and hence were able to hold their shape for significant periods of time.

The development of rubber and rubber elasticity theory can be traced through several stages. Perhaps the first scientific investigation of rubber was by Gough in 1805 (13). Working with unvulcanized rubber, Gough reached three conclusions of far-reaching thermodynamic impact:

1. A strip of rubber warms on stretching and cools on being allowed to contract. (This experiment can easily be confirmed by a student using a rubber band. The rubber is brought into contact with the lips and stretched rapidly, constituting an adiabatic extension. The warming is easily perceived by the temperature-sensitive lips.)

2. Under conditions of constant load, the stretched length decreases on heating and increases on cooling. Thus it has more retractive strength at higher temperatures. This is the opposite of that observed for most other materials.

3. On stretching a strip of rubber and putting it in cold water, the rubber loses some of its retractile power, and its relative density increases. On warming, however, the rubber regains its original shape. In the light of present-day knowledge, this last set of experiments involved the phenomenon known as strain-induced crystallization, since unvulcanized natural rubber crystallizes easily under these conditions.

In 1844 Goodyear vulcanized rubber by heating it with sulfur (14). In modern terminology, he cross-linked the rubber. (Other terms meaning cross-linking include "tanning" of leather, "drying" of oil-based paints, and "curing" of inks.) Vulcanization introduced dimensional stability, reduced creep and flow, and permitted the manufacture of a wide range of rubber articles, where before only limited uses, such as waterproofing, were available (15). (The "MacIntosh" raincoat of that day consisted of a sandwich of two layers of fabric held together by a layer of unvulcanized natural rubber.)

Using the newly vulcanized materials, Gough's line of research was continued by Kelvin (16). He tested the newly established second law of thermodynamics with rubber and calculated temperature changes for adiabatic stretching. The early history of rubber research has been widely reviewed (17,18).

All of the applications above, of course, were accomplished without an understanding of the molecular structure of polymers or of rubber in particular. Beginning in 1920, Staudinger developed his theory of the long-chain structure of polymers (19,20). [Interestingly Staudinger's view was repeatedly challenged by many investigators tenaciously adhering to ring formulas or colloid structures held together by partial valences (21).] See Appendix 5.1.

In the early days the only elastomer was natural rubber. Starting around 1914, a polymer of 2,3-dimethylbutadiene known as methyl rubber was made in Germany. This was replaced by a styrene-butadiene copolymer called Buna-S (butadiene–natrium–styrene), where natrium is, of course, sodium. This sodium-catalyzed copolymer, as manufactured in Germany in the period from about 1936 to 1945 had about 32% styrene monomer (22).

In 1939 the U.S. government started a crash program to develop a manufactured elastomer, called the Synthetic Rubber Program (23). The new material was called GR-S (government rubber-styrene). GR-S was made by emulsion polymerization. While the Bunas was catalyzed by sodium, the latter was catalyzed by potassium persulfate. Incidentally, the emulsifier in those days was ordinary soap flakes. Both materials played crucial roles in World War II, a story told many times (23–25).

While Buna-S had about 32% styrene monomer (22), the GR-S material, started a few years later with the benefit of the German recipe, had about 25% styrene (26). This difference in the composition was important for lowering the glass transition temperature. Using the Fox equation, equation (8.73), with a T_g of polystyrene of 373 K, and that of polybutadiene of 188 K, Table 8.7, values of T_g for Buna-S and GR-S are estimated at $-47°$ and $-58°C$, respectively. Noting that winter temperatures in European Russia reach $-40°$ to $-51°C$ (27), lore has it that use of Buna-S seriously influenced the winter Russian campaigns.

Today, SBR elastomers are widely manufactured with only minor improvements in the GR-S recipe. One such is the use of synthetic surfactants as emulsifiers. These and other improvements allow the production of more uniform latex particles.

9.3 RUBBER NETWORK STRUCTURE

Once the macromolecular hypothesis of Staudinger was accepted, a basic understanding of the molecular structure was possible. Before cross-linking, rubber (natural rubber in those days) consists of linear chains of high molecular weight. With no molecular bonds between the chains, the polymer may flow under stress if it is above T_g.

The original method of cross-linking rubber, via sulfur vulcanization, results in many reactions. One such may be written

$$2 \sim CH_2 - \overset{\overset{\displaystyle CH_3}{|}}{C} = CH - CH_2 \sim + \text{ sulfur}$$

$$\rightarrow \sim CH - \overset{\overset{\displaystyle CH_3}{|}}{C} = CH - CH_2 \sim$$

$$\underset{\underset{\displaystyle \sim CH_2 - CH - CH = \overset{\overset{\displaystyle CH_3}{|}}{C} - \underset{\underset{\displaystyle S-S-R}{|}}{CH} \sim}{\overset{\overset{\displaystyle |}{S}}{\underset{\displaystyle S}{|}}}{} \tag{9.1a}$$

where R represents other rubber chains.

Two other methods of cross-linking polymers must be mentioned here. One is radiation cross-linking, with an electron beam or gamma irradiation. Using polyethylene as an example,

$$2(\sim CH_2-CH_2-CH_2-CH_2\sim)$$

$$\xrightarrow{h\nu} \begin{array}{c} \sim CH_2-CH-CH_2-CH_2\sim \\ | \\ \sim CH_2-CH-CH_2-CH_2\sim \end{array} +H_2 \qquad (9.2)$$

Another method involves the use of a multifunctional monomer in the simultaneous polymerization and cross-linking of polymers. Taking poly(ethyl acrylate) as an example, with divinyl benzene as cross-linker,

$$(9.3)$$

where the upper and lower reactions taken place independently in time.

After cross-linking, flow of one molecule past another (viscoelastic behavior) is suppressed. Excluding minor impurities, an object such as a rubber band can be considered as one huge molecule. [It fulfills the two basic requirements of the definition of a molecule: (a) every atom is covalently bonded to every other atom, and (b) it is the smallest unit of matter with the characteristic properties of rubber bands.]

The structure of a cross-linked polymer may be idealized (Figure 9.3). The primary chains are cross-linked at many points along their length. For materials such as rubber bands, tires, and gaskets, the primary chains may have molecular weights of the order of 1×10^5 g/mol and be cross-linked (randomly) every 5 to 10×10^3 g/mol along the chain, producing 10 to 20 cross-links per primary molecule. It is convenient to define the average molecular weight between cross-links as M_c and to call chain portions bound at both ends by cross-link junctions active network chain segments.

In the most general sense, an elastomer may be defined as an amorphous, cross-linked polymer above its glass transition temperature (see Section 8.12). The two terms "rubber" and "elastomer" mean nearly the same thing. The term rubber comes from the "rubbing out" action of an eraser. Originally, of course, rubber was natural rubber, *cis*-polyisoprene. The term elastomer is more general and refers to the elastic-bearing properties of the materials.

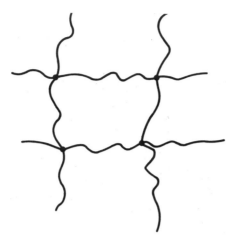

Figure 9.3   Idealized structure of a cross-linked polymer. Wavy lines, polymer chains; circles, cross-links.

9.4 RUBBER ELASTICITY CONCEPTS

The first relationships between macroscopic sample deformation, chain extension, and entropy reduction were expressed by Guth and Mark (28) and by Kuhn (29,30) (see Section 5.3). Mark and Kuhn proposed the model of a random coil polymer chain (Figure 9.4) which forms an active network chain segment in the cross-linked polymer. When the sample was stretched, the chain had extended in proportion, now called an affine deformation. When the sample is relaxed, the chain has an average end-to-end distance, r_0 (Figure 9.4), which increases to r when the sample is stretched. (Obviously, if the sample is compressed or otherwise deformed, other chain dimensional changes will occur.)

Through the research of Guth and James (31–35), Treloar (36), Wall (37), and Flory (38), the quantitative relations between chain extension and entropy reduction were clarified. In brief, the number of conformations that a polymer chain can assume in space were calculated. As the chain is extended, the number of such conformations diminishes. (A fully extended chain, in the shape of a rod, has only one conformation, and its conformational entropy is zero.)

The idea was developed, in accordance with the second law of thermodynamics, that the retractive stress of an elastomer arises through the reduction of entropy rather than through changes in enthalpy. Thus long-chain molecules, capable of reasonably free rotation about their backbone, and joined together in a continuous, monolithic network are required for rubber elasticity.

In brief, the basic equation relating the retractive stress, σ, of an elastomer in simple extension to its extension ratio, α, is given by

$$\sigma = nRT\left(\alpha - \frac{1}{\alpha^2}\right)$$

(9.4)

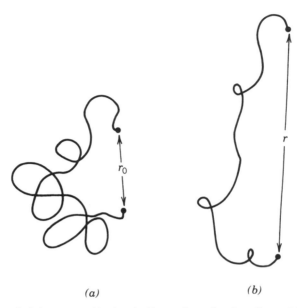

(a) (b)

Figure 9.4 A network chain segment: (a) relaxed, with a random coil conformation, and (b) extended, owing to an external stress.

where the original length, L_0, is increased to L ($\alpha = L/L_0$), and RT is the gas constant[†] times the absolute temperature. The quantity n represents the number of active network chain segments per unit volume (39–42). The quantity n equals ρ/M_c, where ρ is the density and M_c the molecular weight between cross-links.

Several quantities must be distinguished here:

1. The quantity M_c in a polymer network (the present case) is the number-average molecular weight between cross-links.
2. For linear polymers the entanglement molecular weight, M_c', is the value given later in this book by the elbows in Figure 10.16. (Note that some texts list this as M_c, leading to confusion.)
3. The quantity M_e is the molecular weight between the entanglements. Wool (43) defines $M_e = (4/9)M_c'$ for linear polymers but points out that experimentally $M_e \cong (1/2)M_c'$.

As described in Section 9.10.5, real polymer networks contain both chemical and physical cross-links, the latter being the various kinds of entanglements. It will be observed that equation (9.4) is nonlinear; that is, the Hookean simple proportionality between stress and strain does not hold. Young's modulus is often close to 2×10^6 Pa.

Equation (9.4) is compared to experiment in Figure 9.5 (44,45). The theoretical value of M_c was chosen for the best fit a low extensions. The sharp upturn of the experimental

---

[†] Convenient values of R for calculation purposes are 8.31×10^7 dynes·cm/mol·K, when the stress has units of dynes/cm^2, and 8.31 Pa·m^3/mol·K, when stress has units of Pa.

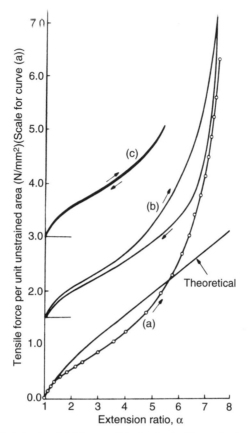

Figure 9.5 Stress–strain behavior of lightly cross-linked natural rubber at 50°C. Curve (*a*), experimental. Theoretical is equation (9.4). Curve (*c*) illustrates the reversible nature of the extension up to $\alpha = 5.5$. At higher elongations, curve (*b*), hysteresis effects become important. The theoretical curve has been fitted to the experimental data in the region of small extensions, with $nRT = 0.39$ N/mm² (47,48).

data above $\alpha = 7$ is due to the limited extensibility of the chains themselves,[†] which can be explained in part by more advanced theories (46).

In the following sections the principal equations of the theory of rubber elasticity are derived, emphasizing the relationships between molecular chain characteristics, stress, and strain.

9.5 THERMODYNAMIC EQUATION OF STATE

As a first approach to the equation of state for rubber elasticity, we analyze the problem via classical thermodynamics. The Helmholtz free energy, F, is given by

$$F = U - TS \qquad (9.5)$$

where U is the internal energy and S is the entropy.

[†] Part of the effect is usually strain-induced crystallinity especially for natural rubber and *cis*-polybutadiene.

The retractive force, f, exerted by the elastomer depends on the change in free energy with length:

$$f = \left(\frac{\partial F}{\partial L}\right)_{T,V} = \left(\frac{\partial U}{\partial L}\right)_{T,V} - T\left(\frac{\partial S}{\partial L}\right)_{T,V} \tag{9.6}$$

For an elastomer, Poisson's ratio is nearly 0.5, so the extension is nearly isovolume. When the experiment is done isothermally, the analysis becomes significantly simplified; note the subscripts to equation (9.6).

According to the statistical thermodynamic approach to be developed below, each conformation that a network chain segment may take is equally probable. The number of such conformations depends on the end-to-end distance, r, of the chain, reaching a rather sharp maximum at r_0. The retractive force of an elastomer is developed by the thermal motions of the chains, statistically driven toward their most probable end-to-end distance, r_0.

The changes in numbers of chain conformations can be expressed as an entropic effect. Thus, for an ideal elastomer, $(\partial U/\partial L)_{T,V} = 0$.

By contrast, most other materials develop internal energy-driven retractive forces. For example, on extension the iron atoms in a steel bar are forced farther apart than normal, calling into play energy well effects and concomitant increased atomic attractive forces. Such a model assumes the opposite effect, $(\partial S/\partial L)_{T,V} = 0$.

As derived by Wall (46), there is a perfect differential mathematical relationship between the entropy and the retractive force:

$$-\left(\frac{\partial S}{\partial L}\right)_{T,V} = \left(\frac{\partial f}{\partial T}\right)_{L,V} \tag{9.7}$$

Equation (9.6) can then be expressed as

$$f = \left(\frac{\partial U}{\partial L}\right)_{T,V} + T\left(\frac{\partial f}{\partial T}\right)_{L,V} \tag{9.8}$$

which is sometimes called the thermodynamic equation of state for rubber elasticity.

Equation (9.8) can be analyzed with the aid of a construction due to Flory (18b) (see Figure 9.6), which illustrates a general experimental curve of force versus temperature. This line is extended back to 0 K. For an ideal elastomer, the quantity $(\partial U/\partial L)_{T,V}$ is zero, and the entropic portion (tangent) goes through the origin. Of course, the experimental line is straight in the ideal case, the slope being proportional to $-(\partial S/\partial L)_{T,V}$ or $(\partial f/\partial T)_{L,V}$.

The first term on the right of equation (9.8) expresses the energetic portion of the retractive force, f_e, and the second term on the right expresses the entropic portion of the force, f_s. Thus

$$f = f_e + f_s \tag{9.9}$$

Equations (9.8) and (9.9) call for stress–temperature (isometric) experiments (47).[†] While a detailed analysis of such experiments will be presented below in Section 9.10, Figure 9.7 (47) shows the results of such an isometric study. The quantity f_s accounts for

---

[†] Of course, the term "stress" refers to the force per unit initial cross section (see Section 8.1). Much of the early literature talks about force but measures stress. However, f refers to force.

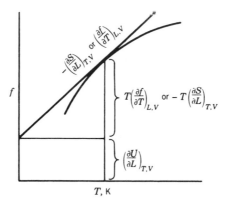

Figure 9.6 An analysis of the thermodynamic equations of state for rubber elasticity (18b).

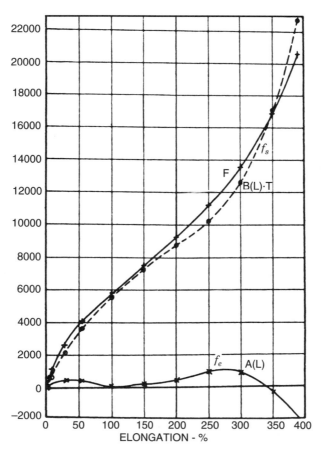

Figure 9.7 The total retractive force and its entropic, f_s, and energetic components, f_e, as a function of elongation. Natural rubber vulcanized with 8% sulfur; values at 20°C (47).

more than 90% of the stress, whereas f_e hovers near zero. The turndown of f_e above 300% elongation may be due to incipient crystallization.

Incidentally, equation (9.4) fits Figure 9.7 much better than it does Figure 9.5. There are two reasons: (a) Figure 9.7 represents a much closer approach to an equilibrium stress–strain curve, and (b) the much higher level of sulfur used in the vulcanization (8% vs. 2%) reduces the quantity of crystallization at high elongations.

9.6 EQUATION OF STATE FOR GASES

The equation of state of rubber elasticity will now be calculated via statistical thermodynamics, rather than the classical thermodynamics of Section 9.5. Statistical thermodynamics makes use of the probability of finding an atom, segment, or molecule in any one place as a means of computing the entropy. Thus tremendous insight is obtained into the molecular processes of entropic phenomena, although classical thermodynamics illustrates energetic phenomena adequately.

However, most students not broadly exposed to statistical thermodynamics find such calculations difficult to follow at first. For this reason we first derive the ideal gas law via the very same principles that are employed in calculating the stress–elongation relationships in rubber elasticity.

Consider a gas of ν molecules in an original volume of V_0 (18c). Let us calculate the probability Ω that all the gas molecules will move spontaneously to a smaller volume, V (see Figure 9.8).

The probability of finding one molecule in the volume V is given by

$$p_1 = \frac{V}{V_0} \tag{9.10}$$

Neglecting the volume actually occupied by the molecules (an ideal gas assumption), the probability of finding two molecules in the volume V is given by

$$p_2 = \left(\frac{V}{V_0}\right)^2 \tag{9.11}$$

and the probability of finding all the molecules (spontaneously) in the volume V is

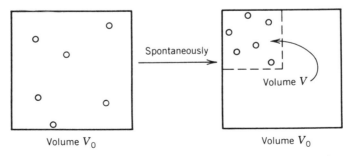

Figure 9.8 Ideal gas molecules spontaneously moving from a volume V_0 to a volume V.

given by

$$\Omega = \left(\frac{V}{V_0}\right)^v \tag{9.12}$$

The change in entropy, ΔS, is given by the Boltzmann relation[†]

$$\Delta S = +k \ln \Omega \tag{9.13}$$

which yields

$$\Delta S = kv \ln\left(\frac{V}{V_0}\right) \tag{9.14}$$

The pressure of the gas is given by

$$P = -\left(\frac{\partial F}{\partial V}\right)_T = -\left(\frac{\partial U}{\partial V}\right)_T + T\left(\frac{\partial S}{\partial V}\right)_T \tag{9.15}$$

where F is the Helmholtz free energy. For a perfect gas, $(\partial U/\partial V)_T = 0$, and

$$P = T\left(\frac{\partial S}{\partial V}\right)_T = \frac{kvT}{V} \tag{9.16}$$

If moles instead of molecules are considered, k becomes R and v becomes n, yielding the familiar

$$PV = nRT \tag{9.17}$$

If the internal energy is not required to be zero, more complex equations arise. For example, van der Waal's law per mole gives

$$P = -\frac{a}{V^2} + \frac{nRT}{V - b} \tag{9.18}$$

where the first term on the right indicates an energetic (attractive force) term and the second term on the right is the entropic term corrected for the molar volume of the gas (47).

Returning to Figure 9.8, if any large number of molecules are involved (e.g., 1 mol), the probability of the gas molecules spontaneously moving to a much smaller volume is very small. Of course, if they are arbitrarily so moved, a pressure P is required to keep them there.

In an analogous strip of rubber, the corresponding situation would be the spontaneous elongation of the strip (a rubber band stretching itself). On a molecular level, such motions are spontaneous, but because of the low probabilities involved, are but momentary. Again, the phenomenon is possible but unlikely. Instead of a pressure P to

---

[†] The Boltzmann hypothesis *assumes* rather than derives a logarithmic relationship between the probability of a state (or the number of such states, the inverse) and the entropy. Boltzmann's constant, k, is defined as the constant of proportionality.

Table 9.1 Corresponding concepts in ideal gases and ideal elastomers

$PV = nRT$	$G = nRT$
Entropy calculated from probabilities of finding n molecules in a given volume	Entropy calculated from probability of finding end-to-end distance r at r_0
Probability of the gas volume spontaneously decreasing $(\partial U/\partial V)_T = 0$; internal energy assumed zero	Probability of an elastomer strip spontaneously elongating $(\partial U/\partial L)_{T,V} = 0$; internal energy assumed constant
Molar volume of gas assumed zero	Elastomer assumed incompressible (molar volume is constant)
Pressure P given by $-(\partial F/\partial V)_T$	Retractive force f given by $-(\partial F/\partial L)_{T,V}$

hold the gas in the volume V, a stress σ (force per unit area) will be required to keep the elastomer stretched from L_0 to L ($\alpha = L/L_0$).

In both problems, to the first approximation, the internal energy component can be assumed to be zero. Table 9.1 compares the concepts of an ideal gas with those of an ideal elastomer, to be developed below.

9.7 STATISTICAL THERMODYNAMICS OF RUBBER ELASTICITY

It is useful to consider again the freely jointed chain (Section 5.3.1). In this case the root-mean-square end-to-end distance is given by $(\overline{r_f^2})^{1/2} = ln^{1/2}$. In a real random coil, with fixed bond angles, the quantity $(\overline{r_f^2})^{1/2}$ is larger but still obeys the $n^{1/2}$ relationship. For a given value of n, however, the root-mean-square end-to-end distance can vary widely, from zero, where the ends touch, to nl, the length of the equivalent rod. The probability of finding particular values of r underlies the following subsections.

9.7.1 The Equation of State for a Single Chain

It is convenient to divide the derivation of equations such as (9.4) into two parts. First, the equation of state for a single chain in space is derived. Then we show how a network of such chains behaves.

It is convenient to start again with the general equation for the Helmholtz free energy, equation (9.5):

$$F = U - TS$$

This can be rewritten in statistical thermodynamic notation:

$$F = \text{constant} - kT \ln \Omega(r, T) \tag{9.19}$$

where the quantity $\Omega(r, T)$ [see equations (9.12) and (9.13)] now refers to the probability that a polymer molecule with end-to-end distance r at temperature T will adopt a given conformation.[†]

[†] The term "conformation" refers to those arrangements of a molecule that can be attained by rotating about single bonds. Configurations refer to tacticity, steric arrangement, *cis* and *trans*, and so on; see Chapter 2.

From a quantitative point of view, at each particular end-to-end distance, all possible conformations of the chain need to be counted, holding the ends fixed in space. Then the sum of all such conformations as the end-to-end distance is varied needs to be calculated. (Later the sum of all conformations of a distribution of molecular weights is considered.)

The retractive force is given by

$$f = \left(\frac{\partial F}{\partial r}\right)_{T,V} = -kT\left(\frac{\partial \ln \Omega(r,T)}{\partial r}\right)_{T,V} \tag{9.20}$$

As before, the quantity U, assumed to be constant (or zero), drops out of the calculation, leaving only the entropic contribution. In this case, for a single chain, the quantity f for force must be used. The cross section of the individual chain, necessary for a determination of the stress, remains undefined.

A particular direction in space is selected first. Then, using vector notation, the probability that r lies between r and $r + dr$ in that direction is given by

$$W(r)\,dr = \frac{\Omega(r,T)\,dr}{\int_0^\infty \Omega(r,T)\,dr} \tag{9.21}$$

where the denominator serves as a normalizing factor. Of course, the integral does not need to extend to infinity; in reality different conformations only go to the fully extended, rodlike chain.

Removing the directional restrictions on r,

$$W(r)\,dr = \frac{\Omega(r,T)\,dr}{\int_0^\infty \Omega(r,T)\,dr}\,4\pi r^2 \tag{9.22}$$

A spherical shell between r and $r + dr$ is generated (see Figure 9.9), depicted in Cartesian coordinates.

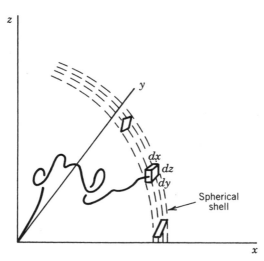

Figure 9.9 A spherical shell at a distance r (inner surface) and $r + dr$ (outer surface), defining all conformations in space having that range of r.

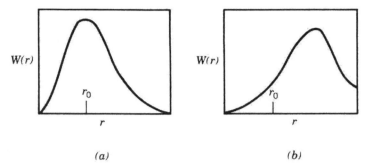

Figure 9.10 The radial distribution function $W(r)$ as a function of r for (a) a relaxed chain, where the most probable end-to-end distance is r_0, and (b) for a chain under an extensive force, f.

Rearranging equation (9.22) and taking logarithms,

$$\ln \Omega(r, T) = \ln W(r) + \ln \int_0^\infty \Omega(r, T)\, dr - \ln 4\pi - \ln r^2 \tag{9.23}$$

Differentiating with respect to r,

$$\left(\frac{\partial \ln \Omega(r, T)}{\partial r} \right)_{T, V} = \left(\frac{\partial \ln W(r)}{\partial r} \right)_{T, V} - \frac{2}{r} \tag{9.24}$$

since $\int_0^\infty \Omega(r, T)\, dr$ is independent of r.

The quantity $W(r)$ can be expressed as a Gaussian distribution (18):

$$W(r) = \left(\frac{\beta}{\pi^{1/2}} \right)^3 e^{-\beta^2 r^2} 4\pi r^2 \tag{9.25}$$

In molecular terms, $\beta^2 = 3/(2\overline{r_0^2})$, where $\overline{r_0^2}$ represents the average of the squares of the relaxed end-to-end distances.

The quantity $W(r)$ is shown as a function of r in Figure 9.10. The radial distribution function $W(r)$ is shown for a relaxed chain and a chain extended by a force f.

Equation (9.24) can now be expressed as

$$\left(\frac{\partial \ln \Omega(r, T)}{\partial r} \right)_{T, V} = -2\beta^2 r \tag{9.26}$$

Substituting equation (9.26) into equation (9.20),

$$f = \frac{3kTr}{\overline{r_0^2}} \tag{9.27}$$

and the equation of state for a single chain is obtained. The force appears to be zero at $r = 0$, because of the spherical shell geometry assumed (Figure 9.9). Again, the quantity

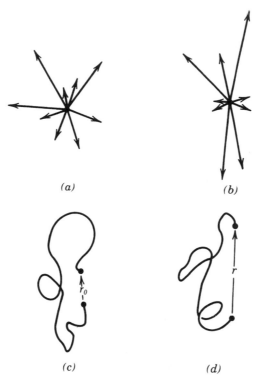

Figure 9.11 Illustration of an affine deformation, end-to-end vectors drawn from a central point. (*a*) End-to-end vectors have spherical symmetry. (*b*) After extending the macroscopic sample, end-to-end vectors have ellipsoidal symmetry. Parts (*c*) and (*d*) illustrate the corresponding effects on a single chain.

r_0 is the isotropic end-to-end distance for a free chain in space. This approximates the end-to-end distance expected both in θ-solvents and in the bulk state, for linear chains.

9.7.2 The Equation of State for a Macroscopic Network

Equation (9.27) expresses the retractive force of a single chain on extension. Since the retractive force is proportional to the quantity r, the chain behaves as a Hookean spring on extension. The problem now is to link a large number, n, of these chains together per unit volume to form a macroscopic network.

The assumption of affine deformation is required; that is, the junction points between chains move on deformation as if they were embedded in an elastic continuum (Figure 9.11). The mean square length of a chain in the strained state is given by

$$\overline{r^2} = \tfrac{1}{3}(\alpha_x^2 + \alpha_y^2 + \alpha_z^2)\overline{r_i^2} \tag{9.28}$$

The alphas represent the fractional change in shape in the three directions, equal to $(L/L_0)_i$, where $i = x, y,$ or z.

The quantity $\overline{r_i^2}$ represents the isotropic, unstrained end-to-end distance in the network. The two quantities $\overline{r_i^2}$ and $\overline{r_0^2}$ [see equation (9.27)] bear exact comparison. They represent the same chain in the network and un-cross-linked states, respectively. Under many circumstances the quantity r_i^2/r_0^2 approximately equals unity. In fact the simpler derivations of the equation of state for rubber elasticity do not treat this quantity, implicitly assuming it to be unity (42). However, deviations from unity may be caused by swelling, cross-linking while in tension, changes in temperature, and so on, and play an important role in the development of modern theory.

The difference between r_i and r_0 may be illustrated by way of an example. Assume a cube of dimensions $1 \times 1 \times 1$ cm. Its volume, of course, is 1 cm³. If it is swelled to 10 cm³ volume, then each linear dimension in the sample is increased by $(10)^{1/3}$, the new length of the sides, in this instance. Assuming an affine deformation, the end-to-end distance of the chains will also increase by $(10)^{1/3}$; that is, $r_i = (10)^{1/3} r_0$. The value of r_0 does not change, because it is the end-to-end distance of the equivalent free chain. The value of r_i is determined by the distances between the cross-link sites binding the chain. Again, neither r_i nor r_0 represents the end-to-end distance of the whole primary chain but rather the end-to-end distance between cross-link junctions.

For the work done by the n network chains,

$$-W = \Delta F_{el} \tag{9.29}$$

where ΔF_{el} represents the change in the Helmholtz free energy due to elastic deformation. Returning to equation (9.27), for n chains,

$$\Delta F_{el} = \frac{3nRT}{r_0^2} \int_{(r_i^2)^{1/2}}^{(\overline{r^2})^{1/2}} r\, dr \tag{9.30}$$

Integrating and substituting equation (9.28) yields

$$\Delta F_{el} = \frac{nRT}{2} \frac{\overline{r_i^2}}{r_0^2} (\alpha_x^2 + \alpha_y^2 + \alpha_z^2 - \alpha_{x_0}^2 - \alpha_{y_0}^2 - \alpha_{z_0}^2) \tag{9.31}$$

By definition, $\alpha_{x_0}^2 = \alpha_{y_0}^2 = \alpha_{z_0}^2 = 1$, since these terms deal with the unstrained state.

Since it is assumed that there is no volume change on deformation (Poisson's ratio very nearly equals 0.5—i.e., an incompressible solid),

$$\alpha_x \alpha_y \alpha_z = 1 \tag{9.32}$$

If α_x is taken simply as α, then $\alpha_z = \alpha_y = 1/\alpha^{1/2}$, at constant volume.

After making the above substitutions, the work on elongation equation may be written, per unit volume,

$$-W = \Delta F_{el} = \frac{nRT}{2} \frac{\overline{r_i^2}}{r_0^2} \left(\alpha^2 + \frac{2}{\alpha} - 3 \right) \tag{9.33}$$

The stress is given by

$$\sigma = \left(\frac{\partial F}{\partial \alpha} \right)_{T,V} = nRT \frac{\overline{r_i^2}}{r_0^2} \left(\alpha - \frac{1}{\alpha^2} \right) \tag{9.34}$$

which is the equation of state for rubber elasticity. Equation (9.34) bears comparison with equation (9.4). The quantity $\overline{r_i^2}/\overline{r_0^2}$ is known as the "front factor."

The quantity n in the above represents the number of active network chains per unit volume, sometimes called the network or cross-link density. The number of cross-links per unit volume is also of interest. For a tetrafunctional cross-link (see Figure 9.3), the number of cross-links is one-half the number of chains. (see Section 9.9.2) Equation (9.34) holds for both extension and compression.

Several other relationships may be derived immediately. Young's modulus can be written

$$E = L\left(\frac{\partial \sigma}{\partial L}\right)_{T,V} \tag{9.35}$$

which yields

$$E = nRT\frac{\overline{r_i^2}}{\overline{r_0^2}}\left(2\alpha^2 + \frac{1}{\alpha}\right) \cong 3n\frac{\overline{r_i^2}}{\overline{r_0^2}}RT \tag{9.36}$$

for small strains. This is the engineering modulus, which utilizes the actual cross section at α rather than the relaxed value (i.e., $\alpha = 1$). At low extensions, the two moduli are nearly identical. The shear modulus may be written

$$G = E/2(1+v) \tag{9.37}$$

Poisson's ratio, v, for rubber is approximately 0.5 (incompressibility assumption), so

$$G = n\frac{\overline{r_i^2}}{\overline{r_0^2}}RT \tag{9.38}$$

Then, to a good approximation

$$\sigma = G\left(\alpha - \frac{1}{\alpha^2}\right) \tag{9.39}$$

thus defining the work to stretch, the stress–strain relationships, and the modulus of an ideal elastomer. As equations (9.34) and (9.39) illustrate, the stress–strain relationships are non-Hookean; that is, the strain is not proportional to the stress. Again, these equations yield curves of the type illustrated in Figures 9.5 and 9.7.

Another relationship treats biaxial extension (49). If equibiaxial extension is assumed, such as in a spherical rubber balloon, then

$$\sigma = nRT(\alpha^2 - \alpha^{-4}) \tag{9.40}$$

assuming $\overline{r_i^2}/\overline{r_0^2} \cong 1$, and the volume changes of the elastomer on biaxial extension are nil.

Other moduli are occasionally used to characterize polymeric materials. Appendix 9.1 describes the use of the ball indentation method to characterize the cross-link density of gelatin. In general, this method can be used for sheet rubber and other large sample types.

9.7.3   Some Example Calculations

9.7.3.1   An Example of Rubber Elasticity Calculations   Suppose that an elastomer of 0.1 cm × 0.1 cm × 10 cm is stretched to 25 cm length at 35°C, a stress of 2×10^7 dynes/cm^2 being required. What is the concentration of active network chain segments?

Use equation (9.34), assuming that $\overline{r_i^2} = \overline{r_0^2}$.

$$n = \frac{\sigma}{RT(\alpha - 1/\alpha^2)}$$

$$= \frac{2 \times 10^7 \text{ dynes/cm}^2}{8.31 \times 10^7 (\text{dynes} \cdot \text{cm/mol} \cdot \text{K}) \times 308 \text{ K} \times (2.5 - 1/2.5^2)}$$

noting that $\alpha = L/L_0 = 25/10 = 2.5$.

$$n = 3.34 \times 10^{-4} \text{ mol/cm}^3$$

Note that the rubber band in Appendix 9.2 has $n = 1.9 \times 10^{-4}$ mol/cm^3.

9.7.3.2   An Example of Work Done during Stretching   The elastomer strip in Section 9.7.3.1 was stretched to 45 cm length at 25°C. How much work was required?

Equation (9.33) provides the basis. The quantity $\overline{r_i^2}/\overline{r_0^2}$ is taken as substantially unity. The quantity $n = 3.34 \times 10^{-4}$ mol/cm^3.

$$-W = 3.34 \times 10^{-4} \text{ (mol/cm}^3) \times 8.31 \times 10^7 (\text{dyn} \cdot \text{cm/mol} \cdot \text{K}) \times 298 \text{ K} (4.5^2 + 2/4.5 - 3).$$
$$-W = 1.46 \times 10^8 \text{ erg/cm}^3, \text{ noting that dyn} \cdot \text{cm} \equiv \text{erg}.$$

But noting that there are 10 cm^3 of elastomer,

$$-W = 1.46 \times 10^9 \text{ erg or } 1.46 \times 10^2 \text{ J}.$$

This calculation ignores the mass of the elastomer itself.

Note that work, as given in equation (9.33), must have units that match the units of the gas constant R.

9.7.4   Comparison to Metal Springs

As described above, the elasticity in elastomers arises through the entropic straightening and recoiling of the polymer chains. This is substantially an isovolume phenomenon, Poisson's ratio being close to 0.5.

By contrast, the elasticity in metals arises through increases in the distances between atoms, removing the atoms from their equilibrium positions in the energy well. This is an enthalpic effect. The volume of such materials increases substantially on stretching, Poisson's ratio being about 0.35 for many metals.

Thus elasticity in solid materials can arise through two general mechanisms: reduction in entropy for elastomers and increases in internal energy for most other materials. As described in Section 9.2, a strip of rubber warms on stretching; a metal spring cools on stretching.

9.8 THE "CARNOT CYCLE" FOR ELASTOMERS

In elementary thermodynamics, the Carnot cycle illustrates the production of useful work by a gas in a heat engine. This section outlines the corresponding thermodynamic concepts for an elastomer and illustrates a demonstration experiment.

The conservation of energy for a system may be written

$$dU = V\,dp + T\,dS + \sigma\,dL + \cdots \tag{9.41}$$

where the internal energy, U, is equated to as many variables as exist in the system. For an ideal gas (Section 9.6), $P-V-T$ variables are selected. The corresponding variables for an ideal elastomer are $\sigma-L-T$ [see equation (9.34)]. Since Poisson's ratio is nearly 0.5 for elastomers, the volume is substantially constant on elongation.

By carrying a gas, elastomer, or any material through the appropriate closed loop with a high- and low-temperature portion, they may be made to perform work proportional to the area enclosed by the loop. A system undergoing such a cycle is called a heat engine.

9.8.1 The Carnot Cycle for a Gas

In the Carnot heat engine, a gas is subjected to two isothermal steps, which alternate with two adiabatic steps, all of which are reversible (see Figure 9.12) (48). Briefly, the gas undergoes a reversible adiabatic compression from state 1 to state 2. The temperature is increased from T_1 to T_2. During this step the surroundings do work $|\omega_{12}|$ on the gas. The absolute signs are used because conventions require that the signs on some of the algebraic quantities herein be negative.

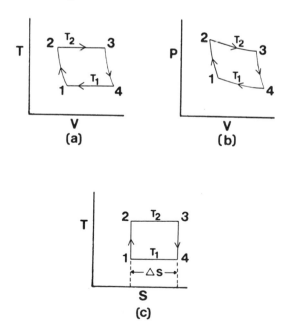

Figure 9.12 Carnot cycle for a gas (48).

Next the gas undergoes a reversible isothermal expansion from state 2 to state 3. While expanding, the gas does work $|\omega_{23}|$ on the surroundings while absorbing heat $|q_2|$. Then there follows a reversible adiabatic expansion of the gas from state 3 to state 4, the temperature dropping from T_2 to T_1. During this step, the gas does work $|\omega_{34}|$ on the surroundings.

Last, there is an isothermal compression of the gas from state 4 to state 1 at T_1. Work $|\omega_{41}|$ is performed on the gas, and heat $|q_1|$ flows from the gas to the surroundings.

9.8.2 The Carnot Cycle for an Elastomer

For an elastomer, the rubber goes through a series of stress–length steps, two adiabatically and two isothermally, as in the Carnot cycle (see Figure 9.13) (50). Beginning at length L_1 and temperature T^I, a stress, σ, is applied stretching the elastomer adiabatically to L_2. The elastomer heats up to T^{II}. The quantity σ is related to the length by the nonlinear equation

$$\sigma = nRT\left[\frac{L}{L_0} - \left(\frac{L_0}{L}\right)^2\right] \tag{9.42}$$

[see equation (9.34)]. In this step work is done on the elastomer.

At T^{II}, the elastomer is allowed to contract isothermally to L_3. It absorbs heat from its surroundings in this step and does work. As the length decreases, its entropy increases by ΔS (see Figure 9.13c). The elastomer then is allowed to contract adiabatically to L_4, doing work, and its temperature falls to T^I again. The length of the sample is then

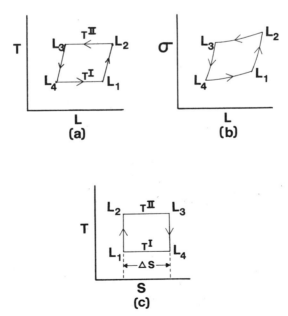

Figure 9.13   Thermal cycle for an elastomer (50).

increased isothermally from L_1 to L_1, work being done on the sample, and heat is given off to its surroundings. This step completes the cycle.

An increase in the volume of the gas, however, corresponds to a decrease in the length of a stretched elastomer. It is important to note that at no time does the elastomer come to its rest length, L_0. Interestingly the corresponding "rest volume" of a gas is infinitely large.

9.8.3 Work and Efficiency

The equations governing the work done during the two cycles may also be compared. For a gas,

$$\omega_g = -\oint P \, dV \tag{9.43}$$

For an elastomer,

$$\omega_e = -\oint \sigma \, dL \tag{9.44}$$

In both cases the cyclic integral measures the area enclosed by the four steps in Figures 9.12 and 9.13.

The efficiencies, $\bar{\eta}$, of the two systems may also be compared. For a gas,

$$\bar{\eta}_g = \frac{q_1 + q_2}{q_2} \tag{9.45}$$

where q_1 and q_2 are the heat absorbed and released (opposite signs), as above. For the elastomer,

$$\bar{\eta}_e = \frac{\oint \sigma \, dL}{Q_{\mathrm{II}}} = \frac{(T^{\mathrm{II}} - T^{\mathrm{I}})\Delta S}{Q_{\mathrm{II}}} = \frac{Q_{\mathrm{I}} + Q_{\mathrm{II}}}{Q_{\mathrm{II}}} \tag{9.46}$$

or in a different form,

$$\bar{\eta}_e = \frac{T^{\mathrm{II}} - T^{\mathrm{I}}}{T^{\mathrm{II}}} \tag{9.47}$$

where Q_{I} and Q_{II} are the amounts of heat released to the low-temperature reservoir (T^{I}) and absorbed from the high-temperature reservoir (T^{II}), respectively.

While the entropy change is zero for either system during the reversible adiabatic steps (see Figures 9.12c and 9.13c), it must be emphasized that the entropy change is greater than zero for an irreversible adiabatic process. An example for an elastomer is "letting go" of a stretched rubber band.

9.8.4 An Elastomer Thermal Cycle Demonstration

The elastomer thermal cycle is demonstrated in Figure 9.14 (50). A bicycle wheel is mounted on a stand, with a source of heat on one side only. Stretched rubber bands

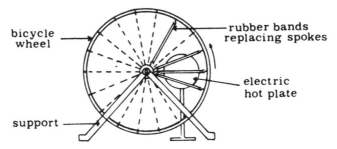

Figure 9.14   A thermally rotated wheel, employing an elastomer as the working substance (50).

replace the spokes. On heating, the stress that the stretched rubber bands exert is increased so that the center of gravity of the wheel is displaced toward 9 o'clock in the drawing. The wheel then rotates counterclockwise (51).

Each of the steps in Figure 9.13 may be traced in Figure 9.14, although none of the steps in Figure 9.13 are purely isothermal or adiabatic, and then of course they are not strictly reversible. Steps 1 to 2 in Figure 9.13 occur at 6 o'clock in Figure 9.14, where there is a (near) adiabatic length increase due to gravity. At 3 o'clock, at T^{II}, heat is absorbed (nearly) isothermally, and the length decreases, doing work. At 12 o'clock, corresponding to steps 3 to 4, there is an adiabatic length decrease due to gravity. Last, at 9 o'clock, steps 4 to 1, there is a (nearly) isothermal length increase, and heat is given off to the surroundings at T^{I}, and work is done on the elastomer.

9.9   CONTINUUM THEORIES OF RUBBER ELASTICITY

9.9.1   The Mooney–Rivlin Equation

The statistical theory of rubber elasticity is based on the concepts of random chain motion and the restraining power of cross-links; that is, it is a molecular theory. Amazingly, similar equations can be derived strictly from phenomenological approaches, considering the elastomer as a continuum. The best known such equation is the Mooney–Rivlin equation (52–55),

$$\sigma = 2C_1\left(\alpha - \frac{1}{\alpha^2}\right) + 2C_2\left(1 - \frac{1}{\alpha^3}\right) \qquad (9.48)$$

which is sometimes written in the algebraically identical form,

$$\sigma = \left(2C_1 + \frac{2C_2}{\alpha}\right)\left(\alpha - \frac{1}{\alpha^2}\right) \qquad (9.49)$$

Equations (9.48) and (9.49) appear to be corrections to equation (9.34), with an additional term being added. However, they are derived from quite different principles.

According to equation (9.34), the quantity $\sigma/(\alpha - 1/\alpha^2)$ should be a constant. Equation (9.49), on the other hand, predicts that this quantity depends on α:

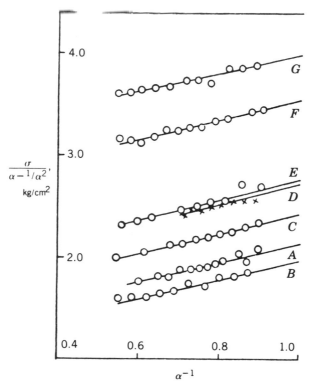

Figure 9.15 Plot of $\sigma/(\alpha - 1/\alpha^2)$ versus α^{-1} for a range of natural rubber vulcanizates. Sulfur content increases from 3 to 4%, with time of vulcanization and other quantities as variables (56).

$$\frac{\sigma}{(\alpha - 1/\alpha^2)} = 2C_1 + \frac{2C_2}{\alpha} \tag{9.50}$$

Plots of $\sigma/(\alpha - 1/\alpha^2)$ versus $1/\alpha$ are found to be linear, especially at low elongation (see Figure 9.15) (56). The intercept on the $\alpha^{-1} = 0$ axis yields $2C_1$, and the slope yields $2C_2$. The value of $2C_1$ varies from 2 to 6 kg/cm^2, but the value of $2C_2$, interestingly, remains constant near 2 kg/cm^2. Appendix 9.2 describes a demonstration experiment that illustrates both rubber elasticity [see equation (9.34)] and the nonideality expressed by equation (9.50).

On swelling, the value of $2C_2$ drops rapidly (see Figure 9.16) (56), reaching a value of zero near v_2 (volume fraction of polymer) equal to 0.2. This same dependence is observed for the same polymer in different solvents, different levels of cross-linking for the same polymer, or (as shown) different polymers entirely (56).

The interpretation of the constants $2C_1$ and $2C_2$ has absorbed much time; the results are inconclusive (42). It is tempting but generally considered incorrect to equate $2C_1$ and $nRT(\overline{r_i^2}/\overline{r_0^2})$. The original derivation of Mooney (46) shows that $2C_2$ has to be finite, but it does not indicate its value relative to $2C_1$. According to Flory (41), the ratio $2C_2/2C_1$ is related to the looseness with which the cross-links are embedded within the structure.

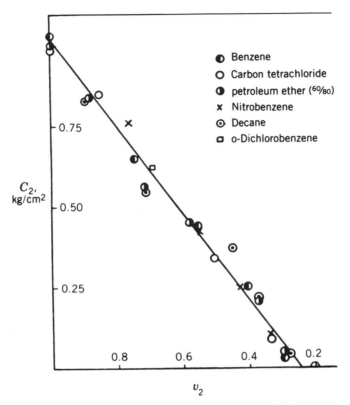

Figure 9.16 Dependence of C_2 on v_2 for synthetic rubber vulcanizates (56). Polymers: ○, butadiene–styrene, (95/5); ◑, butadiene–styrene, (90/10); ◐, butadiene–styrene, (85/15); ●, butadiene–styrene, (75/25); ⊙, butadiene–styrene, (70/30); X, butadiene–acrylonitrile, (75/25).

Trifunctional cross-links have larger values of $2C_2/2C_1$ than tetrafunctional cross-links, for example (57).

As indicated above, $2C_2$ decreases with the degree of swelling. Furthermore Gee (58) showed that during stress relaxation, swelling increased the rate of approach to equilibrium. Ciferri and Flory (59) showed that $2C_2$ is markedly reduced by swelling and deswelling the sample at each elongation. The samples, actually measured dry, had $2C_2$ values about half as large after the swelling–deswelling operation as those measured before. These results suggest that the magnitude of $2C_2$ is caused by nonequilibrium phenomena. Gumbrell et al. (56) stated it in terms of the reduced numbers of conformations available in the dry state versus the swollen state.

Other possible explanations include non-Gaussian chain or network statistics (see Section 9.10.6) and internal energy effects (42). The latter, bearing on the front factor, will be treated in Section 9.10.

9.9.2 Generalized Strain–Energy Functions

Following the work of Mooney, more generalized theories of the stress–strain relationships in elastomers were sought. The central problem was how to calculate the work, W, stored in the body as strain energy.

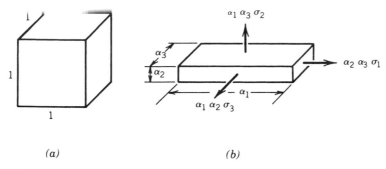

Figure 9.17 An elastomeric cube. (a) Undeformed and (b) deformed states, showing principal stresses and strains.

Rivlin (55) considered the most general form that such strain–energy functions could assume. As basic assumptions, he took the elastomer to be incompressible and isotropic in the unstrained state. Symmetry conditions required that the three principal extension ratios, α_1, α_2, and α_3, depend only on even powers of the α's. In three dimensions (see Figure 9.17), the simplest functions that satisfy these requirements are

$$I_1 = \alpha_1^2 + \alpha_2^2 + \alpha_3^2 \tag{9.51}$$

$$I_2 = \alpha_1^2\alpha_2^2 + \alpha_2^2\alpha_3^2 + \alpha_3^2\alpha_1^2 \tag{9.52}$$

$$I_3 = \alpha_1^2\alpha_2^2\alpha_3^2 \tag{9.53}$$

where I_1, I_2, and I_3 are termed strain invariants.

The third strain invariant is equal to the square of the volume change,

$$I_3 = \left(\frac{V}{V_0}\right)^2 = 1 \tag{9.54}$$

which under the assumption of incompressibility equals unity. Alternate formulations have been proposed by Valanis and Landel (60) and by Ogden (61), which have been reviewed by Treloar (42).

Consider the deformation of a cube (Figure 9.17). The work that is stored in the body as strain energy can be written (62).

$$W(\alpha) = \int \sigma_1 \, d\alpha_1 + \int \sigma_2 \, d\alpha_2 + \int \sigma_3 \, d\alpha_3 \tag{9.55}$$

where the σ's are the stresses.

The work, in a more general form, can be expressed as a power series (55):

$$W = \sum_{i,j,k=0}^{\infty} C_{ijk}(I_1 - 3)^i (I_2 - 3)^j (I_3 - 1)^k \tag{9.56}$$

Equation (9.56) is written so that the strain energy term in question vanishes at zero strain.

For the lowest member of the series, $i = 1$, $j = 0$, and $k = 0$,

$$W = C_{100}(I_1 - 3) \tag{9.57}$$

which is functionally identical to the free energy of deformation expressed in equation (9.31). For the case of uniaxial extension,

$$\alpha_1 = \alpha \tag{9.58}$$

and noting equation (9.32) and equation (9.54),

$$\alpha_2 = \alpha_3 = \left(\frac{1}{\alpha}\right)^{1/2}$$

Equation (9.57) can now be written

$$W = C_{100}\left(\alpha^2 + \frac{2}{\alpha}\right) \tag{9.59}$$

and the stress can be written [see equation (9.54)]

$$\sigma = \frac{\partial W}{\partial \alpha} = 2C_{100}\left(\alpha - \frac{1}{\alpha^2}\right) \tag{9.60}$$

This equation is readily identified with equation (9.34), suggesting (for this case only) that

$$2C_{100} = nRT\frac{\overline{r_i^2}}{\overline{r_0^2}} \tag{9.61}$$

On retention of an additional term in equation (9.55), with $i = 0$, $j = 1$, and $k = 0$,

$$W = C_{100}(I_1 - 3) + C_{010}(I_2 - 3) \tag{9.62}$$

which leads directly to the Mooney–Rivlin equation, equation (9.49).

Interestingly, if we retain one more term, C_{200}, an equation of the form (62)

$$\sigma = \left(C + \frac{C'}{\alpha} + C''\alpha^2\right)\left(\alpha - \frac{1}{\alpha^2}\right) \tag{9.63}$$

can be written, where

$$C = 2(C_{100} - 6C_{200}) \tag{9.64}$$
$$C' = 2(4C_{200} + C_{010}) \tag{9.65}$$

and

$$C'' = 4C_{200} \tag{9.66}$$

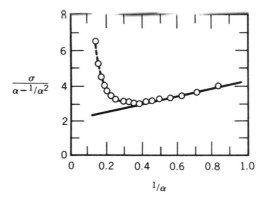

Figure 9.18 Mooney–Rivlin plot for sulfur-vulcanized natural rubber. Solid line, equation (9.48); dotted line, equation (9.62).

Equation (9.63), with two additional terms over the statistical theory of rubber elasticity, fits the data quite well (see Figure 9.18) (63).

Because no particular molecular model was assumed, theoretical values cannot be assigned to C, C', and C'', nor can any molecular mechanisms be assigned. These phenomenological equations of state, however, accurately express the form of the experimental stress–strain data.

9.10 SOME REFINEMENTS TO RUBBER ELASTICITY

The statistical theory of rubber elasticity has undergone significant and continuous refinement, resulting in a series of correction terms. These are sometimes omitted and sometimes included in scientific and engineering research, as the need for them arises. In this section we briefly consider some of these.

9.10.1 The Inverse Langevin Function

The Gaussian statistics leading to equation (9.34) are valid only for relatively small strains—that is, under conditions where the contour length of the chain is much more than its end-to-end distance. In the region of high strains, where the ratio of the two parameters approaches $\frac{1}{3}$ to $\frac{1}{2}$, this limit is exceeded.

Kuhn and Grün (64) derived a distribution function based on the inverse Langevin function. The Langevin function itself can be written

$$L(x) = \coth x - \frac{1}{x} \tag{9.67}$$

and was first applied to magnetic problems. In this case

$$L(\beta') = \frac{r}{n'l} \tag{9.68}$$

where n' is the number of links of length l. (It must be pointed out that the quantity n' in this case need not be identical with the number of mers in the chains.) Thus the quantity $n'l$ represents a measure of the contour length of the chain, and $r/n'l$ is the fractional chain extension. Of course, for the inverse Langevin function of interest here,

$$L^{-1}\left(\frac{r}{n'l}\right) = \beta' \tag{9.69}$$

The stress of an elastomer obeying inverse Langevin statistics can be written (42,62)

$$\sigma = nRT(n')^{1/2}\left[L^{-1}\left(\frac{\alpha}{(n')^{1/2}}\right) - \alpha^{-3/2}L^{-1}\left(\frac{1}{\alpha^{1/2}(n')^{1/2}}\right)\right] \tag{9.70}$$

At intermediate values of α (and hence of $r/n'l$), equation (9.70) predicts a sharp upturn in the stress at α's greater than 4, as observed in experiments. Because of the complexity of equation (9.70), the Gaussian-based (9.34) is preferred where possible.

9.10.2 Cross-link Functionality

In order to form a network, at least some of the mers need to have a functionality greater than 2; that is, more than two chain portions must emanate from those mers. In the structure depicted in Figure 9.3, the functionality of each cross-link is 4. When divinyl benzene or sulfur is used as a cross-linker, the functionality will indeed be 4.

Suppose, however, that glycerol is used as the cross-linker in the synthesis of a polyester. Then the functionality of the cross-link site will be 3. Use of trimethylol propane trimethacrylate or pentaerythritol tetramethacrylate results in functionalities of 6 and 8, respectively (65).

Duiser and Staverman (66) and Graessley (69) have shown that the front factor depends on the functionality of the network. Representing the network functionality as f^*, equation (9.34) can be written

$$\sigma = \left(\frac{f^* - 2}{f^*}\right)nRT\frac{\overline{r_i^2}}{\overline{r_0^2}}\left(\alpha - \frac{1}{\alpha^2}\right) \tag{9.71}$$

For tetrafunctional cross-links, defined as four chain segments emanating from each cross-link site [the same type as obtained with the use of divinyl benzene (see Figure 9.3)], $f^* = 4$, equation (9.71) predicts one-half the stress that equation (9.34) predicts.

Another way of writing the correction for cross-link functionality is (41,68)

$$\sigma = (n - \mu)RT\frac{\overline{r_i^2}}{\overline{r_0^2}}\left(\alpha - \frac{1}{\alpha^2}\right) \tag{9.72}$$

where n and μ are the number densities of elastically active strands and junctions. A junction is elastically active if at least three paths leading away from it are independently attached to the network. A strand, meaning a polymer chain segment, is elastically active if it is bound at each end by elastically active junctions (69). Equation (9.72) also predicts a front-factor correction of $\frac{1}{2}$ for a tetrafunctional network, since there are half as many cross-links as there are chain segments.

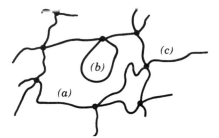

Figure 9.19 Network structure and defects: (*a*) elastically active chain, (*b*) loop, and (*c*) dangling chain end.

9.10.3 Network Defects

There are two major types of network defects: (a) the formation of inactive rings or loops, where the two ends of the chain segment are connected to the same cross-link junction, and (b) loose, dangling chain ends, attached to the network by only one end (70–73) (see Figure 9.19).

Both of these defects tend to decrease the retractive stress, because they are not part of the network. The equation in use to correct for dangling ends may be written

$$\sigma = nRT\left(1 - \frac{2M_c}{M}\right)\frac{\overline{r_i^2}}{\overline{r_0^2}}\left(\alpha - \frac{1}{\alpha^2}\right) \tag{9.73}$$

where M_c is the molecular weight between cross-links and M is the primary-chain molecular weight. Where $M \gg M_c$, the correction becomes negligible.

9.10.4 Volume Changes and Swelling

If a polymer network is swollen with a "solvent" (it does not dissolve), equation (9.32) may be rewritten

$$\alpha_x \alpha_y \alpha_z = \frac{1}{v_2} \tag{9.74}$$

where v_2 is the volume fraction of polymer in the swollen material. Of course, v_2 is less than unity, and $\alpha_x \alpha_y \alpha_z$ is larger than unity, as is commonly experienced.

The detailed effect has two parts:

1. Effect on the front factor, $\overline{r_i^2}/\overline{r_0^2}$. The quantity $\overline{r_i^2}$ increases with the volume V to the two-thirds power, since r_i itself is a linear quantity. Of course, $\overline{r_0^2}$ remains constant. Thus

$$\left(\frac{\overline{r_i^2}}{\overline{r_0^2}}\right)_s = \left(\frac{V}{V_0}\right)^{2/3}\left(\frac{\overline{r_i^2}}{\overline{r_0^2}}\right) = \frac{1}{v_2^{2/3}}\left(\frac{\overline{r_i^2}}{\overline{r_0^2}}\right) \tag{9.75}$$

where the subscript s refers to the swollen state and where V_0 is the volume of the unswollen polymer.

2. Effect on the number of network chain segments concentration, n. The quantity n decreases with volume:

$$\left(\frac{V_0}{V}\right)n = n_s \qquad (9.76)$$

where n_s is the chain segment concentration in the swollen state.

$$\left(\frac{V_0}{V}\right)n = v_2 n \qquad (9.77)$$

Incorporating the right-hand sides of equations (9.75) and (9.77) into equation (9.34) leads to an equation of the form (74,75)

$$\sigma = nRTv_2^{1/3}\frac{\overline{r_i^2}}{\overline{r_0^2}}\left(\alpha - \frac{1}{\alpha^2}\right) \qquad (9.78)$$

The stress, defined as force per unit actual cross section, is decreased by $v_2^{1/3}$, since the number of chains occupying a given volume has decreased.

When volume change caused by deformation alone is considered (usually less than 1%), the equation of state can be written (76–79)

$$\sigma = nRT\left(\frac{V_0}{V}\right)^{2/3}\frac{\overline{r_i^2}}{\overline{r_0^2}}\left(\alpha - \frac{1}{\alpha^2}\frac{V}{V_0}\right) \qquad (9.79)$$

9.10.5 Physical Cross-links

9.10.5.1 *Trapped Entanglements* So far the discussion has been restricted to ordinary covalent cross-links. There are, however, several types of physical cross-links that exist as permanent loops or entanglements existing in the network structure. (They may slide, however, yielding a mode of stress relaxation also.)

Three types of trapped entanglements are shown in Figure 9.20 (80–85). They each portray the same phenomenon, but with increasing rigor of definition.

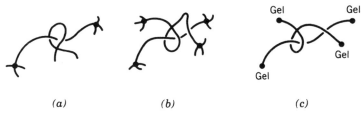

(a)   *(b)*   *(c)*

Figure 9.20 Three types of trapped entanglements: (a) The Bueche trap (80–85), (b) the Ferry trap (83), and (c) the Langley trap (76). The black circles are chemical cross-link sites. After Ferry (85).

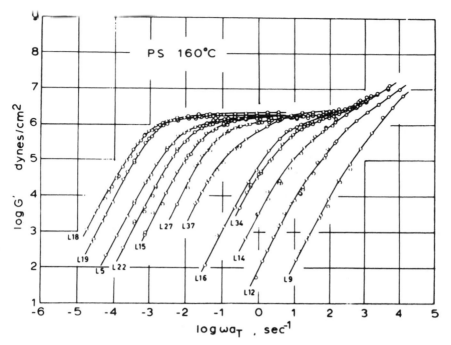

Figure 9.21 Shear storage modulus versus frequency for narrow molecular-weight polystyrenes at 160°C. Molecular weights range from $M_w = 8900$ g/mol (L9) to $M_w = 581,000$ g/mol (L18) (87).

Early works referred to the chemical and physical cross-links in a simple manner,

$$\sigma = (n_c + n_p)RT\frac{\overline{r_i^2}}{\overline{r_0^2}}\left(\alpha - \frac{1}{\alpha^2}\right) \tag{9.80}$$

where n_c and n_p are the concentration of chains bound by chemical and physical cross-links, respectively. It had been established early, for example, that the retractive stress was higher than expected by nearly a constant amount; indeed, for short relaxation times even linear polymers above T_g behaved as if they had some type of cross-linking (86–87) (see also Section 8.2).

Figure 9.21 illustrates the rubbery plateau (see Section 8.2) for a dynamic mechanical study of polystyrene as a function of frequency. The plateau shear modulus, near 3×10^6 dynes/cm^2, corresponds to a number of active network chains of near 1×10^{-4} mol/cm^3, nearly independent of the molecular weight of the polymer.

A more recent approach uses the concept of the potential entanglements that have been trapped by the cross-linking process. Langley (84) defines the quantity T_e as the fraction (or probability) that an entanglement is trapped in this manner.

Two theories, developed by Flory (88) and Scanlan (89), yield the calculation of chemical cross-links (57,58). For Flory's theory, the total number of effective cross-links is

$$n_{\text{tot}} = n_c W_g T_e^{1/2} + n_e T_e \tag{9.81}$$

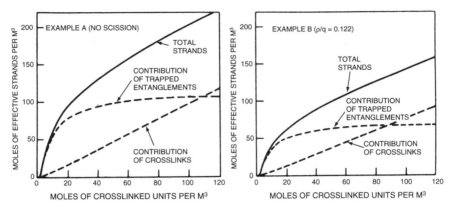

Figure 9.22 Contributions of chemical cross-links and trapped entanglements to the total cross-link level, for polystyrene, $M_n = 5 \times 10^5$ g/mol, $M_w = 1 \times 10^6$ g/mol, based on equation (9.81) (90).

where n_e is the concentration of potential entanglement strands. For $W_g = 1$, equation (9.81) reduces to $n_c + n_p$ [see equation (9.79)].

With the Scanlan criterion,

$$n_{\text{tot}} = \tfrac{1}{2} n_c T_e^{1/2} (3W_g - T_e^{1/2}) + n_e T_e \tag{9.82}$$

which also reduces to $n_c + n_p$ for $W_g = 1$. The value of these relationships, of course, is that for real networks, $W_g < 1$, and the way is open to evaluate n_c and n_e. Some calculations are shown in Figure 9.22. As they illustrate, the effective (permanent) physical cross-links start out at zero when the system is linear and increase rapidly to a plateau level (90).

9.10.5.2 The Phantom Network
It must be remarked that considerable controversy exists over the existence of physical cross-links (40,41,86,91). A theory has been proposed by Flory (40,41) using mathematics of a simplified network, called the "phantom network."

This model consists of a network of Gaussian chains connected in any arbitrary manner. The physical effect of the chains is assumed to be confined exclusively to the forces they exert on the junctions to which they are attached.

For a perfect phantom network of functionality f^*, the front factor contains the term $(f^* - 2)/f^*$, leading to equation (9.71), which considers chemical cross-links of arbitrary functionality, but no physical cross-links.

Flory (40) argues that the presence or absence of the term $(f^* - 2)/f^*$ may depend on the magnitude of the strain. At small deformations the displacement of junctions may conform more nearly to the older assumptions (i.e., affine in the macroscopic sense), and hence the term $(f^* - 2)/f^*$ might not appear for real chains that do have entanglements.

9.10.6 Small-Angle Neutron Scattering

In the preceding text the chains were assumed to deform in an affine manner when the networks were swelled or stretched. Until recently there was no way to approach this ııı/ılıₑııı ₑᵥ ₚₑₜₜₘₒₐₜₐₗₗⱼ. ᵂⁱⁱⁱₗₗ ⱼₕₑ ₐₜᵢᵥₑₙₜ ₒₜ small angle ₙₑᵤₜᵣₒₙ scattering (SANS), the conformation of the chains in the bulk state could be investigated (see Section 5.2.2.1).

On swelling, chain dimensions increase, usually isotropically. When SANS studies are done on the stretched elastomers, the scattering pattern yields greater dimensions in one direction than the other, because the chains are anisotropic (see Section 5.2) (93–105).

The question arose whether polymer chains in a network had the same conformation as in the melt before cross-linking. Beltzung et al. (92) prepared well-defined poly(dimethyl siloxane) (PDMS) chains containing Si — H linkages in the α and ω positions. Blends of PDMS(H) and PDMS(D) were prepared, where H and D, of course, represent the protonated and deuterated analogues. These blends were end-linked by tetrafunctional or hexafunctional cross-linkers under stoichiometric conditions.

Neutron scattering was carried out on both the PDMS melts and the corresponding networks. The principal results were that (a) the Gaussian character of the network chains in the undeformed state was confirmed, and (b) the chain dimensions were not changed by the cross-linking process.

Benoit et al. (93–97) prepared two types of tagged polystyrene networks: (a) type "A" networks containing labeled (deuterated) cross-link sites. This permitted a characterization of the spatial distribution of the cross-link points. (b) type "B" networks containing a few percent of perdeuterated polystyrene chains (see Figure 9.23) (93). Cross-linking utilized divinyl benzene (DVB).

Benoit et al. (93) studied these polystyrene networks as is, swollen in several solvents, and stretched. For the latter, stretching was done above T_g, followed by cooling in the stretched state. They found a maximum in the angular scattering curves of type "A" networks.

The mean pair separation distance between chain ends, h, was found to be proportional to $M_c^{0.5}$ both in the dry state and in the swollen state. On extension, $h_\parallel$ and $h_\perp$ values followed the expected affine deformation.

The "B" network, on the other hand (93), appeared to deviate significantly from the affine. As illustrated in Figure 9.24 (93), the chain radius of gyration increased on swelling far less than predicted by the affine deformation mechanism. The values of R_g were also less than predicted by the "end-to-end pulling mechanism" (96), which accentuates the extension of the end portions of the chains rather than the central section. One might imagine that entanglements prevent the motion of the central portions of long

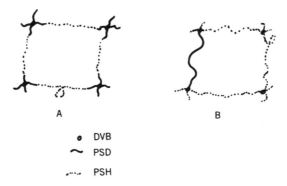

A   B

● DVB
∼ PSD
▪⋯▪ PSH

Figure 9.23 Schematic representation of labeled polystyrene networks. (*A*) Cross-linking points labeled. (*B*) Random labeled chains added (95).

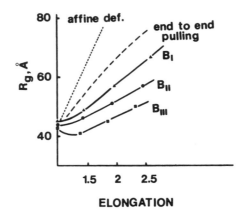

Figure 9.24 Variation of the radius of gyration of type B networks of different functionalities. Dotted line, theoretical behavior for affine deformation; dashed line, theoretical behavior for the end-to-end pulling mechanism (95), for polystyrene.

chains. Another possible explanation (see below) is that the chain's cross-link junction points rearrange to yield the system with the lowest free energy. This rearrangement minimizes the actual extension of the chain.

More recent SANS experiments on stretched networks were performed by Hinkley et al. (98) and by Clough et al. (99,100). Hinkley et al. (98) prepared blends of polybutadiene and polybutadiene-d_6. Both polymers were made by the "living polymer" technique, end-capped with ethylene oxide, and water-washed to yield the dihydroxy liquid prepolymer. Uniform networks were prepared by reacting the prepolymers with stoichiometric amounts of triphenyl methane triisocyanate. The value of using polybutadiene over polystyrene, of course, is that the networks are elastomeric at ambient temperatures.

These networks (98) were extended up to $\alpha = 1.6$ and characterized by SANS. Owing to large experimental error, no definitive conclusion could be reached, although the data fit the junction affine model better than either the chain affine model or the phantom network model (Section 9.10.5.2).

Random types of cross-linking are of special interest for real systems. Clough et al. (99,100) blended anionically polymerized polystyrene with PS-d_8 and cross-linked the mutual solution with ^{60}Co γ-radiation. Bars of the cross-linked polystyrene were elongated at 145°C and cooled. Specimens were cut in both the longitudinal and transverse directions and characterized by SANS. The quantity

$$R = \frac{R_g \text{ (stretched)}}{R_g \text{ (unstretched)}} \tag{9.83}$$

was plotted versus α, as illustrated in Figure 9.25 (99,100). The transverse measurements are divided into "anisotropic" and "end-on." The anisotropic measurements refer to the case where the beam was perpendicular to the direction of orientation (100), and the end-on measurements refer to the case where the beam was parallel to the stretch direction. In neither of these cases was affine chain deformation followed. Both of these

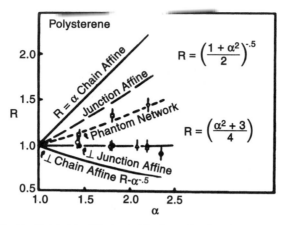

Figure 9.25 The ratio $R = R_g$ (stretched)$/R_g$ (unstretched) as a function of sample elongation (99): □, longitudinal; ▽, longitudinal; ○, transverse, anisotropic; ●, transverse, "end-on"; △, transverse, anisotropic; ▲, transverse, "end-on."

experiments, however, yielded identical results within experimental error, confirming important macromolecular hypotheses.

In Figure 9.25 the quantity $R = [(\alpha^2 + 3)/4]^{0.5}$ was obtained from the dependence predicted for a tetrafunctional phantom network (100). The quantity $R = [(\alpha^2 + 1)/2]^{0.5}$ represents the affine junction case.

The constant value of $R_\perp$ up to $\alpha = 2$ suggests that the chains are deforming far less than the junctions. These results follow neither the $R = \alpha$ (parallel) nor the $R = \alpha^{-0.5}$ (perpendicular) prediction but rather support Benoit et al. (93) that affine chain behavior is not followed.

In a series of theoretical papers, Ullman (102–104) reexamined the phantom network theory of rubber elasticity, especially in the light of the new SANS experiments. He developed a semiempirical equation for expressing the lower than expected chain deformation on extension:

$$\lambda^{*2} = \lambda^2(1 - \alpha') + \alpha' \tag{9.84}$$

as the basis of a network unfolding model. The quantity λ^* is defined as the ensemble average of the deformation of junction pairs connected by a single submolecule in the network, and λ is the corresponding quantity calculated for the phantom network model. The quantity α' expresses the fractional deviation from ideality. The phantom network corresponds to $\alpha' = 0$. If $\alpha' = 1$, the chain does not deform at all on network stretching or swelling. From the data of Clough et al. (99–100), Ullman (102,103) concluded that α' was in the range of 0.36 to 0.53.

In the above, Hinkley et al. (98), Clough et al. (99,100), and Benoit et al. (93,94) utilized end-linked networks. Ullman (102–103) delineated the differences between the two types of network. He pointed out that randomly cross-linked chains deform to a greater extent than end-linked chains, that sensitivity to network functionality is much greater for end-linked chains, and that for high cross-linking levels, the randomly cross-linked chain approaches the macroscopic deformation of the sample. Ullman (104)

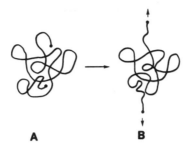

Figure 9.26 Model of the end-pulling mechanism, showing how $R_{\parallel}$ increases, while $R_{\perp}$ remains nearly constant: (A) relaxed and (B) stretched.

recently reviewed these and other SANS experiments on the deformation of polymer networks.

Recently Hadziioannou et al. (105) prepared amorphous polystyrene with extrusion ratios up to 10, using a solid-state coextrusion technique. Their polystyrene had a molecular weight of 5×10^5 g/mol. The anisotropy of the R_g values agreed with those predicted on the basis of a chain affine model, in accord with point 6 above.

While general conclusions appear to be premature, it appears that the cross-link sites rearrange themselves during deformation to achieve their lowest free-energy states; thus the chains deform less than the affine mechanism predicts. A modified end-pulling mechanism is also possible. A possible molecular mechanism, which results in minimal changes in $R_{\perp}$, is illustrated in Figure 9.26 (106). The debate over the exact molecular mechanism of deformation is sure to continue.

9.11 INTERNAL ENERGY EFFECTS

9.11.1 Thermoelastic Behavior of Rubber

In Section 9.5 some of the basic classical thermodynamic relationships for rubber elasticity were examined. Now the classical and statistical formulations are combined (107,108).

Rearranging equation (9.8),

$$f_e = f - T\left(\frac{\partial f}{\partial T}\right)_{L,V} \tag{9.85}$$

Dividing through by f and rearranging,

$$\frac{f_e}{f} = 1 - \left(\frac{\partial \ln f}{\partial \ln T}\right)_{L,V} \tag{9.86}$$

Rewriting equation (9.79) in terms of force, and substituting equation (9.38), we find that

$$f = GA_0\left(\alpha - \frac{V}{V_0}\frac{1}{\alpha^2}\right) \tag{9.87}$$

where A_0 is the initial cross-sectional area; substituting equation (9.87) into the right-hand side of equation (9.86) and carrying out the partial derivative yields

$$\left(\frac{\partial \ln f}{\partial \ln T}\right)_{L,V} = \frac{d \ln G}{d \ln T} + \frac{\beta T}{3} \tag{9.88}$$

where β is the isobaric coefficient of bulk thermal expansion, $(1/V)(\partial V/\partial T)_{L,P}$. Substituting equation (9.88) into equation (9.86),

$$\frac{f_e}{f} = 1 - \frac{d \ln G}{d \ln T} - \frac{\beta T}{3} \tag{9.89}$$

Returning to equation (9.38), and differentiating the natural logarithm of the network end-to-end distance with respect to the natural logarithm of the temperature obtains

$$\frac{d \ln \overline{r_0^2}}{d \ln T} = 1 - \frac{d \ln G}{d \ln T} - \frac{\beta T}{3} \tag{9.90}$$

Noting that the right-hand sides of equation (9.89) and (9.90) are identical,

$$\frac{f_e}{f} = \frac{d \ln \overline{r_0^2}}{d \ln T} = \frac{1}{T}\frac{d \ln \overline{r_0^2}}{dT} \tag{9.91}$$

which expresses the fractional force due to internal energy considerations in terms of the temperature coefficient of the free chains end-to-end distance.

9.11.2 Experimental Values

Values of f_e/f are usually derived by applying the equations above to force–temperature data of the type presented in Figure 9.27 (109). These data, carefully taken after extensive relaxation at elevated temperatures, are reversible within experimental error; that is, the same result is obtained whether the temperature is being lowered (usually first) or raised.

Some values of f_e/f are shown in Table 9.2. For most simple elastomers, f_e/f is a small fraction, near ± 0.20 or less. This indicates that some 80% or more of the retractive force is entropic in nature, as illustrated from early data in Figure 9.7. These same values, of course, lead to temperature coefficients of polymer chain expansion [equation (9.91)].

9.12 THE FLORY–REHNER EQUATION

9.12.1 Causes of Swelling

The equilibrium swelling theory of Flory and Rehner (75) treats simple polymer networks in the presence of small molecules. The theory considers forces arising from three sources:

1. The entropy change caused by mixing polymer and solvent. The entropy change from this source is positive and favors swelling.

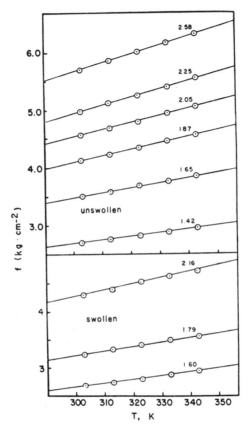

Figure 9.27 Force–temperature relationships for natural rubber. Extension ratios, α, are indicated by the numbers associated with the lines (109).

Table 9.2 Thermoelastic behavior of various polymers

Polymer	f_e/f	Reference
Natural rubber	0.12	(a)
trans-Polyisoprene	0.17	(b)
cis-Polybutadiene	0.10	(c)
Polyethylene	−0.42	(d)
Poly(ethyl acrylate)	−0.16	(e)
Poly(dimethyl siloxane)	0.15	(e)

References: (a) G. Allen, M. J. Kirkham, J. Padget, and C. Price, *Trans. Faraday Soc.*, **67**, 1278 (1971). (b) J. A. Barrie and J. Standen, *Polymer*, **8**, 97 (1967). (c) M. Shen, T. Y. Chem. E. H. Cirlin, and H. M. Gebhard, in *Polymer Networks, Structure, and Mechanical Properties*, A. J. Chompff and S. Newman, eds., Plenum Press, New York, 1978. (d) A. Ciferri, C. A. J. Hoeve, and P. J. Flory, *J. Am. Chem. Soc.*, **83**, 1015 (1961). (e) L. H. Sperling and A. V. Tobolsky, *J. Macromol. Chem.*, **1**, 799 (1966).

2. The entropy change caused by reduction in numbers of possible chain conformations on swelling. The entropy change from this source is negative and opposes swelling.

3. The heat of mixing of polymer and solvent, which may be positive, negative, or zero. Usually it is slightly positive, opposing mixing.

The Flory–Rehner equation may be written

$$-[\ln(1 - v_2) + v_2 + \chi_1 v_2^2] = V_1 n \left[v_2^{1/3} - \frac{v_2}{2} \right] \tag{9.92}$$

where v_2 is the volume fraction of polymer in the swollen mass, V_1 is the molar volume of the solvent, and χ_1 is the Flory–Huggins polymer–solvent dimensionless interaction term. Appendix 9.3 describes the application of the Flory–Rehner theory. This theory, of course, is also related to the thermodynamics of solutions (see Section 3.2). As a rubber elasticity phenomenon, it is an extension in three dimensions.

The value of equation (9.92) here lies in its complementary determination of the quantity n [see equation (9.4) for simplicity]. Both equations (9.4) and (9.92) determine the number of elastically active chains per unit volume (containing, implicitly, corrections for front factor changes). By measuring the equilibrium swelling behavior of an elastomer (χ_1 values are known for many polymer–solvent pairs), its modulus may be predicted. Vice versa, by measuring its modulus, the swelling behavior in any solvent may be predicted.

Generally, values from modulus determinations are somewhat higher, because physical cross-links tend to count more in the generally less relaxed mechanical measurements than in the closer-to-equilibrium swelling data. However, agreement is usually within a factor of 2, providing significant interplay between swelling and modulus calculations (110–112).

Simple elastomers may swell a factor of 4 or 5 or so, leading to a quantitative determination of n. However, two factors need to be considered before the final numerical results is accepted:

1. The front-factor, not explicitly stated in the Flory–Rehner equation, may be significantly different from unity (113).

2. While step polymerization methods lead to more or less statistical networks and good agreement with theory, addition polymerization and vulcanization nonuniformities lead to networks that may swell as much as 20% less than theoretically predicted (114,115).

9.12.2 Example Calculation of Young's Modulus from Swelling Data

At equilibrium, a sample of poly(butadiene–*stat*–styrene) swelled 4.8 times its volume in toluene at 25°C. What is Young's modulus at 25°C?

This material is a typical elastomer, widely used for rubber bands, gaskets, and rubber tires. Table 3.4 gives $\chi_1 = 0.39$. The molar volume of toluene can be calculated from its density, 0.8669 g/cm^3, Table 3.1. A molecular weight of 92 g/mol for toluene yields a molar volume of 106 cm^3/mol. The quantity $v_2 = 1/4.8 = 0.208$. Substituting into equation (9.92) yields

$$n = \frac{-\ln(1 - 0.208) + 0.208 + 0.39 \times 0.208^2}{(106 \text{ cm}^3/\text{mol})\{0.208^{1/3} - 0.208/2\}}$$

$$n = 1.5 \times 10^{-4} \text{ mol/cm}^3$$

Young's modulus is given by equation (9.36),

$$E = 3 \times 1.5 \times 10^{-4}(\text{mol/cm}^3) \times 8.31 \times 10^7 (\text{dyn} \cdot \text{cm/mol} \cdot \text{K}) \times 298(\text{K})$$

$$E = 1.1 \times 10^7 \text{ dyn/cm}^2 \quad \text{or} \quad 1.1 \text{ MPa}$$

This is a typical Young's modulus for such elastomers. Of course, the reverse calculation can be performed, starting with the modulus, and estimating the equilibrium swelling volume.

9.13   GELATION PHENOMENA IN POLYMERS

Gelation in polymers may be brought about in several ways: temperature changes, particularly important in protein gelation formation; polymerization with cross-links; phase separation in block copolymers; ionomer formation; or even crystallization. Such materials are usually thermoreversible for physical crosslinks, or thermoset through the advent of chemical cross-links. Of course, there must be at least two cross-link sites per chain to induce gelation. A major task in polymer science centers on obtaining unambiguous measures of the gelation point.

9.13.1   Gelation in Solution

Let us consider a polymeric network that contains solvent, usually called a polymeric gel. There are several types of gels. A previously cross-linked polymer subsequently swollen in a solvent follows the Flory–Rehner equation (Section 9.12). If the network was formed in the solvent so that the chains are relaxed, the Flory–Rehner equation will not be followed, but rubber elasticity theory can still be used to count the active network segments.

Gels may be prepared using either chemical or physical cross-links. Physical cross-links in gels may involve dipole–dipole interactions, traces of crystallinity, multiple helices, and so on, and thus vary greatly with the number and strength of the bonds. The number of physical cross-links present in a given system depends on time, pressure, and temperature. Many such gels are thermoreversible; that is, the bonds break at elevated temperature and reform at lower temperatures.

A common thermoreversible gel is gelatin in water. Gelatin is made by hydrolytic degradation, boiling collagen in water. Collagen forms a major constituent of the skin and bones of animals. The gelatin made from it forms the basis for many foods and desserts. (The latter is usually made by dissolving gelatin plus sugar in hot water, then refrigerating to cause gelation.)

Thermoreversible gels may be bonded at single points, called point cross-links; junction zones, where the chains interact over a portion of their length; or in the form of fringe micelles—see Figure 9.28 (114). Gelatin has the junction zone type of cross-links, where the chains form multiple helices. Natural collagen is in the form of a triple helix. In hot aqueous solution, the denatured protein forms random coils. The helix–coil

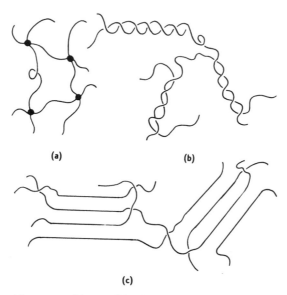

Figure 9.28 Types of thermoreversible cross-links: (*a*) point cross-links, (*b*) junction zones, and (*c*) fringed micelles.

transition is at about 40°C at about 0.5% concentration, as commonly used in foods. Under these conditions gelatin forms either double or triple helices, a subject of current research. The critical nucleation length is about 20 to 50 peptide (mer) units, the most important interactions being among proline, hydroxyproline, and glycine.

Gels can be made from a number of polymers, including poly[acrylamide–*stat*–(acrylic acid)], poly(vinyl acetate), poly(dimethyl siloxane), and polyisocyanurates. The poly[acrylamide–*stat*–(acrylic acid)] hydrogels can be made to swell up to 20,000 times, v/v. Thermorevesible gels can be prepared in organic solvents from polyethylene and *i*-polystyrene, both of which crystallize on cooling but go back in solution on heating.

A type of gel formation involves the globular proteins. Egg white is a typical example. On cooking, it goes form a sol to a gel irreversibly, becoming denatured. Current models (116) show the hydrophobic groups partially flipping out in the range of 75 to 80°C, making contact with similar groups on other globules. Above 85°C, the —S—S— intramolecular cross-links in cysteine become labile, partially interchanging to form intermolecular crosslinks. As is well known, hard-boiled egg white is highly elastic. The white appearance is caused by the nonuniform distribution of protein and water, causing intense light-scattering.

Gels sometimes undergo the phenomenon known as syneresis, where the solvent is exuded from the gel. Two types of syneresis are distinguishable (117): (a) The χ-type, where the polymer phase separates from the solvent due to poor thermodynamics of mixing. Spinodal decomposition is common in such circumstances. Such gels may be turbid in appearance. (b) The *n*-type, which exudes solvent because of increasing cross-link density. The polymeric gel still forms one phase with the solvent, but its equilibrium swelling level decreases. Such gels remain clear. In both cases, various amounts of fluid surround the gel.

Fully formed gels may be highly elastic; see Appendix 9.1.

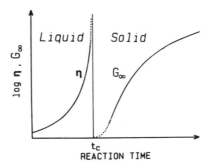

Figure 9.29 Behavior of melt viscosity and shear modulus near the gel point in a crosslinking polymerization as a function of reaction time. The steady shear viscosity goes to infinity, while the equilibrium shear modulus rises from zero, eventually reaching a plateau for the fully reacted material.

9.13.2 Gelation during Polymerization

If cross-linkers are present during a polymerization, the material may gel at a certain point, its viscosity going to infinity; see Figure 9.29 (118). Before gelation the material is fluid, not having an equilibrium modulus. Thus the definition of a gel point requires an infinite steady-state shear viscosity and an equilibrium modulus of zero. At least one of the molecules of the cross-linking polymer must be very large, having grown to the dimensions of the order of the macroscopic sample.

During a polymerization with cross-links, the storage and loss shear moduli, G' and G'', respectively, may cross. While this phenomenon is not entirely universal, it has been frequently taken as the gelation point, see Figure 9.30 (119). The viscous behavior of the oligomeric material dominates in the initial part of the polymerization. With increasing

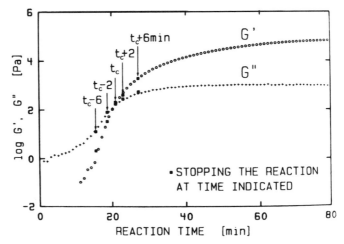

Figure 9.30 Dynamic mechanical experiments of gelling systems such as illustrated in Figure 9.29 often show a crossing of the storage and loss shear modulus. The crossing point is then taken as the gelation point. Illustrated is the behavior of a crosslinking of poly(dimethyl siloxane) in an oscillatory shear experiment at constant frequency, ω.

molecular weight, the loss modulus increases while the storage modulus rises sharply until it intersects, then exceeds the loss modulus. By this definition, the gelation point is the time where the sample exhibits congruent $G' = G''$ behavior, independent of frequency and temperature.

9.13.3 Hydrogels

Hydrogels, or water-containing gels are polymeric materials characterized by both hydrophilicity and insolubility in water (120). Hydrogels that are capable of absorbing very large amounts of aqueous fluids are sometimes referred to as superwater adsorbents.

Their hydrophilicity arises from the presence of water-solubilizing groups such as $-OH$, $-COOH$, $-CONH_2$, $-CONH-$, $-SO_3H$, and so on. Alternately, the main chain of the polymer may be water soluble, such as poly(ethylene oxide) derivatives (121), or based on acrylic, acrylamide, N-vinyl-2-pyrrolidinone, poly(vinyl alcohol), ionomers or glycopolymers (122). The insolubility arises from the presence of a three-dimensional network. The cross-links may be covalent, electrostatic, hydrophobic, or dipole–dipole in character. The extent of hydration may exceed a factor of 100, based on dry gel weight.

Many ionic hydrogels exhibit a first-order volume-phase transition (123). In such a transition the equilibrium extent of swelling can change dramatically with only a small change in conditions. The transition may be brought about by changing a range of variables, see Figure 9.31 (123). A characteristic manifestation of the transition involves the coexistence of two gel phases, and hence the presence of a first-order transition between them. Owing to the high values of hydrogel swelling, the inverse Langevin function (Section 9.10) is used to describe the chain conformations at near full extension (124). The Debye–Hückel theory of ionic osmotic pressures is utilized. Of course, thermodynamics ultimately governs the equilibrium swelling (125).

Applications for hydrogels depending on their high-swelling ability include soft contact lenses based on poly(2-hydroxyethyl methacrylate), polyHEMA, and a number of drug delivery systems of various composition (120). While simple entrapment of the medicament in the hydrogel works, controlled drug-release systems include eroding reservoir devices and other strategies. Hydrogels based on poly(sodium acrylate) (126,127) form the absorbent portion of disposable diapers (123). Another application of this technology involves switching materials, where *on* and *off* are two extents of swelling, as described above.

9.14 GELS AND GELATION

Most of the above cross-linked polymers were considered in the dry state, although the Flory–Rehner theory (Section 9.12) made use of equilibrium swollen gels in the evaluation of the cross-link density. Generally, a polymeric gel is defined as a system consisting of a polymer network swollen with solvent. It must be understood that the solvent is dissolved in the polymer, not the other way around.

Polymeric gels may be categorized into two major classes (128): thermoreversible gels and permanent gels. The thermoreversible gels undergo a transition from a solid-like form to a liquid-like form at a certain characteristic temperature. The links between the polymeric chains are transient in nature and support a stable polymeric network only below a characteristic "melting" point.

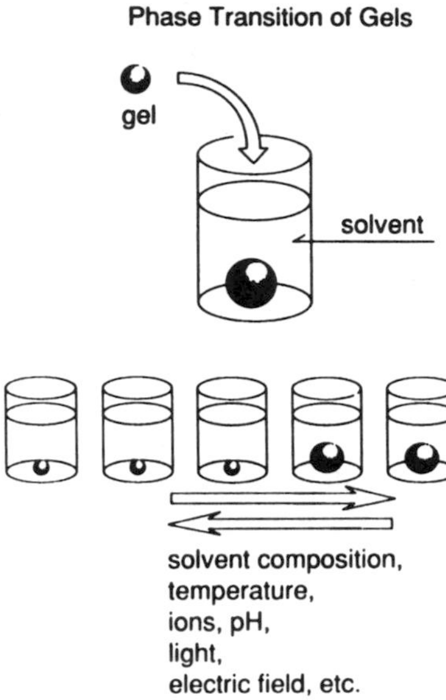

Phase Transition of Gels

Figure 9.31 Illustration of the first-order phase transition in ionic hydrogels. The extent of swelling may change dramatically with environmental factors, as shown.

The permanent gels consist of solvent-logged covalently bonded polymer networks. One family of such networks is formed by cross-linking preexisting polymer chains, such as by vulcanization. Another family makes use of simultaneous polymerization and cross-linking. Some of the more important types of gels are delineated in Table 9.3 some specific examples include Jello® (comprised of approximately 3% of collagen-derived protein gelatin plus colored, flavored, and sweetened water), vitreous humor that fills the interior of the eye, membranes (both natural and synthetic), and soft-contact lenses (129).

Taking gelatin as an example, the thermal naturation and denaturation transition occurs near 40°C in water for native soluble collagen, and near 25°C for partially rena- tured gelatin. After setting, the initial growth rate of the modulus is nearly third order in gelatin concentration. The mechanism of gelation thus is thought to involve slow asso- ciation of three rapidly formed single-helix segments (Figure 9.32) (130). a demonstra-

Table 9.3 Types of gels

Class	Bonding Mode
Thermoreversible	Hydrogen bonds
	Microcrystals
	Entwined helical structures
	Specific interactions
Permanent gels	Vulcanization
	Multifunctional monomers
	(a) Chain polymerizable
	(b) Stepwise polymerizable

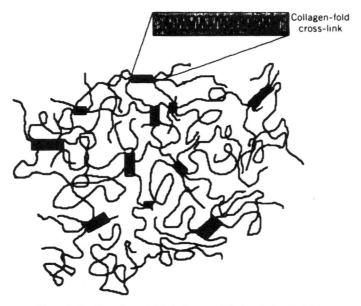

Collagen-fold cross-link

Figure 9.32 Formation of triple-helix cross-links in gelatin gels (130).

tion experiment measuring the modulus of gelatin as a function of concentration is given in Appendix 9.1.

The gelatin-type of triple helix may be considered a type of crystal; indeed, the denaturing and renaturing of these gels is a first-order transition. However, it has been modeled as a nucleated, one-dimensional crystallization, as opposed to the ordinary three-dimensional crystallization of bulk meterials.

9.15 EFFECT OF STRAIN ON THE MELTING TEMPERATURE

Most elastomers are amorphous in use. Indeed, significant crystallinity deprives the polymer of its rubbery behavior. However, some elastomers crystallize during strains such as extension. The most important of these are *cis*-polybutadiene, *cis*-polyisoprene, and *cis*-polychloroprene. Crystallization on extension can be responsible for a rapid upturn in the stress–strain curves at high elongation; see Figure 9.5.

Such crystallization can result in significant engineering advantage. Consider, for example, the wear mechanisms in automotive tires. It turns out that abrasion is the most important mode of loss of tread. Tiny shreds of rubber are torn loose, hanging at one end. In contact with the road, these shreds are strained at each revolution of the tire, gradually tearing off more rubber. If the rubber crystallizes during extension, it becomes self-reinforcing when it is needed most, thus slowing the failure process. When the strain is released, the crystallites melt, returning the rubber to its amorphous state reversibly.

The elevation in melting temperature can be expressed (131) as

$$\frac{1}{T_f} = \frac{1}{T_f^0} - \frac{R}{2n\Delta H_f}\left(\alpha^2 + \frac{2}{\alpha} - 3\right) \tag{9.93}$$

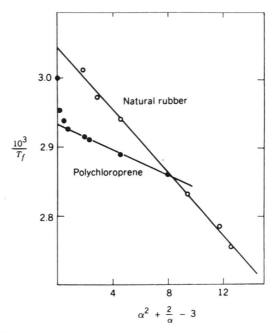

Figure 9.33   Fit of equation (9.93) to melting point elevation (128).

where T_f and T_f^0 are the thermodynamic melting temperatures for the strained and unstrained polymers, respectively, n is the number of repeating units per network chain, and ΔH_f is the heat of fusion per mer. The student will note the relationship to equations (6.38) to (6.43), albeit that these equations treat melting point depressions, and the present case treats melting point elevation. Also note equation (9.33).

Equation (9.93) was derived from the assumption that the system can be treated as being composed of two independent phases, T_f obtained directly by equating the chemical potentials of the strained amorphous and crystalline phases. The function of the extension ratio, given in equation (9.33) yields the dependence of the change in free energy on elongation.

Figure 9.33 (131) illustrates the application of the equation to natural rubber and polychloroprene (see Section 9.15 for structures). The equation fits the data for natural rubber somewhat better than for polychloroprene, although both polymers are fit for reasonable extensions. It must be pointed out that while the pure polymers both crystallize somewhat above room temperature, ordinary vulcanization and compounding lower the melting temperatures significantly, shifting the data in Figure 9.33 upward. Typical numerical values for equation (9.93) are as follows:

Polymer	ΔH_f (cal/mol)	n
Natural rubber	1000	73
Polychloroprene	2000	37

9.16   ELASTOMERS IN CURRENT USE

The foregoing sections outline the theory of rubber elasticity. This section describes the classes of elastomers in current use. While many of these materials exhibit low modulus, high elongation, and rapid recovery from deformation and obey the theory of rubber elasticity, some materials deviate significantly, have limited extensibility or poorly defined rubbery plateaus, but are considered elastomers.

9.16.1   Classes of Elastomers

9.16.1.1   Diene Types   The diene elastomers are based on polymers prepared from butadiene, isoprene, their derivatives and copolymers. The oldest elastomer, natural rubber (polyisoprene), is in this class (see Section 9.2). Polybutadiene, polychloroprene, styrene–butadiene rubber (SBR), and acrylonitrile–butadiene rubber (NBR) are also in this class.

The general polymerization scheme may be written

$$n\text{CH}_2\!\!=\!\!\overset{\displaystyle \text{X}}{\underset{\displaystyle |}{\text{C}}}\!\!-\!\!\text{CH}\!\!=\!\!\text{CH}_2 \;\rightarrow\; \left(\!\!\text{CH}_2\!\!-\!\!\overset{\displaystyle \text{X}}{\underset{\displaystyle |}{\text{C}}}\!\!=\!\!\text{CH}\!\!-\!\!\text{CH}_2\!\!\right)_{\!n} \tag{9.94}$$

where X— may be H—, CH_3—, Cl—, and so on (see Table 9.4).

The diene double bond in equation (9.94) may be either *cis* or *trans*. The *cis* products all have lower glass transition temperatures and/or reduced crystallinity, and they make superior elastomers. A random copolymer of butadiene and styrene is polymerized to form SBR (styrene–butadiene rubber). This copolymer forms the basis for tire rubber (see below). The *trans* materials, such as the balata and gutta percha polyisoprenes (12), are highly crystalline and make excellent materials such as golf ball covers.

Natural rubber is widely used in truck and aircraft tires, which require heavy duty. They are self-reinforcing because the rubber crystallizes when stretched.

9.16.1.2   Saturated Elastomers   The polyacrylates exemplify these materials:

$$n\text{CH}_2\!\!=\!\!\underset{\displaystyle \underset{\displaystyle \text{O}\!-\!\text{X}}{\overset{\displaystyle |}{\text{C}\!\!=\!\!\text{O}}}}{\underset{\displaystyle |}{\text{CH}}} \;\rightarrow\; \left(\!\!\text{CH}_2\!\!-\!\!\underset{\displaystyle \underset{\displaystyle \text{O}\!-\!\text{X}}{\overset{\displaystyle |}{\text{C}\!\!=\!\!\text{O}}}}{\underset{\displaystyle |}{\text{CH}}}\!\!\right)_{\!n} \tag{9.95}$$

where X— may be CH_3 —, CH_3CH_2 —, and so on (see Table 9.4). Ethyl and butyl are the two most important derivatives, with glass transition temperatures in the range of −22°C and −50°C, respectively. The main advantage of the saturated elastomers is resistance to oxygen, water, and ultraviolet light, which attack the diene elastomers in outdoor conditions (132).

An important saturated elastomer is based on a random copolymer of ethylene and propylene, EPDM, or ethylene–propylene–diene monomer. The diene is often a bicyclic

Table 9.4 Structures of elastomeric materials

Name	Structure
A. Diene elastomers	$\begin{array}{c} X \\ \| \\ -(CH_2-C{=}CH-CH_2)_n \end{array}$
Polybutadiene	$X--=H-$
Polyisoprene	$X--=CH_3-$
Polychloroprene	$X--=Cl-$
B. Acrylics	$\begin{array}{c} -(CH_2-CH)_n \\ \| \\ O{=}C-O-X \end{array}$
Poly(ethyl acrylate)	$X--=CH_3CH_2-$
C. EPDM[a]	$\begin{array}{c} CH_3 \\ \| \\ -(CH_2-CH_2)_n-(CH_2-CH)_m \end{array}$
D. Thermoplastic elastomers	ABA
Poly(styrene–*block*–butadiene–*block*–styrene)	A = polystyrene B = polybutadiene
Segmented polyurethanes	$-(AB)_n$ A = polyether (soft block)
	B = aromatic urethane (hard block)
	A = poly(butylene oxide)
	B = poly(terephthalic acid–ethylene glycol)
E. Inorganic elastomers Silicone rubber	$\begin{array}{c} CH_3 \\ \| \\ -(Si-O)_n \\ \| \\ CH_3 \end{array}$
Polyphosphazenes	$\begin{array}{c} R \quad\ R' \\ \backslash\ / \\ -(N{=}P)_n \end{array}$

[a] Ethylene-propylene diene monomer.

compound introduced at the 2% level to provide cross-linking sites:

Dicyclopentadiene   5-Ethylidene-2-norbornene   (9.96)

It must be pointed out that although both polyethylene and polypropylene are crystalline polymers, the random polymer at midrange compositions is totally amorphous. These materials, with a glass transition of $-50°C$, make especially good elastomers for toughening polypropylene plastics (133–135).

9.16.1.3 Thermoplastic Elastomers These new materials contain physical cross-links rather than chemical cross-links. A physical cross-link can be defined as a noncovalent bond that is stable under one condition but not under another. Thermal stability is the most important case. These materials behave like cross-linked elastomers at ambient temperatures but as linear polymers at elevated temperatures, having reversible properties as the temperature is raised or lowered.

The most important method of introducing physical crosslinks is through block copolymer formation (136–139). At least three blocks are required. The simplest structure contains two hard blocks (with a T_g or T_f above ambient temperature) and a soft block (with a low T_g) in the middle (see Figure 4.16). The soft block is amorphous and above T_g under application temperatures, and the hard block is glassy or crystalline.

The thermoplastic elastomers depend on phase separation of one block from the other (see Chapter 13), which in turn depends on the very low entropy gained on mixing the blocks. The elastomeric phase must form the continuous phase to produce rubbery properties; thus the center block has a higher molecular weight than the two end blocks combined.

Examples of the thermoplastic elastomers include polystyrene–*block*–polybutadiene–*block*–polystyrene (SBS) or the saturated center block counterpart (SEBS). In the latter, the EB stands for ethylene–butylene, where a combination of 1,2 and 1,4 copolymerization of butadiene on hydrogenation presents the appearance of a random copolymer of ethylene and butylene (see Table 9.4).

When the polymer illustrated in Figure 4.16 is of the SBS or SEBS type, it is sold under the trademark Kraton.®

Important applications of the triblock and its cousins, the starblock copolymer thermoplastic elastomers, include the rubber soles of running shoes and sneakers (139,140) and hot melt adhesives. In the former application, sliding friction generated heat momentarily turns the elastomer into an adhesive (see Section 8.2), reducing slips and falls. When sliding stops, the sole surface cools again, regenerating the rubber.

The so-called segmented polyurethanes form thermoplastic elastomers of the $+AB+_n$ type, where A is usually a polyether such as poly(ethylene oxide) and B contains aromatic urethane groups (141,142). These polyurethanes make excellent elastic fibers that stretch about 30% and are widely used in undergarments under the trade names Spandex® and Lycra®.

The literature distinguishes various kinds of polyurethanes. There are those that are not thermoplastic elastomers, often densely cross-linked. These need not be elastomeric at all.

The thermoplastic elastomer polyurethanes, TPU, may be of two general types: partly crystalline elastic fibers (see preceding paragraph) or softer elastomers, depending on the relative length of the soft segment. Those polyurethanes based on polyester soft segments tend to be more resistant to hydrocarbons, while the polyether types are more resistant to hydrolysis but tend to swell more in aqueous environments (143). The block copolymer characteristics of polyurethanes are discussed further in Chapter 13.

A newer type of $(AB)_n$ block copolymer, known as the poly(ether–ester) elastomers and sold as Hytrel®, contains alternating blocks of poly(butylene oxide) and butylene terephthalate as the soft and hard blocks, respectively (143–145):

$$\left[\begin{matrix}\overset{O}{\overset{\|}{C}}-\hspace{-4pt}\langle\hspace{-2pt}\bigcirc\hspace{-2pt}\rangle\hspace{-4pt}-\overset{O}{\overset{\|}{C}}-O+CH_2\!\!+_4-O\end{matrix}\right]_m \left[\begin{matrix}\overset{O}{\overset{\|}{C}}-\hspace{-4pt}\langle\hspace{-2pt}\bigcirc\hspace{-2pt}\rangle\hspace{-4pt}-\overset{O}{\overset{\|}{C}}-O+(CH_2)_4-O\end{matrix}\right]_n \quad (9.97)$$

Hard segment   Soft segment

Usually $m = 1$ or 2, and $n = 40$–60; thus the soft segment is much longer than the hard segment (144). The hard blocks crystallize in this case.

9.16.1.4 Inorganic Elastomers

The major commercial inorganic elastomer is poly(dimethyl siloxane), known widely as silicone rubber (see Table 9.4). This specialty elastomer has the lowest known glass transition temperature, $T_g = -130°C$ (146); it also serves as a high-temperature elastomer. A common application of this elastomer is as a caulking material. It cross-links on exposure to air.

Another covalently bonded inorganic elastomer class is the polyphosphazenes (146,147),

$$+N=\underset{\underset{R'}{|}}{\overset{\overset{R}{|}}{P}}+_n \tag{9.98}$$

In elastomeric compositions R and R' are mixed substituent fluoroalkoxy groups. The current technological applications depend on the oil resistance and nonflammability of these elastomers; low T_g's are also important. Gaskets, fuel lines, and O-rings are made from this class of elastomer.

Another interesting inorganic elastomer is polymeric sulfur (148). Under equilibrium conditions at room temperature, rhombic sulfur consists mainly of eight-membered rings. On heating, it melts at 113°C to a relatively low viscosity, reddish-yellow liquid. Above approximately 160°C, the viscosity increases suddenly as the eight-membered rings open into long, linear chains of about 1×10^5 degree of polymerization. On further heating, the viscosity declines, as the polymer depolymerizes again. The sudden polymerization on heating is called a *floor temperature*, below which the free energy of polymerization is positive, and above which the free energy of polymerization is negative. (This is the opposite of many organic polymers, which exhibit *ceiling temperatures*, above which they depolymerize back to the monomer.)

If the polymerized form of elemental sulfur is quenched, it becomes highly elastomeric in the vicinity of room temperature. The situation is complicated by the presence of unpolymerized S_8 rings, which behave as plasticizers. In addition the polymer is only metastable, reverting back to S_8 rings in a relatively short time.

The above-mentioned heating and subsequent quenching of sulfur has long been a favorite laboratory demonstration both in high school and college chemistry (149,150). Since the sulfur exhibits both simple liquid and polymeric properties at laboratory temperatures without complex equipment or other chemicals, it illustrates several principles of science easily.

9.16.2 Reinforcing Fillers and Other Additives

Natural rubber has a certain degree of self-reinforcement, since it crystallizes on elongation (151). The thermoplastic elastomers also gain by the presence of hard blocks (139). However, nearly all elastomeric materials have some type of reinforcing filler, usually finely divided carbon black or silicas (136).

These reinforcing fillers, with dimensions of the order of 100 to 200 Å, form a variety of physical and chemical bonds with the polymer chains. Tensile and tear strength are increased, and the modulus is raised (152–156). The reinforcement can be understood

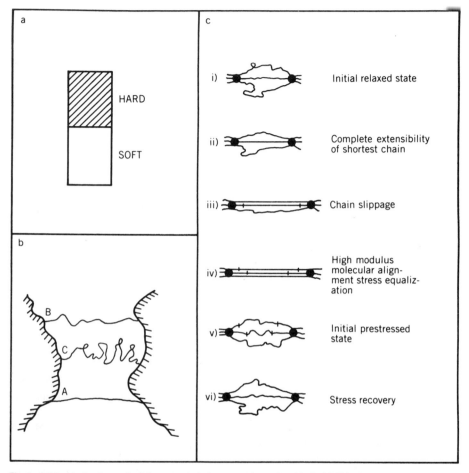

Figure 9.34 Mechanisms of reinforcement of elastomers by carbon black: (*a*) Takayanagi model approach; (*b*) on stretching, some chains (A) will become taut before others (B, C); (*c*) chain slippage on filler surface maintains polymer–filler bonds. From J. A. Harwood, L. Mullins, and A. R. Payne, *J. IRI*, **1**, 17 (1967).

through chain slippage mechanisms (see Figure 9.34) (136). The filler permits local chain segment motion but restricts actual flow.

While it is beyond the scope of this book to treat all the other components of commercial elastomers, a few must be mentioned. Frequently, an extender is added. This is a high-molecular-weight oil that reduces the melt viscosity of the polymer while in the linear state, easing processing, and lowers the price of the final product.

Antioxidants or antiozonites are added, especially to diene elastomers, to slow attack on the double bond. Ultraviolet screens are used for outdoor application (132).

The major application of carbon-black-reinforced elastomers is in the manufacture of automotive tires. Table 9.5 illustrates an automotive tire recipe involving synthetic rubber. The synthetic-rubber recipes usually contain two or more elastomers that are blended together with the other ingredients and then covulcanized.

Table 9.5   A modern tire tread recipe

Ingredient	phr[a]	Function
Styrene-butadiene rubber	50	Elastomer
Natural rubber	50	Elastomer
Carbon black	45	Reinforcing filler
Aromatic oil	9	Extender
Paraffin wax	1	Processing aid and finish
Antiozonant	2	Reduce chain scission
Aryl diamine	1	Antioxidant
Stearic Acid	3	Accelerator-activator
Zinc oxide	3	Accelerator-activator
Sulfur	1.6	Vulcanizing agent
N-Oxydiethylene benzothiazo-ε-2-sulfanamide)	0.8	Accelerator, delayed action type
N,N-Diphenyl guanidine	0.4	A secondary accelerator

Source: The Vanderbilt Rubber Handbook, R. F. Ohm, Ed., R. T. Vanderbilt Co., Norwalk, CT, 14 Ed., 1995 (CD ROM version).

[a] Parts per hundred parts of rubber, by weight.

Current problems in tire usage involve rolling resistance, traction, and skid resistance. High values of tan δ can cause heat generation and fatigue problems. Rolling resistance involves a whole tire frequency of 10 to 100 Hz at 50 to 80°C, the average running temperature of tires.

In the case of skid resistance, stress is generated by friction with the road surface and concomitant motions of the rubber at the surface of the tire tread. The equivalent frequency of this movement is higher, depending on the roughness of the road surface, but is around 10^4 to 10^7 Hz at room temperature. High values of tan δ are desired for gripping the road under wet or icing conditions. By way of the WLF equation (Section 8.6.1.2), a reduced temperature for different tire properties at 1 Hz can be used as the criterion for polymer and filler development for tire compounds; see Figure 9.35 (157).

From a viscoelastic point of view, a high-performance tire should have a low tan δ value at 50 to 80°C to reduce rolling resistance and save energy. The ideal material should also possess high hysteresis from −20 to 0°C, for high skid resistance and wet grip.

While Figure 9.35 reads on the glass transition behavior of the elastomer, the filler also greatly influences the result. Today, carbon black is the principal reinforcing filler added to tire rubber; see item three in Table 9.5. However, newer commercial materials, such as a carbon/silica dual phase filler (158,159), improve the behavior especially with regard to the properties needed in Figure 9.35. The new filler consists of two phases, a carbon black phase, with a finely divided silica phase dispersed therein. The new filler causes stronger filler-polymer interactions in comparison with either conventional carbon black or silica alone having comparable surface areas. This results in a higher *bound rubber* content, which is the fraction of rubber bound to the filler tightly enough that it cannot be easily extracted by solvents. The effect is related to the introduction of greater numbers of surface defects in the graphitic crystal lattice of the carbon phase. The tighter binding of the rubber does not affect the value of the tan δ maximum at low temperatures, but the reduced chain slippage causes a lower value of tan δ at higher temperatures.

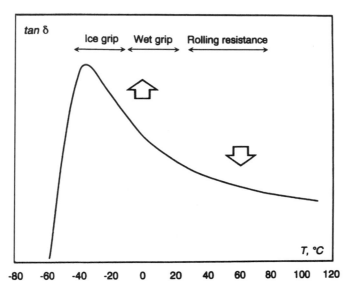

Figure 9.35 Desired values of tan δ as a function of temperature for tire performance. Frequencies reduced to 1 Hz. While a high value of tan δ is need at low temperatures to reduce skidding on ice and water, low values of tan δ at higher temperatures reduce fuel consumption, tire overheating, and tire wear.

9.17 SUMMARY OF RUBBER ELASTICITY BEHAVIOR

When an amorphous cross-linked polymer above T_g is deformed and released, it "snaps back" with rubbery characteristics. The dependence of the stress necessary to deform the elastomer depends on the cross-link density, elongation, and temperature in a way defined by statistical thermodynamics.

The theory of rubber elasticity explains the relationships between stress and deformation in terms of numbers of active network chains and temperature but cannot correctly predict the behavior on extension. The Mooney–Rivlin equation is able to do the latter but not the former. While neither theory covers all aspects of rubber deformation, the theory of rubber elasticity is more satisfying because of its basis in molecular structure.

The theory of rubber elasticity is one of the oldest theories in polymer science and has played a central role in its development. Now one of the key assumptions of rubber elasticity theory, the concept of affine deformation, is under question in the light of small-angle neutron scattering experiments. The SANS experiments suggest a nonaffine chain deformation via an end-pulling mechanism.

If these new results stand the test of time, they may help answer one of the important challenges to the Mark–Flory random coil model (see Section 5.3). In this challenge the considerable discrepancy between rubber elasticity theory and experiment is blamed on the presumed nonrandom coiling of the polymer chain in space. These discrepancies may, however, lie rather in the realm of the mode of chain disentanglement on extension.

An understanding of how an elastomer works, however, has led to many new materials and new types of elastomers. A leading new type of elastomer is based on physical cross-links rather than chemical cross-links. The new materials are known as thermoplastic elastomers.

REFERENCES

1. D. Adolf, J. E. Martin, and J. P. Wilcoxon, *Macromolecules*, **23**, 527 (1990).
2. D. Adolf and J. E. Martin, *Macromolecules*, **23**, 3700 (1990).
3. J. E. Martin, D. Adolf, and J. P. Wilcoxon, *Phys. Rev. A*, **39**, 1325 (1989).
4. P. G. de Gennes, *J. Phys. Lett.*, **37**, 61 (1976).
5. D. J. Stauffer, *J. Chem. Soc. Faraday Trans. 2*, **72**, 1354 (1976).
6. D. Stauffer, A. Coniglio, and M. Adam, *Adv. Poly. Sci.*, **44**, 103 (1982). [*Polymer Networks*, K. Dusek, ed., Springer, New York, 1982).]
7. B. Gutte and R. B. Merrifield, *J. Am Chem. Soc.*, **91**, 501 (1969).
8. F. Rodriguez, *Principles of Polymer Systems*, 3rd ed., Hemisphere, New York, 1988, Chap. 14.
9. K. Dusek, in *Advances in Polymerization. 3*, R. N. Haward, ed., Applied Science Publishers, London, 1982.
10. K. Dusek and W. J. MacKnight, in *Cross-Linked Polymers: Chemistry, Properties, and Applications*, R. A. Dickie, S. S. Labana, and R. S. Bauer, eds., ACS Symposium Series 367, ACS Books, Washington, DC, 1988.
11. H. F. Mark, *Giant Molecules*, Time Inc., New York, 1966, p. 124.
12. H. J. Stern, *Rubber: Natural and Synthetic*, 2nd ed., Plamerton, New York, Chap. 1.
13. J. Gough, *Mem. Lit. Phil. Soc. Manchester*, **1**, 28 (1805).
14. C. Goodyear, U.S. Patent 3, 633 (1844).
15. J. L. White, *Rubber Industry*, August 1974, p. 148.
16. Lord Kelvin (W. Thompson), *J. Pure Appl. Math.*, **1**, 57 (1857).
17. C. Price, *Proc. R. Soc. Lond.*, **A351**, 331 (1976).
18. P. J. Flory, *Principles of Polymer Chemistry*, Cornell University Press, Ithaca, NY, 1953, (a) pp. 434–440, (b) p. 442, (c) p. 464.
19. H. Staudinger, *Ber. Otsch. Chem. Ges.*, **53**, 1073 (1920).
20. H. Staudinger, *Die hochmolekularen organische Verbindungen*, Verlag von Julius Springer, Berlin, 1932.
21. Reference 18, Chap. 1.
22. H. J. Stern, in *Rubber Technology and Manufacture*, C. M. Blow and C. Hepburn, eds., Butterworth Scientific, London, 1982.
23. Anonymous, *Chemical Heritage*, **17**(1), 4 (1999).
24. Y. Furukawa, *Inventing Polymer Science*, University of Pennsylvania Press, Philadelphia, 1998.
25. P. J. T. Morris, in *The American Synthetic Rubber Research Program*, University of Pennsylvania Press, Philadelphia, 1989.
26. W. M. Saltman, in *Rubber Technology*, M. Morton, ed., Van Nostrand Reinhold, New York, 1973.
27. R. A. French, in *Encyclopedia Britannica*, Vol. 22, William Benton, Publisher, Chicago, 1968, p. 506.
28. E. Guth and H. Mark, *Monatsh. Chem.* **65**, 93 (1934).
29. W. Kuhn, *Angew. Chem.*, **51**, 640 (1938).
30. W. Kuhn, *J. Polym. Sci.*, **1**, 380 (1946).

31. E. Guth and H. M. James, *Ind. Eng. Chem.*, **33**, 624 (1941).
32. E. Guth and H. M. James, *J. Chem. Phys.*, **11**, 455 (1943).
33. H. M. James, *J. Chem. Phys.*, **15**, 651 (1947).
34. H. M. James and E. Guth, *J. Chem. Phys.*, **15**, 669 (1947).
35. H. M. James and E. Guth, *J. Polym. Sci.*, **4**, 153 (1949).
36. L. R. G. Treloar, *Trans. Faraday Soc.*, **39**, 36, 241 (1943).
37. F. T. Wall, *J. Chem. Phys.*, **11**, 527 (1943).
38. P. J. Flory, *Trans. Faraday Soc.*, **57**, 829 (1961).
39. G. Allen, *Proc. R. Soc. Lond.*, **A351**, 381 (1976).
40. P. J. Flory, *Polymer*, **20**, 1317 (1979).
41. P. J. Flory, *Proc. R. Soc. Lond.*, **A351**, 351 (1976).
42. L. R. G. Treloar, *The Physics of Rubber Elasticity*, 3rd ed., Clarendon Press, Oxford, 1975.
43. R. P. Wool, *Polymer Interfaces: Structure and Strength*, Hanser, Munich, 1995.
44. L. R. G. Treloar, *Trans. Faraday Soc.*, **40**, 49 (1944).
45. L. R. G. Treloar, *Trans. Faraday Soc.*, **42**, 83 (1946).
46. F. T. Wall, *Chemical Thermodynamics*, 2nd ed., Freeman, San Francisco, 1965, p. 314.
47. R. L. Anthony, R. H. Caston, and E. Guth, *J. Phys. Chem.*, **46**, 826 (1942).
48. K. J. Laidler, *The World of Physical Chemistry*, Oxford University Press, Oxford, 1995.
49. J. E. Mark and B. Erman, *Rubberlike Elasticity: A Molecular Primer*, Wiley, New York, 1988.
50. E. Pines, K. L. Wun, and W. Prins, *J. Chem. Ed.*, **50**, 753 (1973).
51. K. J. Mysels, *Introduction to Colloid Chemistry*, Wiley-Interscience, New York, 1965.
52. M. Mooney, *J. Appl. Phys.*, **11**, 582 (1940).
53. M. Mooney, *J. Appl. Phys.*, **19**, 434 (1948).
54. R. S. Rivlin, *Trans. R. Soc. (Lond.)*, **A240**, 459, 491, 509 (1948).
55. R. S. Rivlin, *Trans. R. Soc. (Lond.)*, **A241**, 379 (1948).
56. S. M. Gumbrell, L. Mullins, and R. S. Rivlin, *Trans. Faraday Soc.*, **49**, 1495 (1953).
57. J. E. Mark, R. R. Rahalkar, and J. L. Sullivan, *J. Chem. Phys.*, **70** 1794 (1979).
58. G. Gee, *Trans. Faraday Soc.*, **42**, 585 (1946).
59. A. Ciferri and P. J. Flory, *J. Appl. Phys.*, **30**, 1498 (1959).
60. K. C. Valanis and R. F. Landel, *J. Appl. Phys.*, **38**, 2997 (1967).
61. R. W. Ogden, *Proc. R. Soc.*, **A326**, 565 (1972).
62. M. Shen, in *Science and Technology of Rubber*, F. R. Eirich, ed., Academic Press, Orlando, 1978.
63. Y. Sato, *Rep. Prog. Polym. Phys. Jpn.*, **9**, 369 (1969).
64. W. Kuhn and F. Grün, *Kolloid Z.*, **101**, 248 (1942).
65. J. K. Yeo, L. H. Sperling, and D. A. Thomas, *J. Appl. Polym. Sci.*, **26**, 3977 (1981).
66. J. A. Duiser and J. A. Staverman, in *Physics of Noncrystalline Solids*, J. A. Prins, Ed., North-Holland, Amsterdam, 1965.
67. W. W. Graessley, *Macromolecules*, **8**, 186 (1975).
68. K. J. Smith Jr., and R. J. Gaylord, *J. Polym. Sci. Polym. Phys. Ed.*, **13**, 2069 (1975).
69. D. S. Pearson and W. W. Graessley, *Macromolecules*, **13**, 1001 (1980).
70. P. J. Flory, *Ind. Eng. Chem.*, **38**, 417 (1946).
71. L. Mullins and A. G. Thomas, *J. Polym. Sci.*, **43** 13 (1960).
72. A. V. Tobolsky, D. J. Metz, and R. B. Mesrobian, *J. Am. Chem. Soc.*, **72**, 1942 (1950).
73. J. Scanlan, *J. Polym. Sci.*, **43**, 501 (1960).
74. H. M. James and E. Guth, *J. Chem. Phys.*, **11**, 455 (1943).

75. P. J. Flory and J. Rehner, *J. Chem. Phys.*, **11**, 521 (1943).

76. H. M. James and E. Guth, *J. Polym. Sci.*, **4**, 153 (1949).

77. P. J. Flory, *Trans. Faraday Soc.*, **57**, 829 (1961).

78. A. V. Tobolsky and M. C. Shen, *J. Appl. Phys.*, **37**, 1952 (1966).

79. L. H. Sperling and A. V. Tobolsky, *J. Macromol. Chem.*, **1**, 799 (1966).

80. A. M. Bueche, *J. Polym. Sci.*, **19**, 297 (1956).

81. L. Mullins, *J. Appl. Polym. Sci.*, **2**, 1 (1959).

82. G. Kraus, *J. Appl. Polym. Sci.*, **7**, 1257 (1963).

83. R. G. Mancke, R. A. Dickie, and J. O. Ferry, *J. Polym. Sci.*, **A-2** (6), 1783 (1968).

84. N. R. Langley, *Macromolecules*, **1**, 348 (1968).

85. J. D. Ferry, *Viscoelastic Properties of Polymers*, 3rd ed., Wiley, New York, 1980, pp. 408–411.

86. O. Kramer, *Polymer*, **20**, 1336 (1979).

87. S. Onogi, T. Masuda, and K. Kitagawa, *Macromolecules*, **3**, 111 (1970).

88. P. J. Flory, *Chem. Rev.*, **35**, 51 (1944).

89. J. Scanlan, *J. Polym. Sci.*, **43**, 501 (1960).

90. N. R. Langley and K. E. Polmanter, *J. Polym. Sci. Polym. Phys. Ed.*, **12**, 1023 (1974).

91. J. D. Ferry, *Polymer*, **20**, 1343 (1979).

92. M. Beltzung, C. Picot, P. Rempp, and J. Hertz, *Macromolecules*, **15**, 1594 (1982).

93. H. Benoit, D. Decker, R. Duplessix, C. Picot, P. Remp, J. P. Cotton, B. Farnoux, G. Jannick, and R. Ober, *J. Polym. Sci. Polym. Phys. Ed.*, **14**, 2119 (1976).

94. H. Benoit, R. Duplessix, R. Ober, M. Daoud, J. P. Cotton, B. Farnoux, and G. Jannick, *Macromolecules*, **8** 451 (1975).

95. L. H. Sperling, *Polym. Eng. Sci.*, **24**, 1 (1984).

96. L. H. Sperling, A. M. Fernandez, and G. D. Wignall, in *Characterization of Highly Cross-linked Polymers*, S. S. Labana and R. A. Dickie, eds., ACS Symposium Series No. 243, American Chemical Society, Washington, 1984.

97. C. Picot, R. Duplessix, D. Decker, H. Benoit, F. Boue, J. P. Cotton, and P. Pincus, *Macromolecules*, **10**, 436 (1977).

98. J. A. Hinkley, C. C. Han, B. Mozer, and H. Yu, *Macromolecules*, **11**, 837 (1978).

99. S. B. Clough, A. Maconnachie, and G. Allen, *Macromolecules*, **13**, 774 (1980).

100. S. B. Clough, Private communication, December 29, 1982.

101. D. S. Pearson, *Macromolecules*, **10**, 696 (1977).

102. R. Ullman, *Macromolecules*, **15**, 1395 (1982).

103. R. Ullman, *Macromolecules*, **15**, 582 (1982).

104. R. Ullman, in *Elastomers and Rubber Elasticity*, J. E. Mark and J. Lal, eds., ACS Symposium Series No. 193, American Chemical Society, Washington, DC, 1982, Chap. 13.

105. G. Hadziiannou, L. H. Wang, R. S. Stein, and R. S. Porter, *Macromolecules*, **15**, 880 (1982).

106. F. Boue, M. Nierlich, G. Jannick, and R. C. Ball, *J. Physiol. (Paris)*, **43**, 137 (1982).

107. M. Shen and P. J. Blatz, *J. Appl. Physiol.*, **39**, 4937 (1968).

108. M. Shen, *Macromolecules*, **2**, 358 (1969).

109. A. Ciferri, *Macromol. Chem.*, **43**, 152 (1961).

110. K. Dusek, ed., *Polymer Networks*, Springer, New York, 1982.

111. J. E. Mark, B. Erman, and F. R. Eirich, *Science and Technology of Rubber*, 2nd ed., Academic Press, San Diego, 1994.

112. J. E. Mark and J. Lal, eds., *Elastomers and Rubber Elasticity*, American Chemical Society, Washington, DC, 1982.

113. A. V. Galanti and L. H. Sperling, *Polym. Eng. Sci.*, **10**, 177 (1970).

114. B. A. Rozenberg, Presented at *Networks 91*, Moscow, April 1991.

115. B. A. Rozenberg, *Epoxy Resins and Composites II*, Advances in Polymer Science Vol. 75, Springer, Berlin, 1986.

116. S. B. Ross-Murphy, Polymer, **33**, 2622 (1992).

117. L. Z Rogovina and V. G. Vasil'yev, *Networks 91*, Moscow, April 1991.

118. H. H. Winter and F. Chambon, *J. Rheol.*, **30**, 367 (1986).

119. F. Chambon and H. H. Winter, *Polym. Bull.*, **13**, 499 (1985).

120. V. Kudela, *Encyclopedia of Polymer Science and Engineering*, J. I. Kroschwitz, ed., Wiley, New York, 1987.

121. R. M. Ottenbrite, in *Encyclopedia of Polymer Science and Engineering*, J. I. Kroschwitz, ed., Wiley, New York, 1989.

122. T. Miyata and K. Nakamae, *Trends in Polym. Sci. (TRIP)*, **5**, 198 (1997).

123. M. Shibayama and T. Tanaka, in *Responsive Gels: Volume Transitions I*, K. Dusek, ed., Springer-Verlag, Berlin, 1993.

124. M. Ilavsky, in *Responsive Gels: Volume Transitions I*, K. Dusek, ed., Springer, Berlin, 1993.

125. A. R. Khokhlov, S. G. Starodubtzev, and V. V. Vasilevskaya, in *Responsive Gels: Volume Transitions I*, K. Dusek, ed., Springer, Berlin, 1993.

126. W. C. Buzanowski, S. S. Cutie, R. Howell, R. Papenfuss, and C. G. Smith, *J. Chromatography*, **677**, 355 (1994).

127. D. Benda, J. Snuparek, and V. Cermak, *J. Dispersion Sci. Tech.*, **18**, 115 (1997).

128. S. M. Aharoni and S. F. Edwards, *Macromolecules*, **22**, 3361 (1989).

129. T. Tanaka, in *Encyclopedia of Polymer Science and Engineering*, 2nd ed., Vol., 7, Wiley, New York, 1987, p. 514.

130. P. I. Rose, in *Encyclopedia of Polymer Science and Engineering*, 2nd ed., Vol. 7, Wiley, New York, 1987, p. 488.

131. W. R. Krigbaum, J. V. Dawkins, G. H. Via, and Y. I. Balta, *J. Polym. Sci. A-2*, **4**, 475 (1966).

132. F. Rodriguez, *Principles of Polymer Systems*, 4nd ed., Taylor & Francis, Washington, DC, 1996.

133. J. Karger-Kocis, A. Kallo, A. Szafner, G. Bodor, and Z. Senyei, *Polymer*, **20**, 37 (1979).

134. D. W. Bartlett, J. W. Barlow, and D. R. Paul, *J. Appl. Polym. Sci.*, **27**, 2351 (1982).

135. E. Martuscelli, M. Pracela, M. Avella, R. Greco, and G. Ragosta, *Makromol. Chem.*, **181**, 957 (1980).

136. L. H. Sperling, *Polymeric Multicomponent Materials: An Introduction*, Wiley, New York, 1997.

137. J. E. McGrath, *J. Chem. Ed.*, **58** (11), 914 (1981).

138. A. Noshay and J. E. McGrath, *Block Copolymers—Overview and Critical Survey*, Academic Press, Orlando, 1977.

139. G. Holden, *Understanding Thermoplastic Elastomers*, Hanser, Munich, 2000.

140. G. Holden, in *Recent Advances in Polymer Blends, Grafts, and Blocks*, L. H. Sperling, ed., Plenum, New York, 1974.

141. E. Pechhold, G. Pruckmayr, and I. M. Robinson, *Rubber Chem. Technol.*, **53**, 1032 (1980).

142. J. Blackwell and K. H. Gardner, *Polymer*, **20**, 13 (1979).

143. R. P. Brentin, *Rubber World*, **208** (1), 22 (1993).

143. L. L. Zhu, G. Wegner, and U. Bandara, *Makromol. Chem.*, **182**, 3639 (1981).

144. A. Lilaonitkul and S. L. Cooper, *Rubber Chem. Technol.*, **50**, 1 (1977).

145. P. C. Mody, G. L. Wilkes, and K. B. Wagener, *J. Appl. Polym. Sci.*, **26**, 2853 (1981).

146. J. E. Mark, H. R. Allcock, and R. West, *Inorganic Polymers*, Prentice-Hall, Englewood Cliffs, NJ, 1992.

147. D. P. Tate, *J. Polym. Sci. Polym. Symp.*, **48**, 33 (1974).

148. J. E. Mark, H. R. Allcock, and R. West, *Inorganic Polymers*, Prentice-Hall, Englewood Cliffs, NJ, 1992.

149. H. R. Allcock, *Sci. Am.*, **230** (3), 66 (1974).

150. K. R. Birdwhistell and J. W. Long, *J. Chem. Educ.*, **72**, 56 (1995).

151. J. E. Mark, A. Eisenberg, W. W. Graessley, E. T. Samulski, J. L. Koenig, and G. D. Wignall, *Physical Properties of Polymers*, 2nd Ed., ACS Books, Washington, DC, 1993.

152. A. V. Galanti and L. H. Sperling, *Polym. Eng. Sci.*, **10**, 177 (1970).

153. G. Kraus, *Rubber Chem. Technol.*, **51**, 297 (1978).

154. B. B. Boonstra, *Polymer*, **20**, 691 (1979).

155. Z. Rigbi, *Adv. Polym. Sci.*, **36**, 21 (1980).

156. K. E. Polmanteer and C. W. Lentz, *Rubber Chem. Technol.*, **48**, 795 (1975).

157. M.-J. Wang, *Rubber Chem. Technol.*, **71**, 520 (1998).

158. L. J. Murphy, M.-J. Wang, and K. Mahmud, *Rubber Chem. Technol.*, **71**, 998 (1998).

159. L. J. Murphy, E. Khmelnitskaia, M.-J. Wang, and K. Mahmud, *Rubber Chem. Technol.*, **71**, 1015 (1998).

GENERAL READING

M. F. Bukhina, *Technical Physics of Elastomers*, Chemi, Moscow, 1984.

W. Burchard and S. B. Ross-Murphy, *Physical Networks: Polymers and Gels*, Elsevier, New York, 1990.

K. Dusek, Ed., *Polymer Networks*, Springer, New York, 1982.

R. N. Haward, ed., *Developments in Polymerization-3, Network Formation and Cyclization in Polymer Reactions*, Applied Science Publishers, London, 1982.

O. Kramer, Ed., *Biological and Synthetic Polymer Networks*, Elsevier, London, 1988.

J. E. Mark, B. Erman, and F. R. Eirich, eds., *Science and Technology of Rabber*, 2nd ed., academic Press, SanDiego, 1994.

M. Morton, ed., *Rubber Technology*, 3rd ed., Van Nostrand–Reinhold, New York, 1987.

K. te Nijenhuis, *Thermoreversible Networks: Viscoelastic Properties and Structure of Gels*, Advances in Polymer Science Vol. 130, Springer, Berlin, 1997.

J. P. Queslel and J. E. Mark, *Encyclopedia of Polymer Science and Engineering*, 2nd ed., Vol. 5, Wiley, New York, 1986 p. 365.

L. H. Sperling, *Interpenetrating Polymer Networks and Related Materials*, Plenum, New York, 1981.

B. Erman and J. E. Mark, *Structures and Properties of Rubberlike Networks*, Oxford University Press, Oxford, 1997.

STUDY PROBLEMS

1. Why does a rubber band snap back when stretched and released? An explanation including both thermodynamic and molecular aspects is required. (Equations/diagrams/figures and as few words as possible will be appreciated.)

2. A strip of elastomer 1 cm $\times$ 1 cm $\times$ 10 cm is stretched to 25-cm length at 25°C, a stress of 1.5×10^7 dynes/cm^2 being required.

   (a) Assuming a tetrafunctional cross-linking mode, how many moles of network chains are there per cubic centimeter?

   (b) What stress is required to stretch the sample to only 15 cm, at 25°C?

   (c) What stress is required to stretch the sample to 25-cm length at 100°C?

3. The theory of rubber elasticity and the theory of ideal gas dynamics show that the two equations, $G = nRT$ and $PV = n'RT$, share certain common thermodynamic ideas. What are they?

4. Write the chemical structure for polybutadiene, and show its vulcanization reaction with sulfur.

5. For a swollen elastomer, the equation of state can be written

$$\sigma = nRTv_2^{1/3} \frac{\overline{r_i^2}}{r_0^2} \left(\alpha - \frac{1}{\alpha^2} \right)$$

Explain, qualitatively and very briefly, where the term $v_2^{1/3}$ originates. A derivation is not required.

6. Read any paper, 1999 or more recent, on rubber elasticity and write a 200-word report on it *in your own words*. (Give the reference!) Key figures, tables, or equations may be photocopied. How is the science or engineering of elastomers advanced beyond this book?

7. Show how equations (9.81) and (9.82) reduce to $n = n_c + n_p$ for $W_g = 1$. What do you get when $W_g = 0$?

8. Recent experimental evidence using SANS instrumentation suggests that the ends of a network segment deform affinely, yet the chain itself barely extends in the direction of the stress and contracts in the transverse direction even less. Develop a model to explain the results, and comment on how you think the theory of rubber elasticity ought to be modified to accommodate the new finding.

9. We just had Halloween. Do you believe in "phantom networks"? Why?

10. A sample of vulcanized natural rubber, *cis*-polyisoprene, swells to five times its volume in toluene. What is Young's modulus of the unswollen elastomer at 25°C? (The interaction parameter, χ_1, is 0.39 for the system *cis*-polyisoprene–toluene.) [*Hint:* See equation (9.92); the molar volume of toluene is 106.3 cm^3/mol.]

11. Young's modulus for an elastomer at 25°C is 3×10^7 dynes/cm^2. What is its shear modulus? What is the retractive stress if a sample 1 cm $\times$ 1 cm $\times$ 10 cm is stretched to 25 cm length at 100°C?

12. A sample of elastomer, cross-linked at room temperature, is swollen afterward to 10 times its original volume. Then it is stretched. What value of the "front factor" should be used in the calculation of the stress?

13. Two identical 10-cm rubber bands, A and B, are tied together at their ends and stretched to a total of 40 cm length and held in that position. Rubber band A is at 25°C, and rubber band B is at 150°C. How far from the B end is the knot?

14. In the rubber heat engine described in Figure 9.14, the wheel is heated and turns accordingly. Does this experiment have the equivalent of all four steps illustrated in Figure 9.13a? [*Hint:* Don't forget gravity.]

15. A rubber ball is dropped from a height of 1 yard and bounces back 18 in. Assuming a perfectly elastic floor, approximately how much did the ball heat up? The heat capacity, C_p, of SBR rubber is about 1.83 kJ kg^{-1} · K^{-1}.

16. According to recent papers published in the *J. Theor. Hypothet. Polym. Sci.*, wheels go round, plastics break, and balls don't bounce. In one paper of recent vintage, two identical rubber bands, A and B, were dissolved in identical baths A' and B'. The solvent de-cross-linked the rubber bands but otherwise was a simple solvent. Rubber band A was dropped in unstretched. Rubber band B was rapidly stretched from its initial length of 10 cm to 25 cm and instantly placed in the bath while held stretched by a holder of no physical properties during the solution process. Right after the solution process was completed, the poor investigator found he had mixed up the baths. Quick, how would you help him identify the baths? What basic and simple experiment would you perform? Assuming you found the difference, how does this difference change, algebraically, with the extent of stretch of rubber band B? If the two baths are identical, and no difference should exist, write a brief paragraph giving the correct reasons for full credit.

17. A sample of rubber was mixed with sulfur and other curing agents, molded into a sheet, and heated briefly in the relaxed state. A total of n_1 cross-links were introduced. Then the sheet was stretched α times its original length, and the heating continued, introducing n_2 new cross-links. The sample was then released and cooled. In terms of α, to what extent will the sample remain stretched?

18. You are shipwrecked on a desert island. You find some bushes that have a sticky sap. You also find the island has all kinds of minerals. How do you get off the island?

APPENDIX 9.1 GELATIN AS A PHYSICALLY CROSS-LINKED ELASTOMER[†]

Introduction

Ordinary gelatin is made from the skins of animals by a partial hydrolysis of their collagen, an important type of protein (A1,A2). At home, a crude type of gelatin can be prepared from the broth of cooked meats and fowl; this material also frequently gels on cooling.

When dissolved in hot water, the gelatin protein has a random coil type of conformation. On cooling, a conformational change takes place to a partial helical arrangement. At the same time, intermolecular hydrogen bonds form, probably involving the N—H linkage. On long standing, such gels may also crystallize locally. The bonds that form in gelatin are known not to be permanent, but rather they relax in the time frame of 10^3 to 10^6 s (A3–A5). The amount of bonding also decreases as the temperature is raised. The purpose of this appendix is to demonstrate the counting of these bonds via modulus measurements.

[†] Reproduced in part from the G. V. Henderson, D. O. Cambell, V. Kuzmicz, and L. H. Sperling, *J. Chem. Ed.*, **62**, 269 (1985).

Theory

By observing the depth of indentation of a sphere into the surface of gelatin, "indentation" modulus is easily determined. The indentation modulus yields its close relative, Young's modulus. The cross-link density and thus the number of hydrogen bonds (simple physical cross-links) are readily determined by treating the gelatin as a hydrogen-bonded elastomer.

Young's modulus may be determined by indentation using the Hertz (A6) equation:

$$E = \frac{3(1 - v^2)F}{4h^{3/2}r^{1/2}} \tag{A9.1.1}$$

where F represents the force of sphere against the gelatin surface $= mg$ (dynes), h represents the depth of indentation of sphere (cm), r is the radius of sphere (cm), g represents the gravity constant, and v is Poisson's ratio.

The ball indentation experiment is the scientific analogue of pressing on an object with one's thumb to determine hardness. The less the indentation, the higher the modulus.

Young's modulus is related to the cross-link density through rubber elasticity theory; see equation (9.36):

$$E = 3nRT \tag{A9.1.2}$$

Assuming a tetrafunctional cross-linking mode (four chain segments emanating from the locus of the hydrogen bond):

$$E = 6\mu RT \tag{A9.1.3}$$

where n represents the number of active chain segments in network and μ is the cross-link density (moles of cross-links per unit volume). For this experiment, the gelatin was at 278.0 K, the temperature of the refrigerator employed.

Experimental

Time: About 30 minutes, the gelatin prepared previously.
Principles Illustrated:

1. Helix formation and physical cross-linking in gelatin.
2. Rubber elasticity in elastomers.
3. Physical behavior of proteins.

Equipment and Supplies:

Five 150 × 75 mm Pyrex® crystallizing dishes or soup dishes
Five 2-cup packets of flavored Jello® brand gelatin (8 g protein per packet)
Eighteen 2-cup packets of unflavored Knox® brand gelatin (6 g protein per packet)
One metric ruler
One steel bearing (1.5-in. diameter and 0.226 kg—or any similar spherical object)

Table A9.1.1   Gelatin concentrations

Dish	1	2	3	4	5
Concentration[a]	3.0	2.0	1.0	0.75	0.50
Jello[b]	1	1	1	1	1
Gelatin[c]	8	5	2	1.25	0.5

[a] Concentration = number of times the normal gelatin concentration (each dish contains 600 ml of water).

[b] Jello = number of 2-cup packets of Jello® brand black raspberry flavored gelatin.

[c] Gelatin = number of 2-cup packets of Knox® brand unflavored gelatin.

One lab bench

One knife

Five different concentrations (see Table A9.1.1) of gelatin were prepared, each in 600 ml of water, and allowed to set overnight in a refrigerator at 5.0°C. Then indentation measurements were made by placing the steel bearing in the center of the gelatin samples and measuring the depth of indentation, h (see Figure A9.1.1). As it is difficult to see through the gelatin to observe this depth, it is desirable to measure the height of the bearing from the level surface of the gelatin and subtract this quantity from the diameter of the bearing (Figure A9.1.1).

The measured depth of indentation, the radius of the bearing, and the force due to the bearing are algebraically substituted into equation (A9.1.1). This value of Young's modulus is substituted into equation A9.1.3 to yield hydrogen bond cross-link density.

Results

A plot of E as a function of gelatin concentration (Figure A9.1.2) demonstrates a linear increase in Young's modulus at low concentrations. The slight upward curvature at high concentrations is caused by the increasing efficiency of the network. However, the line should go through the origin.

Physical cross-link concentrations were determined using equation (A9.1.3), and the results are shown in Table A9.1.2. Assuming a molecular weight of about 65,000 g/mol for the gelatin, there is about 1.2 physical bonds per molecule (see Table A9.1.2).

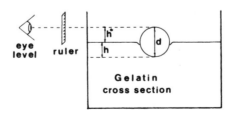

DEPTH OF INDENTATION.

$$h(cm) = d - h^*$$

Figure A9.1.1 Schematic of experiment, measuring the indentations of the heavy ball in the gelatin.

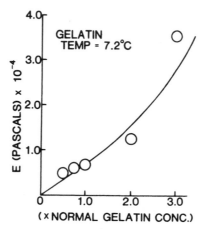

Figure A9.1.2 Modulus of gelatin samples versus concentration. By comparison, a rubber band has a Young's modulus of about 1×10^6 pascals (1 pascal = 10 dynes/cm^2).

Using gelatin as a model cross-linked elastomer, its rubber elasticity can also be demonstrated by a simple stretching experiment. Thin slices of the more concentrated gelatin samples were cut and stretched by hand. On release from stretches up to about 50%, the sample snaps back, illustrating the rubberlike elasticity of these materials. At greater elongation the sample breaks, however. The material is weak because the gelatin protein chains are much diluted with water.

Discussion

For rubbery materials, Young's modulus is related to the number of cross-links in the system. In this case the cross-links are of a physical nature, caused by hydrogen bonding. Measurement of the modulus via ball indentation techniques allows a rapid, inexpensive method of counting these bonds. Table A9.1.2 shows that the number of these bonds is of the order of 10^{-7} mol/cm^3. The number of these bonds also was shown to increase linearly with concentration, except at the highest concentrations.

Table A9.1.2 also demonstrates that at each gelatin concentration, the number of bonds per gelatin molecule is relatively constant. This number, of course, is the number of bonds taking part in three-dimensional network formation. Not all the gelatin chains

Table A9.1.2 Gelatin indentations yield bond numbers

Concentration	h (cm)	μ (mol/cm^3)	N
0.5	1.50	3.5×10^{-7}	1.20
0.75	1.30	4.4×10^{-7}	1.00
1.0	1.20	5.0×10^{-7}	1.00
2.0	0.80	9.1×10^{-7}	0.86
3.0	0.40	2.6×10^{-6}	1.70

Key: h, indentation measured at gelatin temperature of 5°C; C_x, number of hydrogen bonds; N, number of bonds per molecule.

are bound in a true tetrafunctionally cross-linked network. Many dangling chain ends exist at these low concentrations, and the network must be very imperfect.

The gelation molecule is basically composed of short α-helical segments in the form of a triple helix with numerous intramolecular bonds at room temperature; see Section 9.13. The α-helical segments are interrupted by proline and hydroxy proline functional groups. These groups disrupt the helical structure, yielding intervening portions of chain that behave like random coils, and which may be relatively free to develop inter-molecular bonds. The subject has been reviewed by Djabourov (A7) and Mel'nichenko et al. (A8).

In this experiment the concentration of sugar was kept constant so as to minimize its effect on the modulus. In concluding, it must be pointed out that if sanitary measures are maintained, the final product may be eaten at the end of the experiment. If gelation five times normal or higher is included in the study, the student should be prepared for his or her jaws springing open after biting down!

REFERENCES

A1. A. Veis, *Macromolecular Chemistry of Gelatin*, Academic Press, Orlando, 1964.

A2. E. M. Marks, in *Encyclopedia of Chemical Technology*, Kirk-Othmer, Interscience, New York, 1966, Vol. 10, p. 499.

A3. J. L. Laurent, P. A. Janmey, and J. D. Ferry, *J. Rheol.*, **24**, 87 (1980).

A4. M. Miller, J. D. Ferry, F. W. Schremp, and J. E. Eldridge, *J. Phys. Colloid Chem.*, **55**, 1387 (1951).

A5. J. D. Ferry, *Viscoelastic Properties of Polymers*, 3rd ed., Wiley, New York, 1980, pp. 529–539.

A6. L. H. Sperling, *Interpenetrating Polymer Networks and Related Materials*, Plenum Press, New York, 1981, p. 177.

A7. M. Djabourov, *Contemp. Phys.*, **29** (3), 273 (1988).

A8. Yu. Mel'nichenko, Yu. P. Gomza, V. V. Shilov, and S. I. Osipov, *Polym. Intern. (Brit. Polym. J.)*, **25**(3), 153 (1991).

APPENDIX 9.2   ELASTIC BEHAVIOR OF A RUBBER BAND†

Stretching a rubber band makes a good demonstration of the stress–strain relationships of cross-linked elastomers. The time required is about 30 minutes. The equipment includes a large rubber band (Star® band size 107, E. Faber, Inc., Wilkes-Barre, PA, is suitable), a set of weights up to 25 kg, and a meter stick. Also required are hooks to attach the weights and a high place from which to hang the rubber band.

First, the rubber band is measured, both in length and cross section, and the hooks are weighed. Increasing weight is hung from the rubber band, its length being recorded at each step. When it nears its breaking length, caution is advised.

A plot of stress (using initial cross-sectional area) as a function of α, Figure A9.2.1, demonstrates the nonlinearity of the stress–strain relationship. Initial values of the slope of the curve yield Young's modulus, E. The sharp upturn of the experimental curve at

---

†Reproduced in part from A. J. Etzel, S. J. Goldstein, H. J. Panabaker, D. G. Fradkin, and L. H. Sperling, *J. Chem. Ed.*, **63**, 731 (1986).

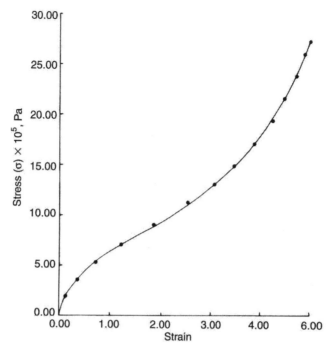

Figure A9.2.1 Simple rubber–elastic behavior of a rubber band under increasing load.

high elongations is due to the limited extensibility of the chains themselves. The number of active network chains per unit volume can be calculated from equation (9.34) as 1.9×10^2 mol/m^3.

A Mooney–Rivlin plot according to equation (9.50) yields a curve that rapidly increases for values of $1/\alpha$ greater than 0.25; see Figure 9.18. The constants $2C_1$ and $2C_2$ are calculated from the intercept and slope, respectively. Values of 2.3×10^5 Pa and 2.8×10^5 Pa were obtained, respectively.

APPENDIX 9.3 DETERMINATION OF THE CROSS-LINK DENSITY OF RUBBER BY SWELLING TO EQUILIBRIUM[†]

The present experiment is based on the rapid swelling of elastomers by organic solvents. Application of the Flory–Rehner equation yields the number of active network chain segments per unit volume, a measure of the extent of vulcanization (C1,C2).

Experiment

Time: About 1 hour

Level: Physical Chemistry

Principles Illustrated:

[†] Reprinted in part from L. H. Sperling and T. C. Michael, *J. Chem. Ed.*, **59**, 651 (1982).

1. The cross-linked nature of rubber
2. Diffusion of a solvent into a solid

Equipment and Supplies:

One large rubber band
One 600-ml beaker (containing 300 ml toluene)
One ruler or yardstick
One long tweezers to remove swollen rubber band
Paper towels to blot wet swollen rubber band
One clock or watch
One lab bench

First, cut the rubber band in one place to make a long rubber strip. Measure and record its length in the relaxed state. Place in the 600-ml beaker with toluene, making sure the rubber band is completely covered. Remove after 5 to 10 minutes. Blot dry. **Caution: toluene is toxic and can be absorbed through the skin.** Again, measure and record length. Repeat for about 1 hour. Optional: Cover and store overnight. Measure the length of the band the next day.

Expected Results

The rubber band swells to about twice its original length, but then it remains stable. Note that swelling to twice its length means a volume increase of about a factor of 8. Also, *note that the swollen rubber band is much weaker than the dry material and may break if not treated gently.*

Chemically most rubber bands and similar materials are composed of a random copolymer of butadiene and styrene, written poly(butadiene–*stat*–styrene), meaning that the placement of the monomer units is statistical along the chain length. Usually this product is made via emulsion polymerization.

Swelling of a Rubber Band with Time

Length, cm	Time, min
16.5	0
24.0	14
26.0	25
27.0	36
28.0	70

Typical results are shown in the table. Over a period of 70 minutes, the length of the rubber band increased from 16.5 to 28.0 cm, for a volume increase of about 4.9. This is sufficiently visible to be seen at the back of an ordinary classroom. The rubber band would continue to swell slowly for some hours, or even days, but for the purposes of demonstrations and classroom calculations, the swelling can be considered nearly complete.

Calculations

For the system poly(butadiene–*stat*–styrene) and toluene, χ_1 is 0.39. Assuming that additivity of volumes v_2 is found from the swelling data to be 0.205. The quantity V_1 is 106.3 cm^3/mol for toluene. Algebraic substitution into equation (9.92) yields n equal to 1.55×10^{-4} mol/cm^3. (Compare result with Appendix 9.2.)

Extra Credit

Two experiments (or demonstrations) can be done easily for extra credit.

1. Obtain some unvulcanized rubber. Most tire and chemical companies can supply this. Put a piece of this material into toluene overnight and observe the results. It should dissolve to form a uniform solution.
2. The quantity n can be used also to predict Young's modulus (the stiffness) of the rubber band. The equation is

$$E = 3nRT \tag{A9.2.1}$$

where E represents Young's modulus, and R in these units is 8.31×10^7 dynes $\cdot$ cm/ mol $\cdot$ K. For the present experiment, E is calculated to be 1.1×10^7 dynes/cm^2, typical of such rubbery products.

REFERENCES

C1. P. J. Flory, and J. Rehner, *J. Chem. Phys.*, **11**, 521 (1943).
C2. J. E. Mark, A. Eisenberg, W. W. Graessley, L. Mandelkern, E. T. Samulski, J. E. Koenig, and G. D. Wignall, *Physical Properties of Polymers*, 2nd ed., American Chemical Society, Washington, DC, 1993, Chap. 1.

10

POLYMER VISCOELASTICITY AND RHEOLOGY

The study of polymer viscoelasticity treats the interrelationships among elasticity, flow, and molecular motion. In reality, no liquid exhibits pure Newtonian viscosity, and no solid exhibits pure elastic behavior, although it is convenient to assume so for some simple problems. Rather, all deformation of real bodies includes some elements of both flow and elasticity. Because of the long-chain nature of polymeric materials, their viscoelastic characteristics come to the forefront. This is especially true when the times for molecular relaxation are of the same order of magnitude as an imposed mechanical stresses.

Chapters 8 and 9 have introduced the concepts of the glass transition and rubber elasticity. In particular, Section 8.2 outlined the five regions of viscoelasticity, and Section 8.6.1.2 derived the WLF equation. This chapter treats the subjects of stress relaxation and creep, the time–temperature superposition principle, and melt flow. Parts of this topic are commonly called rheology, the science of deformation and flow of matter.

10.1 STRESS RELAXATION AND CREEP

In a stress relaxation experiment the sample is rapidly stretched to the required length, and the stress is recorded as a function of time. The length of the sample remains constant, so there is no macroscopic movement of the body during the experiment. Usually the temperature remains constant also (1).

Creep experiments are conducted in the inverse manner. A constant stress is applied to a sample, and the dimensions are recorded as a function of time. Of course, these experiments can be generalized to include shear motions, compression, and so on.

Section 8.1.6 defined the modulus of a material as a measure of its stiffness, and compliance as a measure of its softness. Under conditions far from a transition, $E \cong 1/J$. Frequently stress relaxation experiments are reported as the time-dependent modulus, $E(t)$, whereas creep experiments are reported as the time-dependent compliance, $J(t)$.

433

10.1.1   Molecular Bases of Stress Relaxation and Creep

While the exact molecular causes of stress relaxation and creep are varied, they can be grouped into five general categories (2):

1. *Chain scission.* Oxidative degradation and hydrolysis are the primary causes. The reduction in modulus caused by chain scission during stress relaxation can be illustrated by a model where three chains are bearing a load and one is cut:

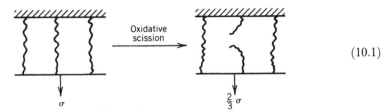

$$(10.1)$$

   Inversely, this causes an increase in elongation during creep.

2. *Bond interchange.* While this is not a degradation in the sense that the molecular weight is decreased, chain portions changing partners cause a release of stress. Examples of stress relaxation by bond interchange include polyesters and polysiloxanes. Equation (10.2) provides a simple example (3).

   Bond interchange is going on constantly in polysiloxanes, with or without stress. In the presence of a stress, however, the statistical rearrangements tend to reform the chains so that the stress is reduced.

3. *Viscous flow.* Caused by linear chains slipping past one another, this mechanism is responsible for viscous flow in pipes and elongational flow under stress. An example is the pulling out of Silly Putty®.

4. *Thirion relaxation* (4). This is a reversible relaxation of the physical cross-links or trapped entanglements in elastomeric networks. Figure 10.1 illustrates the motions involved. Usually an elastomeric network will relax about 5% by this mechanism, most of it in a few seconds. It must be emphasized that the chains are in constant motion of the reptation type (5) (see Section 5.4).

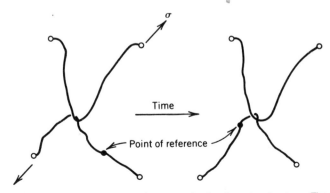

Figure 10.1   An illustration of reversible motion in a trapped entanglement under stress. The marked point of reference moves with time. When the stress is released, entropic forces return the chains to near their original positions.

```
     CH3      CH3    |  CH3
      |        |     |   |
~~Si—O—Si—O—|—Si—O~~
      |        |     |
     CH3      CH3    |  CH3
                               → σ = constant

        CH3   H3C   |    CH3
         |     |    |     |
~O—Si—O—Si—|—O—Si~
         |     |    |     |
        CH3   H3C   |    CH3
```

interchange along dotted line
————————————————→   (10.2)

```
     CH3      CH3     |        CH3
      |        |      |         |
~~~~Si—O—Si—CH3 | CH3—Si—O~~
      |                |         |
     CH3               |         O
                       |
     CH3     O         |         O
      |      |         |         |
~O—Si—O—Si—CH3 | CH3—Si~~
      |      |         |         |
     CH3    CH3        |        CH3
                               → σ = 0
```

5. *Molecular relaxation, especially near T_g.* This topic was the major subject of Chapter 8, where it was pointed out that near T_g the chains relax at about the same rate as the time frame of the experiment. If the chains are under stress during the experiment, the motions will tend to relieve the stress.

It must be emphasized that more than one of the relaxation modes above may be operative during any real experiment.

10.1.2 Models for Analyzing Stress Relaxation and Creep

To permit a mathematical analysis of the creep and relaxation phenomenon, spring and dashpot elements are frequently used (see Figure 10.2). A spring behaves exactly like a metal spring, stretching instantly under stress and holding that stress indefinitely. A dashpot is full of a purely viscous fluid. Under stress, the plunger moves through the fluid at a rate proportional to the stress. On removing the stress, there is no recovery. Both elements may be deformed indefinitely.

10.1.2.1 *The Maxwell and Kelvin Elements* The springs and dashpots can be put together to develop mathematically amenable models of viscoelastic behavior. Figure 10.3 illustrates the simplest two such arrangements, the Maxwell and the Kelvin (sometimes called the Voigt) elements. While the spring and the dashpot are in series in the Maxwell element, they are in parallel in the Kelvin element. In such arrangements it is convenient to assign moduli E to the various springs, and viscosities η to the dashpots.

In the Maxwell element both the spring and the dashpot are subjected to the same stress but are permitted independent strains. The inverse is true for the Kelvin element,

Spring

Dashpot

Figure 10.2 Springs and dashpots are the basic elements in modeling stress relaxation and creep phenomena.

which is equivalent to saying that the horizontal connecting portions on the right-hand side of Figure 10.3 are constrained to remain parallel.

As examples of the behavior of combinations of springs and dashpots, the Maxwell and Kelvin elements will be subjected to creep experiments. In such an experiment a stress, σ, is applied to the ends of the elements, and the strain, ε, is recorded as a function of time. The results are illustrated in Figure 10.4.

On application of the stress to the Maxwell element, the spring instantly responds, as illustrated by the vertical line in Figure 10.4. The height of the line is given by $\varepsilon = \sigma/E$. The spring term remains extended, as the dashpot gradually pulls out, yielding the slanted upward line. This model illustrates elasticity plus flow.

The spring and the dashpot of the Kelvin element undergo concerted motions, since the top and bottom bars (see Figure 10.3) are constrained to remain parallel. The dashpot responds slowly to the stress, bearing all of it initially and gradually transferring it to the spring as the latter becomes extended.

The rate of strain of the dashpot is given by

$$\frac{d\varepsilon}{dt} = \frac{\sigma}{\eta} \tag{10.3}$$

When the spring bears all the stress, both the spring and the dashpot stop deforming together, and creep stops. Thus, at long times, the Kelvin element exhibits the asymp-

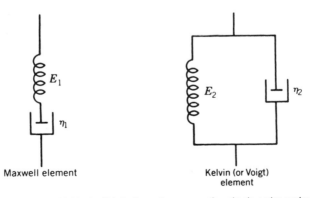

Maxwell element

Kelvin (or Voigt) element

Figure 10.3 The Maxwell and Kelvin (or Voigt) elements, representing simple series and parallel arrays of springs and dashpots.

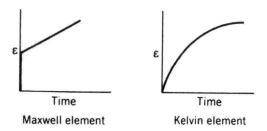

Figure 10.4 Creep behavior of the Maxwell and Kelvin elements. The Maxwell element exhibits viscous flow throughout the time of deformation, whereas the Kelvin element reaches an asymptotic limit to deformation.

totic behavior illustrated in Figure 10.4. In more complex arrangements of springs and dashpots, the Kelvin element contributes a retarded elastic effect.

10.1.2.2 The Four-Element Model

While a few problems in viscoelasticity can be solved with the Maxwell or Kelvin elements alone, more often they are used together or in other combinations. Figure 10.5 illustrates the combination of the Maxwell element and the Kelvin element in series, known as the four-element model. It is the simplest model that exhibits all the essential features of viscoelasticity.

On the application of a stress, σ, the model (Figure 10.5a) undergoes an elastic deformation, followed by creep (Figure 10.5b). The deformation due to η_3, true flow, is nonrecoverable. Thus, on removal of the stress, the model undergoes a partial recovery.

The four-element model exhibits some familiar behavior patterns. Consider the effects of stretching a rubber band around a book. Initially E_1 stretching takes place. As time passes, $E_2 + \eta_2 + \eta_3$ relaxations take place. On removing the rubber band at a later time, the remaining E_1 recovers. Usually the rubber band circle is larger than it was initially. This permanent stretch is due to η_3. Although less obvious, the Kelvin element motions

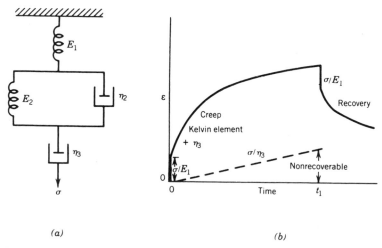

(a)

(b)

Figure 10.5 (a) The four-element model. (b) Creep behavior as predicted by this model. At t_1, the stress is relaxed, and the model makes a partial recovery.

can also be observed by measuring the rubber band dimensions immediately after removal and again at a later time.

The quantities E and η of the models shown above are not, of course, simple values of modulus and viscosity. However, as shown below, they can be used in numerous calculations to provide excellent predictions or understanding of viscoelastic creep and stress relaxation. It must be emphasized that E and η themsevles can be governed by theoretical equations. For example, if the polymer is above T_g, the theory of rubber elasticity can be used. Likewise the WLF equation can be used to represent that portion of the deformation due to viscous flow, or for the viscous portion of the Kelvin element.

More complex arrangements of elements are often used, especially if multiple relaxations are involved or if accurate representations of engineering data are required. The Maxwell–Weichert model consists of a very large (or infinite) number of Maxwell elements in parallel (2). The generalized Voigt–Kelvin model places a number of Kelvin elements in series. In each of these models, a spring or a dashpot may be placed alone, indicating elastic or viscous contributions.

10.1.2.3 The Takayanagi Models
Most polymer blends, blocks, grafts, and interpenetrating polymer networks are phase-separated. Frequently one phase is elastomeric, and the other is plastic. The mechanical behavior of such a system can be represented by the Takayanagi models (6). Instead of the arrays of springs and dashpots, arrays of rubbery (R) and plastic (P) phases are indicated (see Figure 10.6) (7). The quantities λ and φ or their indicated multiplications indicate volume fractions of the materials.

As with springs and dashpots, the Takayanagi models may also be expressed analytically. For parallel model Figure 10.6a, the horizontal bars connecting the two elements must remain parallel and horizontal, yielding an isostrain condition ($\varepsilon_P = \varepsilon_R$). Then

$$\sigma = \sigma_R + \sigma_P \tag{10.4}$$

$$\sigma_i = \varepsilon E_i, \qquad i = P, R \tag{10.5}$$

$$E = (1 - \lambda)E_P + \lambda E_R \tag{10.6}$$

Figure 10.6a represents an upper bound model, meaning that the modulus predicted is the highest achievable.

For example, take a 50/50 blend of a plastic and an elastomer. Typical Young's moduli are $E_P = 3 \times 10^{10}$ dynes/cm^2, and $E_R = 2 \times 10^7$ dynes/cm^2. The quantity λ equals 0.5, and equation (10.6) yields 1.5×10^{10} dynes/cm^2. To an excellent approximation, this is half the modulus of the plastic.

For the series model Figure 10.6b, an isostress strain condition exists. The strains are additive,

$$\varepsilon = \varepsilon_P + \varepsilon_R \tag{10.7}$$

yielding a Young's modulus of

$$E = \left(\frac{\varphi}{E_R} + \frac{1 - \varphi}{E_P} \right)^{-1} \tag{10.8}$$

This is a lower bound modulus, meaning that equation (10.8) yields the lowest modulus

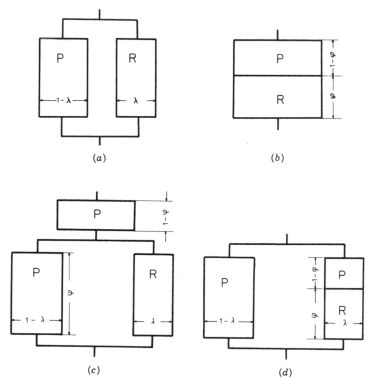

Figure 10.6 The Takayanagi models for two-phase systems: (*a*) an isostrain model; (*b*) isostress model; (*c, d*) combinations. The area of each diagram is proportional to a volume fraction of the phase.

any material may have. If the example above is used, with $\varphi = 0.5$, equation (10.8) yields $E = 4 \times 10^7$ dynes/cm^2, or approximately twice the elastomer value.

Surprisingly, it is possible to make two different 50/50 blends of plastics and elastomers and obtain very nearly both of the results predicted by the above examples. This depends on the morphology, and particularly on phase continuity; see Figure 4.3. When the plastic phase is continuous in space and the elastomer dispersed (Figure 10.6*a*), the material will be stiff, and substantially exhibit upper bound behavior. On the other hand, when the elastomer is continuous and the plastic discontinuous (Figure 10.6*b*), a much softer material results and a lower bound situation prevails. The rubber phase is much more obviously discontinuous in the models represented in Figures 10.6*c* and 10.6*d*.

While the examples above used Young's modulus, many other parameters may be substituted. These include other moduli, rheological functions such as creep, stress relaxation, melt viscosity, and rubber elasticity. Each element may itself be expressed by temperature-, time-, or frequency-dependent quantities. It must also be noted that these models find application in composite problems as well. For example, a composite of continuous fibers in a plastic matrix can be described by Figure 10.6*a* if deformed in the direction of the fibers, and by Figure 10.6*b* if deformed in the transverse direction.

Figures 10.6*c* and 10.6*d* represent a higher level of sophistication, allowing for more precise calculations involving continuous and discontinuous phases. For Figures 10.6*c* and 10.6*d*, respectively, Young's moduli are given by

$$E = \left[\frac{\varphi}{\lambda E_R + (1 - \lambda)E_P} + \frac{1 - \varphi}{E_P} \right]^{-1} \tag{10.9}$$

$$E = \lambda \left(\frac{\varphi}{E_R} + \frac{1 - \varphi}{E_P} \right)^{-1} + (1 - \lambda)E_P \tag{10.10}$$

For these latter models, the volume fraction of each phase is given by a function of λ multiplied by a function of φ. The volume fraction of the upper right-hand portion of Figure 10.6*d* is given by $\lambda(1 - \varphi)$. Equations (10.9) and (10.10) simplify for numerical analysis if φ is taken equal to λ.

10.2   RELAXATION AND RETARDATION TIMES

The various models were invented explicitly to provide a method of mathematical analysis of polymeric viscoelastic behavior. The Maxwell element expresses a combination of Hooke's and Newton's laws. For the spring,

$$\sigma = E\varepsilon \tag{10.11}$$

the time dependence of the strain may be expressed as

$$\frac{d\varepsilon}{dt} = \frac{1}{E}\frac{d\sigma}{dt} \tag{10.12}$$

The time dependence of the strain on the dashpot is given by

$$\frac{d\varepsilon}{dt} = \frac{\sigma}{\eta} \tag{10.13}$$

Since the Maxwell model has a spring and dashpot in series, the strain on the model is the sum of the strains of its components:

$$\frac{d\varepsilon}{dt} = \frac{1}{E}\frac{d\sigma}{dt} + \frac{\sigma}{\eta} \tag{10.14}$$

10.2.1   The Relaxation Time

For a Maxwell element the relaxation time is defined by

$$\tau_1 = \frac{\eta}{E} \tag{10.15}$$

The viscosity, η, has the units of dynes $\cdot$ sec/cm^2, and the modulus, E, has the units of dynes/cm^2, so τ_1 has the units of time. Thus τ_1 relates modulus and viscosity.

On a molecular scale the relaxation time of a polymer indicates the order of magnitude of time required for a certain proportion of the polymer chains to relax—that is, to respond to the external stress by thermal motion. It should be noted that the chains are in constant thermal motion whether there is an external stress or not. The stress tends to be relieved, however, when the chains happen to move in the right direction, degrade, and so on.

Alternatively, τ_1 can be a measure of the time required for a chemical reaction to take place. Common reactions that can be measured in this way include bond interchange, degradation, hydrolysis, and oxidation. Combining equations (10.14) and (10.15) leads to

$$\frac{d\varepsilon}{dt} = \frac{1}{E}\frac{d\sigma}{dt} + \frac{\sigma}{\tau_1 E} \tag{10.16}$$

where the first term on the right is important for short-time changes in strain, and the second term on the right controls the longer-time changes.

The stress relaxation experiment requires that $d\varepsilon/dt = 0$; that is, the length does not change, and the strain is constant. Integrating equation (10.16) under these conditions leads to

$$\sigma = \varepsilon_0 E e^{-(t/\tau_1)} \tag{10.17}$$

or

$$\sigma = \sigma_0 e^{-(t/\tau_1)} \tag{10.18}$$

Equations (10.17) and (10.18) predict a straight-line relationship between $\ln \sigma$ or $\log \sigma$ and linear time if a single mechanism controls the relaxation process. If experiments other than simple elongation are done (i.e., relaxation in shear), the appropriate modulus replaces E in equation (10.17).

10.2.2 Applications of Relaxation Times to Chemical Reactions

10.2.2.1 Chemical Stress Relaxation
As indicated in Section 10.1.1, stress relaxation can be caused by either physical or chemical phenomena. Examples will be given of each.

Figure 10.7 (8) illustrates the stress relaxation of a poly(dimethyl siloxane) network, silicone rubber, in the presence of dry nitrogen. The reduced stress, $\sigma(t)/\sigma(0)$, is plotted, so that under the initial conditions its value is always unity. Since the theory of rubber elasticity holds (Chapter 9), what is really measured is the fractional decrease in effective network chain segments. The bond interchange reaction of equation (10.2) provides the chemical basis of the process. While the rate of the relaxation increases with temperature, the lines remain straight, suggesting that equation (10.2) can be treated as the sole reaction of importance.

The relaxation times may be estimated from the time necessary for $\sigma(t)/\sigma(0)$ to drop to $1/e = 0.368$. The results are shown in Table 10.1 (8).

Appendix 10.1 derives a relationship between the relaxation time and the energy of activation, ΔE_{act},

$$\tau_1 = \text{constant} \times e^{\Delta E_{act}/RT} \tag{10.19}$$

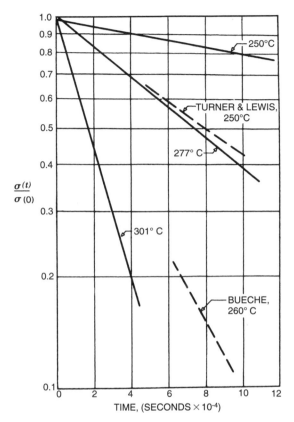

Figure 10.7 Stress relaxation of silicone rubber, poly(dimethyl siloxane). The rate of stress relaxation increases with temperature, but the lines remain straight (8). Equation (10.18) suggests that a logarithmic y-axis should produce straight lines.

A plot of ln τ_1 versus $1/T$ yields $\Delta E_{act}/R$ for the slope. An apparent energy of activation of 35 kcal/mol was estimated from such a plot. Thus a purely chemical quantity can be deduced from a mechanical experiment.

Stress relaxation experiments were used to determine the mechanism of degradation in synthetic polymers (9) during World War II, when these materials were first being made. Tobolsky later described the results of these famous experiments (2):

Table 10.1   Chemical stress relaxation times for silicone rubber (8)

Temperature, °C	$\tau_1 \times 10^{-4}$, s
250	48
277	10.5
301	2.45

It was found that in the temperature range of 100 to 150°C, these vulcanized rubbers showed a fairly rapid decay to zero stress at constant extension. Since in principle a cross-linked rubber network in the rubbery range of behavior should show little stress relaxation, and certainly no decay to zero stress, the phenomenon was attributed to a chemical rupture of the rubber network. This rupture was specifically ascribed to the effect of molecular oxygen since under conditions of *very low* oxygen pressures ($<10^{-4}$ atm) the stress–relaxation rate was markedly diminished. However at moderately low oxygen pressures the rate of chemical stress relaxation was the same as at atmospheric conditions. This result parallels the very long established fact that in the liquid phase the rate of reaction of hydrocarbons with oxygen is independent of oxygen pressure down to fairly low pressures.

10.2.2.2 *Procedure X*

Stress relaxation can also be used to separate and identify two or more reactions causing relaxation, provided the rates are sufficiently different. Consider reactions *a* and *b* going on simultaneously:

$$\sigma(t) = \sigma_a(0)e^{-t/\tau_{1a}} + \sigma_b(0)e^{-t/\tau_{1b}} \tag{10.20}$$

If the two relaxation times, chemical or physical, are sufficiently different, two straight lines may be obtained by algebraic analysis. Tobolsky named this method of analysis "procedure X" (2).

10.2.2.3 *Continuous and Intermittent Stress Relaxation*

Another "trick" to separate two reactions involves continuous and intermittent stress relaxation measurements. Figure 10.8 illustrates the separation of degradation and cross-linking in cis-1,4-polybutadiene. In an intermittent stress relaxation experiment, the sample is maintained in a relaxed, unstretched condition at a constant temperature. At suitably spaced time intervals the rubber is rapidly stretched to a fixed elongation, and the stress is measured. Then the sample is returned to its unstretched length. Of course, in the continuous stress relaxation experiment, the strain is maintained continuously.

At 130°C, the temperature of the experiment, oxidative scission and cross-linking are both going on all the time. The reactions happen, however, in the condition of the network at the time. This means that the continuous stress relaxation experiment measures only the degradation step, because the new cross-links form in the stretched chains; that is, the second network develops in the extended state, at equilibrium. There is no significant change in conformational entropy on formation of these cross-links, and the change in stress is near zero.

On the other hand, if the oxidative cross-links form when the sample is unstretched, then they can be measured afterward by using the theory of rubber elasticity. The intermittent experiment measures the total number of active chain segments at any given instant of time. As illustrated in Figure 10.8 the cross-linking reaction predominates after 10 hours, and the sample actually gets harder. This last is often observed at home, where old rubber materials tend to stiffen up.

10.2.3 The Retardation Time

As the relaxation time, τ_1, is defined for the Maxwell elements, the retardation time, τ_2, is defined for the Kelvin element. The equation for the Kelvin element under stress can be written

$$\sigma = \eta\frac{d\varepsilon}{dt} + E\varepsilon \tag{10.21}$$

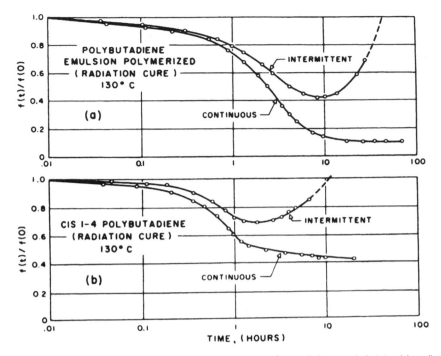

Figure 10.8 Continuous and intermittent stress relaxation at 130°C on radiation cured *cis*-1,4-polybutadiene in the presence of air. The experiment separates oxidative cross-linking and chain scission. The quantity $f(t)$ is the time-dependent retractive force. From A. V. Tobolsky, *Properties and Structure of Polymers*, Wiley, New York, 1960.

Under conditions of constant stress, equation (10.21) can be integrated to

$$\varepsilon = \frac{\sigma}{E}(1 - e^{-(E/\eta)t}) \tag{10.22}$$

which can be rewritten in terms of the retardation time,

$$\varepsilon = \frac{\sigma}{E}(1 - e^{-t/\tau_2}) \tag{10.23}$$

The equation for the four-element model (Figure 10.5a) may now be written for the condition of constant stress:

$$\varepsilon = \varepsilon_1 + \varepsilon_2 + \varepsilon_3 \tag{10.24}$$

$$\varepsilon = \frac{\sigma}{E_1} + \frac{\sigma}{E_2}(1 - e^{-t/\tau_2}) + \frac{\sigma}{\eta_3}t \tag{10.25}$$

The first term on the right of equation (10.25) represents an elastic term, the second term expresses the viscoelastic effect, and the third term expresses the viscous effect.

It is of interest to compare the retardation time with the relaxation time. The retardation time is the time required for E_2 and η_2 in the Kelvin element to deform to $1 - 1/e$, or 63.21% of the total expected creep. The relaxation time is the time required for E_1 and η_3 to stress relax to $1/e$ or 0.368 of σ_0, at constant strain. Both τ_1 and τ_2, to a first approximation, yield a measure of the time frame to complete about half of the indicated phenomenon, chemical or physical. A classroom demonstration experiment showing the determination of the constants in the four-element model is shown in Appendix 10.2.

10.2.4 Dynamic Mechanical Behavior of Springs and Dashpots

In addition to stress relaxation and creep, springs and dashpots can model the loss and storage characteristics of polymers undergoing cyclic motions. Since a principal application of such modeling is for noise and vibration damping analysis (see Section 8.12) where the motions are of a shearing nature, the equations for shear are emphasized below. The viscoelastic motions of a Maxwell element in shear may be written for an angular frequency ω (rad/s) (10):

$$J(t) = J + \frac{t}{\eta} \tag{10.26}$$

$$G(t) = Ge^{-t/\tau_1} \tag{10.27}$$

$$G'(\omega) = \frac{G\omega^2\tau_1^2}{1 + \omega^2\tau_1^2} \tag{10.28}$$

$$G''(\omega) = \frac{G\omega\tau_1}{1 + \omega^2\tau_1^2} \tag{10.29}$$

$$J'(\omega) = J \tag{10.30}$$

$$J''(\omega) = \frac{J}{\omega\tau_1} = \frac{1}{\omega\eta_1} \tag{10.31}$$

$$\tan\delta = \frac{1}{\omega\tau_1} \tag{10.32}$$

The scaling of time in relaxation processes is achieved by means of the "Deborah number," which is defined as

$$D_e = \tau_e/t \tag{10.33}$$

where t is a characteristic time of the deformation process being observed and τ_e is a characteristic time of the polymer. For a Hookian elastic solid, τ_e is infinite, and for a Newtonian viscous liquid, τ_e is zero; see Section 10.2.1. For polymer melts, τ_e is often of the order of a few seconds. Of course, τ_e can be the relaxation time, or the retardation time of the polymer. The Deborah number is widely used in engineering as a single-number approximation to the several molecular phenomena involved in the deformation and flow of many materials, including polymers.

The loss tangent is seen to be a maximum when $\tau_1 = 1/\omega$—that is, when the time required for one cycle of the experiment equals the relaxation time. Of course, $G = 1/J$ far from a transition (Chapter 8).

The corresponding quantities for the Kelvin element are as follows (10):

$$J(t) = J(1 - e^{-t/\tau_2}) \tag{10.34}$$

$$G(t) = G \tag{10.35}$$

$$G'(\omega) = G \tag{10.36}$$

$$G''(\omega) = G\omega\tau_2 = \omega\eta_2 \tag{10.37}$$

$$\eta'(\omega) = \eta \tag{10.38}$$

$$J'(\omega) = \frac{J}{1 + \omega^2\tau_2^2} \tag{10.39}$$

$$J''(\omega) = \frac{J\omega\tau_2}{1 + \omega^2\tau_2^2} \tag{10.40}$$

$$\tan \delta = \omega\tau_2 \tag{10.41}$$

If springs are equated to capacities and resistances to dashpots, the storage and dissipative units are seen to correspond to time-dependent electrical behavior. However, the topology is backward; that is, series electrical connections correspond to parallel mechanical connections (10).

10.2.5 Molecular Relaxation Processes

The preceding sections describe the mechanical behavior of a polymeric sample in terms of creep and stress relaxation. Both elastomeric and plastic materials can be modeled by combinations of springs and dashpots. However, these are only models. Ultimately stress relaxation and creep derive from molecular origins, and it is in this area that more recent studies have been concentrated (11–15).

The most important method of characterizing the molecular relaxation processes has been through the use of small-angle neutron scattering (see Figure 10.9) (16). Boue et al. (15) investigated linear polystyrene of medium and high molecular weight. In each case, blends of deuterated and protonated polystyrenes were prepared, and the samples stretched up to an α value of 3 at temperatures above T_g in the range of 113 to 134°C. The samples were held for various periods of time at that extension, the stress being recorded (15), and then the samples were quickly quenched to the glassy state at room temperature. Then SANS measurements were made in both the perpendicular and parallel directions.

Figure 10.10 (15) shows the changes in the transverse radius of gyration with time, presented in the form of a master curve (see next section). On an absolute scale, R_g for the isotropic sample was 280 Å, and R_g for the samples undergoing relaxation ranged from 161 to 210 Å. Boue et al. (15) pointed out that immediately after stretching their sample, the radius of gyration showed affine deformation. This can be interpreted as indicating that the experiment did not fail to capture any major coil relaxation process.

Figure 10.10 (15) also applies the molecular diffusion theories of de Gennes (17), Doi and Edwards (18), and Daoudi (19), which assume that the major mode of molecular relaxation is by reptation. In this way, the chains move back and forth within a hypothetical tube. Relaxation occurs by the chain disengaging itself from the tube, only at the ends, in a backward-and-forward reptation.

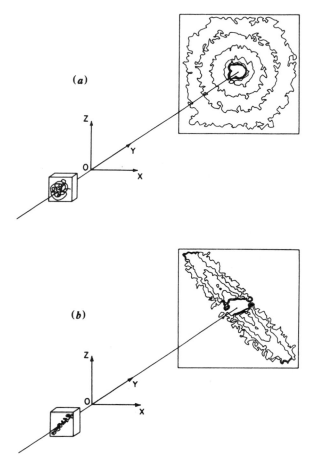

Figure 10.9 Geometry of the SANS experiment and resulting intensity contour plots: (a) unoriented amorphous polymer and (b) oriented polymer (16).

Two characteristic times are employed: (a) T_d, defined as the longest of the relaxation times for defect equilibration, proportional to M^2, and (b) T_r, defined as the renewal time for chain conformation, proportional to M^3. This latter time is the time to form a new isotropic tube. For polystyrene of $M_w = 650,000$ g/mol at 117°C, they found that

$$3 \times 10^3 \text{ s} \leq T_d \leq 4 \times 10^4 \text{ s} \tag{10.42}$$

$$6 \times 10^5 \text{ s} \leq T_r \leq 1 \times 10^7 \text{ s} \tag{10.43}$$

The data in Figure 10.10 are in good agreement with the value of T_r. The authors state that more data are required to determine T_d decisively. The reader is referred to Section 5.4.2 and Appendix 5.2.

By way of contrast, Maconnachie et al. (12), using polystyrene of $M_w = 144,000$ g/mol, found that most of the relaxation was complete in about 5 minutes at 120°C. However, the chains never quite returned to their initial dimensions, both $R_\parallel$ and $R_\perp$

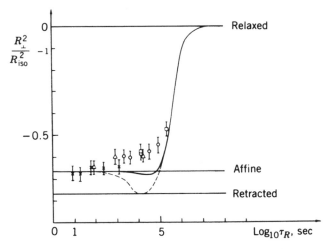

Figure 10.10 Molecular relaxation of polystyrene at different relaxation times and temperatures, as followed by SANS data taken in the transverse direction (15). Data are reduced to 117°C (just above T_g) via the time–temperature superposition principle.

remaining slightly higher than the unstretched polymer dimensions even after 1×10^4 s. The deformation appeared to behave as if the affine deformation theory held only for distances separating effective cross-links.

For better comparisons, these data are in the rubbery flow portion of the spectrum. The SANS data in Figure 10.10 correspond to the earlier stress relaxation studies by Tobolsky (20) on almost identical polystyrenes.

More recently Wool (21) described three relaxation times. Referring to Figure 10.11, the relaxation time is given by

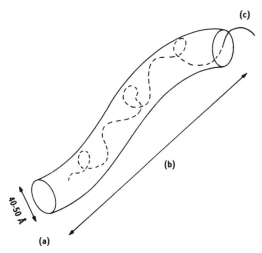

Figure 10.11 Tube model illustrating three relaxation times, each with a specific motion: (a) across the tube, (b) along the tube, and (c) in and out of the tube.

$$T_e \sim M_e^2 \qquad (10.44)$$

which is constant for a given polymer. The quantity M_e refers to the segment molecular weight involved in movement about the diameter of the equivalent tube. It is also the molecular weight between entanglements. The Rouse relaxation time of the whole chain is given by

$$T_R \sim M^2 \qquad (10.45)$$

which expresses the time for one end of the chain to respond to or communicate with the other end. If one end of the chain is stretched, this is the characteristic time for the other end to retract. The reptation time is given by

$$T_r \sim M^3 \qquad (10.46)$$

similar to the de Gennes theory, where M is the molecular weight of the whole chain.

10.2.6 Diffusion at Short Times

Modern theory differentiates two cases for the time dependence of diffusion. The steady state condition first described by de Gennes (Section 5.4.2) holds for times longer than T_r,

$$X(t) \sim t^{1/2} M^{-1} \qquad (10.47)$$

where $X(t)$ represents the average mer interpenetration depth into new territory. For times shorter than T_r,

$$X(t) \sim t^{1/4} M^{-1/4} \qquad (10.48)$$

Equation (10.48) is particularly useful for studies of crack healing, welding time in molding operations, and film formation from latexes (21). Alternately, equations (10.47) and (10.48) can be cast in the form,

$$X(t) = X_\infty \left(\frac{t}{T_r}\right)^{1/2} \qquad (10.49)$$

and

$$X(t) = X_\infty \left(\frac{t}{T_r}\right)^{1/4} \qquad (10.50)$$

respectively. Equations (10.48) and (10.50) provide a nonclassical mode of diffusion, called minor chain reptation (21), before the chain completely escapes its tube.

The value of X_∞ is approximately $0.4R_g$ (22). The actual break point between $t^{1/4}$ and $t^{1/2}$ depends on the model (23). If center-of-mass motion from one side of the interface is examined, where the center-of-mass of the chains at the interface are about $0.4R_g$ from the interface, then Wool (21) calculates that the interface is completely healed at about $0.8R_g$ total interdiffusion distance. However, interdiffusion is measured from the surface, rather than from $0.4R_g$ inside, leading to the experimental value of $0.4R_g$ (24). With symmetrical interfaces, interdiffusion from both sides simultaneously leads to a model of $0.4R_g + 0.4R_g = 0.8R_g$ total motion. From a physical point of view, after about $0.8R_g$ total interdiffusion, the chains overlap such that the interface is healed.

Therefore the interdiffusion action described by equations (10.48) or (10.50) is substantially complete when the chains have diffused a distance of approximately $0.8R_g$ (21). At that point, the material has gained substantially all of its mechanical properties, such as tensile strength, impact resistance, and so on, which is to say, the material is healed; see Section 11.5. Further interdiffusion merely results in chains wandering randomly in the material.

10.2.7 Relationships among Molecular Parameters

Fetters et al. (25) summarized several molecular relationships, which broadly express values for random coil polymers,

$$G^0 = 10.52 MPa(\mathring{A})^3 \left(\frac{r^2}{M}\rho N_a\right)^3 \tag{10.51}$$

$$M_e = 2225.8 \frac{cm^3}{(\mathring{A})^3\,mol} \left(\frac{r^2}{M}\right)^{-3} \rho^{-2} N_a^{-3} \tag{10.52}$$

$$d_t = 19.36 \left(\frac{r^2}{M}\right)^{-1} \rho^{-1} N_a^{-1} \tag{10.53}$$

$$\frac{d_t^2}{M_e} = \frac{r^2}{M} \tag{10.54}$$

where G^0 represents the shear plateau modulus, ρ the polymer density, M_e the molecular weight between entanglements, d_t the reptation tube diameter (see Figure 10.11), r the end-to-end distance of the unperturbed (relaxed) chain, M the molecular weight of the polymer chain, and N_a is Avogadro's number.

The ratio r^2/M is a constant, see Table 5.4, where $r^2/M = 6R_g^2/M$. The constants shown in equations (10.51) to (10.54) are universal for random coil polymers of high enough molecular weight. Typical values for M_e and d_t are shown in Table 10.2 (4).

Table 10.2 Molecular characteristics of linear polymers at 413 K

Polymer	M_e, g/mol	d_t, Å
Polyethylene	828	32.8
Poly(ethylene oxide)	1624	37.5
Poly(methyl methacrylate)	10013	67.0
Polystyrene	13309	76.5
Poly(dimethyl siloxane)	12293	78.6
1,4-Polybutadiene	1815	44.4
1,4-Polyisoprene	6147	62.0
a-Polypropylene	4623	60.7

Source: Values taken from Table 1, L. J. Fetters, D. J. Lohse, D. Richter, T. A. Witten, and A. Zirkel, *Macromolecules*, **27**, 4639 (1994).

10.2.7.1 *Example Problem for Polystyrene*

Values for polystyrene are $r^2/M = 0.275^2 \times 6 = 0.434\,\text{Å}^2\text{mol/g}$ (from Table 5.4), and $\rho = 1.06$ g/cm^3 (Table 3.2). Then $G^0 = 2.24 \times 10^5$ Pa, $M_e = 1.13 \times 10^4$ g/mol, and $d_t = 70.0\,\text{Å}$. The value of G^0 should be compared with the plateau value in Figure 9.21, where G^0 equals about 2×10^5 Pa, providing excellent agreement with experiment. The idea of a plateau modulus implies that the shear rate or frequency is higher than the relaxation rate of the chains, which means that they do not become disentangled during the experiment. This differs from the concept of an equilibrium, infinite time modulus, which is zero for linear amorphous polymers above their glass transition temperature, or cross-linked polymers short of their gel point.

Thus the conformational statistics of chain molecules are now increasingly able to provide a basis for estimating the rheological and viscoelastic behavior of linear amorphous polymers above their glass transition temperatures.

10.2.7.2 *Example Problem Relating to Adhesives*

Many modern adhesives, particularly pressure sensitive adhesives (see Section 12.8.4), deliberately utilize partly crosslinked, partly linear polymers (26). An important factor relating to the strength of the adhesive is whether single chains are able to diffuse into the cross-linked portions, thus forming a linear chain-cross-linked chain entanglement. It is known that strong contact adhesives require $M_c > M_e$, where the first term reads on the cross-linked portion, and the second term reads on the linear polymer (26). (Usually the linear polymer and the crosslinked polymer have similar compositions.)

As a simplified model, M_c will assume monodisperse values in a regular tetrafunctionally cross-linked network. The tube diameter of the linear polymer, d_t, must fit into the net. One side of the net then must be larger than d_t to permit interdiffusion. If as a minimum requirement d_t is taken equal to the equivalent radius of gyration, R_g, then

$$d_t = R_g$$

$$R_g = (KM_c)^{1/2}$$

values of K are shown in Table 5.4; multiply by $6^{1/2}$ for end-to-end distances, and

$$d_t = (M_e K)^{1/2}$$

yielding

$$M_e = M_c$$

for this simplified model.

Taking poly(methyl methacrylate) as an example, $M_e = 10{,}013$ g/mol (Table 10.2), M_c should be equal or greater than 10,013 g/mol. Experiments by Zosel and Ley (27) on crosslinked latex films show that values of $M_c \geq M_e$ are required for good mechanical properties, the oversimplified model above providing the minimum net size. Aspects of adhesion are discussed further in Section 12.8.

Figure 10.12 (22) illustrates stress relaxation in poly(methyl methacrylate) resulting from molecular relaxation processes. Here, the logarithm of the relaxation Young's modulus is plotted against the logarithm of time. At low temperatures, the polymer is glassy, and only slow relaxation is observed. As the glass transition temperature is ap-

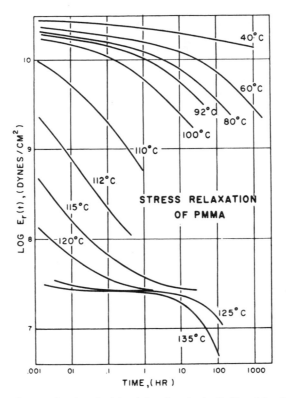

Figure 10.12   Stress–relaxation of poly(methyl methacrylate) with $M_v = 3.6 \times 10^6$ g/mol (20).

proached (106°C at 10 s), the relaxation rate increases, reaching a maximum just above the classical glass transition temperature range. Then the rate of relaxation decreases as the rubbery plateau is approached. This is exemplified by the data at 125 and 135°C in the range of 0.01 to 1 h. At higher temperatures or longer times, the polymer begins to flow.

The data in Figure 10.12 thus show two distinct relaxation phenomena: first, chain portions corresponding to 10 to 50 carbon atoms are relaxing, which corresponds to the glass–rubber transition; then, at higher temperatures, whole chains are able to slide past one another.

10.2.8 Physical Aging in the Glassy State

Whereas the molecular motion in the rubbery and liquid states involves 10 to 50 carbon atoms, molecular motion in the glassy state is restricted to vibrations, rotations, and motions by relatively short segments of the chains.

The extent of molecular motion depends on the free volume. In the glassy state, the free volume depends on the thermal history of the polymer. When a sample is cooled from the melt to some temperature below T_g and held at constant temperature, its volume will decrease (see Figure 8.18). Because of the lower free volume, the rate of stress relaxation, creep, and related properties will decrease (29–33). This phenomenon is sometimes called physical aging (29,32), although the sample ages in the sense not of degradation or oxidation but rather of an approach to the equilibrium state in the glass.

The effect of physical aging can be illustrated through a programmed series of creep studies. First, the sample is heated to a temperature T_0, about 10 to 15°C above T_g. A period of 10 to 20 minutes suffices to establish thermodynamic equilibrium. Then the sample is quenched to a temperature T_1 below T_g, and kept at this temperature. At a certain elapsed time after the quench, a creep experiment is started. The sample is subjected to a constant stress, σ_0; the resulting strain, ε, is determined as a function of time. The sample is then allowed to undergo creep recovery. Each creep period is short in comparison with the previous aging time as well as the last recovery period preceding it. The student should note the difference between the two time scales: t_e is the aging time, beginning at the time of quenching, whereas t is the creep time, which begins at the moment of loading for each run.

Figure 10.13 (29) illustrates typical creep compliance results for poly(vinyl chloride), which has a glass transition temperature of about 80°C. The most important conclusion obtained from such studies is that the rate of creep slows down as the sample ages. Note that in the first and last curves in Figure 10.13, the polymer reaches a tensile creep compliance of 5×10^{-10} m^2/N after about 10^3 s after being aged for 0.03 days, but after an aging period of 1000 days, 10^7 s are required to reach the same tensile creep compliance level.

The problem of determining the theoretical behavior of polymers in the glassy rates is treated by Curro et al. (34,35). The time dependence of the volume in the glassy state is accounted for by allowing the fraction of unoccupied volume sites to depend on time.

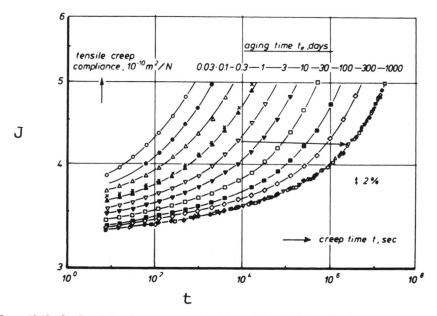

Figure 10.13 Small-strain tensile creep curves for glassy poly(vinyl chloride) quenched from 90°C (about 10°C above T_g) to 40°C and stored at that temperature for four years. The different curves were measured for various times t_e elapsed after the quench. The master curve gives the result of a superposition by shifts that were almost horizontal; the shifting direction is shown by the arrow. The crosses refer to another sample quenched in the same way, but only measured for creep at t_e of 1 day (29).

This permits the application of the Doolittle equation to predict the shift in viscoelastic relaxation times,

$$\tilde{D} = \tilde{D}_r \exp[-B(f^{-1} - f_r^{-1})] \tag{10.55}$$

where $\tilde{D}$ is the diffusion constant for holes, and the subscript r represents the reference state, taken for convenience at the glass transition temperature, and f is the fractional free volume; see Section 8.6.12.

10.3 THE TIME–TEMPERATURE SUPERPOSITION PRINCIPLE

10.3.1 The Master Curve

As indicated above, relaxation and creep occur by molecular diffusional motions which becomes more rapid as the temperature is increased. Temperature is a measure of molecular motion. At higher temperatures, time moves faster for the molecules. The WLF equation, derived in Section 8.6.1.2, expresses a logarithmic relationship between time and temperature. Building on these ideas, the time–temperature superposition principle states that with viscoelastic materials, time and temperature are equivalent to the extent that data at one temperature can be superimposed on data at another temperature by shifting the curves along the log time axis (2).

The importance of these ideas becomes clear when one considers that data can be obtained conveniently over only a narrow time scale, say, from 1 to 10^5 s (see Figure 10.12). The time–temperature superposition principle allows an estimation of the relaxation modulus and other properties over many decades of time.

Figure 10.14 (36–38) illustrates the time–temperature superposition principle using polyisobutylene data. The reference temperature of the master curve is 25°C. The reference temperature is the temperature to which all the data are converted by shifting the curves to overlap the original 25°C curve. Other equivalent curves can be made at other temperatures. The shift factor shown in the inset corresponds to the WLF shift factor, especially at the lower temperatures. Thus the quantitative shift of the data in the range T_g to $T_g + 50$°C is governed by the WLF equation, and this equation can be used both to check the data and to estimate the shift where data are missing.

Most interestingly, the master curve shown in Figure 10.14 looks much like the modulus temperature curves (see Figures 1.6 and 8.2). This likeness derives, of course, from the equivalence of log time with temperature.

In multicomponent polymeric systems such as polymer blends or blocks, each phase stress relaxes independently (39–41). Thus each phase will show a glass–rubber transition relaxation. While each phase follows the simple superposition rules illustrated above, combining them in a single equation must take into account the continuity of each phase in space. Attempts to do so have been made using the Takayanagi models (41), but the results are not simple.

10.3.2 The Reduced Frequency Nomograph

As illustrated in Figure 10.14, the master curve allows the extrapolation of data over broad temperature and time ranges. Similar master curves can be constructed with frequency as the variable, instead of time. More elegant still is the reduced frequency

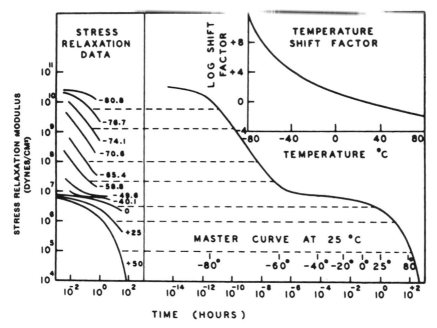

Figure 10.14   The making of a master curve, illustrated with polyisobutylene data. The classical T_g at 10 s of this polymer is $-70°C$ (36–38).

nomograph, which permits both reduced frequency and temperature simultaneously; see Figure 10.15 (42).

The reduced frequency nomograph (42,43) is constructed as follows. First, the storage and loss modulus (or tan δ) are plotted versus reduced frequency, making use of the reduced variables shift factor, A_T. The reduced frequency is defined as $f_j A_{T_i}$, where f_j is the frequency and A_{T_i} is the value of A_T at temperature T_i. Then an auxiliary frequency scale is constructed as the ordinate on the right side of the graph; see Figure 10.15. The values of $f_j A_{T_i}$ and f form a set of corresponding oblique lines representing temperature. Weissman and Chartoff (42) point out that computer software has been developed to speed the generation of such nomographs.

Simulated data are used in Figure 10.15 for ease of instruction. Assume that the value of E' and tan δ at T_{-1} and some frequency f are needed, illustrated by point C in Figure 10.15. The intersection of the horizontal line CX, with $f_j A_{T_i}$ (point X) defines a value of $f A_T$ at point D of approximately 4×10^2. From this value of $f A_T$ it follows from the plots of E' and tan δ that the quantity E' equals 10^{-3} N/m^2, point B, and tan δ equals 1.2, point A.

A critical point of general information in Figure 10.15 is that, at high enough frequencies, the storage modulus increases into the glassy range. At low enough frequencies, any linear (or lightly cross-linked polymer) will flow (or be rubbery). As the frequency increases beyond the capability of the polymer chains to respond, the polymer glassifies. Since the reduced frequency is usually plotted with increasing values to the right, this figure appears backward to figures where log time is the x-axis variable. While stress relaxation studies employ time as the variable, dynamic experiments employ frequency.

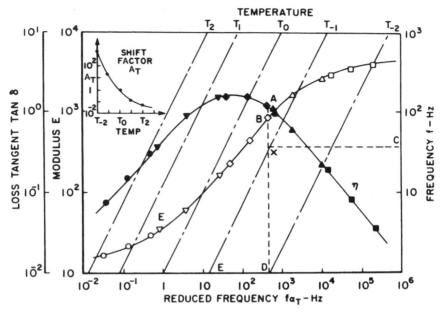

Figure 10.15 The reduced frequency nomograph. Here, the quantity $\eta \equiv \tan \delta$ (42).

Applications of the reduced frequency nomograph include sound and vibration damping (see Section 8.12) and earthquake damage control, interdisciplinary fields involving polymer scientists and mechanical engineers, and other disciplines. Frequencies of interest range from about 0.01 Hz (tall buildings) to 10^5 Hz, in the ultrasonic range.

10.4 POLYMER MELT VISCOSITY

10.4.1 The WLF Constants

In regions 4 and 5 of the modulus–temperature curve, linear amorphous polymers are capable of flow if they are subjected to a shear stress. While the melt viscosity, η, of low-molecular-weight substances may be Newtonian [see equation (8.3)], the flow behavior of polymers always contains some elements of viscoelasticity. In Section 8.6.1, the WLF equation was derived:

$$\log\left(\frac{\eta}{\eta_{T_g}}\right) = \frac{-C_1'(T - T_g)}{C_2' + (T - T_g)} \tag{10.56}$$

where C_1' and C_2' are constants, and η_{T_g} is the melt viscosity at the glass transition temperature. If data are not available on the polymer of interest, values of $C_1' = 17.44$ and $C_2' = 51.6$ may be used. However, these constants vary significantly if conditions other than η_{T_g} and T_g are used. Selected values of C_1' and C_2' are tabulated in Table 10.3 (44–46). The universal constants are widely used, as it is believed that the values for the

Table 10.3   WLF parameters

Polymer	C_1'	C_2'	T_g, K
Polyisobutylene	16.6	104	202
Natural rubber (Hevea)	16.7	53.6	200
Polyurethane elastomer	15.6	32.6	238
Polystyrene	14.5	50.4	373
Poly(ethyl methacrylate)	17.6	65.5	335
"Universal constants"	17.4	51.6	

Source: J. J. Aklonis and W. J. MacKnight, *Introduction to Polymer Viscoelasticity,* Wiley-Interscience, New York, 1983, Table 3-2, p. 48.

individual polymers differ only by experimental error. Frequently $\eta_{T_g} \cong 1 \times 10^{13}$ poises $= 1 \times 10^{12}$ Pa · s.

10.4.2   The Molecular-Weight Dependence of the Melt Viscosity

10.4.2.1   *Critical Entanglement Chain Length* Viscoelasticity in polymers ultimately relates back to a few basic molecular characteristics involving the rates of chain molecular motion and chain entanglement. The increasing ability of chains to slip past one another as the temperature is increased governs the temperature dependence of the melt viscosity. One embodiment of this concept is the WLF equation (10.56). The increased resistance to flow caused by entanglement governs the molecular-weight dependence.

Basic parameters in discussing the molecular weight dependence of the melt viscosity are the degree of polymerization (DP), which represents the number of monomer units linked together, and the number of atoms along the polymer chain's backbone (Z). For styrenics, acrylics and vinyl polymers, $Z = 2DP$, and for diene polymers, $Z = 4DP$. The point is that melt viscosity characteristics depend more on the number of backbone atoms than on the side foliage. It was also found very early that the melt viscosity depends on the weight-average degree of polymerization.

The molecular weight dependence of the melt viscosity exhibits two distinct regions (47), depending on whether the chains are long enough to be significantly entangled (see Figure 10.16). A critical entanglement chain length, $Z_{c,w}$ is defined as the weight-average number of chain atoms in the polymer molecules to cause intermolecular entanglement. Below $Z_{c,w}$ the melt viscosity is given by

$$\eta = K_L Z_w^{1.0} \tag{10.57}$$

and above $Z_{c,w}$ the melt viscosity is given by

$$\eta = K_H Z_w^{3.4} \tag{10.58}$$

where K_L and K_H are constants for low and high degrees of polymerization.

The 1.0 power dependence in equation (10.57) represents the simple increase in viscosity as the chains get longer. The dependence of the viscosity on the 3.4 power of the degree of polymerization as shown in equation (10.58) arises from entanglement and diffusion considerations. The algebraic power 3.0 was derived by de Gennes (48), using

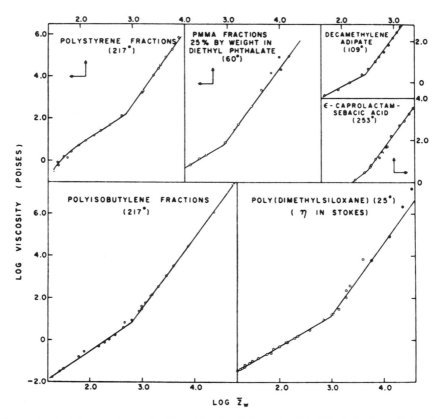

Figure 10.16 A plot of melt viscosity, log η, versus log Z_w for several polymers of importance. Below about $Z_w = 600$, the melt viscosity goes as the first power of Z_w; above about 600, the melt viscosity depends on the 3.4 power of Z_w.

scaling concepts. According to the reptation model (see Section 5.4), there is a maximum relaxation time of the chains, τ_{max}, which is the time required for complete renewal of the tube that constrains the chain. That is, τ_{max} represents the time necessary for the polymer chain to diffuse out of its original tube and assume a new conformation.

10.4.2.2 An Example Calculation of Melt Viscosity You are the engineer in charge of a polystyrene drinking cup manufacturing unit. Normally you process the material at 160°C, where the viscosity is 1.5×10^3 poises at $Z_w = 800$. Today, your polystyrene has $Z_w = 950$. What changes in processing temperature will bring the viscosity down to 1.5×10^3 again?

T_g for polystyrene is 100°C. The melt viscosity relationships are given by equations (10.58) and (8.48):

$$\eta = K_H Z_w^{3.4}$$

$$\log\left(\frac{\eta}{\eta_{T_g}}\right) = -\frac{17.44(T - T_g)}{51.6 + (T - T_g)}$$

Solving for K_H obtains

$$K_H = \frac{1.5 \times 10^3}{(800)^{3.4}} = 2.02 \times 10^{-7}$$

At 160°C the new viscosity is

$$\eta = 2.02 \times 10^{-7} \times (950)^{3.4} = 2.69 \times 10^3 \text{ P}$$

Solving the WLF equation for η_{T_g} obtains

$$\log\left(\frac{2.69 \times 10^3}{\eta_{T_g}}\right) = -\frac{17.44(433 - 373)}{51.6 + (433 - 373)}$$

$$\eta_{T_g} = 6.39 \times 10^{12} \text{ P}$$

Now the new temperature must be calculated, using the WLF equation:

$$\log\left(\frac{1.5 \times 10^3}{6.39 \times 10^{12}}\right) = -\frac{17.44(T - 373)}{51.6 + (T - 373)}$$

$$T = 436.6 \ K \quad \text{or} \quad 163.6°C$$

Thus the temperature should be raised 3.6°C, to return to the same melt viscosity. It is often important in operating polymer melt machinery to work at the same melt viscosity. These two equations, used in tandem, can be used to solve many problems involving simultaneous changes in temperature and molecular weight.

The quantity Z_w may cross 600, necessitating a change from equation (10.57) to equation (10.58), and vice versa. The required new constant (K_L or K_H) can be obtained by noting that the viscosity for both equations is the same at Z_w equals 600; see Figure 10.16.

10.4.2.3 *Analysis of* $Z_{c,w}$ Most of the numerical values of $Z_{c,w}$ are in the range of a few to several hundred; see Table 10.4. This suggested a theoretical analysis, to find if a

Table 10.4 Selected values of $Z_{c,w}$

Polymer	$Z_{c,w}$
Polyisobutylene	460
Poly(dimethyl siloxane)	630
Polystyrene	600
Poly(vinyl acetate)	570
Polyethylene	275
Poly(decamethylene sebacate)	290
Poly(ε-caproamide) (Polyamide-6)	324
Poly(propylene oxide)	400
Poly(decamethylene adipate)	280
Poly(methyl methacrylate)	210

Source: Data taken from T. G. Fox and V. R. Allen, *J. Chem. Phys.*, **41**, 344 (1964).

universal quantity to express the breakpoint in the melt-viscosity *versus* molecular weight could be found. The basic relationship was expressed by Fox and Allen (49):

$$X_c = \left(\frac{R_g^2}{M}\right) Z_{c,w} \rho v_2 \tag{10.59}$$

where R_g^2/M is a constant (see Table 5.4), ρ the density, and v_2 the volume fraction of the polymer. In the bulk melt state, v_2 equals unity.

The universal value of X_c is $47.0\,\text{Å}^2\text{mol/cm}^3$, $\pm 20\%$ (50). Where values of $Z_{c,w}$ or X_c are known, these values should be used for calculating the molecular weight dependence of the melt viscosity. Where the values are unknown, such as with a new polymer, the value of $Z_{c,w} = 600$ is still sometimes used in melt-viscosity–molecular weight estimations.

10.4.2.4 The Tube Diffusion Coefficient   Under a steady force, f, the chain moves with a velocity, v, in the tube. Then the "tube mobility" of the chain, μ_{tube}, is given by (48)

$$\mu_{\text{tube}} = \frac{v}{f} \tag{10.60}$$

If long-range backflow effects are assumed negligible, then the friction force v/μ_{tube} is essentially proportional to the number of atoms in the chain,

$$\mu_{\text{tube}} = \frac{\mu_1}{Z} \tag{10.61}$$

where μ_1 is independent of Z. Similarly the tube diffusion coefficient, D_{tube}, is related to μ_{tube} and Z through an Einstein relationship $D_{\text{tube}} = \mu_1 T/Z$

$$D_{\text{tube}} = \frac{D_1}{Z} \tag{10.62}$$

Since the time necessary to diffuse a certain distance depends on the square of the distance, the time necessary to completely renew the tube depends on the length of the tube, L, squared. Then

$$\tau_{\text{max}} \cong \frac{L^2}{D_{\text{tube}}} \tag{10.63}$$

and

$$\tau_{\text{max}} \cong \frac{ZL^2}{D_1} \tag{10.64}$$

since L is proportional to Z,

$$\tau_{\text{max}} = \tau_1 Z^3 \tag{10.65}$$

Table 10.5   Viscosities of some common materials (51)

Composition	Viscosity Pa · s	Consistency
Air	10^{-5}	Gaseous
Water	10^{-3}	Fluid
Polymer latexes	10^{-2}	Fluid
Olive oil	10^{-1}	Liquid
Glycerine	10^{0}	Liquid
Golden Syrup	10^{2}	Thick liquid
Polymer melts	10^{2}–10^{6}	Toffee-like
Pitch	10^{9}	Stiff
Plastics	10^{12}	Glassy
Glass	10^{21}	Rigid

The viscosity of the system is given by equation (10.15), $\eta = \tau E$. According to the reptation model, the modulus E depends on the distance between obstacles and does not depend on the chain length. Therefore

$$\eta \propto Z^{3} \tag{10.66}$$

Equation (10.66) should be compared with equation (10.58). While the power dependence is not quite correct in this simple derivation, it illustrates the principal molecular-weight dependence of the viscosity (47).

10.5 POLYMER RHEOLOGY

Rheology is the study of the deformation and flow of matter. As such, the reader will recognize that a significant fraction of this book already involves rheological concepts. Some important areas not yet considered include shear rate dependence of flow and the effect of normal stress differences (51,52).

The range of melt viscosities ordinarily encountered in materials is given in Table 10.5 (51). Polymer latexes and suspensions are aqueous dispersions with viscosities dependent on solid content and additives. Polymer solutions may be much more viscous, depending on the concentration, molecular weight, and temperature.

The temperature dependence of the viscosity is most easily expressed according to the Arrhenius relationship:

$$\eta = Ae^{-B/T} \tag{10.67}$$

where T represents the absolute temperature and A and B are constants of the liquid. The Arrhenius equation may easily be shown to be an approximation of the WLF equation far above the glass transition temperature.

The rate of energy dissipation per unit volume of the sheared polymer may be expressed as either the product of the shear stress and the shear rate or, equivalently, the product of the viscosity and the square of the shear rate. The rate of heat generated during viscous flow may be significant. Thus the heat generated may actually reduce the viscosity of the material. From an engineering point of view, these quantities provide an important measure of the power necessary to maintain a given flow rate.

All real liquids have both viscous and elastic components, although one or the other may predominate. For example, water behaves as a nearly perfect viscous medium, while a rubber band is a nearly perfect elastomer. A polymer solution of, for example, polyacrylamide in water may exhibit various ranges of viscoelasticity, depending on the concentration and temperature. According to Weissenberg (53), when an elastic liquid is subjected to simple shear flow, there are two forces to be considered:

1. The shear stress, characteristic of ordinary viscosity.
2. A normal force, observed as a pull along the lines of flow.

The several aspects of polymer rheology are developed below.

10.5.1 Shear Dependence of Viscosity

Any liquid showing a deviation from Newtonian behavior is considered non-Newtonian. As soon as reliable viscometers became available, workers found departures from Newtonian behavior for polymer solutions, dispersions, and melts. In the vast majority of cases, the viscosity decreases with increasing shear rate, giving rise to what is often called "shear-thinning."

An example for xanthan gum solutions is given in Figure 10.17 (54). Note especially the wide range of shear rates obtainable. Xanthan gum is a high-molecular-weight polysaccharide (MW $= 7.6 \times 10^6$ g/mol in Figure 10.17), a biopolymer used in food applications, oil recovery, and textile printing. Less than 1% of xanthan increases water's viscosity by a factor of 10^5 at low shear rates, yet only by a factor of 10 at high shear rates. The viscosity reduction is caused by the reduced number of chain entanglements as

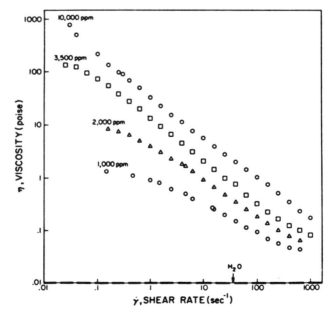

Figure 10.17 Viscosity of polymer melts and solutions usually decreases with increasing shear rate (54).

Table 10.6 Typical power-law parameters of a selection of well-known materials for a particular range of shear rates (52)

Material	$K_2(Pa \cdot s^n)$	n	Shear rate range (s^{-1})
Ball-point pen ink	10	0.85	$10^0 - 10^3$
Fabric conditioner	10	0.6	$10^0 - 10^2$
Polymer melt	10,000	0.6	$10^2 - 10^4$
Molten chocolate	50	0.5	$10^{-1} - 10$
Synovial fluid	0.5	0.4	$10^{-1} - 10^2$
Toothpaste	300	0.3	$10^0 - 10^3$
Skin cream	250	0.1	$10^0 - 10^2$
Lubricating grease	1000	0.1	$10^{-1} - 10^2$

the chains orient along the lines of flow. Note that water has 0.01 poise (1×10^{-3} Pa·s), shown as the base line in Figure 10.17.

The general case of shear rate dependence includes a limit of constant viscosity at very low shear rates, and a lower constant viscosity at the limit of very high shear rates. While this is not often observed with polymeric materials, a suggestion of both limits can be seen in Figure 10.17. The reason is that, even at the lowest shear rates, the chains begin to orient, and entanglements slip easier. At the limit of very high shear rates, polymers degrade. An extremely useful relationship is the well known "power-law" model,

$$\eta = K_2 \dot{\gamma}^{n-1} \qquad (10.68)$$

where $\dot{\gamma}$ is the shear rate, n is the power-law index, and K_2 is called the "consistency." When n equals zero, a form of Newton's law is generated; see Section 8.2. The quantity $\dot{\gamma}$ is identical to ds/dt, equation (8.3), with units of reciprocal seconds. Typical values of n and K_2 for a number of materials are given in Table 10.6 (52). Many practical materials, such as skin creams and inks, contain polymers.

10.5.2 Normal Stress Differences

Consider a small plane surface of area ΔA drawn in a deforming medium; see Figure 10.18. The material flows or deforms through ΔA. For analytic purposes, the vector direction of motion is divided among x, y, and z. The direction of motion is not necessarily normal to the plane of ΔA.

The stress components may be written f_{nx}, f_{ny}, and f_{nz}, the first index referring to the orientation of plane surface, and the second to the direction of the stress. Then the quantity f_{xx} means that a stress perpendicular to the plane ΔA is under consideration. There are three quantities that are rheologically relevant (55)—the shear stress, f_{yx}, and the first and second normal stress differences, N_1 and N_2, respectively:

$$N_1 = f_{xx} - f_{yy} \qquad (10.69)$$

$$N_2 = f_{yy} - f_{zz} \qquad (10.70)$$

Both N_1 and N_2 are shear rate dependent. From a physical point of view, the generation of unequal normal stress components arises from the anisotropic structure of a polymer fluid undergoing flow; that is, the polymer becomes oriented. The largest of the three normal stress components is always f_{xx}, the component in the direction of flow. How-

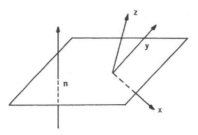

Figure 10.18   Arbitrary infinitesimal plane of area ΔA; n is the direction perpendicular to the plane. The quantities x, y, and z are orthogonal coordinates.

ever, since the molecular structures undergoing flow are anisotropic, the forces are also anisotropic. The deformed chains often have the rough shape of ellipsoids, which have their major axis tilted toward the direction of flow. Thus the restoring force is greater in this direction than in the two orthogonal directions. It is these restoring forces that give rise to the normal forces (52).

The observable consequences of the normal forces are quite dramatic. Perhaps the best known is the so-called Weissenberg effect; see Figure 10.19 (52). It is produced when

Figure 10.19   The Weissenberg effect illustrated by rotating a rod in a solution of polyisobutylene in polybutene (52).

Figure 10.20 Molecular forces involved in the Weissenberg effect. Note polymer orientation and entanglements.

a rotating rod is placed into a vessel containing a viscoelastic fluid such as a concentrated polymer solution. Whereas a Newtonian liquid would be forced toward the rim of the vessel by inertia, the viscoelastic fluid moves *toward* the rod, climbing it. The chains are extended yet remain partly entangled, which causes a rubber elasticity type of retractive force; see Figure 10.20. This "hoop stress" around the rod forms the basis for the Weissenberg effect.

Another phenomenon, seen in fiber and plastic extruding operations, is "die swell"; see Figure 10.21 (56). Here a viscoelastic fluid is oriented in a tube during flow. When it is extruded, it flows from the exit of the tube and retracts. This results in swelling to a much greater diameter than that of the hole or slit. Die swell becomes much greater when the polymer chains are long enough to contain significant numbers of entanglements, being a direct consequence of the elastic energy stored by the polymer in the tube.

10.5.3 Dynamic Viscosity

Periodic or oscillatory viscosity experiments provide a powerful rheological analogue to dynamic mechanical experiments. The theoretical relations are, in fact, closely interlocked. The complex viscosity, η^*, is defined as (10)

$$\eta^* = \eta' - i\eta'' \tag{10.71}$$

The complex viscosity can be determined from dynamic mechanical data,

$$\eta^* = \frac{[(G')^2 + (G'')^2]^{1/2}}{\omega} \tag{10.72}$$

where ω is the angular frequency. The individual components are given by

$$\eta' = \frac{G''}{\omega} \tag{10.73}$$

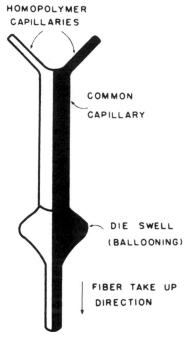

HOMOPOLYMER
CAPILLARIES

COMMON
CAPILLARY

DIE SWELL
(BALLOONING)

FIBER TAKE UP
DIRECTION

Figure 10.21 Bicomponent fibers are often used in the manufacture of rugs, because a mismatch in the expansion coefficients causes curling on cooling. However, the second normal forces must match, so that the die swell on both sides is the same (56).

and

$$\eta'' = \frac{G'}{\omega} \tag{10.74}$$

Thus the phase relations are the opposite of those for G^*. The in-phase, or real component, η' for a viscoelastic liquid approaches the steady-state flow viscosity as the frequency approaches zero.

10.5.4 Instruments and Experiments

There are several different types of viscometers, of varying complexity, good for specific purposes and/or ranges of viscosity (57,58). The three major classes are described below.

Falling-Ball Viscometers. This is the simplest type of viscometer, useful for determining relatively low viscosities. In one simple case, a graduated cylinder is filled with the fluid of interest (perhaps a concentrated polymer solution), and a steel ball is dropped in. The time to fall a given distance is recorded. According to Stokes, the terminal velocity is given by

$$\bar{u} = \frac{2}{9} \frac{R^2}{\eta} g(\rho - \rho_0) \tag{10.75}$$

where R is the sphere radius, ρ and ρ_0 are the densities of the sphere and the medium, respectively, and g is the gravitational constant. Within its range of operation, it is inexpensive, easily used, and the viscosity is absolute.

Capillary Viscometers. These instruments are used for intrinsic viscosities and also for more viscous melts, solutions, and dispersions. The viscosity is given by the Hagen–Poiseuille expression,

$$\eta = \frac{\pi' r^4 \Delta p t}{8 V L} \tag{10.76}$$

where r is the capillary radius, Δp represents the pressure drop, and V is the volume of liquid that flows through the capillary of length L in time t. For a given viscometer with similar liquids,

$$\eta = Kt \quad \text{or} \quad \bar{v} = \eta/\rho_0 = K't \tag{10.77}$$

where $\bar{v}$ is the kinematic viscosity.

Rotational Viscometers. These relatively complex instruments can be used in the steady state or in an oscillatory, dynamic mode. Some are useful up to the glassy state of the polymer. The working mechanism, in all cases, is one part that moves past another. Designs include concentric cylinders (cup and bob), cone-and-plate, parallel-plate, and disk, paddle, or rotor in a cylinder.

The most important device is the cone-and-plate viscometer; see Figure 10.22 (52). The advantage of the cone-and-plate geometry is that the shear rate is very nearly the same everywhere in the fluid, provided the gap angle, θ_0, is small. The shear rate in the fluid is given by

$$\dot{\gamma} = \frac{\Omega_1}{\theta_0} \tag{10.78}$$

where Ω_1 represents the angular velocity of the rotating platter. The viscosity is given by

$$\eta = \frac{3 f \theta_0}{2\pi a^3 \Omega_1} \tag{10.79}$$

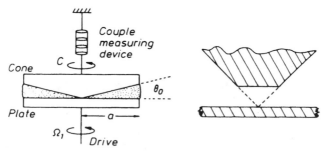

Figure 10.22 The cone-and-plate viscometer, showing the cone on top, a rotating plate, and couple attached to the cone. The inset shows a form of truncation employed in many instruments (52).

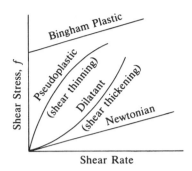

Figure 10.23   Several shear-rate-dependent rheological phenomena.

where the stress f is measured by the couple C on the cone, and a is the radius of the cone-and-plate. An important factor in deciding on a viscometer, of course, is the viscosity of the fluid and the information desired.

10.5.5   General Definitions and Terms

Several shear-rate-dependent phenomena are illustrated in Figure 10.23. A Bingham plastic has a linear relationship between shear stress and shear strain but does not go through the origin. It must be realized that this is an idealization. Pseudoplastic materials, which include most polymer solutions and melts, exhibit shear thinning. Dilatant materials shear thicken.

There are several terms that apply to time-dependent behavior. Thixotropic fluids possess a structure that breaks down as a function of time and shear rate. Thus the viscosity is lowered. A famous example of this reversible phenomenon is the ubiquitous catsup bottle, yielding its contents only after sharp blows. The opposite effect, although rarely observed, is called antithixotropic, or rheopectic behavior, where materials "set up" as a function of time and shear rate; that is, the viscosity increases or the material gels.

Viscoelastic fluids exhibit elastic recovery from deformations that occur during flow. The Weissenberg effect and die swell have been discussed previously. Another very simple example is the stirring of concentrated polymer solutions rapidly, then stopping. Elastic effects make the fluid move backward for a time.

Several flow equations are summarized in Table 10.7 (58) for various models. Here, f_0 is the yield stress, η_0 the viscosity at low shear rates and η_∞ at high shear rates, and α and n are constants.

10.6   OVERVIEW OF VISCOELASTICITY AND RHEOLOGY

This chapter has illustrated how stress relaxation, creep, and rheology in polymers depend on the rate of molecular motion of the chains and on the presence of entanglements. It must be remembered that all macroscopic deformations of matter depend ultimately on molecular motion. In the case of high polymers, the chain's radius of gyration is changed during initial deformation or flow. Thermal motions tend to return the polymer to its

Table 10.7   Flow models and equations (58)

Model	Equation
Newtonian	$f = \eta \dot{\gamma}$
Bingham plastic	$f - f_0 = \eta \dot{\gamma}$
Power law	$f = k \dot{\gamma}^n$
Power law with yield value	$f - f_0 = k \dot{\gamma}^n$
Casson fluid	$f^{1/2} - f_0^{1/2} = \eta_\infty^{1/2} \dot{\gamma}^{1/2}$
Williamson	$\eta - \eta_\infty = \dfrac{(\eta_0 - \eta_\infty)}{1 + f/f_m}$
Cross	$\eta - \eta_\infty = \dfrac{(\eta_0 - \eta_\infty)}{1 + \alpha \dot{\gamma}^n}$

initial conformation, thus raising its entropy. Clearly, there is a direct relationship between the mechanical or viscous behavior of polymeric materials and their molecular behavior.

REFERENCES

1. J. J. Aklonis, *J. Chem. Ed.*, **58**, 893 (1981).
2. A. V. Tobolsky, *Properties and Structure of Polymers*, Wiley, New York, 1960.
3. L. H. Sperling, S. L. Cooper, and A. V. Tobolsky, *J. Appl. Polym. Sci.*, **10**, 1735 (1966).
4. P. Thirion and R. Chasset, *Proceedings of the 4th Rubber Technology Conference*, London, 1962, p. 338.
5. P. G. de Gennes, *J. Chem. Phys.*, **55**, 572 (1971).
6. M. Takayanagi, *Mem. Fac. Eng. Kyushu Univ.*, **23**, 11 (1963).
7. L. H. Sperling, *Polymeric Multicomponent Materials: An Introduction*, Wiley, New York, 1997.
8. T. C. P. Lee, L. H. Sperling, and A. V. Tobolsky, *J. Appl. Polym. Sci.*, **10**, 1831 (1966).
9. A. V. Tobolsky, I. B. Prettyman, and J. H. Dillon, *J. Appl. Phys.*, **15**, 380 (1944).
10. J. D. Ferry, *Viscoelastic Properties of Polymers*, 3rd ed., Wiley, New York, 1980.
11. C. Picot, R. Duplessix, D. Decker, H. Benoît, F. Boue, J. P. Cotton, M. Daoud, B. Farnoux, G. Jannick, M. Nierloch, A. J. deVries, and P. Pincus, *Macromolecules*, **10**, 957 (1977).
12. A. Maconnachie, G. Allen, and R. W. Richards, *Polymer*, **22**, 1157 (1981).
13. F. Boue and G. Jannink, *J. Phys. Colloq.*, **39**, C2-183 (1978).
14. F. Boue, M. Nierlich, G. Jannink, and R. C. Ball, *J. Phys. Lett.* (*Paris*), **43**, 593 (1982).
15. F. Boue, M. Nierlich, G. Jannink, and R. C. Ball, *J. Phys.* (*Paris*), **43**, 137 (1982).
16. G. Hadziiannou, L. H. Wang, R. S. Stein, and R. S. Porter, *Macromolecules*, **15**, 880 (1982).
17. P. G. de Gennes, *J. Chem. Phys.*, **55**, 572 (1971).
18. M. Doi and J. F. Edwards, *J. Chem. Soc. Faraday Trans.*, **74**, 1789, 1802, 1818 (1978).
19. S. Daoudi, *J. Phys.*, **38**, 731 (1977).
20. A. V. Tobolsky, *J. Polym. Sci. Lett.*, **2**, 103 (1964).
21. R. P. Wool, *Polymer Interfaces: Structure and Strength*, Hanser, Munich, 1995.
22. K. A. Welp, R. P. Wool, G. Agrawal, S. K. Satija, S. Pispas, and J. Mays, *Macromolecules*, **32**, 5127 (1999).
23. S. D. Kim, A. Klein and L. H. Sperling, *Macromolecules*, **33**, 8334 (2000).
24. K. D. Kim, L. H. Sperling, A. Klein and B. Hammouda, *Macromolecules*, **27**, 6841 (1994).

25. L. J. Fetters, D. J. Lohse, D. Richter, T. A. Witten, and A. Zirkel, *Macromolecules*, **27**, 4639 (1994).

26. S. Tobing, A. Klein, L. H. Sperling, and B. Petrasko, accepted, J. Appl. Polym. Sci., 2000.

27. A. Zosel and G. Ley, in *Film Formation in Water Borne Coatings*, T. Provder, M. Winnik and M. Urban, eds., ACS Books, Washington, DC, 1996.

28. J. R. McLaughlin and A. V. Tobolsky, *J. Colloid Sci.*, **7**, 555 (1952).

29. L. C. E. Struik, *Physical Aging in Amorphous Polymers and Other Materials*, Elsevier, New York, 1978.

30. H. C. Booij and J. H. M. Palmen, *Polym. Eng. Sci.*, **18**, 781 (1978).

31. S. Matsuoka, H. E. Bair, S. S. Bearder, H. E. Kern, and J. T. Ryan, *Polym. Eng. Sci.*, **18**, 1073 (1978).

32. F. H. J. Maurer, J. H. M. Palmen, and H. C. Booij, *Polym. Mater. Sci. Eng. Prepr.*, **51**, 614 (1984).

33. F. H. J. Maurer, in *Rheology, Vol. 3: Applications*, G. Astarita, G. Marrucci, and L. Nicolais, Eds., Plenum, New York, 1980.

34. J. G. Curro, R. R. Lagasse, and R. Simha, *J. Appl. Phys.*, **52**, 5892 (1981).

35. J. G. Curro, R. R. Lagasse, and R. Simha, *Macromolecules*, **15**, 1621 (1982).

36. E. Castiff and A. V. Tobolsky, *J. Colloid Sci.*, **10**, 375 (1955).

37. E. Castiff and A. V. Tobolsky, *J. Polym. Sci.*, **19**, 111 (1956).

38. L. E. Nielsen, *Mechanical Properties of Polymers*, Reinhold, New York, 1962.

39. T. Horino, Y. Ogawa, T. Soen, and H. Kawai, *J. Appl. Polym. Sci.*, **9**, 2261 (1965).

40. R. E. Cohen and N. W. Tschoegl, *Int. J. Polym. Mater.*, **2**, 205 (1973).

41. D. Kaplan and N. W. Tschoegl, *Polym. Eng. Sci.*, **14**, 43 (1974).

42. P. T. Weissman and R. P. Chartoff, in *Sound and Vibration Damping with Polymers*, R. D. Corsaro and L. H. Sperling, eds., ACS Symposium Series No. 424, American Chemical Society, Washington, DC, 1991.

43. D. I. G. Jones, *Shock Vibration Bull.*, **48**(2), 13 (1978).

44. M. L. Williams, R. F. Landel, and J. D. Ferry, *J. Am. Chem. Soc.*, **77**, 3701 (1955).

45. J. D. Ferry, *Viscoelastic Properties of Polymers*, 3rd ed., Wiley, New York, 1980, Chap. 11.

46. J. J. Aklonis and W. J. MacKnight, *Introduction to Polymer Viscoelasticity*, 2nd ed., Wiley, New York, 1983, Chap. 3.

47. T. G. Fox, S. Gratch, and S. Loshaek, in *Rheology*, F. R. Eirich, ed., Academic Press, Orlando, 1956, Vol. 1, Chap. 12.

48. P. G. de Gennes, *Scaling Concepts in Polymer Physics*, Cornell University Press, Ithaca, NY, 1979, Chap. 8.

49. T. G. Fox and V. R. Allen, *J. Chem. Phys.*, **41**, 344 (1964).

50. T. G. Fox, *J. Polym. Sci.*, **9C**, 35 (1965).

51. F. N. Cogswell, *Polymer Melt Rheology*, G. Godwin, ed., Wiley, New York, 1981.

52. H. A. Barnes, J. F. Hutton, and K. Walters, *An Introduction to Rheology*, Elsevier, New York, 1989.

53. K. Weissenberg, *Nature*, **159**, 310 (1947).

54. P. J. Whitcomb and C. W. Macosko, *J. Rheol.*, **22**, 493 (1978).

55. J. Meissner, R. W. Garbella, and J. Hostettler, *J. Rheol.*, **33**, 843 (1989).

56. J. A. Manson and L. H. Sperling, *Polymer Blends and Composites*, Plenum, New York, 1976.

57. Perry's *Chemical Engineer's Handbook*, 7th ed., R. H. Perry and D. W. Green, eds., McGraw-Hill, New York, 1997.

58. C. K. Schoff, in *Encyclopedia of Polymer Science and Engineering*, Vol. 14, Wiley, New York, 1988, p. 454.

GENERAL READING

J. J. Aklonis and W. J. MacKnight, *Introduction to Polymer Viscoelasticity*, 2nd ed., Wiley, New York, 1983.

H. A. Barnes, J. F. Hutton, and K. Walters, *An Introduction to Rheology*, Elsevier, Amsterdam, 1989.

J. D. Ferry, *Viscoelastic Properties of Polymers*, 3rd ed., Wiley, New York, 1980.

R. K. Gupta, *Polymer and Composite Rheology*, 2nd ed., Dekker, New York, 2000.

L. E. Nielsen and R. F. Landel, *Mechanical Properties of Polymers and Composites*, 2nd ed., Dekker, New York, 1994.

L. C. E. Struik, *Physical Aging in Amorphous Polymers and Other Materials*, Elsevier, Amsterdam, 1978.

C. D. Craver and C. E. Carraher, Jr., eds., *Applied Polymer Science: 21st Century*, Elsevier, Amsterdam, 2000.

N. W. Tschoegl, *The Theory of Linear Viscoelastic Behavior*, Academic Press, Orlando, 1981.

STUDY PROBLEMS

1. Draw the creep and creep recovery curves for the three-element model consisting of a Kelvin element and a spring in series.

2. If the modulus of the *cis*-1,4-polybutadiene in Figure 10.8 was 2.5×10^7 dynes/cm^2, plot the total number of remaining active chain segments from the original network as a function of time. Also plot the number of active chain segments formed by the oxidative cross-linking.

3. At 200°C, how long would it take for the silicone rubber in Figure 10.7 to relax 50%?

4. Derive equations to express Young's modulus as a function of rubber and plastic composition using the Takayanagi models (*c*) and (*d*) in Figure 10.6.

5. Draw stress relaxation curves for the four-element model.

6. Derive an equation for the creep recovery of a Kelvin element, beginning after a creep experiment extending it to t_1, a later time.

7. A poly(methyl methacrylate) bridge is to be placed across a river in the tropics, where the average temperature is 40°C. The bridge is a simple platform 100 ft long and 5 ft thick and 10 ft wide. The bridge will fail when creep slopes the sides of the bridge more than a 30° angle, so cars get stuck in the middle. How long will the bridge last? [*Hint:* As a simple approximation, the simple beam, center-loaded bending distance Y is given by $Y = fL^3/4a^3bE$, where a is the thickness, b is the width, f is the force, L is the length, and E is Young's modulus. See Figure 10.12.]

8. Prepare a master curve at 110°C for poly(methyl methacrylate), using the data in Figure 10.12.

9. The melt viscosity of a fraction of natural rubber is 2×10^3 Pa · s at 240 K. What is the melt viscosity of this fraction at 250 K?

10. A new polymer with a mer weight of 211 g/mol and five atoms in the chain was found to have a weight-average molecular weight of 300,000 g/mol. Its melt viscos-

ity is 1500 poises. What is the viscosity of the polymer if its molecular weight is doubled?

11. A polymer with a Z_w of 200 was found to have a melt viscosity of 100 Pa · s. What is the viscosity of this polymer when $Z_w = 800$?

12. A polymer with a T_g of 110°C and a Z_w of 400 was found to have a melt viscosity of 5000 Pa·s at 160°C. What is its melt viscosity at 140°C when $Z_w = 900$? [*Hint:* Combine the WLF equation with the DP dependence.]

13. The three-element springs and dashpot model shown is subject to a creep experiment. Show how the length (or strain) increases with time. At time $= t$, the stress is removed. Show how the sample recovers.

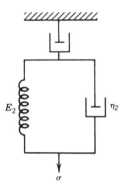

Figure P10.13 The three-element model.

14. A new polymer follows the Kelvin model. The quantity η obeys the WLF equation, and E obeys rubber elasticity theory. The glass transition temperature of the polymer is 5°C, where it has a viscosity of 1×10^{13} poises. The concentration of active chain segments is 1×10^{-4} mol/cm^3. The temperature of the experiment is 30°C.

(a) How does this polymer creep with time under a stress of 1×10^7 dynes/cm^2? A plot of strain versus time is required.

(b) Briefly discuss two ways to slow down or reduce the rate of creep in part (a).

15. A certain extruder for plastics was found to work best at a melt viscosity of 2×10^4 Pa · s. The polymer of choice had this viscosity at 145°C when its DP$_w$ was 700. This polymer has a T_g of 75°C. Because of a polymerization kinetics miscalculation by someone who did not take polymer science, today's polymer has a DP$_w$ of 500. At what temperature should the extruder be run so that the viscosity will remain at optimum conditions? (Assume two carbon atoms per mer backbone.)

16. In the spinning of rayon fibers, the viscosity of the viscose solution must be carefully controlled. In a falling ball experiment on a viscose solution, it took a 2 mm lead ball 324 s to fall 10 cm. The density of the viscose was 1.2 g/cm^3. What is its viscosity?

17. A cone-and-plate experiment was performed on a polymer exhibiting first and second normal forces. Will the cone and plate be forced together, pushed apart, pushed sideways, or what? Discuss your reasoning.

18. Based on the concepts of Section 10.2.7, what is the value of G^0 for poly(methyl methacrylate)? How does it compare with the value in Figure 10.12?

APPENDIX 10.1   ENERGY OF ACTIVATION FROM CHEMICAL STRESS RELAXATION TIMES

From first-order chemical kinetics,

$$\frac{-dc_A}{dt} = kC_A \tag{A10.1.1}$$

where C_A is the concentration of species A. Then on integration,

$$-\ln C_A = kt + \text{constant} \tag{A10.1.2}$$

$$C_A = C_{A_0} e^{-kt} \tag{A10.1.3}$$

Note that k has the units of inverse time. If $k = 1/\tau_1$, an immediate relationship with equation (10.18) is noted.

From the Arrhenius equation,

$$k = \frac{RT}{Nh} e^{\Delta S/R} e^{-\Delta H/RT} \tag{A10.1.4}$$

For the present purposes,

$$k = se^{-\Delta E_{act}/RT} \tag{A10.1.5}$$

which may be rewritten

$$\tau_1 = s^{-1} e^{\Delta E_{act}/RT} \tag{A10.1.6}$$

where ΔE_{act} is the activation energy of the process. Then

$$\ln \tau_1 = \text{constant}^{-1} + \frac{\Delta E_{act}}{RT} \tag{A10.1.7}$$

A plot of $\ln \tau_1$ versus $1/T$ yields $\Delta E_{act}/R$ as the slope.

APPENDIX 10.2   VISCOELASTICITY OF CHEESE[†]

Cheese is made from milk. While the composition of cheeses varies greatly, ordinary hard cheeses have approximately 31% fat, mostly triglycerides, 25% protein, 1.7% carbohydrates, mostly lactose, about 2.2% ash, mostly sodium salts, and about 40% moisture (B1).

[†] Reprinted in part from Y. S. Chang, J. S. Guo, Y. P. Lee, and L. H. Sperling, *J. Chem. Educat.*, **63**, 1077 (1986).

Pasteurized, prepared cheese products such as Velveeta® have about 21% fat, 11% carbohydrates, and about 18% protein. Most of the remaining material is water with some salt. While the morphology of cheese is complex, the water tends to plasticize the protein. The lower protein content is largely responsible for prepared cheese products which are softer (lower modulus) and have greater viscoelasticity, making the material suitable for the present demonstration.

The proteins are largely the biopolymer casein. There are four types of casein present, $\alpha_{s1}, \beta, \alpha_{s2}$, and κ. These are present in the molecular ratio of $4:4:1:1$, and have 199, 209, 208, and 169 amino acid residues (mers) per chain, respectively (B2). Of course, the structures of all proteins are basically polyamide-2 copolymers,

$$\left(\!\!\begin{array}{c} \text{R} \quad \text{O} \\ | \quad\;\; \| \\ \text{NH} - \text{CH} - \text{C} \end{array}\!\!\right)_{\!\!n}$$

where R may take any of 20 structures; see Appendix 2.1. The spatial structure of the protein chain is maintained by hydrogen bonding and internal cross-links (see Figure 9.2), but α_{s1}-casein is known to have a relatively open structure (B3) contributing significantly to the viscoelastic characteristics of the cheese.

The objective of the experiment is to measure the viscoelastic characteristics of cheese, and interpret the data in terms of the four-element model.

The time required is one hour. The level is junior or senior standing. The principles to be investigated are the applicability of the four-element model to polymers in general, and cheese in particular.

Equipment and Supplies:

One 1-pound block processed cheese (Velveeta® or similar), at room temperature

Cut the cheese inhalf, place one-half on top of the other

One clock (stopwatch is fine)

One meter stick

One small, flat plate (insert between cheese and weight)

One flat, hard surface (most desks are suitable)

Weights of about 1 kg (rectangular weights of the same cross-sectional area as the cheese are fine)

The original height of the two layers of cheese is recorded. Then the plate and weight are placed on top of the cheese. Its new height is recorded immediately and each succeeding 5 minutes thereafter. This experiment measures creep in compression.

Replace the weight and deformed cheese with another weight and new cheese, and repeat.

Results

The initial dimensions of the 1-pound block of cheese were 4 cm × 6 cm × 15 cm. Weights of 500 and 700 g were employed. The resulting creep curves are illustrated in Figure A10.2.1. An immediate elastic compression is noted, a measure of the modulus E_1. This is followed by a curved line of strain versus time, and a straight-line portion lasting at least 2 hours; see Figure A10.2.1.

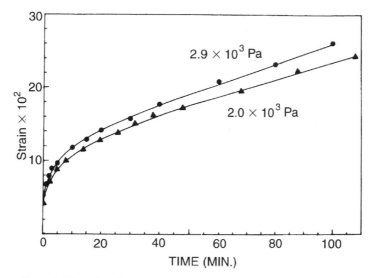

Figure A10.2.1 Experimental creep curves for Velveeta® cheese under two loads.

The data were analyzed in Figure A10.2.2 according to the four-element model. First, at zero time, the strain yields E_1. The straight-line portion at long times, separated as curve 1, yields η_3. Then, by subtraction, curve 2 was obtained. By simple curve fittings E_2, η_2, and τ_2 can be determined. The retardation time for Velveeta cheese was found to be about 7 to 9 minutes (see Table A10.2.1).

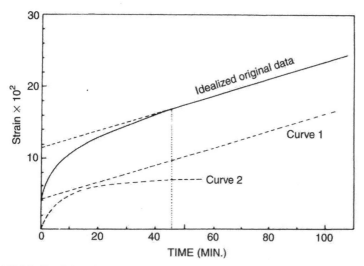

Figure A10.2.2 Resolution of the experimental data according to the four-element model. Curve 1 represents flow according to η_3. Curve 2 illustrates the behavior of the middle element, which is really a Kelvin element. The strain at zero time up to the interception of curve 1 represents the instantaneous deformation according to the spring E_1.

Table A2.10.1   Characteristics of Velveeta cheese at room temperature according to the four-element model

Variable	Applied Weight	
	500 g	700 g
σ (Pa)	2.00×10^3	2.88×10^3
E_1 (Pa)	4.88×10^4	5.18×10^4
E_2 (Pa)	2.82×10^4	4.24×10^4
η_2 (Pa·s)	1.52×10^7	1.78×10^7
η_3 (Pa·s)	1.00×10^8	1.21×10^8
τ_2 (min)	9	7

The value of E_1, near 5×10^4 Pa, places it in the fourth region of viscoelasticity, rubbery flow. The results should be compared to known values of modulus (A4).

REFERENCES

B1. A. R. Hill, *Chemistry of Structure-Function Relationships in Cheese*, E. L. Malin and M. H. Tunick, eds., Plenum, New York, 1995.

B2. W. N. Eigel, J. E. Butler, C. A. Ernstrom, J. H. M. Farrell, V. R. Harwalkar, R. Jenness, and R. M. Whitney, *J. Dairy Sci.*, **67**, 1599 (1984).

B3. E. L. Malin and E. M. Brown, *Int. Dairy J.*, **9**, 207 (1999).

B4. J. H. Prentice, in *Encyclopedia of Food Sci. and Technol.*, Vol. 1, Y. H. Hui, ed., Wiley, 1992, p. 348.

11

MECHANICAL BEHAVIOR
OF POLYMERS

Up until not too long ago, problems of fatigue, fracture, and failure in polymers were treated empirically. The old adage, "a chain is as strong as its weakest link," prevailed. To an increasing extent, fortunately, ideas of thermodynamics, viscoelasticity, and molecular bonding are now being used to explain what happens when polymers undergo fracture.

Fracture in polymers presents two important aspects. On the one hand, a knowledge of the strengths (and weaknesses) of polymers is of enormous engineering importance. Second, it is becoming increasingly popular as a field of fundamental research, combining the skill of the theoretician with that of the experimenter.

11.1 AN ENERGY BALANCE FOR DEFORMATION AND FRACTURE

11.1.1 The First Law of Thermodynamics

Consider a closed system, into which an increment of energy, δU_1, is transferred. While the nature of the energy can and should remain general, for the most part δU_1 is intended to consist of mechanical work; see Figure 11.1 (1). Inside the closed system, δU_1 is divided into three types of energy: δU_2, the change in irreversibly dissipated energy; δU_3, the change in stored or potential energy; and δU_4, the change in the kinetic energy.

The quantity δU_2 represents that energy dissipated by plastic or viscous flow being converted to heat. Two subcases must be considered—adiabatic systems and isothermal systems. In the adiabatic case, there is a temperature rise in the system. In the isothermal case, the thermal energy is all transferred to the surroundings across the closed system's boundary.

The quantity δU_3 indicates the stored elastic energy in the system (note the springs in Chapter 10) but also includes stored thermal energy. The quantity δU_4 indicates changes in energy associated with changes in linear or angular velocity.

For the isothermal case where the stored energy is solely elastic, the net input of energy can be written

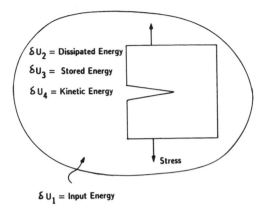

Figure 11.1   Mechanical behavior of a closed system, with energy input, causing crack growth (1).

$$\delta U_1 - \delta U_2 \qquad (11.1)$$

The first law of thermodynamics yields the conservation of energy,

$$\delta U_1 - \delta U_2 = \delta U_3 + \delta U_4 \qquad (11.2)$$

Now consider that the body undergoes a displacement δu. Then

$$\frac{\delta U_1}{\delta u} - \frac{\delta U_2}{\delta u} = \frac{\delta U_3}{\delta u} + \frac{\delta U_4}{\delta u} \qquad (11.3)$$

An incremental increase in energy in the system may be written

$$\delta U = \sigma \delta u \qquad (11.4)$$

where σ represents the stress in the direction of the deformation or displacement u.

In the simplest case, that of rigid body dynamics, U_3 and U_2 are both zero. Then

$$\frac{\delta U_1}{\delta u} = \frac{\delta U_4}{\delta u} \qquad (11.5)$$

11.1.2   Relations for Springs and Dashpots

By way of review, if a mass M moves with a velocity v,

$$U_4 = (\tfrac{1}{2}) M v^2 \qquad (11.6)$$

The derivative of U_4 with respect to u yields the acceleration, a:

$$\frac{dU_4}{du} = Ma \qquad (11.7)$$

Since the external force f equals dU_1/du [see equation (11.5)], Newton's second law of motion is obtained:

$$f = Ma \qquad (11.8)$$

The springs and dashpots of Chapter 10 clearly have both a thermodynamic and a mechanical basis. For a spring of modulus E attached to a mass M, the stored energy on stretching is given by

$$U_3 = \tfrac{1}{2}E\varepsilon^2 \qquad (11.9)$$

where ε is the strain, and as a function of stretch,

$$\frac{dU_3}{du} = E\varepsilon \qquad (11.10)$$

The dashpot, of course, always dissipates its energy, according to its viscosity,

$$\frac{dU_2}{du} = \eta\frac{d\varepsilon}{dt} \qquad (11.11)$$

Substitution of these quantities into equation (11.3) yields the equation of forced vibration of a viscously damped system:

$$\sigma = E\varepsilon + \eta\frac{d\varepsilon}{dt} + M\frac{d^2\varepsilon}{dt^2} \qquad (11.12)$$

This is the motion of a Maxwell element, including both mass and acceleration concepts.

The important points here are that the mechanical deformation of a body, including that of a polymer, is governed by simple thermodynamic laws applied to a mechanical system. Two additional quantities will be defined before leaving this section.

Consider the case of the body fracturing, where the area of the crack increases by δA. Then the energy changes can be written

$$\frac{\delta U_1}{\delta A} - \frac{\delta U_2}{\delta A} = \frac{\delta U_3}{\delta A} + \frac{\delta U_4}{\delta A} \qquad (11.13)$$

The term $\delta U_2/\delta A$ is called the fracture resistance, $\bar{R}$:

$$\bar{R} = \frac{\partial U_2}{\partial A} \qquad (11.14)$$

It indicates the energy dissipated in propagating a fracture over an increment of crack area δA. This is the work required to create new surfaces. Clearly, a large value of $\bar{R}$ indicates a tougher material.

As will be amplified below, contributions to $\bar{R}$ arise from the energy actually required to create the new surface (a quantity related to the surface tension), the orientation of chains near the surface, the breaking of chains that spanned the cracking region, and rubber elasticity energy storage effects.

A closely related quantity is the strain energy release rate, $\mathcal{G}$,

$$\mathcal{G} = \frac{dU_1}{dA} - \frac{dU_3}{dA} \tag{11.15}$$

If $\mathcal{G} > \bar{R}$, then the system is unstable, and the crack velocity increases. These terms define the conditions of crack growth and stability, and energy storage within the body (1). The quantity $\mathcal{G}$ is also known as the work of fracture, or the fracture energy per unit area of crack created. The quantity $\mathcal{G}_c$ represents the critical energy of crack growth, per unit area of crack created.

11.2 DEFORMATION AND FRACTURE IN POLYMERS

The previous section described the thermodynamics of deformation and fracture in terms of the energy required to elongate and break the sample. This section describes two major experiments to evaluate the deformation and fracture energy: stress–strain and impact resistance. In a tensile stress–strain experiment, the sample is elongated until it breaks. The stress is recorded as a function of extension. Stress–strain studies are usually relatively slow, of the order of mm/s. Impact strength measures the material's resistance to a sharp blow, typically m/s (2). In both stress–strain and impact studies, energy is absorbed within the sample by viscoelastic deformation of the polymer chains, and finally by the creation of new surface areas (3). Energy may be absorbed by shear yielding, crazing, or cracking.

It is important to distinguish between a craze and a crack. When a stress is applied to a polymer, the first deformation involves shear flow of the polymer molecules past one another if it is above T_g, or bond bending and stretching for glassy polymers. Eventually a crack will begin to form, presumably at a flaw of some kind, and then propagate at high speed, causing catastrophic failure. A craze is not an open fissure but is spanned top to bottom by fibrils that are composed of highly oriented polymer chains (see Figure 11.2) (4). Although the volume of the sample is increased by the void space, these fibrils hold the material together. A crazed material may appear whitened because of light-scattering effects. Then the material is said to be stress-whitened.

A third mechanism of energy absorption is called shear yielding. In shear yielding, oriented regions are formed parallel to the planes of maximum shear (>45°) for pressure sensitive materials such as polymers. Shear bands may be seen as birefringent entities; no void space is produced.

Another fundamental point relates to the actual mechanism(s) of fracture. Consider a sample composed of long polymer chains capable of viscoelastic motion. Does the crack grow through the polymer by breaking the chains, by viscoelastic flow of one chain past another, or by some combination of these factors? See Figure 11.3 (3). This will be considered further in Section 11.5. Crack initiation and ultimately failure in all materials begins at some type of flaw or defect. This can be a scratch or cut, even microscopic, a foreign particle, a bubble, a thin, degraded, or swollen region of the specimen, and so on.

In order to cause crack growth at a predefined position, many fracture researchers deliberately introduce notches of various kinds before testing. While this presupposes that the early stages of fatigue (Section 11.4) or other types of internal damage can be ignored, notching does produce better controlled experimental conditions.

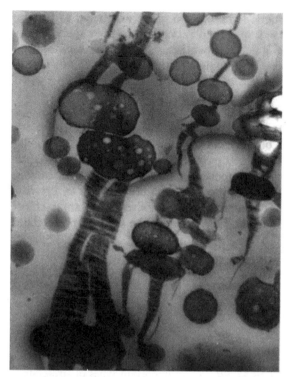

Figure 11.2 Crazing in ABS plastic (4). Electron micrograph of an ultrathin section cut parallel to the stress-whitened surface. The dark portions have been stained with OsO_4. The white plastic phase is poly(styrene–*stat*–acrylonitrile), and the stained phase is poly(butadiene–*stat*–acrylonitrile).

Some of the more important methods of failure studies include stress–strain, impact loading, and fatigue. Creep and stress relaxation (Chapter 10) may cause serious damage to engineering materials, but they normally do not result in fracture per se except for creep rupture. Emphasis in this chapter will be on the study of fracture energy, kinetics

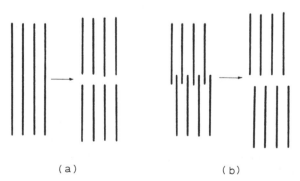

(a)   (b)

Figure 11.3 Schematic of fracture mechanisms in polymeric materials: (a) bond breakage and (b) chain slippage (3). In reality, both chain scission and slippage (pull-out) are governed by entanglements and polymer molecular weight, see Figure 1.3b.

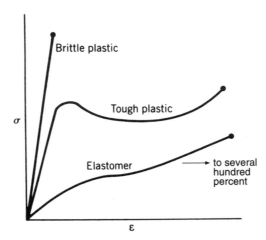

Figure 11.4 Stress–strain behavior of three types of polymeric materials. Young's modulus of brittle plastics is often close to 3×10^9 Pa, σ/ε.

of crack growth, and molecular mechanisms. The reader is directed to Chapter 13 for a fuller discussion of plastic toughening.

11.2.1 Stress–Strain Behavior of Polymers

11.2.1.1 General Behavior Three important types of stress–strain curves are illustrated in Figure 11.4. The brittle plastic stress–strain curve is linear up to fracture at about 1% to 2% elongation. Typical stresses at break are of the order of 10,000 psi, or about 6×10^7 Pa. Ordinary polystyrene behaves this way.

An example of a tough plastic is polyethylene, which is semicrystalline, with the amorphous portions above T_g. Its Young's modulus, given by the initial slope of the stress–strain curve, is somewhat lower than that of the brittle, glassy plastic (see Figure 11.4). Typically this class of polymer exhibits a yield point, followed by extensive elongation at almost constant stress. This is called the plastic flow region and is clearly a region of nonlinear viscoelasticity. Stresses in the range of 2 to 5×10^7 Pa are commonly exhibited in this range. Extension at constant stress in the plastic flow region is often referred to as cold drawing (see Section 11.4). Finally, the polymer strain hardens, then ruptures. Many tough plastics break at about 50% elongation.

The third type of stress–strain curve is that exhibited by elastomers. The equation of state for rubber elasticity governs here, with its peculiar nonlinear curve. The student will remember that $\varepsilon = \alpha - 1$. Elongation to break for the elastomer may be of the order of several hundred percent and is indicated by the dot at the end of the curve. For crystallizing elastomers such as natural rubber, the curve swings upward rather sharply at the point of crystallization, and tensile strengths of 2 to 5×10^3 psi are common. Noncrystallizing elastomers such as SBR have much lower tensile strengths, often below 1000 psi. However, with the addition of reinforcing fillers such as finely divided carbon black, the tensile strength is much increased. Of course, this last material is widely used for automobile tires, one of the toughest materials known.

Toughness as such is measured by the area under the stress–strain curve. This area has the units of energy per unit volume and is the work expended in deforming the material (see Section 11.1). The deformation may be elastic, and recoverable, or permanent (irreversible deformation). Elastic energy is stored in the sample in terms of energy per unit volume. Because of the development of crazes within the strained material, which are microscopic voids, the volume of the sample may increase, sometimes by several percent. This volume increase should not be equated with volume increases related to Poisson's ratio, which is recoverable.

There are several terms in use that describe the "strength" of a polymer. The tensile strength describes the stress to break the material, usually in elongation, for example. The tensile strength is certainly important, but in engineering practice a polymer is rarely stressed so greatly that it breaks immediately. The toughness of the polymer is frequently a more useful parameter.

11.2.1.2 Tensile Strength of Plastics

A major test of the mechanical behavior of polymers, especially those plastics below their glass transition temperature, involves the measurement of tensile strength. While it can be argued that tensile strength is not the best quantity to characterize engineering behavior, it is simple, inexpensive, and very widely reported.

Table 11.1 (5) summarizes the tensile yield strength of the three subclasses of plastics: amorphous, crystalline, and thermoset. Often crystalline plastics have higher elongations to break than amorphous materials because the crystalline regions act as reinforcement. The thermosets are almost always amorphous.

The values shown in Table 11.1 usually increase with the molecular weight of the polymer up to about $8M_c$ (see Section 11.5), decrease with branching, increase with increasing crystallinity, and for the thermosetting materials, decrease with very high levels of cross-linking. Orientation, where present, usually increases tensile strength, note Table

Table 11.1 Tensile strength of selected plastics

Plastic	Tensile Strength, MPa	Elongation to Break, %
A. Amorphous plastics		
Polystyrene	50	2.5
Poly(methyl methacrylate)	65	10
Poly(vinyl chloride)	50	30
Poly(bisphenol carbonate)	60	125
B. Semicrystalline plastics		
Polyethylene (HD)	30	600
Polypropylene	33	400
Polytetrafluoroethylene	25	200
Polyamide 66	80	200
Poly(ethylene terephthalate)	54	275
C. Thermosets		
Phenol-formaldehyde resin	55	1
Epoxy resin	90	2.4
Unsaturated polyester resin	60	3

Source: D. W. van Krevelen, *Properties of Polymers*, 3rd ed., Elsevier, Amsterdam, 1990, Table 13.11.
Note: MPa×145 = PSI.

Table 11.2   Mechanical behavior of fibers

Polymer	Modulus, GPa	Tensile Strength, GPa	Elongation to Break, %	T_f, °C	Reference
Polystyrene	4	0.08	2	100, T_g	(a)
Polyethylene	1	0.05	50	140	—
Polyethylene, ultradrawn	172	3.5	5	149	(a)
Rayon	3	0.3	20	180 decomp.	
Polyamide	3.8	0.8	25	250	(b)
Poly(1,4-benzamide)	100	3.0	6	500 decomp.	(b)

References: (a) S. Kavesh and D. C. Prevorsek, U.S. Pat. No. 4,413,110 (1983), basis for Spectra.® (b) J. R. Schaefgen, T. I. Bair, J. W. Ballou, S. L. Kwolek, P. W. Morgan, M. Panar, and J. Zimmerman, both in *Ultra-High Modulus Polymers*, A. Ciferri and I. M. Ward, eds., Applied Science, London, 1979.

Note: 1 GPa $= 10^9$ N/m^2 $= 10^{10}$ dynes/cm^2 $= 1.1 \times 10^4$ kg/cm^2 $= 1.45 \times 10^5$ lb/in.^{2}.

11.2 for fibers. Increasing strain rate usually decreases elongation to failure but increases tensile strength, because the polymer chains cannot relax as well. Polymers that are above their brittle–ductile transition temperature (see Section 11.2.3) often exhibit yield points, usually at several percent elongation, and have higher elongation to break. Poly(bisphenol carbonate), polycarbonate, is above its brittle–ductile transition temperature, while polystyrene and poly(methyl methacrylate) are not, for example. Of course, the energy to fracture is proportional to the area under the stress–strain curve, so that higher elongations generally mean tougher materials. No fillers, rubber, or plasticizers are assumed present in these materials. For all of these reasons, the values shown in Table 11.1 are only approximate, average values.

11.2.1.3   Time and Temperature Effects

The viscoelastic characteristics of stress–strain behavior of glassy amorphous polymers can be approached through considering that the stress at any time t is the sum of all the little stresses, $\delta\sigma$, each of which is the result of many incremental relaxing stresses each started at a progressively different time, t'. Each $\delta\sigma$ produces an incremental strain, $\delta\varepsilon$, and

$$\delta\sigma(t) = E(t')\delta\varepsilon \tag{11.16}$$

Thus the polymer can be considered as undergoing stress relaxation during the actual deformation process. On dividing the time interval from 0 to t into equal time increments, δx, we find the total stress, $\sigma(t)$, given by the convolution integral (6),

$$\sigma(t) = \int_0^t E(t - x)\frac{d\varepsilon}{dx}\,dx \tag{11.17}$$

While the stress relaxation studies in Chapter 10 emphasized the presence of a single relaxation time, or at most a few relaxation times, real plastics undergoing extensive deformation display a broad distribution of relaxation times. Eyring's model of holes and flow, Section 8.6, can also be used to predict the time and temperature effects of shear yielding.

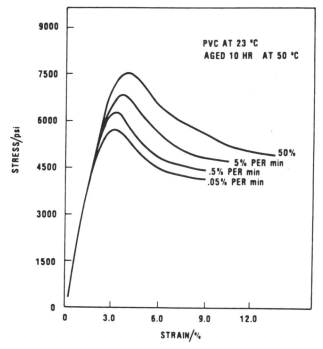

Figure 11.5 Stress–strain curves for poly(vinyl chloride) at various strain rates. Note the well-defined yield point at about 5% strain. (1 psi = 6.9×10^3 N/m²) (6).

Figures 11.5 and 11.6 illustrate the effects of time and temperature (6). Both sets of data show yield stresses, followed by strains nearly independent of the stress. Note the well-defined yield points. Many people use the yield strength for design rather than the tensile strength, because the yield point represents the onset of permanent damage.

11.2.1.4 Semicrystalline Polymers

There are two cases to be considered in the stress–strain relationships of semicrystalline plastics. If the amorphous portion is rubbery, then the plastic will tend to have a lower modulus, and the extension to break will be very large. If the amorphous portion is glassy, however, then the effect will be much more like that of the glassy amorphous polymers. Orientation of semicrystalline polymers is also much more important than for the amorphous polymers. A special case involves fibers, where tensile strength is a direct function of the orientation of the chains in the fiber direction.

Typical data for low density polyethylene are shown in Figure 11.7 (6). At the lowest temperature shown, −60°C, the polymer is just above its glass transition temperature, −80°C. As the temperature is reduced, the modulus increases, and a yield point is developed at about 5% strain. At higher elongations, polyethylene exhibits cold drawing; see Section 11.2.2.

Some semicrystalline polymers such as poly(ethylene terephthalate) and polycarbonates based on bisphenol A are used commercially at very low molecular weights, down

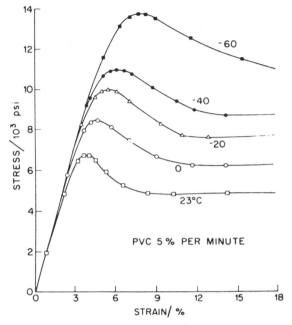

Figure 11.6   For poly(vinyl chloride) in the glassy region, Young's modulus stays substantially constant as the temperature is lowered from 23 to −60°C, but the stress needed to stretch the sample beyond the yield point increases (6).

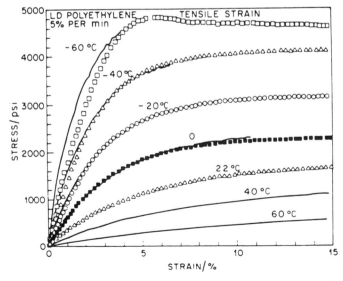

Figure 11.7   Low-density polyethylene's Young's modulus increases greatly as the temperature is lowered. Fit curves were generated theoretically using a scaling rule for crystalline polymers (6).

to $M_v = 13,000$ g/mol. Under these conditions the polymer may be extremely brittle, with low tensile strength, because its molecular weight is below $8M_c$. The chains do not form Gaussian coils at such low molecular weights, but rather are stiff, semicoils. Such polymers are most often used as matrix resins with glass fibers, useful because of their very low melt viscosities.

Polycarbonates and other polymers are considered by some to be toughened by the presence of low-temperature secondary transitions, such as the β-transition. The theory is that the flexibility imparted to the chain by the secondary chain motion contributes to the chain's ability to respond to stress.

11.2.1.5 Liquid Crystalline Polymers Fibers made from polymers in the liquid crystalline state exhibit both very high moduli and very high tensile strengths. As described in Chapter 7, these fibers are actually very highly crystalline, being made up of rigid-rod polymer chains, and hence highly aligned. Table 11.2 compares the mechanical behavior of several plastics and fibers. As noted previously, polystyrene is a brittle plastic, with a high modulus but breaking at only 2% elongation. Polyethylene is softer but much more extensible. Ultradrawn polyethylene fibers are made from very high molecular weight polymer 3 to 6×10^6 g/mol, by drawing the material 50 to 100 times its original spun length, thus introducing a very high orientation. Note that ultradrawn polyethylene has a modulus 70 times higher than the ordinary plastic.

The liquid crystalline-forming polymer, poly(1,4-benzamide), has a modulus and tensile strength still higher than the ultradrawn polyethylene. The properties of poly(1,4-benzamide) are compared with the well-known fibers nylon and rayon. The latter are much softer but are more extensible.

The fibers made from poly(1,4-benzamide) and other liquid crystalline forming polymers are used in the manufacture of bullet-proof jackets. Here, large and rapid energy-absorbing capabilities are essential.

The ultradrawn polyethylenes are the simplest fiber possible, with every carbon atom contributing to its load-bearing characteristics. Even though ultradrawn polyethylenes is clearly not a rigid-rod polymer, through ultradrawing virtually all the crystalline chain folds are eliminated, and the finished material mimics those actually made from rigid rods. The material is also used for bullet-proof jackets.

11.2.2 Cold Drawing in Crystalline Polymers

The tough plastic in Figure 11.4 is shown to have a yield point, followed by a region of cold drawing at almost constant stress. There are two basic causes for this phenomenon. First, for rubber-toughened amorphous plastics, the region of cold drawing is where extensive orientation of the chains takes place, accompanied by significant viscoelastic flow.

Second, for semicrystalline polymers with amorphous portions above T_g, cold drawing rearrangement of the chains takes place in a characteristic, complex manner, beginning with necking. A neck is a narrowing down of a portion of the stressed material to a smaller cross section. This is a form of shear yielding. The neck grows, at the expense of the material at either end, eventually consuming the entire specimen.

In the region of the neck, a very extensive reorganization of the polymer is taking place. Spherulites are broken up, and the polymer becomes oriented in the direction of stretch (see Figure 11.8) (7). The number of chain folds decreases, and the number of tie molecules between the new fibrils is increased. The crystallization is usually enhanced by

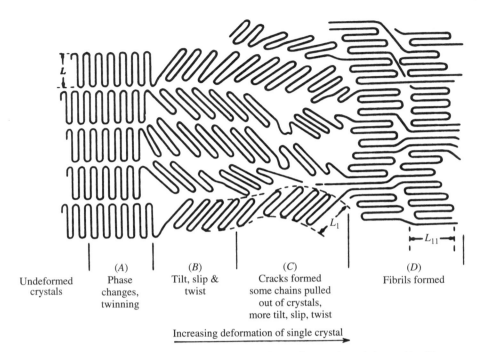

(A) (B) (C) (D)

| Undeformed crystals | Phase changes, twinning | Tilt, slip & twist | Cracks formed some chains pulled out of crystals, more tilt, slip, twist | Fibrils formed |

Increasing deformation of single crystal

Figure 11.8 Proposed mechanism of reorientation of crystalline structures during necking (7).

the chain alignment. At the end of the reorganization, a much longer, thinner, and stronger fiber or film is formed.

This phenomenon is easily demonstrated at home or in the classroom. A strip of polyethylene film, perhaps with print on it, is satisfactory. The strip is slowly stretched between the hands at room temperature, and the neck will form after several percent elongation. Such a material can then be elongated 2 to 5 times its original length. Draw ratios can be significantly higher under more scientific conditions.

The molecular reorganization illustrated in Figure 11.8 has attracted the attention of the theoreticians (8,9). Does the polymer actually melt under stress, and then reorganize, or do the crystals themselves rotate? At this point the evidence seems to be divided between the two possibilities.

It must be pointed out that drawing operations are critically important in the manufacture of fibers such as polyamide and rayon. In the case of polyamide, the polymer is spun in the melt state. The fiber is then cooled until it crystallizes, and finally stretched with a draw ratio of 4 to 8.

The case of rayon is slightly complicated because actually an alkaline solution of sodium cellulose xanthate is spun. The nascent fiber is spun into an acid bath, which removes the xanthate groups, precipitating the polymer. As it precipitates, it crystallizes. The fiber is then stretched wet, approximately four to seven times its original length. In both cases the fibers are highly oriented, as shown by X-ray studies. The high degree of orientation contributes to the high strength of such fibers.

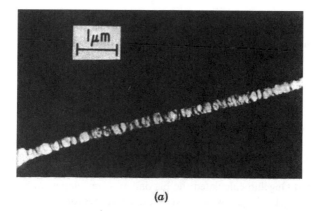

(a)

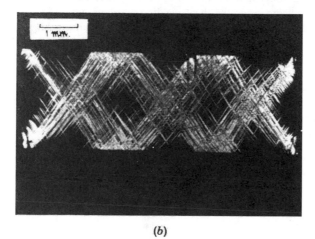

(b)

Figure 11.9 (a) Transmission electron micrograph of a craze in polystyrene. Courtesy of R. P. Kambour, General Electric Co. Note fibrillar structures spanning the craze. (b) Section of polystyrene deformed at 60°C, viewed between crossed polarizers. These shear bands started to form at a compressive strain of about 4%. P. B. Bowden and S. Raha, *Philos. Mag.*, **22**, 463 (1970).

11.2.3   The Brittle–Ductile Transition

As illustrated in Figure 11.9 (10) crazing involves the formation of an intricate network of fibrils connecting the upper and lower surfaces of the craze. The craze itself forms at right angles to the applied stress. Shear yielding, on the other hand, involves molecular slip, usually at 45° to the applied stress, assuming uniaxial stress–strain relationships. If the polymer forms crazes, it usually fails after 1% or 2% extension, and is said to be brittle. If it undergoes yielding, the polymer usually can be stretched at least 10% or 20% and is said to be ductile.

What criteria determine which mechanism will prevail? Polymers sometimes change from brittle to ductile on raising the temperature. Similarly, if the polymer is tested under compression, or under hydrostatic pressure, it tends to fail by shear yielding. Sternstein and Ongchin (11) studied the yield criteria for plastic deformation of glassy polymers in general stress fields, because the fracture resistance of a polymer is determined by its ability to develop a yield zone in the region of a crack tip, where it is usually in a state of triaxial stress. Sternstein and Ongchin used the so-called von Mises criterion (12–15),

$$(\sigma_1 - \sigma_2)^2 + (\sigma_2 - \sigma_3)^2 + (\sigma_3 - \sigma_1)^2 = 6C^2 \tag{11.18}$$

where σ_1, σ_2, and σ_3 are the triaxial stresses. For polymers, the quantity C is pressure, temperature, and strain rate dependent. For pressure, a linear dependence has been used. If the left-hand side exceeds $6C^2$, then shear yielding occurs, according to the pressure-modified von Mises criterion.

Sternstein and Ongchin calculated the biaxial stress envelopes for poly(methyl methacrylate); Figure 11.10 (11) shows separately the envelopes for craze initiation and shear yielding predicted from equation (11.18). In the first quadrant of stress space, the crazing envelope is everywhere inside the shear yielding envelope, which implies that all combinations of tensile biaxial stress produce crazes prior to shear yielding. When the sample is under compression, third quadrant yielding will always take place. In the second and fourth quadrants, the two envelopes intersect each other. Generally, the failure mechanism that takes place is the one requiring the lower stress. The 45° line running through the second and fourth quadrants represents pure shear deformation, marking the boundary between hydrostatic compression and hydrostatic tension. Crazing does not take place below this line because the pressure component of the stress tends to reduce rather than increase the volume.

While it can be argued that the intersection points in the second and fourth quadrants of Figure 11.10 constitute a brittle–ductile transition, both yielding mechanisms being

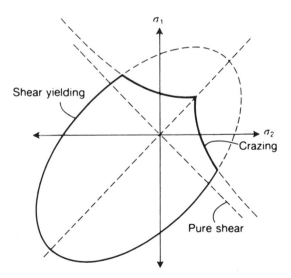

Figure 11.10   Biaxial stress envelope for poly(methyl methacrylate) at room temperature; $\sigma_3 = 0$. The curves indicate onset of crazing and shear yielding, as indicated. Note the intersection of crazing and shear yielding envelopes, and concomitant heavy continuous line (11).

coexistent. The brittle–ductile "transition" is a transition in the sense of which mode is predisposed to happen first in a given stress field (14).

11.2.4   Impact Resistance

Impact resistance is defined in terms of the energy required to fracture a sample when struck with a sharp blow.

11.2.4.1   *Instrumentation*   Resistance to impact loading may be measured by using specially designed universal testing machines, which permit very high rates of loading, or, more commonly, by using one of two types of instruments: the Izod and the Charpy impact test machines. Each of these strikes the specimen with a calibrated pendulum. The Charpy instrument is illustrated in Figure 11.11 (2). In both cases, the sample is often notched to provide a standardized weak point for the initiation of fracture. In the Charpy experiment the specimen is supported on both ends and struck in the middle. The notch is on the side away from the striker. In the Izod experiment, on the other hand, the sample is supported at one end only, cantilever style. The notch is placed on the same side as the striker. Izod testing is usually reported in units of J/m (Joules per meter of crack), based on samples of 12.7 mm wide, 3.2 mm thickness, and 65 mm length. Charpy testing is usually reported in units of J/m^2, or energy required to create a unit area of crack.

The interpretation of the results may be complicated even if the experiment is one of the simplest. The striker is released from an elevated position. On striking the sample, part of the momentum of the striker goes into the creation of new surface area. The broken portion(s) of the sample also leaves with a certain momentum, which must be accounted for during analysis. For certain kinds of standardized width samples, the impact strength is reported in terms of J/m^2. When multiplied by the width, the work required to create a unit surface is determined. Typical values for Izod impact strengths are summarized in Table 11.3 (3). Ideally impact strength ought to measure the work required to create new surfaces, especially for brittle, glassy plastics. However, viscoelastic effects cause the numerical results to be much larger (3). The student will note that two surfaces are created in each fracture.

11.2.4.2   *Toughening by Rubber Addition*   The impact resistance of glassy plastics may be increased by the addition of small quantities of rubber in the form of a polymer blend or graft copolymer (3,4,16–20). The rubber promotes crazing and shear yielding, which absorbs the energy locally. This may be seen in everyday life as the white spot that results after a plastic has been hit accidently. This is called stress whitening.

The characteristics of the rubber additive are of critical importance in determining the toughness of the final product. Several important factors include the size of the rubber droplets, the phase structure within the rubber particles (see Figures 4.14 and 11.2), grafting of the rubber to the plastic, and the glass transition of the rubber.

The rubber droplets must be at least as large as the crack tip radius. This puts the minimum size at several hundred angstroms, and a maximum size at about 3000 to 5000 Å. A size greater than 400 Å is required for cavitation.

The glass transition effect arises from a consideration of the speed of crack growth, which in turn depends on the velocity of sound in the material (3). In a plastic matrix

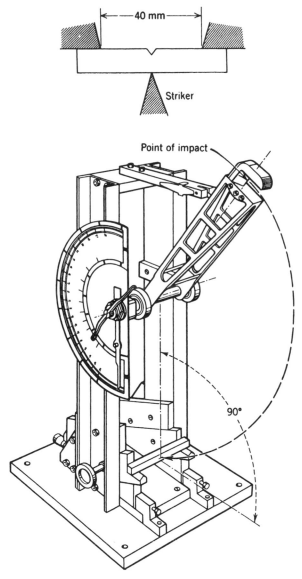

Figure 11.11   Charpy-type impact instrument. *Inset:* Charpy test piece showing notch and point of striker (2).

having Young's modulus equal to 3 GPa, the velocity of a crack under impact conditions is about 620 m/s. If the crack radius is about 1000 Å, an equivalent cyclic deformation frequency of about 10^9 Hz can be calculated. If the glass transition temperature increases at about 6 to 7°C per decade of frequency, the effective glass transition temperature of the rubbery phase is increased about 60°C above values measured at low frequencies, about 10^{-1} Hz. This calculation, while grossly oversimplified, suggests that the T_g of the elastomer phase must be about 60°C below the test temperature, which

Table 11.3 Mechanical strength of several plastics

Polymer	Izod Impact Strength (J/m)	$\mathcal{G}_{1c}$ (kJ/m^2)[a]	K_{1c} (MPa-m$^{1/2}$)
Polystyrene	20–30	0.4	1.1[a]
High-impact polystyrene, HIPS	80–150	3.5	1–2[c]
ABS plastics	200–400	—	2.0[c]
Epoxy resin (no filler)	75	0.1	0.5[c]
Epoxy resin (glass-fiber-filled)	—	7	7[c]
Cellulose acetate	120	—	—
Poly(methyl methacrylate)	40–60	0.5	1.1[c]
Phenol-formaldehyde plastics	20	—	—
Poly(vinyl chloride)	40–70	1.23[d]	2–4[c]
Polycarbonate	600–800	—	2.4

[a] A. J. Kinloch and R. J. Young, *Fracture Behavior of Polymers*, Applied Science Publishers, London, 1983, table 3.4. MNm$^{-3/2}$ = MPa-m$^{1/2}$.

[b] *Modern Plastics Encyclopedia*, W. A. Kaplan, Ed., McGraw-Hill, New York (1998).

[c] J. G. Williams, *Fracture Mechanics of Polymers*, Ellis Horwood Ltd., Chichester, 1984, table 6.4.

[d] Williams, ibid., table 8.1.

correlates well with the experimental evidence (see Table 11.4) (18). Thus, if the impact experiments are done at about 20°C, the glass transition of the rubber must be below about −40°C in order to attain significant improvement in impact resistance (21).

One of the ways of obtaining the greatest overall toughness in a plastic is by combining shear yielding, crazing, and cracking in the proper order to absorb the highest total energy. Usually it is desirable to have the sample yield in shear first. This absorbs energy without serious damage to the plastic. Then crazes should be encouraged to form within the shear-banded areas. The shear bands tend to stop the propagation of the crazes, limiting their growth. Only lastly does the "molecular engineer" want an open crack to form, because its propagation leads to failure.

A certain amount of viscoelastic molecular motion is required for shear yielding to occur first. The rubber domains actually cause crazes to multiply. This is a valuable occurrence, because as the crazes multiply they absorb energy locally in great quantities.

For example, polystyrene homopolymer is brittle, and under impact loading it tends to crack. High-impact polystyrene, HiPS, the graft copolymer of polystyrene with 5% to 10% polybutadiene, crazes first. A blend of this graft copolymer with poly(2,6-dimethyl-

Table 11.4 Impact resistance of ABS polymers at 30°C (16)

Sample Number	Composition of Rubber Component		T_g of Rubber Component, °C	Charpy Impact Strength, kJ/m^2
	BD[a]	ST[b]		
1	35	65	40	0.75
2	55	45	−20	18
3	65	35	−35	30
4	100	0	−85	40

[a] Polybutadiene.

[b] Poly(styrene–*stat*–acrylonitrile).

1,4-phenylene oxide), PPO, undergoes shear yielding first, then crazes. The ABS plastics craze, but with some shear yielding.

Similarly homopolymers and copolymers with a flexible bond in their backbone tend to be tougher than those without. Examples include the polycarbonates and poly(methyl methacrylate) containing small amounts of acrylate comonomers. Flexible side chains seem not to be as effective, probably because the mechanical energy is not absorbed where it is needed.

Semicrystalline plastics such as polyamide 610 and polyethylene are particularly tough. In this case the amorphous portions are well above T_g, the $-CH_2-$ groups being highly flexible. The crystalline portions are hard and act to hold the entire material together.

11.2.5 Mechanical Behavior of Elastomers

11.2.5.1 Stress–Strain Behavior In Chapter 9 the theory of rubber elasticity was developed. Young's modulus was given as $E = 3nRT$. Indeed, the modulus, a direct measure of the stiffness, increases with cross-link density. Ordinary cross-linked networks have a distribution of active chain lengths. Assuming a random cross-linking process, then $(M_c)_w/(M_c)_n$ should be about 2.

Mark and co-workers (22–26) examined the influence of the distribution of M_c, using end-linked poly(dimethyl siloxane) networks. Networks were formed from polymers having various ratios of low-molecular-weight component (660 g/mol) and high-molecular-weight component (21.3×10^3 g/mol), both number averages. The stress–strain behavior of the bimodal networks is illustrated in Figure 11.12 (22). A maximum in the stress to break as well as the toughness (as measured by the area under the curves) was attained at about 95 mol% of short chains. However, because of the difference in chain length between the two species, this is actually close to a 50/50 weight ratio. The over-all mechanical advantages of the bimodal network are illustrated in Figure 11.12 (22).

At high elongations the curves in Figure 11.13 turn upward owing to the limited extensibility of the short network chains. The elongations at rupture increase with decreasing mol% of short chains, causing the energy to rupture and the stress to break to go through maxima as the composition is changed.

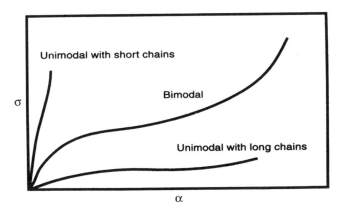

Figure 11.12 Dependence of stress versus elongation for two unimodal networks having either all short chains or all long chains, and a bimodal network having some of both. Note that the bimodal network exhibits both higher stress and elongation at failure than either unimodal network.

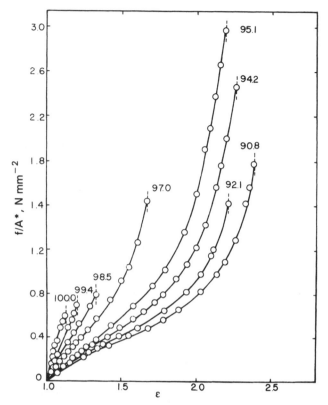

Figure 11.13 Stress–strain behavior of bimodal poly(dimethyl siloxane) networks. Each curve is labeled with the mol% of the short chains. The area under each curve represents the energy required for rupture (22).

The theory of rubber elasticity (Section 9.7) assumes a monodisperse distribution of chain lengths. Earlier, the *weakest link* theory of elastomer rupture postulated that a typical elastomeric network with a broad distribution of chain lengths would have the shortest chains break first, the cause of failure. This was attributed to the limited extensibility presumably associated with such chains, causing breakage at relatively small deformations. The flaw in the weakest link theory involves the implicit assumption that all parts of the network deform affinely (24), whereas chain deformation is markedly nonaffine; see Section 9.10.6. Also it is commonly observed that stress-strain experiments are nearly reversible right up to the point of rupture.

Apparently the short chains act primarily to increase the ultimate strength through their limited deformability, while the long chains thwart the spread of rupture nuclei that would otherwise lead to catastrophic failure. Mark (24) notes that this is similar to a *delegation of responsibilities*. In general, of course, if energy can be absorbed by a material by more than one mode, it is likely to be stronger. The conclusion from the research of Mark and his co-workers is that a broad distribution of M_c values may actually be beneficial in attaining good elastomer mechanical properties.

This surprising conclusion runs counter to the findings for the mechanical behavior of plastics and fibers. In that case high molecular weights of narrow polydispersity indexes usually yield superior mechanical behavior.

Another feature of some elastomeric networks is their capability of crystallization. This too brings about an anomalous increase in the modulus at high elongations (25). Natural rubber has a melting temperature near room temperature in the relaxed state. (The melting temperature is slightly depressed by the vulcanization process.) At high elongations the melting temperature climbs, so that at $\alpha = 4.5$, T_f is about 75°C.

Strain-induced crystallization also has a pronounced reinforcing effect within the network, and this increases the ultimate strength and maximum extensibility (26). The capability of crystallizing with extension while remaining elastomeric at low elongations provides an important self-reinforcement mode for natural rubber. The only other elastomer with these properties is *cis*-polybutadiene. These materials are the elastomers of choice for many heavy-duty applications.

11.2.5.2 *Viscoelastic Rupture of Elastomers*   Failure in highly stretched elastomers is by no means instantaneous (27–29). A common home example involves putting a rubber band around some objects and setting them aside. Some time later the rubber band may be seen broken. Interestingly the time frame of the failure can be predicted through the application of ordinary principles of viscoelasticity.

In the laboratory highly stretched elastomers undergo a period of smooth relaxation followed by sudden failure. The time frame of failure decreases with increasing initial stretch (see Figure 11.14) (27). The dashed line illustrates the locus of failure. The data in Figure 11.14 are for one temperature. When the experiment is repeated at different temperatures, a family of such curves will be generated, as shown in Figure 11.15 (27).

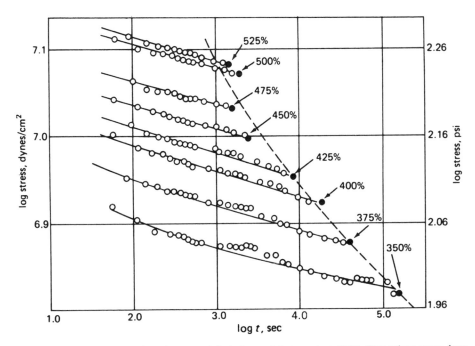

Figure 11.14 Stress relaxation of *cross*–poly(butadiene–*stat*–styrene) at 1.7°C. Elongations range from 350% to 525%. Solid points indicate rupture. The dashed line gives the ultimate stress as a function of log time (27).

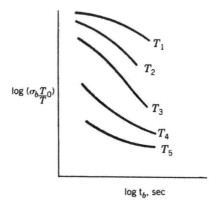

Figure 11.15 The locus of failure points obtained at different temperatures (27).

The quantity σ_b represents the stress to break. The quantity T_0/T multiplying the stress to break applies the theory of rubber elasticity, reducing the stress to the temperature of interest.

Through the application of the time–temperature superposition principle (Section 10.3), these curves can be shifted to yield a master curve at a particular temperature; see Figure 11.16 (27). Note that the quantity A_T appearing in Figure 11.16 is the shift factor appearing in the WLF equation (Section 8.6.12).

It is customary to record the strain to break, ε_b, as well as the stress to break. When $\log \varepsilon_b$ is plotted against $\log(\sigma_b T_0/T)$, a special curve called the failure envelope is generated; see Figure 11.17 (27). The failure envelope is widely used to predict the stability of elastomers under stress. If a specimen falls in region A of Figure 11.17, it will not fail. On the other hand, a specimen in region B is subject to eventual failure.

Elastomers exposed to wear and tear, such as automotive tires, are always reinforced. The usual reinforcing substance is either finely divided silicas or carbon blacks (see Sec-

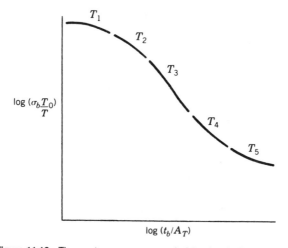

Figure 11.16 The master curve composed of the data in Figure 11.15.

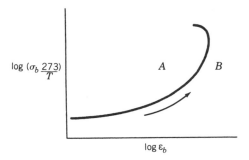

Figure 11.17   The failure envelope prepared from the master curve in Figure 11.16, obtained by plotting the reduced stress to break versus the strain to break, on a log–log scale. The specimen will not fail in region *A*, but will in region *B*. The arrow indicates the direction of lower temperatures or higher strain rates, the latter for dynamic experiments.

tion 9.16.2). Tires contain up to 60% of the latter (see Table 9.5), raising their tensile strength several times over. It must be pointed out that the major cause of wear in tires is abrasion. Tiny shreads of rubber are torn loose and stretched on the road, gradually ripping them off (30). While tear strength and tensile strength are not the same, they are closely related.

11.2.6   Fiber-Reinforced Plastics

In the manufacture of engineering materials based on plastics, much emphasis has been placed on high modulus and great strength. The practice of polyblending small quantities of rubber into plastics has already been mentioned as a way to improve impact resistance and strain to break. The addition of carbon black to rubber increases tensile strength and abrasion resistance.

A third way to toughen plastics is to add fibers, to make composite materials (3). Fibers in use today include glass, boron, graphite, or other polymers. These may be chopped or added as continuous filaments. With continuous filaments, the fibers carry most of the mechanical load, while the polymeric matrix serves to transfer stresses to the load-bearing fibers and to protect them against damage (31–34). Thus a controlled amount of adhesion between the fibers and the matrix is required.

11.2.6.1   *Matrix Materials*   Polymers used for matrix materials are usually thermosets, with epoxies being the most common (35). Others are polyesters and phenolics. Sometimes thermoplastics are used, principally polyimides, polysulfones, and polyetheretherketone (PEEK). The matrix has several roles in polymer–fiber composites. These include maintaining desired fiber spacing, transmitting shear loads between layers of fiber, and reducing the tendency to transmit stress concentration from broken fibers to intact fibers.

The role of the fiber is to support all main loads and limit deformations. Thus the fibers increase strength, stiffness, and toughness, and decrease corrosion, creep, and fatigue. Figure 11.18 (36) illustrates a scanning electron micrograph of the fracture surface of an injection-molded PEEK composite reinforced with 30 wt% of short carbon fibers. The sample was static loaded, with crack growth being from lower right to upper left. Minimum pullout of fibers indicates good adhesion between fiber and matrix.

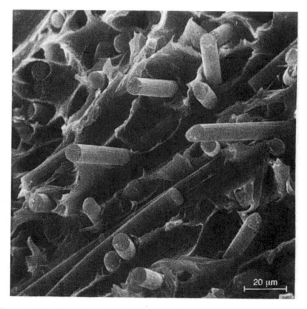

Figure 11.18   Fracture surface of short carbon fiber–PEEK composite (36).

Because of their light weight and great strength and stiffness, fiber–polymer composites are among the strongest materials known, especially on specific modulus and strength basis (the property is divided by the density). Table 11.5 (35) illustrates the increase in properties that can be achieved. Thus the modulus is increased up to 80 times, and the tensile strength about 20 times on the addition of various fibers. On the other hand, the densities are less than double. Chapter 13 explores composites further.

Table 11.5   Comparative fiber and unidirectional composite properties (35)

Fiber/Composite	Elastic Modulus, GPa[a]	Tensile Strength, GPa[a]	Density, g/cm³	Specific Stiffness, MJ/kg	Specific Strength, MJ/kg
Epoxy matrix resin	3.5	0.09	1.20	—	—
E-glass fiber	72.4	2.4	2.54	28.5	0.95
Epoxy composite	45	1.1	2.1	21.4	0.52
S-glass fiber	85.5	4.5	2.49	34.3	1.8
Epoxy composite	55	2.0	2.0	27.5	1.0
Boron fiber	400	3.5	2.45	163	1.43
Epoxy composite	207	1.6	2.1	99	0.76
High-strength graphite fiber	253	4.5	1.8	140	2.5
Epoxy composite	145	2.3	1.6	90.6	1.42
High-modulus graphite fiber	520	2.4	1.85	281	1.3
Epoxy composite	290	1.0	1.63	178	0.61
Aramid[b] fiber	124	3.6	1.44	86	2.5
Epoxy composite	80	2.0	1.38	58	1.45

[a] To convert GPa to psi, multiply by 145,000.

[b] Aromatic polyamide.

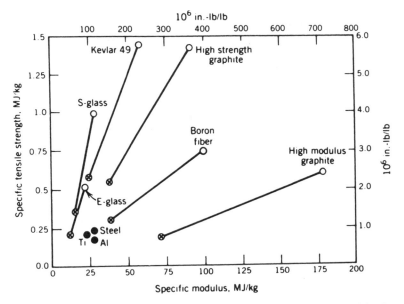

Figure 11.19 Specific strength versus specific modulus for fiber-reinforced epoxy materials: ○, unidirectional composite; ⊗, quasi-isotropic material. Properties are compared to some common metallic materials (●). To convert MJ/kg to 10^6 in.·lb/lb, multiply by 4.023 (35).

Figure 11.19 (35) illustrates the specific modulus and strength of these materials, compared to common metals. It is especially important to note that for many applications, particularly aerospace, weight is the important quantity, not volume of material.

11.2.6.2 Silane Interfacial Bonding Agents   A critically important parameter involves the bonding of the fiber to the matrix. If the fiber is not well bonded to the matrix, it will pull out under load. If it is too well bonded, no energy can be absorbed by the debonding process. Ideally the matrix should fracture first, followed by the debonding of the fiber–matrix interface, followed by the fracture of the fiber itself for maximum absorption of energy; see Figure 11.18. For glass fibers particularly, the extent of bonding is controlled by the use of silanes as interfacial bonding agents. The silane molecule has one end that will react with the glass surface, and the other end with the matrix. For an epoxy–glass system, a silane such as γ-aminopropyltriethoxysilane is used,

$$NH_2-CH_2-CH_2-CH_2-Si\begin{subarray}{l} \diagup O-C_2H_5 \\ -O-C_2H_5 \\ \diagdown O-C_2H_5 \end{subarray} \qquad (11.19)$$

The amino group reacts with the oxirane of the epoxy, and the silane bonds to the glass as illustrated in Figure 11.20 (37).

11.2.6.3 Molecular Composites   Among the newest experimental materials are the so-called molecular composites that make use of rod-shaped polymer chains dispersed or

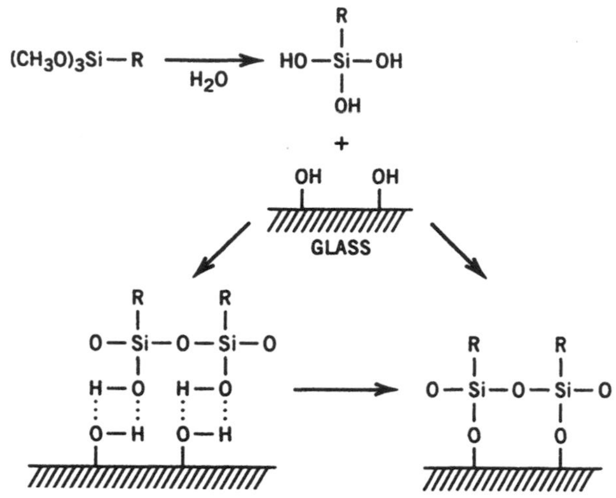

$$(CH_3O)_3Si-R \xrightarrow{H_2O} HO-\underset{\underset{OH}{|}}{\overset{\overset{R}{|}}{Si}}-OH$$

+

OH   OH

GLASS

$$O-\underset{\underset{H-O}{|}}{\overset{\overset{R}{|}}{Si}}-O-\underset{\underset{H-O}{|}}{\overset{\overset{R}{|}}{Si}}-O$$

O—H   O—H $\longrightarrow$

$$O-\underset{\underset{O}{|}}{\overset{\overset{R}{|}}{Si}}-O-\underset{\underset{O}{|}}{\overset{\overset{R}{|}}{Si}}-O$$

Figure 11.20   Illustration of the process by which silane coupling agents are adsorbed onto silica surfaces (37). The group R may be a vinyl-bearing moiety or may contain another reactive group that will bond it to the polymeric matrix.

dissolved in a random coil matrix; see Figure 11.21 (38). Since the extent of reinforcement increases with the ratio of length to width of the fiber (called the aspect ratio), ideally very strong materials should result. A major experimental problem, however, has been a proper dispersal of the rod-shaped molecules in the matrix, caused by the very low entropy of mixing two kinds of polymer molecules; see Section 4.3. Many scientists and engineers think that a rapid coagulation followed by orientation, such as used in fiber technology, may yield a breakthrough for the anticipated excellent properties.

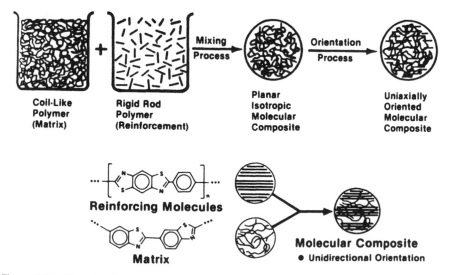

Figure 11.21   The molecular composites concept utilizes a rod-shaped polymer dispersed in a random-coil-shaped polymer. When the rod-shaped chains are oriented, reinforced fibers can be made (38).

11.3 CRACK GROWTH

11.3.1 The Griffith Equation

As a material is strained, energy is stored internally by chain extension, bond bending, or bond stretching modes (39–42). This energy will be dissipated if bond breakage or viscoelastic flow occurs. The first analysis of the balance between the energy applied and the energy released in bond breakage as a crack propagates was due to Griffith (39). He showed that when the release of strain energy per unit area of the crack surface exceeds the energy required to break the bonds associated with the unit area of surface, the surface tension, sometimes called intrinsic surface energy being designated as γ_s, a crack would propagate. Several designs of specimens containing various cracks, or notches, are shown in Figure 11.22 (42).

When σ is the gross applied stress, E is Young's modulus, and a is half the crack length, Griffith showed that the critical point is defined by the relation

$$\sigma = \left(\frac{2E\gamma_s}{\pi' a}\right)^{1/2} \tag{11.20}$$

This equation was specifically derived for the center-notched panel (Figure 11.22b). Experimentally determined values for the surface energy for polymers are 10^2 to 10^3 times values calculated on the basis of bond breakage alone (40,41). The increase is largely due to viscoelastic flow, which absorbs large amounts of energy.

11.3.2 The Stress Intensity Factor

A more general equation was derived by Orowan (43), who replaced γ_s with the term $\gamma_s + \gamma_p$, where γ_p accounts for the energy involved in plastic deformation. Irwin (44)

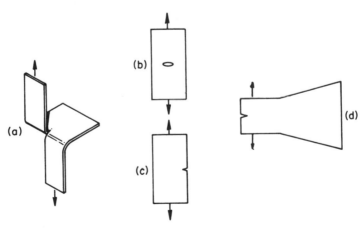

Figure 11.22 Important specimen shapes for the determination of the characteristic energy and stress intensity parameters for fracture mechanics (40): (*a*) Trouser-leg design, often used with elastomers. (*b*) Center-notched panel. (*c*) Edge-notched panel. (*d*) Cantilever-beam type. Types (*b*)–(*d*) are used for plastics.

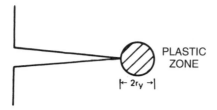

Figure 11.23 Schematic illustration of the plastic zone at the crack tip (45).

considered the fracture of solids from a thermodynamic point of view and arrived at the equation

$$\sigma = \left(\frac{E\mathscr{G}}{\pi'a}\right)^{1/2} \tag{11.21}$$

where $\mathscr{G}$ represents the fracture energy, $\partial U/\partial a$ (see Section 11.1). Then

$$\mathscr{G} = 2(\gamma_s + \gamma_p) \tag{11.22}$$

A key development in fracture mechanics was the analysis of the stress field at a crack tip (45). If σ_{ys} is the material's yield strength and r_y the crack-tip plastic zone radius (see Figure 11.23),

$$r_y = \frac{1}{2\pi'}\frac{K^2}{\sigma_{ys}^2} \tag{11.23}$$

where the stress intensity factor, K, is given by

$$K^2 = E\mathscr{G} \tag{11.24}$$

for the case of plane stress, or $K^2 = E\mathscr{G}/(1 - v^2)$ for plane strain, where v is Poisson's ratio (45).

For the simple case represented by Figure 11.21b the stress intensity factor is given by

$$K = \sigma(\pi'a)^{1/2} \tag{11.25}$$

At the critical conditions of crack spreading, the critical stress intensity factor, K_c, is given by

$$K_c = \sigma_c(\pi'a)^{1/2} \tag{11.25a}$$

Similarly the work done per unit area of new crack surface is given by $\mathscr{G}_c$. It must be remembered that a crack will spread only if the total energy of the system is decreased. A more general case may be written

$$K_c = cy\sigma(\pi'a)^{1/2} \tag{11.26}$$

where Y equals 1.00 for a center-crack plate of infinite width. Graphically, K represents the number of times that the stress is magnified at the crack tip.

The stress intensity factor forms the central part of a great deal of the fracture and fatigue literature. Once the stress intensity factor is known, it is possible to determine the maximum value that would cause failure. This critical value, K_c, is known as the fracture toughness of the material. Under plane strain (tensile) test conditions, the fracture toughness is given the special designation K_{1c} (46). The corresponding notation for shear strain is K_{2c}, and the notation for tearing (antiplane) test conditions is K_{3c}. These tests refer to conditions where the crack is pulled open, where the portions of the sample above and below the crack are pulled in opposite directions parallel to the crack, and where portions of the sample above and below the crack are pulled in opposite directions normal to the direction of the crack, respectively. Values of the critical fracture energy and critical stress intensity factor are shown in Table 11.3 and discussed further in Section 13.8.

11.3.3   A Worked Example

A center-notched panel of polystyrene at 25°C has a crack of 1.0 cm width; see Figure 11.22b. The critical stress intensity factor, K_c, was determined by other experiments to be 6 MPa $\cdot$ m$^{1/2}$. Will the sample fail under a stress of 10 MPa?

From equation (11.25), we have

$$K = \sigma(\pi'a)^{1/2}$$

$$K = 10 \text{ MPa } (3.14 \times 0.5 \text{ cm})^{1/2}$$

since a is half the width of the crack.

$$K = 1.25 \text{ MPa} \cdot \text{m}^{1/2}$$

The sample will not fail because K is smaller than K_c.

What is the critical strain energy release rate, $\mathscr{G}_c$? From equation (11.24),

$$\mathscr{G}_c = \frac{K_c^2}{E}$$

Young's modulus for polystyrene at room temperature is approximately 3×10^9 Pa; see Chapter 8.

$$\mathscr{G}_c = (6 \text{ MPa} \cdot \text{m}^{1/2})^2/3 \times 10^3 \text{ MPa}$$

$$= 1.2 \times 10^{-2} \text{ MJ/m}^2$$

Note the units of work per unit area for $\mathscr{G}_c$.

11.4   CYCLIC DEFORMATIONS

In cyclic deformation, stress and strain patterns are repeated over and over again, perhaps tens of thousands of times. While the stresses are usually well below the ultimate

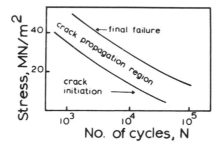

Figure 11.24 The fatigue response of a polymer showing the relationship between applied stress, σ, and the logarithm of the number of cycles to failure, N, for both initiation of fatigue cracks and final failure. Note that fatigue cracks may form on the interior of the material, and not be visible on cursory inspection.

failure stress measured in simple extension, cracks may grow and the sample may fail. This experiment simulates the mechanical vibrations that engineering plastics are subjected to in service; indeed, such plastics are frequently more likely to fail because of repeated stressing than because of a single excessive deformation or hard blow.

Material failure because of cyclic stressing is called fatigue. Phenomenologically there are three regions in the stress versus number of cycles to failure (S–N curves); see Figure 11.24 (45): For a sample undergoing cyclic stress–strain exercises, first there is a region where there may be no externally apparent damage. This is followed by a region of crack propagation, where the crack grows steadily with each cycle. The last region involves catastrophic failure, where one or two cycles more causes total material fracture. Fatigue response in this way is commonly observed: A material in service appears to be fine, and then breaks with little or no warning. Of course, the cracks may be propagating internally, not easily seen without instrumental examination. In terms of molecular processes, fatigue failure involves two principal mechanisms: (a) the sample may get hot owing to adiabatic heating (see Section 11.1)—this is especially true at the crack tip, and (b) growth of the crack itself (47,48). The number of cycles to failure depends on the stress level, as illustrated in Figures 11.25 (48) and 11.26 (49). In some cases a so-called endurance limit is observed below which fatigue failure does not occur in realistic lifetimes of experiments.

The fatigue phenomenon is not an all-or-nothing situation. First, it is assumed by many investigators that all materials are flawed as made, even if the flaws are too small to be ordinarily noticed. On repeated stressing, these tiny flaws initiate cracks. In the next stage, these cracks propagate slowly, perhaps growing a bit with every cycle, or sometimes intermittently. Finally, the material fails, the crack growing through the sample in one stress cycle.

11.4.1 Mechanisms of Crack Growth

Hertzberg and Manson (45) have identified several important molecular aspects of the fatigue process:

1. High molecular weights and narrow moleclar-weight distributions are generally more fatigue resistant. Wool (50) has shown that plastics having molecular weights greater than about $8M_c$ arrive at a plateau of maximum mechanical strength.

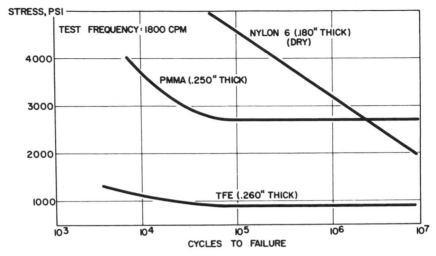

Figure 11.25 Fatigue lifetimes versus stress level for nylon 6, poly(methyl methacrylate), and polytetra-fluoroethylene (46). This plot is known as an S–N plot, for stress and number of cycles to failure.

2. Chemical changes such as bond breakage are to be avoided or minimized.*
3. Viscoelastic deformation of the chains is desired, as these motions absorb energy and tend to prevent crack growth, provided the sample does not heat up excessively.
4. Morphological changes such as drawing, orientation, and crystallization also absorb energy and are desired.
5. Samples undergoing cyclic stressing at frequencies and temperatures near the glass transition temperature or secondary transitions will tend to heat up more than otherwise, causing softening or degradation. If the starting temperature is just below T_g, the heating rate may increase as time progresses.
6. Adiabatic heating caused by such hysteretic effects is clearly also undesirable.
7. Inhomogeneous deformations, such as crazing and shear banding, absorb energy and are desired. It is especially desired to have shear banding occur before crazing; that is, the former mechanism should have a lower free energy of initiation.

Of course, several of these factors may be operative simultaneously.

11.4.2 Fatigue Crack Propagation

Fatigue crack propagation relates to the slow growth of cracks caused by cyclic stressing. In many fatigue crack propagation experiments, a notch is deliberately introduced in the specimen at a convenient location (51). The rate of growth of this crack and the morphology of the crack surface outline areas of current interest.

For a crack of length a, the fatigue crack growth rate is da/dn, where n represents the number of stress cycles. The stress intensity factor range is given by

$$\Delta K = K_{max} - K_{min} \tag{11.27}$$

---

*However, chain scission is now thought to be a major failure mechanism of many plastics; see Section 11.5.2.

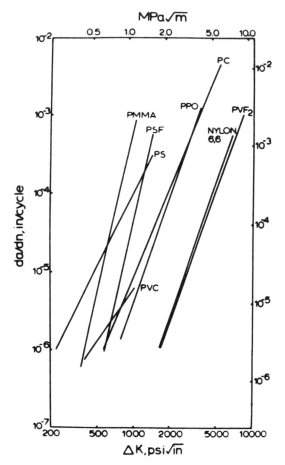

Figure 11.26 The relationship between ΔK and fatigue crack growth rates for poly(methyl methacrylate), PMMA; polysulfone, PSF; polystyrene, PS; poly(vinyl chloride), PVC; poly(2,6-dimethyl-1,4-phenylene oxide), PPO; polycarbonate, PC; nylon 66; and poly(vinylidene fluoride), PVF$_2$ (49). Samples with data to the right are more fatigue resistant than on the right, because they require higher levels of stress at the crack tip to propagate the crack.

where $K_{\max}$ and $K_{\min}$ are the maximum and minimum values of the stress intensity factor during the cyclic loading. The crack growth rate is related to the stress intensity factor range by the equation (Paris' law)

$$\frac{da}{dn} = A\Delta K^m \qquad (11.28)$$

where A and m are parameters dependent on material variables, mean stress, environment, and frequency. While the rate of crack growth increases with crack length, equation (11.28) implies a straight line, which is frequently observed over much of the range in ΔK (see Figure 11.27) (52). The actual rates of crack growth depend on several factors. First is the temperature of the experiment. For low frequencies, the temperature of

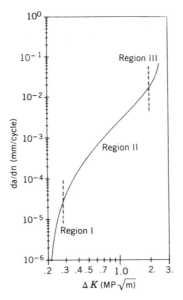

Figure 11.27 Fatigue propagation in plastics. Region I, substantially nil crack growth. Region II, figure crack propagation proceeds with thousands of cycles required to produce failure. Region III, failure after only a few cycles.

the crack tip may approximate the ambient conditions. At higher frequencies, the crack tip may get quite hot. This is important because viscoelastic motions depend strongly on the temperature and the frequency of the loading. Either the glass transition or secondary transitions such as the β transition may be involved (51), but the latter is usually the more important because most engineering plastics are well below their glass transition temperatures. The frequency sensitivity, defined as the multiple by which the fatigue crack propagation rate changes per decade of frequency, goes through a maximum at the temperature where the β transition segmental jump frequency is comparable to the test machine loading frequency (51).

Second, the rates of crack growth depend on the value of ΔK. If ΔK is low enough, the rate of crack growth may be essentially zero. If ΔK is high enough, one cycle may produce fracture. This results in a sigmoidal character to the fatigue crack growth curve (see Figure 11.27) (52). The data in Figure 11.26 are essentially in the middle range.

Third, fatigue propagation depends inversely on the weight-average molecular weight (52). Below a critical value of M_w, the sample is brittle. At higher molecular weight ranges, the fatigue crack propagation rate decreases. Indeed, higher molecular weight samples are stronger in a variety of experiments.

11.5 MOLECULAR ASPECTS OF FRACTURE AND HEALING IN POLYMERS

From a kinetics point of view, formation of new surfaces, fracture, and subsequent crack healing have a certain symmetry, with chains variously being pulled out or diffusing into the fracture surface. The symmetry is far from complete, however, because of chain breakage, the chemical and physical state of the surfaces at the time of crack healing, and so on (53–60).

In general, the two surfaces being "healed" are from different sources. Examples of this include molding operations, film formation from latexes, and adhesion. This section examines the micromechanisms involved in fracture and healing in polymeric materials. The effect of chemical bonding and adhesion between a polymer and a substrate surface is considered in Section 11.6.

11.5.1 Interdiffusion at a Polymer–Polymer Interface

When two otherwise identical polymer surfaces are brought into juxtaposition at $T > T_g$ and annealed, a healing process may take place. As a result the interface gradually disappears. Basic additional requirements are that the polymer be linear (or branched) and amorphous. The major mechanism involves the interdiffusion of polymer chain segments across the interface. For polymer chain interdiffusion, the reptation theory of de Gennes (54) and Edwards (55) is employed. In this model the chain is confined to a tube, as described in Chapter 5. The equations of motion of this chain in the bulk state are described in Section 10.4.2. In the following, a linear monodisperse molecular weight polymer whose chain ends are uniformly distributed in the bulk melt material is considered. Assuming instantaneous wetting of the two surfaces, Wool and co-workers (56–59) and Tirrell (60) analyzed the molecular aspects of the interface as interdiffusion proceeds. Wool developed the "minor chain model"; see Figure 11.28 (57). At time equals

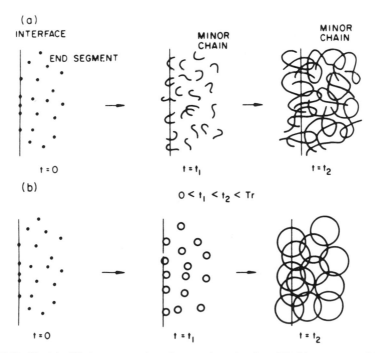

Figure 11.28 The interdiffusion process at a polymer–polymer interface (57): (a) appearance of the minor chains and (b) the minor chain spherical envelopes.

zero, molecular contact is achieved at the interface. The chains begin to reptate via their chain ends, which are shown initially as dots randomly distributed throughout the polymer. As time proceeds, only that portion of the chain that has escaped its original conformation (i.e., the minor chain) can contribute to interdiffusion across the interface. Wool does not consider the exact location of each segment, but only the spherical envelope to which it belongs; see Figure 11.28.

11.5.2 Molecular Basis of Fracture in Glassy Plastics

While chain scission is important in the deformation and fracture of many polymers, this micromechanism normally consumes only a small fraction of the fracture energy. However, its appearance limits the extent of viscoelastic energy dissipation, the major energy-consuming factor in most polymer fractures. Thus, for tougher materials, chain scission should be delayed or avoided in favor of molecular relaxation.

11.5.2.1 Energy Well Activation In the relaxed amorphous state (Chapter 5), polymer chains exist in the form of random coils. The initial effect of an external stress is to elongate those coil segments residing in the fracture plane, see Figure 11.29 (53). This

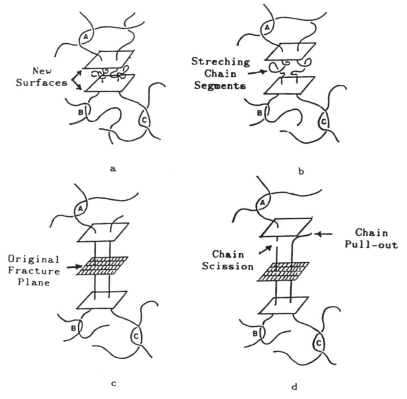

Figure 11.29 Micromechanisms of fracture: (*a*) new surfaces appear under stress; (*b*) chain segments elongate; (*c*) maximum elongation; and (*d*) chain scission or pull-out (51).

is followed by chain scission or pullout. The total energy consumed in fracturing a glassy polymer is usually large compared to simple chain scission consuming only approximately 83 kcal/mol (4.9×10^{-19} J/bond) per C—C bond broken. The extra energy is required because according to the Lake–Thomas theory (61), all of the bonds between two entanglement points have their energy wells activated. This continues until all such bonds are stressed almost to the breaking point (62,63). For polystyrene, this represents some 300 bonds per adjacent pair of entanglement points, M_e. When all of these bonds are excited to the quantum state just below bond separation, the next quantum of energy added causes one of the bonds to break. All of the energy from the remaining bonds is then lost as heat. Experimentally the total energy to fracture a unit volume (or unit area of material) was obtained via a dental burr grinding instrument (64). The energy was divided into chain scission versus pullout components on the basis of molecular weight reduction to count the number of scissions. The pullout energy was assumed to be the remainder after chain scission calculations. Chain straightening and actual creation of new surfaces together were estimated at less than 1% of the total energy.

Two types of deformation may be imagined to excite the energy wells. First, the 109° bond angle is opened (ideally) to 180°. Second, the carbon–carbon bond distance is elongated from 1.54 Å to some larger value. Mark (65) assumed that stretching the C—C bond to 2.54 Å was required to cause the bond to fail.

The question arises as to the shape of the energy well, and the spacing between quantum states. In order for all the bonds between entanglement points to be excited to the scission point, it has to be that after the first quantum of energy raises one bond to the first excited state, the next quantum added will go to some other bond, and so on, until all the bonds are excited to the first excited state level. Then the second and succeeding quantum levels must behave similarly. Thus the quantum energy spacings must get larger with each excited state. This is the opposite behavior known for the hydrogen–hydrogen bond, where the energy spacings get closer together as the molecule gets more and more excited (62). Increasing evidence now points to the stretching and orientation action of craze fibril formation as the actual site of chain scission (64a,64b,64c), rather than the growing tip of the craze. Both models support the same energy of chain scission, etc.

11.5.2.2 Chain Pullout

If the chain portion is bound by one entanglement only, namely either because of very low molecular weight or because of a chain end, the chain end pulls out of its surroundings rather than scissioning. Prentice (66) showed that chain pullout energy leads directly from a consideration of the velocity of a chain moving through its de Gennes's tube. This leads to the molecular frictional coefficient per unit length of mer,

$$\mu_0 = \frac{2E_v}{vnL^2} \tag{11.29}$$

where E_v represents the energy for chain pullout, v the velocity of the chain being pulled out, $n = N_0 - N_a'$, where N_0 is the total number of chains per unit volume, and N_a' is the number of chain scissions per unit volume, and L is the length of the segment being pulled out. For polystyrene, E_v is about 280×10^6 J/m^3 for a molecular weight of 151,000 g/mol, v was taken as the velocity of rotation of the burr, 8.3×10^{-3} m/s, $N_a' = 13 \times 10^{23}$ scissions/m^3, $N_0 = 4.2 \times 10^{24}$ chains/m^3 (for 151,000 g/mol), and the quantity L is given by 75 mers $\times$ 2.534 Å/mer = 190 Å. Thus μ_0 is found to be

Table 11.6   Viscoelastic energy of pullout: Theoretical and experimental values

Pullout Energy (J/m^3)	Basis
3.1×10^8	H. Mark approach to pull out 75 mers
0.5×10^8	Evans's equation for 75 mers between entanglements
2.6×10^8	Experimental for $M_n = 1.5 \times 10^5$ g/mol

64.4 J-s/m^3. This frictional value measures the energy expended in moving one chain relative to its neighbors in the bulk, and it can be used (below) to estimate the temperature of the chain in motion.

Further Evans (67) examined the energy required to pull a chain out of a tube. The energy for pullout of a chain end, E_p, is given by

$$E_p = \frac{kTN_e^2}{2} \tag{11.30}$$

where N_e represents the number of mers between entanglements. For polystyrene, N_e is about 75.

An alternate, but simpler approach was presented earlier by Mark (65). His reasoning was that on average, half of the number of mers between entanglements would need to be pulled out, since that was the average number appearing in any given chain end. Thus, about one-half of the work to break one of the bonds between entanglements would be required for one chain end to be pulled out. The results of experiment *vs.* theory is shown in Table 11.6 (64) for polystyrene. While both the Evans and Mark approach are the same order of magnitude as the experimental value, the Mark approach provides somewhat better agreement with experiment.

11.5.2.3 *Temperature of Chains Being Pulled Out*   Through an estimate of the frictional forces involved in the chain pullout, the temperature of the polystyrene chain pullout was found to be between 150 and 250°C. This is thought to be the temperature of the chain being pulled out, not of the surroundings.

Independently Wool (50), quoting Kramer's (68) analysis of the Saffman–Taylor (69) meniscus instability mechanism, Figure 11.30 (68) arrived at 170°C for the craze-tip temperature in polystyrene via a WLF equation analysis of the melt viscosity. Thus, although the sample being fractured is substantially at room temperature, the chain portions being pulled out at the craze tip are above their T_g (64). There is a growing body of evidence that the actual fracture process of plastics takes place in the melt state, creating a revolution in the thinking of plastic failure processes. Note that the growing crack tip in fatigue is actually quite hot.

One approach to obtain the energy to break single chains involves placing them across a polymer blend interface. Polymer interfaces between two polymers, however well annealed, are often weak; see Chapter 12. One way to strengthen an interface is to put small quantities of block copolymers at the interface. Usually the two blocks are identical to or at least soluble in their respective homopolymers. Thus a covalently bonded chain crosses the interface. In the case of the block copolymer {composed of polystyrene and poly(2-vinyl pyridine)} while the interface was substantially saturated, the number of chains crossing the interface was small compared to the homopolymer

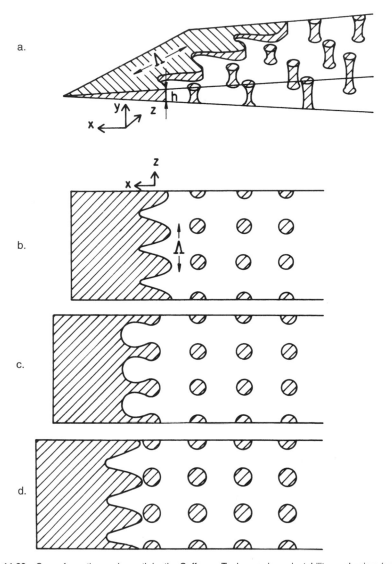

Figure 11.30 Craze formation and growth by the Saffman–Taylor meniscus instability mechanism (courtesy of E. J. Kramer). An analysis of the viscous motions with the WLF equation allows an estimation of the temperature of the crazes on formation.

case; see Table 11.7 (70). The energy per chain to fracture for both cases is within experimental error of the theoretical value, 1.79×10^{-16} J.

11.5.2.4 *Molecular Weight Requirements* The effect of molecular weight on fracture behavior has been examined by Wool (50). He estimates that polymers reach their maximum tensile strength and energy to break at about $8M_c'$; that is, about eight en-

Table 11.7   Molecular basis of fracture at block copolymer and homopolymer interfaces

Property	Block Copolymer	Reference	Homopolymer	Reference
Force to break a single chain	2×10^{-9} N/bond	(a)	5.7×10^{-9} N/bond	(d)
Static friction per mer	6.3×10^{-12} N/mer	(b)	1.2×10^{-10} N/mer	(d)
Fracture toughness, G	15 J/m^2	(b)	1.2×10^2 J/m^2	(d)
Area/chain density at saturation	0.11 chains/nm^2	(b)	1.93 chains/nm^2	(e)
Interfacial tension	3.2 ergs/cm^2	(c)	0	—
Energy per chain	1.36×10^{-16} J	(c)	1.55×10^{-16} J	—

References: (a) C. Creton, E. J. Kramer, C. Y. Hui, and H. R. Brown, *Macromolecules*, **25**, 3075 (1992). (b) J. Washiyama, E. J. Kramer, and C. Y. Hui, *Macromolecules*, **26**, 2928 (1993). (c) S. Wu, *J. Phys. Chem.*, **74**, 632 (1970). (d) M. Sambasivam, A. Klein, and L. H. Sperling, *Macromolecules*, **28**, 152 (1995). (e) N. Mohammadi, A. Klein, and L. H. Sperling, *Macromolecules*, **26**, 1019 (1993).

tanglements per chain. Quantitatively

$$\mathscr{G}_{1c} \sim \left(\frac{M}{M_c'} \right) \left\{ 1 - \left(\frac{M_c'}{M} \right)^2 \right\} \tag{11.31}$$

in the range $M_c' < M < 8M_c'$. This puts the behavior shown in Figure 1.4 on a molecular basis. Actually mechanical properties tend to drift up a bit beyond $8M_c'$ before reaching a plateau. However, because of the large increases in melt viscosity on increasing the molecular weight [see equation (10.58)], many industrial polymers tend to have molecular weights approximated by $7M_c'$, providing the best trade-off. Thus, while the molecular basis of fracture is still not completely understood, the roles of chain scission and chain pullout, combined with known chain entanglement densities, provides a unique approach to calculating the mechanical behavior of glassy plastics.

11.5.3   Scaling Laws for a Polymer–Polymer Interface

Kim and Wool (55) derived a number of functions that describe the molecular state at the interface as a function of time. For example, the number of random-coil chains intersecting unit area of the interface as a function of contact time, t, and molecular weight, M, can be written

$$n(t) \sim t^{1/4} M^{-5/4} \qquad (t < \tau_r) \tag{11.32}$$

$$n_\infty \sim M^{-1/2} \qquad (t > \tau_r) \tag{11.33}$$

where τ_r is the relaxation time. Note that the static or virgin state solution at $t > \tau_r$ is obtained by substituting $t = \tau_r \sim M^3$ in the dynamic solution. This condition constitutes the state when the crack is completely healed, and the sample behaves as if the crack never existed. A number of analytical aspects of interdiffusion at polymer–polymer interfaces are shown in Table 11.8 (59). The general relations may be written

$$H(t) = H_\infty (t/\tau_r)^{r/4} \tag{11.34}$$

$$H_\infty = M^{(3r-s)/4} \tag{11.35}$$

Table 11.8 Molecular aspects of interdiffusion at a polymer–polymer interface (57)

Molecular Aspect	Symbol	Dynamic Relation $H(t)$	Static Relation H_∞	r	s
General property	$H(t)$	$t^{r/4}M^{-s/4}$	$M^{(3r-s)/4}$	r	s
Number of chains	$n(t)$	$t^{1/4}M^{-5/4}$	$M^{-1/2}$	1	5
Number of bridge	$p(t)$	$t^{1/2}M^{-3/2}$	M^0	2	6
Average monomer depth	$X(t)$	$t^{1/4}M^{-1/4}$	$M^{1/2}$	1	1
Total monomer depth	$X_0(t)$	$t^{1/2}M^{-3/2}$	M^0	2	6
Average contour length	$\ell(t)$	$t^{1/2}M^{-1/2}$	M	2	2
Total contour length, number of monomers, N	$L_0(t)$	$t^{3/4}M^{-7/4}$	$M^{1/2}$	3	7
Average bridge length	$\ell_p(t)$	$t^{1/4}M^{-1/4}$	$M^{1/2}$	1	1
Center of mass depth[a]	X_{cm}	$t^{1/2}M^{-1}$	$M^{1/2}$	2	5
Diffusion front length	N_f	$t^{1/2}M^{-3/2}$	M^0	2	6

[a]This equation applies to chains in the bulk and does not apply to chains whose center of mass is within a radius of gyration of the surface.

Values for r and s for individual properties are shown in Table 11.8 (59). Related fractal geometry (71–73) aspects are treated in Section 11.5.5 and in Chapter 12.

11.5.4 Tensile Strength Dependence on Interdiffusion Distance

One of the major conclusions of the Wool theory (56,57) is that mechanical strength buildup during interface healing should have a one-fourth power time dependence. Note that the time dependence of equation (11.29) and others are $t^{1/4}$ rather than $t^{1/2}$ as expected for usual atomic diffusion processes.

Yoo et al. (74,75) confirmed this dependence using uniform polystyrene latex films. (See Section 5.4.4.) The process of film formation from a latex can be divided into three stages: (a) evaporation of the water containing the latex, (b) coalescence and deformation of the latex particles, and (c) interdiffusion of the polymer chains into adjacent latex particles.[†] The time for complete healing is given by the reptation time, $T_r = r^2/(3\pi'^2 D)$ where r is the end-to-end distance of the chain, D represents the diffusion coefficient, and $\pi' = 3.14$.

Figure 11.31 (74) shows the tensile strength buildup as a function of time to the one-fourth power. There is an apparent induction period, followed by a portion that agrees with Wool's theory. At long times the fully developed tensile strength remains nearly constant. Concomitant small-angle neutron scattering experiments showed that the level-off conditions correspond to an interdiffusion distance of 110 to 120 Å, which corresponds roughly to the radius of gyration of the polystyrene chains. In other words, at full tensile strength, the lead interdiffusing chains were half in the old latex particle and half in the new. Thus no further increase in tensile strength could be expected, because the interface between the latex particles was substantially healed.

[†] These are the stages of film formation in ordinary latex paint, such as used on buildings. In this case the glass transition temperature of the polymer is usually just below room temperature to facilitate interdiffusion. The latex dispersion contains filler to increase modulus and decrease scratch resistance, as well as colorants, and so on.

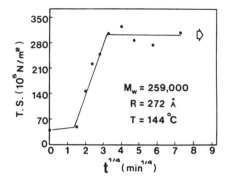

Figure 11.31   Wool plot, showing tensile strength buildup from polystyrene latex (74).

11.6 SUMMARY

Deformation, fracture, and fatigue in polymeric materials are clearly established as being controlled by viscoelastic quantities as well as the thermodynamics of energy storage and the energy required to create new surfaces. There are several ways of modifying polymers so that they can withstand greater stresses. The most important include the incorporation of finely divided fillers, which reinforce polymers, particularly elastomers; the use of fiber composites, which bear the load; and the blending in of a rubbery phase to toughen an otherwise brittle plastic.

Under some well-defined conditions, both plastics and elastomers may be used almost indefinitely without significant fear of failure. This is specified by the failure envelope for elastomers and by the endurance limit in plastics. The glass transition temperature and the secondary transitions are important because their associated molecular motions absorb energy, heating the sample. Thus both the temperature and frequency of the experiment are important.

Two major mechanisms prevail in determining the fracture resistance of polymeric materials (76). One primary contribution comes from the energy required to extend polymer chains, or some subsection thereof to the point of rupture. This can be divided into two subparts: (a) extensional energy is required to deform a polymer chain to a fully extended conformation, and (b) each of the main-chain bonds along this fully extended chain will store its rupture energy before the chain breaks. This is a consequence of the shape of the energy well.

The second major contribution arises from viscous dissipation. Viscoelastic loss associated with chain deformation in regions around the crack tip is significant in determining fracture resistance in polymers.

Keeping in mind that thermodynamic conservation of energy holds everywhere, the energy to propagate a crack, supplied by an outside source, is consumed by the deformation of the chains, their breaking, and slipping. New surface formation also consumes energy, as well as a host of other mechanisms. If they work well, the polymer is strong or tough. If not, the polymer is weak or brittle, and easy to break.

In concluding, it must be emphasized that research in the areas of fracture and fatigue in polymers is in its infancy. Indeed, research in the whole area of polymers is developing rapidly and at an increasing rate.

Although natural polymers were used since biblical times and before for clothing, building, and transportation [note that Noah's Ark was painted with pitch, a polymeric material (77)], it is only since Staudinger's Macromolecular Hypothesis in the 1920s that significant understanding has begun to be achieved.

REFERENCES

1. J. G. Williams, *Fracture Mechanics of Polymers*, Ellis Horwood, Chichester, England, 1984, Chap. 1.

2. "Methods of Test for Impact Resistance of Plastics and Electrical Insulating Materials," ASTMD 256-54T, ASTM Standards, Part 6, p. 182 (1955). Also ASTM D-256-56, ASTM Standards 1956. American Society for Testing Materials, Baltimore, 1955 and 1956.

3. J. A. Manson and L. H. Sperling, *Polymer Blends and Composites*, Plenum, New York, 1976, Chap. 1.

4. M. Matsuo, *Jpn. Plastics*, **2**, 6 (1968).

5. D. W. van Krevelen, *Properties of Polymers*, 3rd ed., Elsevier, Amsterdam, 1990.

6. S. Matsuoka, in *Failure of Plastics*, W. Brostow and R. D. Corneliussen, eds., Hanser Publishers, New York, 1986.

7. A. Peterlin, *J. Polym. Sci.*, **9C**, 61 (1965).

8. J. D. Hoffman, *Polymer*, **24**, 3 (1983).

9. J. D. Hoffman, *Polymer*, **23**, 656 (1982).

10. P. B. Bowden and S. Raha, *Philos. Mag.*, **22**, 463 (1970).

11. S. S. Sternstein and L. Ongchin, *Polym. Prepr.*, **10(2)**, 1117 (1969).

12. J. C. Bauwens, *J. Polym. Sci. A-2*, **5**, 1145 (1967).

13. J. C. Bauwens, *J. Polym. Sci. A-2*, **8**, 893 (1970).

14. A. J. Kinloch and R. J. Young, *Fracture Behavior of Polymers*, Applied Science Publishers, London, 1983, p. 115.

15. N. G. McCrum, C. P. Buckley, and C. P. Bucknall, *Principles of Polymer Engineering*, Oxford University Press, New York, 1988, p. 117.

16. C. B. Bucknall and D. G. Street, *Soc. Chem. Ind. Monogr.*, **26**, 272 (1967).

17. D. R. Paul and L. H. Sperling, eds., "*Multicomponent Polymer Materials*," ACS Adv. in Chem. Ser. No. 211, American Chemical Society, Washington, DC, 1986.

18. M. Matsuo, *Polym. Eng. Sci.*, **9**, 206 (1969).

19. K. Kato, *Jpn. Plastics*, **2**, 6 (1968).

20. L. H. Sperling, *J. Polym. Sci. Polym. Symp.*, **60**, 175 (1977).

21. C. B. Bucknall and R. R. Smith, *Polymer*, **6**, 437 (1965).

22. M. Y. Tang and J. E. Mark, *Macromolecules*, **17**, 2616 (1984).

23. J. G. Curro and J. E. Mark, *J. Chem. Phys.*, **80**, 4521 (1984).

24. J. E. Mark, *Rubber Chem. Technol.*, **72**, 465 (1999).

25. J. E. Mark, *Polym. Eng. Sci.*, **19**, 254 (1979).

26. J. E. Mark, *Polym. Eng. Sci.*, **19**, 409 (1979).

27. K. W. Scott, *Polym. Eng. Sci.*, **7**, 158 (1967).

28. T. L. Smith and P. J. Stedry, *J. Appl. Phys.*, **31**, 1892 (1960).

29. T. L. Smith, *J. Appl. Phys.*, **35**, 27 (1964).

30. J. C. Halpin and F. Bueche, *J. Appl. Phys.*, **35**, 3142 (1964).

31. L. J. Broutman and R. H. Krock, *Modern Composite Materials*, 8 Vols., Academic Press, Orlando, FL, 1974.

32. L. Nicholais, *Polym. Eng. Sci.*, **15**, 137 (1975).

33. S. Tsai, J. C. Halpin, and N. J. Pagano, eds., *Composite Materials Workshop*, Technomic, Stamford, CT, 1969.

34. J. C. Halpin, *Polym. Eng. Sci.*, **15**, 132 (1975).

35. F. P. Gerstle Jr., *Encyclopedia of Polymer Science and Engineering*, Vol. 3, J. I. Kroschwitz, ed., Wiley, New York, 1985, p. 776.

36. K. Friedrich, R. Walter, H. Voss, and J. Karger-Kocsis, *Composites*, **17**, 205 (1986).

37. J. C. Seferis and L. Nicolais, in *The Role of the Polymeric Matrix in the Processing and Structural Properties of Composite Materials*, J. L. Koenig and C. Chiang, eds., Plenum, New York, 1983.

38. D. R. Ulrich, *Polymer*, **28**, 533 (1987).

39. A. A. Griffith, *Philos. Trans. R. Soc.*, **A221**, 163 (1921).

40. J. P. Berry, *J. Polym. Sci.*, **50**, 107 (1961).

41. B. Rosen, ed., *Fracture Processes in Polymeric Solids, Phenomena and Theory*, Interscience, New York, 1964.

42. R. S. Rivlin and A. G. Thomas, *J. Polym. Sci.*, **10**, 291 (1953).

43. E. Orowan, *Phys. Soc. Rep. Prog. Phys.*, **12**, 186 (1948).

44. G. R. Irwin, in *Fracture, Handbuch der Physik*, Vol. VI, S. Flugge, ed., Springer, Berlin, 1958, p. 551.

45. R. W. Hertzberg and J. A. Manson, *Fatigue of Engineering Plastics*, Academic Press, Orlando, FL, 1980.

46. ASTM Standard E399-78, "Test for Plane-Strain Fracture Touchness in Metallic Materials," Part 10, American Society for Testing Materials, Philadelphia, 1979.

47. J. A. Manson and R. W. Hertzberg, *CRC Crit. Rev. Macromol. Sci.*, **1**, 433 (1973).

48. N. M. Riddell, G. P. Koo, and J. L. O'Tool, *Polym. Eng. Sci.*, **6**, 363 (1966).

49. R. W. Hertzberg, J. A. Manson and M. D. Skibo, *Polym. Eng. Sci.*, **15**, 252 (1975).

50. R. P. Wool, *Polymer Interfaces: Structure and Strength*, Hanser, Munich, 1995.

51. R. W. Hertzberg, J. A. Manson, and M. D. Skibo, *Polymer*, **19**, 359 (1978).

52. J. Michel, R. W. Hertzberg, and J. A. Manson, *Polymer*, **25**, 1657 (1984).

53. N. Mohammadi, J. N. Yoo, A. Klein, and L. H. Sperling, submitted *J. Polym. Sci. Polym. Phys. Ed.*, **30**, 1311 (1992).

54. P. G. de Gennes, *J. Chem. Phys.*, **55**, 572 (1971).

55. S. F. Edwards, *Proc. Phys. Soc. London*, **92**, 9 (1967).

56. R. P. Wool, B. L. Yuan, and O. J. McGarel, *Polym. Eng. Sci.*, **29**, 1340 (1989).

57. Y. H. Kim and R. P. Wool, *Macromolecules*, **16**, 1115 (1983).

58. R. P. Wool, *Rubber Chem. Technol.*, **57**, 307 (1984).

59. H. Zhang and R. P. Wool, *Macromolecules*, **22**, 3018 (1989).

60. M. Tirrell, *Rubber Chem. Technol.*, **57**, 523 (1984).

61. G. J. Lake and A. G. Thomas, *Proc. R. Soc. Lond. Ser. A.*, **A300**, 108 (1967).

62. L. H. Sperling, A. Klein, M. Sambasivam, and K. D. Kim, *Polym. Adv. Technol.*, **5**, 453 (1994).

63. M. Sambasivam, A. Klein, and L. H. Sperling, *Polym. Adv. Technol.*, **7**, 507 (1996).

64. M. Sambasivam, A. Klein, and L. H. Sperling, *Macromolecules*, **28**, 152 (1995).

64a. C. C. Kuo, S. L. Phoenix, and E. J. Kramer, *J. Mater. Sci. Lett.*, **4**, 459 (1985).

64b. L. L. Berger, Macromolecules, **23**, 2926 (1989).

64c. S. D. Kim, P. Suwanmala, A. Klein, and L. H. Sperling, in Press, J. Mater. Sci., 2000.

65. H. Mark, *Cellulose and Cellulose Derivatives, Vol. 5*, E. Ott, ed., Interscience, New York, 1943.

66. P. Prentice, *Polymer*, **24**, 344 (1983).

67. K. E. Evans, *J. Polym. Sci. Polym. Phys. Ed.*, **25**, 353 (1987).

68. E. J. Kramer, in *Advances in Polymer Science: Crazing*, **52/53**, 1, H. H. Kausch, ed., Springer, Berlin, 1983.

69. P. G. Saffman and G. Taylor, *Proc. R. Soc., Ser. A. Math. Phys. Sci.*, **245**, 312 (1958).

70. L. H. Sperling, *Polymeric Multicomponent Materials: An Introduction*, Wiley, New York, 1997.

71. B. B. Mandelbrot, *The Fractal Geometry of Nature*, Freeman, New York, 1983.

72. B. Sapoval, M. Rosso, and J. F. Gouyet, *J. Phys. Lett. (Paris)*, **46**, 149 (1985).

73. R. P. Wool and J. M. Long, *Polym. Prepr.*, **31(2)**, 558 (1990).

74. J. N. Yoo, L. H. Sperling, C. J. Glinka, and A. Klein, *Macromolecules*, **23**, 3962 (1990).

75. L. H. Sperling, A. Klein, J. N. Yoo, K. D. Kim, and N. Mohammadi, *Polym. Adv. Technol.*, **1**, 263 (1990).

76. K. A. Mazch, M. A. Samus, C. A. Smith, and G. Rossi, *Macromolecules*, **24**, 2766 (1991).

77. The Bible, *Genesis* 6: 14.

GENERAL READING

J. A. Brydson, *Plastics Materials*, 6th ed., Butterworth-Heinemann, Oxford, 1995.

W. Brostow and R. D. Corneliussen, eds., *Failure of Plastics*, Hanser Publishers, New York, 1986.

A. Ciferri and I. M. Ward, *Ultra-High Modulus Polymers*, Applied Science, London, 1979.

W. J. Feast and H. S. Monroe, *Polymer Surfaces and Interfaces*, Wiley, New York, 1987.

C. T. Herakovich, *Mechanics of Fibrous Composites*, Wiley, New York, 1988.

R. W. Hertzberg, *Deformation and Fracture Mechanics of Engineering Materials*, 4th ed., Wiley, New York, 1996.

R. W. Hertzberg and J. A. Manson, *Fatigue of Engineering Plastics*, Academic Press, Orlando, FL, 1980.

H. H. Kausch, *Polymer Fracture*, 2nd ed., Springer, Berlin, 1987.

A. J. Kinloch and R. J. Young, *Fracture Behavior of Polymers*, Applied Science, New York, 1983.

N. G. McCrum, C. P. Buckley, and C. B. Bucknall, *Principles of Polymer Engineering*, 2nd ed., Oxford University Press, Oxford, 1997.

L. E. Nielsen and R. L. Landel, *Mechanical Properties of Polymers and Composites*, 2nd ed., Vols. I and II, Dekker, New York, 1994.

L. C. Sawyer and D. T. Grubb, *Polymer Microscopy*, 2nd ed., Chapman and Hall, London, 1996.

L. H. Sperling, *Polymeric Multicomponent Materials: An Introduction*, Wiley, New York, 1997.

L. G. Williams, *Fracture Mechanics of Polymers*, Horwood, Chichester, England, 1984.

A. E. Woodard, *Atlas of Polymer Morphology*, Hanser Publishers, New York, 1989.

R. P. Wool, *Polymer Interfaces: Structure and Strength*, Hanser, Munich, 1995.

S. Wu, *Polymer Interface and Adhesion*, Dekker, New York, 1982.

STUDY PROBLEMS

1. An engineering plastic is stretched until it breaks. Briefly, discuss all the possible places that the energy might have gone.

2. What are the definitions of the following terms: impact resistance; tensile strength; failure envelope; fatigue crack propagation; craze?

3. Theoretically, polyethylene can be perfectly oriented and crystallized 100%. At perfect orientation,
   (a) What is the theoretical work to break of this material?
   (b) What is its theoretical tensile strength?
   (c) What is its theoretical modulus in the direction of orientation?

4. Read any scientific or engineering paper in the field of mechanical behavior of polymers written since 2000, and show how it advances the understanding of polymer science beyond when this book was finished. Provide exact reference: Author, journal, volume, page, year.

5. A panel of poly(methyl methacrylate) is bolted in the long direction only at 50°C between two other components with zero coefficients of expansion. What is the stress on the polymer at 0°C? Will the polymer break?

6. You are designing new plastics and elastomers for space and planetary exploration. Some of the planets to be visited are very cold, and some of them have very dense atmospheres.
   (a) What will the effect of hydrostatic pressure and temperature be on the relaxation behavior of polystyrene? Of *cis*-polyisoprene?
   (b) What will be the effects of hydrostatic pressure and temperature on the stress–strain behavior of these polymers?

7. A sample of poly(methyl methacrylate) at room temperature has a center notched crack of 2.0 cm in diameter. If K_c is 0.80 MPa $\cdot$ m$^{1/2}$, what is the critical stress to propagate the crack?

8. A sample of a new plastic 100 cm in length 1.0 cm in width, and 0.1 cm in thickness was subjected to a force of 1×10^6 dynes. If it is stretched to 100.1 cm long, what is its Young's modulus?

9. Note in Figure 11.5 the 5% per minute strain rate. How much work would it take to break a $1 \times 1 \times 10$ cm^3 strip of this poly(vinyl chloride) under these conditions?

10. Your boss decides to lower the molecular weight of the polystyrene manufactured from 225,000 g/mol to 150,000 g/mol. "Our customers will save lots of energy in processing the material," she exclaimed. As chief engineer in charge of mechanical properties, how will this affect the fracture energy? Do you agree with her decision?

12

POLYMER SURFACES
AND INTERFACES

Polymer surfaces and interfaces have long been important in engineering and manufacturing, but until recently remained a largely empirical science. Today, with the advent of many new instruments and important new theories, this aspect of polymer science and engineering is among the most rapidly growing areas with greatly improved understanding at the molecular level.

Five basic classes of polymer surfaces and interfaces can be distinguished:

1. The polymeric surface commonly exposed to air that people see and touch. Strictly speaking, a surface (or *free* surface) refers to that part of a pure condensed substance in contact with a vacuum, while in reality most surfaces may be in contact with air, oxidized, oily, or dirty; see Figure 12.1 (1). For purposes of this chapter, the polymer making up the surface will be assumed to be pure, unless stated otherwise.

2. The dilute polymer solution–solid interface, where most often the solid is dispersed as a colloid. Because of the relatively huge surface area presented by such surfaces, polymer adsorption becomes a highly significant effect. A single polymer chain may be adsorbed at one or many sites, the remaining portion of the chain sticking out into the solution and remaining solvated (2).

3. A symmetric polymer interface, where two more or less identical polymers are in contact with each other. Important factors include chain interdiffusion, healing, and fracture (3).

4. An asymmetric polymer interface, involving two different polymers. These constitute the ordinary polymer blends and related materials of commerce. If the polymers are immiscible, the usual situation, the interface may remain indefinitely. Interpenetration at the interface ranges from a depth of a few to several nanometers, depending on the statistical segment length and χ_1. The term *interphase* describes the interpenetration zone, and the interphase may have physical properties distinctly different from either polymer. The two polymers comprising the

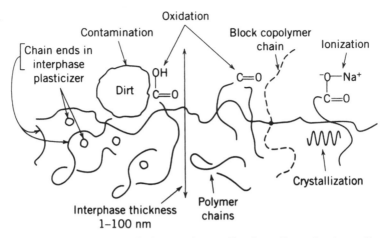

Figure 12.1 Features characteristic of polymer surfaces and interfaces. Note various types of impurities.

interphase may be bonded, as in block or graft copolymers, or contain physical bonds such as hydrogen bonding. Such bonds strengthen the surface, making the material much tougher (1).

5. A composite interface, composed of a polymer and a nonpolymer. Usually the nonpolymer is a solid phase such as glass, carbon, or boron fibers, steel, or particulates such as calcium carbonate or titanium dioxide. In such a case the polymer cannot interdiffuse into the nonpolymer surface but may bond and adhere to it (4). While a foam constitutes a special case of a composite material, for the purposes of this chapter it will be considered along with the air interfaces.

12.1 POLYMER SURFACES

The surface of any material is different from the interior or bulk of the material. Differences can be chemical or physical, or both. An example is rust on iron. In polymers, surface oxidation can also exist. Special effects on polymer surfaces include contamination, plasticization, grafting, and a host of other chemical changes or modifications, some wanted, and some adventitious; see Figure 12.1.

There are numerous physical changes as well. The orientation of the polymer at the surface is almost always different from the interior. The polymer chains may be lying flat, oriented in the surface plane, or if some special group (especially a group at the end of the chain) is attracted to the surface, the orientation of adjacent mers may be normal to the surface plane. In such cases, of course, the concentration of the special group will be higher at the surface than in the interior.

According to Dee and Sauer (5,6), the interfacial width between a bulk polymer melt and pure air (assuming the vapor pressure of the polymer is negligible) is of the order of 10 to 15 Å. There are two aspects to be considered. First, the depth of the surface of interest depends on the nature of the quantity being studied. Second, the measurable thickness depends on the instrument being used. Both of these aspects will be discussed in the following sections.

12.2 THERMODYNAMICS OF SURFACES AND INTERFACES

12.2.1 Surface Tension

The inhomogeneous organization of the atoms at the surface of a condensed phase causes the phenomenon known as surface tension, γ. The surface tension is the reversible work needed to create a unit surface area in a substance. The idea of surface tension goes back to the concept that the surface of a liquid has some kind of contractile skin. The surface tension is sometimes called the specific surface energy, the intrinsic surface energy, or the true surface energy. Surface tension has the cgs units of ergs/cm^2, or SI units of J/m^2. However, since work can be expressed as force times distance, the units are sometimes expressed as dyn/cm or N/m.

Table 12.1 (7) summarizes the surface tensions of selected polymers. While the surface tension of most polymers varies from about 20 to 50 erg/cm^2, these values are low compared to the surface tension of water, 72.94 erg/cm^2, or mercury, 486.8 erg/cm^2. Thus water beads up on polymers with low surface tensions, not wetting them. For example, low surface tension underlies the utility of the Teflon® polytetrafluoroethylene frying pan.

Just as gases, liquids, and solids have thermodynamic properties, so do surfaces and interfaces. The work, W, required to create a unit of surface area, dA, is expressed by

$$W = \gamma \, dA \tag{12.1}$$

The change in the free energy of a surface, dG, on incrementing the area by dA is also given by

$$dG = \gamma \, dA \tag{12.2}$$

Table 12.1 Surface tensions of selected polymers

Polymer	Surface Tension, γ (erg/cm^2)		$-(d\gamma/dT)$ (erg/cm^2-deg)
	20°C	140°C	
Polyethylene	35.7	28.8	0.057
Polystyrene	40.7	32.1	0.072
Poly(methyl methacrylate)	41.1	32.0	0.076
Poly(dimethyl siloxane)	19.8	14.0	0.048
Poly(ethylene oxide)	42.9	33.8	0.076
Polycarbonate	49.2	35.1	0.060
Polytetrafluoroethylene	23.9	16.9	0.058
Poly(n-butyl methacrylate)	31.2	24.1	0.059
Poly(vinyl acetate)	36.5	28.6	0.066
Polychloroprene	43.6	33.2	0.086
Poly(ethylene terephthalate)	44.6	28.3	0.065

Source: S. Wu, *Polymer Interface and Adhesion*, Dekker, New York, 1982, table 3.7, p. 88.

The surface free energy, G^s, is given by

$$G^s = \left(\frac{\partial G}{\partial A}\right)_{T,P} = \gamma \qquad (12.3)$$

where the superscript s indicates a surface property. Following classical thermodynamics the surface entropy can be expressed,

$$S^s = -\left(\frac{\partial G^s}{\partial T}\right)_P \qquad (12.4)$$

and

$$S^s = -\frac{d\gamma}{dT} \qquad (12.5)$$

Thus the quantity $-d\gamma/dT$ shown in Table 12.1 can be interpreted in terms of the surface entropy of the system. The total enthalpy per square centimeter,

$$H^s = G^s + TS^s \qquad (12.6)$$

is larger than G^s itself (8). Since $-d\gamma/dT$ is positive, the change in free energy on forming the surface is positive. This provides evidence for the orientation of chains at surfaces and interfaces, or at least some form of organization.

The temperature dependence of γ, Table 12.1, can be expressed (5)

$$\gamma = c_1 + c_2 T + c_3 T^2 \qquad (12.7)$$

where known values of c_2 are negative. The variation of γ with molecular weight can be expressed,

$$\gamma(M_n) = \gamma_\infty - k/M_n \qquad (12.8)$$

where γ_∞ represents the value of $\gamma(M_n)$ at infinite molecular weight, and k is a constant. Some low molecular weight polymers correlate well with an $M_n^{-2/3}$ dependence (7,9). At reasonably high molecular weights, the molecular weight dependence is small according to both theories.

12.2.2 Example Calculation of Surface Tensions

What is the surface entropy of polystyrene? Using data from Table 12.1 in equation (12.5),

$$S^s = \frac{-(32.1 - 40.7)}{140 - 20} \text{ erg/cm}^2\text{-deg} = +0.0716 \text{ erg/cm-deg}$$

which rounds off to the same value as in Table 12.1.

Table 12.2   Polymer blend interfacial properties

Polymer Pairs, $_{b/a}$[a]	γ_{ab} (erg/cm^2, 140°C)	$S_{b/a}$ (erg/cm^2, 140°C)	$-d\gamma_{ab}/dT$ (erg/cm^2-deg)	W_a (erg/cm^2, 140°C)	Shear Strength (dyn/cm^2)
PMMA/PS	1.7[b]	−1.6	0.013	62.4	6.5×10^7
PE/PMMA	9.7	−6.5	0.018	51.1	5.2×10^7
PnBM/PVA	2.9	1.6	0.010	49.8	9.6×10^7
PDMS/PCLP	6.5	12.0	0.0050	40.8	1.3×10^8
PB/PS[c]	3 (100°C)	—	~0	—	—

[a] *Abbreviations:* PMMA, poly(methyl methacrylate); PDMS, poly(dimethyl siloxane); PS, polystyrene; PCLP, polychloroprene; PE, polyethylene; PnBM, poly(*n*-butyl methacrylate); PVA, poly(vinyl acetate); PB, polybutadiene.

[b] From P. C. Ellington, D. A. Strand, A. Cohen, R. L. Sammler, and C. J. Carrier, *Macromolecules*, 1643 (1994) report γ(PMMA/PS), 190°C) = 1.2 erg/cm^2.

[c] From U. Bianchi, E. Pedemonte, and A. Turturro, *Polymer*, **11**, 268 (1970).

From S. Wu, *Polymer Interface and Adhesion*, Dekker, New York, 1982, table 11.1, p. 362, and table 3.20, p. 126.

12.2.3   Polymer Blend Interfaces

When any two condensed phases are in contact, it is common to talk about the interfacial tension of the system. A case of special interest involves two polymers in contact, Table 12.2 (7). Numerically the values of the interfacial tension, γ_{ab}, are significantly smaller than the values of any of these polymers facing air. This results from the more gradual and less total change in composition at a blend interface. The orientation may be different also. While in many simple homopolymer surfaces the polymer chains tend to lie parallel to the surface, there is some theoretical evidence that polymer chain ends tend to protrude vertically into the interphase between the two homopolymers (10,11). A smaller interfacial tension suggests the system is closer to miscibility, see Section 12.3.7.2. Other things being equal, a larger interfacial entropy value suggests more disorder, and hence more mixing.

A temperature of 140°C is shown in Table 12.2 because all of the polymers must be above their glass transitions if amorphous, and above their melting temperatures if semicrystalline for the determination of interfacial tensions and related properties. Values of surface tension sometimes listed for glassy or semicrystalline polymers are usually extrapolated from the melt.

A quantity of interest is the spreading coefficient, $S_{b/a}$, which determines whether liquid b will spread over liquid a, or bead up. (Note: Subscript b/a, S superscript s represents surface entropy.) Consider a drop of polymer b placed on the liquid surface of polymer a. The surface free energy change on increasing the contact area by dA is given by

$$dG = \left(\frac{\partial G}{\partial A_a}\right) dA_a + \left(\frac{\partial G}{\partial A_b}\right) dA_b + \left(\frac{\partial G}{\partial A_{ab}}\right) dA_{ab} \tag{12.9}$$

Since the same areas are involved,

$$dA_b = -dA_a = dA_{ab} \tag{12.10}$$

and noting that $\partial G/\partial A_a = \gamma_a$, and so forth. Then

$$S_{b/a} = -\partial G/\partial A_b = \gamma_a - \gamma_b - \gamma_{ab} \tag{12.11}$$

since the coefficient, $-(\partial G/\partial A_b)$ represents the free energy for the spreading of the polymer b over polymer a.

If $S_{b/a}$ is positive, spreading is accompanied by a decrease in the free energy, and the process is spontaneous. If

$$\gamma_{ab} > |\gamma_a - \gamma_b| \tag{12.12}$$

then $S_{b/a}$ must always be negative, and neither polymer liquid spreads on the other. Table 12.2 shows both positive and negative values of $S_{b/a}$.

12.2.4 Crystalline and Glassy Polymers

The surface tension of a crystalline polymer can be approximately calculated from the amorphous or melt surface tension, and the difference in densities (7):

$$\frac{\gamma_c}{\gamma_a} = \left(\frac{\rho_c}{\rho_a}\right)^n \tag{12.13}$$

where n is about 4, ρ represents the density, and c and a represent the crystalline and amorphous states, respectively. Since the density of crystalline polymers is always higher than the corresponding amorphous state, the surface tension is higher.

For polymers undergoing the glass–rubber transition (7),

$$\left(\frac{d\gamma}{dT}\right)_G = \frac{\alpha_G}{\alpha_R}\left(\frac{d\gamma}{dT}\right)_R \tag{12.14}$$

where α represents the expansion coefficient. Since $\alpha_G < \alpha_R$, then

$$-\left(\frac{d\gamma}{dT}\right)_G < -\left(\frac{d\gamma}{dT}\right)_R \tag{12.15}$$

While the surface tensions themselves are continuous across the glass–rubber transition, the temperature derivatives are not.

12.3 INSTRUMENTAL METHODS OF CHARACTERIZATION

Many new instruments are now available which can be used to characterize various depths of a specimen, as well as a number of older ones; see Table 12.3 (1,12,13). In the following, descriptions of several instruments methods will be developed, and methods of application explored.

Table 12.3   Methods of characterizing surfaces and interfaces

Technique	Advantages	Disadvantages
Surface tension and contact angle methods	Inexpensive, rapid, provides information related to surface energetics, and hydrated samples can be observed	Artifact-prone: swelling of substrate, leaching, etc.; liquid purity is of critical importance; only surface energetics information is obtained
Brewster angle reflectivity	Measures interfacial mass and thickness	Multiple surfaces and reflects provides complications
Dynamic light-scattering	Yields thickness of adsorbed polymer layers on latexes and particles	Limited by the diffusion coefficients of the particles
Force-balance apparatus (surfaces forces apparatus)	Determines adsorbed layer thickness on solid–liquid surfaces	Electrostatic or other forces must be absent
Electron spectroscopy for chemical analysis (ESCA), also called XPS	Sampling depth is relevant to surface interactions (~ 10–100 Å)	Expensive; artifact-prone, high vacuum required
Atomic force microscopy	Inexpensive; sample may be insulator	Atomic resolution difficult
Scanning electron microscopy (SEM)	Rapid, visually rich, three-dimensional perspective	Differences in secondary electron emission can be confused with topographic features
Attenuated total reflection (ATR) infrared	Readily available, spectra rich in chemical information, orientation information can be obtained	Analysis to 1-μm depth studies both bulk and surface, some samples are not amenable to ATR analysis
Secondary ion mass spectrometry (SIMS)	Look at a highly surface localized region (~ 10 Å), rich in chemical information, depth profiling (destructive) is possible	Expensive, artifact-prone, theory, for polymers not developed
Evanescent wave method	Provides information on adsorbed layers	Adsorbate must interact with radiation
Auger electron spectroscopy (AES) or scanning Auger microprobe (SAM)	Rapid, good spatial resolution (2.5×10^{-3} μm^2), all elements detected	Destructive to organics, limited chemical information, quantitative analysis is difficult
Small-angle neutron scattering (SANS)	Interior interfaces can be characterized	Expensive, deuterated probe molecules often required
Scanning tunneling microscopy	Atomic resolution, relatively inexpensive	Resolution limited with insulating materials
Light-scattering	Determines interior surface areas	Secondary scattering must be eliminated
Ion beam analysis (forward recoil spectrometry)	Surface composition and diffusion coefficients can be determined	Length scale resolution 150 Å at best
Chaudhury-Whitesides apparatus	Measures adhesion forces nondestructively	Polymer must be in the rubber-elastic state

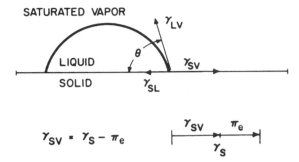

Figure 12.2 Vector forces controlling the contact angle of a liquid drop on a smooth, homogeneous, planar, rigid surface. The quantity π_e relates to an adsorbed film pressure caused by the saturated vapor.

12.3.1 Surface Tension and Contact Angles

If a drop of liquid is placed on a surface, either it wets the surface, spreading, in principle, over the entire surface, or else it beads up. For drops that bead up, an important concept in understanding surface tension involves the contact angle, θ; see Figure 12.2 (7). The subscripts S, L, and V, stand for solid, liquid, and vapor, respectively. The angle θ is controlled by three vector forces: γ_{LV}, which causes the droplet to bead up, γ_{SV}, which if dominating, leads the droplet to cover the surface, and γ_{SL}, which wants to clear or avoid the surface. This assumes that the vapor from the liquid phase is saturated, and that the system has reached equilibrium.

A balance of the three relations leads to Young's equation, one of the oldest in surface science (14):

$$\gamma_{SV} = \gamma_{SL} + \gamma_{LV} \cos \theta \qquad (12.16)$$

Young's equation has two measurable quantities, $\cos \theta$ and γ_{LV}. Unfortunately, determination of the remaining two quantities still presents a problem.

While Figure 12.2 relates to the equilibrium situation, more can be learned by a study of such droplets in motion on the surface. Consider the droplet on an inclined plane. Then there are two contact angles, the advancing contact angle and the receding contact angle. There are three main causes of the hysteresis between the two angles:

1. Contamination of either the solid or liquid phase; see Figure 12.1.
2. Rough surfaces.
3. Low mobility, causing either/or slow desorption or hindered spreading.

Selected advancing contact angles are shown in Table 12.4. Confirming common experience, water beads up on polytetrafluoroethylene, while n-octane, a much less polar material, spreads but does not completely wet polytetrafluoroethylene. The decreases in contact angle with increasing temperature suggest increasing tendency to wet the surface.

Modern methods of measuring the surface tension include the pendant drop method, the sessile drop method, and others (7,8,15). These methods depend on the shape of a drop of the polymer or a bubble in it, and on the balance of surface tension and gravitational forces; see Figure 12.3 (8).

Table 12.4   Selected contact angle data

Liquid (γ, ergs/cm^2)	Solid	Advancing θ (deg)	$d\theta/dT$	References
Water (72)	Polytetrafluoroethylene	108	—	(a)
Water (72)	Polyethylene	96	−0.11	(b)
CH$_2$I$_2$ (67)	Polyethylene	46	—	(c)
n-Octane (21.6)	Polytetrafluoroethylene	30	−0.12	(d)

[a]W. A. Zisman, *Adv. Chem. Ser. No.* 43, 1964.

[b]F. D. Petke and B. R. Ray, *J. Coll. Interf. Sci.*, 31, 216 (1969).

[c]A. El-Shimi and E. D. Goddard, *J. Coll. Interf. Sci.*, 48, 242 (1974).

[d]C. L. Sutula, R. Haritala, R. A. Dalla Betta, and L. A. Mitchel, *Abstracts*, 153rd Meeting, American Chemical Society, April 1967.

12.3.2 ESCA (XPS)

Another important method for characterizing polymer surfaces involves electron spectroscopy for chemical analysis, ESCA, also known as X-ray photoelectron spectroscopy (XPS); see Table 12.3. This method is based on the observation that electrons are emitted by atoms under X-ray irradiation. The energy of the emitted electrons yields the binding energy of the electron to the particular atom, the primary information. While the energies measured are characteristic of the elements, they are sensitive to the electronic environment of the atom, which of course, yields information about surrounding atoms and molecules. A monoenergetic beam of soft X-rays, 3 to 30 Å in wavelength bombards the surface in question. Energies of interest range from near 0 to 1500 eV. A good excitation source utilizes Al Kα X-rays at 1486.6 eV. Since the electrons are emitted with low energy, they are easily scattered unless produced very close to the surface (16). Thus the maximum sampling depth is 10 to 100 Å. ESCA can be used to study the oxidation of polymers, surface fluorination (used to improve the chemical resistance of polyethylene and other polymers), and copolymer composition, which may differ from the interior substance.

For example, ESCA shows that a poly(dimethyl siloxane)–*block*–polycarbonate co-

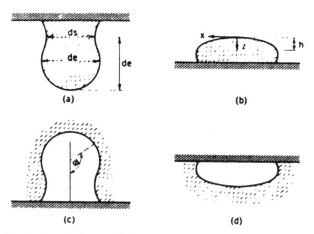

Figure 12.3 Surface tension of polymer and other fluids can be calculated via (*a*) hanging drop, (*b*) sessile drop, (*c*) hanging bubble, and (*d*) sessile bubble shapes.

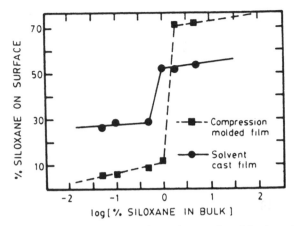

Figure 12.4 Poly(dimethyl siloxane) rises to the surface preferentially in polycarbonate–*blend*–polycarbonate–*block*–poly(dimethyl siloxane) segmented copolymer blends, as characterized by ESCA instrumentation.

polymer covers 50% to 70% of the surface of a blend with polycarbonate at concentrations of only about 1%; see Figure 12.4 (17). Noting that poly(dimethyl siloxane) has a much lower surface tension, Table 12.1, this rise to the surface is to be expected. LeGrand and Gaines (18) found that the contact angles of a number of blends of polycarbonate and poly(dimethyl siloxane) displayed contact angles very close to that of pure poly(dimethyl siloxane), providing important supporting information.

ESCA can also be used to analyze fracture surfaces, particularly those involving immiscible (asymmetric) interphases. While finely divided polymer blends offer some difficulties, plates of two different polymers welded together provide valuable model materials. Foster and Wool (19) and Willett and Wool (20) studied the polystyrene/poly(methyl methacrylate) interface welded at 125 and 140°C; see Figure 12.5 (20). As described above, ESCA makes use of the photoelectric effect, with low-energy X-rays bombarding the sample surface, causing electrons to be ejected. The kinetic energy of the electrons was analyzed, leading to an identification of the type of atoms lying on the fractured surfaces.

The question of interest was whether adhesive or cohesive fracture was taking place. Adhesive fracture refers to cracking in the interphase, while cohesive fracture refers to cracking on one side or the other of the interphase, in one of the homopolymers; see Figure 12.5a.

The temperatures shown in Figure 12.5b are significantly above the glass transition temperatures of polystyrene, 100°C, and poly(methyl methacrylate), 106°C, so polymer chain motion, reptation, and interdiffusion take place. However, polystyrene and poly(methyl methacrylate) are immiscible, interdiffusing only 20 to 50 Å, as determined via neutron reflection techniques (21–23); see Section 12.3.7. Theoretical calculations (see Section 12.3.7.2) yield an interphase thickness of 27 Å (3).

For purposes of ESCA studies, some typical atoms and energies include

O_{1s} 532 eV

C_{1s} 285 eV

N_{1s} 400 eV

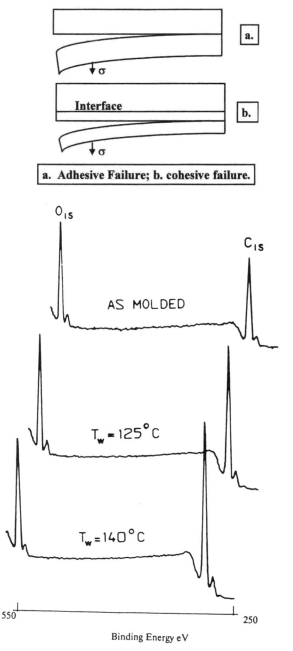

Figure 12.5 (*a*) Illustration of adhesive and cohesive modes of failure. (*b*) XPS spectra of poly(methyl methacrylate) fracture surfaces welded to polystyrene. Range is 250 to 550 eV. The ratio of the oxygen-to-carbon peaks are important in deciding whether adhesive or cohesive failure took place. Top portion, as molded poly(methyl methacrylate).

The analysis of the results utilized the following reasoning: If, during the fracture process, poly(methyl methacrylate) were transferred to the polystyrene side, a strong O_{1s} signal should develop on the polystyrene side, whereas if polystyrene were transferred to the poly(methyl methacrylate) side, a decrease in the O_{1s} signal should be observed on the poly(methyl methacrylate) side. Since poly(methyl methacrylate) has two oxygens for every five carbons, a theoretical ratio of 0.40 for the two peaks, respectively, would indicate that the poly(methyl methacrylate) side had no polystyrene transferred.

The experimental results shown in Figure 12.5 yield O/C ratios of 0.26 and 0.28 for the 125 and 140°C molding temperatures, respectively, indicating a decrease in the oxygen content.

By contrast, the poly(methyl methacrylate) homopolymer (also shown in Figure 12.5) yields a ratio of 0.38, in good agreement with the theoretical value of 0.40. The ratio 0.28/0.38 suggests that the fracture process is more adhesive than cohesive by a ratio of about 3 : 1 for this polystyrene/poly(methyl methacrylate) interface.

The O/C ratios observed on the corresponding polystyrene fracture surfaces were 0.01 or lower, indicating that crack growth at the polystyrene/poly(methyl methacrylate) interface contains both cohesive and adhesive components, with cohesive fracture only occurring on the polystyrene side.

12.3.3 Scanning Electron Microscopy

Scanning electron microscopy (SEM) constitutes one of the older, and one of the most widely used, instruments for surface analysis. Because SEM provides a three-dimensional visual image, the qualitative analysis is relatively straightforward. SEM forms its image by scanning a focused electron beam probe across the specimen of interest (24). The probe interacts with a thin surface layer of the specimen, of a few micrometers or less in thickness. Commonly secondary electrons emitted from the sample surface are used to form a TV-type of image. These secondary electrons are produced by the interaction of the primary electron beam with the specimen. Usually back-scattered primary beam electrons are to be avoided. Topography is promoted by where the beam falls. If the beam falls on a tilted surface, or onto an edge, more secondary electrons will escape. If the beam falls into a valley or pit, fewer secondaries will escape. Nonconducting materials, such as most polymers, require conductive coatings or the use of low accelerating voltages to prevent them from charging in the electron beam. Atomic number contrast can be increased by inclusion of some backscattered primary beam electrons.

Returning to the PS sample examined above by ESCA, SEM on the fractured PS surface revealed a series of ridges of highly deformed materials, with a well-defined spacing of about 50 μm, and hundreds of micrometers in length as shown in Figure 12.6 (20). Such features indicate an unstable crack growth pattern commonly called *slip-stick*. An examination of the corresponding poly(methyl methacrylate) fracture surfaces show both polystyrene residue and rather featureless portions. Thus the two sides of the fracture can be matched up to a significant extent.

Interestingly the fracture energy, $\mathscr{G}$, of the polystyrene/poly(methyl methacrylate) interface is about 50 J/m², while that of polystyrene is about 500 J/m². Since the interface fracture energy is thought to be about 0.1 J/m², this corresponds to 10% ridges and 90% interface (adhesive) fracture observed in Figure 12.6, also roughly matching the values calculated from Figure 12.5.

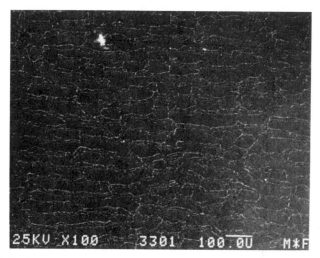

Figure 12.6 SEM micrograph of polystyrene fracture surfaces welded to poly(methyl methacrylate), T_w at 140°C.

12.3.4 Scanning Probe Microscopy

12.3.4.1 Near Atomically Sharp Rasters The basic ideas of scanning probe microscopy (SPM) date back to the work of Binnig and Rohrer in 1982 (25,26), involving a near atomically sharp raster which is moved over conducting surfaces (27). The original instrument lead to the field called scanning tunneling microscopy (STM) because the motion of the electrons between the raster tip and the surface arises because of a quantum mechanical tunneling effect. To build up a three-dimensional topographic surface map, the probe is scanned in a rasterlike fashion above the surface; see Figure 12.7 (28). The tunneling current is maintained at a constant preset value. However, a major limiting factor of its use in polymer science is the requirement of electrical conductivity.

Today, there are many forms of SPM, each with multiple applications; see Table 12.5. The two most important derivative methods are atomic force microscopy (AFM), sometimes called scanning force microscopy (SFM), and friction force microscopy (FFM), sometimes called lateral force microscopy (LFM). In AFM (SFM) the sample base is mounted on a piezoelectric *xyz* translator; see Figure 12.7. Imaging involves a rasterlike scanning of the sample in the *xy* plane, while monitoring localized forces between the sample and the near molecularly sharp stylus attached to a soft microfabricated cantilever (29). The forces are measured by focusing a laser beam near the end of the cantilever and directing the reflected light onto a photodetector array. Vertical cantilever displacements, namely vertical forces, are measured via the voltage difference between the multiple diodes in the array. Cantilever twisting is observed similarly.

12.3.4.2 The Molecular Basis of Friction FFM (LFM) measures lateral force during *xy* scanning. The FFM is a modified AFM with a four-quadrant photodiode, based on a laser beam deflection technique; see Figure 12.8 (30). Although the field of tribology has existed for decades, only with the FFM has progress been made in probing the fundamental mechanisms of friction on a molecular or atomic scale.

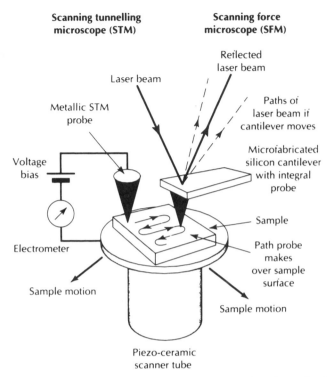

Figure 12.7 Principal features of scanning tunneling microscopy (STM) and scanning force microscopy (SFM). STM is sensitive to local surface conductivity whereas SFM records local topography. A conducting tip is raster-scanned a few Ångstoms above the surface in STM, a small voltage being applied across the gap. In SFM, the probe is mounted on a "soft" spring (cantilever) such that it experiences a very small interaction force with the surface, causing bending in the supporting cantilever.

In general, both energy dissipative and energy conservative lateral forces have been found to exist. The molecular mechanisms underlying frictional processes involve the same nonequilibrium dynamics manifest in the hysteretic measurements surface forces, involving both approach and withdrawal deformation, or loading versus unloading, or contact angle, advancing versus receding (31). Such hysteretic behavior has been categorized as either mechanical or chemical in nature (32). Mechanical hysteresis is similar to that known as stick-slip processes involved in friction. Chemical hysteresis occurs when the conformation or configuration of the molecules changes, such as plastic deformation. In this new field of nanotribology, one of the goals is to connect phenomenological aspects in tribology with a molecular understanding. As such, nanotribological studies follow atomistic and statistical principles, and try to avoid empirical formulas and laws, such as frictional coefficients, that were developed for macroscopic experiments. In the end, of course, it is hoped that nanotribology will be able to make major contributions to the field of tribology by showing what the molecules actually do when rubbed against one another.

12.3.4.3 Elasticity and Modulus Measurements These instruments can make a third type of measurement, that of elasticity, or modulus. Thus, a simultaneous study of

Table 12.5 Scanning probe microscopy: Raster analysis

Technique	Measurements	References
Scanning tunneling microscopy (STM)	Local conductivity; molecular resolution of images	(a) (b)
Scanning force microscopy, SFM, or atomic force microscopy, AFM (same)	Topographic imaging; dynamic real-time crystal growth; fracture surface analysis, nanometer resolution	(c) (d) (e) (f)
Friction force microscopy (FFM) or lateral force microscopy (LFM) (same)	Velocity dependence of frictional forces; two phases with differing friction; anisotropic friction; topography, friction, elasticity	(g) (h) (i) (j) (k)

References: (a) C. J. Roberts, M. C. Davies, K. M. Shakeseff, S. J. B. Tenler, and P. M. Williams, *Trends Polym. Sci. (TRIP)*, 4, 420 (1996). (b) S. F. Bond, A. Howie, and R. H. Friend, *Surf. Sci.*, 196, 331 (1995). (c) G. Haugstad, *Trends Polym. Sci. (TRIP)*, 3, 353 (1995). (d) S. D. Durbin, W. E. Carlson, and M. T. Saros, *J. Phys. D: Appl. Phys.*, 26, B128 (1993). (e) O. L. Shaffer, R. Bagheri, J. Y. Qian, V. Dimonie, R. A. Pearson, and M. S. El-Aasser, *J. Appl. Polym. Sci.*, 58, 465 (1995). (f) B. Drake, C. B. Prater, A. L. Weisenhorn, S. A. Gould, T. R. Albrecht, C. F. Quate, D. S. Cannell, H. G. Hansma, and P. K. Hansma, *Science*, 243, 1586 (1989). (g) G. Haugstad, W. L. Gladfelter, E. B. Weberg, R. T. Weberg, and R. R. Jones, *Langmuir*, 11, 3473 (1995). (h) R. M. Overney, *Trends Polym. Sci. (TRIP)*, 3, 359 (1995). (i) G. Haugstad, W. L. Gladfelter, E. B. Weberg, R. T. Weberg, and T. D. Weatherill, *Langmuir*, 10, 4295 (1994). (j) R. M. Overney, H. Takano, M. Fujihira, W. Paulus, and H. Ringsdorf, *Phys. Rev. Lett.*, 72, 3546 (1994). (k) R. M. Overney, E. Meyer, J. Frommer, H. J. Guntherodt, M. Fujihira, H. Takano, and Y. Gotah, *Langmuir*, 10, 1281 (1994).

topography, friction, and elasticity can be made on different materials comprising the same surface (33). In a study of a surface containing two phases, a 1 : 1 molar mixture of behenic acid, BA, $C_{21}H_{43}COOH$, and partially fluorinated carboxylic acid ether, PFECA, $C_9F_{19}C_2H_4 — O — C_2H_4COOH$, were introduced in organic solvent to an aqueous subphase of polyallylamine (PAA) cationic polymer. The pH of the samples were adjusted to bring out special features.

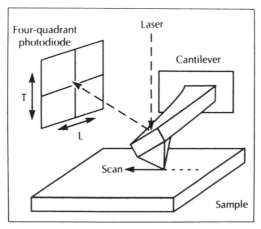

Figure 12.8 Friction force microscopy (FFM) with beam-deflection detection scheme. Topography (T) and lateral forces (L) are measured simultaneously with a four-quadrant photodiode. Irreversible lateral forces are frictional forces, by definition.

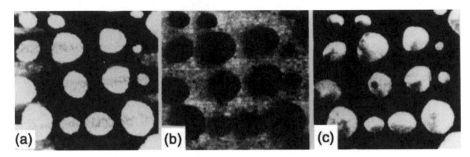

Figure 12.9 SFM studies of topography, friction, and elasticity on the same area, 3 × 3 μm². (*a*) Topography with islandlike hydrocarbon domains (bright) of 300 to 1000 nm diameter on top of a fluorocarbon film (dark). The height of the islands is 2.5 ± 0.5 nm. (*b*) Friction force map shows lower friction (dark) on hydrocarbon islands. (*c*) Elastic compliance (elasticity) map indicates higher Young's modulus (bright) for the hydrocarbon domains.

Figure 12.9 (30) illustrates the complimentary nature of the three types of study for this system. The threefold measurement showed that Young's modulus was higher for the hydrocarbon domains than the fluorocarbon areas, which explained, from a mechanical point of view, why the hydrocarbons showed a lower sliding friction, by a factor of 2.5, than the fluorocarbons. Classically a higher Young's modulus is known to contribute to a lower sliding friction.

Often an AFM instrument of the optical deflection type operates at constant force in the contact (repulsive) mode in which the tip always touches the surface when the feedback loop is on. Such an instrument was used to study the fracture behavior of rubber-toughened epoxies (34). In this case core-shell latexes with styrene-butadiene rubber (SBR) cores and poly(methyl methacrylate–*stat*–acrylonitrile) shells were used to toughen an epoxy matrix, the latter cross-linked with piperidine. The epoxy contained 10 vol-% latex particles. An AFM figure of the stress-whitened zone is shown in Figure 12.10 (34). Examination of the troughs strongly suggests cooperative cavitation. The SBR portion of the latex particles cavitates under triaxial expansive stress. Though the increase in the volume tends to reduce the stress on the plastic, the troughs hence help the plastic resist the externally applied stress. The reader should also note the rims around the cavitated particles. The height of the rims relative to their thickness indicates the extent of plastic deformation, and increased yield strength. Thus the scanning probe microscopy group of instruments, all relatively newcomers to the scene, are providing new and important information about the molecular and supramolecular structure of surfaces.

12.3.5 Secondary Ion Mass Spectrometry, SIMS

SIMS utilizes an impinging ion beam to cause the ejection of charged atoms and molecules (35). The impinging primary ion beams may be Cs^+, O_2^+, O^-, Ar^+, and so on. The secondary ions ejected may be H^+, D^+, C^+, and so on. The sample surface is commonly rastered by the impinging beam to erode a portion of the sample surface (36). The charged species of the secondary beam are filtered and then detected by a mass spectrometer tuned to the element of interest.

Figure 12.11 (35) illustrates the SIMS spectrum of poly(ethylene terephthalate). The protonated repeat unit is seen at 193 g/mol, and peaks at lower masses represent frag-

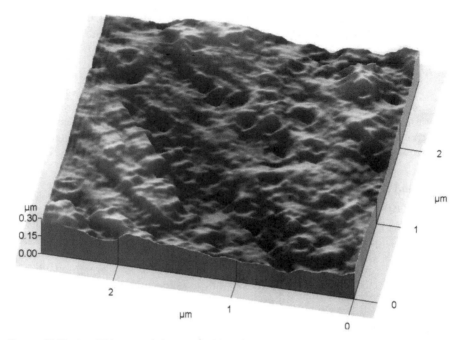

Figure 12.10 An AFM scan of the stress-whitened zone on an epoxy toughened with core-shell SBR/P(MMA–*stat*–AN) latexes. Note appearance of a craze line.

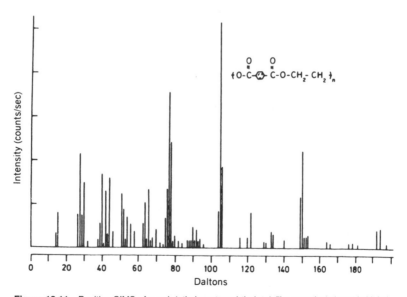

Figure 12.11 Positive SIMS of a poly(ethylene terephthalate) film sample using 4-keV Ar⁺.

mentation with dominant peaks at 149 g/mol ($HOOC—C_6H_4—CO^+$), 104 g/mol ($C_6H_4CO^+$), and 77 g/mol ($C_6H_5^+$). SIMS can be used for microanalysis of polymer surfaces. In one case (35) the experiment was used to detect the presence of dimethyl siloxane contamination on the surface of the polymer.

SIMS has also been used to verify the reptation motion of polymer chains, especially in the short-time region, where diffusion goes as $t^{1/4}$ power (37,38). In this case two thin layers of polystyrene, differing in deuteration mode, were allowed to interdiffuse and were rastered by SIMS as a function of diffusion time. However, since the depth resolution of the SIMS experiment is about 50 to 100 Å, the results are not as conclusive as that carried out by neutron reflection experiments, Section 12.3.7.

12.3.6 Ion Beam Analysis

12.3.6.1 *RBS, FRES, and NRA Instruments*
Ion beam analysis is a generic term that includes Rutherford back-scattering (RBS), forward-recoil spectrometry (FRES), and nuclear reaction analysis (NRA), all of which project helium nuclei on to the polymer surface or interface of interest (39). Typically, the probe helium ions have energies of 1 to 3 MeV. Ion beam analysis provides a depth profile in direct space. (Methods such as neutron reflectometry require some type of integral transform, because they are much more model dependent, so they will be examined in the next section.) The RBS method, the simplest and the oldest, provides an excellent method of depth profiling relatively heavy elements in a matrix of lighter elements. It utilizes a scattering angle, θ, of 180° of the helium nuclei from the nuclei of the heavy atoms. The energy of the helium nuclei are reduced during passage into the matter by numerous collisions with electrons. The helium nuclei then encounter heavy atom nuclei, and are scattered in various directions by the very strong attractive forces between the nuclei. The energy spectrum of those that are back-scattered is recorded. As most polymer problems do not tend to involve heavy atoms, RBS has not been widely used in polymer science.

At very high energies, the so-called Coulomb barrier may be penetrated, and nuclear reactions may take place. The NRA method produces new nuclei of specific energies. As with RBS, these energies are reduced by collisions with electrons, providing for an excellent method of depth analysis.

The most widely used method in polymer science has been FRES, an ion beam analysis technique specifically designed for depth profiling of deuterium and hydrogen. In FRES the deuterium and hydrogen ions recoiling from collisions are observed. Typically FRES is carried out with the beam incident on the sample at a glancing angle of 15°, and with a scattering angle of 30° being measured. In this case values of $0.5E_0$ and $0.67E_0$ are obtained for the recoil energies for hydrogen and deuterium, respectively, where E_0 represents the energy of the helium nuclei (39). In order to obtain improved separations of the hydrogen and deuterium nuclei from those of the helium nuclei also present, newer methods make use of time-of-flight technology. The more slowly moving helium ions are gated out, leaving only the ions of specific velocities to be measured.

Two types of experiments have been carried out using FRES, segregation of various species at surfaces and interfaces, and polymer chain diffusion. In order to carry out depth profiling via FRES, the data are convoluted with an instrumental resolution function, typically Gaussian.

Assume a two-component mixture of some type, in which one of the components is concentrated at either the surface, or at a substrate interface, or both. Then, instead of a sharp break in the concentration at the surface or interface expected in reality, Gaussian

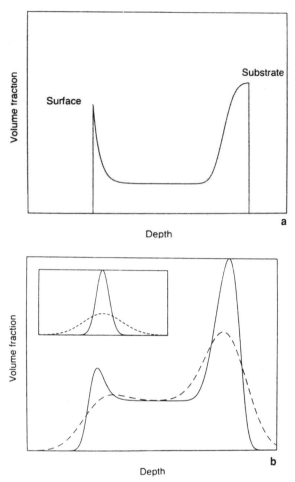

Figure 12.12 Phase domain segregation in thin films deposited on an inorganic substrate. (*a*) A hypothetical concentration profile of one component of a polymer blend, showing segregation both at the surface and at the inorganic interface. (*b*) The concentration profile shown in (*a*) after convolution with two Gaussian instrumental resolution functions of different FWHM, shown in the inset.

resolution functions with a characteristic full width at half maximum (FWHM) are recorded; see Figure 12.12 (39). While this leads to a powerful analysis mode, the composition profile sometimes has unrealistic edges.

12.3.6.2 *Observation of Block Copolymers at Interfaces* The interfacial strength of polymer blends is increased, in general, by lowering the interfacial tension and increasing the interfacial bond strength, as will be further developed in Chapter 13. Both of these quantities can be improved by the addition of block or graft copolymers to the interface. Shull and co-workers (40) carried out FRES experiments on a model polymer blend of layered polystyrene and poly(2-vinylpyridine), each with a chain length of about 6000 mers, containing various amounts of block copolymers of the two. The polystyrene

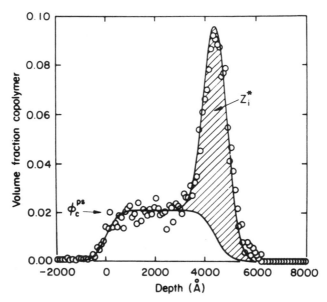

Figure 12.13 Segregation in a tricomponent thin layer on silica system: Polystyrene, $DP = 6000$, poly(2-vinyl pyridine), $DP = 6000$, and a dPS–PVP diblock copolymer, $DP = 391-68$. The polystyrene migrates to the air interface. The polymer–polymer interface copolymer excess, z_i*, corresponds to the hatched area.

block was deuterated. The authors pointed out that block copolymers initially dissolved in one of the homopolymers can go to four places: To the interface between the two polymers, to the air surface, to micellar structures in either or both of the homopolymers, or be dissolved in the homopolymers. Since the value of χ_1 for polystyrene mixing with poly(2-vinyl pyridine) is 0.11, mixing would be highly endothermic, and mutual solution of the two polymers can be neglected. If the total concentration is sufficiently low, little block copolymer migrates to the air surface. However, some block copolymer forms micelles, and some migrates to the blend interface; see Figure 12.13 (40). Here, the deuterated polystyrene block was 391 mers, and the poly(2-vinylpyridine) block was 68 mers in length. The interfacial tension between the two homopolymers was calculated to be 3.04 erg/cm². The volume fraction of copolymer in the polystyrene phase in equilibrium with the segregated layer in Figure 12.13 is 2.1%.

The total excess copolymer at the interface, z_I*, has a corresponding chain dimension of 100 Å. The corresponding calculated dimension based on the thermodynamics of the system yields 126 Å. The corresponding radius of gyration of a polystyrene chain containing 391 mers is 56 Å, based on Table 5.4. This is the smallest of the polymer chain dimensions. The end-to end distance is obtained by multiplying R_g by $6^{1/2}$, yielding 137 Å. These last two values straddle the experimental result, suggesting a unimolecular layer of block copolymer at the interphase. It must be recognized that the block copolymer chain conformations in phase separated systems are not quite random in nature; they tend to be somewhat more extended. The interfacial copolymer excess in chains/ area is obtained from z_I* by dividing by the copolymer molecular volume, yielding 1.52×10^{-3} chains/Å², or 1.52×10^{-1} chains/nm².

12.3.7 Neutron Reflectometry

Neutron reflectometry makes use of grazing angle reflection, of the order of about 1°. It is capable of penetrating to relatively large depths, around 1000 to 5000 Å, and yield a spatial resolution of about 10 Å, about an order of magnitude better than the SIMS (41), or FRES experiments (40). The depth and thickness of an interface can be determined via deconvolution of the multipeaked reflection pattern usually obtained. Such analysis is highly model dependent, so auxiliary experiments are usually necessary. Another significant drawback is the expense of the experiment. However, neutron reflection experiments have been used to analyze a variety of thin solid films, polymers at the air–liquid and solid–liquid interfaces, and polymer electrochemistry problems. Both ion-beam and neutron reflectometry analysis methods share the need to have deuterated polymer in the system since the deuterium nucleus provides the probe species.

One of the major problems in modern polymer science relates to the understanding and verification of the relations governing diffusion in the very short time interval, when polymer chains just leave the de Gennes tube; see Section 5.4.2. Wool and co-workers (3,42) developed the minor chain model, indicating that the diffusion of polymer chains in the melt should scale as $t^{1/4}$; see Sections 5.4.2.2 and 11.5.

12.3.7.1 The Ripple Experiment
Agrawal et al. (37) carried out what they called the "ripple experiment," using neutron reflectometry on selectively deuterated polystyrene block copolymer chains in two layers. In one of the layers, the central 50% of the mers were deuterated, the two end 25% portions were normal, that is, bearing hydrogen. This material was denoted as HDH (1/4H–1/2D–1/4H). In the second layer, the chain labeling was reversed to make DHD (1/4D–1/2H–1/4D). The molecular weights of the two polymers were nearly identical, 2.25×10^5 g/mol, and 2.50×10^5 g/mol, respectively, synthesized by anionic polymerization to have narrow polydispersity indices.

When the temperature of the bilayer film is above the glass transition temperature, interdiffusion begins; see Figure 12.14 (38). During the anisotropic motion of the chains as they first cross the interface, the HDH layer is enriched by deuterated polystyrene, and the DHD layer by hydrogenated (normal) polystyrene. The interdiffusion of the HDH/DHD matching pairs with reptation dynamics creates a ripple in the concentration profile; see Figure 12.15 (37). In Figure 12.15, τ_d represents the reptation time at the temperature of the experiment, 118°C, and τ_{RO} is the Rouse relaxation time, the diffusion time between tube walls. In Figure 12.15 the concentration profiles first show a ripple at the interface, reaching a maximum at about 450 minutes. According to reptation theory, the maximum ripple should be observed at $\tau_d/4$, or 465 minutes, in good agreement with the experiment.

Welp et al. (43) continued the ripple experiment, going to 400,000 g/mol polystyrene to obtain a better definition of the ripple. They concluded that the reptation model proposed by de Gennes (44) with parallel development by Doi and Edwards (45) was the best model to describe the dynamics of polymer interdiffusion. There were six dominant ripple characteristics that were examined:

1. The time at which the peak reached a maximum.
2. The peak maximum amplitude.
3. The peak maximum area.
4. The peak position at maximum amplitude.

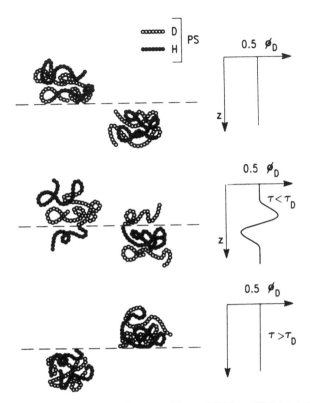

Figure 12.14 Model diagram of the interface between a bilayer of HDH- and DHD-labeled polystyrene. The open circles are the portions of the chains labeled with deuterium, and the filled circles represent the protonated portions. *Top figure:* The initial interface. *Middle figure:* The interface after center-of-mass diffusion over distances less than the radius of gyration. *Bottom figure:* The interface after the molecules have fully diffused across it. The graphs on the right depict the concentration of deuterated polystyrene as a function of distance across the interface.

5. The dependence of the peak position on time.
6. The dependence of the peak position on molecular weight.

All of the characteristics were predicted correctly by the de Gennes theory. These experiments provide significant evidence that polymer chains actually move like snakes through a grass composed of other snakes.

12.3.7.2 Theoretical Polymer Blend Interphase Thickness The structure and composition profile of polymer blend interphases is of current interest because (a) greater degrees of interfacial mixing between components in a binary mixture usually increases the mechanical strength of the material, and (b) observation of the degree of interfacial mixing at a model interface as a function of time provides information about the interdiffusion coefficient of the polymer pair. Both the free energy of mixing and the kinetics of diffusion are important in defining the interface.

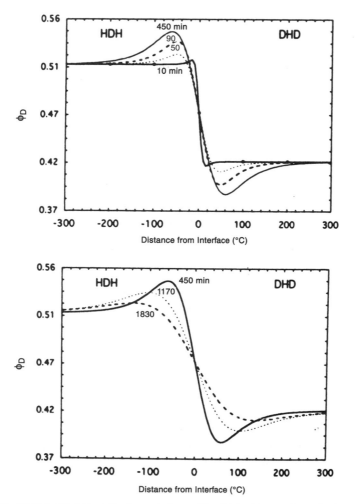

Figure 12.15 Ripple experiment results. Concentration profiles obtained from neutron reflection experiments are shown at several annealing times. *Top:* profiles from 10 to 450 min. *Bottom:* profiles from 450 to 1830 min. Here, $\tau_{RO} \approx 21$ min and $\tau_d \approx 1860$ min.

Helfand and Tagami (10) derived thermodynamically based equations for the interfacial tension at a polymer blend interface,

$$\gamma_{ab} = \left(\frac{\chi}{6}\right)^{1/2} \rho_0 bkT \qquad (12.17)$$

and the interphase surface thickness,

$$S_{th} = \frac{2b}{(6\chi)^{1/2}} \qquad (12.18)$$

Table 12.6   Temperature dependence of χ for selected polymer blends

Blend	Relationship
PS/PMMA	$\chi = 0.0284 + 3.902/T$
SAN/PMMA	$\chi = (624\beta^2 - 727\beta + 193)/T$
SAN/PC	$\chi = (556\beta^2 - 697\beta + 314)/T$
PS/PB	$\chi = 0.083 \ (25°C)$

Source: Based on R. P. Wool, *Polymer Interfaces: Structure and Strength*, Hanser, Munich, 1995, Chap. 9.

Note: β represents the mole fraction of styrene mers. PS = polystyrene; PMMA = poly(methyl methacrylate); SAN = poly(styrene–*stat*–acrylonitrile); PC = polycarbonate; PB = polybutadiene.

where ρ_0 is the density, K is Boltzmann's constant, χ represents the Flory–Huggins interaction parameter, and b is the statistical segment step length, equal to 6.5 Å for both polystyrene and poly(methyl methacrylate). Many other random coil polymers have similar values of b. The values of χ are temperature dependent, important for molding conditions; see Table 12.6. While the polystyrene/poly(methyl methacrylate) system will be immiscible for any reasonable molecular weight and temperature range, the other two blends listed can be made miscible through manipulation of the acrylonitrile comonomer content. If the polymers are in the miscible range, equation (12.18) does not apply.

Another equation relates directly to the critical fracture energy, $\mathscr{G}_{1c}$,

$$\mathscr{G}_{1c} = 2\gamma_{ab}\left(\frac{S_{th}}{b}\right)^2 \tag{12.19}$$

providing a lower bound relationship. The values of $\mathscr{G}_{1c}$ obtained are of the same order of magnitude as that obtained for equation (12.20), but based on quite different concepts.

Another quantity is the work of adhesion, W_a, defined as the reversible work per unit area required to separate the interface between two polymer melts from their equilibrium positions to infinity:

$$W_a = \gamma_a + \gamma_b - \gamma_{ab} \tag{12.20}$$

The quantity W_a is a lower bound value, since chain scission and chain pullout are ignored. In fact, the W_a values shown in Table 12.2 are some 500 times smaller than the interfacial fracture energy, $\mathscr{G}_{1c}$. Thus the experimental value of $\mathscr{G}_{1c}$ for the PMMA/PS interface is about 5×10^4 erg/cm^2 (50 J/m^2) (46). This compares to values of 500 J/m^2 for polystyrene, and about 500 J/m^2 for poly(methyl methacrylate); see Table 13.6. While none of these relationships depend directly on the molecular weight, a real limitation, it is assumed that the molecular weight is high enough that S_{th} has reached its limiting value.

12.3.7.3   *Example Calculation of Interphase Thickness*   What is the theoretical interphase thickness for a polystyrene/poly(methyl methacrylate) material, annealed at 135°C, then quenched to room temperature?

Table 12.6 provides the temperature dependence the Flory–Huggins heat of mixing term,

$$\chi = 0.0284 + \frac{3.902}{T} \tag{12.21}$$

At 135°C, χ equals 0.038. Substituting into equation (12.18) obtains

$$S_{th} = \frac{2 \times 6.5\,\text{Å}}{(6 \times 0.038)^{1/2}}$$

yielding $S_{th} = 27.2\,\text{Å}$.

Of course, rapid quenching substantially freezes in the morphology. At room temperature, both polymers are below T_g, and polymer chain diffusion is essentially zero within the time scale of the following experiments.

12.3.7.4 Experimentally Determined Interphase Thicknesses

Neutron reflection determined values for the interphase thickness for the polystyrene/poly(methyl methacrylate) system were determined by Russell et al. (21) to be $50 \pm 2\,\text{Å}$. Fernandez et al. (23), who also carried out neutron reflection, arrived at $20 \pm 5\,\text{Å}$. By combining equations (12.18) and (12.21), the interphase thickness of this system was determined to be 27 Å (3). Assuming an M_c value of 31,200 g/mol for polystyrene (47), the corresponding R_g value is 48 Å, from Table 5.4. Then the interphase thickness is very close to the critical value of having one entanglement per chain portion present. Physical entanglements are critical to the development of high-fracture energy, about eight per chain necessary for optimum mechanical strength; see Section 11.5.2. Hence, if only one entanglement per chain is in the interface, one would conclude such interphases should be weak, which they are.

12.3.8 Determination of Interior Surface Areas

12.3.8.1 Theory for X-Ray, Neutron, and Light-Scattering

Polymer blends and composites frequently form irregular or even random microscopic or submicroscopic structures. The interfacial areas can be determined by small-angle neutron or X-ray scattering. Debye et al. (48) derived the basic theory for small-angle X-ray scattering (SAXS). Their interest was in the determination of the internal surface areas of porous catalytic particles. Such materials possess very ample differences in electron density, required for good SAXS intensity. For small-angle neutron scattering (SANS), similar contrast can be obtained in polymer blends using deuterated probes (49,50). The method works best for randomly appearing surfaces, whether the minor phase is dispersed or the sample exhibits dual phase continuity. If the interfaces are not randomly structured, for example, dispersed spheres, a mathematical correction must be made to determine the interfacial area of the known geometry.

Debye et al. (48) assumed a randomly structured material in which the probability of encountering a surface at an arbitrary distance could be provided by a correlation function in an exponential form,

$$\gamma(r) = \exp\left(-\frac{r}{a}\right) \tag{12.22}$$

where $\gamma(r)$ represents the correlation function for a characteristic distance r between scattering centers, and the quantity a is the correlation distance defining the size of the

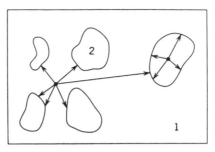

(a) Correlation distance, a

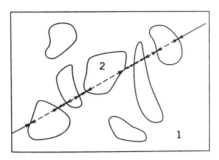

(b) Transverse length, l_1, l_2

Figure 12.16   Specific interface areas and dimensions determined via scattering experiments. (*a*) The correlation distance *a* is determined starting from a random place and traveling in a random direction until an interface is reached. The distance to an interface, on average, is *a*. (*b*) A straight line is drawn through the material. The average distance between two interfaces determines L_1 and L_2.

heterogeneities. The correlation distances are illustrated in Figure 12.16 (1). It does not matter in which phase one starts. After landing, a random direction is selected, and traveled until striking the first surface (or interface). The specific interfacial surface area, S_{sp} is defined as the ratio of the interfacial surface area, A, to the mass, m,

$$S_{sp} = \frac{A}{m} = \frac{4\phi(1 - \phi)}{ad} \qquad (12.23)$$

where ϕ represents the volume fraction of either phase and d represents the density. Of course, whichever phase volume fraction ϕ represents, $1 - \phi$ represents the volume fraction of the remaining phase.

The theory calls for a plot of $I^{-1/2}$ versus K^2 where I represents the excess scattered intensity (after removal of the background scattering, etc.) and $K = 4\pi\lambda^{-1} \sin(\theta/2)$, where λ is the wavelength, and θ represents the scattering angle. The quantity a is then,

$$a = \left(\frac{\lambda}{2\pi}\right)\left(\frac{slope}{intercept}\right)^{1/2} \qquad (12.24)$$

where the slope and intercept refer to the line drawn by plotting $I^{-1/2}$ versus K^2.

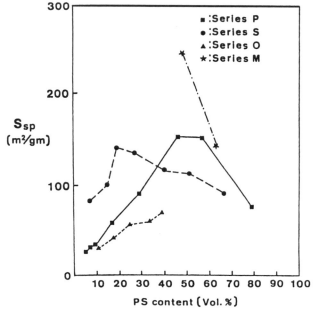

Figure 12.17 The specific interfacial area in an IPN goes through a broad maximum in the midrange compositions.

The Debye et al. theory can also be interpreted in terms of the transverse lengths across the domains L_1 and L_2 of phases 1 and 2,

$$L_1 = \frac{a}{\phi_2}, \quad L_2 = \frac{a}{\phi_1} \tag{12.25}$$

The correlation distances are also illustrated in Figure 12.16. Conceptually the transverse lengths can be obtained by drawing a randomly chosen straight line through the sample. The average distance between any two adjacent interfaces determines L_1 and L_2. The reader should note that in equation (12.25), the subscripts of the L and ϕ quantities are opposite; that is, L_1 goes with ϕ_2, and so on. For larger structures this theory also works with visible light-scattering.

Interpenetrating polymer networks are known to have finely divided morphologies with significantly irregular morphologies; see Figure 4.14. The specific surface areas for a series of *cross*–polybutadiene–*inter*–*cross*–polystyrene IPNs are shown in Figure 12.17 (49). A maximum in S_{sp} occurs in the midrange of compositions. Values of S_{sp} ranged from 50 to 300 m²/g, typical for colloidally dispersed materials and finely divided polymer blends, as well as many IPNs.

12.3.8.2 K^{-4} *Dependence of Interphase Thickness*

A quantity in wide use in small-angle scattering, especially of X-rays and neutrons, is $K = (4\pi/\lambda)\sin(\theta/2)$, where λ and θ represent the wave length and angle between the incident and scattered beam; see Section 5.2. An ideal two-phased system with sharp, nonmixing interfaces was treated by Porod (51), who predicted a dependence of the scattering intensity proportional to K^{-4} at large angles,

$$I(K) = \frac{K_p}{K^4} \tag{12.26}$$

where K_p is the so-called Porod constant.

Deviations from Porod's law lead to estimations the interfacial thickness, S_{th}. One such relationship is (52)

$$I(K) = K'(S_{sp}/K^4) \exp(-K^2 S_{th}^2) \tag{12.27}$$

where K' depends on the differences in electron density or scattering lengths, and thus depends on the type of waves and the atomic characteristics of the two components involved (53). The immediate value is that S_{sp} can be easily estimated.

12.4 CONFORMATION OF POLYMER CHAINS IN A POLYMER BLEND INTERPHASE

By definition, the free energy of mixing of a polymer blend interphase is positive, or else molecular mixing would continue to completion, and at equilibrium the interphase would vanish. Helfand and Tagami (10,11) developed a mean-field theory of polymer interfaces, or interphases, as they are now called. They were particularly interested in the equilibrium composition and interphase density across the interphase, the interfacial tension and thickness (see Section 12.3.7.2), and conformation of the polymer chains making up the interphase.

Briefly, they made four points; see Figure 12.18 (10):

1. The composition would change from substantially all one polymer to substantially all the other polymer in a smooth fashion, the width of the interphase being controlled by the value of χ; see Section 12.3.7.2.

2. Assuming symmetrical polymers with regards to individual density, statistical segment (mer) length, and compressibility, there would be a minimum in the density in the middle of the interphase. This was caused by the repulsive forces between the two chains, namely positive heat of mixing.

3. Small third-molecule components would be preferentially adsorbed in the interphase. The presence of small molecules dilutes the polymers, tending to weaken the interphase mechanically. (If water and/or oxygen were making up some or all of the small molecule composition, clearly polymer degradation might be an increased problem as well.)

4. The chain ends would tend to be slightly more concentrated in the interphase than are segments in the middle of the chain. Thus there is a slight orientation of the chain end portions perpendicular to the plane of the interphase; see Figure 12.19 (1).

More recently Cifra et al. (54) developed a similar theory for composite interfaces involving a polymer and a solid, nonpolymer phase, and showed that the density of chain ends at the interface with the nonpolymer phase was also higher than statistical. However, they also showed that the enhancement of chain ends at the interface was

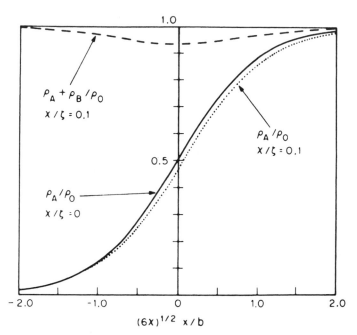

Figure 12.18 The density profile of a polymer blend interface goes through a minimum if the polymers are immiscible. The quantity ζ is a function of the compressibility.

combined with a depletion layer of chain ends just below the enriched layer because the number of chain ends must be conserved within the dimension of a single chain.

Recently there have been several experiments supporting these theories. Kajiyama et al. (55) studied the surface molecular motion of monodisperse polystyrenes via SFM and LFM, finding that the glass transition temperature was significantly depressed. Depth

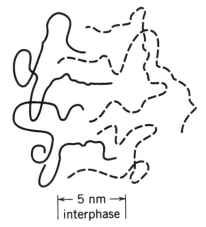

Figure 12.19 An immiscible polymer blend interphase region. The central portions of the chains have a slight tendency to lie parallel to the nominal interface, while the chain ends tend to lie perpendicular.

profiling of deuterated polystyrene films in which the chain end groups were labeled with protonated groups was carried out with SIMS. A significant increase in the end group population on the surface was determined. The increased thermal molecular motion on the polymeric surface was explained by an excess free volume induced by the surface localization of these chain ends (55,56). Goh et al. (57) used AFM to evaluate surface motion in poly(butyl methacrylate) latex film formation, finding that the surface diffusion coefficient was some 10^4 larger than for the bulk polymer homodiffusion coefficient. A glass transition depression of about 10°C could be estimated from the data (1), probably caused by a slightly lower density of the surface material.

Reiter et al. (58) studied molecular weight fractionation in polymer blend interfaces, assuming the normally broad molecular weight distribution present in most polymers. They found that the fraction of shorter chains was larger in the interphase. The shorter chains were also preferentially dissolved in the other polymer. The reason was that the shorter chains are more miscible, and also that in visiting the other phase the gain in entropy is larger relative to the enthalpy loss. (Again, the concentration of short chains in the interphase tends to weaken it, because they have fewer physical entanglements with neighboring chains.)

Shaub et al. (59) put fluorocarbon ends on polystyrene, then formed thin films on silicon. Using neutron reflectivity, significant increases in chain ends in contact with the silicon surface was found, supporting the Cifra (54) theory. Of course, chain end enrichment at the polymer surface or interface will also give rise to an adjacent depletion region over a distance of roughly equivalent to the end-to-end distance of the polymer chain, where the glass transition should approach its bulk value.

12.5   THE DILUTE SOLUTION–SOLID INTERFACE

Another system of great interest relates to the interactions between solid surfaces and dilute polymer solutions. Frequently the solids in question are in the form of colloids, presenting areas of the order of 50 to 300 m^2/g to the exterior solution phase. The polymer chains may be bondable to the solid surfaces at one end, both ends, and/or along the chain, depending on the chemistry.

12.5.1   Nondrip Latex Paint

A common example of such a system is ordinary latex paint, containing both polymer latexes and inorganic fillers and pigments, and dispersed in an aqueous phase. The aqueous phase usually contains a dilute solution of polymers capable of associative bonding, that is, bonding weakly with the colloidal solids and/or each other. Such associative bonding increases the viscosity significantly and, more important, contributes a thixotropic nature to the paint; see Section 10.5.5.

In simplified terms the thixotropic action is as follows: The can of latex paint starts with a high viscosity. Sticking a brush in it produces a high shear rate, lowering the viscosity and allowing the paint to flow easily onto the brush. Raising the brush out of the paint and holding it aloft is a low shear situation for the paint on the brush, and the viscosity increases. This last produces the famous *nondrip* characteristics of such materials. Applying the brush to the wall or ceiling produces a high shear situation again, and concomitant low viscosity, allowing the paint to flow onto the desired surface. When painting a ceiling, properly formulated paint will not run down the brush, because such

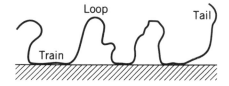

Figure 12.20 A model of a polymer chain partly adsorbed on a surface.

flow would have to take place under low shear conditions. The surface tension of the aqueous phase then acts to *level* the paint on the wall or ceiling, while the now high viscosity retards *sagging*. The result, after drying and film formation, is a smooth coat of paint.

12.5.2 Conformation of Polymer Chains at Solution Interfaces

Consider the simple case of a dilute polymer solution in contact with a solid surface (2). Some of the chains are adsorbed on the surface. Such adsorbed chains have three main parts; see Figure 12.20 (1). The polymer *trains* have all of their mers in contact with the substrate. The trains are separated by *loops* that are unbound and connect the trains, and the *tails*, which are nonadsorbed chain ends. An important parameter in calculations is the train or bound fraction, p. The amount of polymer adsorbed on most such surfaces is of the order of one mg/m^2.

12.5.3 Example Calculation of p

Assume a solution of 200,000 g/mol polystyrene saturating a colloidal surface. Weight increases show that there is 2.5 mg of polymer for each m^2 of surface. What is the bound fraction of polymer?

The molecular weight of a mer is 104 g/mol, and the density of the bulk polystyrene is 1.06 g/cm^3, Table 3.2. Then

$$\frac{104 \text{ g/mol}}{1.06 \text{ g/cm}^3} = 98.1 \text{ cm}^3/\text{mol}$$

or 98.1×10^{-6} m^3/mol. On a per mer basis,

$$\frac{98.1 \times 10^{-6} \text{ m}^3/\text{mol}}{6.02 \times 10^{23} \text{ mers/mol}} = 16.30 \times 10^{-29} \text{ m}^3/\text{mer}$$

Assuming a cubical shape for the mer, the area of one side is

$$(16.30 \times 10^{-29} \text{ m}^3)^{2/3} \text{ per mer} = 3.00 \times 10^{-19} \text{ m}^2/\text{mer}.$$

Inverting, we see that there are 3.35×10^{18} mers/m^2. The weight of the mers saturating a m^2 of the surface is

$$\frac{3.35 \times 10^{18} \text{ mers/m}^2}{6.02 \times 10^{23} \text{ mers/mol}} \times 104 \text{ g/mol} = 5.79 \times 10^{-4} \text{ g/m}^2$$

or 5.79×10^{-1} mg/m^2. This the bound material at saturation, assuming that the surface completely covered. Since there are 2.5 mg of polystyrene per m^2, the maximum bound fraction is

$$p = \frac{5.79 \times 10^{-1} \text{ mg/m}^2}{2.5 \text{ mg/m}^2} = 0.23$$

This is the bound fraction.

If the assumption is further made of one long train and no loops and one long tail per polystyrene chain (unrealistic), then the length of the average train is about

$$200{,}000 \text{ g/mol} \times 0.23 = 46{,}000 \text{ g/mol}$$

and the length of the tail is

$$200{,}000 \text{ g/mol} - 46{,}000 \text{ g/mol} = 154{,}000 \text{ g/mol}$$

So, the polymer is mostly tail, with more than three times the amount of tail as bound material. Of course, there may be loops as well.

12.5.4 Conformation of the Bound Chains with Increasing Solution Concentration

At a very low solution concentration, the surface concentration is also very low. Under these conditions the chains have a tendency to flatten out, or *pancake*; see Figure 12.21. In this concentration range much of the surface may be vacant. As the concentration increases, the competition for surface space increases. The chains may form a conformation known as *mushrooms*, where the chains are either bound at one end, or by a few mers at one end. At still higher solution concentrations, a *brush* structure appears. Also in this case each chain attached to the surface may only be bound at one end or by a few mers, the remainder of the chain sticking nearly straight up into the solution to avoid contact with neighboring chains. In these latter cases, of course, most of the chains remain entirely in solution, there being no space at the surface.

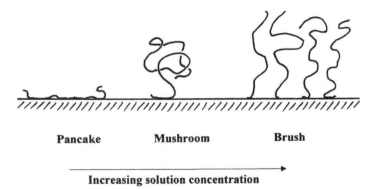

Pancake   **Mushroom**   **Brush**

Increasing solution concentration

Figure 12.21   Models of three types of bonding a chain and a surface may have. With increasing concentration of polymer on the surface, the chains have less room per chain, and tend to stand up.

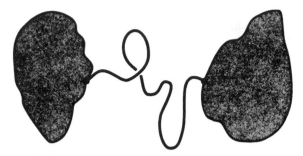

Figure 12.22 Illustration of two particles bound by a single chain. How far apart can the particles be before one end of the chain "lets go?"

Once on the surface, the chains are often more difficult to remove than they were to attach in the first place. The cause is related to multiple points of attachment. If one considers hydrogen bonding, for example, at 5 kcal/mol, then one such bond is rather easy to break. To break two such bonds is more complicated than just having 10 kcal/mol available, because both bonds need to be broken simultaneously, or, that the second bond needs to be broken before the first one reforms. In the case of several such bonds, it often proves difficult to debond all of them at the same time. Hence the kinetics of debonding is much slower than the kinetics of bonding in the first place.

12.5.5 Example Calculation: A Single Chain Bound to Two Colloid Particles

Above, an example of the power of an associative polymer to cause thixotropic behavior was given in terms of the use of latex paint. Assume an associative poly(ethylene oxide), PEO, chain bonded two different colloidal particles; see Figure 12.22 (1). Poly(ethylene oxide) is soluble in aqueous media, being near Flory-θ conditions in water at room temperature. The molecular weight of the PEO chain is 1×10^5 g/mol. The bonding energy to each of the two particles is 5.9 kcal/mol, or 10 kT, approximately equal to one hydrogen bond. A shearing force is applied to the system, tending to separate the two particles. The question is: To what distance must the two particles be separated before the PEO debonds from one of them?

At rest the PEO chain has a radius of gyration governed by the relation (2)

$$R_{g0}(nm) = 343 \times 10^{-4}\ M^{0.5} \tag{12.28}$$

Thus

$$R_{g0}(nm) = 343 \times 10^{-4} \times (1 \times 10^5)^{1/2} = 10.8\ nm$$

The average distance between the bound surfaces with no external forces applied is given by the end-to-end distance, r ($r^2 = 6R_g^2$), or 26.4 nm.

The theory of rubber elasticity for single chains (Section 9.7.1) relates the force f necessary to hold a chain at an end-to-end distance r,

$$f = \frac{3kTr}{r_0^2} \tag{12.29}$$

where r_0 is the value of r under Flory θ-temperature conditions, substantially the present case. The force here is purely entropic in nature, generated by the reduction in the number of conformations possible as the end-to-end distance is increased.

Of course, work, W, equals force times distance,

$$W = \int_0^r \left(\frac{3kTr}{r_0^2}\right) dr \tag{12.30}$$

$$W = \left(\frac{3}{2}\right)kT\left(\frac{r^2}{r_0^2}\right) = \left(\frac{3}{2}\right)kT\alpha^2 \tag{12.31}$$

where α is the ratio of the actual end-to-end distance to that in a θ solvent. Noting that 5.9 kcal/mol translates into 4.11×10^{-20} J/chain at 25°C,

$$4.11 \times 10^{-20} \text{ J/chain} = (\tfrac{3}{2})(1.38 \times 10^{-23} \text{ J/K}) \times 298 \text{ K}\alpha^2$$

Solving for α, $\alpha^2 = 6.66$, or $\alpha = 2.58$. Then $2.58 \times 6^{1/2} \times R_{g0} = 68.3$ nm for the distance between the particles, the entropic retraction forces will be sufficient to break one of the bonds, and the two particles will separate.

Of course, the chain ends are usually buried somewhere within the polymer coil. In the calculation above, the chain ends are attached to the colloidal particles, and hence forced outward. There are two limiting cases:

1. If the colloidal particles are large with respect to the R_g of the polymer chain, and the surface is smooth, the particle can be approximated by a flat plane.
2. If the particles are small relative to the R_g of the polymer chain, then a point approximation may be valid, and the conformation of the polymer chain may be substantially random.

In reality most colloid particle sizes lie between these two extremes. The conformations of polymer chains have been discussed by de Gennes (60); see below. In many cases, the polymer chain tries to avoid the wall at points other than the attachment point.

Milner (61) addressed the energetics of polymer brushes. In this case neighboring polymer chains need to avoid one another. This volume exclusion effect provides the driving force for the chains to extend out into the solution. However, the problem is also different from the one chain bonding two particles, above, because one end of the chain is free. Milner also included a correction term for surface tension effects. He concluded that excess energies in the range of 8 to 10 kT were reasonable for many brush conformations.

12.5.6 Depletion Zones

Consider the case where there are no attractive forces between the polymer chain in solution and the surface. Then, as a polymer chain approaches such a surface, the required conformational change causes a loss of entropy. This may result in a depletion zone next to the surface, where the polymer concentration is lower than in the solution.

The free (air) surface of a dilute polymer solution may also suffer a depletion zone because the polymer, effectively, is forbidden by loss of free energy to have significant numbers of loops or ends sticking in the air. This causes vapor pressure and surface tension to be reduced less than expected, for example (62).

Another example of the depletion zone relates to the core-shell effects in the synthesis of latexes (63). Consider a partly polymerized latex. The remaining monomer may be the same or different from the already formed polymer. Many such systems have most of the polymer in the interior, and most of the monomer near the surface, because the polymer avoids the aqueous interface. A requirement is that the latex particle be larger than the end-to-end distance of the polymer chains already formed. This requirement is usually met. This has two consequences:

1. When polymerizing a monomer II in the presence of polymer I, a core-shell structure may develop with the shell making up polymer II.
2. The major theories of polymerization kinetics (64), which assume the remaining monomer is uniformly distributed in the latex, may need a correction term.

The depletion zone also reduces the effective pore size in gel permeation chromatography particles. The reaction rates of polymers at surfaces may be reduced below expected, ignoring the depletion zone.

12.6 INSTRUMENTAL METHODS FOR ANALYZING POLYMER SOLUTION INTERFACES

A series of modern instruments have been designed to study the conformations and other aspects of polymer chains bound to a surface.

12.6.1 The Force Balance Apparatus (Surface Forces Apparatus)

Consider a polymer chain adsorbed on a surface, all immersed in a liquid phase. In general, the polymer chains are partly dissolved, with those portions composed of loops and tails. These may have conformations similar to that of random coils. When two such adsorbed polymer chains approach one another, there will be an entropic repulsive force between them, arising from the excluded volume effect, as well as other forces (65,66). Sometimes the repulsive forces are called steric or overlap repulsion.

Chains attached to colloidal surfaces provide powerful forces for stabilization. Colloidal particles that normally coagulate from a solvent dispersion can thus be stabilized by adding a small amount of polymer to the dispersion. Such polymer additives are sometimes known as protective colloids, leading to *steric stabilization*. Both synthetic polymers and biopolymers such as proteins and gelatin are commonly used in both nonpolar and polar solvents. Industrially they are used in paints, toners, emulsions, suspensions, cosmetics, pharmaceuticals, processed foods, and lubricants.

The surface forces apparatus (SFA) was designed to measure these forces; see Figure 12.23 (67,68). In the SFA the force between two surfaces in controlled vapors or liquid immersion can be directly measured with a resolution of about 0.1 nm, and a force sensitivity of about 1×10^{-8} N.

Typically the SFA features two curved, molecularly smooth surfaces of mica of radius of about 1 cm. The interaction forces are measured via force-measuring springs. Usually, the sheets of mica are placed in a crossed cylinder configuration, locally equivalent to a sphere near a flat surface or to two spheres close together. An optical tech-

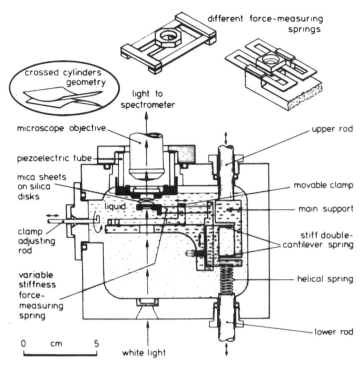

Figure 12.23 Illustration of the surface forces apparatus, useful for direct measurement of the forces between surfaces in liquids or vapors at the Ångstrom level. Both normal and shear stresses may be applied to the crossed cylinders.

nique involving interference fringes provides a measure of the distance between the two cylinders. The motion of the two cylindrical surfaces can be toward each other, resulting in the measure of entropic interference between the two layers of chains, or laterally, providing a measure of the friction between the chains, and so on.

Several experiments of interest can be carried out with the SFA. One possibility is to have the chains attached at their ends to both surfaces, thus providing an attractive *bridging* effect (69,70). Bridging effects are extremely important in flocculation of colloidal dispersions, the opposite of stabilization. Clarification of waste water provides an important application.

Another possibility involves measuring the thickness of the layers, if the polymer chains are bound at one end. Taunton et al. (71) studied polystyrene chains containing either an ionic end group or a short end block of poly(ethylene oxide), both of which bond to mica. The polystyrenes were dissolved in toluene were allowed to contact the mica, and bond to equilibrium. Figure 12.24 (71) illustrates two cases: The forces generated when 150,000 g/mol polystyrene occupied both sheets of mica, and when one of the sheets was replaced by a clean mica sheet of the same thickness. When both sheets contain the polymer, the repulsive forces between the chains constitute the result, but when the polymer was on one sheet only, it was possible to measure the forces on compression of a single layer of the terminally anchored, nonadsorbing polymer against a rigid wall.

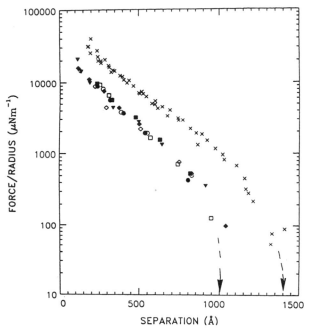

Figure 12.24 Using the Surface Forces Apparatus, the repulsive forces between polymer chains on both crossed cylinders, left, and one coated surface and one clean mica sheet, right, can be measured. Poly(ethylene oxide) anchors the nonadsorbing polystyrene chains to the mica, with toluene as the solvent.

For the polymer adsorbed on both sheets of mica, a parameter of interest was $2L_o$, defined as the separation at which repulsion could just be detected, providing an effective layer thickness. Values of $2L_o$ plotted against molecular weight of the polystyrene chain are shown in Table 12.7 (71); see the arrows in Figure 12.24. The values follow the relationship

$$L_o \propto M^{0.6} \tag{12.32}$$

(except for the highest molecular weight sample), and they are thus related to the radius

Table 12.7 Values of $2L_0$ for polystyrenes end bound to mica

Molecular Weight, g/mol	$2L_0$, Å
26,000	440 ± 50
58,000	800 ± 80
140,000	1350 ± 100
150,000	1450 ± 100
184,000	1500 ± 100
375,000	2400 ± 100
666,000	2100 ± 100

Source: Based on H. J. Taunton et al., *Macromolecules*, **23**, 571 (1990).

of gyration of the chains (Section 3.6.2). However, L_o values found were approximately twice that found for more strongly adsorbing chains of the same molecular weight at comparable surface coverages. This latter arises from the partial replacement of loops and tails by end-bound brushes for the larger L_o values.

In Figure 12.24 the value of L_o for the one coated/one clean sheet of mica is seen to be 1000 Å, significantly more than half of the distance for the case where both sheets contain polymer. A possible interpretation is that if two irregular, soft surfaces meet, a repulsive force develops only after the chains significantly begin to overlap; many chains will have passed by each other without interacting. However, with a hard surface facing the chains on one side, the first chains encountering the uncoated side will produce a measurable repulsive force. This provides some insight as to the topology of the brush formation.

12.6.2   Total Internal Reflectance Fluorescence, TIRF

While the SFA provided direct evidence as to the thickness of a polymer layer adhering to a surface, TIRF provides a measure of the surface area concentration, Γ. TIRF spectroscopy makes use of the total internal reflection of light at the interface between a solid adsorption substrate of relatively high refractive index and a polymer solution of lower refractive index. The total internal reflection, however, generates a standing *evanescent wave*, a nonclassical penetration of the light into the lower refractive index phase. The evanescent wave may penetrate some 60 to 65 nm into the lower refractive index phase, be absorbed by it, fluoresce, and so on (72,73).

If the molecules do not naturally fluoresce, frequently a fluorophore such as fluorescein is added to the polymer. This special synthesis may tag one end of the chain, for example. As fluorescently labeled molecules from a dilute polymer solution diffuse to the interface and adsorb, their fluorophores become excited by the evanescent wave. To the extent that the quantum yield of the label is uninfluenced by the chemical differences between the solution and the interfacial regions, the fluorescence is proportional to the changing surface excess, in real time.

12.6.3   Brewster Angle Reflectivity

Another optical method involves Brewster angle reflectivity, which depends on light passing from a material of lower refractive index, n_0, into a medium of higher refractive index, n_1. Zero-plane-polarized reflected light will occur at the Brewster angle defined by

$$\tan \theta_B = \frac{n_1}{n_0} \tag{12.33}$$

The reflectivity around the Brewster angle depends on the angle of incidence, θ_0, and the angle of refraction, θ_1, which are related through Snell's law,

$$n_0 \sin \theta_0 = n_1 \sin \theta_1 \tag{12.34}$$

The addition of a thin film to the interface creates multiple reflections at the air–film and film–water interfaces (8), which change the properties of the Brewster angle and provide information about its thickness. Frequently the assumption of a constant layer density is made for data analysis, although density variations clearly occur physically.

The two methods, TIRF and Brewster angle reflectivity measurements, were combined (73) to examine the competitive adsorption of poly(ethylene oxide) of different molecular

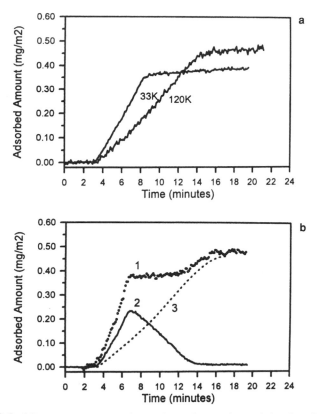

Figure 12.25 Reflectivity measurements can be used to estimate polymer chain adsorption kinetics. (*a*) Noncompetitive poly(ethylene oxide) adsorption kinetics for chains of 33,000 and 120,000 g/mol. (*b*) Competitive adsorption kinetics for a 50–50 wt-% mixture of 33,000 and 120,000 g/mol poly(ethylene oxide).

weights onto silica surfaces. To facilitate the TIRF experiments, the poly(ethylene oxide) had coumarin end tags.

Figure 12.25*a* (73) shows the adsorption kinetics for 33,000 and 120,000 g/mol poly(ethylene oxide)s at 2.5 ppm each in noncompetitive adsorption. The kinetics of adsorption of the lower molecular weight species is clearly faster.

Figure 12.25*b* (73) shows the different facets of a competitive adsorption of a 50/50 w/w% mixture of the long and short chains from a total concentration of 5 ppm. Curve 1 illustrates the reflectivity results, presenting a stepwise appearance. The initial signal rise was observed to be the sum of the two signal rises in Figure 12.25*a*. This suggests that at low coverage, both species adsorb independently at their mass transport limited rates. This persists until the total coverage reaches a level corresponding to the equilibrium coverage of the short chains shown in Figure 12.25*a*, determined by near-Brewster angle reflectivity. At this point, the surface appears full to the short chains. Next, a period of nearly constant interfacial mass ensues, followed by a second rise to an ultimate coverage equal to the adsorbed amount of the long chains in Figure 12.25*a*.

Figure 12.25*b* also illustrates the TIRF signal for the coumarin-tagged short chains, curve 2, as they adsorb from the mixture. Initially the short chains from the mixture are uninfluenced by the presence of the long chains, as the rise in curve 2 is compared to the rise in the 33,000 g/mol adsorption in Figure 12.25a. After this point the short chains are displaced from the surface by the incoming long chains, which continues until substantially all of the short chains are removed.

In Figure 12.25b, curve 3 is determined by subtracting curve 2 from curve 1. A comparison of curve 3 with that in Figure 12.25a, 120,000 g/mol sample, shows that the long chain adsorption during competition is not influenced by the adsorbed short chains. The arrival of the long chains from solution constitutes the rate limiting step. The free energy of having the long chains in contact with the silica surface is lower, the competitive adsorption interchange being a transport-limited process.

12.6.4   Dynamic Light-Scattering

Dynamic light-scattering, sometimes called quasi-elastic light scattering or photon correlation spectroscopy, can be used to measure the diffusion coefficients of polymer chains in solution and colloids, a kind of Doppler effect; see Section 3.6.6. In a dilute dispersion of spherical particles, the diffusion coefficient D is related to the particle radius, a, through the Stokes–Einstein equation,

$$D = \frac{kT}{6\pi\eta a} \tag{12.35}$$

where η represents the viscosity of the fluid. Today, there are computer programs to convert the wavelength distribution via Fourier transform to the needed diffusion coefficient.

When polymers dissolved in the fluid adsorb onto the colloidal particles, the particles' radii increase, and vice versa for desorption. Thus the change in the apparent particle radius supplies information about the thickness of the adsorbed layer.

The kinetics of exchange of short and long poly(ethylene oxide) associative polymers was examined by Gao and Ou-Yang (74) by dynamic light-scattering. The poly(ethylene oxide) chains were capped symmetrically on both ends with $C_{16}H_{33}$ or $C_{20}H_{41}$ aliphatic hydrophobes and placed in the presence of uniform polystyrene latexes in dispersion. In one experiment (74), 17,000 or 100,000 g/mol end capped poly(ethylene oxide) was placed in a dispersion of polystyrene latexes containing the other component, and the rates of exchange were observed; see Figure 12.26. Equilibrium was attained in both cases after about 8 hours under the conditions employed. The authors concluded that the long chains replaced the short chains more effectively than the reverse case. While the result is similar to that observed in Figure 12.25b, in that data there were no hydrophobes attached, and portions of the whole chain adsorbed, producing a distribution of trains, loops, and ends, as opposed to that in Figure 12.26, involving morphologies more like brushes.

Sometimes, however, it has been noted that the short chains are preferentially adsorbed. Several competing factors come into play in each case; see Table 12.8 (1).

12.7   THEORETICAL ASPECTS OF THE ORGANIZATION OF CHAINS AT WALLS

The theoretician de Gennes pointed out that both attractive walls and repulsive walls must be considered (75). Figure 12.27 (75) illustrates the conditions at attractive versus repulsive walls. For the attractive wall situation, Figure 12.27a, the concentration profile

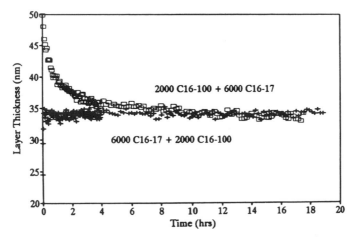

Figure 12.26 A study in both equilibrium and reversibility in polymer chains adhering to surfaces. Poly(ethylene oxides) containing C_{16} hydrocarbon tails on both ends, adhering to polystyrene latex particles in aqueous dispersion. Note that poly(ethylene oxide) is water soluble. Samples of first 100,000 g/mol ($C_{16}-100$) and then 17,000 g/mol ($C_{16}-17$) were placed on the latex, and the other allowed to diffuse in with partial replacement.

exhibits three regions: the proximal regions, which is very sensitive to the details of the interactions; the central regions, which exhibits self-similarity (see below); and the distal region, which is controlled by a few large loops and tails. Scaling analysis shows that the concentration, $\varphi(z)$, in the central region follows the relationship,

$$\varphi(z) = \left(\frac{a}{z}\right)^{4/3} \tag{12.36}$$

where a represents the proximal distance (about the length of a mer), and z is the distance from the wall. The total thickness of the interface is governed by the size of the polymer chains and may be tens of nanometers thick.

The repulsive wall relationships, Figure 12.27*b*, are governed by the polymer chain correlation length, ξ. Because of the decrease in entropy if the chain lies near the wall, the region next to the wall may contain nearly all solvent. This is called the depletion layer; see Section 12.5.5. In certain concentration ranges, the two effects combine to produce the mushroom structures, see Figure 12.21.

Table 12.8   Factors affecting the molecular weight dependence of adsorption

Favoring high molecular weight	Favoring low molecular weight
More adsorption sites per molecule: harder to *unstick* all bound sites simultaneously	Smaller conformational entropy losses on approaching the surface
May bind to multiple particles or surfaces, becoming protected from the exterior	Faster diffusion to the surface
Chemistry/charge distribution/side chains/ polarity often more complex	Less subject to shearing effects such as chain extension and degradation
	Fits into smaller pores

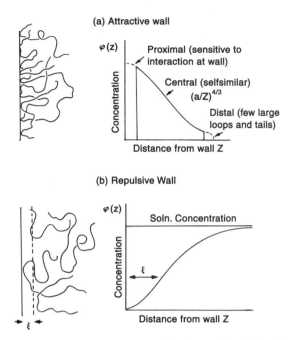

Figure 12.27 Concentration profiles at a wall are governed by the size of the chain and its bonding to the wall.

12.7.1 The Self-similarity of Polymer Chains near Walls

The central region of the polymer layer adsorbed onto an attractive wall may be modeled as a self-similar grid; see Figure 12.28 (75). The mesh size is determined by the overlap distance of the polymer chains, borrowing the concept from an analysis of the semidilute

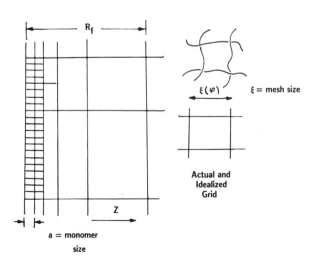

Figure 12.28 A self-similar polymer grid at an interface.

Table 12.9   Factors controlling polymers at interfaces

Thermodynamic	Kinetic
Equilibria	Diffusion of single chains
Competitive adsorption	Evolution of structural rearrangements
Extent of spatial entanglement	Segment-surface contact life
Number of bound segments	Lateral diffusion
Adsorbed layer thickness	Diffusion through adsorbed layers
Patchiness and bare spots	Replacement of one species by another
	Effect of dynamic entanglement

region; see Section 4.2. The mesh size is a decreasing function of the concentration,

$$\xi = a\varphi^{-3/4} \tag{12.37}$$

At a distance z from the wall, the local mesh size equals z, increasing with distance,

$$\xi(\varphi(z)) = z \tag{12.38}$$

There are a number of factors, both thermodynamic and kinetic, which control the concentration at the interface; see Table 12.9 (76). One of the more complicated kinetic aspects involves diffusion of a polymer chain through already adsorbed polymer layers. It is assumed diffusion follows reptation motions. However, the fraction of remaining available bonding sites becomes important when the chain actually reaches the surface.

While the discussion above assumes that each mer of the chain has an equal probability of bonding with the surface, this need not be so. The end groups may be more active, yielding an end-bonded chain. Or the chain may be a block copolymer, where one block is attracted to the wall and the other block is repulsed. Actual chemical bonds may be involved; in this case the surface is described as being grafted.

12.7.2   Fractionation of Polymer Chains near Walls

If the polymer in solution has a broad molecular weight distribution, a question arises as to whether there is any preferential motion of one molecular species or another to the surface. Fleer et al. (2) examined two cases theoretically: a 1000 mer polymer dissolved in its own monomer and a 1000 mer polymer dissolved in a 100 mer polymer; see Figure 12.29 (1). In both cases the longer chains avoided the surface region because of loss of conformational entropy. Hence the high-molecular weight polymer is depleted from near the surface at equilibrium. The thickness of the depletion zone is proportional to $M^{1/2}$, extending several mer-distances into the fluid.

This result is important in several ways: First, the kinetics of adsorption from dilute polymer solutions of high polydispersity polymers is complicated by the depletion zone; second, if a polymerization is carried out in the presence of a nonbonding filler, the first formed polymer may actually form a depletion zone, avoiding the filler in favor of the remaining monomer; and third, in bulk composite materials the shortest chains will tend to be in contact with the filler. Since lower molecular weight materials, in general, have poorer mechanical properties, this may reduce the strength of the composite. The Fleer et al. theory is particularly important because corresponding research on the actual

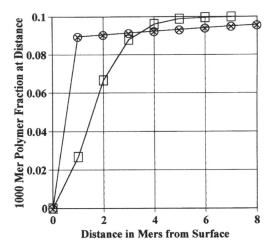

Figure 12.29 Effect of molecular weight on the depletion zone. Ten percent of 1000-mer polymer dissolved in: □, its monomer; ○, in 100-mer polymer. The longer polymer chains tend to avoid a neutral surface.

distribution of polymer chains adjacent to fillers in the bulk case have yet to be carried out. The general conclusion, however, confirms engineering results of long standing: melt blending of fillers with narrower molecular weight distribution polymers produces stronger products.

12.7.3 Fractal-like Interfaces

Fractal geometry was made popular by Mandelbrot (77), with the solution of his famous problem, *How long is the coast of Britain?* The problem revolves around the length of the measurement unit: As the measurement unit becomes shorter and shorter, the length of the coastline seems to increase without limit! The solution lay in the development of fractal dimensions, noting that many real geometry problems may have dimensions lying between the Euclidean 1, 2, and 3 dimensions. For the length of the coastline, the fractal dimension was found to be 1.2618, resulting in a finite coast length, as might be intuitively imagined.

In Section 9.1 the concept of percolation was developed to examine the spatial continuity of polymer networks. The concept can be extended to clusters of molecules or particles in space. Percolation clusters are fractal objects which do not fill space in a dense manner, but fill only a fraction of space. Thus the mass, M, versus distance, R, is given by

$$M \sim R^{D'} \qquad (12.39)$$

where D' represents the fractal dimension, which is typically less than the embedding dimension, d.

As an example, polymer chains in the melt or in Flory θ-solvents form random coils, see Section 5.3. The radius of gyration, R_g, of a random coil depends on its molecular weight, M, to the $\frac{1}{2}$ power. Thus $M \sim R^2$, where the fractal dimension is 2. For self-

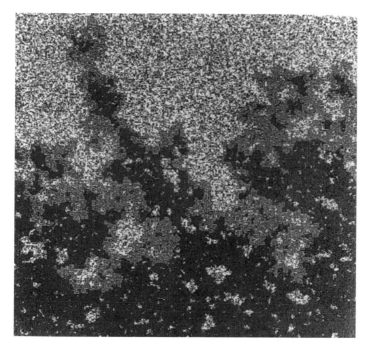

Figure 12.30 Computer generated gradient percolation of atoms with a diffusion length L_d of 10,240 on a 512^2 lattice. The fractal dimension was estimated to be 1.76 (see Appendix 12.1).

avoiding walks, as observed in thermodynamically good solvents, the fractal dimension is somewhat smaller.

Gradient percolation arises when considering diffusion fronts (3), for example, two polymers diffusing across an interface. A diffusion front can be characterized by three quantities: its position, its width, and the total number of particles in the front.

Such a front can also be generated theoretically; see Figure 12.30 (78). The diffusion profile was generated using a Fickian profile together with an error function. At the bottom, $x = 0$, initially the fraction of particles to be diffused was unity. At the top, initially there were other atoms. At any interdiffusion depth x, a site has a probability p of containing an atom of interest. The diffusion front (white) becomes more extensive and ramified with increasing diffusion depth. The atoms connected to the $x = 0$ plane are shown in black. Atom clusters shown in gray are not connected to the source at $x = 0$. If the black atoms were metal, they could conduct electricity across the system. Several black pendent regions are also noted in Figure 12.30. These represent regions connected through the front via isthmus or similar connections. The fractal dimension of the black material in Figure 12.30 was estimated by Wool and Long (78) to be 1.76.

Fractal dimensions can be estimated from peak height and spacing such as provided by atomic force microscopy micrographs, yielding much information about connectivity and organization. As an example calculation, the fractal dimension of the two-dimensional figure shown in Figure 12.30 will be estimated in Appendix 12.1.

In general, the larger the fractal dimension, the more connected is the structure. For example, consider a three-dimensional object containing some disconnected kind of structures. A fractal dimension of 2.8 would mean greater connectivity than 2.5.

Besides diffusion problems, fractal analysis can be applied to such problems as lamination of composites, powder and pellet resin coalescence, latex paint structure, internal weld line shape, and tack of uncured elastomers. In this area, theory is ahead of experiment.

The student should note that there are a number of new theoretical approaches to solving all kinds of problems, including polymer problems. Among others, scaling, self-similarity, percolation, and now fractals have been introduced.

12.8   ADHESION AT INTERFACES

12.8.1   Forms of Adhesives

Adhesives and glues were known to antiquity. Early materials were based on proteins from boiled-down fish, bones, albumin, and so on, making gelatinous water solutions. Many animal glues form bonds stronger than the wood used in furniture construction. However, most such materials were restricted to indoor uses, for the polymers rapidly dissolved in rainwater. Later, rubber dissolved in solvents formed the so-called rubber cement, all still in service.

Adhesives may be used to bond two different structures together to form a composite, as in polymer–polymer welding. The problem of adhesion also extends to the bonding of an engineering polymer directly to a nonpolymer substrate.

In Chapter 8 an adhesive was defined as a linear amorphous polymer above its glass transition temperature. While this definition holds in its simplest form for pressure-sensitive adhesives, many real adhesives are more complex in application. For example, the adhesive may be applied as a monomer, as in epoxy glues and the *instantaneous* adhesives based on cyanoacrylates, such as Krazy Glue® or Super Glue.5®. Similarly an adhesive may be applied as a prepolymer (low-molecular-weight reactive polymeric species), such as is the case for urethane types. Many of these are thermosetting on polymerization. Other types of adhesives include block copolymers, suspensions, and latexes. A common house hold glue, sometimes called "white glue," is an aqueous suspension of poly(ethylene-co-vinyl acetate). Of course, a polymer can be its own adhesive, bonding directly to a structure on one side, while filling another purpose on the other side. Coatings such as house paints serve in this manner. All of these types, however, work by providing either chemical or physical bonding to the material on the other side of the interface.

There are several types of adhesive bonds:

1. Mechanical adhesion is defined as when the adhesive flows around the substrate surface roughness. An interlocking action takes place, like the pieces of a puzzle.
2. Specific adhesion is defined as the case where secondary bonds, such as hydrogen bonds, are formed between the adhesive and the substrate.
3. A special case of specific adhesion is when there are primary chemical bonds present. For example, graft or block copolymers may bond the different phases of a multicomponent material together; see Chapter 13. Direct bonding to substrates is also encouraged in many systems. For example, maleic anhydride comonomers are used to bond to metallic surfaces.

4. Surface interpenetration between two polymers is important, and depends on diffusion rates as well as thermodynamic parameters; see Section 12.3.7.2. If one or both polymers are subsequently cross-linked, the bond strength may be improved.

5. There are various nonspecific forces, such as the van der Waals forces, which contribute to bonding.

Of course, several of these adhesive bonds may be active simultaneously in real systems.

12.8.2 Strength of Adhesion with Cross-linking

The science of adhesion recognizes two types of failure. As already mentioned in Section 12.3.2, *adhesive failure* occurs when the bond between the adhesive and the adherend breaks on stressing. *Cohesive failure* is when the failure takes place within either the substrate or the adherend. Many cases are known where the adhesive bond between two substances is stronger than the substances themselves.

Gent and co-workers (79,80) and Lake and Thomas (81) examined the molecular aspects of both cohesive and adhesive failure in elastomeric networks. One of the most interesting results is that the work of fracture per unit area in the cohesive failure of elastomer, $\mathscr{G}_c$, is about 30 to 100 J/m^2, which compares with only a few J/m^2 expected for the theoretical value of a plane of C—C covalent bonds.

According to Lake and Thomas, this difference has been attributed to the polymeric nature of the molecular chains comprising the network; many bonds must be stressed in order to break any one bond. As defined in Chapter 9, M_c is the molecular weight of an active network chain segment bound on either side by cross-links. The equation $\mathscr{G} \sim M_c^{1/2}$ predicts that lower cross-link densities will produce stronger materials. (This is true as long as each chain has at least two cross-links, thus maintaining a reasonable network structure.) This equation is also useful for calculations involving the molecular basis of fracture in linear plastics, where then M_e is the molecular weight between entanglements; see Sections 9.4 and 11.5.

The adhesive strength of elastomers was examined by Chang and Gent (82). Two partially cross-linked layers of elastomer were pressed together, and the gelation reaction then taken to completion. By decreasing the extent of initial gelation, the degree of chemical bonding could be increased between the two layers. This was true both when the elastomers were identical (82), a special case of cohesive failure, and when the elastomers were different (83), a more obvious case of adhesive failure. This latter is important in automotive tires, where co-cross-linked rubber blends are used; see Section 9.16.2. Gent and co-workers also concluded that the equation $\mathscr{G}_c = KM_c^{1/2}$ was valid, as the extent of cross-linking was varied. This relation has also been found to be valid for a number of cross-linked plastics such as epoxies as well as elastomers. Large values of M_c or M_e contribute to increased viscoelastic motions and increased numbers of bonds stressed on deformation, and hence greater toughness.

However, densely cross-linked plastics (thermosets) are sometimes deliberately used. Reasons tend to be nonmechanical: increased solvent and swelling resistance, high-temperature resistance, and reduced flammability are cited. The last arises because the polymer chars rather than degrades with concomitant reduction of volatile (and flammable) components. Examples include epoxies, phenol-formaldehyde resins, and some polyurethanes; see Section 14.2.

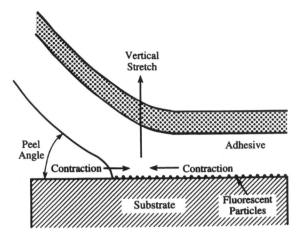

Figure 12.31 Details of a peel test involving an adhesive tape on a substrate. Note the strains and motions of the adhesive, a linear amorphous polymer above T_g. Fluorescent particles helped determine the motions of the adhesive.

12.8.3 Experimental Measurement of Adhesion Forces

The experimental observation of adhesion frequently employs peel and lap shear tests (84,85); see Figure 12.31 (86). The fracture energy, $\mathscr{G}$, is obtained from the peel force, P, and the peel angle, θ through the relation

$$\mathscr{G} = P(1 - \cos \theta) \tag{12.40}$$

When the quantity P is given in N/m (force per unit adhesive width), $\mathscr{G}$ is given in J/m^2. Common values of $\mathscr{G}$ range from 10^2 to 10^4 J/m^2 (87) and depend not only on the angle of peel, but on the temperature and pull rate; see Section 12.8.4. In addition the advancing crack tip may exhibit various wavy patterns, and portions of the adhesive may be pulled up in the form of fibrillar structures and/or undergo slip-stick motions. All of these usually serve to increase the fracture energy.

12.8.4 Pressure-Sensitive Adhesives

Pressure-sensitive adhesives form physical bonds with other materials upon brief contact and with light pressure. Examples include self-stick stamps, packaging tape, double-sided tapes, paper labels, and the ubiquitous Post-it™ notes. Bond formation results from the polymeric material being able to flow under light pressure, thereby establishing good contact area with the substrate. The debonding step involves deformation of the polymeric material under stress (see Section 12.8.3), followed by separation from the substrate. The adhesives most often involve triblock copolymers such as poly(styrene–*block*–isoprene–*block*–styrene) or poly(styrene–*block*–butadiene–*block*–styrene), SBR elastomers, natural rubber, or acrylic copolymers (88).

The acrylic-type pressure sensitive adhesives are often partly cross-linked, partly linear in composition. Theory and data available suggest that the cross-linked component should be only very lightly cross-linked, so that the R_g calculated from M_c of the cross-linked portion is larger than the R_g based on M_e of the linear portion; see Section 10.2 (89,90), especially Figure 10.11. Then the linear polymer can enter the network structure through reptation, developing physical bonds.

Pressure-sensitive adhesives must be very sticky, that is, exhibit high *tack*. The tack of an adhesive usually reaches a maximum in the range of 40 to 70°C above T_g. Frequently the polymer by itself is not sufficiently tacky for commercial purposes. To solve the problem, people dissolve *tackifiers* in the adhesive. A tackifier is a compound that increases the T_g of the material while lowering the modulus. By contrast, plasticizers decrease T_g, as well as lower the modulus. Tackifiers are often based on natural product derived *rosins*, obtained from ground-up pine tree stumps and related materials (91). These rosins are multicyclic steroid-like ring structures.

The dynamic shear modulus of pressure-sensitive adhesives at one second must be less than 3×10^6 dyn/cm^2, known as the Dahlquist criterion (88,92,93). The Dahlquist criterion is best met by employing a tackifier on an adhesive at 40 to 70°C above T_g.

12.8.5 Surface Modification

Surface modification of polymers involves altering the characteristics of the material. A polymer may be given a low surface tension by fluorinating the surface, or a reactive surface may be introduced. For example, the surface may be exposed to plasma or corona treatments. A plasma is defined as a partially ionized gas, with equal number densities of electrons and positive ions (94). A corona involves a controlled discharge of electrons. Depending on the atmosphere and/or ions used, the surface of a polymer may be oxidized or specific moieties placed on the surface. Thus the bondability of polymers to a surface can be altered significantly.

12.9 OVERVIEW OF POLYMER SURFACE AND INTERFACE SCIENCE

No material can come into being without creating a surface or interface! An examination of the literature shows that modern polymer interface science dates only from about 1989, although clearly there are many older papers. In that year several key studies of both theoretical and experimental basis were published. Instruments capable of measuring surface properties down to the atomic scale were available. The roles of nonrandom chain conformation, entanglements between different kinds of chains, interfacial bonding, and the role of compatibilizers, tackifiers, and plasticizers are becoming better understood, although clearly interface science is just beginning, and far from mature.

Figure 12.32 (1) summarizes the state of the art of polymeric and multicomponent polymer materials from the point of view of the role of surfaces and interfaces. The three main types of surfaces and interfaces are (a) free surfaces, (b) polymer blend interfaces, and (c) polymer composite interfaces. While the dilute solution–colloid interfaces are not *free* in the ordinary sense, the fluid phase exhibits a low viscosity allowing rapid diffusion similar in some ways to the free air surface, and is classified as such for the present purposes. The concepts of polymer blends and composites will be further developed in Chapter 13.

Very many extremely useful applications of polymer surface and interface science exist. First, consider applications in the dilute polymer solution–colloid interface field (Sections 12.5–12.7):

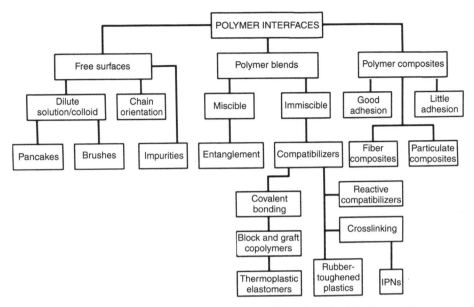

Figure 12.32 Classification of multicomponent polymer materials on the basis of interfaces and composition of the phases.

1. *Enhanced oil recovery.* Underground oil droplets are freed by using water soluble polymers as dispersants.
2. *Lubrication.* Polymer adsorb onto engine parts. Also, they coat and lubricate hard drives in computers, etc.
3. *Wastewater treatment.* Coagulate and precipitate particulates when polymer chains attach themselves to multiple particles.
4. *Papermaking operations.* Coating of inorganic particulates, sometimes known as opacifiers.
5. *Drug delivery.* Facilitates dispersal of colloidal particles bearing drugs.
6. *Nondrip latex paints.* Polymer chains weakly bond to multiple particles; bonds are broken by shear (as described in Section 12.5.1).

An equal number of applications have been developed for bulk interfaces between two polymers:

1. *Polymer molding.* Healing of internal weld lines. For plastics made from pellets, the sintering together of the pellets. For latex paints, the interdiffusion of polymer chains during film formation.
2. *Polymer recycling.* Improving compatibility between immiscible polymers, thus improving mechanical strength.
3. *Microelectronic packaging.* Adhesion and protection of computer and related parts.
4. *Manufacture of thermoplastic composites.* Bonding at the polymer/filler interfaces. For fiber glass reinforced plastics, bonding between the glass and the polymer.

5. *Development and manufacture of rubber-toughened plastics by increased bonding between the phases.*

6. *Automotive tire manufacture.* Optimizing the bonding between the carbon black and the rubber.

Then, there are some general applications, such as studies of porous polymer gels, important in gel permeation chromatography and other types of separations. The preceding lists are obviously incomplete. It is the hope of the author that you, the student, will discover yet unknown applications.

REFERENCES

1. L. H. Sperling, *Polymeric Multicomponent Materials: An Introduction*, Wiley, New York, 1997.

2. G. J. Fleer, M. A. C. Stuart, J. M. H. M. Scheutjens, T. Cosgrove, and B. Vincent, *Polymers at Interfaces*, Chapman and Hall, London, 1993.

3. R. P. Wool, *Polymer Interfaces: Structure and Strength*, Hanser, Munich, 1995.

4. A. Kelly, ed., *Concise Encyclopedia of Composite Materials*, rev. ed., Pergamon, Oxford, England, 1994.

5. G. T. Dee and B. B. Sauer, *J. Colloid Interface Sci.*, **152**, 85 (1992).

6. G. T. Dee and B. B. Sauer, *Trends Polym. Sci. (TRIP)*, **5**, 230 (1997).

7. S. Wu, *Polymer Interface and Adhesion*, Dekker, New York, 1982.

8. A. W. Adamson and A. P. Gast, *Physical Chemistry of Surfaces*, 6th ed., Wiley, New York, 1997.

9. D. G. Legrand and G. L. Gaines, *J. Colloid Interface Sci.*, **31**, 162 (1969).

10. E. Helfand and Y. Tagami, *J. Chem. Phys.*, **56**, 3592 (1972).

11. E. Helfand and Y. Tagami, *J. Polym. Sci.*, **B, 9**, 741 (1971).

12. B. D. Ratner, S. C. Yoon, and N. B. Mateo, *Polymer Surfaces and Interfaces*, W. J. Feast and H. S. Munro, eds., Wiley, Chichester, England, 1987.

13. B. D. Ratner, *Ann. Biomed. Eng.*, **11**, 313 (1983).

14. T. Young, *Phil. Trans. R. Soc. London*, **95**, 65, 84 (1805).

15. G. L. Gaines, *Polym. Eng. Sci.*, **12**, 1 (1972).

16. D. Campbell and J. R. White, *Polymer Characterization: Physical Techniques*, Chapman and Hall, New York, 1989.

17. I. Yilgor and J. E. McGrath, *Adv. Polym. Sci.*, **86**, 1 (1988).

18. D. G. LeGrand and J. G. L. Gaines, *Polym. Prepr.*, **11**, 442 (1970).

19. K. L. Foster and R. P. Wool, *Macromolecules*, **24**, 1397 (1991).

20. J. L. Willett and R. P. Wool, *Macromolecules*, **26**, 5336 (1993).

21. T. P. Russell, A. Menelle, W. A. Hamilton, G. S. Smith, S. K. Satija, and C. A. Majkrzak, *Macromolecules*, **24**, 5721 (1991).

22. T. P. Russell, *Mater. Sci. Rep.*, **5**, 171 (1990).

23. M. L. Fernandez, J. S. Higgins, J. Penfold, R. C. Ward, C. Shackleton, and D. Walsh, *Polymer*, **29**, 1923 (1988).

24. L. C. Sawyer and D. T. Grubb, *Polymer Microscopy*, 2nd ed., Chapman and Hall, London, 1996.

25. G. Binnig and H. Rohrer, *Helv. Phys. Acta*, **55**, 726 (1982).

26. G. Binnig, H. Rohrer, C. Gerber and E. Weibel, *Phys. Rev. Lett.*, **49**, 57 (1982).

27. B. Hoffmann-Millack, C. J. Roberts and W. S. Steer, *J. Appl. Phys.*, **67**, 1749 (1990).

28. C. J. Roberts, M. C. Davies, K. M. Shakesheff, S. J. B. Tendler, and P. M. Williams, *Trends Polym. Sci. (TRIP)*, **4**, 420 (1996).

29. G. Haugstad, *Trends Polym. Sci. (TRIP)*, **3**, 353 (1995).

30. R. M. Overney, *Trends Polym Sci. (TRIP)*, **3**, 359 (1995).

31. G. Haugstad, W. L. Gladfelter, E. B. Weberg, R. T. Weberg, and T. D. Weatherill, *Langmuir*, **10**, 4295 (1994).

32. J. N. Israelachvili, *Surf. Sci. Rep.*, **14**, 109 (1992).

33. R. M. Overney, E. Meyer, J. Frommer, H. H. Guntherodt, M. Fujihira, H. Takano and Y. Gotoh, *Langmuir*, **10**, 1281 (1994).

34. O. L. Shaffer, R. Bagheri, J. Y. Qian, V. Dimonie, R. A. Pearson, and M. S. El-Aasser, *J. Appl. Polym. Sci.*, **58**, 465 (1995).

35. D. Briggs, *Polymer Surfaces and Interfaces*, W. J. Feast and H. S. Munro, eds., Wiley, Chichester, England, 1987.

36. A. Benninghoven, F. G. Rudenaur, and H. W. Werner, eds., *Secondary Ion Mass Spectrometry*, Wiley, New York, 1987.

37. G. Agrawal, R. P. Wool, W. D. Dozier, G. P. Felcher, T. P. Russell, and J. W. Mays, *Macromolecules*, **27**, 4407 (1994).

38. T. P. Russell, V. R. Deline, W. D. Dozier, G. P. Felcher, G. Agrawal, and R. P. Wool, *Nature*, **365**, 235 (1993).

39. R. A. L. Jones, in *Polymer Surfaces and Interfaces II*, W. J. Feast, H. S. Munro and R. W. Richards, eds., Wiley, Chichester, 1993.

40. K. R. Shull, E. J. Kramer, G. Hadziioannou, and W. Tang, *Macromolecules*, **23**, 4780 (1990).

41. R. W. Richards and J. Penfold, *Trends Polym. Sci. (TRIP)*, **2**, 5 (1994).

42. Y. H. Kim and R. P. Wool, *Macromolecules*, **16**, 1115 (1983).

43. K. A. Welp, R. P. Wool, S. K. Satija, S. Pispas, and J. Mays, *Macromolecules*, **31**, 4915 (1998).

44. P. G. de Gennes, *J. Chem. Phys.*, **55**, 572 (1971).

45. M. Doi and S. F. Edwards, *J. Chem. Soc., Faraday Trans. II*, **74**, 1789, 1802, 1818 (1978).

46. K. L. Foster and R. P. Wool, *Macromolecules*, **24**, 1397 (1991).

47. R. P. Wool, *Macromolecules*, **26**, 1564 (1993).

48. P. Debye, J. H. R. Anderson and H. Brumberger, *J. Appl. Phys.*, **28**, 679 (1957).

49. J. H. An and L. H. Sperling, in *Cross-Linked Polymers: Chemistry, Properties, and Applications*, R. A. Dickie, S. S. Labana and R. S. Bauer, eds., American Chemical Society, Washington, DC, 1988.

50. J. H. An, A. M. Fernandez, and L. H. Sperling, *Macromolecules*, **20**, 191 (1987).

51. G. Porod, *Kolloid Zeitschrift*, **124**, 83 (1951).

52. I. W. Hamley, *The Physics of Block Copolymers*, Oxford University Press, Oxford (1998).

53. J. T. Koberstein, B. Morra, and R. S. Stein, *J. Appl. Cryst.*, **13**, 34 (1980).

54. P. Cifra, E. Nies, and F. E. Karasz, *Macromolecules*, **27**, 1166 (1994).

55. T. Kajiyama, K. Tanaka, and A. Takahara, *Macromolecules*, **30**, 280 (1997).

56. A. M. Mayes, *Macromolecules*, **27**, 3114 (1994).

57. M. C. Goh, D. Juhue, O. M. Leung, Y. Wang, and M. A. Winnik, *Langmuir*, **9**, 1319 (1993).

58. J. Reiter, G. Zifferer, and O. F. Olaj, *Macromolecules*, **23**, 224 (1990).

59. T. F. Schaub, G. J. Kellog, and A. M. Mayes, *Macromolecules*, **29**, 3982 (1996).

60. P. G. de Gennes, *Macromolecules*, **13**, 1075 (1980).

61. S. T. Milner, *Science*, **251**, 905 (1991).

62. D. Ausserre, H. Hervet, and F. Rondelez, *Phys. Rev. Lett.*, **54**, 1948 (1985).

63. S. E. Yang, A. Klein, L. H. Sperling, and E. F. Casassa, *Macromolecules*, **23**, 4852 (1990).

64. P. A. Lovell and M. S. El-Aasser, eds., *Emulsion Polymerization and Emulsion Polymers*, Wiley, Chichester, England, 1997.

65. J. N. Israelachvili and D. Tabor, *Proc. R. Soc. Lond.*, **A331**, 19 (1972).

66. J. N. Israelachvili, *Intermolecular and Surface Forces*, 2nd ed., Academic Press, London (1992).

67. J. N. Israelachvili and G. E. Adams, *J. Chem. Soc. Faraday Trans. I*, **74**, 975 (1978).

68. J. N. Israelachvili, *Accounts Chem. Res.*, **20**, 415 (1987).

69. J. Klein and P. F. Luckham, *Nature*, **308**, 836 (1984).

70. Y. Almog and J. Klein, *J. Colloid Interface Sci.*, **106**, 33 (1985).

71. H. J. Taunton, C. Toprakcioglu, L. J. Fetters, and J. Klein, *Macromolecules*, **23**, 571 (1990).

72. M. S. Kelly and M. M. Santore, *Colloids Surfaces A: Physicochem. Eng. Aspects*, **96**, 199 (1995).

73. M. Santore and Z. Fu, *Macromolecules*, **30**, 8516 (1997).

74. Z. Gao and H. D. Ou-Yang, in *Complex Fluids*, E. B. Sirota, D. Weitz, T. Witten, and J. Israelachvili, eds., MRS, Pittsburgh, 1992.

75. P. G. de Gennes, *Adv. Colloid Interface Sci.*, **27**, 189 (1987).

76. Based on M. Santore, Seminar, Lehigh University, April 1991.

77. B. B. Mandelbrot, *The Fractal Geometry of Nature*, Freeman, New York (1983).

78. R. P. Wool and J. M. Long, *Macromolecules*, **26**, 5227 (1993).

79. A. Ahagon and A. G. Thomas, *J. Polym. Sci. Polym. Phys. Ed.*, **13**, 1903 (1975).

80. A. N. Gent and R. H. Tobias, *J. Polym. Sci. Polym. Phys. Ed.*, **22**, 1483 (1984).

81. G. J. Lake and A. G. Thomas, *Proc. R. Soc. Lond. Ser. A.*, **300**, 108 (1967).

82. R. J. Chang and A. N. Gent, *J. Polym. Sci. Polym. Phys. Ed.*, **19**, 1619 (1981).

83. R. J. Chang and A. N. Gent, *J. Polym. Sci. Polym. Phys. Ed.*, **19**, 1635 (1981).

84. F. Garbassi, M. Morra, and E. Occhiello, *Polymer Surfaces: From Physics to Technology*, Wiley, Chichester (1994).

85. M. B. Kaczinski and D. W. Dwight, *J. Adhesion Sci. Technol.*, **7**, 165 (1993).

86. B. M. Z. Newby and M. K. Chaudhury, *Langmuir*, **13**, 1805 (1997).

87. A. M. Gent and G. R. Hamed, *Encyclopedia of Polymer Science and Engineering*, J. I. Kroschwitz, ed., Wiley, New York, 1985.

88. H. W. H. Yang and E. P. Chang, *TRIP (Trends Polym. Sci.)*, **5**, 380 (1997).

89. H. Mohd. Gazahly, The Effect of Crosslinking on Latex Film Formation, Ph.D. Dissertation, Lehigh University, Bethlehem, PA, 1999.

90. S. Tobing, A. Klein, L. H. Sperling, and B. Petrasko, accepted, *J. Appl. Polym. Sci.*, 2000.

91. R. J. Stokes and D. F. Evans, *Fundamentals of Interfacial Engineering*, Wiley-VCH, New York, 1997.

92. C. A. Dahlquist, *Adhesion, Fundamentals and Practice*, Ministry of Technology, eds., Gordon and Breach, New York, 1966.

93. L. Clemens, Personal communication, 1998.

94. J. E. Klemberg-Sapieha, L. Marinu, S. Sapieha, and M. R. Wertheimer, *The Interfacial Interactions in Polymeric Composites*, G. Akovali, ed., Kluwer, Dordrecht, The Netherlands, 1993.

GENERAL READING

A. W. Adamson and A. P. Gast, *Physical Chemistry of Surfaces*, 6th ed., Wiley-Interscience, New York, 1997.

T. L. Barr, *Modern ESCA*, CRC, Boca Raton, FL, 1994.

G. Beamson and D. Briggs, *High Resolution XPS of Organic Polymers*, Wiley, New York, 1992.

D. Briggs, Chap. 9 in *Practical Surface Analysis*, 2nd ed., Wiley, New York, 1990.

G. J. Fleer, M. A. Cohen Stuart, J. M. H. M. Scheutjens, T. Cosgrove, and B. Vincent, *Polymers at Interfaces*, Chapman and Hall, London, 1993.

F. Garbassi, M. Morra, and E. Occhiello, *Polymer Surfaces: From Physics to Technology*, 2nd ed., Wiley, Chichester, 1998.

J. Israelachvili, *Intermolecular and Surface Forces*, 2nd ed., Academic Press, London, 1991.

D. J. Lohse, T. P. Russell, and L. H. Sperling, eds., *Interfacial Aspects of Multicomponent Polymer Materials*, Plenum, New York, 1997.

R. W. Richards and S. K. Peace, *Polymer Surfaces and Interfaces III*, Wiley, Chichester, 1999.

I. C. Sanchez, ed., *Physics of Polymer Surfaces and Interfaces*, Butterworth-Heinemann, Boston, 1992.

I. Skeist, ed., *Handbook of Adhesives*, 3rd ed., Chapman and Hall, New York, 1990.

L. H. Sperling, *Polymeric Multicomponent Materials: An Introduction*, Wiley, New York, 1997.

R. J. Stokes and D. F. Evans, *Fundamentals of Interfacial Engineering*, Wiley-VCH, New York, 1997.

T. L. Vigo and B. J. Kinzig, eds., *Composite Applications: The Role of Matrix, Fiber, and Interface*, VCH, New York, 1992.

R. P. Wool, *Polymer Interfaces: Structure and Strength*, Hanser, Munich, 1995.

S. Wu, *Polymer Interfaces and Adhesion*, Dekker, New York, 1982.

STUDY PROBLEMS

1. Assuming the values in Section 12.5.3, what is the maximum distance above the surface can the loose ends of the chains be, assuming a brush conformation? What is your estimate of a realistic average distance?

2. What is the surface entropy of polychloroprene?

3. Although poly(phenylene sulfide) and poly(ethylene terephthalate) are immiscible, you suspect there is some mixing at the interphase of the blends. In one typed page (double-spaced, 12-point type) or less, write a proposal to your university or college describing how you would measure the extent of mixing, if any. Assume funding, instrumentation, and manpower are unlimited.

4. Look up a 2000 or more recent paper or patent in the field of polymer surfaces and interfaces. How does it advance the field beyond this chapter? Provide the full reference: Authors, journal, volume number, page number, year.

5. What is the theoretical thickness of a SAN/PC interphase at 50°C, where the SAN contains 0.25 mol fraction acrylonitrile?

6. Noting Table 12.10, what force is required to carry out a 90° peel test of a 1 cm strip of O_2 plasma treated LDPE, to separate it from the epoxy?

7. Consider polystyrene and poly(methyl methacrylate) of normal molecular weight, at 140°C.
   (a) How thick is the interface? (Calculate.)
   (b) What is the interfacial tension? (Calculate.)

(c) What is the equilibrium energy to fracture of the interface after cooling to room temperature?

8. Polybutadiene with carboxylic acid chain ends is blended with poly(ethylene terephthalate). Assuming no chemical bonding, what can be said about the location of the polybutadiene chain ends?

9. How does the minor chain reptation model work with reference to the ripple experiment? (Please limit description to 100 words and a drawing.)

10. With your own drawings and calculations, what is your independent estimation of the fractal dimension in Figure 12.30?

11. Propose a new application for any of the materials or subjects discussed in this chapter. *Limit:* One page of double-spaced words, plus one page of illustrations and related material. *Remember:* People have gotten rich by finding new applications for polymers.

12. A new polymer, poly(whatcha micallit) was thought by some not to obey either the de Gennes reptation laws or the scaling laws for short-time intervals. Devise experiment(s) to allow the best analysis possible. What composition, molecular weight ranges and temperatures would you recommend? *Remember:* The *last* Nobel prize has not yet been won! *Length limit:* One page of double-spaced words, and one page of equations, diagrams, and the like.

13. Assuming an adhesive containing a partly crosslinked portion with tetrafunctional cross-linking, show that R_g based on M_c of the cross-linked portion must be at least as large as R_g based on M_e, as discussed in Section 10.2.7.2, in order to have diffusion of the linear chains into the cross-linked portion.

14. Using the block copolymer information in Section 12.3.6.2, compare the interfacial concentration with the critical overlap concentration, Section 4.2.

15. You suspect that a polymer blend has phase separated *via* spinodal decomposition, resulting in cylindrical domains. For an 80/20 blend, a specific surface area of 150 m^2/g was measured. Assuming unit density, what is the diameter of the cylinders?

16. One million molecular weight polybutadiene chains are bound to a surface at one end only, packed in at a density of one chain every 6400 Å^2. What is the minimum bonding energy between the polymer and the surface to hold the chains in place at 25°C? [*Hint:* Use rubber elasticity theory for single chains.]

APPENDIX 12.1 ESTIMATION OF FRACTAL DIMENSIONS

Assume that the white and gray portions of Figure 12.30 can be washed away, leaving a black, insoluble fractal surface. What fractal number characterizes this surface? Approaches to the problem are discussed by Maletsky et al. (A1), Liebovitch (A2), and Avnir (A3).

First, Figure 12.30 was enlarged to 13×14 cm, and three series of horizontal and vertical lines are inserted, dividing the figure into squares, which were subsequently divided into squares one-fourth the size, and then one-fourth the size again. Those squares containing the dividing line between black and another color were counted. The

Table A12.1.1 Fractal analysis of figure 12.30

Linear Cell Size, S (cm)	Log S	Counts, P	Log P
2.0	0.30	18	1.25
1.0	0.0	50	1.70
0.5	−0.30	140	2.15

basic equation is given by

$$P_{box} \propto S^{-D'} \tag{12.41}$$

where P represents the number of nonempty cells (containing the dividing line), and S is the cell size. The results shown in Table A12.1.1 were obtained.

A plot of log P against log S (not shown) yields $D = 1.6$. More precise calculations performed by Wool (78) yield $D = 1.76$. For other types of data, circles, spheres, and the like, can be used instead of the squares of the present case.

REFERENCES

A1. H. O. Peitgen, H. Jurgens, and D. Saupe, *Fractals for the Classroom. Part One: Introduction to Fractals and Chaos*, Springer, New York, 1992.

A2. L. S. Liebovitch, *Fractals and Chaos Simplified for the Life Sciences*, Oxford University Press, New York, 1998.

A3. D. Avnir, ed., *The Fractal Approach to Heterogeneous Chemistry: Surfaces, Colloids, Polymers*, Wiley, Chichester, 1989.

13

MULTICOMPONENT POLYMERIC MATERIALS

Multicomponent polymeric materials consist of polymer blends, composites, or combinations of both. A polymer blend has two definitions: The broad definition includes any finely divided combination of two or more polymers. The narrow definition specifies that there be no chemical bonding between the various polymers making up the blend. Table 2.5 and Section 2.7 summarize the basic types of polymer blends based on the broad definition; primarily these are the block, graft, star, starblock, and AB-cross-linked copolymers (conterminously grafted copolymers), interpenetrating polymer networks, as well as the narrow definition of polymer blends. More complex arrangements of polymer chains in space can be shown to be combinations of these several topologies.

Basically a polymer composite contains a polymer and a nonpolymer. While polymer composites include such compositions as foams and some types of gels, this chapter will be restricted to compositions of one or more polymers and one or more nonpolymers in the bulk state. There are a few points of overlap between blends and composites: polymer-impregnated wood (where wood itself is a natural polymer blend), and organic fiber (e.g., polyester) reinforced plastics constitute examples. Compositions of special interest to this chapter include glass fiber reinforced plastics, carbon black reinforced rubber, and mineral-pigmented coatings.

Since no multicomponent polymeric material can exist without interfaces, the reader's attention is called to Chapter 12. In brief, the nature and extent of bonding between the various phases critically determines the mechanical strength and other properties of the material.

Types of overall properties include mechanical behavior, electrical conductivity, diffusion coefficients, and biomedical compatibility, among many others, not the least of which is price. While this last will not be emphasized in this text, many blends and composites are manufactured because they contain one or more very low price components. If a product can be produced with 90% of the properties with 50% of the cost, that product has a very important advantage in today's world.

13.1 CLASSIFICATION SCHEMES FOR MULTICOMPONENT POLYMERIC MATERIALS

While this brief chapter cannot cover all of the types of polymer blends and composites known in the literature (1–3), Figures 13.1 (4) and 13.2 (5) categorize the more important classes of polymer blends and composites, respectively. Figure 12.30 already summarized the classification of polymer interfaces.

13.1.1 Combinations of Two Kinds of Mers

Figure 13.1 provides a classification scheme for combinations of two kinds of mers. (With only slight modification, three or more kinds of mers can be accommodated.) First are the random (statistical) copolymers and the alternating copolymers, which usually form one phase. While of great interest both theoretically and commercially, they will not be treated further in this chapter. Most of the remaining materials shown in Figure 13.1 normally exhibit phase separation.

The polymer blend materials stem from the narrow definition of a polymer blend: no chemical bonding between the polymers in question. The graft copolymers contain a side chain bonded to a backbone chain. In most such materials the extent of grafting is limited and irregular, but frequently plays important roles in compatibilizing the polymers in question and/or in cross-linking a rubbery phase.

The block copolymers, especially those made via anionic polymerization, tend to be much more regular than the corresponding graft copolymers. The interpenetrating polymer networks are thermoset materials. While nominally not grafted, some grafts often appear. The IPN structures in Figure 4.14 show typical morphologies of sequential IPNs.

The classification scheme illustrated in Figure 13.1 contains many possible connections, not shown. For example, the latex core/shell materials and the latex IPNs differ principally by the presence of crosslinks in the latter, but significant morphological and mechanical differences may also occur. The mechanochemical blends are synthesized when a polymer 1/monomer 2 mix is masticated with sufficient shear to degrade polymer 1 to a certain extent. The various free radicals, anions and cations formed then initiate the polymerization of monomer 2 with various extents with both graft and block copolymer formation.

The mechanochemical blends differ from the solution graft copolymers in that the shear rate in the latter is much lower. However, in the solution graft copolymers, polymer 1 is usually a polydiene such as polybutadiene or polyisoprene, which grafts in connection with the remaining double bonds. The HIPS structure illustrated in Figure 4.14 shows a typical morphology of a solution graft copolymer.

The IENs (interpenetrating elastomer networks) are blends of two kinds of latexes, both of which are pre-cross-linked. The surface modification grafts involve grafting of polymer 2 onto the surface of a polymer 1, often to alter the surface characteristics of the material such as contact angle, polarity, or bondability.

The bicomponent and biconstituent polymer materials refer to fibrous materials. A bicomponent fiber has one part (perhaps one hemisphere, in cross section) consisting of one polymer, the other part of the fiber consisting of the other polymer. A biconstituent fiber has little fibrils of one polymer dispersed in the other polymer. The bicomponent fibers are often used for rugs, since a difference in expansion coefficients on cooling from the melt causes a curlicue, springy structure. A higher glass transition of the dispersed

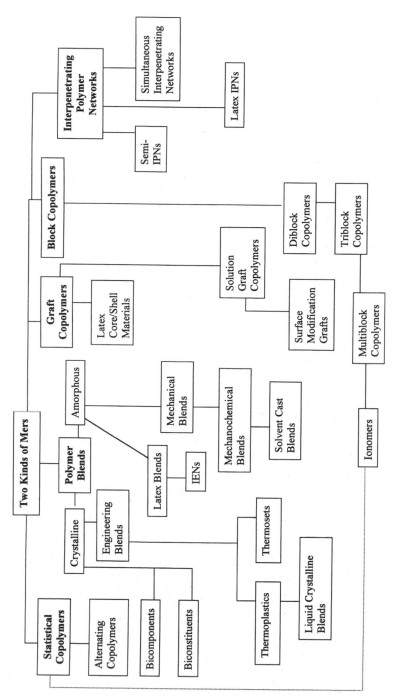

Figure 13.1 Classification scheme for multicomponent polymer materials.

fibrils in a biconstituent material usually raises the overall modulus of the fiber, stiffening the product particularly at higher temperatures.

13.1.2 Polymer-Based Composites

The two major classes of composites, Figure 13.2, differ in which phase is continuous: the polymer or the nonpolymer. In the filled porous materials, ceramic, concrete, or wood (considered here as a nonpolymer) are usually filled with monomers under pressure, followed by polymerization. The resulting materials are tougher, or wear longer, and in the case of polymer-filled concrete are also more freeze–thaw resistant.

In reinforced plastics various inorganic materials are dispersed in the polymer. Carbon black reinforced elastomers have already been considered; see Section 9.16.2. For fiber composites, two subtypes are important, the short fiber-containing materials, which are thermoplastic, and the continuous filament types, which cannot flow. While short fibers can be melt blended with thermoplastics, they are often embedded in monomeric mixes, followed by polymerization in situ. Continuous fibers are always processed via monomeric mixes which can flow over the beds of fibers. Of course, these monomeric mixes may have polymers or prepolymers dissolved in them, raising the viscosity, and reducing shrinkage on polymerization. An example of the continuous filament type is a tape composite, familiar as the strapping tape used for packaging.

Strapping tape itself is a very complex material. Continuous glass fibers run though a plastic phase, providing high strength in the long direction. Adhered onto the plastic phase is a pressure-sensitive adhesive; see Section 12.8.4.

The macroscopic composites are those in which one or both phases tend to have structures large enough to be easily seen with the naked eye. These include paint films (paint itself is usually a particulate composite), adhesive joints, and foams (e.g., the familiar polystyrene foams used for coffee cups because of their low heat transfer coefficient). The fabric laminates alternate layers of inorganic cloth and plastic. An important subclass usually has multiple layers of glass cloth or chopped fiber interspersed with layers of unsaturated polyester-styrene AB-cross-linked materials; see Section 2.7.3. These sheet molding compounds are widely used for the hulls of pleasure boats, because they are light weight and very puncture resistant, as well as for automotive bodies; see Section 13.9.1.

13.2 MISCIBLE AND IMMISCIBLE POLYMER PAIRS

Most polymer pairs are immiscible; that is, mixtures of the two remain phase separated. This is caused by the (usually) positive heat of mixing, and very small entropy of mixing, as developed in Section 4.3. A list of selected immiscible polymer pairs is shown in Table 13.1 (3). A corresponding list of miscible polymer pairs is shown in Table 13.2 (3).

Most of the polymer pairs in both tables are commercial materials. For example, in Table 13.1 the addition of several percent of polybutadiene to polystyrene results in high-impact strength polystyrene, HIPS. Similarly the addition of CTBN to epoxy materials results in a much tougher product. Blends of *cis*-polybutadiene and SBR are widely used in automotive tires. The polybutadiene, while more expensive, crystallizes on extension, providing reinforcement when it is needed most (a smart polymer!).

Polystyrene is miscible with poly(2,6-dimethylphenylene oxide), PPO. The commercial composition actually uses HIPS blended with PPO, again resulting in a very tough

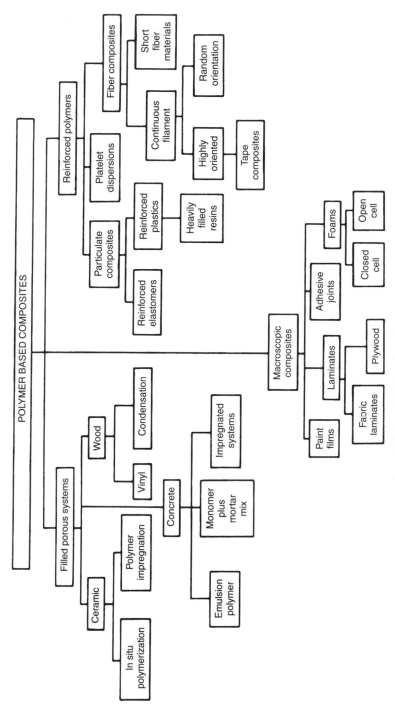

Figure 13.2 Classification scheme for polymer composites.

Table 13.1 Selected pairs of immiscible polymers

Polymer I	Polymer II
Polystyrene	Polybutadiene
Polystyrene	Poly(methyl methacrylate)
Polystyrene	Poly(dimethyl siloxane)
Polypropylene	EPDM
Epoxies	CTBN
Polycarbonate	ABS
Polyamide 6,6	EPDM
Polyamide 6	Poly(ethylene terephthalate)
Polybutadiene	SBR

Abbreviations: EPDM, ethylene-propylene-diene monomer (a copolymer elastomer); CTBN, carboxy-terminated butadiene nitrile (an elastomer prepolymer); ABS, acrylonitrile-butadiene-styrene (a complex latex structure used both independently and to toughen other polymers); SBR, styrene-butadiene rubber (an elastomer).

product. A blend of poly(vinylidene fluoride) with poly(methyl methacrylate) imparts both light stability and gasoline resistance, making it useful for coating automotive parts.

While polymer blends are used for a wide variety of purposes, those related to toughness or impact resistance constitute the largest tonnage. Similar results are seen with composites: note the effect of carbon black on rubber, and glass fibers on plastics. However, specific applications depend on physical properties and morphology, treated next.

13.3 THE GLASS TRANSITION BEHAVIOR OF MULTICOMPONENT POLYMER MATERIALS

13.3.1 The Glass Transitions of Polymer Blends

Nearly every polymer exhibits a glass transition temperature. When two polymers are blended together, broad definition, one of several different events may happen; see Figure 13.3 (3). First of all, the polymers may be miscible and one-phased. In this case, the Fox equation (8.78) can be used to estimate the glass transition of the blend. The opposite extreme is that the two polymers are totally immiscible. Then the two polymers retain their original glass transition temperatures.

Table 13.2 Selected pairs of miscible polymers

Polymer I	Polymer II
Polystyrene	Poly(2,6-dimethylphenylene oxide)
Polystyrene	Poly(vinyl methyl ether)
Poly(vinyl chloride)	Poly(butylene terephthalate)
Poly(methyl methacrylate)	Poly(vinylidene fluoride)
Poly(ethylene oxide)	Poly(acrylic acid)

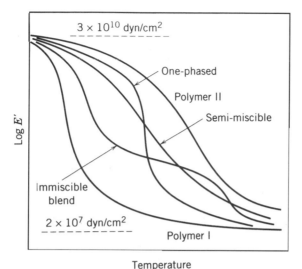

Figure 13.3 Possible glass transition behavior of polymer blends. The two polymers may be immiscible, showing two distinct glass transitions; miscible, with one glass transition defined by the Fox equation; or semi-miscible, usually with one very broad glass transition.

Actually one could argue successfully that some molecular mixing occurs in *all* mixtures of two or more chemicals, inorganic, organic, or polymeric. However, for practical purposes, glass is totally immiscible with organic polymers, and many polymer pairs exhibit very little mixing, and the like.

A more complex case arises when there is partial mixing; see Figure 13.3. There might be two glass transitions, each moved inward; see Section 8.8.2. There might be microheterogeneous phase structures formed, where the composition varies from location to location. This condition, generally nearly miscible, is sometimes called semimiscible. Clearly, a new term is required to express this exact and important condition. The Fox equation can be used within each phase in phase separated systems, but now more information is required.

13.3.2 An Example Calculation of Polymer Blend Glass Transition Temperatures

Assume a blend of an epoxy polymer with an acrylate polymer, 60/40 w/w overall composition (3). The loss moduli peaks of both the pure components and the blend are shown in Table 13.3.

Figure 13.4 illustrates the physical situation. To be calculated are the questions: (a) What is the composition of each phase? (b) What is the volume fraction of each phase?

What may be assumed is as follows:

1. The system is in thermodynamic equilibrium.
2. The kinetics of polymer chain interdiffusion are fast relative to the time span of setting up the experiment.

Table 13.3 Model data for glass transition temperature calculations

Composition	T_g, °C	T_g, K
Pure acrylate	−40	233
Acrylate rich blend phase	−10	263
Pure epoxy	120	393
Epoxy rich phase	95	368

3. The phase composition temperature dependence is small between the glass transition temperatures and the temperature of interest (room temperature), or changes in composition are slow relative to the time required for temperature changes during the experiment.

4. That both phases are homogeneous on a molecular scale, allowing the Fox equation to be applied to each phase separately.

The phase diagram, emphasizing a lower critical solution temperature (true for most blends), is shown in Figure 13.5. For simplicity, assume that the densities of the two polymers and their blends are all equal. Note the relationship between weight fractions w_1 and w_2 for the composition within each phase.

For the epoxy-rich phase, the Fox equation yields

$$\frac{1}{368} = \frac{w_1}{233} + \frac{1 - w_1}{393}$$

noting that $w_1 + w_2 = 1$. On calculation,

$$w_1 = 0.0989 \text{ acrylate}$$

$$w_2 = 0.901 \text{ epoxy}$$

For the acrylate-rich phase,

$$\frac{1}{263} = \frac{w_1}{233} + \frac{1 - w_1}{393}$$

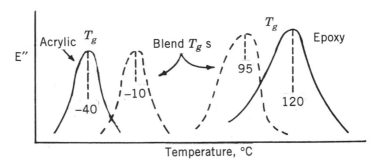

Figure 13.4 When two polymers are slightly miscible, their glass transitions are shifted inward.

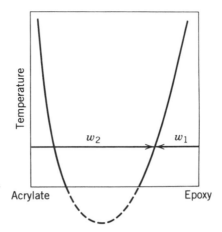

Figure 13.5 A lower critical solution phase diagram, lowest portion often not accessible, illustrating a calculation method for determining the compositions of the phases.

(Note that the two equations are identical except for the composition-dependent glass transition temperature on the left.) Then

$$w_1 = 0.72 \text{ acrylate}$$
$$w_2 = 0.28 \text{ epoxy}$$

thus providing the compositions of the two phases.

For the volume fraction of the acrylate-rich phase, X is assumed to be the acrylate phase volume. Since the entire blend was 40% acrylate,

$$0.72X + 0.0989(1 - X) = 0.40$$

yielding $X = 0.48$. The volume fraction Y is obtained similarly for the epoxy-rich phase:

$$0.901Y + 0.28(1 - Y) = 0.60$$

with $Y = 0.52$. Note that $X + Y = 1$. Thus both the compositions within each phase and the volume fraction of each phase can be easily estimated.

13.3.3 Glass Transitions of Polymer Composites

The glass transition of a polymer in the vicinity of a composite interface depends on the extent of mixing; see Chapter 12. If the polymer is attracted to or bonded to the filler interface, then such bonding may reduce the free volume of the polymer in the vicinity of the filler, thus raising its glass transition temperature. Frequently the glass transition appears to be broadened on the high side for carbon black reinforced elastomers.

For example, Kraus and Gruver (6) found such effects studying thermal expansion, free volume, and molecular mobility of HAF carbon black reinforced styrene-butadiene rubber (SBR); see Section 9.16.2. The SBR elastomer bonds to carbon black through a

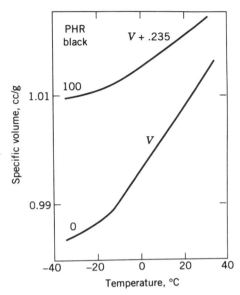

Figure 13.6 On addition of carbon black to SBR rubber, the glass transition is shifted upward slightly and broadened on the high side.

variety of secondary bonding modes. Thus, the train portion of the polymer chains in contact with the carbon black is relatively large and held to the surface rather firmly through multiple secondary bonds.

Kraus and Gruver chose dilatometry as their main instrument; see Figure 13.6 (6). Here, the glass transition appears as an increase in the expansion coefficient at the elbows in the data. The width of the glass transition region increased from 9 to 20°C for the elastomer, principally on the upper side of the transition.

The analysis of the data is shown in Table 13.4. Here, α_g and α_r represent the expansion coefficients in the glassy and rubbery regions, respectively. The calculated

Table 13.4   Interfacial properties of SBR at carbon black interfaces

Property	Neat	100 phr HAF
A. Basic properties		
T_g, °C	−11.4	−9.3
$\alpha_g \times 10^4$ cm^3/g-°C	2.10	1.35
$\alpha_r \times 10^4$ cm^3/g-°C	5.70	5.56
Surface area, m^2/g	—	80
Width of T_g, °C	9	20
B. Derived properties		
Thickness of restricted segmental motion, nm	—	3
Increase in T_g in restricted region, °C	—	10
Fraction of polymer restricted	—	0.25

Source: Based on G. Kraus and J. T. Gruver, *J. Polym. Sci.*, **A-2**(8), 571 (1970).

width of the restricted segmental motion region was 3 nm, based on the known HAF surface area and volume of rubber. This is the interphase thickness, in fact. Thus, for 100 phr (parts per hundred), the fraction of polymer with restricted motion amounted to approximately one-fourth.

13.4 THE MODULUS OF MULTICOMPONENT POLYMERIC MATERIALS

Qualitatively, the modulus of polymer blends and composites is expected to be intermediate between the modulus of the materials involved. Quantitatively, the picture is more complex, depending on phase morphology and continuity. The Takayanagi models (Section 10.1.2.3) explore the basic methods of calculating not only the modulus, but many other viscoelastic quantities. While the original models were developed with glassy and rubbery polymers in mind, they are quite general and useful for composite systems as well.

13.4.1 Moduli of Particulate Composites

Particles that are dispersed in plastics are usually harder than the plastic itself, thus the modulus of the particulate composite is increased; see Figure 13.7 (7). The increase in modulus depends not only on the modulus and concentration of the filler, as modeled by Takayanagi, but also on the particle shape. For example, in Figure 13.7, the modulus of 20% mica is higher than 20% calcium carbonate, because the mica forms platelets while the calcium carbonate forms more or less isodimensional structures. The 20% asbestos modulus is still higher, being fibrillar in shape. Thus, all other aspects being equal, the modulus increases with the shape factor of dispersed particles as follows: E(spheroidal) < E(planar) < E(fibrillar). The reason lies in the increase in spatial continuity (higher fractal dimension), even though the structure itself forms a discontinuous phase. Another factor relates to the specific surface area of the material.

In Figure 13.7 the glass transition of the polymer appears to increase because at the same modulus, such as $G = 6 \times 10^9$ dyn/cm^2, the data are shifted to higher temperatures. However, the T_g analysis is better made at the maximum rate of down-turn of the modulus, where the glass transitions are seen to be closer to the homopolymer transition, but still apparently slightly higher. This may be due to polymer filler bonding, with concomitant local reduction in free volume.

13.4.2 Example Calculation of Composite Moduli

Today, thermosets such as epoxy resins are frequently reinforced with either glass fibers or carbon fibers (also known as graphite fibers) (8,9). As an example calculation, the modulus of a glass fiber–epoxy composite will be calculated in both the longitudinal and transverse directions. The properties of the materials are

Young's modulus, epoxy	4.0 GPa
Young's modulus, glass fiber	80 GPa
Fiber volume fraction	62%

First, assume the longitudinal direction. In response to a unidirectional stress in the longitudinal direction, both the fibers and the matrix are continuous. The Takayanagi

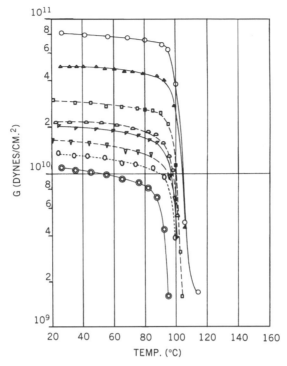

Figure 13.7 A series of filled polystyrenes, illustrating the increases in glassy shear modulus, and apparent or slight increases in the glass transition temperature. Materials: ◎, control; ▽, 20% mica; △, 40% mica; ○, 20% calcium carbonate; ◰, 40% asbestos; ○, 60% mica; ▷, 20% asbestos; □, 60% asbestos.

model shown in Figure 10.6a will be assumed, with λ equal to 0.62. The basic relation is given by equation (10.6). Then

$$E = 0.38 \times 4 + 0.62 \times 80 \ (\text{GPa})$$

$$E = 51.1 \ \text{GPa}$$

Thus, to a first approximation, the modulus is given by the modulus of the fiber times its volume fraction. However, in the transverse direction the fibers are discontinuous, while the matrix retains its continuity. The Takayanagi model Figure 10.6b and equation (10.8) hold. Then

$$E = \left(\frac{0.62}{80} + \frac{0.38}{4} \right)^{-1} \ \text{GPa}$$

$$E = 9.73 \ \text{GPa}$$

This value is much lower than the longitudinal values. The experimental values (8), of the Takayanagi model, and computer-calculated values (9) are compared in Table 13.5.

Thus the Takayanagi longitudinal modulus prediction is seen to be nearly correct, but the Takayanagi transverse modulus prediction is seen to be low. However, the Takaya-

Table 13.5 Estimation of glass fiber–epoxy Young's moduli, GPa

Property	Experimental	Takayanagi Model	Computer Simulation
Longitudinal moduli	53.48	51.1	53.37
Transverse moduli	17.7	9.73	17.57

nagi model is intended for use when the discontinuous phase portions are separated. At 62% fiber, the fibers have to be almost or actually touching in many places.

The computer simulation (9) employed a program called ICAN (Integrated Composite Analyzer). While it can calculate simple moduli as above, its primary use is to estimate fracture properties, not only of unidirectional composites as above, but for various laminates; see Figure 13.8 (9). Cross-ply laminates have the advantage of having good mechanical behavior both in the *x*- and *y*-directions, providing a balance of high modulus and high tensile strength.

13.4.3 Other Mathematical Relationships for Modulus Calculations

The literature abounds with relationships to calculate the modulus of polymer blends and composites, besides the Takayanagi equations. While most of them are older, they remain easy to apply and still enjoy wide use both for scientific and engineering applications, providing insight as to the organization of multiphase matter. Three such relationships will be provided below.

13.4.3.1 The Kerner Equation Perhaps the most widely used relationship is the Kerner equation (10):

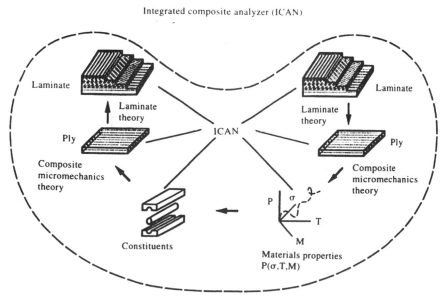

Figure 13.8 Illustration of how the ICAN computer program analyzes a laminate composite for mechanical behavior.

$$\frac{G_c}{G_p} = \frac{G_f v_f / [(7 - 5v)G_p + (8 - 10v)G_f] + v_p / [15(1 - v)]}{G_p v_f / [(7 - 5v)G_p + (8 - 10v)G_f] + v_p / [15(1 - v)]} \tag{13.1}$$

where G_c represents the shear modulus of the composite, G_p and G_f are the shear moduli of the polymer and the filler, the continuous and discontinuous phases, respectively, v is Poisson's ratio of the polymer, v_f and v_p represent the volume fraction of the filler and polymer, respectively. The equation works best when the polymer is glassy or semicrystalline, and the filler is spheroidal in shape. For most plastics Poisson's ratio is in the range of 0.30 to 0.35; see Section 8.1.3. The Kerner equation predicts a lower bound increase in the modulus, that is the smallest increase possible, assuming that the filler has a higher modulus than the polymer. If Young's modulus is desired, equation (8.10) can be applied.

13.4.3.2 The Guth–Smallwood Equation

The increase in modulus in particulate reinforced elastomers, and particularly those containing carbon black or silica (11,12), can be predicted by the equation

$$\frac{G_c}{G_p} = 1 + 2.5v_f + 14.1v_f{}^2 \tag{13.2}$$

The coefficient 2.5 of the first term on the right arises from the Einstein viscosity equation (see Section 3.8.2), the origin of the Guth–Smallwood equation. The second term on the right is important for more concentrated dispersions of the fillers, taking into account their interaction. This equation also assumes the filler is dispersed in the polymer, and also predicts a lower bound increase in the modulus.

13.4.3.3 The Davies Equation

Davies (13) derived an equation useful for dual phase continuity in blends of plastics and elastomers:

$$G_c{}^{1/5} = v_1 G_1{}^{1/5} + v_2 G_2{}^{1/5} \tag{13.3}$$

where G_1 and G_2 represent the shear moduli of the two polymers. Equation (13.3) predicts the shape of the curve as the concentration of the blend is changed from all elastomer to all plastic, being more accurate on the plastic end. Often experimental data have an S-shaped curve as the system undergoes phase inversion. These and other equations have been recently summarized (3).

13.4.3.4 Example Calculations with the Davies Equation

A blend of 30% rubber, $G = 3 \times 10^5$ Pa, and 70% of a plastic, $G = 1 \times 10^9$ Pa, is prepared. What shear modulus is predicted for the blend? Applying the Davies equation,

$$G_c = \{0.30 \times (3 \times 10^5)^{1/5} + 0.70 \times (1 \times 10^9)^{1/5}\}^5$$

$$G_c = 2.5 \times 10^8 \text{ Pa}$$

The value predicted is lower than that predicted by the Takayanagi discontinuous rubber phase model, 7×10^8 Pa. Young's modulus can be obtained from the Davies equation by the approximation $E = 3G$, or $E_c = 7.5 \times 10^8$ Pa.

13.5   THE MORPHOLOGY OF MULTIPHASE POLYMERIC MATERIALS

The morphology of polymer blends and composites have been widely explored by both X-ray and neutron scattering techniques, and by electron microscopy. Figure 4.17 illustrated the morphology of block copolymers, in this case with planar oriented cylinders. A series of graft copolymers and IPNs morphologies was shown in Figure 4.15, with varying degrees of phase continuity.

Graft copolymers such as HIPS are only slightly grafted; however, sufficient polystyrene is grafted from polybutadiene chain to polybutadiene chain to produce an AB-cross-linked copolymer in the rubber-rich domains. This grafting tends to lower the interfacial tension, and bonds the two phases together. Commercially this material is produced by dissolving 5% to 10% of polybutadiene in styrene monomer, and polymerized *with stirring*. At first, the polybutadiene is soluble in the styrene-polystyrene polymerizing mix. However, after several percent of conversion, the two polymers phase separate, each phase highly swollen with styrene monomer. Next, the stirring induces a *phase inversion*, that is, the earlier continuous rubber-rich phase becomes discontinuous, and vice versa. Another reason for stirring is heat dissipation.

On continued polymerization, the styrene in the polystyrene-rich phase just produces more polystyrene, but the styrene in the rubber-rich phase produces occluded cylindrical or spherical cellular domains; see Figure 13.9. Note the phase-within-a-phase-within-phase morphology.

This complex morphology results primarily from spinodal decomposition within the rubber domains, although nucleation and growth kinetics are sometimes important. As described in Section 4.3.5, nucleation and growth results in spheroidal domains, while spinodal decomposition often results in interconnected cylinders. The actual morphology is not always obvious in thin section transmission electron microscopy.

The rubber cellular domain structures are sometimes called *salami structures*, after their appearance. The toughness obtained in such materials is related to the *rubber phase volume*, which is the rubber volume plus the occluded polystyrene cellular domain volume.

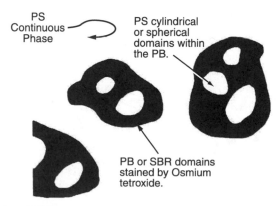

PS Continuous Phase

PS cylindrical or spherical domains within the PB.

PB or SBR domains stained by Osmium tetroxide.

Figure 13.9 Illustration of HIPS morphology. Note that this is a phase-within-a-phase-within-a-phase organization, with polystyrene being the continuous phase.

13.5.1 Phase Inversion

Phase inversion itself usually requires some type of shearing action on the polymers, or else metastable morphologies may result. During a polymerization the timing of the phase inversion is controlled by the volume fraction and the viscosity of each phase (14,15),

$$\frac{V_1}{V_2} \cdot \frac{\eta_2}{\eta_1} = X \tag{13.4}$$

where if $X > 1$, phase 1 is continuous, and if $X < 1$, phase 2 is continuous, and for $X \cong 1$, dual phase continuity exists, or a phase inversion may be in progress. In equation (13.4), V represents the volume fraction, and η the viscosity, with phases indicated by the subscripts. Equation (13.4) holds in the limiting case of zero shear. Utracki (16) has developed somewhat more complex relationships for finite shear rates.

Figure 13.10 (15) illustrates the regions of phase continuity expressed by equation (13.4), together with some supporting experimental data. Phase inversion results from passage across the dual phase continuity curve. Stirring or shearing is often a subsidiary requirement for phase inversion, because diffusion across phases of limited miscibility is very slow.

The path of the ratio of the volume fractions to the inverse viscosity ratios in equation (13.4) may be tortuous in the polymerization of monomer 2 in the presence of polymer 1. As the polymerization of monomer 2 proceeds, the viscosity of both phases increases but not necessarily at the same rate. However, the volume fraction of the polymer-2 rich phase increases at the expense of the polymer-1 rich phase.

Consider another case for the application of equation (13.4), this one involving the crosslinking of a polymer, which increases its viscosity. Blends of isotactic polypropylene

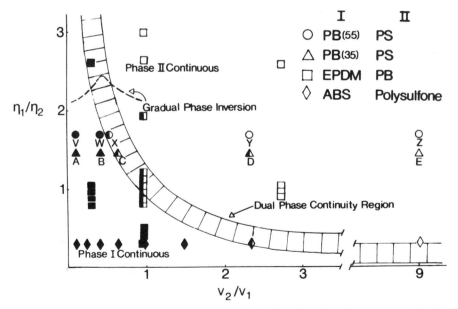

Figure 13.10 Phase continuity and inversion diagram. Filled points, phase I continuous, open points, phase II continuous, and half-filled points, dual phase continuity.

and EPDM are widely used in energy-absorbing applications such as automotive bumpers. (EPDM is an elastomer made from ethylene, propylene, and a crosslinking monomer.) The *i*-PP and the EPDM are blended under shear, while the EPDM undergoes the cross-linking. During crosslinking, the viscosity of the EPDM-rich phase increases, leading to phase inversion if carried to completion. However, the most desired final morphology involves is dual phase continuity, where long cylinders of partly crosslinked EPDM traverse the *i*-PP material.

13.5.2 Example Calculation of Phase Inversion

A batch of 70/30 *i*-PP and EPDM is blended. The viscosity of the *i*-PP is 5×10^4 Pa at the blending temperature of 180°C. The initial viscosity of the EPDM at 180°C is 2×10^3 Pa, and increases 1×10^3 Pa every minute after the peroxide initiator is added. How long should the reaction be continued in the blender to obtain dual phase continuity?

The quantity X in equation (13.4) should equal unity. Assume that phase 1 represents the *i*-PP. Since the overall composition is 70/30,

$$\left(\frac{70}{30}\right) \times \left(\frac{\eta_2}{5 \times 10^4} \text{ Pa}\right) = 1$$

The final value of η_2 should be 2.1×10^4 Pa. At the reaction rate indicated, the cross-linking should be stopped after 11 minutes. Usually it is extruded directly into the application mold. Clearly, any such actual study should be accompanied by microscopy to follow the development of the morphology.

13.5.3 Morphology of ABS Plastics

ABS (acrylonitrile-butadiene-styrene) plastics are actually a type of partially grafted copolymer, similar to HIPS but more oil resistant because of the polar acrylonitrile, and with significantly higher impact resistance. It can be made by several methods, through emulsion polymerization, suspension polymerization, or bulk polymerization, but the most important method utilizes emulsion polymerization. In this case a *seed* latex of cross-linked polybutadiene is made, which constitutes up the *core* of the latex. This is followed by the addition of a mix of styrene and acrylonitrile monomers, usually 72/28 or similar in weight, respectively, followed by continued polymerization to form the *shell* of the latex particle.

Figure 13.11 (17) shows that the osmium tetroxide-stained polybutadiene has occluded SAN inside the core latex portion, with SAN also forming the shell. On film formation, the SAN shell component makes up the continuous phase. A straight SAN latex may be added to increase the separation of the polybutadiene rubber domains. The rubber portion must be cross-linked to minimize morphological damage to the core during processing.

13.5.4 Phase Domain Size in Sequential IPNs

In the case of sequential IPNs, the phase domain size of the second polymerized polymer is governed primarily by the cross-link density of the first polymerized polymer and the overall composition. The disperse phase diameter of the second polymerized polymer,

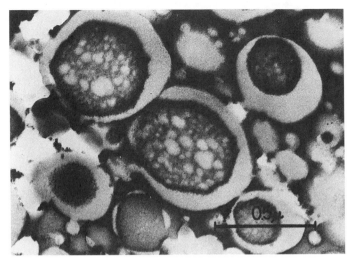

Figure 13.11 Morphology of a ABS latex particle, osmium tetroxide stained. Note the core–shell structure, with AS copolymer constituting the shell, and also AS domains inside of the polybutadiene rubber core.

D_2, assuming spheres may be written (18)

$$D_2 = \frac{4\gamma}{RT(An_1 + Bn_2)} \tag{13.5}$$

where

$$A = \frac{3v_1^{1/3} - 3v_1^{4/3} - v_1 \ln v_1}{2v_2} \tag{13.6}$$

and

$$B = \frac{\ln v_2 - 3v_2^{2/3} + 3}{2} \tag{13.7}$$

where γ represents the interfacial tension, RT the gas constant times the absolute temperature, n_1 and n_2 are the concentration of the effective network chains of polymers I and II per unit volume as a measure of crosslink density (see Chapter 9), respectively, and v_1 and v_2 represent the volume fractions of polymers I and II, respectively.

13.6 PHASE DIAGRAMS IN POLYMER BLENDS (BROAD DEFINITION)

Section 4.3 delineated some of the basic notions of polymer blend phase diagrams. Principally due to the very small entropy of mixing and usually positive heats of mixing, two polymers will be immiscible unless some strong interaction such as hydrogen bonding exists between them. Since the number of mer-sized holes in a polymer depends on the interactions between the attractive forces and molecular motion of the polymer chains, lower critical solution temperatures are usually observed.

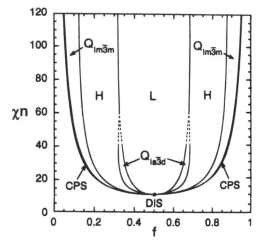

Figure 13.12 Mean-field phase diagram for conformationally symmetric diblock copolymer melts. Phases are labeled L, lamellar; H, hexagonal cylinders; Q_{Ia3d}, bicontinuous Ia3d cubic; Q_{Im3m}, bcc spheres; CPS, close-packed spheres; and DIS, disordered region. The dashed lines denote extrapolated phase boundaries, and the dot denotes the mean-field critical ODT.

13.6.1 The Order–Disorder Transition in Block Copolymers

Classical composition–temperature phase diagrams cannot, however, be obtained for block copolymers because the chain segments (blocks) are attached to each other. Hence the volume fraction and the molecular weights of the blocks cannot be independently varied. However, people use the term *order-disorder transition* (ODT), sometimes called the *microphase separation transition* (MST) to describe the critical phase separation conditions.

The theoretical ODT is shown in Figure 13.12 (19), including crystallographic space group designations. The location of the order–disorder phase boundary can be calculated using mean-field theories. Simple mean-field theory enables the calculation of the size and number of chains in a micelle and its free energy of formation. The domain shape with the lowest free energy of formation is the one predicted for equilibrium conditions.

Leibler (20) found that $\chi_{12}n = 10.5$, where n represents the copolymer total degree of polymerization, and χ_{12} is the Flory interaction parameter between the two polymeric blocks to be the critical condition for phase separation. At smaller values of $\chi_{12}n$, miscible materials are predicted. Thus the y axis increases with total molecular weight, assuming that χ_{12} remains constant. For Figure 13.12 the two kinds of mers are assumed symmetrical in size and shape, resulting in the minimum in χ_{12} appearing at $f = 0.5$. {Incidentally, $\chi_{12}n = 2$ for the critical point of polymer blends of equal mer lengths, where n represents the combined degrees of polymerization, as above (21).}

In a more detailed calculation the self-consistent mean field theory reduces the problem of calculating the interactions among polymer chains to that of a single non-interacting polymer chain placed in an external field self-consistent with the composition profiles (22). Again, the primary objective is to compute the free energy and polymer distribution functions near the order–disorder transition.

Originally three main phase morphologies were known: spheres, cylinders, and alternating lamellae; see Section 4.3.9. More recently several other morphologies were discovered for block copolymers: *close-packed spheres, ordered bicontinuous double diamond* (OBDD) (23,24) which occurs in star block copolymers [but recently questioned (25)], *gyroid* phases, *hexagonally perforated layers* (HPL), now thought not to be thermodynamically stable, but rather a long-lived transient structure (26), and a host of exotic phase structures for ABC block copolymers bearing three or more chemically distinct blocks (27,28); see below. Figure 13.12 illustrates the location of the close-packed spheres, spherical, cylindrical, gyroid, and lamellar phases, based on self-consistent mean-field theory (19,29). Note the location of the order–disorder phase boundary at $\chi_{12}n = 10.5$. The diagram is perfectly symmetrical because the mers were presumed to have identical sizes and shapes.

Experimental phase diagrams for amorphous block copolymers were explored by Khandpur and co-workers (25). First, low-frequency isochronal shear modulus–temperature curves were developed on a series of polyisoprene–*block*–polystyrene polymers to guide the selection of temperatures for the transmission electron microscopy and SAXS experiments to follow; see Figure 13.13 (25). Both order–order (OOT) and ODT

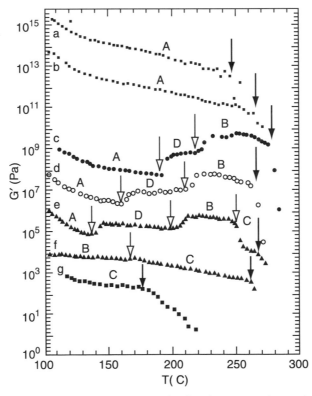

Figure 13.13 Dynamic storage shear modulus as a function of temperature for a series of polyisoprene–polystyrene diblock copolymers. Open and filled arrows indicate order–order and order–disorder phase transitions, respectively. The *y*-axis data have been multiplied for clarity; sample *g* has the actual values of *G'*. Note that all samples are above the glass transition temperatures of both polymers, and hence in the melt state.

transitions were identified. The OOT are marked by open arrows, while the ODT are shown by filled arrows. Since the ODT occurs as the temperature is raised, an upper critical solution temperature is indicated, much more frequent with block copolymers than with polymer blends. The regions marked *A*, *B*, *C*, and *D* denote lamellar, bicontinuous, cylindrical, and perforated layered microstructures, respectively. The changes in morphology are driven by the temperature dependence of χ_{12},

$$\chi_{12} = \frac{71.4}{T} - 0.0857 \tag{13.8}$$

where χ_{12} is assumed to be independent of composition (30). The changes in χ_{12} with temperature also tend to make many block copolymers more mutually soluble at higher temperatures, in contrast to the reverse for most polymer blends.

The experimental phase diagram illustrated in Figure 13.14 (25) shows the regions of several morphologies,

Im $\bar{3}$ m space group (body-centered cubic spheres)

HEX (hexagonally packed cylinders)

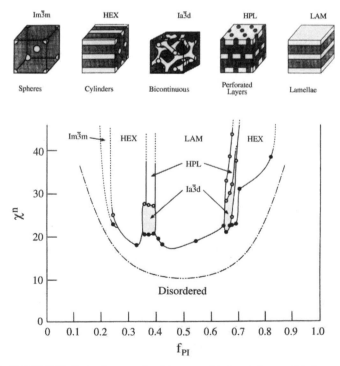

Figure 13.14 Experimental phase diagram of polyisoprene–*block*–polystyrene diblock copolymers. The quantity f_{PI} represents the volume fraction of polyisoprene. Open and filled circles represent the OOT and ODT transitions, respectively. The dash–dot curve represents the mean-field prediction for the ODT. Solid curves delineate the different phases observed. Five different ordered microstructures have been observed for this system.

Ia $\bar{3}$ d space group (bicontinuous gyroid)

HPL (hexagonally perforated layers)

LAM (lamellae).

These morphologies are depicted in the upper portion of Figure 13.14.

While the overall topology of Figure 13.14 is similar to that shown in Figure 13.12, some significant differences exist. First, there is an overall asymmetry in Figure 13.14 with respect to $f = 0.5$. This occurs partially because styrene and isoprene mers have different sizes and shapes, something not included in the calculations resulting in Figure 13.12. Also some of the asymmetry results because the styrene–isoprene interactions are not accurately represented by a single χ parameter: the free-energy cost of moving a styrene mer from pure polystyrene surroundings to pure polyisoprene surroundings is not exactly the same as moving an isoprene mer from pure polyisoprene to pure polystyrene.

13.6.2 Self-assembly in ABC Block Copolymers

When three or more *different* blocks make up the polymer chain, there will be three or more phases if the polymers are immiscible. This results in a very large number of possible phase domain structures, noting that *ABC*, *ACB*, and *BAC* sequences all result in different morphologies. If the blocks are adjacent in the chain, then the phase domains are required to be adjacent in space.

Mogi et al. (31) investigated the morphologies of *ABC* triblock copolymers as well as some *ABCB* tetrablock copolymers; see Figure 13.15. Osmium tetroxide staining produced black, white, and gray images that denote polyisoprene, polystyrene, and poly(2-vinyl pyridine) domains, respectively. Figure 13.15a shows a lamellar structure of polyisoprene–*block*–polystyrene–*block*–poly(2-vinyl pyridine)–*block*–polystyrene. Figure 13.15b illustrates a morphology consisting of two kinds of mutually interpenetrated frameworks formed by the two end blocks, embedded in a matrix composed of the middle block. Figure 13.15c shows a cylindrical morphology of a triblock copolymer with a composition of 1:4:1. This morphology consists of two kinds of cylindrical domains formed by the two end blocks, both embedded in a continuous phase of

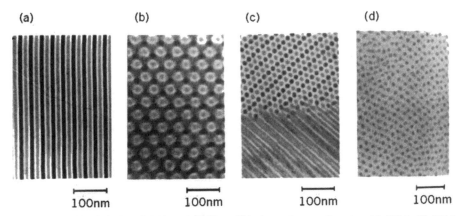

Figure 13.15 Morphologies of ABC and ABCB multiblock copolymers. Samples: (a) ISP-4, (b) ISP-3, (c) ISP-18, (d) ISP-12. Exact block sequences are described in the text.

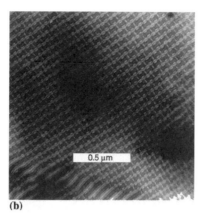

(b)

Figure 13.16 Polystyrene–*block*–poly(ethylene–*stat*–butylene)–*block*–poly(methyl methacrylate) bright field transmission electron micrograph, sample stained with RuO₄. Note knitting pattern.

the center block. Figure 13.15*d* shows a micrograph of spherical domains of two end blocks in a polystyrene matrix form from a composition of 1 : 8 : 1.

Breiner et al. (32) studied the so-called *knitting pattern* found for the triblock copolymer polystyrene–*block*–poly(ethylene–*stat*–butylene)–*block*–poly(methyl methacrylate); see Figure 13.16. [The poly(ethylene–*stat*–butylene) component was formed by hydrogenating the original polybutadiene block, which contained some 1,2-placements as well as the more usual 1,4-placements.] In this morphology, poly(methyl methacrylate) forms peristaltic lamellae in which opposite maxima and minima are spanned by ellipsoidal-shaped cylindrical poly(ethylene–*co*–butylene) (hydrogenated 1,4- and 1,2-polybutadiene domains. Of course, such complexity arises from the combined need of the chains to be able to wander from domain to domain, and to have uniform density within a domain all while maintaining the chains as nearly randomly coiled as possible; see Figure 13.17 (32).

13.6.2.1 *Domain Periods* The domain sizes as a function of block molecular weight for simple block copolymers was given in Section 4.3.9. The student will remember that the radius of a spherical block is given by $R = 1.33\alpha K M^{1/2}$. In the more general case, minimization of the total free energy with respect to the domain period, d, leads to the scaling relation

$$d \sim bn^{2/3}\chi^{1/6} \qquad (13.9)$$

where the quantity b is the statistical segment length [approximately 0.65 nm for polystyrene and poly(methyl methacrylate)], n represents the block copolymer degree of polymerization, and χ is the Flory–Huggins interaction parameter (33). This relation was derived based on the strong segregation limit theory, namely, that the two polymers are very immiscible. The weak segregation limit theory, the composition fluctuation theory, and the self-consistent field theory each yield somewhat different relations, as discussed by Hamley (22). A major point, however, is that the repeat domain spacing, where the morphology repeats itself, depends primarily on the total molecular weight of

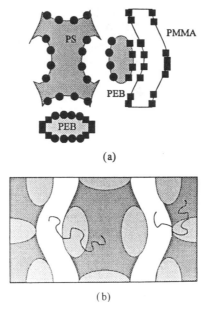

(a)

(b)

Figure 13.17 Triphase knitting pattern morphology, illustrating the chain conformation within each phase domain.

the block copolymer and the (positive) heat of mixing between the polymers. Again, this is a direct consequence of the fact that the blocks are attached to one another.

Zheng and Wang (34) modeled some possible ABC triblock copolymer morphologies; see Figure 13.18. While 11 possible morphologies are illustrated, neither the knitting pattern nor the gyroid morphology are included. On the theoretical side, the proof of thermodynamic stability provides great challenges because the calculations are numerically intensive. A great deal of computational effort is involved in mapping phase diagrams in a large parameter space with numerous competing microphase structures. Such calculations run the risk of overlooking complex three-dimensional microphases not previously identified experimentally. At the time of this writing, numerous advances are being made in this field.

13.6.2.2 Crystallizable Block Copolymer Morphologies   While the largest part of the block copolymer literature describes totally amorphous materials, one or more of the blocks may form semicrystalline regions. Examples include polyester-polyether block copolymers (35), where the poly(tetramethylene terephthalate) polyester blocks crystallize, and the thermoplastic polyurethane elastomers, where the polyurethane hard blocks crystallize (36).

The morphology of the product may depend on the time order of the events: phase separation, crystallization of the hard block, and/or vitrification, if any (37).

13.6.3 The Development of IPN Morphologies

13.6.3.1 The TTT Cure Diagram   Evidence to date suggests that interpenetrating polymer networks, like most polymer blends, exhibit lower critical solution temperatures

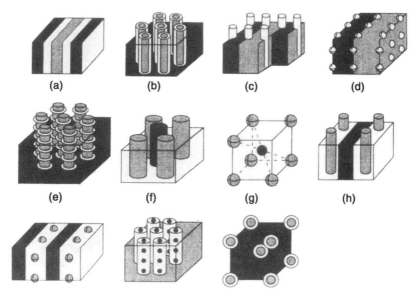

Figure 13.18 Schematic representation of eleven possible triphase morphologies. Note black, white, and gray shadings. (*a*) lamellar, (*b*) coaxial cylinder, (*c*) lamella-cylinder, (*d*) lamella-sphere, (*e*) cylinder-ring, (*f*) cylindrical domains in a square lattice structure, (*g*) spherical domains in a CsCl–salt lattice structure, (*h*) lamella-cylinder-II, (*i*) lamella-sphere-II, (*j*) cylinder-sphere, (*k*) concentric spherical domains in the bcc structure.

(38). However, the development of morphology with polymerization of one or more of its components is complicated by the presence of cross-linking. Thus the materials cannot be stirred beyond a certain point, as can the HIPS materials, and cannot be made to flow at elevated temperatures; they are thermoset materials.

Kim et al. (39) studied the time–temperature-transformation (TTT), diagram (see Section 8.7.2) for the system poly(ether sulfone)–*inter–net*–epoxy semi-II IPN; see Figure 13.19. They identified five general steps in the development of IPN morphology in the figure:

1. Onset of phase separation
2. Gelation
3. Fixation of the phase-separated morphology (the domain size becomes fixed)
4. End of phase separation
5. Vitrification

Of course, not every system has all of these, and the time order of events may vary. Figure 13.19 was for a semi-IPN. In the case of a full IPN, both components undergo gelation. In the case of simultaneous interpenetrating network polymerizations, it may not be known in general which phase will gel first. However, it is a useful generalization that the first phase to gel tends to be more continuous in space.

13.6.3.2 The Metastable Phase Diagram

The time order of the above-mentioned five events depends on such factors as miscibility of components, cross-linking level, and

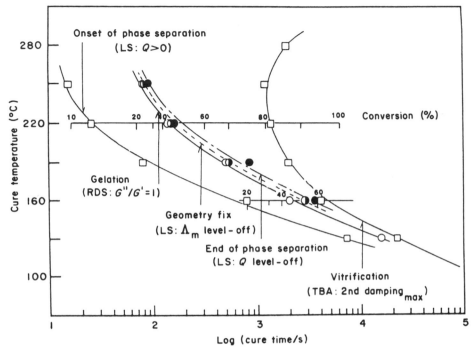

Figure 13.19   Time-temperature-transformation (TTT) cure diagram for poly(ether sulfone)–*inter–net*–epoxy semi-II IPN. Data gathered via light-scattering, LS, dynamical mechanical spectroscopy, RDS, and torsional braid analysis, TBA; see Section 8.3.

overall composition. The time order of these events controls the morphology of the material, and hence physical properties such as modulus and mechanical strength. To a great extent, however, the time order of these events can be controlled with a knowledge of the metastable phase diagram.

Consider the case of poly(methyl methacrylate) and polyurethane simultaneous polymerizations; see Figures 13.20 and 13.21 (40). The four corners of the tetrahedrons indicate methyl methacrylate monomer, poly(methyl methacrylate), the urethane prepolymer, "U", and the fully polymerized polyurethane. If one were polymerizing pure methyl methacrylate, for example, the MMA–PMMA edge would be followed.

The floor triangle composed of MMA–"U"–PU is all one phased, since methyl methacrylate monomer and the urethane prepolymer are both soluble in polyurethane. A simultaneous interpenetrating network, SIN, polymerization starts somewhere along the line MMA–"U", depending on composition. The individual rates of polymerization depend on initiator or catalyst levels, temperature, and so on, and in general, polymerization paths will not be straight lines. However, a polymerization beginning at point M must have a nominal end at point P, Figure 13.20. However, because of phase separation the final product will have phase compositions near the PU and PMMA corners.

In the specific system examined, the poly(methyl methacrylate) contained 0.5% tetraethylene glycol dimethacrylate as a cross-linker, causing gelation after about 8% conversion, as indicated by the G_1–U–PU plane. The phase separation curve for the ternary

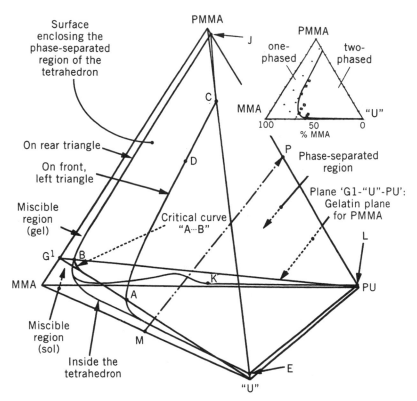

Figure 13.20 Metastable phase diagram for the poly(methyl methacrylate)–polyurethane SIN polymerization. The four corners of the tetrahedron represent the pure monomers and polymers indicated.

system MMA–PMMA–"U" is indicated by the line C–D–A–E. The inset in Figure 13.20 shows the actual experimentally determined curve, with data points just behind the MMA–PMMA–"U" face of the tetrahedron. Similarly the phase separation curve for the MMA–PMMA–PU system, rear triangle (Figure 13.20) is illustrated by the points J–B–K–L. Thus the entire tetrahedron volume is divided into phase separated and single phased regions, separated by C–D–A–E–L–K–B–J.

The curvilinear line A–B, indicating the intersection of the poly(methyl methacrylate) gelation plane with that of the phase separation region represents the critical line along which there is simultaneous gelation of poly(methyl methacrylate) and phase separation of the polyurethane from the poly(methyl methacrylate). Thus reactions moving to the left of this curve will have the poly(methyl methacrylate) gel before phase separation, while reactions to the right of the line A–B will phase separate before gelation.

The plane G_2–MMA–PMMA in Figure 13.21 demarcates the onset of gelation of the polyurethane. The intersection of the two planes, G_1–G_2 indicates the line of simultaneous gelation of both polymers. Thus, reactions passing to one side or the other of this line will have one polymer or the other gel first. The line G_1–G_2 of Figure 13.21 also intersects the line A–B of Figure 13.20, not shown. This intersection represents a triple critical point, where both polymers simultaneously gel and phase separate. Passing through this point could be used to define an ideal SIN polymerization. These meta-

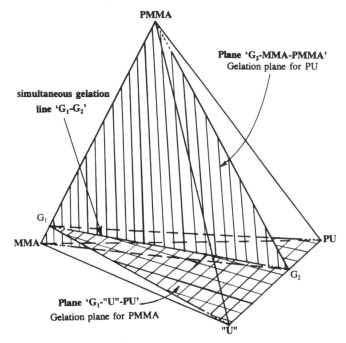

Figure 13.21   Figure 13.20 redrawn, to illustrate the gelation planes of both poly(methyl methacrylate), G_1, and polyurethane, G_2. The intersection of the two planes denotes the line of simultaneous gelation. In reality, neither G_1 nor G_2 are perfectly planar.

stable phase diagrams can be generalized to very many IPN and SIN polymerizations. For example, in the case of a semi-IPN the appropriate gelation plane will be omitted.

13.7 MORPHOLOGY OF COMPOSITE MATERIALS

13.7.1 Reinforcing Carbon Blacks

The carbon blacks used in toughening elastomers such as tire rubber are extremely divided colloidal structures, usually about 10 nm in size, with 50 to 150 m^2/g of surface area. These carbon blacks have extremely active surfaces, containing hydrogen, hydroxyl, carboxyl, ester, and aldehyde groups, all capable of interacting with the elastomer to produce a series of weak bonds; see Section 9.16.2.

Figure 13.22 (41) shows the aggregated structure of a graphitized medium thermal (MT) carbon black by three methods: transmission electron microscopy, scanning electron microscopy, and scanning tunneling microscopy. The graphitization process tends to remove the active groups from the surface, emphasizing the crystalline structure of the carbon.

In tire manufacture a certain amount of aggregation produces optimum toughness. A major problem is keeping the carbon black properly dispersed among the several elastomers generally blended for the purpose.

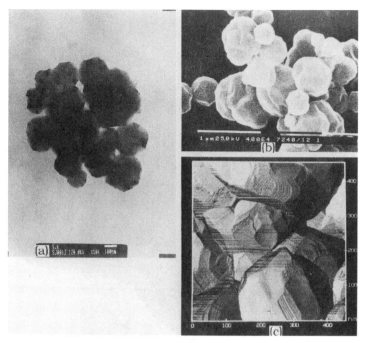

Figure 13.22   Three experimental views of graphitized MT carbon black: (a) TEM, (b) SEM, (c) STM images. The last shows the carbon black as polyhedrons covered with both smooth and terraced facets.

13.7.2   Fiber-Reinforced Plastics

A short carbon fiber reinforced poly(ether ether ketone), PEEK, is illustrated in Figure 13.23 (42). The structure of PEEK is

$$\left(\text{O} - \bigcirc - \overset{\overset{\displaystyle O}{\|}}{C} - \bigcirc - \text{O} - \bigcirc \right)_n$$

serving as a semicrystalline high-temperature and solvent-resistant thermoplastic.

In Figure 13.23 the fibers are nearly randomly oriented. Note the residual polymer still bonded to the fibers in the lower figure, indicating better bonding between the components than shown in the upper figure. Better bonding usually increases fracture strength.

For maximum energy absorption during fracture, the bonds between the polymer and the reinforcing filler should break first absorbing as much energy as possible, followed by fracture of the fiber. This suggests that the bonding energy between the polymer and the filler and the polymer should be just below the fracture energy of the filler (3). As observed in Figure 13.23, cohesive fracture of the polymer adjacent to the interface can replace adhesive fracture.

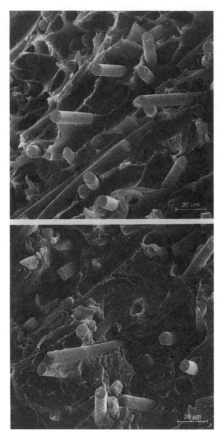

Figure 13.23 Scanning electron micrographs of short carbon fiber reinforced poly(ether ether ketone), emphasizing the fracture surface. Note the polymer still bonded to the fibers (especially in the lower picture), which suggests good bonding.

13.8 FRACTURE BEHAVIOR OF MULTIPHASE POLYMERIC MATERIALS

A major objective in synthesizing polymer blends and composites has been to make tough materials. Mechanically speaking, the word *toughness* refers to the area under the stress–strain curve. This area equals the work done to fracture the material. However, it is also common to speak of toughness under impact loading, fatigue cycling, and compression stresses as well as simple stress strain behavior.

13.8.1 Rubber-Toughened Plastics

Polystyrene and poly(methyl methacrylate) homopolymers, while inexpensive, are relatively brittle, failing in stress–strain experiments after only 1% to 3% extension, and so on. These materials can be toughened very greatly by the addition of 5% to 10% of an elastomer; see HIPS in Figure 4.15. On stressing, the energy is absorbed by multiple

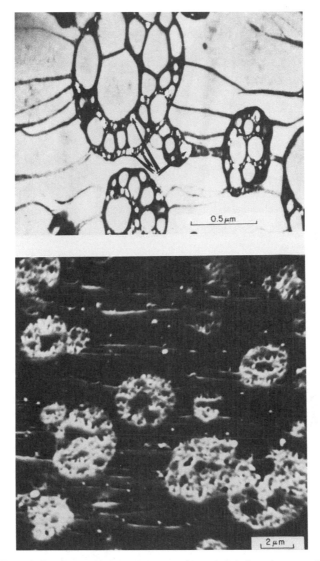

Figure 13.24 Transmission electron microscopy, upper, and scanning electron microscopy, lower, of crazing and fracture in HIPS. Note the presents of cavities in the rubber domains, indicated by the arrows, upper figure.

crazes and the formation of cavities within the rubber domains, among others; see Figure 13.24 (43).

These phenomena are modeled in Figure 13.25. A *crack* is an open fissure in the material. However, a *craze* constitutes a relatively stable structure, with highly oriented *fibrillar* or *microfibrillar structures* running from the roof to the floor of the craze. The rubber domains may cavitate, forming internal voids, thus relieving triaxial stresses inside the plastic.

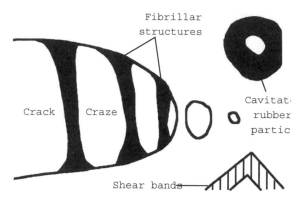

Figure 13.25 Illustration of several structures found in the fracture of rubber-toughened plastics. A preferred order of appearance for maximum energy absorption is shear bands, followed by cavitation of the rubber particles, followed by crazing, because cracks and crazes cannot easily cross shear bands, and cavitation relieves triaxial stresses.

Shear yielding involves *shear band* formation, where the chains slip, becoming highly oriented at approximately 45° to the stress direction. Formation of shear bands involves a limited slip or yielding of one portion of the polymer relative to another, leaving a highly oriented band of material connecting the two portions. Ideally shear band formation should precede crazing, forcing the crazes to propagate through the shear bands. This consumes extra energy, providing an additional toughening mechanism.

Using real-time X-ray measurements on HIPS, Bubeck et al. (44) showed that cavitation of the rubber particles actually precedes crazing of the matrix under tensile impact loading conditions. Thus cavities formed within the rubber particles can be seen as nuclei for craze growth. Modeling studies by Lazzeri and Bucknall (45) showed that the rubber particles cavitate when the energy released during cavitation is greater than the void formation energy.

Cheng et al. (46,47), studying MBS (methacrylate–butadiene–styrene, the equivalent of ABS, substituting methyl methacrylate for acrylonitrile) toughened polycarbonates, found that the rubber particles cavitated in arrays, with particles outside of the arrays much less cavitated. While their work also supports the notion that cavitation precedes crazing, they also showed that a somewhat nonrandom distribution of the rubber particles in the plastic was superior to systems mixed more extensively; that is, controlled levels of clumping are desirable. On toughening epoxies with MBS (48) or CTBN (49), the particles also cavitate in the matrix before shear yielding in this case, further supporting the model of early cavitation.

In the case of HIPS, the rubber domains are relatively large. ABS plastics have much smaller rubber domains, being based on latexes. The core–shell type of toughening particle has been carried over into other types of systems, where core–shell latex particles are used to toughen epoxies (48) as well as other plastics.

Rubber particles may also bridge a crack or craze, providing toughening through rubber elasticity forces. Section 11.2.4.2 described how the glass transition of the rubber phase affects the impact resistance of plastics. Again, the T_g must be at least 60°C below ambient to be effective. While the model of toughening outlined in Section 11.2.4.2 points to quite different toughening mechanisms than that described by Lazzeri and Bucknall (45), and others above, both provide important aspects of the overall picture.

Table 13.6 Mechanical values for various materials

Material	Young's Modulus, E (GPa)	$\mathscr{G}_{1c}$ (kJm^{-2})	K_{1c} (MNm$^{-3/2}$)[a]
Rubber	0.001	13	—
Polyethylene	0.15	20 (J_{1c})	—
Polystyrene	3	0.4	1.1
HIPS	2.1	15.8 (J_{1c})	1.5–2[b]
PMMA	2.5	0.5	1.1
Epoxy	2.8	0.1	0.5
Rubber-toughened epoxy	2.4	2	2.2
Glass-reinforced thermoset	7	7	7
Glass	70	0.007	0.7
Wood	2.1	0.12	0.5
Aluminum Alloy	69	20	37
Steel, mild	210	12	50
Steel alloy	210	107	150

[a] MNm$^{-3/2}$ = MPa-m$^{1/2}$.
[b] M. Parvin and J. G. Williams, *J. Mater. Sci.*, 11, 2045 (1976).

Generally speaking, the more ways a material under stress can absorb energy, the tougher it will be.

The energy absorbing capabilities of several materials are summarized in Table 13.6 (50), which compares typical values for epoxy homopolymers and rubber-toughened materials. Wood, glass, and metals are shown for comparison. Impact strengths are shown in Table 11.3. Both rubber-toughened plastics and composites alike are seen to exhibit larger values of fracture energy, $\mathscr{G}_{1c}$, and the mode I critical stress intensity factor, K_{1c}.

13.8.2 Toughening Mechanisms in Composite Materials

In the case of composite materials, several additional toughening mechanisms may also be present: *crack pinning* by several closely spaced particles may retard crack passage, *microcracking via bifurcation* and *crack path deflection*, where the particles behave as a mirror to the growing crack thus extending its path, all contributing to toughness.

Laminate materials, especially those containing cross-ply features, introduce several special features. First of all, the polymer usually behaves primarily as a binder, holding the fibers together. The majority phase by weight and volume is the fibers.

On the application of an external stress, some local fiber fractures usually occur earlier than the final material fracture, owing to the misalignment and scatter during the manufacture of the fibers, and minor material faults in the fibers themselves (51). The stresses of the broken fibers are transferred to intact fibers through the matrix polymer. Such transfers continue until the remaining intact fibers cannot maintain the load, leading to gross failure. Along the way, however, further increases in stress may result in debonding of the fiber from the matrix. This is useful, since more energy is absorbed before final failure. Since actual crack propagation on a large area is chaotic and may propagate on several surfaces simultaneously, the exact origin of the crack leading to final failure often cannot be easily ascertained.

The experimental longitudinal (fiber direction) mechanical properties of epoxy matrices and E-glass reinforced unidirectional materials (8) are compared in Table 13.7.

Table 13.7 Longitudinal mechanical properties of epoxy and unidirectional E-glass–epoxy laminates, 60% fibers

Property	Epoxy	Laminate
Fiber volume fraction	—	60
Modulus (E) (GPa)	3.35	45.6
Tensile strength (MPa)	80	1280
Tensile failure strain (%)	5	2.9[a]
Thermal coefficient of expansion $(10^{-6}/°C)$	58	8.6
Poisson's ratio	0.35	0.28

[a] Of the fibers.

Computer calculations (9) were within a few percent of the experimental, except for the longitudinal strain to failure of the fibers, not calculated. For 60% fiber volume content, the modulus is raised a factor of 15, while the tensile strength is raised more than a factor of ten. The longitudinal failure strain is reduced, because the fibers cannot be extended as much as the polymer. The longitudinal (linear) thermal coefficient of expansion is also seen to decrease more than a factor of 10, contributing greatly to dimensional stability. Poisson's ratio is seen to decline slightly, suggesting a somewhat greater volume increase on extension.

For both polymer blends and composites, one must distinguish among various loading modes: static, slowly applied loads, impact loading, shearing, bending, and compression. In reality, several of these can be present simultaneously, called mixed mode stressing. Again, the more different ways in which mechanical energy can be degraded into chemical changes such as chain scission or debonding, cavitation, crazing, and fiber fracture, the more difficult it will be to cause failure in the overall material.

13.8.3 Matrix and Fiber Roles in Composites

The polymer and the fibers have distinct roles in composites (52). The roles of the matrix polymer include the following:

1. Maintain desired fiber spacing.
2. Transfer loads from the polymer to the fiber, and transmit shear loads between layers of fiber.
3. Broaden the transmission of stress concentration from broken fibers to intact fibers.
4. Controlled interfacial bond failure between the polymer and the fiber ahead of the crack (and normal to the crack) blunts the crack, absorbing additional energy.
5. Brittle fibers and brittle matrix polymers can be synergistic, leading to tough materials, operating by a combination of mechanisms to keep cracks small, isolated, blunted, and dissipate energy.

The fibers also have specific roles:

1. Support all main loads, and limit deformations.
2. Increase overall strength, stiffness, toughness, and decrease corrosion, creep, and fatigue.

Thus fiber-based polymer composites make up a superior family of engineering plastics, involving many applications.

13.9 PROCESSING AND APPLICATIONS OF POLYMER BLENDS AND COMPOSITES

13.9.1 Processing of Sheet Molding Compounds

Polymeric materials have very many applications in today's world. These applications frequently depend on the details of the processing procedures as well as the physical and mechanical properties of the materials involved. As an example, the processing of sheet molding compounds (SMC) will be developed.

The most common polyesters used are based on polypropylene fumarate polymers (called the *resins* in the trade), prepared from maleic anhydride and propylene glycol. Usually about 40% of styrene monomer is added. It first serves as a solvent and then polymerizes by reaction through some of the fumarate double bonds of the polyester to form a thermoset AB-cross-linked copolymer. The polystyrene and the polyester microphase separate during the polymerization. The final product is commonly called polyester, although both glass fibers and polystyrene constitute integral portions of the material.

Beginning in 1955, the plastic body of the Corvette automobile introduced fiber-reinforced thermoset polyesters as the preeminent plastic for cars (53). The original process was called preform molding, where glass mats or preforms were laid over a hot mold and peroxide catalyzed mixes of unsaturated polyesters and styrene were poured over them. When the mold was closed, the resin spread throughout, displacing the air, and was cross-linked (called *cured* in the trade).

Today, about 30% of 2.5 cm chopped rovings of E-glass (electrical grade of glass fibers), usually containing silane or other coupling agents (54), replaces the glass mats.

Important early problems included handling problems of the *green* uncured sheets, and resin shrinkage on polymerization that left ripple patterns between the fibers, creating an aesthetic problem. This was significantly solved by two developments in the industry. The first of these problems was fixed by the inclusion of metal oxides such as MgO or CaO in finely divided form, causing slow but enormous viscosity increases in polyesters containing carboxylic acid end groups. This constitutes a type of ionic cross-linking. After a few days of aging, the material becomes tack-free, and can be cut to the required dimension, and manually or robotically placed in the desired mold.

The second breakthrough was the addition of thermoplastics such as poly(methyl methacrylate), poly(vinyl acetate) (55), polyester urethanes and/or long-chain fatty acid capped terephthalic acid-based polyester oligomers (56), called low-profile additives (LPAs). (Oligomers are low molecular weight polymers, generally a few thousand g/mol; see Section 3.1.1.) These new LPAs enabled molding of SMC parts having substantially zero shrinkage and a very high surface quality. The added thermoplastics are initially soluble in the mix, but precipitate during the polymerization of the polystyrene. The newly formed interfaces are weak. A dendritic void structure then develops in the thermoplastic phase in response to the shrinkage stresses. These microvoid structures constitute a type of controlled cracking (57), relieving the triaxial stresses and preventing shrinkage. Presumably the void volume approximately compensates for the polymerization shrinkage. Since the strength of the finished sheet lies primarily in the fiber structure, the mechanical properties are relatively unaffected.

Table 13.8 Relationships among selected physical quantities and engineering properties

Physical Quantity	Engineering Property
Glass transition temperature, crystallinity, cross-linking, fibers fillers, plasticizers	Hardness (modulus)
Molecular weight, chain orientation, reinforcement, rubber additions	Tensile strength, toughness, impact resistance, fatigue resistance
Molecular motion, reactive groups	Adhesion
Free motion of electrons	Electrical conductivity, thermal conductivity, color/absorbance
Concentration of electrons	Refractive index, reflectance
Atomic nuclei concentration, atomic number	Density/specific gravity
Polymer chain motion, glass transition temperature	Damping, creep and relaxation, dielectric constant
Glass transition temperature, crystallinity, fillers	Coefficient of expansion
Aromatic structures, ladder polymers, cross-linking, fillers	Heat resistance, chemical resistance
Free radical absorbers	Flammability resistance

Besides the need for high mechanical strength, low shrinkage, dimensional stability, and ease of processing, the final formulations must also be resistant to the elements such as water and ultraviolet light (54). For automotive use, the materials also must be paintable. Thus the SMC technology involves not only aspects of polymer blends and composites but controlled fracture, adhesion, and environmental resistance as well.

13.9.2 General Considerations in Application Science

For the following, the reader is reminded of the definitions of elastomers, plastics, adhesives, coatings, and fibers in Section 8.13. Successful industrial and commercial application of polymers requires consideration of many properties; see Table 13.8. In general, increases in the listed physical quantity brings about increases in the corresponding engineering property.

Mechanical strength properties such as tensile strength, impact resistance and fatigue resistance are among those most often cited in the patent literature (16). Other quantities of great importance include compressive and bending strength and shear resistance. While tensile strengths are more frequently reported in the literature, engineering applications more frequently place load-bearing polymer materials in compression. In fact most such materials are far stronger in compression than in tension.

Critical factors in developing mechanical strength include high molecular weight. Wool (58) cites $8M_c$ as the value required to achieve nearly full strength. Actually a number of polymers are used at about $7M_c$, primarily because of melt viscosity considerations, since the viscosity increases as $M^{3.4}$ power in this range. At $7M_c$ about 80% to 90% of most mechanical properties are achieved. Another important factor is chain orientation, especially important in fiber manufacture. Addition of fillers, especially fibers for plastics and carbon black or silicas for elastomers, makes for significant improvements in mechanical properties.

The art of compounding (59) rises to the fore in the development of polymeric materials. Thus, fillers can be either reinforcing or inert and low price, or serve as pigments.

Plasticizers and lubricants need to be added on occasion. Antioxidants and ultraviolet light absorbers lengthen the life of many polymers. Curatives such as sulfur for vulcanization or peroxides reduce flow, producing cross-linking.

Good adhesion may require significant molecular motion in order to flow onto the surface in question, or monomers capable of polymerizing rapidly on surfaces with bonding. Generally, a good adhesive will have reactive sites such as carboxyl or carbonyl groups. Often latex paints for the home consumer have small concentrations of many types of reactive groups, so that they will adhere reasonably well to a variety of surfaces: wood, metal, old paint, and so on.

In order to conduct electricity, greater or lesser free motion of electrons in the material is required. However, electrical behavior involves questions not only of conduction and resistance but of static charges. Most polymers are insulators, except if they have alternating double and single bonds; see Section 14.7.3. The motion of electrons in polymers, and in materials generally, also affects thermal conductivity. Thus good electrical insulation and good thermal insulation (low coefficient of heat transfer) are closely related physically. That is why pots are made out of metal, but pot handles are made out of plastic. Lack of electron-free motion also contributes to transparency in pure, amorphous plastics such as polystyrene and poly(methyl methacrylate).

The number of electrons per unit volume however, controls the refractive index of a polymer and its reflectance. For many applications a high refractive index clear material is esthetically pleasing.

Refractive index and density are also linked, because while the former measures the number of electrons, the latter measures the mass of atomic nuclei per unit volume.

Polymer chain motion contributes to damping and dielectric constants, via conversion of mechanical or electrical energy to molecular motion.

The coefficient of expansion of a polymer increases above its glass transition temperature; see Section 8.6. The presence of crystallinity or fillers generally decreases the coefficient of expansion. For the production of complex parts, low coefficients of thermal expansion are strongly preferred. Generally, plastics have higher coefficients of thermal expansion than metals, important in automotive manufacture, for example. Inappropriate joining of different classes of materials may lead to parts literally torn apart because of the relatively large stresses brought about through temperature changes.

While heat resistance and chemical resistance are not closely linked, the presence of aromatic groups, special structures such as liquid crystalline polymers, dense cross-linking, or fillers are often beneficial. When examining chemical resistance, it must be in context to specific chemicals: aqueous acids or bases, organic solvents, oxidizing agents, and so on. Flammability resistance is usually brought about by the incorporation of chemical capable of absorbing free radicals, hence quenching the chain reaction phenomena associated with fire. Such chemicals are required for baby clothes where a soft, fuzzy texture is desired but such texture also provides ready access to oxygen. Similarly rugs are often treated for fire resistance.

13.9.3 Applications of Polymers, Polymer Blends, and Composites

Polymers are used in nearly all aspects of modern living. Clothing is composed of either natural or synthetic polymers in fiber form. Paper is based on cellulose, a natural polymer, but with both inorganic opacifiers and synthetic polymers added for purposes of adhesion and writing ease. Rubber, plastics, adhesives, coatings, and fibers are everywhere. Polymer-based products are one of the fastest growing industries in the world.

Table 13.9 Selected homopolymer applications

Polymer	Application	Physical Properties
Poly(methyl methacrylate)	Artificial glass	Toughness, clarity
Polystyrene	Coffee cups	Low thermal coefficient as foam
Polyamide 66, and others	Fibers	Semicrystalline, abrasion resistant, degradation resistant
cis-Polyisoprene	Elastomers	Self-reinforcing on extension (crystallizes)
Polyethylene	Films and containers	Waterproof, clarity in thin section
Polyolefins	Wire and cable insulation	Low dielectric constants
Poly(ethylene terephthalate)	Soft drink containers	Low permeability to carbon dioxide
Epoxies	Adhesives	Good adhesion, high strength

Selected applications for polymers, polymer blends, and composites are shown in Tables 13.9, 13.10, and 13.11, respectively. In Table 13.9, poly(ethylene terephthalate) (PET) used for soft drink containers incorporates a bit of comonomer to reduce crystallinity just to the incipient point, rendering it optically clear for esthetic purposes. Most important, PET has exceedingly low permeability to carbon dioxide (see Section 4.4), allowing for relatively long shelf life of the sodas. The crystalline homopolymer makes the fiber widely known as polyester.

Table 13.10 first emphasizes rubber-toughened plastics. In each of these cases a low T_g elastomer is dispersed in a plastic. Examples are given of amorphous (HIPS and ABS) and crystalline (nylon) thermoplastics, and a thermoset (epoxy). Usually it is desirable to have some limited grafting between the elastomer and plastic phases, to increase interfacial strength.

Table 13.10 also delineates triblock copolymers based on polystyrene–*block*–polybutadiene–*block*–polystyrene. Polyisoprene may be substituted for the polybutadiene, and/or the center block may be hydrogenated. The hydrogenation reduces environmental degradation due to light, heat, or water. These triblock copolymers are widely used as shoe and sneaker soles.

The triblock copolymers exhibit increasing friction with increasing slide rate (60). Under high-slide rates, frictional work heats the surface of the shoe soles up to the glass transition of the hard block, about 75°C as it appears in commercial materials. Only the surface 40 to 50 nm or so need to be heated. Above the glass transition temperature of the hard block, the elastomer temporarily becomes an adhesive; note Section 8.13. By contrast, chemically cross-linked homopolymer elastomers may shred at high-slide rates, reducing their coefficient of friction.

Table 13.10 also describes a homo-IPN of poly(methyl methacrylate), densely cross-linked, that provides a basis for artificial (false) teeth (61,62). These teeth have two superior properties: Because of the suspension polymerization route used, the dentist can grind the teeth to a fine powder, which is useful when fitting them to the opposing set of teeth. Second, because they are densely cross-linked, they do not swell significantly in salad oils, margarine, and butter, which plasticize poly(methyl methacrylate).

There are very many new applications of polymers in the biomedical field. In Table 13.11, the reaction product of glycidyl methacrylate with bisphenol A, known widely as *bis*-GMA filled with fumed silicas forms the basis for dental composites for filling cavities. Such products are widely used for front teeth because of ease of color matching

Table 13.10 Selected applications of polymer blends (broad definition)

Material	Application	Physical Properties
A. Rubber-toughened plastics		
HIPS	Automotive, appliances, packaging, lids	Impact resistance, low price
ABS	Appliances (ice cube trays), business machine housings	Greater impact resistance, low temperature capabilities
Nylon/EPDM	Automotive, gears	Very tough, wear resistant
Epoxy/CTBN	Printed circuit boards	Tough, dimensionally stable
PP/EPDM	Automotive bumpers	Energy absorbing
PC/ABS	Automotive and transportation industries	Tough, wide temperature range of application
B. Block copolymers		
SBS or SIS	Shoe soles	Increasing friction coefficient with slide velocity
SBS or SIS (plasticized)	Pressure sensitive adhesives	Rapid flow, bonding
SEBS	Wire and cable insulation	Nonconductor, flexible
PU	Undergarment elastics and swim suit clothing	Short-range elasticity
C. IPNs		
SEBS/Polyester	Under hood insulation	Wide temperature range of constant, leathery modulus
PU/Polyester–styrene	Sheet molding compounds (automotive bodies)	High green strength, toughness
Anionic/cationic	Ion exchange resin	Positive and negative charges in juxtaposition
Acrylic based	Artificial teeth	Low swelling in salad oils, good grindability by dentist
Vinyl/phenolics	Sound and vibration damping	Broad T_g, loss modulus
PDMS/PTFE	Burn dressing	High moisture permeability, clarity of film

Abbreviations: HIPS, high-impact polystyrene; ABS, acrylonitrile–butadiene–styrene graft copolymer; EPDM, ethylene–propylene–diene copolymer; CTBN, carboxyl terminated butadiene nitrile telomer ($M \cong 5000$ g/mol); PC, polycarbonate; SBS, styrene–butadiene–styrene triblock copolymer; SIS, styrene–isoprene–styrene triblock copolymer; SEBS, SBS with hydrogenated center block; PU, segmented (block copolymer) polyurethanes; PDMS, poly(dimethyl siloxane); PTFE, polytetrafluoroethylene.

to natural teeth, an esthetic objective (63). The structure of *bis*-GMA is

Note the presence of two double bonds, leading to dense crosslinking. From a chemical point of view, the *bis*-GMA exhibits low shrinkage on polymerization (cure), so as to

Table 13.11 Selected composite applications

Material	Application	Physical Properties
Unsaturated polyester/styrene–glass fiber	Aerospace and transport industries, sporting goods	High strength, light weight
Rubber blends/carbon black	Automotive tires	Abrasion resistance
Epoxy/mica/core–shell latexes	Electronic component packaging	Adhesion to substrate, temperature cycling resistant, impact resistant
Acrylic latexes/TiO_2 and other fillers	Coatings (wall paint)	Adhesion, weather resistant, scratch resistant
bis-GMA/fumed silica	Dental composites	Low shrinkage on cure, high strength, wear resistant, esthetics
UHMWPE/carbon fibers	Artificial knee and hip joints	Low friction, low wear, good biocompatibility
PMMA/metal wire or carbon fiber or various fillers	Bone cements	Good compressive and shear strength

reduce the probability of unwanted bacteria, and so on, entering between the filling and the remaining natural tooth material (64). [Unfortunately, an effective *expanding* polymerization for dental applications has yet to be developed; see (65).]

Bone cements (Table 13.11) provide another example of composite materials being used in the biomedical field. Poly(methyl methacrylate) compositions can be used for tissue implants, dental materials, bone grafts, pins, screws, plates, stents, and so on. A PMMA semi-IPN composition containing biodegradable polymers (66) as well as proteins and inorganic salts exhibits properties desired for tissue or bone regrowth. The inorganic salts have diameters of 100 to 250 μm. These are dissolved out in the body, providing relatively evenly spaced interconnected interstitial spaces or pores into which living cells can migrate, attach, and proliferate. The porosity of the matrix can be as high as 60% to 90%, depending on the medical requirements.

Good sound and vibration damping requires high values of the loss modulus or tan δ; see Section 8.12. If damping over a wide temperature range is desired, the glass transition must cover the required temperature span. IPNs or polymer blends with χ_{12} nearly zero (but still positive) often exhibit a microheterogeneous morphology, where the composition varies from one microscopic region to another, each region having its own T_g (67). With some control over structure or cross-linking level, materials having iso-G'' or iso-tan δ over the required temperature range can be prepared.

Thus, while the structure, morphology, and properties of polymeric materials are interesting intellectually, their usefulness in today's society has led to huge volumes of research, development, and concomitant applications.

While this section emphasized tough materials a wide range of applications of soft polymers and gels also exist. These range from soft contact lenses and the absorbing portion of diapers (see Section 9.13.3) to food thickeners and a host of biomedical applications.

REFERENCES

1. L. E. Nielsen and R. L. Landel, *Mechanical Properties of Polymers and Composites*, 2nd ed., Dekker, New York, 1994.
2. J. A. Manson and L. H. Sperling, *Polymer Blends and Composites*, Plenum, New York, 1976.
3. L. H. Sperling, *Polymeric Multicomponent Materials: An Introduction*, Wiley, New York, 1997.
4. L. H. Sperling, in *Recent Advances in Polymer Blends, Grafts, and Blocks*, L. H. Sperling, ed., Plenum, New York, 1974.
5. L. H. Sperling, *Polym. Prepr.*, **14**, 431 (1973).
6. G. Kraus and J. T. Gruver, *J. Polym. Sci.*, **A-2**, **8**, 571 (1970).
7. L. E. Nielsen, R. A. Wall, and P. G. Richmond, *Soc. Plastics Eng. J.*, **11** (Sept.), 22 (1955).
8. P. D. Soden, M. J. Hinton, and A. S. Kaddour, *Compos. Sci. Technol.*, **58**, 1011 (1998).
9. P. K. Gotsis, C. C. Chamis, and L. Minnetyan, *Compos. Sci. Technol.*, **58**, 1137 (1998).
10. E. H. Kerner, *Proc. Phys. Soc. London*, **69B**, 802, 808 (1956).
11. E. Guth and H. Smallwood, *J. Appl. Phys.*, **15**, 758 (1944).
12. E. Guth, *J. Appl. Phys.*, **16**, 20 (1945).
13. W. E. A. Davies, *J. Phys. (D)*, **4**, 318 (1971).
14. D. R. Paul and J. W. Barlow, *J. Macromol. Sci. Rev. Macromol. Chem.*, **C18**, 109 (1980).
15. G. M. Jordhamo, J. A. Manson, and L. H. Sperling, *Polym. Eng. Sci.*, **26**, 518 (1986).
16. L. A. Utracki, *Polymer Alloys and Blends: Thermodynamics and Rheology*, Hanser, Munich (1990).
17. K. Kato, *Jpn. Plastics*, **2** (Apr.), 6 (1968).
18. J. K. Yeo, L. H. Sperling, and D. A. Thomas, *Polymer*, **24**, 307 (1983).
19. M. W. Matsen and F. S. Bates, *Macromolecules*, **29**, 1091 (1996).
20. L. Leibler, *Macromolecules*, **13**, 1602 (1980).
21. R. L. Scott, *J. Chem. Phys.*, **17**, 279j (1949).
22. I. W. Hamley, *The Physics of Block Copolymers*, Oxford University Press, Oxford, England (1998).
23. D. B. Alward, D. J. Kinning, E. L. Thomas, and L. J. Fetters, *Macromolecules*, **19**, 215 (1986).
24. E. L. Thomas, D. B. Alward, D. J. Kinning, J. D. L. Handlin, and L. J. Fetters, *Macromolecules*, **19**, 2197 (1986).
25. A. K. Khandpur, S. Forster, F. S. Bates, I. W. Hamley, A. J. Ryan, W. Bras, K. Almdal, and K. Mortensem, *Macromolecules*, **28**, 8796 (1995).
26. D. A. Hajduk, H. Takenouchi, M. A. Hillmyer, F. S. Bates, M. E. Vigild, and K. Almdal, *Macromolecules*, **30**, 3788 (1997).
27. S. Sakurai, *TRIP (Trends in Polymer Science)*, **3**, 90 (1995).
28. F. S. Bates and G. H. Fredrickson, *Physics Today*, **52(2)**, 32 (1999).
29. M. W. Matsen and M. Schick, *Phys. Rev. Lett.*, **72**, 2660 (1994).
30. N. A. Rounds, *Block Copolymers*, Ph.D. Dissertation, University of Akron, 1970.
31. Y. Mogi, H. Kotsuji, Y. Kaneko, K. Mori, Y. Matsushita, and I. Noda, *Macromolecules*, **25**, 5408 (1992).
32. U. Breiner, U. Krappe, E. L. Thomas, and R. Stadler, *Macromolecules*, **31**, 135 (1998).
33. A. N. Seminov, *Soviet Physics JETP*, **61**, 733 (1985).
34. W. Zheng and Z. G. Wang, *Macromolecules*, **28**, 7215 (1995).
35. J. J. R. Wolfe, *Block Copolymers: Science and Technology*, D. J. Meier, ed., Hanser, Munich, 1983.

36. G. Holden, *Understanding Thermoplastic Elastomers*, Hanser, Munich, 2000.

37. Z. D. Cheng, *Polymer Processing Society Proceedings, PPS-16*, Shanghai, June 2000, p. 5.

38. D. Sophiea, D. Klempner, V. Senjijarevic, B. Suthar, and K. C. Frisch, *Interpenetrating Polymer Networks*, D. Klempner, L. H. Sperling and L. A. Utracki, eds., ACS Books, Washington, DC, 1994.

39. B. S. Kim, T. Chiba, and T. Inoue, *Polymer*, **34**, 2809 (1993).

40. V. Mishra, F. E. DuPrez, E. Gosen, E. J. Goethals, and L. H. Sperling, *J. Appl. Polym. Sci.*, **58**, 331 (1995).

41. S. J. Kim and D. H. Reneker, *Rubber Chem. Technol.*, **66**, 559 (1993).

42. K. Friedrich, R. Walter, H. Voss, and J. Karger-Kocis, *Composites*, **17**, 205 (1986).

43. H. Keskkula, M. Schwarz, and D. R. Paul, *Polymer*, **27**, 211 (1986).

44. R. A. Bubeck, D. J. Buckley, E. J. Kramer, and H. Brown, *J. Mater. Sci.*, **26**, 6249 (1991).

45. A. Lazzeri and C. B. Bucknall, *J. Mater. Sci.*, **28**, 6799 (1993).

46. C. Cheng, A. Hiltner, E. Baer, P. R. Soskey, and S. G. Mylonakis, *J. Mater. Sci.*, **30**, 587 (1995).

47. C. Cheng, A. Hiltner, E. Baer, P. R. Soskey, and S. G. Mylonakis, *J. Appl. Polym. Sci.*, **55**, 1691 (1995).

48. H. R. Azimi, R. A. Pearson, and R. W. Hertzberg, *J. Mat. Sci.*, **31**, 3777 (1996).

49. A. F. Yee, D. Li, and X. Li, *J. Mater. Sci.*, **28**, 6392 (1993).

50. A. J. Kinloch and R. J. Young, *Fracture Behavior of Polymers*, Applied Science, London, 1983.

51. H. Thom, *Composites Part A*, **29A**, 869 (1998).

52. F. P. Gerstle Jr., *Encyclopedia of Polymer Science and Engineering*, Vol. 3, J. Kroschwitz, ed., Wiley, New York, 1985, p. 776.

53. J. M. Castro, E. G. Melby, and R. M. Griffith, *Automotive Polymers & Design*, **28** (Oct. 1988).

54. M. Avella, E. Martuscelli, G. Orsello, M. Cocci, G. Caramaschi, M. Leonardo, and S. Sanchioni, *J. Mat. Sci.*, **31**, 5135 (1996).

55. C. B. Bucknall, I. K. Partridge, and M. J. Phillips, *Polymer*, **32**, 786 (1991).

56. D. H. Fisher, T. A. Tufts, and T. Moss, U.S. 5,504,151, to Ashland, Inc., 1996.

57. L. Suspene, D. Fourquier, and Y. S. Yang, *SAMPE Quarterly*, 18 (1990).

58. R. P. Wool, *Polymer Interfaces: Structure and Strength*, Hanser, Munich (1995).

59. F. Rodriguez, *Principles of Polymer Systems*, 4th ed., Taylor and Francis, Washington, DC (1996).

60. G. Holden, in *Recent Advances in Polymer Blends, Grafts, and Blocks*, L. H. Sperling, ed., Plenum, New York, 1974.

61. F. D. Roemer and L. H. Tateosian, US 4,396, 476, 1983.

62. F. D. Roemer and L. H. Tateosian, US 4,396,377, 1984.

63. R. L. Bowen, U.S. 3,066,112, 1962.

64. D. Holter, H. Frey, and F. Mulhaupt, *Polym. Prepr.*, **38**(2), 84 (1997).

65. N. Moszner, T. Volkel, F. Zeuner, and V. Rheinberger, *Polym. Prepr.*, **38**(2), 86 (1997).

66. V. R. Shastri, R. S. Langer, and P. J. Tarcha, U.S. 5,837,752, to Massachusetts Institute of Technology, 1998.

67. R. D. Corsaro and L. H. Sperling, eds., *Sound and Vibration Damping with Polymers*, American Chemical Society, Washington, DC, 1990.

GENERAL READING

T. Arai, Z. Tran-Cong, and M. Shibayama, eds., *Structure and Properties of Multiphase Polymeric Materials*, Dekker, New York, 1998.

C. B. Arends, ed., *Polymer Toughening*, Dekker, New York, 1996.

E. J. Bardero, *Introduction to Composite Materials Design*, Taylor and Francis, Philadelphia, PA, 1999.

M. M. Coleman, J. F. Graf, and P. C. Painter, *Specific Interactions and the Miscibility of Polymer Blends*, Technomic, Lancaster, PA, 1991.

A. A. Collyer, ed., *Rubber Toughened Engineering Plastics*, Chapman and Hall, London, 1994.

C. F. Dangelmajer, *Chemical Heritage*, **18**(2), 8 (2000).

A. Eisenberg and J. S. Kim, *Introduction to Ionomers*, Wiley-Interscience, New York, 1998.

M. J. Folks and P. S. Hope, eds., *Polymer Blends and Alloys*, Blackie Academic & Professional, London, 1993.

R. K. Gupta, *Polymer and Composite Rheology*, 2nd ed. Dekker, New York, 2000.

I. W. Hamley, *Block Copolymers*, Oxford University Press, Oxford, England, 1998.

G. Holden, *Understanding Thermoplastic Elastomers*, Hanser, Munich, 2000.

G. Holden, N. R. Legge, R. Quirk, and H. E. Schroeder, eds., *Thermoplastic Elastomers*, 2nd Ed., Hanser, Munich, 1996.

A. D. Jenkins, E. Martuscelli, P. Musto, and G. Ragosta, eds., *Advanced Routes for Polymer Toughening*, Elsevier, Amsterdam, 1996.

R. M. Jones, *Mechanics of Composite Materials*, 2nd ed., Taylor and Francis, Philadelphia, PA, 1999.

D. Klempner, L. H. Sperling, and L. A. Utracki, eds., *Interpenetrating Polymer Networks*, American Chemical Society, Washington, DC. 1994.

A. Kelly, ed., *Concise Encyclopedia of Composite Materials*, rev. ed., Elsevier, Amsterdam, 1994.

N. M. K. Lamba, K. A. Woodhouse, and S. L. Cooper, *Polyurethanes in Biomedical Applications*, CRC Press, Boca Raton, FL, 1998.

P. A. Lovell, *Trends in Polym. Sci. (TRIP)*, **4**, 264 (1996).

G. M. Michler, *Trends Polym. Sci. (TRIP)*, **3**(4), 124 (1995).

D. R. Paul and C. B. Bucknall, eds., *Polymer Blends: Formulation and Performance*, Vols. I and II, Wiley, New York, 2000.

R. A. Pearson, H.-J. Sue, and A. F. Yee, eds., *Toughening of Plastics*, ACS Symp. Ser. No. 759, ACS Books, American Chemical Society, Washington, DC, 2000.

C. K. Riew and A. J. Kinloch, *Toughened Plastics II: Novel Approaches in Science and Engineering*, American Chemical Society, Washington, DC, 1996.

L. H. Sperling, *Polymeric Multicomponent Materials: An Introduction*, Wiley, New York, 1997.

L. A. Utracki, *Commercial Polymer Blends*, Chapman and Hall, London, 1998.

L. A. Utracki, *Polymer Alloys and Blends*, Hanser, Munich, 1990.

T. L. Vigo and B. J. Kinzig, eds., *Composite Applications: The Role of Matrix, Fiber, and Interface*, VCH, New York, 1992.

B. C. Wendle, *What Every Engineer Should Know About Developing Plastics Products*, Dekker, New York, 1991.

STUDY PROBLEMS

1. The glass transition temperature of an acrylic copolymer was $-30°C$, and that of a styrenic copolymer was $+90°C$. A 60/40 acrylic/styrenic composition was mixed to equilibrium at 120°C, the two phases having glass transition temperatures of −10°C and +70°C. On mixing at 150°C, two phases had glass transition temperatures of −20°C and +80°C.

(a) What are the compositions of the two phases at 150°C? (Assume that the approach to equilibrium is much slower than the rate of temperature change during the measurements.)

(b) What does the resultant phase diagram look like? Sketch in any missing data portions.

2. Can Figures 13.20 and 13.21 be generalized to analyze a sequential IPN synthesis? Assume that the urethane is polymerized first, then methyl methacrylate is swollen in, and then polymerized. What path is followed? What are the requirements to cause phase separation first, and then gelation of the poly(methyl methacrylate)?

3. A diblock copolymer having 720 mers of styrene in one block and 2000 mers of butadiene in the other block forms spherical domains of polystyrene. How many polystyrene chains are there in one such domain?

4. A blend of plastic ($E = 3 \times 10^9$ Pa) and rubber ($E = 1.2 \times 10^6$ Pa), containing 25% of rubber was prepared, and properties measured at 25°C. Transmission electron microscopy showed that the rubber phase was discontinuous.

(a) What is your estimate of Young's modulus of the blend?

(b) If both phases were continuous, what Young's modulus would you now calculate?

5. A 75/25 polybutadiene/polystyrene sequential IPN has 1% cross-linking in the polybutadiene and 1.5% cross-linking in the polystyrene, using divinyl benzene as the cross-linker. What is the diameter of the polystyrene domains at 25°C?

6. Assuming an $A–B$ type of diblock copolymer with amorphous blocks and spherical domains for the A blocks, what is a general equation expressing the relationship among block A molecular weights, the average number of chains in each spherical domain, and the radius of the spherical domains? Derive said equation. [*Hint*: Use of the relationship $R = 1.33\alpha K M^{1/2}$ may lead one astray. A new derivation is required.]

7. If the blend in Problem 4 is mixed at 150°C, above the T_g of both polymers, and the elastomer has a melt viscosity of 6×10^3 Pa·s, what are the viscosity requirements of the plastic such that it be the continuous phase?

8. A center notched rubber-toughened epoxy sample has a crack of 0.015 m width. If a stress of 1×10^8 Pa is imposed, will the sample fail?

9. Read any scientific or engineering paper on polymer blends or composites, 1999 or later. What are the main points? Please provide the full reference. How does the research update this textbook?

10. Describe or invent three possible morphologies for ABC triblock copolymers *not* mentioned in the text. How would you search for them experimentally?

11. A sound and vibration damping material effective over a very broad temperature range with iso-E'' is required. What kinds of morphologies might work best?

12. A crossply two-layer glass fiber laminate containing 30% glass in an epoxy resin is to be evaluated for mechanical properties in the longitudinal direction to one of the

plies. To provide your technicians with working estimates, what values of Young's modulus, tensile strength, strain to failure, thermal coefficient of expansion, and Poisson's ratio can you calculate for them?

13. Ten percent of glass spheres are dispersed in a plastic of $G = 1 \times 10^9$ Pa. What is the shear modulus of the composite?

14. How would you propose to quantify the relative values of crack pinning, microcracking, and crack path deflection toughening mechanisms in a mica-reinforced epoxy resin? Draw pictures of possible morphological changes. [*Hint*: Mica forms thin platelets under these conditions.]

15. Invent or describe a new application for a polymer, polymer blend, or composite. Be fanciful if you must. What processes will you need to prepare the material?

16. A particular polymer blend was found to have *negative* values for $-(d\gamma_{ab}/dT)$ for its interfacial tension. Does this blend exhibit an upper or lower critical solution temperature? What is your reasoning?

14

MODERN POLYMER
TOPICS

The preceding chapters provide an introduction to physical polymer science. However, there are many more topics of great interest not yet covered. While some of the topics to be discussed have been known for some time, a significant part of this chapter describes a series of relatively new developments, often still in their infancy.

14.1 POLYOLEFINS

The polyolefins are those polymers based only on carbon and hydrogen, originating from monomers containing a double bond in the 1-position, sometimes called α-olefins. Principally, these include polyethylene, polypropylene, copolymers of polyethylene containing various comonomers such as 1-butene, 1-hexene, and 1-octene, ethylene–propylene monomer (EPM), and ethylene–propylene–diene–monomer (EPDM). All of these are plastics except EPM and EPDM, which are elastomers.

14.1.1 The Global Picture

Among the synthetic polymers, polyethylene and polypropylene are the largest tonnage synthetic polymers; for comparison, cellulose constitutes the largest tonnage natural polymer; see Table 14.1. These materials are relatively easy to manufacture, have a range of useful properties, and are low priced.

14.1.2 Polyethylene Properties

The melting temperature, extent of crystallinity, modulus, and mechanical behavior depend on the method of manufacture and the addition of comonomers, as well as overall molecular weight. Polyethylene is manufactured by several major processes (1,2): The high-pressure, free-radical polymerization, the Ziegler process, and the newer metallocene-catalyzed polymers and the metallocene–Ziegler processes; see Table 14.2.

Table 14.1   Production of key global polymers

Polymer	Annual Tonnage, Millions (Late 1990s Quantities)	Reference
Polyethylene	45	(a)
Polypropylene	22	(b)
Cellulose[a]	24	(c)
Poly(vinyl chloride)	25	(b)
Polystyrene	12	(b)

References: (a) *Modern Plastics Encyclopedia '98*, **A-15** (1998). (b) M. S. Reisch, *C&E News*, **75** (21), 14 (1997). (c) G. Stanley, *TAPPI*, **82** (1), 40 (1999).
[a] Quantity for paper pulp only. Rayon, cellophane, and cellulose esters and ethers not included.

While the largest tonnages by far are based on the high-pressure process and the Ziegler process, the newest method, via metallocene catalysis, offers great promise for controlled properties. Note that only the high-pressure process is based on free-radical synthesis. The reader is referred to Brydson (1) and Benedikt and Goodall (2) for catalyst and synthesis details.

The various polymers are named based on their density (Table 14.3). The densities of the polyethylenes decrease with increased side group mole fraction, such as obtained via copolymerization with small amounts of propylene or *n*-butene. In LDPE, short side chains are caused by a *back-biting* phenomenon during polymerization; in addition there are long side chains caused by hydrogen abstraction and subsequent branching. HDPE contains substantially no branches, long or short. In Table 14.3 the LLDPE and *m*-LLDPE are linear polyethylene (no long branches) with controlled quantities of comonomer such as 1-butene added to reduce the crystallinity of the product. The *m*-represents metallocene. The presence of short chain branches in LDPE, or their synthetic equivalent added in the form of comonomers in LLDPE disrupt the sequence of ethylene mers; therefore the crystallinity of the ethylene copolymer is reduced. These various compositions are modeled in Figure 14.1.

Below the LLDPEs, there are designations such as ULDPE for still more side groups. The densities in Table 14.3 should be compared to that of 100% crystalline polyethylene, 1.00 g/cm^3; see Table 6.2 and Section 6.3.1. By extrapolation of data from above the melting point, the density of amorphous polyethylene at room temperature is reported

Table 14.2   Synthetic methods for polyethylene manufacture

Method	Polyethylene Properties
High pressure (free radical)	Broad molecular weight distribution, both short and long branches along chain, low melting, low density
Ziegler process (coordination catalysts, titanium tetrachloride/triethyl aluminum)	Broad molecular weight distribution, few branches, high density, linear polymers, high melting, comonomers control crystallinity levels
Metallocene catalysis (bis-cyclopentadienyl–metal complexes)	Relatively narrow molecular weight distributions, controlled levels of branching, improved control of comonomer distribution
Metallocene–Ziegler	High comonomer incorporation

Table 14.3 Polyethylene Properties

Polymer	Designation	Degree of Branching, CH$_3$/100C	Density Range, g/cm^3	Melting Temperature Range, °C
Low-density polyethylene	LDPE	2–7	0.915–0.94	100–129
High-density polyethylene	HDPE	0.1–2	0.94–0.97	108–129
Linear low-density polyethylene	LLDPE	2–6	0.91–0.94	99–108
Metallocene linear low-density polyethylene	*m*-LLDPE	3–7	0.90–0.92	83–102
Ultra-low-density polyethylene	ULDPE	~7	0.86–0.90	~80–85

Source: Data collated from J. Brandrup and E. H. Immergut, eds., *Polymer Handbook*, 3rd ed., Wiley-Interscience, New York, 1989, and G. M. Benedikt and B. L. Goodall, eds., *Metallocene-Catalyzed Polymers: Materials, Properties, Processing, and Markets*, Plastics Design Library, Norwich, NY, 1998.

as 0.855 g/cm^3 (3), so that the 0.86 g/cm^3 end of the ultra-low-density polyethylene is essentially amorphous.

For some applications, HDPE has too high a modulus; however, the rheological and mechanical properties are improved by having a linear chain. LLDPE and *m*-LLDPE offer the best of both worlds for many applications. Frequently polymers with various extents of comonomers are blended to tailor materials for such properties as improved puncture resistance.

14.1.3 Polypropylene Properties

The major manufacturing process utilizes Ziegler-type catalysts. However, the newer metallocene-based materials offer improved control and specialty properties. The major product is isotactic polypropylene. With the Ziegler-type catalysts, various quantities of atactic polypropylene are included as by-products. The degree of isotacticity of the polypropylene also varies with the exact process. In general, the higher the isotacticity index, the higher the modulus and yield stress, and the lower the elongation to break.

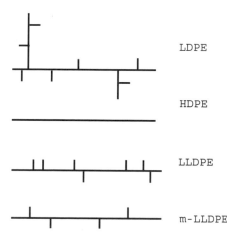

Figure 14.1 Model structures of polyethylenes, illustrating the various types of regularity of side chains.

Scheme 1

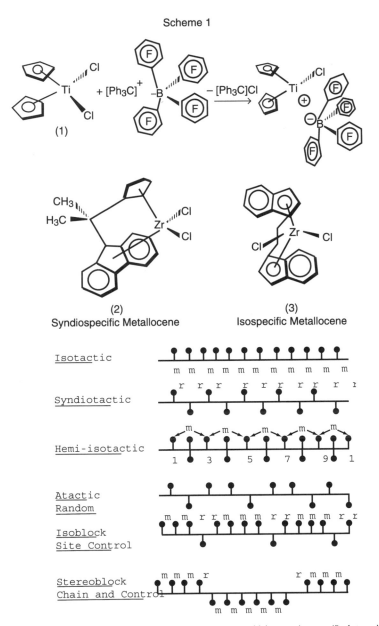

Figure 14.2   *Upper:* Illustration of metallocene catalyst structure, which must be specific for each stereo-specific polymer structure. *Lower:* Microstructures of polypropylene; most can be made via metallocene catalysis.

With the advent of metallocene catalysis, a range of tacticities and structures are possible; see Figure 14.2 (4). Thus a range of compositions not easily prepared before, such as syndiotactic, hemi-isotactic, and isoblock copolymers, are now possible.

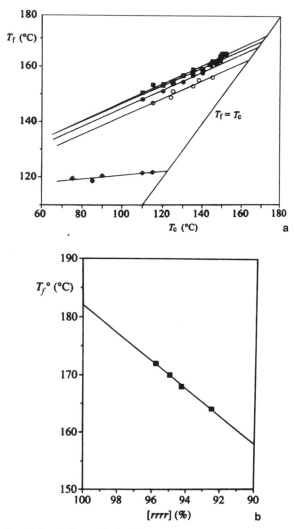

Figure 14.3 Utilization of the Hoffman–Weeks plot to establish the theoretical melting temperature of *rrrr* syndiotactic polypropylene. (a) Melting temperatures T_f of melt-crystallized *syn*-polypropylene, taken from DSC scans at 2.5°C/min, as a function of the crystallization temperature, T_c, extrapolated to $T_f = T_c$. (b) Extrapolation of the $T_f = T_c$ values in (a) to 100% pentad content, *rrrr*.

The melting temperatures of syndiotactic polypropylene were studied as a function of syndiotactic pentad content, *rrrr* (see Section 2.3.4) values determined via ^{13}C NMR spectra; see Figure 14.3 (5). The upper figure utilizes the Hoffman–Weeks extrapolation procedure; see Section 6.8.6. The lower figure shows a second extrapolation of the data to 100% *rrrr* content, arriving at a melting temperature of nearly 182°C. By comparison, an isotactic polypropylene with >99% *mmmm* composition has a melting temperature of 170°C (6).

14.1.4 Polyolefin Elastomers

Two elastomers are made based on statistical copolymers of ethylene and propylene: EPDM, in which the diene portion serves as a cross-linking site, and its non-cross-linking counterpart, EPM. Since these materials have few reactive sites, they are relatively impervious to oxidation or hydrolysis. Frequently they are blended with either polyamides or polypropylene to form impact-resistant plastics; see Section 13.8.

14.2 THERMOSET POLYMER MATERIALS

A number of plastic, adhesive, and coating materials are utilized in a densely crosslinked form known as thermosets. Some of the more important of these include the phenol–formaldehyde resins, urea–formaldehydes, polyimides, epoxies, amino resins, and the alkyds, among others (7). Because these materials are densely cross-linked, they are amorphous, usually do not exhibit any rubber elasticity behavior, and even their glass transition may be suppressed. This last arises because of insufficient chain mobility. The thermosets are synthesized through step-polymerization kinetics, often through condensation routes.

14.2.1 Phenol–Formaldehyde Resins

These materials are produced by heating phenol and formaldehyde in the presence of either an acid or a base catalyst. With a base catalyst and an excess of formaldehyde, a composition called resole resins are prepared. With an acid catalyst and an excess of phenol, a two-stage resin called novolacs are made. In both cases high-melting or viscous oligomers are made, which react further at elevated temperatures to produce high modulus but brittle materials (8).

The basic structure of these materials in the fully cured (cross-linked) state may be written

The open bonds indicate points of attachment to other regions of the network. Of course, this structure is only an illustration of the very many possibilities of a very irregular network. Applications include electrical connectors, pot handles, and so on.

14.2.2 Epoxy Resins

One of the most important monomers in epoxy chemistry is the diglycidyl ether of bisphenol A:

$$H_2C\overset{O}{-}CH-CH_2-O-\left\langle\bigcirc\right\rangle-\underset{\underset{CH_3}{|}}{\overset{\overset{CH_3}{|}}{C}}-\left\langle\bigcirc\right\rangle-O-CH_2\cdot CH\overset{O}{-}CH_2$$

A simple way to polymerize these materials involves reacting them with primary or secondary amines. In such reactions the epoxy group his a functionality of one, resulting in ether bonds of the type:

$$-CH_2-CH_2$$
$$|$$
$$O$$
$$|$$
$$-CH-CH_2-NH-R-$$

Many epoxides have structures more complex than the diglycidyl ether of bisphenol A, having longer linear structures or more epoxy groups per monomer unit, or both. For some adhesives the amine might be replaced by a low-molecular-weight polyamide. Applications include marine coatings, electronic packaging, and adhesives.

This last is the two-part adhesive sold in hardware stores. One part is based on a solution of a dilgycidyl ether derivative, and the other part is based on a solution of a polyamine or polyamide. On mixing, they react with a *pot life* of from 5 to 30 minutes, usually given with the instructions. The pot life is the length of time before gelation, after which it does not flow, and if applied too late may be weak. (See the TTT diagram, Figure 8.27.) The adhesive itself bonds to the two surfaces involved, as well as flowing into the (usually) irregular surface, creating mechanical interlocks.

14.2.3 Mechanical Behavior

The thermoset resins are important because they have high moduli, and a relatively high range of useful temperatures. Table 14.4 compares the properties of selected thermosets

Table 14.4 Mechanical behavior of selected thermosets and thermoplastics

Composition	Density (g/cm^3)	Young's Modulus, (Gpa)	Tensile Yield Strength, MPa	Heat Distortion Temperature, °C
Thermosetting compositions				
Urea/Formaldehyde	1.56	10.0	43	—
Phenol/Formaldehyde	1.36	8.6	50	121
Epoxy resin	1.20	3.6	72	>110
Polyimide	1.40	5.0	72	>243
Thermoplastic polymers				
Polystyrene	1.05	3.2	46	73
High-density Polyethylene	0.96	1.1	32	49
Polyamide 6,6	1.10	2.9	65	75

Source: Based on H. G. Elias, *Macromolecules*, 2nd ed., Plenum, New York, 1984.

with the corresponding properties of common thermoplastics. The thermosets have higher moduli because of their dense cross-linking. The heat distortion temperature, given as a certain deflection under a load of 1.85 MPa, is somewhat lower than the glass temperature or melting temperature, where they exist. The properties of the thermosets in Table 14.4 are only approximate, as these are representative of classes of the indicated materials.

In commercial use a common application of all of these materials is in the form of glass fiber reinforced polymer composites; see Sections 13.7 and 13.8. The glass fiber provides a significant toughening for these materials. For other applications, a variety of fillers are commonly used.

14.3 POLYMER AND POLYMER BLEND ASPECTS OF BREAD DOUGHS

14.3.1 Polymer Blend Aspects of Bread Making

Bread is made from water plasticized wheat flour, which in turn consists of ground-up wheat seeds. Dry flour consists of approximately 12% of protein, 87% of starch, and approximately 1% of everything else, such as minerals and salts. Since both the protein and the starch are polymeric, bread dough and bread making can be considered from the polymer blend point of view. As with most polymer blends, the starch and protein are immiscible, see Chapter 13.

For simplicity, the following discussion will consider bread making from just the recipe of wheat flour, water, and heat; only yeast is added. The yeast, of course, causes bubbles of carbon dioxide to form in the bread, hence the finished product is also a foam. Components such as rye flour, eggs, caraway seeds, and the like, will be considered absent, except where noted.

Wheat starch is composed of two components itself: amylose, a linear, amorphous polymer; and amylopectin, a branched, semicrystalline polymer (9). The protein is known as gluten, also composed of two polymers, gliadin, a low molecular weight, soluble polymer; and glutenin, a high-molecular-weight, cross-linked, elastic polymer primarily responsible for the viscoelastic properties of bread doughs.

14.3.2 Morphology Changes during Baking

The morphology of flour consists of a continuous gluten phase with a dispersed phase of starch granules. During the formation of bread dough from flour, water is added—about 34% to 37% water being present in fresh baked breads. During dough formation at room temperature, the water diffuses into the protein phase, the starch remaining essentially unchanged and dispersed. In Figure 14.4 (10) as viewed with an environmental scanning electron microscope (ESEM) the starch granules were large, plump, and round in shape. The hydrated gluten matrix appeared not to constrict the structure significantly. During baking, starch gelatinizes, meaning that it becomes plasticized with the water, between 52 and 66°C.

The gelatinization temperature appears related to the melting temperature of the amylopectin portion of the starch, which drops rapidly with increasing moisture content. Under normal atmospheric conditions most starches contain 10% to 17% moisture. In particular, starch with 11% water content melts at 65°C (11). Thus, on heating, the water is free to swell the starch after it melts. Much of the water then migrates from the minor protein phase into the major starch phase.

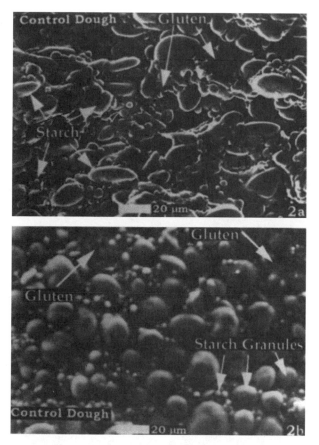

Figure 14.4 Scanning electron microscope studies of bread dough morphology. (*a*) Dried dough showing the starch granules embedded in the gluten matrix. (*b*) Environmental SEM (ESEM) of fresh dough showing hydrated starch granules and the thick gluten matrix that holds the dough together.

Figure 14.5 (10) shows the starch granules to be highly distorted in shape and variable in size, indicating extreme damage due to gelatinization; see Figure 14.5*b* (inset) arrows. The lines between the hydrated, gelatinized starch granules and the surrounding gluten matrix become indistinct, possibly indicating that amylose leached out of the granules and interacted with the gluten matrix. Thus a degree of dual phase continuity is brought about.

The protein portion of the bread also undergoes a transformation on heating. It forms a thermoset network via disulfide cross-linking involving a rearrangement of the cysteine–cysteine amino acids, similar to the chemistry of the hard-boiling of egg whites. This process has sometimes been called a denaturization. (See, e.g., Figure 9.2.)

On cooling back to room temperature, the whole material is transformed from a viscoelastic dough to a soft solid as the amylopectin begins to recrystallize. Of course, the yeast-caused foam bubbles play a critical role in the final texture.

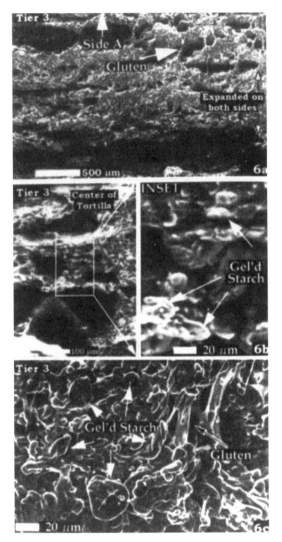

Figure 14.5 Fully baked tortilla. (*a*) SEM of cross section of the fully baked disk showing the considerable expansion of the crumb. (*b*) ESEM illustrating the gelatinization of the starch granules. (*c*) SEM of the crumb of a dried fully cooked tortilla showing the extensive gluten structure and gelatinized starch.

14.3.3 Mechanical Treatment (Kneading)

Kneading constitutes a form of shearing action. Kneading of the bread dough (12) greatly affects the structure–property relationships by establishing an equilibrium between the co-existing phases and orientation of polypeptide chains of the gluten phase. Thus basic mixing of doughs includes the deformation and breaking down of the liquid and gas-dispersed particles, orientation of the gluten, decreasing the size of the liquid and gas-dispersed particles and the thickness of the gluten component, and twisting the starch granules in the shear flow, which provides a higher fluidity due to a "ball-bearing" effect. The starch granules migrate toward the higher-shear regions, providing a decrease in water content of the central layers and formation of substantially starch-empty surface layers of lower stickiness. Kneading also reduces the size of the gas bubbles.

14.3.4 Bread Staling

Increases in amylopectin crystallinity, underlies part of the phenomenon known as bread staling (9). However, it has been recently shown that a rise in the glass transition temperature of the gluten, probably due to moisture migration, is more important (13–15).

The glass transition temperature of fresh baked crumb is just below room temperature (16), and the starch phase substantially amorphous, so the bread is soft. Thus, when bread is first baked, the combination of moisture migration and heat destroy much of the crystallinity in the amylopectin. However, on standing at room temperature, the starch undergoes retrogradation; that is, moisture leaves, and crystallinity sets in again. On re-heating such breads, as in common experience, the bread may become soft again as the crystallites remelt. This assumes that the staling was not primarily due to moisture evaporation, which can also play a role, and may be substantially irreversible.

Thus large bodies of polymer theory, rheology of dispersed phases, polymer blend and polymer interface interactions, as well as excluded volume effects of polymer solutions are seen to be true for bread doughs, as indeed they must.

14.3.5 Bread and Meatlike Textures

The *texture* or feel of bread-type foods arises through having the starch as a continuous phase. If the protein constitutes the continuous phase, the texture becomes meaty to the mouth. While oversimplified, this is the idea behind soy-bean-based texturized products, resembling bacon, chicken, ham, frankfurters, and the like (17).

Soy protein concentrates typically contain 70% to 72% crude protein. For example, bacon strips can be made by a texturization processes involving twin screw extrusion (18). The high-protein content yields the protein as the continuous phase.

14.4 SILK FIBER SPINNING

Nature produces many fine and useful polymeric materials. Important fibers include wool, cotton, and silk. Silk fiber is a natural product protein, produced by the silkworm, *Bombyx mori*, a species of moth, and other insect and arachnid species such as wasps, bees, butterflies, and spiders. The *Bombyx mori* silkworm is a caterpillar that primarily eats mulberry leaves. Silkworms construct the silk cocoons to protect themselves during metamorphosis. The fiber itself is a continuous double monofilament of 1000 to 1600 m in length (19).

Two proteins are synthesized in the silk glands of the silkworm, fibroin and sericin. The double-fibroin monofilaments constitute the main fibrous material. These are covered with sericin, an adhesive. In the solid state, the fibroin may have one of three chain conformations: a random coil, an α-form (silk fibroin I), and an antiparallel-chain pleated-sheet β-form (silk fibroin II). The α-form is a crankshaft pleated structure. The β-

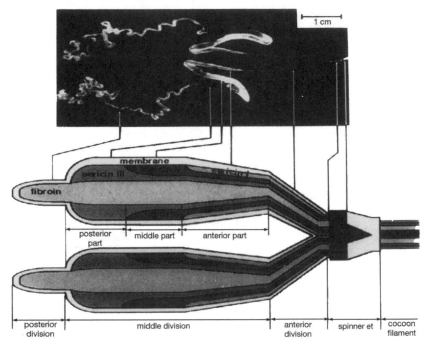

Figure 14.6 Photomicrograph (upper) and schematic diagram (lower) of the silk gland of the larvae of the silkworm, *Bombyx mori*. This is Mother Nature's original solution spinning process.

form is the important structure observed in nature. The fibroin consists primarily of glycine, alanine, and serine, with important parts of the crystalline regions being the 59-mer sequence Gly–Ala–Gly–Ala–Gly–Ser–Gly–Ala–Ala–Gly–{Ser–Gly–(Ala–Gly)$_2$}$_8$Tyr (20). Of course, while proteins are copolymers, the positioning of each mer is specific and determined genetically.

The silkworm has two glands, one on each side of the larva head. Each gland has three divisions: posterior, middle, and anterior, as well as a spinneret region, Figure 14.6 (19). The filament flows from left to right in the figure. The motions of the silk worm's head provides a significant drawing action for the orientation of the silk filaments.

The silkworm forms its fibers from the liquid crystalline nematic state (21,22), which is subsequently transformed into a gel state during spinning. The liquid crystalline state is indicated by the streaming birefringence observed in liquid silk solutions flowing out of the anterior division of the spinning gland. Because of the liquid crystalline organization of the silk solution, its viscosity is thought to be significantly lower than otherwise expected (23).

The fibers are lustrous and strong, with 0.3 to 0.4 GPa tensile strength and about 22% elongation (24); see Figure 14.7 (25). These stress-strain curves are in the same range as many synthetic fibers, see Table 11.2.

Research on silk fiber spinning continues, even intensifying on several bases:

1. Silk fibers remain one of the more important articles of commerce on the planet. The silkworm is being improved genetically, and more efficient production methods are under development,

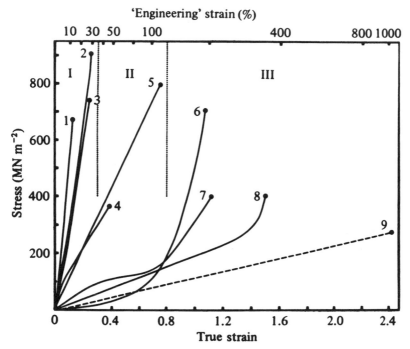

Figure 14.7 Silks from different sources exhibit a wide range of stress–strain properties. The examples shown are divided into three groups on the basis of their extensibility. (1) *Anaphe moloneyi*; (2) *Araneus sericatus* dragline; (3) *Bombyx mori* cocoon; (4) *A. diadematus* cocoon; (5) *Galleria mellonella* cocoon; (6) *A. sericatus* viscid. (7) *Apis mellifora* larval; (8) *Chrysopa carnea* egg stalk; (9) *Meta reticulata* viscid. Data collected from various sources.

2. Other sources of silk, particularly from spiders, continues to draw interest.
3. Modern polymer science is trying to understand and emulate Mother Nature in the liquid crystal manufacture of highly oriented fibers. The natural silk fibers are spun via aqueous solutions (ecologically *green* by today's standards); the resultant materials combine high strength, luster, and light weight.

14.5 DENDRITIC POLYMERS AND OTHER NOVEL POLYMERIC STRUCTURES

14.5.1 Aspects of Self-Assembly

Classically polymers may be linear, cross-linked, block or graft copolymers, and so on, as delineated throughout this text. However, there are a host of novel structures now synthesized, with some beginning to play important roles in polymer science and engineering. Many fall under the category called *self-assembling* polymers.

Many microorganisms, such as viruses, appear to have important aspects of self-assembly in their structure. Fyfe and Stoddart (26) point out that for the synthetic chemist to be able to build nanosystems, similar to those in the natural world, they must learn to control the intermolecular, noncovalent bond. This has lead to the field of

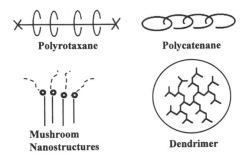

Figure 14.8   Novel polymer structures. These four structures are being examined for applications in various engineering and biomedical areas.

supramolecular chemistry, as well as a host of self-assembling polymers. These include the dendritic polymers, polycantananes, polyrotaxanes, and others; see Figure 14.8.

14.5.1.1   *Polyrotaxanes*   A polyrotaxane is a polymer chain with rings on it (26). The rings are unable to slip off because the ends are blocked with bulky stopper groups. Since the rings are not chemically attached to the polymer, the rings may be said to be self-assembled onto the polymer chain backbone. A closely related material is the polypseudorotaxane, defined as a molecular thread encircled by one or more macrorings. At least one of the thread's extremities must not possess a bulky stopper group.

There are several methods of synthesizing polyrotaxanes; see Figure 14.9 (27). Synthesis method *a*, by polymerization of a chain in the presence of macrocycles, constitutes one of the more important ways to make a polyrotaxane. For the polypseudorotaxanes, synthesis method *d* leads to an equilibrium between threaded and dethreaded rings. Synthesis method *e* leads to freezing in of the structure. Controlled dethreading may be useful for pharmaceutical drug delivery, for example. As a class, the polyrotaxanes and polypseudorotaxanes are also being considered as molecular switches and sensors.

The catenanes resemble a series of chain links. Each link is bound to its neighbors by physical forces only. Since each one of the links constitutes a separate molecule in the chemical sense, the links also bear some of the features of self-assembly (28,29).

14.5.1.2   *Mushroom Nanostructures*   Stupp et al. (30) have evolved strategies to create supramolecular units of various sizes and shapes via self-assembly. They found that rod–coil block copolymers can self-assemble into long striplike aggregates measuring 1 μm or more in length and a few nanometers in other dimensions. The mushroom nanostructures in Figure 14.8 constitute yet another example of a self-assembling nanostructure. These are triblock copolymers of polystyrene–*block*–polyisoprene–*block*–diphenylester, the latter exhibiting a rodlike structure. These materials have molecular weights in the range of 200,000 g/mol. They exhibit strong second harmonic generation nonlinear optical activity because of their lack of a center of inversion; see Section 14.8.

14.5.1.3   *Dendrimers and Hyperbranched Polymers*   The term *dendrimer* derives from the Greek words *dendron*, meaning tree, and *meros*, meaning part. Dendritic macromolecules are hyperbranched, fractal-like structures emanating from a central core

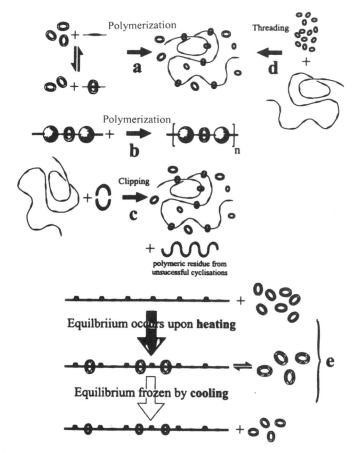

Figure 14.9 Schematic representations of the principal synthetic routes to polyrotaxanes, as illustrated in Figure 14.8 (a) Polymerization in the presence of a macrocycle. (b) Polymerization of a rotaxane. (c) Clipping a macrocycle onto a polymer chain. (d) Treading a macrocycle on to a polymer chain. (e) Slipping macrocycles onto a polymer by heating then *freezing onto* the polymer chain by cooling.

and containing a large number of terminal groups. There are two synthetic approaches, the divergent and the convergent growth approaches (31). In the divergent approach, the synthesis starts with the central core and works outward. In the convergent approach, slices of the "pie" are synthesized separately, and assembled by reaction with the core later. Both kinds of dendrimers are globular in nature, even to the point of mimicking some globular proteins.

During the synthesis of dendrimers, each successive reaction step leads to an additional *generation*, G, of branching (32). Sometimes the generations are numbered $G-1, G-2, G-3, \ldots$. Ideally dendrimers exhibit monodispersity. However, because the synthesis of higher generation materials requires numerous steps, the final products often contain defects. Materials containing up to 10 generations are known (33). Figure 14.10 (31) illustrates the structure of a fourth-generation convergent aromatic polyether dendrimer. With every new generation, the dendrimer doubles the number of terminal

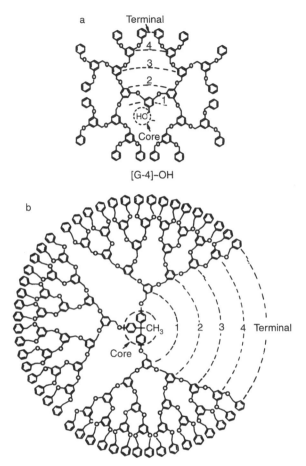

Figure 14.10 Dendrimer structures. (a) {G—4}—OH monodendron obtained by convergent growth. (b) {G—4}$_3$—{C} tridendron obtained by convergent growth.

groups, and also approximately doubles its molecular weight. The mass of a dendrimer increases as 2^G, while the volume available for the mers only increases as G^3. Thus the local density increases with each generation, leading de Gennes and Hervet (34) to predict that a maximum generation number exists, beyond which a perfect growth dendrimer cannot be made.

Besides spherical shapes, dendrimers can also be self-assembled into cylindrical shapes, using tapered monodendrons (35). These materials emulate the protein coats surrounding the nucleic acid in viruses.

In contrast to monodenrons, hyperbranched polymers are synthesized in a single-step reaction. As their name suggests, the products are highly branched and have varying degress of irregularity.

As an example of the broad class of these new materials, the properties of the dendrimers will be briefly explored.

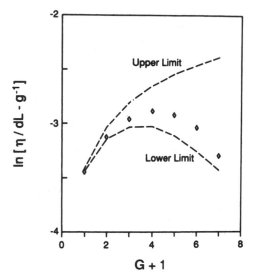

Figure 14.11 Intrinsic viscosity of convergent growth dendrimers. Dashed lines are upper and lower limits calculated from the dimensions of the mers.

14.5.2 Intrinsic Viscosity of Dendrimers

Figure 14.11 (31) shows the intrinsic viscosity of several generations of the aromatic polyether dendrimers. The maximum in viscosity with generation is a fairly common observation (36,37) with dendrimers, but except for the liquid crystalline polymers (Chapter 7), rarely observed in the polymer world. However, the quantity of interest is the numerical value of the intrinsic viscosity. Note that a value of $\ln \eta$ of -3.7 corresponds to an intrinsic viscosity of 0.025 dL/g, the viscosity of Einstein spheres (see Section 3.8.2). These values are only slightly higher, and they might be expected to level off to the Einstein sphere value at about generation 10. In fact Aharoni et al. (38), found this theoretical value for their materials.

14.5.3 Electron Microscopy Studies of Dendrimers

Jackson et al. (33) investigated the transmission electron microscopy images of a series of generations of dendrimers based on a tetrafunctional core of ethylenediamine and successive additions of methyl acrylate and ethylenediamine, called PAMAM; see Figure 14.12. The tenth generation dendrimer, shown in Figure 14.12a, has a mean diameter of 14.7 nm, with a theoretical molecular weight of 934,721 g/mol.

14.5.4 Applications of Dendrimers

There are several applications, actual and proposed, for the dendrimers (39). These include:

1. Drug delivery vehicles. Characteristics include prolonged circulation in the blood, interactions with cell membranes, targeting with antibodies, encapsulation and

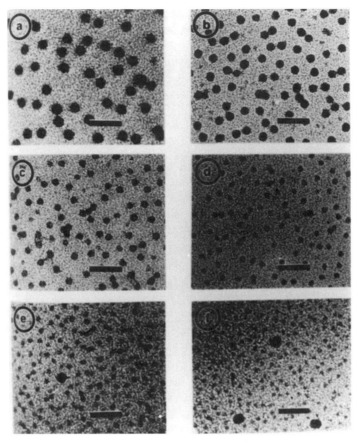

Figure 14.12   PAMAM dendrimers positively stained with a 2% aqueous sodium phosphotunstate imaged by conventional TEM. (a) G10, (b) G9, (c) G8, (d) G7, (e) G6, (f) G5. The scale bars are 50 nm. For G6 and G5, a small amount of G10 was added as a focusing aid.

solubilization of hydrophobic drugs, and so on. The dendritic polymers have two locations for placing drugs: First, the outer shell can be made with reactive species; these can be used to bond drugs. Second, some dendrimers have a lower than average density inside, with high density on the surface. Then, a dendrimer molecule can serve as a host for guest drug molecules, which slowly diffuse out. Similarly such a material can serve as a catalyst carrier, and so on. Working with novel dendrimers, Estfand and co-workers (40) are developing controlled delivery strategies to improve the aqueous solubility of hydrophobic guest molecules via the formation of inclusion complexes. The objective involves tailor-made dendritic hydrophobic cavities for specific drug compounds, resulting in improved and controlled drug delivery.

2. Production of metal and semiconductor nanoparticles in polymer systems. Dendrimers are used for the synthesis of metallic nanoparticles *via* dissolution of a metal precursor in a supercritical fluid followed by reduction.

3. Dendrimer complexes used as ligands for selective binding of toxic metal ions. The system could potentially be used for remediation of contaminated water and soils.

4. Plastics toughened against fracture by increasing the critical energy release rate.

5. Ion-exchange chromatography. Use of specific surface chemistry coupled with the use of different dendrimer generations allows separation selectivity to be modulated.

14.6 POLYMERS IN SUPERCRITICAL FLUIDS

14.6.1 General Properties of Supercritical Fluids

Classically pure substances are solid, liquid, or gaseous. A supercritical fluid is a pure substance compressed and heated above its critical point; see Figure 14.13 (41). The major characteristics of a supercritical fluid are liquidlike density, gaslike diffusivity and viscosity, and zero surface tension.

The solvent power of a supercritical fluid can usually be increased by increasing the pressure. The decaffination of coffee provides a simple commercial example (42,43). Here, supercritical carbon dioxide is contacted with ground coffee. The caffeine preferentially dissolves in the carbon dioxide, which is subsequently moved to another location. Then, when the pressure is lowered, the caffeine precipitates out, and the cycle repeated.

Table 14.5 shows some supercritical fluids used to dissolve polymers. Polymers can be synthesized in supercritical fluids in a manner similar to that in ordinary aqueous or organic fluids (44,45). Polymers in supercritical fluids exhibit the same solution properties as discussed in Chapter 3: The molecular weight can be determined by light-scattering or neutron scattering, for example. However, values of χ or A_2 can now be varied by

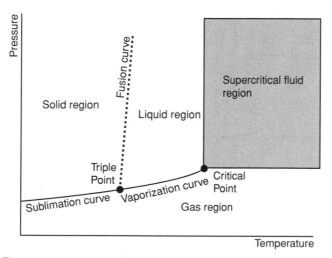

Figure 14.13 The pressure–temperature phase diagram of a pure substance, emphasizing the supercritical fluid region. The critical point is the highest pressure and temperature at which a pure substance can exist in a vapor–liquid equilibrium.

Table 14.5 Supercritical solvents for polymers

Solvent	T_c, C	P_c, Bar	Polymers Dissolved	Reference
Carbon dioxide	31.0	73.8	Poly(dimethyl siloxane), fluoropolymers	(a)
Propane	96.6	42.5	Polypropylene, EPDM	(b)
1,1,1,2-tetrafluoroethylene	101.1	40.6	Polystyrene	(c)

References: (a) E. Buhler, A. V. Dobrynin, J. M. DeSimone, and M. Rubinstein, *Macromolecules*, **31**, 7347 (1998). (b) S. J. Han, D. J. Lohse, M. Radosz, and L. H. Sperling, *Macromolecules*, **31**, 5407 (1998). (c) C. Kwag, C. W. Manke, and E. Gulari, *J. Polym. Sci.: Part B: Polym. Phys.*, **37**, 2771 (1999).

altering the pressure as well as the temperature. For many systems, changing the pressure is far easier than changing the temperature.

14.6.2 Dispersion Polymerization via Supercritical Fluids

Dispersion polymerization starts as a one-phase, homogeneous system such that both the monomer and the polymerization initiator are soluble in the polymerization medium (the supercritical fluid here), but the resulting polymer is not. The polymerization is initiated homogeneously with the resulting polymer, being insoluble, phase separating into primary particles. Once nucleated, these primary particles become stabilized by added amphipathic polymer molecules that prevent particle flocculation and aggregation. The particles, forming in the colloid size range, are usually stabilized by a *steric* mechanism as opposed to the electrostatic mechanism common in aqueous environments; see Section 4.5. The amphipathic polymer molecules impart steric stabilization because part of the chain becomes adsorbed onto the surface of the dispersed phase, becoming the anchoring segment, and part of the polymer chain projects into the continuous supercritical phase. This projecting moiety prevents flocculation by mutual excluded volume repulsion during Brownian collisions, thereby imparting stability to the colloidal particles.

In a carbon dioxide supercritical fluid, polystyrene can be polymerized by such dispersion a polymerization, using polystyrene–*block*–poly(1,1-dihydroperfluorooctyl acrylate), FOA, as the stabilizer. The perfluorinated moiety, soluble in supercritical carbon dioxide, provides the steric stabilization. Figure 14.14 (46) illustrates the particles of polymer produced in this way. With increasing molecular weight of the stabilizing block copolymer, the particles become smaller and more uniform.

14.6.3 Applications of Polymers in Supercritical Fluids

There are several applications for this new technology:

1. Environmentally green solvents. Supercritical carbon dioxide presents an environmentally benign medium for polymerizations (and other chemical operations), minimizing pollution from organic solvents and facilitating the isolation of the polymeric product.
2. Assisting in blending by lowering the melt viscosity of the polymers (47).
3. Extraction of dyes and other molecules from water into liquid or supercritical carbon dioxide (48) using dendritic surfactants.
4. Serving as a polymerization medium.

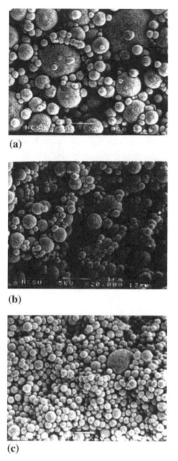

(a)

(b)

(c)

Figure 14.14 SEM of polystyrene particles synthesized via dispersion polymerization utilizing (a) 3.7 K/17 K, (b) 4.5 K/25 K, and (c) 6.6 K/35 K polystyrene–*block*–poly(FOA) stabilizer.

14.7 ELECTRICAL BEHAVIOR OF POLYMERS

Most polymers are insulators. In fact, polyethylene and poly(vinyl chloride) are widely used as the insulating materials for electrical wiring, because they are highly insulating and weather resistant. However, to say a polymer is an insulator is a qualitative statement. To what extent does it conduct electricity? Recently, families of conducting polymers were invented, resulting in the 2000 Nobel Prize in Chemistry.

14.7.1 Basic Electrical Relationships

There are several basis relationships governing the electrical behavior of all materials. Ohm's law can be written

$$I = \frac{V}{R} \tag{14.1}$$

where I is the current in amperes (A), V is the voltage in volts, and R is the resistance in ohms (Ω). The concept of the capacitance, C, involves storage of electrical charge,

$$C = \frac{Q}{V} \qquad (14.2)$$

where Q is the charge in coulombs (C) (49).

When an alternating voltage is applied to an imperfectly conducting material, a current flows that is displaced in time in such a way that it is out of phase with the voltage by an angle δ. The tangent of this angle plays a role similar to that described in Chapter 8 for mechanical behavior and is referred to as the dissipation factor. The power loss is given by

$$W_{\text{loss}} = 2\pi' f C_p (\tan \delta) V^2 \qquad (14.3)$$

where W_{loss} is loss in watts, f is the frequency of the applied voltage, and C_p represents the parallel capacitance of the material. The relative dielectric constant (permittivity) is given by

$$\varepsilon' = \frac{C_p}{C_v} \qquad (14.4)$$

where C_v is the capacitance of vacuum. The dielectric loss constant, ε'', sometimes called the loss factor, is given by

$$W_{\text{loss}} \cong \varepsilon' \tan \delta = \varepsilon'' \qquad (14.5)$$

These quantities can be used to describe the onset of molecular motion at the glass transition temperature; see Section 8.3. In the present setting, we are interested in determining the electrical behavior of polymers.

Two other quantities of interest are the resistivity, the resistance per unit distance, with units of Ω/cm, and its inverse, conductivity, with units of $\Omega^{-1}/\text{cm}^{-1}$, or siemens per centimeter (S/cm).

14.7.2   Range of Polymer Electrical Behavior

The range of conductivities available for a range of polymers and other materials is given in Figure 14.5 (50). An insulator has conductivities in the range of 10^{-18} to 10^{-5} S/cm. A semiconductor is in the range of 10^{-7} to 10^{-3} S/cm, and a conductor is usually given as 10^{-3} to 10^6 S/cm. As seen in Figure 14.15, polystyrene, polyethylene, nylon, and a host of other ordinary polymers are insulators.

On the other end of the scale are the conducting and semiconducting polymers. Several types exist. For example, conducting fillers may be added, such as short metallic fibers that touch each other or carbon black. Alternately, an ionic polymer may be employed, or a salt may be added to the polymer. Conductivity in the latter systems depends on the moisture content of the polymer.

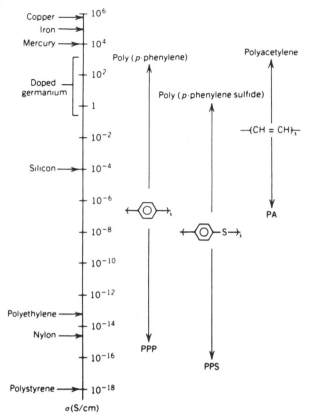

Figure 14.15 Conductivity scale comparing several polymers doped with AsF_5 with conventional materials (50).

14.7.3 Conducting Polymers

In the fall of 2000, Drs. A. J. Heeger, A. G. MacDiarmid, and H. Shirakawa (50a, 50b, 50c) were awarded the Nobel Prize in Chemistry for the discovery and development of conducting polymers. These materials, based on doped polyacetylene and other con-jugated polymers, are sometimes called *synthetic metals*. These materials are widely used as anti-static agents, shields for computer screens against unwanted electromagnetic radiation, and for "smart" windows that can exclude sunlight, light-emitting diodes, solar cells, and electronic displays. Some doped alternating double bond systems are shown in Table 14.6 and Figure 14.15 (51). However, conjugated organic polymers in their pure state are still electrical insulators. Chemical dopants include oxidative mate-rials such as AsF_5 and I_2. Some newer doping materials include SbF_5, $AlCl_3$, Br_2, O_2 and a host of others. The identities of the anionic counterions derived from these dopants show a variety of structures such as $Sb_2F_{11}^-$, AsF_6^-, and I_3^-.

The basic mechanism of electronic conduction is illustrated in Figure 14.16 (50). The importance of π-bonds in electronic conduction must be emphasized, as the overlapping

Table 14.6 Structures and conductivity of doped conjugated polymers (51)

Polymer	Structure	Typical Methods of Doping	Typical Conductivity, S/cm
Polyacetylene		Electrochemical, chemical (AsF_5, I_2, Li, K)	$500–1.5 \times 10^3$
Polyphenylene		Chemical (AsF_5, Li, K)	500
Poly(phenylene sulfide)		Chemical (AsF_5)	1
Polypyrrole		Electrochemical	600
Polythiophene		Electrochemical	100
Poly(phenyl quinoline)		Electrochemical, chemical (sodium naphthalide)	50

electronic clouds contribute to the conduction. Strong interactions among the π-electrons of the conjugated backbone are indicative of a highly delocalized electronic structure and a large valence bandwidth. For good conductivity, the ionization potential, IP, must be small. The electron affinity, EA, reflects the ease of addition of an electron to the polymer, especially through polymer doping. The bandgap, E_g, correlates with the

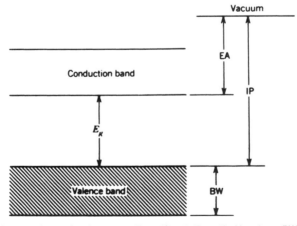

Figure 14.16 Polymer π-electron band structure. E_g represents the optical bandgap; *BW* is the bandwidth of the fully occupied valence band; *EA* is the electron affinity; and *IP* is the ionization potential (50).

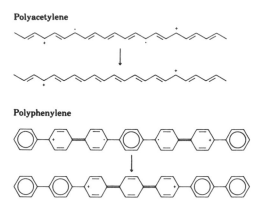

Polyacetylene

Polyphenylene

Figure 14.17 Structures of *trans*-polyacetylene and poly(*p*-phenylene) showing reactions of two polarons (radical cations) to produce a bipolaron (dication) (50).

optical-absorption threshold. The bandwidth, BW, correlates with carrier mobility; a large bandwidth suggests a high intrachain mobility, which favors high conductivity.

The mechanisms by which these polymers conduct electricity have been a source of controversy ever since conducting polymers were first discovered. At first, doping was assumed to remove electrons from the top of the valence band, a form of oxidation, or to add electrons to the bottom of the conduction band, a form of reduction. This model associates charge carriers with free spins, unpaired electrons. This results in theoretical calculations of conduction that are much too small (51). To account for spinless conductivity, the concept of transport via structural defects in the polymer chain was introduced. From a chemical viewpoint, defects of this nature include a radical cation for oxidation effects, or radical anion for the case of reduction. This is referred to as a polaron. Further oxidation or reduction results in the formation of a bipolaron. This can take place by the reaction of two polarons on the same chain to produce the bipolaron, a reaction calculated to be exothermic; see Figure 14.17 (50). In the bulk doped polymer, both intrachain and interchain electronic transport are important.

14.8 POLYMERS FOR NONLINEAR OPTICS

The development of nonlinear optic, NLO, materials constitutes one of the newest and most exciting interdisciplinary branches of research, demanding cooperation among polymer scientists, chemists, physicists, and materials scientists. Nonlinear optics is basically concerned with the interaction of optical frequency electromagnetic fields with materials, resulting in the alternation of the phase, frequency, or other propagation characteristics of the incident light. Some of the more interesting embodiments of second-order NLO include second harmonic generation, involving the doubling of the frequency of the incoming light; frequency mixing, where the frequency of two beams is either added or subtracted; and electrooptic effects, involving rotation of polarization and frequency and amplitude modulation (52). Third-order NLO involves three photons, for similar effects.

Nonlinear optical effects have been explored in a range of materials—inorganic, organic, and polymeric. The basic structural requirement for second-order NLO activity is a lack of an inversion center, which is derived principally from asymmetric molecular configurations and/or poling. Poling, the application of an electric field of high voltage across the material, orients all or a portion of the molecule, usually in the direction of the field. Third-order NLO behavior does not require asymmetry. Active materials include inorganic crystals such as lithium niobate and gallium arsenide (53,54), organic crystals such as 2-methyl-4-nitroaniline, MNA (55,56), polymers such as polydiacetylenes (57), and a plethora of polymers with liquid crystal side chains (52,58–61). The organics and polymers with mesogenic side chains were found to be many times more active than lithium niobate (56), one of the best inorganic materials.

14.8.1 Theoretical Relationships

The starting point for nonlinear optics involves the constitutive relationship between the polarization induced in a molecule, P, and the electric field components, E, of the applied constant, low frequency, or optical fields (52). Ignoring magnetic dipoles and higher order multipoles, the induced polarization yields the approximation

$$P_i = \alpha_{ij}E_j + \beta_{ijk}E_jE_k + \gamma_{ijkl}E_jE_kE_l + \cdots \qquad (14.6)$$

The quantities P and E are vectors, and α, β, and γ are tensors. The subscripts arise through the introduction of Cartesian coordinates.

A similar expression can be written for the polarization induced in an ensemble of molecules, whether in the gas, liquid, or solid state. For this case, the polarization P can be written

$$P_i = \chi_{ij}^{(1)}E_j + \chi_{ijk}^{(2)}E_jE_k + \chi_{ijkl}E_jE_kE_l + \cdots \qquad (14.7)$$

where the coefficients $\chi_{ij}^{(1)}, \ldots,$ are tensors describing the polarization induced in the ensemble. The quantity $\chi_{ijk}^{(2)}$ will yield a zero contribution to nonlinear polarization if the system is centrosymmetric. For nonzero values, some type of asymmetry must be induced in the medium. For polymeric materials, asymmetry is generally induced by the generation of a polar axis in the medium through electric field poling.

Several of the effects occurring through $\chi_{ijk}^{(2)}$ are shown in Figure 14.18 (52). The combination of two photons of frequency ω produces a single photon at 2ω in second harmonic generation. Basically, for second harmonic generation, two photons must arrive at nearly the same time at a point in a nonsymmetric sample capable of absorbing them. Then the combined energy is released as one photon. The other effects in Figure 14.18 depend on other relationships. For example, the linear electrooptic effect follows from the interaction between an optical field and a direct current field in the nonlinear medium. This changes the propagation characteristics of the electromagnetic waves in the medium.

14.8.2 Experimental NLO Studies

When a polymer is subject to an intense sinusoidal electric field such as that due to an intense laser pulse, Fourier analysis of the polarization response can be shown to contain not only terms in the original frequency ω, but also terms in 2ω and 3ω (52). The intensity of the nonlinear response depends on the square of the the intensity of the incident beam for 2ω, and the third power for 3ω. For the second-order effects, the system must have

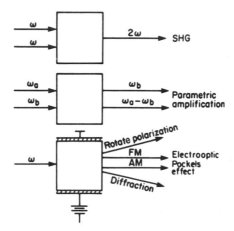

Figure 14.18 Schematic portrayal of some of the more exciting experiments that compose the field of nonlinear optics. Note the frequency doubling in second harmonic generation, *SHG* (52).

some asymmetry, as discussed previously. For poling, this means both high voltage and a chemical organization that will retain the resulting polarization for extended periods of time. Polymeric systems investigated have been of three basic types:

1. Molecules exhibiting large NLO effects are dissolved or dispersed in the polymer, which then merely acts as a carrier (61). An example is 2-methyl-4-nitroaniline dissolved or dispersed in vinylidene fluoride copolymers (62).

2. A segment of the polymer chain backbone is the active portion. This method is clearly the most efficient, taking into account the weight of the system. If the polymer is poled above T_g, then cooled under poling, the orientation will have temporal longevity (63).

3. A special side-chain group can be attached to the polymer backbone. The advantage of side groups is their relatively easy orientation under electric fields. Some types of side groups can be bonded not only to the main chain but at a second point, providing cross-linking. This latter then holds the active portion of the chain in position for temporal longevity (64).

Typical responses for second-order effects are values in the range of $0.2\text{--}20 \times 10^{-9}$ esu. Williams (52) remarked that uniquely useful applications for organic and polymeric materials will require a $\chi^{(2)}$ of 10^{-7} esu or greater. In addition, a variety of additional application-dependent properties and attributes must be present. These include uniform birefringence, minimized scattering losses, transparency, stability, and processibility. The relaxation time of the poled structures must be increased significantly.

NLO materials based on lithium niobate, for example are now in service for devices for second-harmonic generators, optical switches and routing components. One of the great advantages of electromagnetic waves instead of electrons in communications is the reduction in cross-talk. Effects in the picosecond time range have been investigated by Prasad and co-workers (65,66) and Garito (67). With better polymer NLO materials, perhaps they will compete with the inorganic materials.

14.9 MODERN ENGINEERING PLASTICS

One definition of an engineering plastic is one that has a high glass transition or a high melting temperature, for use at elevated temperatures. Other points of importance mention impact resistance, toughness, tensile strength, heat deflection temperature under load, creep, dimensional stability, and mold shrinkage.

Very few polymers are used commercially in the "pure" state. Some polyethylenes and polystyrenes are sold as homopolymers without any additives. However, several types of materials are added to most polymers to improve their properties:

1. Finely divided rubber is added to brittle plastics to toughen them.
2. Composites with glass, carbon, or boron fibers are made for high modulus and high strength.
3. Carbon black or silicas are added to rubber formulations to improve tear resistance and raise the modulus.
4. Various plasticizers may be added to lower the glass transition or reduce the amount of crystallinity. The main effect is to soften the final product.
5. Properties may be improved by adding silanes or other bonding agents to composites, to improve the bonding between polymer and other solid phases such as glass fibers.
6. Many types of polymers, both glassy and rubbery, are cross-linked to improve elastomer behavior or to control swelling.
7. Other items include fire retardants, colorants, and fillers to reduce price.

In each case, an understanding of how polymers behave and the effect of the various interactions allows the scientist and engineer to approach individual cases scientifically, rather than empirically.

14.10 MAJOR ADVANCES IN POLYMER SCIENCE AND ENGINEERING

While the use of natural polymers dates back to the dawn of civilization, polymer science developed rather recently. Tables 14.7, 14.8, and 14.9 delineate three distinct stages in the discovery, invention, and theory. The vulcanization of natural rubber and the synthesis of phenol-formaldehyde plastics (Table 14.7) were marked by trial and error, the basic idea of a polymer molecule not yet existing. Staudinger's Macromolecular Hypothesis (see Section 1.6) marked the first theoretical understanding of polymer structure.

In the classical period (Table 14.8) a basic understanding of polymer synthesis and structure was achieved. Concepts like the random coil, folded-chain crystals, and viscoelasticity involve many of the pages of this text. The synthetic advances play equally important roles in modern books on polymer synthesis, as well as general polymer science books. The invention of high-impact plastics (Chapter 13) played a key role in numerous applications of low-priced plastics.

In the modern era (Table 14.9) includes the discovery of liquid crystalline polymers; see Chapter 7. The newest major scientific development is polymer interface science. While a number of papers on polymer surfaces and interfaces date back into the classical era, the science itself dates from only 1989, when several key papers were published;

Table 14.7 Early polymer discoveries, inventions, and advances

Discovery	Reference
Vulcanization of rubber	C. Goodyear, U.S. 3,633 (1844).
Phenol–formaldehyde thermosets (phenolics)	L. H. Baekeland, U.S. 1,019,406 and U.S. 1,019,407 (1912).
Macromolecular Hypothesis	H. Staudinger, *Ber. Dtsch. Chem. Ges.*, **53**, 1074 (1920); summarized by H. Staudinger, *Die Hochmolekularen Organischen Verbindung*, Springer, Berlin, 1932.
X-ray confirmation of the Macromolecular Hypothesis	K. H. Meyer and H. Mark, *Ber. Dtsch. Chem. Ges.*, **61**, 593 (1928).
Synthesis of nylon (polyamide)	W. H. Carothers, U.S. 2,130,947 (1938).
Mark–Houwink–Sakurada equation	H. Mark, *Der Feste Koerper*, Kirzel, Leipzig, 1938.
Glass transition free volume theory	R. F. Boyer and R. S. Spencer, *J. Appl. Phys.*, **15**, 398 (1944).
Rubber elasticity theory	E. Guth and H. Mark, *Monatschefte Chemie*, **65**, 93 (1934); H. M. James and E. Guth, *J. Chem. Phys.*, **15**, 669 (1947).

Table 14.8 Development of classical polymer science and engineering

Discovery	Reference
Random-coil and polymer solution thermodynamics	P. J. Flory, *J. Chem. Phys.*, **10**, 51 (1942); summarized by P. J. Flory, *Principles of Polymer Chemistry*, Cornell University Press, Ithaca, NY, 1953.
Smith–Ewart theory of emulsion polymerization	W. V. Smith and R. W. Ewart, *J. Chem. Phys.*, **16**, 592 (1948).
High-impact plastics	J. L. Amos, J. L. McCurdy, and O. McIntire, U.S. 2,694,692 (1954).
Folded-chain single crystals	A. Keller, *Philos. Mag.*, **2**, 1171 (1957).
WLF equation and viscoelasticity	M. L. Williams, R. F. Landel, and J. D. Ferry, *J. Am. Chem. Soc.*, **77**, 3701 (1955).
Stereospecific polymerization	K. Ziegler, *Angew. Chem.*, **64**, 323 (1952); **67**, 541 (1955). G. Natta and P. Corradini, *J. Polym. Sci.*, **39**, 29 (1959).

Table 14.9 Modern era of polymer science and engineering

Discovery	Reference
Liquid crystalline polymers	S. L. Kwolek, B.P. 1,198,081 (1966); B.P. 1,283,064 (1968); summarized in S. L. Kwolek, P. W. Morgan, and J. R. Schaefgen, *Encyclopedia of Polymer Science and Engineering*, Vol. 9, Wiley, New York, 1987.
Polymer chain reptation	P. G. de Gennes, *J. Chem. Phys.*, **55**, 572 (1971).
Equation of state theories of polymer blends	P. J. Flory, *J. Am. Chem. Soc.*, **87**, 1833 (1965); I. C. Sanchez, Chap. 3 in *Polymer Blends*, Vol. I, D. R. Paul and S. Newman, Eds, Academic Press, Orlando, FL, 1978.
Polymer interface science	P. G. de Gennes, *Adv. Colloid Interface Sci.*, **27**, 189 (1987); summarized by G. J. Fleer, M. A. Cohen Stuart, J. M. H. M. Scheutjens, T. Cosgrove, and B. Vincent, *Polymers at Interfaces*, Chapman and Hall, London, 1993; and R. P. Wool, *Polymer Interfaces: Structure and Strength*, Hanser, Munich, 1995.

see Chapter 12. These described the shape of the polymer chain at interfaces, and their motion and entanglement across the interfaces.

Polymer science and engineering continues to develop at a rapid and even accelerating pace. The future of the field, and new directions lie with the imagination of the readers.

REFERENCES

1. J. A. Brydson, *Plastics Materials*, 6th ed., Butterworth-Heinemann, Oxford, England, 1995.

2. G. M. Benedikt and B. L. Goodall, *Metallocene-Catalyzed Polymers: Materials, Properties, Processing & Markets*, Plastics Design Library, Norwich, NY, 1998.

3. A. Turner-Jones and A. J. Cobbold, *J. Polym. Sci.*, **B6**, 539 (1968).

4. E. S. Shamshaum and D. Rouscher, in *Metallocene-Catalyzed Polymers: Materials, Properties, Processing and Markets*, G. M. Denedikt and B. L. Goodall, eds., Plastics Design Library, Norwich, NY, 1998.

5. C. D. Rosa, F. Auriemma, V. Vinti, and M. Galimberti, *Macromolecules*, **31**, 6206 (1998).

6. W. Kaminsky, *Pure Appl. Chem.*, **70**, 1229 (1998).

7. S. S. Labana, *Encyclopedia of Polymer Science and Engineering*, J. I. Kroschwitz, ed., Wiley, New York, 1986.

8. J. R. Brown and N. A. S. John, *Trends Polym. Sci. (TRIP)*, **4**, 416 (1996).

9. Y. H. Roos, *Phase Transitions in Foods*, Academic Press, San Diego, 1995.

10. C. M. McDonough, K. Seetharaman, R. D. Waniska, and L. W. Rooney, *J. Food Sci.*, **61**, 995 (1996).

11. S. H. Imam, S. H. Gordon, R. V. Greene, and K. A. Nino, *Polymeric Materials Encyclopedia* (Editor: J. C. Salamone), CRC Press, Boca Raton, FL, 1996.

12. V. Tolstoguzov, *Food Colloids*, **11**, 181 (1997).

13. M. LeMeste, V. T. Huang, J. Panama, G. Anderson, and R. Lentz, *Cereal Foods World*, **37**, 264 (1992).

14. L. M. Hallberg and P. Chinachoti, *J. Food Sci.*, **57**, 1201 (1992).

15. J. H. Jagannath, K. S. Jayaraman and S. S. Arya, *J. Appl. Polym. Sci.*, **71**, 1147 (1999).

16. A. Schiraldi, L. Piazza, O. Brenna, and E. Vittadini, *J. Thermal Analysis*, **47**, 1339 (1996).

17. E. W. Lusas and M. N. Riaz, *J. Nutrition*, **125(3)**, S573 (1995).

18. J. H. Litchfield, *Encyclopedia of Chemical Technology*, J. I. Kroschwitz, ed., Wiley, New York, 1994.

19. J. Magoshi, Y. Magoshi, M. A. Becker, and S. Nakamura, *Polymeric Materials Encyclopedia*, J. C. Salamone, ed., CRC Press, Boca Raton, FL, 1996, p. 669.

20. D. Strydom, T. Haylett, and R. Stead, *Bioch. Biophys. Res. Com.*, **79(3)**, 932 (1977).

21. J. Magoshi, Y. Magoshi, and S. Nakamura, *Repts. Prog. Polym. Phys. Japan*, **23**, 747 (1973).

22. C. Viney, A. E. Huber, D. L. Dunaway, K. Kerkam, and S. T. Case, *Silk Polymers: Materials Science and Biotechnology*, D. Kaplan, W. W. Adams, B. Farmer and C. Viney, eds., American Chemical Society, Washington, DC, 1994.

23. D. L. Kaplan, C. Mello, S. Fossey, and S. Arcidaicono, *Encyclopedia of Chemical Technology*, 4th ed., J. I. Kroschwitz, ed., Wiley, New York, 1997.

24. M. Tsukada, G. Freddi and N. Minoura, *J. Appl. Polym. Sci.*, **51**, 823 (1994).

25. M. W. Denny, in *The Mechanical Properties of Biological Materials*, Cambridge University Press, Cambridge, 1980.

26. M. C. T. Fyfe and J. F. Stoddart, *Acc. Chem. Res.*, **30**, 393 (1997).

27. P. E. Mason, W. S. Bryant, and H. W. Gibson, *Macromolecules*, **32**, 1559 (1999).

28. Y. Geerts, D. Muscat, and K. Mullen, *Macromol. Chem. Phys.*, **196**, 3425 (1995).

29. S. Menzer, A. J. P. White, D. J. Williams, M. Belohradsky, C. Hamers, F. M. Raymo, A. N. Shipway, and J. F. Stoddart, *Macromolecules*, **31**, 295 (1998).

30. S. I. Stupp, V. LeBonheur, K. Walker, L. S. Li, K. E. Huggins, M. Keser, and A. Amstutz, *Science*, **276**, 384 (1997).

31. T. H. Mourey, S. R. Turner, M. Rubinstein, J. M. J. Frechet, C. J. Hawker, and K. L. Wooley, *Macromolecules*, **25**, 2401 (1992).

32. M. Freemantle, *C&E News*, **77**, 27 (1999).

33. C. L. Jackson, H. D. Chanzy, F. P. Booy, B. J. Drake, D. A. Tomalia, B. J. Bauer, and E. J. Amis, *Macromolecules*, **31**, 6259 (1998).

34. P. G. de Gennes and H. Hervet, *J. Phys., Lett.*, **44**, L-351 (1983).

35. V. Percec, C.-H. Ahn, G. Ungar, D. J. P. Yeardley, M. Moller, and S. S. Sheiko, *Nature*, **391**, 161 (1998).

36. J. M. J. Frechet, *Science*, **263**, 1710 (1994).

37. L. J. Hobson and W. J. Feast, *J. Chem. Soc. Chem. Commun.*, 2067 (1997).

38. S. M. Aharoni, C. R. Crosby III and E. K. Walsh, *Macromolecules*, **15**, 1093 (1982).

39. Anonymous, in *Workshop on Properties and Applications of Dendritic Polymers, NIST, Gaithersburg, MD* (July 9–10, 1998).

40. R. Estfand, D. A. Tomalia, E. A. Beezer, J. C. Mitchell, M. Hardy, and C. Orford, *Polym. Prepr.*, **41(2)**, 1324 (2000).

41. S. J. Han, *Ph.D. Dissertation, Processing of Polyolefin Blends in Supercritical Propane Solution,* Lehigh University, 1998.

42. H. Graham, in *Encyclopedia of Food Science and Technology*, Y. H. Hui, ed., Wiley, New York, 1992.

43. S. N. Katz, U.S. 4,820,356, 1989.

44. J. M. DeSimone, Z. Guan, and C. S. Elsbernd, *Science*, **257**, 945 (1992).

45. J. M. DeSimone, E. E. Maury, Y. Z. Menceloglu, J. B. McClean, T. J. Romack, and J. R. Combes, *Science*, **265**, 356 (1994).

46. D. A. Canelas, D. E. Betts, and J. M. DeSimone, *Macromolecules*, **29**, 2818 (1996).

47. M. D. Elkovitch, *CAPCE Newsletter (Ohio State University)*, **2**, 2 (1999).

48. A. I. Cooper, J. D. Londono, G. Wignall, J. B. McClain, E. T. Samulski, J. S. Lin, A. Dobrynin, M. Rubinstein, A. L. C. Burke, J. M. J. Frechet, and J. M. DeSimone, *Nature*, **389**, 368 (1997).

49. K. N. Mathes, in *Encyclopedia of Polymer Science and Engineering*, Vol. 5, J. I. Kroschwitz, ed., Wiley, New York, 1986.

50. J. E. Frommer and R. R. Chance, in *Encyclopedia of Polymer Science and Engineering*, Vol. 5, J. I. Kroschwitz, ed., Wiley, New York, 1986.

50a. H. Shirakawa, E. J. Louis, A. G. MacDiarmid, C. K. Chiang, and A. J. Heeger, J. Chem. Soc. Chem. Comm., 579 (1977).

50b. C. K. Chiang, M. A. Druy, S. C. Gau, A. J. Heeger, E. J. Louis, A. G. MacDiarmid, Y. W. Park, and H. Shirakawa, J. Am. Chem. Soc., **100**, 1013 (1978).

50c. R. B. Kaner and A. C. MacDiarmid, Sci. Am., p. 106, Feb., 1988.

51. M. J. Bowden, in *Electronic and Photonic Applications of Polymers*, M. J. Bowden and S. R. Turner, eds., Advances in Chemistry Series No. 218, American Chemical Society, Washington, DC, 1988.

52. D. Williams, in *Electronic and Photonic Applications of Polymers*, M. J. Bowden and S. R. Turner, eds., Advances in Chemistry Series No. 218, American Chemical Society, Washington, DC, 1988.

53. D. J. Williams, ed., *Nonlinear Optical Properties of Organic and Polymeric Materials*, ACS Symposium Series No. 283, American Chemical Society, Washington, DC, 1983.

54. C. Lee, D. Haas, H. T. Man, and V. Mechensky, *Photonics Spectra*, p. 171, April 1979.

55. B. F. Levine, C. G. Bethea, C. D. Thurmond, R. T. Lynch, and J. L. Bernstein, *J. Appl. Phys.*, **50**, 2523 (1979).

56. B. I. Breene, J. Orenstein, and S. Schmitt-Rink, *Science*, **247**, 679 (1990).

57. C. R. Meredith, J. G. VanDusen, and D. J. Williams, *Macromolecules*, **15** 1385 (1982).

58. R. N. DeMartino, H. N. Yoon, J. R. Stamatoff, and A. Buckley, Eur. Pat. Applic. 0231770 (1987), to Celanese Corporation.

59. D. E. Stuetz, Eur. Pat. Applic. 017212 (1986).

60. N. A. Plate, R. V. Talroze, and V. P. Shibaev, in *Polymer Yearbook 3*, R. A. Pethrick and G. E. Zaikov, eds., Harwood, London, 1986.

61. B. F. Levine, C. G. Bethea, C. D. Thurmond, R. T. Lynch, and J. L. Bernstein, *J. Appl. Phys.*, **50**, 2523 (1979).

62. P. Pantelis and G. J. Davies, U.S. Pat. 4,748,074 (1988), to British Telecommunications.

63. B. I. Breene, J. Orenstein, and S. Schmitt-Rink, *Science*, **247**, 679 (1990).

64. M. Eich, B. Reck, D. Y. Yoon, C. G. Wilson, and G. C. Bjorklund, *J. Appl. Phys.*, **66**, 3241 (1989).

65. D. N. Rao, R. Burzynski, X. Mi, and P. N. Prasad, *Appl. Phys. Lett.*, **48**, 387 (1986).

66. D. N. Rao, J. Swiatkiewicz, P. Chopra, S. K. Chosal, and P. N. Prasad, *Appl. Phys. Lett.*, **48**, 1187 (1986).

67. A. F. Garito, presented at Pacifichem '89, Honolulu, Hawaii, December 1989.

GENERAL READING

A. W. Adamson and A. P. Gast, *Physical Chemistry of Surfaces*, 6th ed., Wiley, New York, 1997.

D. Avnir, *The Fractal Approach to Heterogeneous Chemistry: Surfaces, Colloids, Polymers*, Wiley, Chichester, England, 1989.

G. M. Benedickt and B. L. Goodall, eds., *Metallocene-Catalyzed Polymers: Materials, Processing and Markets*, Plastics Design Library, Norwich, NY, 1998.

M. J. Bowden and S. R. Turner, eds., *Electronic and Photonic Applications of Polymers*, Adv. Chem. Ser. No. 218, American Chemical Society, Washington, DC, 1988.

J. A. Brydson, *Plastic Materials*, Butterworths, London, 6th Ed., 1995.

A. Ciferri, ed., *Supramolecular Polymers*, Dekker, New York, 2000.

H. J. R. Dutton, *Understanding Optical Communications*, Prentice Hall, Englewood Cliffs, 2000.

R. A Hann and D. Bloor, eds., *Organic Materials for Non-Linear Optics*, Royal Society of Chemistry, London, 1989.

A. J. Heeger, J. Orenstein, and D. R. Ulrich, eds., *Nonlinear Optical Properties of Polymers*, Materials Research Society, Pittsburgh, 1988.

D. Kaplan, W. W. Adams, B. Farmer, and C. Viney, eds., *Silk Polymers: Materials Science and Biotechnology*, American Chemical Society, Washington, DC, 1994.

C. C. Ku and R. Liepins, *Electrical Properties of Polymers*, Hanser, Munich, 1987.

J. H. Lai, ed., *Polymers for Electronic Applications*, CRC Press, Boca Raton, FL, 1989.

C. W. Macosko, *RIM Fundamentals of Reaction Engineering*, Hanser, New York, 1989.

M. K. Mishra and S. Kobayashi, eds., *Star and Hyperbranched Polymers*, Dekker, New York, 1999.

H. S. Nalwa and S. Miyata, eds., *Nonlinear Optics of Organic Molecules and Polymers*, CRC Press, Boca Raton, FL, 1997.

P. N. Prasad and D. J. Williams, *Nonlinear Optical Effects in Molecules and Polymers*, Wiley, New York, 1991.

R. W. Siegel, E. Ha, and M. C. Roco, eds., *Nanostructure Science and Technology*, Kluwer Academic, Dordrecht, 1999.

S. Roth, *One-Dimensional Metals*, VCH, Weinheim, 1995.

R. B. Seymour and T. Cheng, eds., *History of Polyolefins*, Reidel, Dordrecht, Holland, 1986.

E. S. Wilks, *Polym. Prepr.*, **40**(2), 6 (1999). (References for catenanes, rotaxanes, and dendritic polymers.)

STUDY PROBLEMS

1. Read any scientific or engineering paper concerned with the topics of Chapter 14 published since 1999. What is the import of the paper? In what ways does it update the text? Please provide the full reference.

2. Assuming a spherical model, what is the theoretical density of the 4th, 6th, and 8th generation dendrimers having the structure illustrated in Figure 14.10? Hint: Assume the bulk material has unit density.

3. In order to improve nutrition, you want to add gelatin proteins to bread doughs. How will you do this? What are the morphological and texture consequences of your actions?

4. Why have polyethylene and polypropylene become the largest tonnage synthetic polymers? (Check the references and/or speculate.)

5. What are the advantages and disadvantages of using increased tonnages of cellulose, rather than polyethylene, polypropylene, polystyrene, and poly(vinyl chloride) for modern plastics? (Check recent literature, and speculate.)

6. Could a good thermoset polymer be made out of copolymers of divinyl benzene and styrene? What would be its modulus, glass transition temperature, and impact strength as a function of copolymer composition? (Draw figures illustrating probable behavior.)

7. You have just joined a silk research institute. Your supervisor asks for your research ideas. Write a one-page double spaced summary of a research proposal dealing with any aspect of silk.

8. For recycling purposes, used consumer plastics need to be separated. Based on superfluid technology, how would you separate polystyrene from polypropylene?

9. Why is polyacetylene such an interesting semiconductor? How does it work?

10. A certain application requires blue light with a wavelength of 400 nm. The only monochromatic light source available is an infrared laser with a wavelength of 800 nm. How will you accomplish this task?

11. If you had listened to Staudinger expound on the Macromolecular Hypothesis in the year 1920, what would you have suggested for a research project to prove or disprove his ideas? (Assume you have all the modern instruments available at your disposal, however.)

INDEX